T5-ASR-120

Frontiers in the
Chemical Sciences

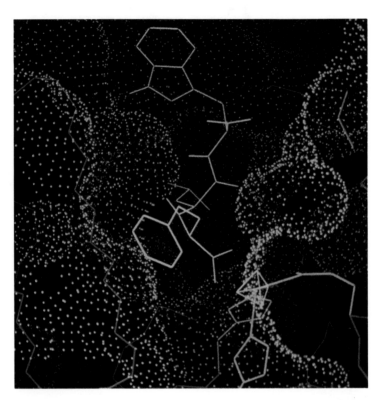

The solvent-accessible surface for thermolysin. The surface of the enzyme is shown by blue dots. Note the hydrophobic cavity that plays a key role in recognizing substrate or inhibitor.

Frontiers in the Chemical Sciences

Edited by William Spindel
and Robert M. Simon

Zhen-dong Chen
1988. 7. 8. in philadelphia.

The American Association for the Advancement of Science

Other titles in this series of *Science* volumes include:

Biotechnology & Biological Frontiers
Edited by Philip H. Abelson

Neuroscience
Edited by Philip H. Abelson, Eleanore Butz, and Solomon H. Snyder

Astronomy & Astrophysics
Edited by Morton S. Roberts

AIDS: Papers from *Science*, 1982–1985
Edited by Ruth Kulstad

Biotechnology: The Renewable Frontier
Edited by Daniel E. Koshland, Jr.

Library of Congress Cataloging-in-Publication Data
Frontiers in the Chemical Sciences
　Reprints from Science magazine originally published
between 1980 and 1985.
　Includes index.
　1. Chemistry.　I. Spindel, William
II. Simon, Robert Michael.　III. Science (Weekly)
QD39.F76　1986　　540　　86–14038
ISBN 0–87168–312–1
ISBN 0–87168–277–X (pbk.)

This material originally appeared in *Science*, the official journal of the American Association for the Advancement of Science.

Distributed to bookstores and overseas through Westview Press, Boulder, Colorado 80301

AAAS Publication No. 85–10

Copyright 1986 by the
American Association for the Advancement of Science
1333 H Street, NW, Washington, DC 20005
Printed in the United States of America

Contents

III. THEORETICAL CHEMISTRY

IV. CHEMICAL CATALYSIS

V. ORGANIC SYNTHESIS

VI. CHEMISTRY OF LIFE PROCESSES

Preface

Cheves Walling, the former editor of the *Journal of the American Chemical Society*, once defined chemistry as the study of matter at the level of organization where everything interesting happens. This wonderful description succinctly states what this book is all about. The collection of scientific articles that follows, selected from the pages of *Science* from 1980 through 1985, shows chemical scientists unraveling the mysteries of molecular reactivity, developing powerful analytical and theoretical tools, exploring and exploiting new pathways for catalysis and synthesis, and contributing to disciplines as diverse as biology and materials science. We have selected and arranged the articles to correspond to these areas of chemists' activities, influenced in no small part by a desire to provide a structure analogous to a recently released report from the National Research Council, *Opportunities in Chemistry*. That report, written by a committee of experts under the chairmanship of George C. Pimentel, provides the first in-depth analysis of chemical research frontiers that has been undertaken by the NAS/NRC in twenty years. This collection of research articles, then, provides discussions, in the words of the researchers themselves, parallel to the major directions in chemical research highlighted by the "Pimentel Report." We hope and expect that most of the readers of this volume will also be readers of *Opportunities in Chemistry*.

Science has recently published many other outstanding articles on cutting-edge chemistry. We are grateful for the advice and assistance of Allen J. Bard, John I. Brauman, Ronald Breslow, and George C. Pimentel as we went about our often difficult task of selecting the articles for this volume.

William Spindel
Robert M. Simon
April 1986
Washington, D.C.

[W]e find ourselves in a time of special opportunity for advances [in chemistry] on all. . .fronts. The opportunity derives from our developing ability to probe and understand the elemental steps of chemical change and, at the same time, to deal with extreme molecular complexity.

Opportunities in Chemistry, page 4.

The seven parts of this book take the reader on a tour of contemporary research frontiers in chemistry. The first five parts focus on the themes of fundamental understanding of chemical change and extreme molecular complexity. The discussion of these themes leads into the final two parts, which contain a series of accounts of frontier chemical research applied to problems in the biological and materials sciences.

Understanding Chemical Reactivity

Our journey through the world of modern chemistry begins with the laser. The availability of precisely tuned, coherent radiation sources has revolutionized the way physical chemists study the dynamics of chemical reactions. Using the laser to carry out detailed studies of small molecules in the gas phase, researchers are uncovering principles applicable to larger molecules and condensed-phase systems.

Stephen R. Leone begins his article on "Laser Probing of Chemical Reaction Dynamics" with a lucid exposition of how laser pulses are used, first to create initial reaction states and then to "interrogate" these states for information as the reaction proceeds. For example, the study of electron transfer from one molecule to another in the gas phase shows which geometric orientations and characteristic bond lengths are necessary to allow for electron switching in simple systems. As another example, high-power ultraviolet lasers can be used to generate large numbers of very fast hydrogen atoms. By studying how these fast atoms collide with simple molecules,

Introduction

Frontiers in the Chemical Sciences

William Spindel and Robert M. Simon

ix

chemists can validate a number of theoretical models for such reactions. They can also learn when a collision will result in the breaking of chemical bonds and the exchange of hydrogen atoms and when a collision will result merely in the vibrational excitement of the target molecule.

While the first article discusses studies of biomolecular reactions, the second article, "Understanding Molecular Dynamics Quantum-State by Quantum-State," written by Warren D. Lawrance, C. Bradley Moore, and Hrvoje Petek, explores the use of laser techniques to study the dynamics of single molecules in the gas phase. Important contributions have been made to our understanding of methylene—the simplest divalent carbon intermediate—to the study of how vibrational energy is exchanged among different vibrational modes in a molecule, and to the study of photochemical fragmentation of simple molecules. The next article in this section takes a close look at the search for a more complete understanding of divalent carbon intermediates. The reader unfamiliar with the chemistry of these reactive species will find "Divalent Carbon Intermediates: Laser Photolysis and Spectroscopy," by Kenneth B. Eisenthal, Robert A. Moss, and Nicholas J. Turro, a helpful introduction to the subject.

The fourth article in this section, "Chemistry and Chemical Intermediates in Supersonic Free Jet Expansions," written by Terry A. Miller, describes an experimental technique that enhances the power of laser techniques in the detection and study of reactive intermediates. A free jet, formed when a high-pressure gas sample passes through a small nozzle to a very low-pressure region, provides a stream of molecules whose random thermal motion has been converted to directed translational motion, resulting in an effective cooling of the molecules to less than 1 degree Kelvin. The result is a nearly ideal source of molecules for study by laser techniques, that is, molecules isolated in the gas phase with only a few populated vibrational or rotational energy levels.

Laser techniques, together with techniques such as ion-cyclotron resonance, ion beams,

and high-pressure and secondary-ion mass spectrometry, have also been an important stimulus to research on ion chemistry in the gas phase. These advances are charted in the paper by Paul B. Comita and John I. Brauman, "Gas-Phase Ion Chemistry."

Lasers—particularly short-pulse lasers—have also played a vital role in advancing the study of chemical reactions. It is appropriate, then, to conclude this section on understanding chemical reactivity with C.V. Shank's look at the frontier of generating ultrashort pulses of coherent light, "Measurement of Ultrafast Phenomena in the Femtosecond Time Domain." The ability to make time-resolved measurements in the femtosecond (10^{-15}) range will enhance chemists' ability to make quantitative observations of dynamic processes and may open doors to the discovery of wholly new phenomena.

Chemical Analysis

The integration of sophisticated instrumentation with computers has introduced a new generation of analytical techniques with unparalleled power. Part-per-trillion and sub-part-per-quadrillion sensitivities are now being achieved. The first article in this section, "Trends in Analytical Instrumentation" by Fred W. McLafferty, likens the ability to detect a part per trillion to the ability to detect a pinhead's area over the entire sweep of a road from New York to California. This basic article sets the stage for two in-depth profiles of advances in mass spectrometry and laser chemical analysis.

"Mass Spectrometry: Analytical Capabilities and Potentials," by R.G. Cooks, K.L. Busch, and G.L. Glish, surveys recent developments in mass spectrometry, including the extension of the mass range of such spectrometers by nearly an order of magnitude, the development of procedures to allow ionic, non-volatile compounds to be examined, and the successful integration of separation and analysis in such developments as tandem mass spectrometry and liquid chromatography-mass spectrometry. These techniques have made possible the application of mass spectrometry in such areas as archeology, biomedicine, botany, chemical kinetics, metabolic studies, and pharmacokinetics. This expanding universe of applications presents a challenge to biological scientists to become more aware of new instrumental capabilities that can be used for structural studies of important biological intermediates as well as for clinical applications.

"Laser Chemical Analysis," by Richard N. Zare, provides a sampling of laser techniques that may be brought to bear on problems of chemical analysis. The use of lasers in multiphoton ionization and laser fluorimetry offers opportunities for ultratrace analysis with unprecedented sensitivity and selectivity, in some cases approaching single-atom or single-molecule detection.

Theoretical Chemistry

The development of new techniques for detecting and studying chemical intermediates, which has been a theme of the previous two sections, has been paralleled by major advances in theoretical chemistry. An excellent and entertaining introduction to this subject is given by William A. Goddard III in "Theoretical Chemistry Comes Alive: Full Partner with Experiment." Several notable examples are given of how predictions of spectroscopic quantities and catalytic pathways, made from theoretical calculations, provided experimental chemists with new insights into studying these systems. As theoretical chemistry comes of age, its practitioners grow more confident of approaching important chemical, biological, and material problems on an equal footing with experimentalists.

One example of the increasing confidence of theoretical chemists is the revival and extension of the van der Waals picture of liquids, a theoretical concept more than a century old. In "Van der Waals Picture of Liquids, Solids, and Phase Transformations," David Chandler, John D. Weeks, and Hans C. Andersen discuss the applicability of van der Waals concepts as a guide to understanding many condensed matter systems, along with the limitations of this picture in systems with strong internal attractive forces. While not a quantitative theory for complex systems, van der Waal concepts are useful in explaining the behavior of systems and phenomena as diverse as molecular liquids, nonequilibrium amorphous and glassy materials, and liquid crystals.

Another example of modern theory interacting with experiment is the paper, "Toward a Coherent Theory of Chemisorption." By combining both experimental and theoretical investigations, Evgeny Shustorovich and Roger C. Baetzold uncover new and unexpected insights into chemisorption, the necessary first step for heterogeneous catalytic processes.

Chemical Catalysis

The contribution that theory has made to experimental chemistry is perhaps best acknowledged by introducing the next two sections of this collection with papers by theoretical chemists whose insights have been recognized in recent years by the Nobel Prize in Chemistry, one indication of how profoundly they have influenced the work of their experimental colleagues.

The seminal work of Henry Taube—winner of the 1983 Nobel Prize—in elucidating the principles of electron transfer between metal complexes, and of Roald Hoffmann—winner of the 1981 Nobel Prize with Kenichi Fukui—in applying molecular orbital theory to organometallic compounds, lies at the heart of modern organometallic chemistry. Their papers lead off this section on chemical catalysis. Taube's Nobel Lecture, "Electron Transfer Between Metal Complexes: Retrospective" is a personal history of the development of his work. "Theoretical Organometallic Chemistry," by Roald Hoffmann, introduces the use of molecular orbital theory in understanding the structure, conformation, and reactivity of organometallic compounds. The article also illustrates the remarkable diversity of structures that organometallic compounds may assume. This diversity is linked to the reason why transition metals play such an important role in catalysis: "There is spatial flexibility—

enough orbitals to provide bonding in just about any direction—and electronic tuning—the possibility of modifying the electronic count, sometimes drastically, by moving across the periodic table Still greater freedom (and control) is provided by the potential of adjusting the orbital size by changing the metal, its oxidation state, or switching from the first transition series to the second or third. No wonder that the variety of geometry, bonding strength, and function in organometallic complexes seems to be infinite."

Historically, though, the first catalytic reactions to be exploited commercially proceeded not by reactions of discrete organometallic complexes, but by reactions occurring at the surface of a bulk metal. For all the research that has been devoted to such reactions, though, it is only in the last twenty years that scientists have been able to study, on the molecular level, the actual steps in these catalytic reactions and the effects of structure, composition, and chemical bonding on chemical events occurring at surfaces and interfaces. In his lucid paper on "Surface Science and Catalysis," Gabor A. Somorjai describes the important atomic and molecular features of an active catalyst. The advent of sophisticated instrumental techniques in surface science has played an important role in transforming the design of catalysts from an art to a science.

Organometallic complexes play the lead role in the next three papers, which explore catalysis by organometallic compounds that are soluble and active in a liquid medium containing both reactants and products. Such "homogeneous" catalysts have proved promising for a number of selective, catalytic transformations of small molecules.

One example of such selectivity is given in Jack Halpern's paper on "Mechanism and Stereoselectivity of Asymmetric Hydrogenation," describing the mechanism of stereoselective organometallic catalysts; that is, complexes capable of effecting chemical transformations that create specific stereochemical relationships in the product molecule. An important commercial applica-

tion of such reactions, the large-scale production of the anti-parkinsonian drug, L-dopa, demonstrates the significance of this achievement.

In "Catalysis by Transition Metals: Metal-Carbon Double and Triple Bonds," Richard R. Schrock discusses a number of catalytic reactions where the catalytically active species contains an organic moiety connected to a metal atom by a double or triple bond. Such species have been implicated in transition-metal-catalyzed metathesis reactions of alkenes and alkynes. The chemistry of these organometallic compounds may also shed an interesting light on the reactivity of commercial Ziegler-Natta polymerization catalysts for ethylene and propylene—Schrock suggests that such reactions may proceed via intermediates whose metal-carbon bonds possess some multiple-bond character.

The search for metal complexes capable of catalyzing a series of reactions to derivitize the carbon-hydrogen bond of alkanes, a plentiful but unreactive class of organic compounds, is one of the Holy Grails of organometallic chemistry. "Activation of Alkanes with Organotransition Metal Complexes," by Robert G. Bergman, is included in this section, not because a new catalytic reaction was discovered, but to give an example of a common roadblock in the search for such reactions. Bergman and his co-workers discovered an organometallic compound of the element iridium that would insert into the carbon-hydrogen bond of an alkane, one necessary step in a catalytic scheme to derivitize alkanes. However, it proved impossible to find a sequence of reactions that would produce a derivitized alkane from this complex *and* regenerate the active organometallic reagent. Since such a regeneration step is needed to form a catalytic cycle, Bergman's intriguing compound merely provided a non-catalytic path to alkyl halides.

A third type of catalysis is driven by the conversion of light energy into electrochemical energy at the interface of semiconductors and solutions. An article on "Photoelectrochemistry," by Allen J. Bard, describes the basic scientific principles underlying photo-driven processes, as well as applications in semiconductor electrochemistry and photo-catalysis.

The collection of papers in this book also includes work on catalysis in biological systems. We will return to this theme in the section on "Chemistry of Life Processes."

Organic Synthesis

Just as the previous section was introduced by the work of Nobelists Henry Taube and Roald Hoffmann, this section starts with an account by Kenichi Fukui, co-winner of the 1981 Nobel Prize in Chemistry, of his theory of the "Role of Frontier Orbitals in Chemical Reactions." One measure of the importance of the concepts first enunciated by Fukui is that these concepts, and important extensions of them such as the Woodward-Hoffmann Rules for the conservation of orbital symmetry, have been part of nearly every textbook of organic chemistry published in the last twenty years.

The current challenges to organic synthesis are introduced by Barry M. Trost in his article on "Sculpting Horizons in Organic Chemistry." One key to future progress is the search for more and better ways of constructing organic molecules with selective structures, and Trost provides a good introduction to the various types of selectivity sought by synthetic organic chemists.

The next three papers pick up this theme of selectivity. Harry S. Mosher and James D. Morrison, in "Current Status of Asymmetric Synthesis," take Trost's general observations one step further, providing an introduction to several notable recent developments in this area, including asymmetric hydrogenation through homogeneous catalysis (mentioned earlier in this introduction), asymmetric epoxidation, asymmetric addition reactions, and the chiral aldol condensation. The latter reaction, a variant of one first described in 1838, has become a powerful method for synthesizing polyfunctional compounds with many asymmetric centers. The developments that made this achievement possible are de-scribed by Clayton H. Heathcock in "Acyclic Stereocontrol Through the Aldol Condensation." A related approach to stereoselectivity, through a photochemical equivalent of *threo*-selective aldol condensation, is described by Stuart L. Schreiber in "[2 + 2] Photocy-cloadditions in the Synthesis of Chiral Molecules." Photochemical [2 + 2] cycloadditions can serve as an attractive starting point for syntheses of highly complex molecules because they allow for rapid construction of the carbon skeleton of a target molecule and because the four-membered rings, which are the product of the cycloadditions, lend themselves to a variety of subsequent stereoselective manipulations. One successful example of this synthetic method, described by Schreiber, is the synthesis of two active components of the sex pheromone of the American cockroach.

Another important trend in organic chemistry is the broad use of organometallic reagents in synthesis. A comprehensive review of the use of organosilicon reagents to facilitate and control the course of organic reactions is given by Leo A. Paquette in "Silicon-Mediated Organic Synthesis."

The Chemistry of Life Processes

In his paper at the beginning of the preceding section, Barry Trost states that "the relationship between chemistry and biology must enter into any discussion of the future [of organic chemistry]. Almost by definition, molecular problems of the life sciences are integral parts of organic chemistry." This section, then, on the "Chemistry of Life Processes" begins with two papers that, at first glance, may seem to be misplaced. But "Cavitands: Organic Hosts with Enforced Cavities," by Donald J. Cram, and "Supramolecular Chemistry: Receptors, Catalysts, and Carriers," by Jean-Marie Lehn, describe crucial steps that are being taken towards the synthesis of artificial enzymes. As Cram notes in his article, "one of the supreme challenges to the organic chemist is to design and synthesize compounds that simulate the working parts of evolutionary chemistry."

The use of organic hosts to influence chemical reactivity is described both by Lehn and by Ronald Breslow in his article on "Artificial Enzymes." Breslow describes several reactions catalyzed by artificial enzymes constructed from cyclodextrins, a class of doughnut-shaped molecules capable of binding small organic molecules, or parts of larger organics, in their cavities. These reactions mimic the selectivity of such well-known enzymatic transformations as transamination of amino acids, and the hydrolytic cleavage of RNA. Modification of the cyclodextrin cavity improves the speed with which reactions with a cyclodextrin proceed. The combination of selectivity with reaction rate acceleration shows the promise of this line of research for designing new, selective synthetic reactions.

A second route to understanding and manipulating enzyme function is exemplified by the use of synthetic methods to construct model compounds, or altered natural compounds, as probes of the influence of structure on biological activity. In "Chemical Mutation of Enzyme Active Sites," E.T. Kaiser and D.S. Lawrence describe their methods for producing "semisynthetic" enzymes, that is, naturally occurring enzymes that have been chemically modified to introduce new active sites. The modified enzymes show reactivities that are dramatically different from their unaltered counterparts. In "Amphiphilic Secondary Structure: Design of Peptide Hormones," Kaiser and F.J. Kézdy study a number of biologically active peptides whose conformation depends entirely on their environment. In an environment like that of many membranes, which are hydrophilic (water-soluble) on one side and hydrophobic on the other, these peptides assume a characteristic structure, with hydrophilic and hydrophobic regions, that seems to be closely related to their biological activity. Indeed, model peptides that are composed of entirely different amino acids, but mimic this combined hydrophilic/hydrophobic (or amphiphilic) behavior, show the same biological activity as the native peptides.

A third approach taken by chemists seeking to unravel the chemistry of life processes is shown by the final two papers in this section. Using powerful instrumental techniques and knowledge gained from model studies on organometallic complexes, chemists are able to unravel the basic reactivity of important biological compounds.

The use of nuclear magnetic resonance (NMR) techniques based on atoms other than hydrogen has become widespread only in the last fifteen years. The power of these techniques is demonstrated in "Studying Enzyme Mechanism by ^{13}C Nuclear Magnetic Resonance," by Neil E. Mackenzie, J. Paul G. Malthouse, and A.I. Scott. Particularly exciting is the use of these researchers of carbon-13 NMR spectroscopy to study the dynamics of enzyme catalysis at sub-zero temperatures. The technique provides a fairly straightforward look at the true intermediates in the reaction, whose structures can be assigned on the basis of the NMR spectrum.

"Mechanisms of Coenzyme B_{12}-Dependent Rearrangements," by Jack Halpern, describes a number of kinetic studies on coenzyme B_{12} and a variety of model studies for probing the electronic and steric environment of the coenzyme and possible pathways for the 1,2-rearrangement seen in the substrate molecules of the enzymatic reaction. These studies establish the likelihood that the only role the coenzyme plays is that of a free radical precursor under the mild conditions of the enzymatic reaction.

Investigations into the nature and sequence of biological processes are by no means limited to chemists. Yet the formidable armamentarium of techniques that chemical scientists have at their disposal guarantees that they will continue to make important contributions to our understanding of the chemistry of life processes.

The Chemistry of New Materials

The final area to be covered by this collection of articles describes a frontier rich with opportunities for chemical scientists—the chemistry of new materials. As the field of material sciences evolves to focus more on molecular and chemical phenomena, the techniques developed by chemical scientists in other areas will provide important insights. The contribution that chemists will make, though, will take place in a highly interdisciplinary context; many of the "chemical" articles in this section are written by physicists, polymer scientists, electrical engineers, and materials engineers.

An example of the crucial importance of chemical structure to material properties is given in an article on "Surface-Active Biomaterials" by Larry L. Hench and June Wilson. The article principally describes the wide range of uses of the surface-active glass, Bioglass, its bonding characteristics, and its mechanism. The response of the body to bioactive glass is determined by the chemical composition of the glass, and maintenance of the ratio of $CaO:SiO_2:Na_2O:P_2O_5$ within certain limits is critical to performance of the glass in living systems.

In spite of all the successful use of Bioglass in animal tests, the field of biocompatible materials has not advanced very far; instrumental techniques for studying interfaces and surfaces in biological systems are nowhere near the level of sophistication of techniques for examining other interfaces. There are important contributions for chemists to make to the study of biocompatibility.

The next three articles discuss current research in conducting organic and organometallic materials. R.L. Greene and G.B. Street, in "Conducting Organic Materials," concisely introduce the two main classes of conducting organic solids—crystalline charge-transfer (CT) complexes and conducting polymers—and the major challenges that now face researchers trying to exploit such systems. An intriguing extension of the CT complexes is described by Tobin J. Marks in "Electrically Conductive Metallomacrocyclic Assemblies." Group IVA phthalocyanines are cofacially linked to form a polymer, e.g., $[Si(Pc)O]_n$, that can be extruded to form strong fibers which, when doped, are both air-stable and electrically conductive. These conductive assemblies of phthalocyanines also provide an easily manipulated system to probe for the connection between structural parameters and electrical behavior.

"Ferroelectric Polymers," by Andrew J. Lovinger, focuses primarily on the extraordinary polymer poly(vinylidene fluoride); a piezoelectric (changes electrical polarization in response to mechanical stress), pyroelectric (changes electrical polarization in response to changes in temperature), and ferroelectric (changes direction of its polar axis by application of an electric field) crystalline polymer. These unique qualities, along with PVF's desirable mechanical properties (light weight, flexibility, and toughness) make it an attractive candidate for a host of electromechanical and acoustic applications.

"Laser-Induced Chemistry for Microelectronics," by applied physicists R.M. Osgood and T.F. Deutsch, exemplifies the interdisciplinary character of current research on materials. Heterogeneous chemistry, involving two different material phases, such as a gas and a solid, now plays a major role in the fabrication of microelectronic circuits, and laser-initiated chemical reactions have been used in novel processing operations. The study of such reactions has, in turn, led to new insights into the nature of light-driven chemical reactions at the surface of solids.

Early research in laser-induced chemistry used laser light to initiate specific reactions in order to synthesize chemicals or separate isotopes. In microelectronics fabrication, the specificity of laser-induced chemistry is important, but equally important is the ability of the laser light to confine reactions to submicrometer-scale regions. The ability to localize reactions on solid surfaces has led to a number of techniques for semiconductor processing. In addition, chemical reactions confined to such small regions of a solid-gas interface are sufficiently different from previously studied reactions that a new form of chemistry, laser-induced microchemistry, is evolving. Some of the techniques, such as laser-assisted etching and excimer-laser photolithography, are immediately applicable to manufacturing. This research has also generated increasing interest in the basic physics and chemistry of light-assisted interface reactions.

"Metalorganic Chemical Vapor Deposition of III-V Semiconductors," by R.D. Dupuis, deals with the preparation of compound semiconductor materials composed of crystalline solid solutions of elements from columns IIIa and Va of the periodic table; these materials are now challenging the monopoly that the semiconductor silicon has enjoyed for the past thirty years in the area of large-scale electronic device applications. These III-V compound semiconductor materials have many unique electronic and optical properties that distinguish them from silicon. For example, many of these alloy semiconductors have a direct band structure. As a result, alloy semiconductors such as gallium arsenide (GaAs) are very efficient emitters of photons, while semiconductors having indirect band structures, such as silicon, are inefficient sources of light.

Several technologies have been developed for growth of the thinfilm epitaxial structures that are required for the realization of such III-V compound semiconductor optoelectronic devices: (i) liquid-phase epitaxy (LPE), (ii) molecular-beam epitaxy (MBE), and (iii) vapor-phase epitaxy (VPE). The VPE growth technique utilizes chemical reactions that occur between the vapors of certain compounds of the IIIa and Va elements when they are heated together. The VPE process that is the subject of this article employs metalorganic compounds as sources of the IIIa elements; the Va elements are derived from either Va hydrides or mixtures of Va hydrides and metalorganics. This metalorganic chemical vapor deposition (MOCVD) process is becoming widely used in the growth of many important III-V compound semiconductors. MOCVD also promises to be an extremely important research tool.

The articles in this book show the chemical sciences to be something quite different from the stereotypical image of test tubes and bunsen burners. As Edward G. Jefferson, Chairman of the Board of the E.I. du Pont de Nemours and Company, wrote in a recent editorial in *Science*, "Chemistry continues to offer major challenges to the 'pure' researcher. At the same time it is the key to meeting many of the needs of society. The chemical sciences have made many vital contributions to the welfare of mankind and have the potential for even greater contributions in the future. . . .[W]e need full support of the chemical sciences by government, industry, and our educational institutions. Society will be repaid manyfold for such an investment of resources."

Part I

Understanding Chemical Reactivity

The advent of lasers has heralded revolutions in almost all fields of technical endeavor (*1*). In chemistry, much of the original excitement centered on the possibilities for laser-driven, bond-selective synthesis and isotope separation (*2*). While a few such schemes for laser-selective chemistry do appear promising, an even greater number of remarkable side benefits have arisen from the intense activity in this field. Tremendous progress in molecular spectroscopy has been made possible by the availability of highly developed tunable laser sources. Another area of rapid progress is in the sophisticated understanding of chemical reaction dynamics which has been obtained through laser studies of state-selective reactions. It is the latter field which is the subject of this article.

Chemical reaction dynamics is a relatively new field (*3, 4*). It is, in essence, a modern-day approach to the study of chemical kinetics, or the rates at which reactions proceed. Rather than merely measuring the rates of chemical reactions under various conditions, chemists are now accustomed to achieving a much finer level of detail. With the availability of laser sources for selective excitation and detection, it is possible to infer many things about the dynamics, that is, the particular motions, that the molecules are likely to undergo in the chemical reaction. It is possible to interrogate, or probe, the specific forms of excitation that best lead to chemical reaction, for example, vibrational, rotational or translational motion, or electronic excitation. Thus, the study of gas-phase, state-resolved dynamics offers many answers to questions of fundamental interest in chemical reactivity.

The laser has played a central role in the development of these studies of

1

Laser Probing of Chemical Reaction Dynamics

Stephen R. Leone

Science 227, 889–895 (22 February 1985)

chemical reaction dynamics. However, laser sources are not the sole means of studying chemical processes with state-selected detail. For many years, complementary techniques of spectrally resolved infrared and visible chemiluminescence and molecular beam reactive scattering have also provided exceedingly detailed results on reaction dynamics (5, 6). Lasers have, however, provided some of the most sensitive methods for detection of molecular states, as well as the means to select specific reagent states with high resolution. Through the inherent polarization of their light, lasers can be used to study numerous chemical reactions as a function of the geometric alignment of reagents or products. With the high resolution of lasers, it is even possible to probe the differences in reactivity of closely spaced spin-orbit states. Lasers even provide the capability to probe the exceedingly short-lived transition states of simple reactions.

Data from experiments on state-selected chemical dynamics now comprise a large body of knowledge from which it is possible to infer many things about the dynamics. It is important to emphasize, though, that even a series of the most cleverly designed experiments together with appropriate theoretical interpretation (7) can only lead to powerful inferences concerning the specific motions that the molecules undergo. Nearly all experiments interrogate the system at a time that is either well before the reaction, by specific reagent state preparation, or well after the transition state of the reaction, by probing the final product states. Most of our simple pictures of the dynamical motions have been derived from this powerful combination of experimental facts, inference, and theory. Experimentalists would like to obtain

even more concrete information, rather than rely so heavily on inference. This is why the possibility of more directly probing the transition states of reactions has intrigued so many investigators.

General Methods of Laser Probing

An apparatus for studying chemical reaction dynamics through either selective laser excitation of reagents or laser detection of product states can have many variations (4, 8). However, a number of key features common to all these devices can be outlined. Figure 1 shows an apparatus that could be used for many types of investigations in reaction dynamics. It consists of a low-pressure reaction chamber, a laser for excitation or dissociation to produce the initial reagent states, a laser for interrogation of product states, and detectors for molecular fluorescence or laser absorption. In some cases the second laser is replaced by a detector for time- and wavelength-resolved emission in either the infrared or visible. Provision can be made to rotate the planes of polarization of the exciting and analyzing lasers. The reagents themselves may be stable molecules or short-lived atomic, radical, or ion species. Thus, many different gas sources are employed to generate the reagents of interest.

In ideal circumstances the reactions are carried out under "single-collision" conditions. These conditions typically require low pressures, so that a product molecule which is born in a particular state is not modified by subsequent collisions during the time of interrogation. Since very short laser pulses are frequently used, the time between excitation and interrogation can be less than

10^{-8} second. In this case, pressures as high as hundreds of pascals (1 Pa = 7.5×10^{-3} torr) can be used with virtually no collisions occurring between the formation of a product and its detection.

It should be immediately apparent that collisional relaxation, which is itself an active area of investigation called molecular energy transfer, is of central importance in the issue of obtaining nascent, or newly born, product state distributions. Moreover, collisional energy transfer pathways can play an underlying role in the dynamics of the reactive event. There are numerous examples of competition between reactive and energy transfer pathways, and the flow of energy between various degrees of freedom during the lifetime of the active transition state can be important in the final outcome of product states. Thus, the field of reaction dynamics requires an implicit understanding of the role that molecular energy transfer plays and the magnitudes of the rates of both energy transfer and reaction.

Laser excitation of selected reagent states most frequently involves a direct absorption process. However, in some cases more subtle methods can be employed. For example, a high-power, po-

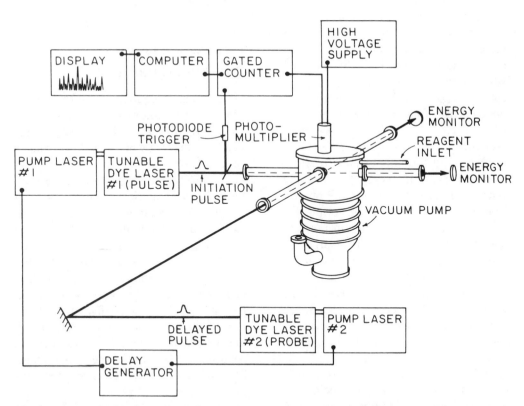

Fig. 1. A typical two-laser, pulse and probe type of apparatus for studies of molecular reaction dynamics. The first laser is used for excitation or dissociation to prepare the reagent states and the second laser is used to interrogate the product states a short time later.

larized laser can be used to photodissociate a collection of molecules, leaving behind a small fraction of molecules with their planes of rotation very purely aligned (9). The laser photolysis method can be used to create high concentrations of reactive free radicals and atoms, and the fact that the laser power density can easily be varied is exploited to control those radical densities (10). Molecules can be prepared in states that are ordinarily inaccessible by pumping with two lasers, first through an intermediate state and then up or down into the state of interest. The latter method of down-pumping is called stimulated emission pumping and can be used to prepare molecules in very high vibrational levels of the ground electronic state (11).

Laser detection schemes are even more varied (12). One of the most common is laser-induced fluorescence, in which a molecule is excited to a fluorescing electronic level and photons are observed as a function of the wavelength of the laser (12). In this way, complete product state distributions can be mapped with vibrational and rotational resolution. Alternatively, the molecules can be ionized with one or more photons from a laser and the electrons or positive ions can be counted to obtain the state information. Transient measurements can be accomplished by absorption of a laser probe beam that is tuned to a particular transition. Various Raman processes, such as coherent anti-Stokes Raman spectroscopy (CARS), can be used as a quantitative probe of molecular-state information (13).

Perhaps the most significant reason for the tremendous variety of techniques now available for interrogating molecules is the extensive development of tunable laser sources that operate throughout the infrared, visible, ultraviolet, and vacuum ultraviolet regions of the spectrum. Techniques have been developed to probe almost any kind of atomic or molecular state, in many cases with sensitivities approaching number densities of 10^5 cm^{-3}, and in special circumstances with detection sensitivity even for single atoms (14). The types of lasers range from solid-state tunable diode lasers in the infrared to liquid-phase organic dye lasers in the visible. By elegant nonlinear optical techniques, the outputs from high-power, pulsed visible dye lasers can be summed and mixed to produce useful tunable ultraviolet and vacuum ultraviolet light, with wavelengths as short as 100 nm (15).

Examples of State-Specific Excitation and Detection

Lasers are frequently used to produce selective excitation of reagent states in an effort to learn about the specific mechanisms that will "drive" a chemical reaction (16, 17). In many cases tremendous enhancements in reactivity are attainable by excitation of certain reagent states. For example, the reactions of calcium and strontium with HF to form CaF and SrF are found to be enhanced by at least 10^4 upon excitation of the HF to the $v = 1$ vibrational state with a chemical laser (18). This rather large increase in the rate coefficient with vibrational excitation also facilitates probing the reactivity as a function of initial rotational state within the $v = 1$ manifold by the same selective laser excitation (19).

Reactions of the type M + HX, where M is a metal and X is a halogen, have many varied characteristics. However,

6

in some of these reactions it is found that rotational excitation may slightly inhibit the reaction (*19*). Vibrational excitation is sometimes more effective than translational energy in promoting the reaction. The mechanism of these reactions is thought to involve an initial electron transfer from the metal atom to the electronegative halogen atom, forming a strong attractive force between the metal and the halogen. Increased vibrational excitation appears to overcome the barriers to this process in an effective way.

Reactions such as the ones described above may involve not only state-specific excitation of the reagents with lasers, but also interrogation of the product states by laser-induced fluorescence probing (*20*). In an elegant series of measurements on the Ca + HF system (*21*), the number of vibrational levels excited in the CaF product was found to increase monotonically with increasing rotational excitation (Fig. 2). The actual population distributions in the vibrational states were very nearly statistical, based on a comparison with calculations in which the total energy available to the products was statistically distributed among all the vibrational, rotational, and translational degrees of freedom. The latter result is suggestive of a long-lived reaction complex.

Insertion reactions. Reactions of electronically excited $O(^1D)$, $C(^1D)$, or $Mg(^1P_1)$ with hydrogen-containing compounds form OH, CH, or MgH as the product. These reactions have been exhaustively studied by laser-induced fluorescence probing of the OH, CH, or MgH product (*22, 23*). The atoms are produced selectively by direct laser excitation or ultraviolet laser photolysis of a molecule, such as ozone. One of the motivations for these extensive studies is

that the reactions can take place through two very different types of chemical mechanisms. One involves direct abstraction of a hydrogen atom from an R–H type molecule, and the other involves insertion into the R–H bond followed by unimolecular decomposition of the highly excited species to form the product.

In principle, the insertion and abstraction mechanisms should be distinguishable by detailed measurements of the vibrational and rotational state distributions of the product. Typically, a direct abstraction reaction yields a very high degree of vibrational excitation (*7*). From 40 to 60 percent of the available reaction energy is channeled into vibration in the diatomic product, which represents a highly nonstatistical fraction of the available energy in vibration. An insertion reaction, on the other hand, may involve a long-lived complex intermediate, and the energy of the reaction may be partitioned in a more statistical manner between all the vibrational, rotational, and translational degrees of freedom. In a large polyatomic intermediate, this would mean that only a small fraction of the available energy would go into the diatomic vibration. However, if the insertion complex breaks up in a time short compared to the time scale for energy randomization, which might especially be the case for a complex with a small number of atoms, then highly nonstatistical distributions in both vibration and rotation can also be observed, making the interpretation ambiguous.

In some cases the ambiguity can be resolved by careful measurements of the rotational state distributions. Bimodal distributions in rotation, in which very different populations and degrees of excitation occur for groupings of low and

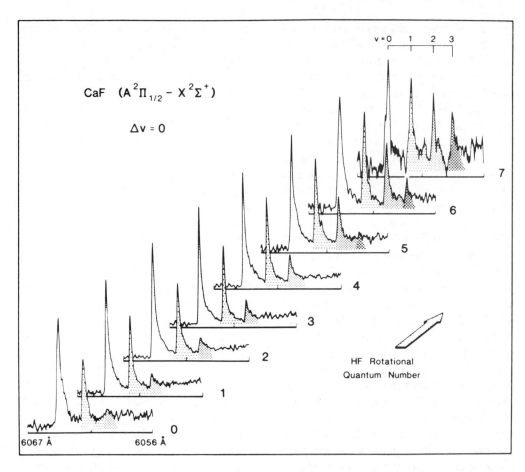

CaF $(A^2\Pi_{1/2} - X^2\Sigma^+)$

$\Delta v = 0$

$v = 0$ 1   2   3

7

6

5

4

3

2

HF Rotational
Quantum Number

1

0

6067 Å   6056 Å

Fig. 2. Laser-induced fluorescence spectra of the CaF product formed in the Ca + HF($v = 1$) reaction as a function of rotational level in the HF reagent. Band heads corresponding to different CaF vibrational levels are shaded differently to indicate the increase in CaF vibrational excitation with reagent rotational excitation. [Reprinted with permission of R. Altcorn, F. E. Bartoszek, J. DeHaven, G. Hancock, D. S. Perry, R. N. Zare, *Chemical Physical Letters* **98**, 212 (1983).]

high rotational states, are typically indicative of multiple pathways or mechanisms. One of the clearest examples of such a bimodal character in rotation comes from the reaction of Mg($^1P_1$) with H_2 (*23*). This reaction is simple enough and the fractions of the insertion and abstraction channels are both large enough that the two pathways are apparent in the rotational data (Fig. 3). There is substantial evidence in this reaction

that the abstraction component is responsible for the low rotational levels and the insertion mechanism is responsible for the high rotational levels.

Several examples of O(^1D) reactions further demonstrate the exceptional level of detail that can be obtained from this type of study but are also illustrative of the frequent ambiguities and uncertainties that occur in the interpretation of data to describe the dynamics. Much of

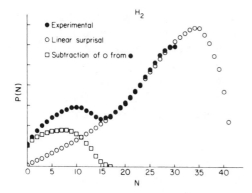

Fig. 3. A plot of the population $P(N)$ versus rotational state N, showing the deconvolution into the low and high rotational components in the reaction of $Mg(^1P_1) + H_2$ to form MgH. A surprisal-type analysis (5), which compares the observed populations to a statistical predicted distribution, is used to estimate the populations of the high rotational components in the blended region. The low-N rotational levels are thought to occur by the direct abstraction mechanism and the high-N levels by insertion. [Reprinted with permission of W. H. Breckenridge and H. Umemoto, *Journal of Chemical Physics* **80**, 4168 (1984). © 1984 American Institute of Physics.]

the experimental evidence indicates that both insertion and abstraction mechanisms are important in the reactions of $O(^1D)$. For example, in the reactions with hydrocarbons, highly nonstatistical distributions in OH vibration and bimodal distributions in rotation of the OH product are interpreted as meaning that both abstraction and insertion are occurring (24). From these measurements it has been possible to estimate the fractions of the abstraction and insertion components, to determine the different extents of vibrational excitation for each component, and to relate these to the statistical and nonstatistical partitioning. Several of these results are shown in Table 1 for the reactions of $O(^1D)$ with H_2, CH_4, C_2H_6, C_3H_8, and $C(CH_3)_4$.

Table 1. Properties of the abstraction and insertion components of $O(^1D) + RH$ reactions (24).

RH	F_A^* $(v = 0)$	$\left(\dfrac{v = 1}{v = 0}\right)_I^\dagger$	$\left(\dfrac{v = 1}{v = 0}\right)_A^\ddagger$
H_2	0.0	0.9	
CH_4	0.05	1.0	1.0
C_2H_6	0.2	0.65	0.3
C_3H_8	0.7	0.55	0.08
$C(CH_3)_4$	0.9		$\leqslant 0.05$

*Fraction of the abstraction component for the OH($v = 0$) channel. †Ratio $(v = 1)/(v = 0)$ for the insertion component. ‡Ratio $(v = 1)/(v = 0)$ for the abstraction component.

Unfortunately, not all the results are so clear-cut concerning the interpretation of $O(^1D)$ mechanisms. For example, in one of the most elegant experimental studies to date, the complete vibrational and rotational partitioning in the ^{16}OD and ^{18}OD products of the reaction between $^{16}O(^1D)$ and $D_2^{18}O$ has been measured (25). The ^{16}OD product, which is the result of a newly formed bond, is found to have 40 percent of the available energy in its vibration. The ^{18}OD molecule, in contrast, has a negligible 2 percent of the energy in its vibration. The rotational excitations in both products are similar, but they also exhibit the typical bimodal pattern, with a very cold component in the low rotational levels and a very hot fraction in the higher rotational levels. In this case, however, the authors suggest that this reaction must be direct, and they favor an abstraction type of mechanism in which the ^{18}O–D bond is a spectator. The interpretation is ambiguous because a very short-lived insertion intermediate might also explain the large difference in vibrational excitation between the two products and still account for the bimodal distributions in rotation.

9

Reactions of fast hydrogen atoms. High-power, pulsed laser outputs at short ultraviolet wavelengths have become available with the advent of excimer lasers. These lasers make it possible to generate high densities of translationally energetic atoms, such as hydrogen atoms, by photolysis of simple precursor molecules such as H_2S, HI, and HBr. The energy of each 193- or 248-nm laser photon in excess of the bond energy is partitioned almost entirely into kinetic energy of the atom of lower mass, due to simple conservation of energy and momentum. This has made possible a number of new and basic studies on the dynamics of reactive and inelastic energy transfer pathways in the collisions of hydrogen atoms with simple molecules.

Recent studies of the reactive collisions of fast hydrogen atoms with D_2 molecules have received wide attention because of the fundamental nature of this simple chemical system and the large number of calculations that have already been carried out to generate accurate potential energy surfaces for the reaction. The vibrational and rotational states of the HD product molecules have been probed by both CARS (*13*) and laser multiphoton ionization (*26*). The hydrogen atoms are produced with high kinetic energy by laser photolysis of HI in the presence of D_2, and the interrogation takes place shortly after some reaction has occurred but before substantial relaxation has taken place. At a collision energy of 1.3 eV, newly born HD product molecules have been observed in the $v = 1$ and $v = 2$ vibrational states and over a range of rotational states from $J = 0$ to 9. The general agreement between experiment and theory for the vibrational and rotational state distributions formed in the reaction is very good (*26*). Further work will help to elucidate

the origins of the small differences and will provide data on other channels, including direct excitation of the D_2 by energy transfer.

Studies have also been made on the competition between vibrational excitation and atom exchange pathways in collisions of laser-generated fast hydrogen and deuterium atoms with HCl and DCl (*27*). Spectrally resolved infrared chemiluminescence is used to determine the amount of vibrational excitation in the HCl and DCl products of collisions between H and DCl. The vibrational population distributions show that the HCl product of the reactive exchange is substantially more excited than the DCl molecules which become vibrationally excited as a result of the direct, inelastic energy transfer process.

As further evidence that the atom exchange and energy transfer processes take place on fundamentally different chemical energy surfaces, the excitation efficiency as a function of the kinetic energy is found to exhibit strikingly different trends (Fig. 4). The efficiency of the pure energy transfer process increases sharply with kinetic energy, whereas the reactive exchange pathway has nearly a constant fraction of energy deposited into vibration over the range of collision energies investigated (1 to 3 eV). The results are interpreted as indicating that quantitatively more vibrational excitation results from the process of breaking and reforming the bond in the atom exchange process, but that once the kinetic energy is high enough to overcome the barrier to atom exchange, very little of the additional excess translational energy is deposited into the diatomic vibration.

Charge transfer reactions. The extremely high sensitivity of the laser-induced fluorescence method allows a

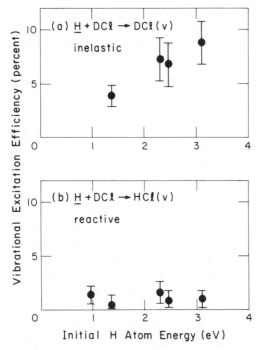

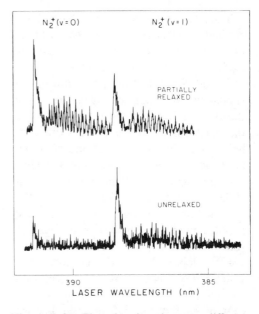

Fig. 4 (left). Plot showing the very different vibrational excitation efficiencies for direct energy transfer and reactive exchange pathways in collisions of H + DCl. Fig. 5 (right). Laser-induced fluorescence spectra of the N_2^+ product in the $Ar^+ + N_2$ charge transfer reaction, showing in the lower trace the nascent (unrelaxed) distribution in which $v = 1$ predominates and in the upper trace a partially relaxed result in which single-collision conditions are not achieved.

number of new kinds of reactions to be studied for the first time. In particular, reactions of ions at near-thermal energies have been difficult to study with product state resolution. The problem of obtaining sufficiently high densities of ions to be able to probe the product states of such reactions under single-collision conditions has been solved by using high-velocity flows of helium to generate the ions (28). One of the most interesting and fundamental classes of reactions involves simple electron transfer, or charge transfer, from one molecule to another in the gas phase. Being free of the complications associated with solvent media, such studies have the potential of showing how the chemical energy of electron transformations can

be channeled into the vibrational, rotational, and translational motions of the relatively heavy nuclei. At thermal collision energies the dynamics will be most sensitive to the strong attractive ion–induced dipole and ion-dipole forces.

The mechanisms of charge transfer can be divided into two extreme models. A Franck-Condon electron transfer would obtain a vibrational distribution that is governed by isolated diatomic overlap factors for the vibrational wave functions of the ion and neutral species; this would look very much like a photoionization process. In contrast, an energy resonance picture would channel the reaction exothermicity as nearly as possible into the vibrational, rotational, or electronic internal degrees of freedom.

Thus far, three reactions have been studied by laser probing of the diatomic product ion vibrational and rotational excitation: $Ar^+ + N_2$ (29), $Ar^+ + CO$ (30), and $N^+ + CO$ (31). In spite of the similarities between the atomic ions and the diatomic molecules in these cases, the reactions exhibit markedly different behaviors. The $Ar^+ + CO$ and N_2 reactions both show an exceptional preference for populating the highest vibrational levels accessible. For example, the N_2^+ is born with ten times as great a population in $v = 1$ as in $v = 0$. Figure 5 shows a spectral scan of the $v = 0$ and $v = 1$ levels of N_2^+ under conditions in which the N_2^+ is in a nascent distribution and partially relaxed by collisions. The CO^+ product of the Ar^+ reaction is born predominantly in the energetically near-resonant states $v = 4$, 5 and 6. In contrast, the $N^+ + CO$ reaction produces 70 percent in $v = 0$, 28 percent in $v = 1$, and 2 percent in $v = 2$, which is in very close agreement with a Franck-Condon distribution of 96, 4, and 0 percent, respectively.

The rotational excitation in these charge transfer reactions further corroborates the differences in the mechanisms. In the $N^+ + CO$ reaction, only 2 percent of the energy goes into rotational excitation, indicating that a substantial fraction of the reactive events occur by long-range electron transfer, which would not impart any angular momentum to the diatomic ion product. In contrast, the reactions of Ar^+ with N_2 and CO both produce substantial rotational excitation of the diatomic ion product, indicating that a close encounter must occur, which would tend to impart angular momentum to the product. It is thought that such a close encounter is necessary to distort the diatomic bond

length before the charge transfer transition state can be reached. Such studies will provide further insights into the geometric orientations and characteristic bond lengths that are necessary to allow the switching of an electron in simple systems.

Spin-orbit effects. The power of the laser to resolve even finer details of molecular reaction dynamics is demonstrated through recent studies of the reactivity of specific spin-orbit states. Spin-orbit states, which arise from the splitting of an atomic energy level by coupling of the nuclear spin with the orbital angular momentum, often involve only minute energy differences, but they can exhibit substantial differences in reactivity. A thorough series of studies has been carried out on the reactions of the $^2P_{1/2}$ and $^2P_{3/2}$ spin-orbit states of fluorine atoms with HBr by using a tunable vacuum ultraviolet laser to probe the corresponding $^2P_{1/2}$, $^2P_{3/2}$ spin-orbit states in the bromine atom product (32). The fluorine atom spin-orbit states could be populated to different extents by thermal heating. The fluorine atom $(^2P_{1/2}/^2P_{3/2})$ spin-orbit state ratio was varied over a factor of 6, while the product bromine atom spin-orbit ratio $(^2P_{1/2}/^2P_{3/2})$ remained constant at 6 percent. These results indicate that there is not a simple direct correlation between the reactant spin-orbit state and product spin-orbit state. Rather, it appears that only the ground $^2P_{3/2}$ electronic state is reactive in this case, most likely because of a high barrier in the potential surface for the excited spin-orbit state. The authors speculate that the constant fraction of bromine atom spin-orbit excitation may be the result of energy transfer in the transition state from the vibrationally excited HF product molecule to the bromine atoms before the products separate.

Despite the small splitting (160 cm^{-1}) of the Ca($^3P_{2,1,0}$) spin-orbit states, selective laser manipulation of the ratios of the three spin-orbit states has shown that they have substantially different reactivities with molecular chlorine to produce electronically excited CaCl products (*33*). The order of reactivity decreases by a factor of 3 to 4 in the series $^3P_2$, $^3P_1$, and $^3P_0$. This selectivity of the reactivity is remarkable in light of the ease with which these states are collisionally transferred from one to the other.

Alignment and orientation. The subject of atomic and molecular alignment and orientation is one of the exciting new areas of chemical reaction dynamics (*34, 35*). The laser allows a number of different effects to be studied by exciting atoms or molecules with polarized light or by probing specific product states which correspond to particular orbital configurations. In the chemical literature the meanings of alignment and orientation are characterized in the following way. Consider a linear molecule with two diffferent ends. Alignment means that the linear axis can be arranged either parallel or perpendicular to the collision partner. Orientation means that either of the two different ends of the molecule can be pointed toward the collision partner. Optical state preparation with lasers can only provide alignment. In a few cases, hexapole electric fields have been used to provide orientation.

In spite of the fact that the degree of alignment cannot always be perfectly controlled, several substantial effects of alignment or reactivity have already been demonstrated. Most remarkable are the reactions of the excited P state of calcium with Cl$_2$, HCl, and CCl$_4$, which

are probed as a function of the *p*-orbital alignment (*36*). The reaction with Cl$_2$ shows a marked preference for perpendicular alignment of the calcium *p* orbital with the incoming reaction partner (Fig. 6). The reaction with HCl shows a variation of the CaCl A and B electronic product states with the orbital alignment, while the reaction with CCl$_4$ shows almost no preferential characteristics associated with alignment.

For molecular reagents, the degree of alignment that can be achieved in a simple absorption process cannot be very

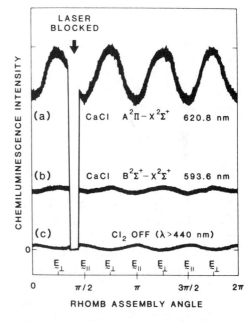

Fig. 6. Variation of the alignment angle between the Ca($^1P_1$) orbital and incoming Cl$_2$ showing the variation of product chemiluminescence for (a) the CaCl (A) state, (b) the CaCl (B) state, and (c) the result with the reagent Cl$_2$ turned off. The latter signal is due to variation of the laser scattered light with rotation of the laser polarization. [Reprinted with permission of T. Rettner and R. N. Zare, *Journal of Chemical Physics* **77**, 2416 (1982). © 1982 American Institute of Physics.]

great. Even so, a polarized laser has been used to align the HF($v = 1$) reagent in the reaction with strontium atoms (*37*). A weak enhancement of the $v = 2$ SrF product state is observed for the broadside collision geometry. If a powerful pulsed laser is used to photodissociate molecules rather than to excite them directly, the remaining undissociated molecules can be prepared with a very high degree of alignment. This method has been used to select the plane of rotation of IBr molecules and to find that the reaction of excited xenon atoms with IBr is preferentially enhanced when the geometry of reagent approach is parallel to the plane of rotation of the IBr molecules (*9*).

Information about the alignment of molecular orbitals in the transition states of chemical reactions can also be obtained by the analysis of certain lambda doublet components in the laser-induced fluorescence product spectra (*22*). This has been particularly applied to the OH products of numerous oxygen atom and hydrogen atom reactions. For high rotational states in the OH product, the lambda doublet components can be uniquely associated with the direction of the orbital containing the lone electron. For example, in the reaction of translationally fast hydrogen atoms with O_2, the energetically lower lambda doublet component is populated by 6:1 preferentially over the energetically higher component (*38*). The preferred component corresponds to the case where the singly occupied orbital lies in the plane of the OH rotation. A similar type of preferential population is observed for many other reactions that produce OH. It is taken to indicate that the OH is released with its plane of rotation rigorously in the plane of the transition state and that the bond that is broken has the lone electron in an orbital within this plane.

Probing the transition state. The laser has been crucial to recent investigations that have brought experimentalists closer to being able to probe the details of the actual transition states of chemical reactions. The high power densities of the light from lasers facilitate the manipulation of even the shortest-lived species, by forcing a competition between the natural rates of the reaction process and the rates of some light absorption or emission processes that can be driven by the laser. There are several reports of the successful manipulation of reactions by absorption of laser light during the lifetime of a transient collision intermediate (*39*). Substantial success has also been achieved in a study of the transition states of chemical reactions by direct emission (*40*). Several of the experiments involve reactions of xenon with Cl_2 to form electronically excited XeCl. Other definitive experiments have been carried out on the laser-assisted associative ionization of two sodium atoms during a collision. In a novel experiment a cluster of Hg with Cl_2 was formed and then excited with a laser, and the prompt formation of electronically excited HgCl product was observed (*41*). As more reports come forth, it will be possible to say much more about the potential of these laser manipulation methods for increasing our understanding of the dynamics of transition states.

Conclusions

The laser provides the means not only to investigate the reactivity of well-specified initial reagent states, but also to quantitatively probe product vibrational, rotational, and electronic states. In several cases, through more subtle features of molecular spectroscopy and optical pumping, it is possible to interrogate the reactivity associated with specific geometries. As new techniques for state detection with lasers become available, the number of possibilities for further detailed study will increase dramatically. Thus the potential of laser methods for the study of chemical reactivity and reaction dynamics is perhaps still largely untapped. The initial goal of many investigations, to demonstrate the possibility of laser isotope separation and bond-selective chemistry, has been partially successful. Far greater rewards have been achieved from the ensuing fundamental studies of the dynamical behavior in state-selected chemical reactions.

References and Notes

1. A. A. Boraiko, *Natl. Geogr.* **165**, 335 (1984).
2. J. Jortner, R. D. Levine, S. A. Rice, Eds., *Photoselective Chemistry* (Wiley, New York, 1981), parts 1 and 2.
3. R. D. Levine and R. B. Bernstein, *Molecular Reaction Dynamics* (Oxford Univ. Press, New York, 1974).
4. M. R. Levy, *Prog. React. Kinet.* **10**, 1 (1979).
5. R. B. Bernstein, *Chemical Dynamics via Molecular Beam and Laser Techniques* (Oxford Univ. Press, New York, 1982).
6. I. W. M. Smith, *Kinetics and Dynamics of Elementary Gas Reactions* (Butterworths, London, 1980).
7. D. Henderson, Ed., *Theoretical Chemistry. Theory of Scattering: Papers in Honor of Henry Eyring* (Academic Press, New York, 1981), vols. 6A and 6B.
8. For an introduction to laser sources, see S. R. Leone and C. B. Moore, in *Chemical and Biochemical Applications of Lasers*, C. B. Moore, Ed. (Academic Press, New York, 1974), pp. 1–27.
9. M. S. deVries, V. I. Srdanov, C. P. Hanrahan, R. P. Martin, *J. Chem. Phys.* **78**, 5582 (1983).
10. D. J. Nesbitt and S. R. Leone, *ibid.* **72**, 1722 (1980).
11. D. E. Reisner, R. W. Field, J. L. Kinsey, H.-L. Dai, *ibid.* **80**, 5968 (1984).
12. W. Demtröder, *Laser Spectroscopy: Basic Concepts and Instrumentation* (Springer-Verlag, New York, 1982).
13. D. P. Gerrity and J. J. Valentini, *J. Chem. Phys.* **79**, 5202 (1983).
14. G. S. Hurst, M. H. Nayfeh, J. P. Young, M. G. Payne, L. W. Grossman, in *Laser Spectroscopy III*, J. L. Hall and J. L. Carlsten, Eds. (Springer-Verlag, New York, 1977), p. 44.
15. R. Hilbig and R. Wallenstein, *Appl. Opt.* **21**, 913 (1982).
16. M. Kneba and J. Wolfrum, *Annu. Rev. Phys. Chem.* **31**, 47 (1980).
17. I. W. M. Smith, in *Physical Chemistry of Fast Reactions*, I. W. M. Smith, Ed. (Plenum, New York, 1980), vol. 2, pp. 1–82.
18. Z. Karny and R. N. Zare, *J. Chem. Phys.* **68**, 3360 (1980).
19. C.-K. Man and R. C. Estler, *ibid.* **75**, 2779 (1981).
20. B. E. Holmes and D. W. Setser, in *Physical Chemistry of Fast Reactions*, I. W. M. Smith, Ed. (Plenum, New York, 1980), vol. 2, pp. 83–214.
21. R. Altcorn, F. E. Bartoszek, J. DeHaven, G. Hancock, D. S. Perry, R. N. Zare, *Chem. Phys. Lett.* **98**, 212 (1983).
22. S. R. Leone, *Annu. Rev. Phys. Chem.* **35**, 109 (1984).
23. W. H. Breckenridge and H. Umemoto, *J. Chem. Phys.* **80**, 4168 (1984).
24. A. C. Luntz, *ibid.* **73**, 1143 (1980).
25. W. A. Guillory, K. H. Gericke, F. J. Comes, *ibid.* **78**, 5993 (1983).
26. E. E. Marinero, C. T. Rettner, R. N. Zare, *ibid.* **80**, 4142 (1984).
27. C. A. Wight, F. Magnotta, S. R. Leone, *ibid.* **81**, 3951 (1984).
28. E. E. Ferguson, F. C. Fehsenfeld, A. L. Schmeltekopf, *Adv. At. Mol. Phys.* **5**, 1 (1969).
29. L. Hüwel, D. R. Guyer, G. H. Lin, S. R. Leone, *J. Chem. Phys.* **81**, 3520 (1984).
30. C. E. Hamilton, V. M. Bierbaum, S. R. Leone, *J. Chem. Phys.* **83**, 2284 (1985); G. H. Lin, J. Maier, and S. R. Leone, *J. Chem. Phys.* **82**, 5527 (1985).
31. D. R. Guyer, L. Hüwel, S. R. Leone, *J. Chem. Phys.* **79**, 1259 (1983); C. E. Hamilton, V. M. Bierbaum, S. R. Leone, *J. Chem. Phys.* **83**, 601 (1985).
32. J. W. Hepburn, K. Liu, R. G. Macdonald, F. J. Northrup, J. C. Polanyi, *J. Chem. Phys.* **75**, 3353 (1981).
33. H.-J. Yuh and P. J. Dagdigian, *ibid.* **79**, 2086 (1983).
34. R. N. Zare, *Ber. Bunsenges. Phys. Chem.* **86**, 422 (1982).
35. S. Stolte, *ibid.*, p. 413.
36. T. Rettner and R. N. Zare, *J. Chem. Phys.* **77**, 2416 (1982).
37. Z. Karny, R. C. Estler, R. N. Zare, *ibid.* **69**, 5199 (1978).
38. K. Kleinermanns and J. Wolfrum, *ibid.* **80**, 1446 (1984).
39. P. R. Brooks, R. F. Curl, T. C. Maguire, *Ber. Bunsenges. Phys. Chem.* **86**, 401 (1982).
40. P. Arrowsmith, S. H. P. Bly, P. E. Charters, J. C. Polanyi, *J. Chem. Phys.* **79**, 283 (1983).
41. C. Jouvet and B. Soep, *Chem. Phys. Lett.* **96** 426 (1983).
42. I am grateful for support from the National Bureau of Standards, National Science Foundation, Air Force Office of Scientific Research, Department of Energy, and Army Research Office.

The process of energy transfer within a molecule or group of molecules is closely related to the making and breaking of chemical bonds. Although it is not possible to photograph step-by-step motions of individual atoms and molecules, it is possible to resolve individual quantum states in the observation of molecular spectra, in the preparation of reactant molecules, and in the analysis of reaction products.

Such information, in combination with theory, reveals a great deal about the dynamics of atomic and molecular motions and about the potential energy surfaces that govern them. In an article in this issue, Leone (1) discusses bimolecular reactions in the gas phase. Our article deals with unimolecular processes, including the nature of excited states and reaction intermediates as revealed by their spectra, the dynamics of intramolecular vibrational energy redistribution as probed by spectroscopy, and the dynamics of photofragmentation as resolved quantum-state by quantum-state. We have selected a few examples to illustrate the power of some of the new types of experiments and of the tools now available, but much equally important and interesting work is not discussed.

Spectroscopy of Transient Molecules

Lasers have made it possible to observe the spectra and study the dynamics and chemical kinetics of many free radicals (2, 3), ions (2, 4), molecular excited states, and other transient species (2). Frequency resolution as high as 1 part in 10^8 is possible. Lasers with pulses as brief as 10^{-14} to 10^{-13} second access the shortest chemically significant time scales, for which the energy uncertainty,

_____ *2*

Understanding Molecular Dynamics Quantum-State by Quantum-State

Warren D. Lawrance,
C. Bradley Moore,
and Hrvoje Petek

Science 227, 895–901 (22 February 1985)

$\Delta E \sim \hbar/\Delta t$, is comparable to chemical bond energies (3, 5). Sensitivities sufficient to detect single molecules have been demonstrated for more modest limits of spectral and temporal resolution (6).

Methylene (CH_2), the prototype for divalent carbon intermediates, has been the focus of many experimental and theoretical studies aimed at determining the structure and the energy separation of the two low-lying electronic states, the "metastable" singlet ($^1CH_2$) and the ground triplet ($^3CH_2$) states. Spectroscopic detection of methylene eluded experimentalists for many years during which the only evidence for the theoretically postulated electronic structure of methylene was the vastly different chemistry exhibited by the two electronic spin states (3). In an article in this issue, Goddard (7) discusses the interaction between experiment and theory that produced accurate determinations of both the singlet and triplet structures, as well as the value of 9 kcal/mol for the energy separation between the singlet and triplet states (Δ_{s-t}).

The pioneering flash-kinetic spectroscopy work of Herzberg provided the first spectra and structure for triplet methylene and showed that it is the ground state (8). Laser magnetic resonance, a technique whereby rotational or vibrational transitions are brought into resonance with a fixed infrared laser frequency by magnetic tuning of fine-structure transitions, has been used to detect spectra of several isotopomers of triplet methylene (9). The $^3CH_2$ structure and potential energy surface derived from these data indicate a bent structure with a barrier to linearity of 1950 cm^{-1} (9), in excellent agreement with state of the art quantum chemical calculations (7).

The analysis of the visible spectrum of $^1CH_2$ by Herzberg and Johns (8) indicates a transition from a bent lower state to a linear upper state. In a recent reinvestigation of this spectrum, ultraviolet laser photolysis of ketene was used to generate a large density of $^1CH_2$, and narrowband tunable infrared or visible lasers were used to probe the absorption spectrum (Fig. 1) (10, 11). This modern version of flash-kinetic spectroscopy, in which traditional photolysis and probe flash lamps have been replaced by lasers, has resulted in significantly greater sensitivity and better wavelength resolution, as well as extension to the infrared—the spectral region from which important structural information for most molecules is readily obtained. The detection of the symmetric and antisymmetric stretch spectra of an extremely reactive molecular transient such as $^1CH_2$ (11) demonstrates the utility of this technique for structural and time-resolved studies of short-lived species. The structure of $^1CH_2$ determined by analysis of the visible and infrared spectra improves on the original work of Herzberg and Johns and agrees well with theoretical work (10, 11).

The detailed analysis of singlet methylene spectra leads to two significant conclusions (10, 11): (i) many calculated transition frequencies are shifted from their observed positions, and (ii) there are many extra transitions not predicted by calculations. This indicates strong coupling between the singlet and triplet manifolds. The simplest spectral manifestation of these perturbations would be a doubling and shifting of optical transitions as a result of the mixing of singlet and triplet quantum states by coupling between electron spin and orbital angular momenta. These mixed states exhibit

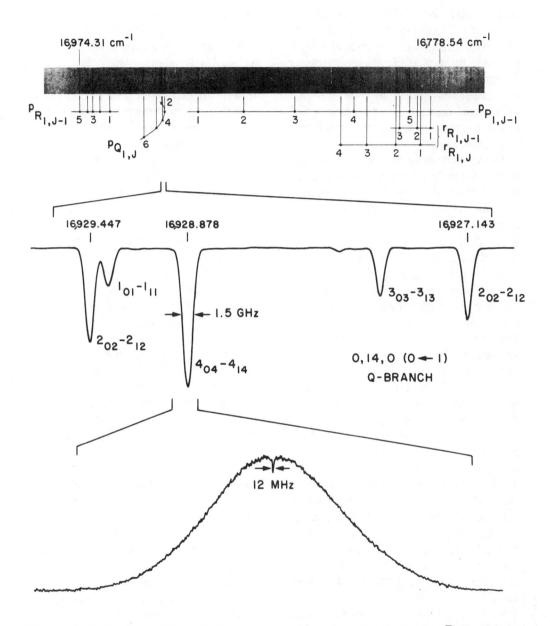

Fig. 1. Flash kinetic spectra of singlet methylene. The electronic transition is $\bar{b}\ {}^1B_1 \leftarrow \bar{a}\ {}^1A_1$. The region presented shows part of the rotational structure of the $(0,14,0) \leftarrow (0,0,0)$ vibrational transition that accompanies this electronic excitation [(n_1,n_2,n_3) refers to the number of quanta in the vibrations ν_1, ν_2, and ν_3, respectively]. From top to bottom: the original photographic plate of Herzberg and Johns showing the P, Q, and R branches; middle, the Q-branch Doppler-limited spectrum; bottom, the sub-Doppler Lamb dip feature on the 4_{04}–4_{14} transition. The width of the Lamb dip corresponds to 5 parts in 10^7 frequency resolutions. [Reprinted with permission of G. Herzberg and J. W. C. Johns in *Proc. R. Soc. London Ser. A* **295**, 106 (1966).]

both singlet and triplet characteristics; the singlet character gives intensity to otherwise forbidden singlet-triplet optical transitions, and the triplet character is expressed by the states having a Zeeman effect arising from the interaction of an external magnetic field with unpaired electron spins of the triplet electronic state component.

These perturbations were first observed by laser magnetic resonance for four rotational transitions of excited bending vibrational states of triplet methylene (9). The analysis of the anomalous Zeeman effect (decrease in magnetic dipole) induced by singlet mixing provided the assignment of these perturbations and the most precise value to date for the singlet-triplet energy difference. The assignment of these perturbations has been confirmed by the visible and infrared spectroscopy of singlet methylene (10, 11). Many other perturbations have been detected by magnetic rotation spectroscopy, an extremely sensitive technique that relies on the rotation of the polarization of the probing light by transitions that are Zeeman-active in an external magnetic field (12). Some of these singlet states couple to the fundamental stretching vibrations of $^3CH_2$. Observation of the CH stretching infrared spectra of $^3CH_2$ can provide a determination of Δ_{s-t} accurate to 0.01 cm^{-1}. Such accuracy should challenge quantum chemical calculations for some time.

Very recently, sub-Doppler resolution has been achieved for flash-kinetic spectroscopy of methylene (13). This will permit an extensive quantitative study of singlet-triplet perturbations. Normally the resolution of molecular spectra is intrinsically limited by the frequency shift due to the thermal motion of the molecules (Doppler effect). Much higher resolution can be achieved by the use of saturation spectroscopy techniques with a narrow-frequency-width laser. The Zeeman effect splits each rotational line into magnetic fine-structure components and gives a quantitative measure of triplet admixture in these states.

These singlet-triplet perturbations have significant consequences for the dynamics of methylene. The collisional quenching of singlet to triplet methylene by inert collision partners occurs efficiently (14). Gelbart and Freed (15) describe collision-induced singlet-triplet quenching in terms of vibrational and rotational energy transfers between mixed singlet and triplet levels, followed by vibrational relaxation in the manifold of the lower-energy electronic state. Thus the intramolecular perturbations provide the pathway for a collisional energy transfer that is forbidden by strict spin conservation rules.

Other dynamic manifestations of singlet-triplet couplings can be seen in the chemistry of methylene. The reaction

$$^3CH_2 + CH_4 \rightarrow \text{products}$$

exhibits an activation energy of 10 ± 2 kcal/mol (16). Since collisional interconversion between $^1CH_2$ and $^3CH_2$ is facile and the reaction of $^1CH_2$ with methane has a nearly gas kinetic rate constant (14), the observed activation energy is approximately equal to the value of Δ_{s-t} required to produce $^1CH_2$ from ground state $^3CH_2$ by collisional excitation.

The observation of strong coupling between the two low-lying electronic states has significantly increased our understanding of methylene structure and dynamics and has made necessary a re-evaluation of previous work on methylene kinetics. The detailed understanding of this coupling and the collision-induced relaxation mediated by the perturbed states are important for understanding similar processes in larger molecules.

Intramolecular Vibrational Energy Redistribution

In any process involving chemical change there is a making or breaking of chemical bonds. Unimolecular reactions are the simplest of chemical reactions because the bond breaking occurs in the isolated molecule. When thermal collisions are used to excite a molecule above its dissociation energy, all of the vibrational motions are excited, even though only one or two bonds are usually broken. If it were possible to excite only those bonds that one wished to break, very specific and interesting new chemistry could be performed. This possibility has proved to be difficult to realize (17).

Vibrational spectroscopy is based on the observation that, at low energy, molecules have distinct vibrations, referred to as normal modes, in which the atomic nuclei have quite different motions that involve the stretching or bending of specific bonds. In this region the potential energy surface can be approximated by a sum of individual parabolas, one for each normal mode. Thus each normal mode is a separate vibrational oscillator. The terms that are omitted in this so-called harmonic approximation are referred to as vibrational anharmonicity and are quite small at low energy. In contrast with this ordered picture at low energy, the almost universally successful theory of thermal unimolecular reactions is based on the assumption that, at the energies of dissociation, vibrational energy is not confined to specific nuclear motions and is free to exchange among different bonds. This exchange of vibrational energy among different vibrational modes, which is referred to as intramolecular vibrational redistribution (IVR), arises when the vibrational modes are no longer isolated oscillators but are coupled together by effects such as anharmonicity. The possibilities for mode-specific chemistry are determined by where and how these two different vibrational regimes merge; to achieve mode-specific chemistry requires that the IVR rate be slower than the reaction rates. Consequently, the study of IVR is central in the study of chemical dynamics.

At low energies there is no time evolution in the vibrational motion because each of the spectral features excited is a molecular eigenstate (that is, a stationary-state solution of Schrödinger's equation). However, at higher energies the molecular eigenstates are a mixture of normal modes because of the coupling referred to above. If all of the states that share the character of a particular normal mode are excited in unison (coherent excitation), the wave functions for the states begin "in phase," so that the sum of all of the atomic displacements corresponds to that normal mode. However, because the energy of each of the states is slightly different, the wave functions each have a slightly different frequency and, with time, shift out of phase. When they are out of phase, the constructive and destructive interference that led to the initial appearance of the normal mode vanishes, and the molecule takes on a new vibrational motion. The IVR rate is the rate at which this dephasing occurs. It is dependent on how strongly the modes are coupled (how mixed the

21

states are) and on the energy spread between the coupled states.

Much of our knowledge of what determines the onset of IVR comes from fluorescence studies at low vibrational energies (typically <6000 cm^{-1}). The fluorescence spectrum for a particular vibrational motion serves as a fingerprint for that motion. Spectral features arising from excitation of vibrations that do not contribute to the intensity of the absorption indicate that IVR is occurring. This technique has been explored in both ground and excited electronic states. There are two experimental approaches. The first involves exciting the same, or similar, vibrations in a number of related molecules in order to see the effect of increasing the vibrational state density by increasing the molecular complexity. Smalley (18) and his co-workers, for example, performed experiments on a series of alkylbenzenes. They found that an increase in the length of the alkyl side chain results in an increase in emission from states with no absorption intensity. Stewart and McDonald (19) studied the CH stretch fundamental in various molecules and showed that there is a definite correlation between the extent of IVR and the number of near-resonant states having the correct symmetry to couple. The second approach is to probe the extent of IVR from different vibrational levels of the same molecule (20). In this case, a steady increase in IVR with the density of vibrational states is particularly obvious. Spectra at low vibrational energy show the structure expected; however, as the vibrational energy increases there is a growth in emission from nearby states until finally all semblance of structure from the excited level vanishes. The conclusion from the large number of different experiments is that

the onset of IVR is a function of the density of states of the correct symmetry surrounding the excited level. However, the range of state densities for which IVR begins will most certainly vary among different classes of molecules and types of vibrations.

Information concerning the coupling mechanisms that are most important in the onset of IVR is beginning to emerge. In this region, mixing of individual states may be observed with high-resolution spectroscopy. Field and Kinsey and their associates (21) have studied the interactions among vibration-rotation levels as a function of vibrational and rotational excitation, in a series of experiments on formaldehyde and acetylene in which they used the technique of stimulated emission pumping. This technique is used to reach the high vibrational levels of the ground electronic state (S_0) in a two-step process: molecules are first excited to a specific vibration-rotation level in the first excited electronic state (S_1) and then transferred to S_0 by stimulated emission. Because a single state is prepared in S_1, only one to six rotational transitions can accompany a given vibrational transition, resulting in a simple and easily assigned spectrum of the S_0 levels. Any lines that are observed but not predicted by the selection rules for optical transitions arise from couplings among the S_0 levels. The formaldehyde data show conclusively that both anharmonic coupling, a purely vibrational effect, and Coriolis coupling, an effect arising from the interaction of rotational and vibrational motion, contribute to the vibrational state mixing. Furthermore, it is seen that Coriolis coupling effects begin to dominate the spectra as the rotational quantum numbers J and K increase. It is clear that rotational excitation can increase the coupling among the vibrational states and hence the rate of IVR.

Studies at energies approaching those of chemical relevance have focused on the behavior of the CH stretching motion, primarily because the combination of high energy and large anharmonicity possessed by these modes is the most favorable for direct excitation. Photoacoustic spectroscopy is used to detect the absorption spectra of these highly excited states. The width of each absorption feature can be interpreted as the time it takes for the excitation to transfer from the CH stretch to other vibrational motions. The broad line widths (of order 10^2 cm^{-1}) observed at high overtones (typically the vibrational energy is in the range of 10,000 to 20,000 cm^{-1}) indicate redistribution times of the order of 10^2 femtoseconds. Berry and his associates (22) have shown that the line widths of CH overtones in benzene first increase with increasing energy as expected on the basis of increasing state density, but then go through a maximum and begin decreasing. Thus it appears that the vibrational redistribution rates do not simply increase monotonically with energy. These results have been interpreted by Sibert *et al.* (23) as arising from the presence of specific pathways for the vibrational energy flow from mode to mode. They suggest that the CH stretch and CCH in-plane wag motions are intimately connected through a strong Fermi resonance (anharmonic coupling) and that excitation of the stretch is quickly transferred to a combination of stretch and wag motions. This transfer requires a proximity of the energy of the vibrations. As the overtone excitation increases, the anharmonicity of the CH overtone alters the energy separation,

first causing an increase in the interaction as the levels approach in energy, and then a decrease as they pass each other.

This kind of mechanical resonance in IVR should be quite general in CH overtone spectra, and for this reason several groups of investigators are gathering precise data on the closely related CH stretch–CH bend interaction (24, 25). For the trihalomethanes (24), the presence of a single CH chromophore and a small number of vibrational modes considerably simplifies the spectra. For example, the strong interaction between the CH stretch and CH bend is clearly demonstrated by the presence of four strong absorption features rather than one for the fifth CH stretch overtone of $CHCl_2F$ (Fig. 2). An analysis of the spectra in terms of a cubic anharmonic coupling between the stretch and bend modes reproduces the observed band positions and intensities. It is clear from this analysis that the interaction between the CH stretch and CH bend is the dominant influence. Furthermore, the coupling constant obtained from the analysis can be broken down into two contributions, the intrinsic anharmonicity of the potential surface for the CH bond, the calculation of which requires extensive ab initio computations, and the more easily calculable kinetic anharmonicity induced by the pendulum-like motion of the CH bend. The kinetic anharmonicity is the largest contribution to the coupling (24), in agreement with the assumption of Sibert *et al.* (23) in their calculations on benzene overtone lineshapes.

It is interesting to compare the information that can be extracted from the overtone spectra of the halomethanes and of benzene. The benzene line widths

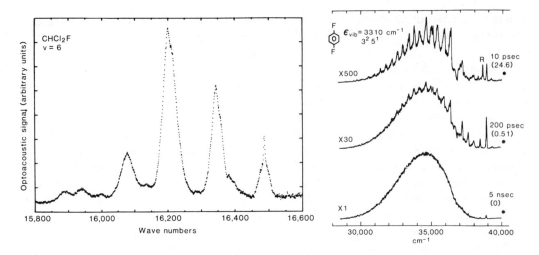

Fig. 2 (left). The $\nu_{CH} = 6 \leftarrow 0$ photoacoustic spectrum for CH stretching in $CHCl_2F$. One band would be observed in the absence of coupling; however, the $\nu_{CH} = 6$ level is strongly coupled to levels with $\nu_{CH} = 5$ and $\nu_{bend} = 2$.   Fig. 3 (right). Fluorescence emission after pumping the $3^2_0 5^1_0$ absorption band of p-difluorobenzene (vibrational energy, 3310 cm^{-1}) with three different pressures of O_2 present. Fluorescence lifetimes and O_2 pressures (kilotorrs, in parentheses) are shown on the right. The feature marked R is Raman scattering from O_2. The laser excitation energy is marked with an asterisk. At short times emission from $3^2 5^1$ is seen; however, at long times this emission is swamped by emission from other levels that are populated by redistribution from $3^2 5^1$.

give the IVR rate for CH stretch to CCH wag, since this effect dominates the line widths; however, information concerning the redistribution to other motions is hidden by this effect. For the halomethanes, the CH stretch–CH bend interaction is separate from other coupling pathways; it appears as resolved structure in the spectra. The IVR rate for CH stretch to CH bend can be calculated from these data. The line widths of the isolated features give the magnitude of the strongest coupling of the CH modes to other vibrational motions. It is clear that the motions of the halogens are much less strongly coupled, and the IVR lifetime is correspondingly much longer than for the CH stretch to CH bend energy transfer.

These overtone spectra, and their in-

terpretation in terms of specific pathways for redistribution, highlight the existence of wide variation in the coupling strengths of different types of nuclear motion. This finding is encouraging for those interested in trying to implement reaction along a particular coordinate.

All of the IVR experiments discussed so far are time-independent. However, the most spectacular view of IVR comes from direct time domain experiments in which the evolution of vibrational motion from the initially prepared state to the surrounding states is monitored. Although experiments on a picosecond time scale are difficult to perform and lie at the forefront of technology, such measurements are now being performed (26, 27). The largest data set for IVR rates, however, resulted from an ingenious method devised by Parmenter (*20, 28*). This method, called "chemical timing," has been used to obtain spectra on a picosecond time scale without recourse to picosecond technology. A high pressure of an electronic state quencher is used to shorten the lifetime of the emitting state to as little as 10 picoseconds, so that only molecules that fluoresce promptly (before undergoing a collision with the quencher) are observed. The spectrum at very short times (that is, at high pressure of the quencher) is just that which is characteristic of the coordinates excited. At longer times (low quencher pressure) there is a relative decrease in this emission and a corresponding increase in a broad emission which results following the transfer of energy from the vibrational mode initially prepared. By varying the pressure of the quencher, one can select molecules with a particular lifetime to obtain a view of the evolution of the vibrational motion. An example of the behavior observed is shown in Fig. 3. Lifetimes for IVR (inverse of the IVR rates) extracted for the low vibrational levels studied (<3500 cm^{-1}) range from nanoseconds to tens of picoseconds. The exact mechanism by which the "chemical timing" process works is unknown, and the absolute accuracy of the extracted rates remains in question. However, it is clear from recent direct picosecond experiments by Hochstrasser and his co-workers (*26*) and by Felker and Zewail (*27*) that the rates are of the right order of magnitude.

Photofragmentation Dynamics

When a diatomic molecule is photoexcited to a repulsive electronic state (potential curve), it dissociates in one-half of a molecular vibration to atomic fragments in states corresponding to that curve and with a translational energy specified by energy conservation. Excitation to a bound excited electronic state results in fluorescence back to vibrational levels of the ground electronic state or in predissociation by crossing (nonradiative transition) to a repulsive electronic curve, or both. The potential surfaces of polyatomic molecules are much more complex; each added atom increases the dimension by three. This increase in complexity opens many new pathways for energy flow from the initial excitation to reaction products. Crossing from a completely bound potential surface to one with an energetically accessible exit valley may occur. A molecule with sufficient energy to dissociate on a given surface may execute many vibrations before passing through an exit valley of the surface to fragments. Molecular fragments may be vibrationally and rotationally excited as well as electronically excited. Thus there are many fragment-state quantum numbers to be determined. The fragment vibration, translation, and both magnitude and direction of angular momentum may be expected to depend on (i) the initial angular momentum and vibrational excitation, (ii) the nature of any transitions between potential surfaces, (iii) coupling among degrees of freedom of the excited molecule, (iv) the geometry of the "transition state," and (v) the shape of the potential surface in the exit valley. Recent results on H_2CO, CH_2CO, and NCNO illustrate how dynamic information may be derived from high-resolution photofragment spectroscopy (*29*).

The formaldehyde molecule (Fig. 4) provides an opportunity for complete spectroscopic resolution and theoretical

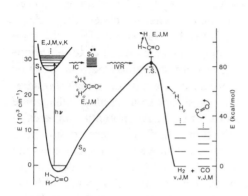

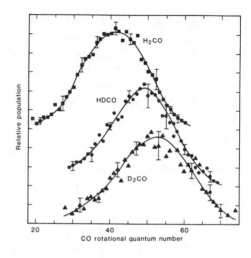

Fig. 4 (left). Energy states along the reaction coordinate for photofragmentation of H_2CO to H_2 and CO. Abbreviations: *IC*, internal conversion; *T.S.*, transition state. *E* refers to the energy; *v* to vibrational quantum numbers; and *J, M,* and *K* to rotational quantum numbers. Fig. 5 (right). A Boltzmann plot of the rotational distribution of the CO photofragment from formaldehyde photolysis near 29,500 cm^{-1}. Curves through the data are drawn by hand. H_2 pushes off from the electron distribution outside the carbon nucleus, imparting a large angular momentum to CO. [Reprinted with permission of D. J. Bamford, S. V. Filseth, M. F. Foltz, J. W. Hepburn, C. B. Moore, *Journal of Chemical Physics* **82**, 3032 (1985). © 1985 American Institute of Physics.]

study of polyatomic molecule dissociation (*30*). The near ultraviolet $\pi^* \leftarrow n$ excitation of the carbonyl of formaldehyde gives a bound singlet state (S_1). The spectrum is completely resolvable, and therefore a pulsed, narrowband laser produces S_1 with all vibrational and rotational quantum numbers defined. This excited state internally converts to a highly vibrationally excited ground singlet state (S_0**) with conservation of total energy and angular momentum, *E* and *J*. This transfer from electronic to vibrational energy corresponds to curve-crossing in diatomics. S_0** then vibrates until nearly all of the energy appears in the dissociation coordinate to bring the molecule to the transition state. The geometry and energy of this transition state

have been established to good accuracy by extensive ab initio calculations (*31*). We describe here experiments that used a total energy just a few kilocalories per mole above this barrier. Some 80 kcal/mol are then released as the H_2 and CO fragments push off from each other.

The vibration-rotation-translation states of the fragments are probed spectroscopically. Molecular-beam time-of-flight spectroscopy shows that most of the energy (65 percent on average) goes into translation (*32*). Laser-excited vacuum ultraviolet fluorescence detection of the rotation-vibration states of the nascent CO product shows that CO is highly rotationally excited (Fig. 5) (*33*). The most probable value of the rotational quantum number *J* for CO is 42 (49 and

53 for HDCO and D_2CO, respectively) independent of the initial rotational excitation of the H_2CO for the two levels studied (for which $J = 3$ and 16). In comparison, the most probable value of J for CO at room temperature is 7. Those few percent of CO molecules formed in vibrational level v = 1 exhibit the same rotational distribution as those in v = 0. H_2 is observed by coherent antistokes Raman spectroscopy (34). The distribution of vibrational energy is found to be v = 0 (23 percent), v = 1 (43 percent), v = 2 (25 percent), v = 3 (9 percent), and v = 4 (≤ 1 percent). The rotational distributions of H_2 peak at $J = 5$ and correspond roughly to temperatures that vary from 1800 K for v = 3 to 3000 K for v = 0. Nuclear spin is found to be conserved in the dissociation process (34); only odd J (that is, ortho) H_2 is produced from ortho-formaldehyde. The overall energy distribution is shown in Table 1.

The distribution of energy is dynamically controlled by the shape of the steep downhill exit valley of the potential surface; very different amounts of energy are deposited in each degree of freedom. The quantum-state distributions for H_2 vibration and CO rotation are far from thermal or statistical. The angular momentum of the CO is much greater than the initial total angular momentum. Conservation of angular momentum thus requires that J of CO be balanced by the

orbital angular momentum (L) of relative motion of H_2 and CO ($J_{CO} \approx L = \mu v b$, where μ is the reduced mass of H_2 with CO, v is the relative velocity, and b the impact parameter). The mean impact parameter corresponding to $J_{CO} = 42$ is 0.8 Å or about 0.2 Å on the outside of the C nucleus (33). H_2 appears to push off from the electron cloud outside the carbon atom. The arrows on the transition state structure indicated are the reaction coordinate vectors from ab initio theory (31).

The fragmentation of ketene to singlet methylene and CO (35)

$$CH_2CO\ (S_1{}^*) \rightarrow CH_2CO\ (S_0{}^{**}) \rightarrow\ {}^1CH_2 + CO$$

occurs after ultraviolet excitation and internal conversion to the ground-state potential surface. The exit valley has no barrier. With excitation of 308 nm (93 kcal/mol) all of the energy except 6.7 kcal/mol is required for breaking the C=C bond (36). The molecule vibrates until enough energy is concentrated in the dissociation coordinate so that fragments move apart. The translational energy distribution of both fragments peaks near zero (36). The rotational distribution of CO is roughly thermal (35): 1300 K for CO (v = 0) and 550 K for CO (v = 1) (Fig. 6). The rotational distributions are exactly statistical as calculated from the assumption that all energetically accessible product states have equal a priori probability. There is just enough energy for CO (v = 1) to be produced. Statistically only 1.3 percent is expected, but 29 percent population of v = 1 is observed. It appears that the soft degrees of freedom, rotation, and bends are strongly coupled and transfer energy rapidly on the time scale of the dissociation and hence are statistically controlled in the product. The CO bond stays intact

Table 1. Average energy in each degree of freedom for photofragmentation of H_2CO.

Frag- ment	Average energy (%)		
	Trans- lation	Rota- tion	Vibra- tion
CO	4	13	1
H_2	61	5	16

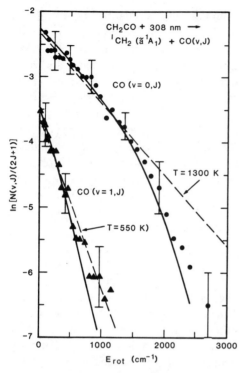

$CH_2CO + 308\ nm \longrightarrow$
$^1CH_2\ (\tilde{a}\ ^1A_1)\ +\ CO(v,J)$

$CO\ (v = 0, J)$

$T = 1300\ K$

$CO\ (v = 1, J)$

$T = 550\ K)$

$E_{rot}\ (cm^{-1})$

Fig. 6. Rotational distribution of CO fragments from ketene photolysis at 308 nm. There is no repulsion between the CH_2 and CO fragments. The solid curves are calculated without any adjustable parameters, on the assumption that all energetically accessible product states have equal probability.

through the dissociation; its behavior is different.

A study of vibration-rotation excitation of CN from

$$NCNO + h\nu \rightarrow CN + NO$$

has been carried out by Wittig and his co-workers (37) for photon energies from the dissociation threshold to 5000 cm^{-1} above threshold. Phase space theory, a statistical theory that is based on the assumption that energy is distributed with equal probability to all of the energetically accessible product states, is successful at matching the data for excitation energies up to 2000 cm^{-1} above

the dissociation threshold. However, as soon as it is energetically possible to produce vibrational excitation of the NO and CN fragments, the phase space theory fails to reproduce the experimental CN vibration-rotation distributions. The vibrational excitation of CN is matched by a statistical calculation that distributes the excess energy among the six vibrations of a very loose NCNO transition state. The experiments give about 50 percent more molecules in v = 1 than is predicted by phase space theory, which includes all nine rotation-vibration degrees of freedom. The analogous calculation for ketone reproduces the dramatic enhancement of CO (v = 1) product (9 percent observed) above the phase space theory value (1.3 percent).

As the relations among potential surface shape, intramolecular dynamics, and photofragment energy distributions become more completely understood, it will become possible to deduce reaction mechanisms from product energy distributions. For example, it should be immediately clear from the v, J states of product CO whether a CO is eliminated in a concerted fashion from a cyclic ketone or in a stepwise process in which the CO equilibrates with the hydrocarbon fragment of a diradical before fragmenting.

So far, we have discussed experiments in which the photodissociation dynamics are studied by the observation of final product vibrational, rotational, and translational excitation. The final product energy distribution is used to infer the nuclear motion on the dissociative potential surface. Imre et al. (38) observed molecules spectroscopically during the half-vibrational period that it takes a molecule to dissociate. They excite a molecule to an unbound elec-

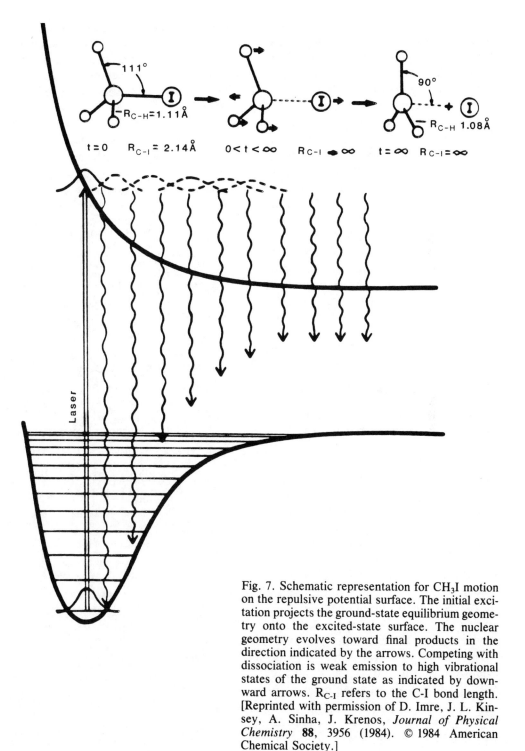

Fig. 7. Schematic representation for CH_3I motion on the repulsive potential surface. The initial excitation projects the ground-state equilibrium geometry onto the excited-state surface. The nuclear geometry evolves toward final products in the direction indicated by the arrows. Competing with dissociation is weak emission to high vibrational states of the ground state as indicated by downward arrows. R_{C-I} refers to the C-I bond length. [Reprinted with permission of D. Imre, J. L. Kinsey, A. Sinha, J. Krenos, *Journal of Physical Chemistry* **88**, 3956 (1984). © 1984 American Chemical Society.]

tronic state. The multidimensional potential energy surface for such an electronic state will have bound motion for all vibrations except the reaction coordinates. Although the molecule is dissociating in half of a vibrational period (10^{-14} second) there is a very small but observable probability of emission. Since electronic motion is much faster than nuclear motion, the light emission has the greatest probability for identical nuclear configuration of the ground and excited states. Thus emission is observed to vibrational levels of the ground electronic state that distort the molecule along the reaction coordinate. Therefore the emission spectrum shows the excitation of those ground-state vibrational modes that reflect the change in nuclear geometry as the excited state evolves into products. The intensity of the fundamental and overtone transitions has been shown by Heller (39) to depend on the exact shape of the potential.

The utility and elegance of this technique have been demonstrated with spectra from CH_3I and O_3 molecules excited to a dissociative continuum (38). The spectrum of CH_3I shows a long progression in the CI stretch (29 quanta) and in the CH_3 umbrella bending mode, reflecting the breaking of the CI bond and transformation of CH_3 from pyramidal to a planar fragment (Fig. 7). Along with the information on the excited-state potential, this technique provides spectra of very highly vibrationally excited levels of the ground electronic state that should be of great value in studying IVR.

The coming years will surely provide completely resolved dynamic studies of a significant number of small molecules. The extension of laser and synchrotron radiation technology in the vacuum ultraviolet will open Rydberg states for study. Many of the principles to be learned from these detailed studies of small gas-phase molecules will be applicable to larger molecules, to gas-surface interactions, and to condensed-phase systems.

References and Notes

1. S. R. Leone, *Science* **227**, 889 (1985).
2. T. A. Miller, *ibid.* **223**, 545 (1984).
3. K. B. Eisenthal, R. A. Moss, N. J. Turro, *ibid.* **225**, 1439 (1984).
4. C. S. Gudeman and R. J. Saykally, *Annu. Rev. Phys. Chem.* **35**, 387 (1984).
5. C. V. Shank, R. L. Fork, R. Yen, R. H. Stolen, W. J. Tomlinson, *Appl. Phys. Lett.* **40**, 761 (1982); A. M. Weiner, J. G. Fujimoto, E. P. Ippen, in *Ultrafast Phenomena IV*, D. H. Auston and K. B. Eisenthal, Eds. (Springer-Verlag, New York, 1984), p. 11.
6. R. N. Zare, *Science* **226**, 298 (1984).
7. W. A. Goddard III, *ibid.* **227**, 916 (1985).
8. G. Herzberg and J. W. C. Johns, *Proc. R. Soc. London Ser. A* **295**, 106 (1966); G. Herzberg, *ibid.* **262**, 291 (1961).
9. A. R. W. McKellar, P. R. Bunker, T. J. Sears, K. M. Evenson, R. J. Saykally, S. R. Langhoff, *J. Chem. Phys.* **79**, 5251 (1983).
10. H. Petek, D. J. Nesbitt, D. C. Darwin, C. B. Moore, in preparation.
11. H. Petek, D. J. Nesbitt, P. R. Ogilby, C. B. Moore, *J. Phys. Chem.* **83**, 5367 (1983).
12. H. Petek, D. J. Nesbitt, C. B. Moore, in preparation.
13. H. Petek, Y. Matsumoto, D. C. Darwin, D. J. Nesbitt, C. B. Moore, in preparation.
14. A. O. Langford, H. Petek, C. B. Moore, *J. Chem. Phys.* **78**, 6650 (1983).
15. W. M. Gelbart and K. F. Freed, *Chem. Phys. Lett.* **18**, 470 (1973).
16. S. Dobe, T. Bohland, F. Temps, H. Gg. Wagner, *Ber. Bunsen Gesellschaft Phys. Chem.* **89**, 432 (1985).
17. F. F. Crim, *Annu. Rev. Phys. Chem.* **35**, 657 (1984).
18. R. E. Smalley, *J. Phys. Chem.* **86**, 3504 (1982).
19. G. M. Stewart and J. D. McDonald, *J. Chem. Phys.* **78**, 3907 (1983).
20. C. S. Parmenter, *J. Phys. Chem.* **86**, 1735 (1982).
21. H. L. Dai, C. L. Korpa, J. L. Kinsey, R. W. Field, *J. Chem. Phys.* **82**, 1688 (1985); D. E. Reisner, P. H. Vaccaro, C. Kittrell, R. W. Field, J. L. Kinsey, H. L. Dai, *ibid.* **77**, 573 (1982); E. Abramson, R. W. Field, D. Imre, K. K. Innes, J. L. Kinsey, *ibid.* **80**, 2298 (1984).
22. K. V. Reddy, D. F. Heller, M. J. Berry, *ibid.* **76**, 2814 (1962).
23. E. L. Sibert, W. P. Reinhardt, J. T. Hynes, *ibid.* **81**, 1115 (1984); *Chem. Phys. Lett.* **92**, 455 (1982).
24. W. H. Green, J. S. Wong, W. D. Lawrance, C. B. Moore, in preparation; J. S. Wong and C. B.

Moore, in *Lasers and Applications*, W. O. N. Guimares, C.-T. Lin, A. Mooradian, Eds. (Proceedings of the Sergio Memorial Symposium, Rio de Janeiro, Brazil, 1980) (Springer-Verlag, New York, 1981), p. 157.
25. S. Peyerimhoff, M. Lewerenz, M. Quack, *Chem. Phys. Lett.* **109**, 563 (1984); H. R. Dubel and M. Quack, *J. Chem. Phys.* **81**, 3779 (1984).
26. R. Moore, F. E. Doany, E. J. Heilweil, R. M. Hochstrasser, *J. Phys. Chem.* **84**, 876 (1984).
27. P. M. Felker and A. H. Zewail, *Chem. Phys. Lett.* **108**, 303 (1984).
28. K. W. Holtzclaw and C. S. Parmenter, *J. Phys. Chem.* **88**, 3182 (1984); R. A. Coveleskie, D. A. Dolson, C. S. Parmenter, *ibid.*, in press.
29. For more comprehensive discussion and references, see S. R. Leone [*Adv. Chem. Phys.* **50**, 255 (1982)] and J. P. Simons [*J. Phys. Chem.* **88**, 1287 (1984)].
30. C. B. Moore and J. C. Weisshaar, *Annu. Rev. Phys. Chem.* **34**, 525 (1983).
31. J. D. Goddard, Y. Yamaguchi, H. F. Schaefer III, *J. Chem. Phys.* **75**, 3459 (1981).
32. P. Ho, D. J. Bamford, R. J. Buss, Y. T. Lee, C. B. Moore, *ibid.* **76**, 3630 (1982).
33. D. J. Bamford, S. V. Filseth, M. F. Foltz, J. W. Hepburn, C. B. Moore, *ibid.*, **82**, 3032 (1985).
34. D. Debarre, M. Lefebvre, M. Pealat, J.-P.

Taran, D. J. Bamford, C. B. Moore, *ibid.* **83**, 4476 (1985); M. Pealat, D. Debarre, J.-M. Marie, J.-P. Taran, A. Tramer, C. B. Moore, *Chem. Phys. Lett.* **98**, 299 (1983); B. Schramm, D. J. Bamford, C. B. Moore, *ibid.*, p. 305.
35. D. J. Nesbitt, H. Petek, M. F. Foltz, S. V. Filseth, D. J. Bamford, C. B. Moore, *J. Chem. Phys.* **83**, 223 (1985).
36. C. C. Hayden, D. M. Neumark, K. Shobatake, R. K. Sparks, Y. T. Lee, *J. Chem. Phys.* **76**, 3607 (1982).
37. I. Nadler, M. Noble, H. Reisler, C. Wittig, *ibid.*, **82**, 2608 (1985).
38. D. Imre, J. L. Kinsey, A. Sinha, J. Krenos, *J. Phys. Chem.* **88**, 3956 (1984).
39. E. J. Heller, *Acc. Chem. Res.* **14**, 368 (1981).
40. We thank C. S. Parmenter for Fig. 3 and J. L. Kinsey for Fig. 7.
41. We are grateful to the U.S. Army Research Office, Research Triangle Park, NC, the Director, Office of Energy Research, Office of Basic Energy Sciences, Chemical Sciences Division of the U.S. Department of Energy under contract No. DE-AC03-76SF00098, the National Science Foundation and the San Francisco Laser Center for support of the works from our own laboratory which have been received here.

In nature and in the laboratory, tetravalent carbon is the normal structural form that underlies the structure of organic molecules. The dominance of tetravalent carbon as a low-energy state of the carbon atom can be traced to the octet rule, which states that first row elements prefer to assume a rare-gas configuration of valence electrons. Since a carbon atom possesses four valence electrons, it can readily obey the octet rule by forming four covalent bonds with atoms capable of donating a total of four electrons.

This article is concerned with abnormal, high-energy, reactive forms of carbon. In contrast to conventional, tetravalent-carbon compounds, which can be put into bottles and whose properties and reactions can be measured on the time scale of minutes or days, we are concerned here with species of carbon that exist only as reactive intermediates and whose properties and reactions must be measured on the time scale of 10^{-6} to 10^{-12} seconds or less. These reactive transients are termed divalent carbon intermediates or, more commonly, carbenes (1–4).

Carbenes possess only six valence electrons and fully lack two valences. The high energy and reactivity implied by the carbene structure has challenged chemists over the past three decades. This article describes advances in carbene chemistry which have resulted mainly from the use of time-resolved laser spectroscopy. This technique has allowed the direct experimental observation of many divalent carbon species in fluid solution at or near room temperature (5–7).

3

Divalent Carbon Intermediates: Laser Photolysis and Spectroscopy

Kenneth B. Eisenthal,
Robert A. Moss,
and Nicholas J. Turro

Science 225, 1439–1445 (28 September 1984)

Reactive Intermediates in Chemical Reactions

To fully understand and appreciate the nature of chemical change, the chemist must know more than the chemical species that exist at the beginning of a reaction and at the end of a reaction, because few chemical reactions proceed by a direct conversion of reactants to products.

If a reaction proceeds in more than one step, intermediates must be postulated in any reaction mechanism. Indeed, it is a standard tactic to postulate reactive intermediates whose existences are fleeting (8, 9).

A crucial task in the investigation of chemical reactions involves the identification and characterization of those intermediates. The scientific fascination and importance of observing species with lifetimes that are vanishingly small relative to conventional molecules has led chemists to a variety of chemical and spectroscopic techniques.

Methylene, the Simplest Carbene

In this article we focus attention on divalent carbon species or carbenes. In order to familiarize the reader with carbene chemistry, we consider some significant structural properties of methylene (CH_2), the parent molecule of the carbenes.

Methylene is a simple three atom system. It can be described in terms of a limiting linear structure possessing a center of symmetry or in terms of a bent structure (Fig. 1) (10–12). The contribution of the central carbon atom to the electronic character of the linear structure may be described in terms of four valence atomic orbitals: a $2s$ orbital and

three $2p$ orbitals. The combination of two orbitals (the s and one p) produces two σ orbitals that are responsible for binding the carbon atom to the two hydrogen atoms. The remaining two p orbitals are labeled as p_x and p_y.

Methylene possesses six valence electrons. The way these six electrons distribute themselves depends on the geometry of the methylene molecule. For linear methylene refer to the molecular orbital scheme in Fig. 2A. The bonding σ_{CH} orbitals are of considerably lower energy than the nonbonding p_x or p_y orbitals. In accordance with both the Aufbau and Pauli Principles, four valence electrons, two from carbon and one from each hydrogen, will be placed pairwise in the two σ_{CH} orbitals. The remaining two electrons will go into the remaining two p orbitals, p_x and p_y. Since the orbital energies of p_x and p_y for the linear methylene are equal by symmetry, one would predict that one electron would occupy p_x and the other p_y, with their spins parallel. This arrangement with parallel spins yields a triplet state; the energy advantage of this arrangement arises from the larger average separation of electrons in the triplet state relative to that of spin-paired electrons in what is called the singlet state (Fig. 2B) (13, 14).

Now let us consider the consequences of bending a linear methylene in the y,z-plane (Fig. 2, C and D). The atomic orbital, p_y, will pick up s character and be lowered in energy. We classify this orbital as a σ orbital in a bent methylene. The atomic p_x orbital will not be significantly perturbed as a result of bending. We classify this orbital as a π orbital in a bent methylene.

How will the electrons distribute themselves in this molecular orbital scheme? Again, four electrons can be

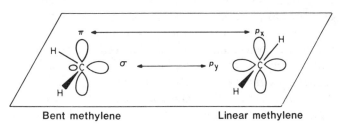

Fig. 1. Orbital representations for bent and linear methylenes. For bent methylene, the hydrogen atoms and the σ orbital are in the indicated plane; the π orbital is above and below it. For linear methylene, the hydrogen atoms and the p_y orbital are in the plane; the p_x orbital is above and below it.

Bent methylene   Linear methylene

immediately placed pairwise in the two low-energy bonding σ_{CH} orbitals. The distribution of the two remaining electrons is determined by achieving the lowest sum of the orbital energies and electron-electron repulsion. To achieve the lowest orbital energy both electrons would occupy the lowest orbital, $(\sigma)^2$, whereas to achieve the smallest electron-electron repulsion the two electrons would occupy different orbitals, $(\sigma)^1(\pi)^1$. However, if placing an electron in the higher energy π orbital gives a lower sum of orbital energies and electron-electron repulsion than when both electrons occupy the σ orbital, the low-energy configuration will be $(\sigma)^1(\pi)^1$, a triplet state (Fig. 2C). Within the $(\sigma)^1(\pi)^1$ configuration, a singlet state is possible (Fig. 2E), but it has a much higher energy than any of the electron states considered here. Additionally, if the energy needed to occupy σ and π is larger than when both electrons occupy the σ orbital, then the electronic configuration will be $(\sigma)^2$, which is a singlet state (Fig. 2D). For methylene, as for many carbenes, the ground state has been found experimentally to be the triplet state, $(\sigma)^1(\pi)^1$, with a low-lying, neighboring singlet state σ^2. The geometries of the triplet and singlet are predicted to be different, with the singlet being more bent. The triplet and singlet geometries and the energy gap between the triplet and singlet will vary

as the groups attached to the central carbon are changed. Despite this variation in bond angles and energies among different carbenes, useful generalizations about this class of chemical intermediates can be made. Now we can explore the key issues that relate the carbene structure and its reaction characteristics to the energies of the triplet and singlet states, to the dynamics of the triplet-singlet interconversion, and to their distinctive chemistries.

Chemistry of Carbenes

Because methylene possesses only six valence electrons, it is expected to seek two electrons for the central carbon atom and thereby achieve the stable octet structure. Reactions of methylene are known in which the CH_2 fragment is inserted into C–H bonds and added across C=C bonds.

Some chemistry of methylene, the prototype of substituted carbenes, is summarized in Fig. 3 in terms of a molecular orbital representation. The forms of methylene in which the two nonbonding electrons occupy different orbitals (symbolized by $\cdot\dot{C}H_2$) are expected to react with C–H bonds as shown in Fig. 3A. The radical character of this state is manifest in the radical-like abstraction of a hydrogen atom to lead to a radical pair,

RP. As a result of the spin selection rules of chemical reactivity, the spin state of the methylene will determine the spin of the radical pair, and thus the reaction path in Fig. 3A (15). For example, the singlet-state radical fragments, ^{1}RP, produced by reaction of singlet-state ·$\dot{C}H_2$ with CH_3OH combine in the solvent cage to form a coupling product from the radical pair. On the other hand, triplet-state radical fragments, ^{3}RP, cannot combine because of spin restrictions; they escape from the solvent cage in which they were born and eventually combine with uncorrelated spin systems to form coupling products such as those shown in Fig. 3A.

The chemistry of methylene in which both nonbonding electrons occupy the same orbital (symbolized by :CH_2) is different from that of the states in which each of the nonbonding electrons occupies a different orbital. The two electrons from the oxygen atom of CH_3OH can overlap nicely with the empty π orbital of :CH_2 to form a C–O bond, the key step in formation of the eventual product, an ether (Fig. 3B). The singlet and triplet states of ·$\dot{C}H_2$ do not possess a low-lying empty orbital, and therefore, do not react with CH_3OH in this fashion.

One of the more fascinating reactions of carbenes is their addition to C=C bonds, that is, a [1 + 2] cycloaddition to form cyclopropanes. Both the singlet and triplet states of ·$\dot{C}H_2$ can be envisioned to yield diradical intermediates. Depending on spin state, the diradical will either close rapidly and with retention of configuration (the path for singlet ·$\dot{C}H_2$) or persist for many bond rotations and eventually close with some loss of configuration (the path for triplet ·$\dot{C}H_2$) (Fig. 3C). The reaction of :CH_2 with a C=C bond is expected to produce an intermediate which has both nonbonding electrons in the same orbital, and which, because further reaction is spin-allowed, should cyclize rapidly to yield a cyclopropane with retention of a configuration (Fig. 3D). Although the reactions of methylene are discussed as proceeding via intermediates, for the singlet states the reaction could be concerted and proceed directly to products without the occurrence of an intermediate (16).

Experimental determination of the low-lying states of divalent carbon. One of the very first questions to be addressed in an investigation of carbenes concerns the nature of the low-lying state of the divalent carbon system under investigation. Prior to 1980 electron spin resonance spectroscopy (17–19), product analyses (20–22), and optical spectroscopy (7, 23–25) were the main experimental methods employed to probe carbene structures and energetics. These methods revealed that methylene possesses a triplet ground state, $(\sigma)^1(\pi)^1$, with a singlet state, $(\sigma)^2$, about 35 kJ/mol higher in energy (26–28). On the other hand, substantial evidence suggests that, although most derivatives of methylene possess triplet ground states, the singlet-triplet energy gap is on the order of only 5 to 20 kJ/mol (1, 29). Furthermore, some derivatives of methylene appear to possess singlet ground states (30).

Measurement of the dynamics of singlet-triplet relaxation and the singlet-triplet energy gap in diphenylcarbene. We were encouraged to study diphenylcarbene by the pioneering theoretical and spectroscopic studies of the dynamics, structure, and chemistry of this aromatic carbene with a triplet ground state (3, 17–19, 23, 24, 31). The spectroscopic information on emission and absorption makes it possible to probe the chemical and physical aspects of carbene formation, intramolecular energy relaxation,

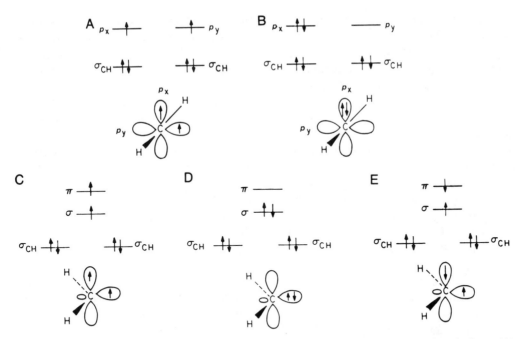

Fig. 2. Energy level and orbital representations for bent and linear geometries of methylene. (A) Linear methylene, the triplet state. (B) Linear methylene, the singlet state. (C) Bent methylene, the triplet state, $(\sigma)^1(\pi)^1$. (D) Bent methylene, the singlet state, $(\sigma)^2$. (E) Bent methylene, the high-energy singlet state, $(\sigma)^1(\pi)^1$.

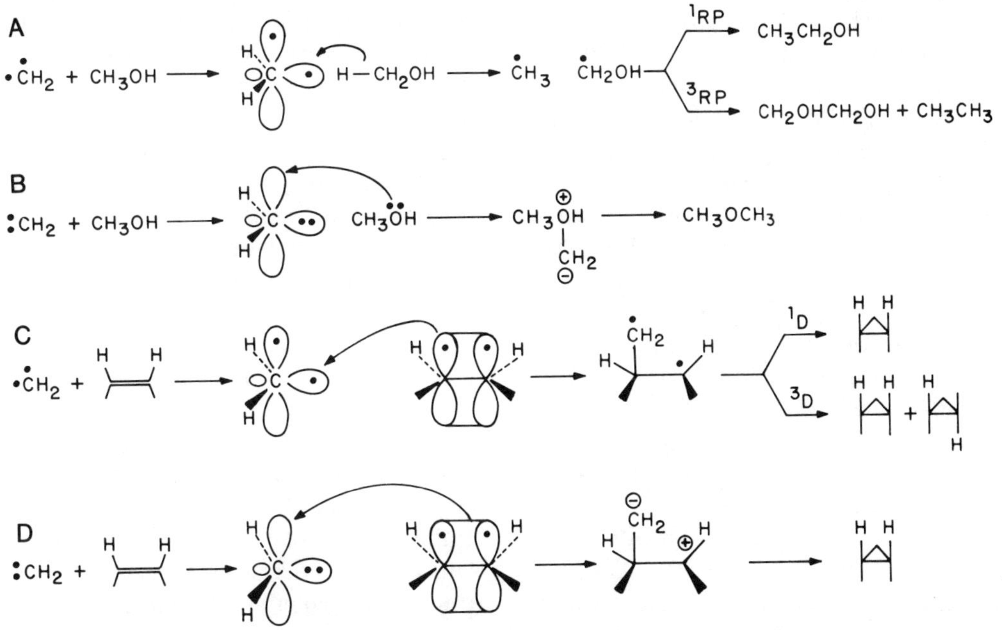

Fig. 3. Some prototypical reactions of divalent carbon with methylene as an example. RP and D represent radical pairs and diradicals, respectively. A superscript 1 stands for a singlet state; a superscript 3, a triplet state.

chemical reactions, and excited electronic-state properties.

The way in which the carbene is generated can be important in unraveling its chemistry. If the carbene is produced in the excited, singlet state, it undergoes singlet reactions that compete with energy relaxation to the triplet ground state where different reactions occur. Similarly, producing the carbene in its triplet state can yield triplet reactions, or the carbene can undergo a thermal excitation to the singlet state, $(\sigma)^2$, where other reactions can occur. The observed chemistry of a carbene thus depends on the difference in energy and interactions between these two states of different spin multiplicity and on their rates of interconversion.

To elucidate the rapid dynamics of carbene formation, energy relaxation within the carbene itself, and reaction of the carbene with other molecules, a picosecond, ultraviolet laser pulse is used to generate the diphenylcarbene (DPC) from diphenyldiazomethane (DPDM) (Eq. 1) (32, 33). Photoexcitation of DPDM to an excited singlet state breaks the $C=N_2$ bond in the diazo compound releasing DPC in a singlet state. The sequence of energy decay steps following the laser photoexcitation of DPDM ulti-

$$\begin{matrix} Ph \\ \\ Ph \end{matrix} C=N_2 + h\nu \longrightarrow \left(\begin{matrix} ^{1*}Ph \\ \\ Ph \end{matrix} C=N_2 \right) \xrightarrow{-N_2}$$

$$\qquad DPDM \qquad\qquad\qquad ^{1*}DPDM$$

$$\left(\begin{matrix} ^1 Ph \\ \\ Ph \end{matrix} C\colon \right) \underset{k_{TS}}{\overset{k_{ST}}{\rightleftarrows}} \left(\begin{matrix} ^3 Ph \\ \\ Ph \end{matrix} \ddot{C}\cdot \right) \qquad (1)$$

$$\qquad\quad ^1DPC \qquad\qquad ^3DPC$$

mately leading to DPC in its triplet ground state is given in Eq. 1, where Ph represents the phenyl group and the su-

perscript * represents excited states. Earlier work showed that ^{3}DPC is generated in a subnanosecond time scale from the excited diazo precursor (25), and the calculated value of k_{ST} (the rate constant for the singlet-triplet transition) was based on the assumed diffusional quenching of the singlet state, ^{1}DPC, by methanol. It was for these reasons picosecond laser methods were necessary to follow the rapid steps of carbene generation and energy relaxation.

Our laser experiments revealed that the bond breaking step in the photoexcited diazo molecule occurred (32) in less than 5 psec. The k_{ST} value was measured using a laser-induced fluorescence technique (Fig. 4) (32). The experiment employs two laser pulses. One pulse is used for the excitation of the diazo compound; a second pulse is delayed with respect to the excitation pulse and is used for determining the formation rate of ^{3}DPC. In acetonitrile, CH_3CN, at room temperature the time it takes for ^{1}DPC to convert into ^{3}DPC, which is the inverse of the rate constant, k_{ST}, is 310 psec.

By measuring the rate constant, k_{TS}, for the reverse process, namely, triplet to singlet transition, in the same solvent we obtain the equilibrium constant of the singlet-triplet interconversion (Eq. 2) (32, 33).

$$K = \frac{k_{ST}}{k_{TS}} \qquad (2)$$

This equilibrium constant, for which an upper limit had been obtained in earlier studies (25), can then be used to calculate the free energy difference between singlet and triplet states (Eq. 3).

$$\Delta G = -RT \ln K \qquad (3)$$

However, the measurement of k_{TS} (^{3}DPC → ^{1}DPC) is not straightforward (25, 33). Clean and rapid production of ^{3}DPC is difficult. As a result, k_{TS} is determined by a procedure that is indirect, because of the required coordination of conventional product analysis, triplet photosensitization, photochemical kinetics, and time resolved methods.

On combining the measured k_{ST} and k_{TS} values for DPC in acetonitrile at room temperature, we obtain an equilibrium constant $K = 290$ and the free energy difference given in Eq. 4.

$$\Delta G = G_T - G_S = -RT \ln K$$
$$= -14 \text{ kJ/mol} \qquad (4)$$

An estimate of the energy difference between the singlet and triplet states can now be made from the thermodynamic relation $\Delta G = \Delta H - T\Delta S$. Since the entropy difference, ΔS, is expected to be primarily due to the multiplicity difference between the singlet and triplet states ($\Delta S = R \ln 3$) and since the enthalpy difference, ΔH, and the energy gap, ΔE, are approximately equal in liquids, we obtain Eq. 5 (25).

$$\Delta G = \Delta H - T\Delta S = \Delta E - RT \ln 3 \quad (5)$$

Therefore, for DPC in acetonitrile at room temperature we obtain the singlet-triplet energy gap in Eq. 6.

$$\Delta E = E_T - E_S$$
$$= -11 \text{ kJ/mol} = -950 \text{ cm}^{-1} \qquad (6)$$

Solvent Effects on Singlet-Triplet

Relaxation

After obtaining these kinetic and thermodynamic results for DPC in acetonitrile, the question of how applicable these findings are to other solvent environments arose. That is, to what extent does the solvent affect the dynamics of the intramolecular spin conversion in carbenes? Prior to this work, the effect of the solvent on intersystem crossing in carbenes was not considered.

We found that the singlet to triplet conversion depends strongly on the choice of solvent (Fig. 4) (34). In isooctane, a hydrocarbon solvent, the time of intersystem crossing, k_{ST}^{-1}, is 95 psec, three times faster than in acetonitrile. There is a clear relation between the speed of singlet to triplet conversion and the polarity of the solvent, the conversion being faster in the less polar solvents. No correlation was seen with other solvent parameters that would be related to possible carbene-solvent complexes or parameters, such as viscosity, that could influence the dynamics of the structural change associated with the singlet to triplet conversion.

To understand how solvent polarity could affect spin conversion, consider the electronic nature of the singlet and triplet states. The singlet state, $(\sigma)^2$, with both unshared electrons in the same orbital, localized on the central carbon, is

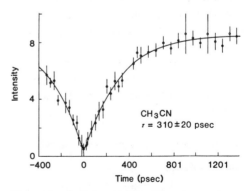

Fig. 4. Laser-induced fluorescence as a function of time delay between excitation and probe pulses yielding ^{1}DPC → ^{3}DPC for diphenylcarbene in acetonitrile.

more polar than the triplet state, $(\sigma)^1(\pi)^1$ (Fig. 2C). Therefore, the polar singlet will be more stabilized in the polar solvents than the triplet. This differential stabilization results in a decrease of the singlet-triplet energy gap as the solvent polarity increases. That the energy gap is smaller in a polar solvent than a nonpolar solvent was confirmed experimentally. We found the energy gap in nonpolar isooctane to be 17 kJ/mol compared to 11 kJ/mol in polar acetonitrile.

Why then does the rate of intersystem crossing increase as the energy gap increases, a seeming contradiction? Upon crossing from the singlet to the triplet state, the excess energy is taken up by the vibrational motions of the triplet. This conversion of electronic energy in the singlet state to vibrational energy in the triplet state is necessary to satisfy energy conservation: the sum of the electronic and vibrational energies in the triplet must equal the electronic energy of the singlet. This is the key. In aromatic carbenes, as the relatively small energy gap increases in magnitude with decreasing solvent polarity, there are more triplet vibrational states approximately equal to the singlet electronic energy, so more triplet vibrations can accept the excess energy from the singlet. Therefore, the carbene chemistry, which is dependent on these relaxation processes, can, in principle, lead to different product distributions in different solvent environments.

Photochemistry of Transient Chemical Intermediates

Although the chemistry of the low-energy electronic states of carbenes—the ground triplet and neighboring singlet—has been extensively studied, the chemistry and general properties of the higher energy electronic states remain unknown. Studies of the photochemistry of carbenes, or more generally the photochemistry of transient chemical intermediates, are just beginning. The transitory nature of reactive ground-state intermediates makes it difficult to generate their excited states in sufficient concentration to observe their chemical properties. Reactions of the thermally accessible states can destroy the chemical intermediate before it is elevated to excited electronic states with light. What is needed is the means to generate the excited intermediate in a time shorter than the reaction time of lower states. One can, of course, extend the reaction time of the lower states by keeping them away from other reactive molecules in the environment. However, if there are few reactive molecules around, then it is not possible to study the reactions of excited state intermediates before they decay back to the ground state, either by the emission of light or by nonradiative processes.

Our approach to this situation is to use picosecond laser methods first to generate the carbene and then to excite the carbene to higher electronic states (35, 36). For the case of diphenylcarbene, a picosecond, ultraviolet laser pulse was used to split the diphenyldiazomethane precursor, as discussed earlier. A second laser pulse arrives shortly after the reaction pulse and excites the triplet ground-state carbene to an excited triplet-state carbene. With these methods we were able to explore the interactions of alcohols with the excited triplet diphenylcarbene, ^{3*}DPC; and the result is a simple model to account for the different reactivity patterns of the singlet, triplet, and excited triplet states of diphenylcarbene.

The model is an orbital filling scheme with nonbonding molecular orbitals, σ and π, and an antibonding orbital, π^*, of the carbene.

The key point in this model is the presence of an empty, low-lying nonbonding orbital for both ^1DPC and $^{3*}DPC$ (Fig. 5). ^3DPC possesses a half-filled, not an empty, π orbital, whereas $^{3*}DPC$ does possess an empty π orbital. Since the ^3DPC has both σ and π orbitals singly occupied, it is expected to behave like a free radical in its reactions. This expectation is borne out experimentally by the hydrogen abstraction reactions ^3DPC undergoes and its stepwise addition to olefins. On the other hand, because ^1DPC has a $(\sigma)^2(\pi)^0$ electron configuration, it can accept a pair of electrons into its empty π orbital. This means the singlet undergoes O–H insertion reactions with alcohols to form ethers (as in Fig. 2B) via the donation of a pair of electrons from the oxygen of the alcohol into the vacant π-orbital of the carbene. In a similar way, $^{3*}DPC$, with a $(\sigma)^1(\pi)^0(\pi^*)^1$ configuration, also has an empty π orbital and could, therefore, undergo reactions resembling ^1DPC. This model not only accounts for the observed difference between the reactions of the ground triplet and lowest singlet state of the carbene but also predicts that the excited triplet should undergo certain reactions that the ground triplet would not.

Reactions of Excited Triplet

Carbenes with Alcohols

By observing the behavior of $^{3*}DPC$ in the presence of a series of alcohols, we can determine whether the mode of reaction for the excited triplet follows the singlet mechanism (attack at the O–H bond), the triplet mechanism (attack at the C–H bond), or some other mechanism. The rate constants, $^{3*}k$, for the reactions of $^{3*}DPC$ with alcohols (Eq. 7)

$$^{3*}DPC + reactant \xrightarrow{^{3*}k} product \quad (7)$$

were obtained from an analysis of the lifetime of $^{3*}DPC$ in the presence of different concentrations of alcohols.

The rate constants for methanol, isopropanol, and t-butanol show that methanol is the most reactive, whereas t-butanol is the least. The observed order of alcohol reactivities closely parallels the relative lability and acidity of the alcohols. This observation supports the mechanism that $^{3*}DPC$ reacts with the O–H bond of the alcohol. If the C–H bond of the alcohol were attacked by $^{3*}DPC$, as in the ground triplet state reaction, then one would not expect to see the observed trend in the rates. The interpretation that $^{3*}DPC$ attacks the O–H bond, rather than the C–H bond, of the alcohols is consistent with its observed kinetic behavior with other electron donors, such as amines and olefins (36). An overall pattern that systematizes the ability of an arbitrary electron donor (an alcohol, amine, or olefin) to quench $^{3*}DPC$ involves two factors: the electron donor's ability to share electrons [a property related to the donor's ionization potential (IP)], and the strength (and reactivity) of the bonds in the donor's structure. When the donor possesses a relatively low ionization potential (IP < 9 eV), for example, amines, a charge transfer quenching mechanism will dominate (Eq. 8).

$$^{3*}DPC + amine \rightarrow (DPC^- \cdots amine^+)$$
$$\rightarrow products \quad (8)$$

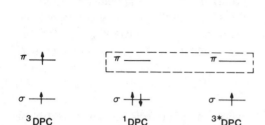

$\pi^* \underline{\qquad}$ $\pi^* \underline{\qquad}$ $\pi^* \underline{\text{—}\uparrow\text{—}}$

$\pi \underline{\text{—}\uparrow\text{—}}$ $[\pi \underline{\qquad}$ $\pi \underline{\qquad}]$

$\sigma \underline{\text{—}\uparrow\text{—}}$ $\sigma \underline{\uparrow\downarrow}$ $\sigma \underline{\text{—}\uparrow\text{—}}$

^{3}DPC ^{1}DPC ^{3*}DPC

Fig. 5. Orbital occupancy for ^{3}DPC (ground state), ^{1}DPC (low-energy singlet state), and ^{3*}DPC (excited state of ^{3}DPC). Note the empty, low-energy π orbital common to ^{1}DPC and ^{3*}DPC.

For donors possessing a relatively high ionization potential (IP > 9 eV), for example, alcohols, the charge transfer quenching mechanism becomes energetically disfavored and the breaking of bonds in the donor becomes the dominant quenching mechanism (Eq. 9).

$$^{3*}\text{DPC} + \text{alcohol} \rightarrow \text{ethers} \quad (9)$$

In comparing the reactivities of ^{3}DPC, ^{1}DPC, and ^{3*}DPC with bimolecular quenchers, the common characteristic of an empty low-energy orbital that can accept a pair of electrons produces a similarity in the reactivity of ^{1}DPC and ^{3*}DPC toward electron donors. Thus, ^{3*}DPC, although formally a triplet, acts more like ^{1}DPC than ^{3}DPC.

Arylhalocarbenes: Reaction Dynamics of Carbenes with Singlet Ground States

As we have said, diaryl carbenes typically possess triplet ground states with low-energy, thermally accessible singlet states. However, appropriate variations of carbene structure lead to such stabilization of the singlet state, $(\sigma)^2$, relative

to the triplet state, $(\sigma)^1(\pi)^1$, that the singlet state becomes the ground state. For example, replacing a phenyl substituent with a halogen atom achieves the switch.

$\underline{\qquad}$ $(\sigma)^2$ $\underline{\qquad}$ $(\sigma)^1 (\pi)^1$

$\underline{\qquad}$ $(\sigma)^1 (\pi)^1$ $\underline{\qquad}$ $(\sigma)^2$

The stabilization of the $(\sigma)^2$ state is so great for phenylhalocarbenes that only the $(\sigma)^2$ state is involved in the chemistry of such species. This simplifies the measurement and interpretation of the reaction dynamics for this class of carbenes, and it presents an opportunity to determine the nature of the relation between reactivity and structure for an exceedingly reactive species.

We have generated phenylchlorocarbene (PCC) by photolysis of phenylchlorodiazirine (PCD) (37–39). When the photolysis is conducted in the presence of olefins, clean production of cyclopropanes is observed (Eq. 10).

$$(10)$$

The absorption spectrum of PCC produced by laser flash photolysis of PCD at room temperature in fluid solution is

42

experimentally identical to the spectrum of PCC in a solid matrix at 77 K (conditions under which the carbene is indefinitely stable). The decay of the absorption of PCC can be monitored by time-resolved spectroscopy. Upon addition of tetramethylethylene (or any olefin), the reaction shown in Eq. 10 occurs, and the PCC absorption decreases more rapidly with time as a result of the reaction of the carbene with the olefin. Analysis of the decay of PCC absorption as a function of olefin concentration allows evaluation of the absolute rate constant (k_{abs}) for reaction of PCC with an olefin.

Relative reactivity experiments have played an important role in developing an understanding of organic reaction mechanisms. The data from kinetic competition experiments have established the electrophilic nature of the selectivity of many carbenes toward reagents such as olefins, and a quantitative selectivity index has been created to standardize relative reactivity parameters (37). Selectivity refers to the ratio of rate constants for two related reactions of a carbene toward two substrates or two different carbenes toward the same substrate. These selectivity parameters have been assumed to imply reactivity parameters: high selectivity implies low reactivity, and low selectivity implies high reactivity. The advent of time-resolved laser spectroscopy allowed a direct test of the implied selectivity-reactivity relationship, because direct measurement of the absolute rate constants for reactions of carbenes became possible by this technique.

The absolute rate constants for the reactions of PCC and some of its derivatives with tetramethylethylene (a relatively reactive olefin) and with 1-hexene (a relatively unreactive olefin) in isooc-

tane have been determined (40–42). For even the most reactive systems, the rate constant ($\sim 1.5 \times 10^9 M^{-1}$ sec^{-1}) is slightly below the diffusion control limit for isooctane ($\sim 1 \times 10^{10} M^{-1}$ sec^{-1}). However, even though carbenes are enormously reactive, their reactivity is a strong function of their structure and of the structure of the other reactant. For example, the rate constant for reaction of 1-hexene and a 4-methoxyl–substituted carbene is on $1.3 \times 10^5 M^{-1}$ sec^{-1}.

Thus, a highly reactive species can be, nonetheless, quite selective. Such a result suggests caution in the application of conventional selectivity-reactivity relationships to the chemistry of carbenes and other highly reactive intermediates.

The activation energy, E_a, for the rate limiting step in a reaction mechanism is conventionally obtained by measuring the absolute rate constant for the reaction (k_{abs}) as a function of temperature and then employing the Arrhenius expression (Eq. 11).

$$\log k_{abs} = \log A - E_a/2.3RT \qquad (11)$$

Chemical reactions are expected to proceed via mechanisms involving an energy barrier for the rate limiting step, and as a result, E_a is generally a positive quantity. This means that a graph of log k_{abs} versus T^{-1} will generally possess a negative slope. For the reaction given in Eq. 10, in toluene or acetonitrile, the slope of the plot is positive for temperatures of 225 K and above (20). A positive slope implies a negative activation energy. Investigation of other carbene reactions have revealed that they fall into three categories (21): (i) those that possess negative activation energies on the order of several kilojoules per mole; (ii) those that possess activation energies equal to zero within experimental error;

and (iii) those that possess positive activation energies on the order of several kilojoules per mole.

There are at least two general (somewhat related) interpretations of the observation of a negative activation energy. Both seek escape from the apparent dilemma that theory does not allow the occurrence of anything other than a zero or positive energy barrier in a single step.

The first solution challenges the assumption that the rate constant, k_{abs}, corresponds to a single reaction step. For example, if the reaction of carbenes and olefins proceeded in two steps, neither of which was uniquely rate determining, the dependence of k_{abs} on temperature could be complex. In the simplest case, a reversibly formed complex between carbene and olefin might form (Eq. 12). As the temperature varies,

$$\text{carbene} + \text{olefin} \rightleftarrows \text{complex}$$
$$\text{complex} \rightarrow \text{product} \qquad (12)$$

the complex can either dissociate to starting materials (slowing down the rate of reaction) or collapse to product (increasing the rate of reaction). Stating that a reaction has a negative activation energy is equivalent to stating that the rate of reaction increases as the temperature decreases. Such would be the case if the carbene complex were stabilized with respect to dissociation as the temperature decreased and, as a result of this stabilization, it proceeded more efficiently to product than it did at higher temperature. Such a condition could exist until the reaction became diffusion controlled; at that point the reaction of the complex would be 100 percent efficient and the reaction rate would be equal to and limited by the diffusion of reactants together. Indeed, below ~250

K in toluene the slope of log k_{abs} versus T^{-1} for the reaction of PCC and tetramethylethylene corresponds to a positive activation energy; that is, the reaction becomes diffusion controlled below 250 K, and the activation energy corresponds to the energy of diffusion.

A second interpretation of negative activation energies focuses on the balance between enthalpy and entropy as the system proceeds along the reaction coordinate, and does not require any complex other than the conventional encounter complex required for bimolecular reactions of freely diffusing species in fluid solutions (43).

Whatever interpretation is applied, it is clear that conventional enthalpy control is not dominant in many of the carbene reactions investigated. Indeed, in all of the carbene reactions studied, the entropies of activation exert a strong influence on k_{abs} (44).

The application of fast spectroscopic and kinetic techniques to carbene chemistry has stimulated important research on reactive intermediates. In spite of the high reactivity of carbenes, which results in a short lifetime, their dynamics and chemistry can be investigated directly via time-resolved laser spectroscopy. In addition to the conventional features of structure-reactivity and the applicability to structure-selectivity relationships, we have been able to use kinetic measurements to obtain information about the singlet-triplet-energy gap, the effects of solvent interactions on the dynamics of singlet-triplet interconversions, and the nature of excited-state carbene reactions from picosecond spectroscopy. This information should be useful to theoretical and experimental chemists as well as to biological chemists who are involved with methods which allow controlled labeling of biological molecules (45).

References and Notes

1. W. Kirmse, *Carbene Chemistry* (Academic Press, New York, ed. 2, 1971).
2. R. A. Moss and M. Jones, Jr., Eds., *Carbenes* (Wiley, New York, 1973), vol. 1.
3. _____, *ibid.*, vol. 2.
4. R. A. Moss, *Chem. Eng. News* 47 (No. 25), 60 (1969).
5. N. J. Turro, M. Aikawa, J. A. Butcher, Jr., *IEEE J. Quantum Electron.* QE-16, 1218 (1980); N. J. Turro, *Tetrahedron* 38, 809 (1972).
6. Y. Wang and K. B. Eisenthal, *J. Chem. Ed.* 59, 482 (1982); K. B. Eisenthal, in *Ultrashort Light Pulses*, S. Shapiro, Ed. (Springer-Verlag, Berlin, 1977), chap. 5.
7. G. L. Closs, in (*3*), chap. 4.
8. M. Jones, Jr., and R. A. Moss, Eds., *Reactive Intermediates* (Wiley, New York, 1978), vol. 1; (1981), vol. 2.
9. R. A. Abramovitch, Ed., *Reactive Intermediates* (Plenum, New York, 1980), vol. 1.
10. R. Hoffmann, G. D. Zeiss, G. W. Dine, *J. Am. Chem. Soc.* 90, 1485 (1968).
11. J. F. Harrison and L. C. Allen, *ibid.* 91, 807 (1969).
12. J. F. Harrison, *Acc. Chem. Res.* 7, 378 (1974).
13. For a qualitative discussion of the basis of singlet-triplet energy differences, see N. J. Turro, *Modern Molecular Photochemistry* (Benjamin/Cummings, Menlo Park, Calif., 1978), p. 28ff.
14. For a quantitative discussion of singlet-triplet energy differences, see S. P. McGlynn, T. Azumi, M. Kinoshita, *Molecular Spectroscopy of the Triplet State* (Prentice-Hall, Englewood Cliffs, N.J., 1969), p. 73.
15. For reviews of the spin states of carbenes, see P. P. Gasper and G. S. Hammond, in W. Kirmse, *Carbene Chemistry* (Academic Press, New York, 1964), chap. 12.
16. R. Hoffmann, *J. Am. Chem. Soc.* 90, 1475 (1968).
17. E. Wasserman and R. S. Hutton, *Acc. Chem. Res.* 10, 27 (1977).
18. G. L. Closs and L. E. Closs, *J. Am. Chem. Soc.* 91, 4549 (1969).
19. D. C. Doetschman and C. A. Hutchinson, Jr., *J. Chem. Phys.* 56, 3964 (1972).
20. P. S. Skell and R. C. Woodworth, *J. Am. Chem. Soc.* 78, 4496 (1956).
21. R. C. Woodworth and P. S. Skell, *ibid.* 81, 3383 (1959).
22. W. von E. Doering and A. K. Hoffman, *ibid.* 76, 6162 (1954).
23. A. M. Trozzolo, *Acc. Chem. Res.* 1, 329 (1968); _____ and W. A. Gibbons, *J. Am. Chem. Soc.* 89, 129 (1961).
24. G. Herzberg, *Proc. R. Soc. London, Ser. A* 262, 291 (1961).
25. G. L. Closs and B. E. Rabinow, *J. Am. Chem. Soc.* 98, 8190 (1976).
26. R. K. Langel and R. N. Zare, *ibid.* 100, 7495 (1978).
27. A. R. W. McKellar, P. R. Bunker, T. J. Sears, K. M. Evenson, R. J. Saykally, S. R. Langhoff, *J. Chem. Phys.* 79, 5251 (1983).
28. C. C. Hayden, D. M. Neumark, K. Shobatke, R. K. Sparks, Y. T. Lee, *ibid.* 76, 3607 (1982).
29. H. Dün and B. Ruge, *Top. Curr. Chem.* 66, 53 (1975).
30. W. V. E. Doering and W. A. Henderson, *J. Am. Chem. Soc.* 80, 5274 (1958).
31. D. Bethell, G. Stevens, P. Tickl, *J. Chem. Soc. Dalton Trans.* (1970), p. 792.
32. C. DuPuy, G. M. Korenowski, M. McCurliffe, W. M. Hetherington, III, K. B. Eisenthal, *Chem. Phys. Lett.* 77, 272 (1981).
33. K. B. Eisenthal, N. J. Turro, M. Aikawa, J. A. Butcher, Jr., C. Dupuy, G. Hefferon, W. Hetherington, G. M. Korenowski, M. J. McAuliffe, *J. Am. Chem. Soc.* 102, 6563 (1980).
34. E. V. Sitzmann, J. Langan, K. B. Eisenthal, *ibid.* 106, 1868 (1984); E. V. Sitzmann and K. B. Eisenthal, in *Applications of Picosecond Spectroscopy to Chemistry*, K. B. Eisenthal, Ed. (Reidel, Dordrecht, Netherlands, 1984), p. 41.
35. Y. Wang, E. V. Sitzmann, F. Novak, C. DuPuy, K. B. Eisenthal, *J. Am. Chem. Soc.* 104, 3238 (1982).
36. E. V. Sitzmann, Y. Wang, K. B. Eisenthal, *J. Phys. Chem.* 87, 2283 (1983); E. V. Sitzmann, J. Langan, K. B. Eisenthal, *Chem. Phys. Lett.* 102, 446 (1983).
37. R. A. Moss, *Acc. Chem. Res.* 13, 58 (1980).
38. N. J. Turro and G. C. Weed, *J. Am. Chem. Soc.* 105, 1861 (1983).
39. T. Koenig and H. Fischer, in *Free Radicals*, J. Kochi, Ed. (Wiley, New York, 1973), vol. 1, p. 157; J. Franck and E. Rabinowich, *Trans. Faraday Soc.* 30, 120 (1934); E. Rabinowich and W. Wood, *ibid.* 32, 1381 (1936); R. M. Noyes, *Prog. React. Kinet.* 1, 130 (1961); T. J. Chang, G. W. Hoffman, K. B. Eisenthal, *Chem. Phys. Lett.* 25, 201 (1974).
40. N. J. Turro *et al.*, *J. Am. Chem. Soc.* 102, 7576 (1980).
41. N. J. Turro, G. F. Lehr, J. A. Butcher, Jr., R. A. Moss, W. Guo, *ibid.* 104, 1754 (1982).
42. D. P. Cox, I. R. Gould, N. P. Hacker, R. A. Moss, N. J. Turro, *Tetrahedron Lett.* 24, 5313 (1983).
43. K. N. Houk, N. G. Rondan, J. Mareda, *Tetrahedron*, in press.
44. For other investigations of absolute rate constants for reactions of other carbenes, see P. B. Grasse, B. E. Brauer, J. J. Zupancic, K. J. Kaufmann, and G. B. Schuster [*J. Am. Chem. Soc.* 105, 6833 (1983)] and D. Griller, N. T. H. Liu, and J. C. Scaino [*ibid.* 104, 5549 (1982)].
45. For reviews of the use of carbenes in photoaffinity labeling, see *Photogenerated Reagents in Biochemistry and Molecular Biology*, H. P. Bayley, Ed. (Elsevier/North-Holland, New York, 1983).
46. R.A.M. thanks the National Science Foundation and K.B.E. and N.J.T. thank the National Science Foundation, the Air Force Office of Scientific Research, and the Joint Services Electronic Program for support of the research described in this article.

The concept of a reactive intermediate, sometimes called a free radical, is an old one in chemistry. To illustrate, a textbook of organic chemistry, first published in 1887, states, "Radicals are groups of atoms that play the part of elements, may combine with these and one another and may be transferred by exchange from one compound to another" (1). Our concept of a chemical intermediate has been refined only a little in the intervening 100 years.

We would now call any molecular or atomic species that is produced during the detailed mechanism of an overall chemical reaction, but is not an isolable end product of that reaction, a chemical intermediate. Generally speaking, such intermediates will be chemically unstable (that is, highly reactive) but physically stable (that is, the unperturbed species will not spontaneously decompose) (2).

Most common chemical intermediates can be classified into several categories that include reactive atoms such as H, O, and N; neutral free radicals such as CH, ClO, C_5H_5 (cyclopentadienyl), and C_7H_7 (benzyl); and charged molecular ions such as N_2^+, Cl_2^+, $C_6H_6^+$ (benzene cation), and $C_6H_3F_3^+$ (sym-trifluorobenzene cation). An important facet of modern physical chemistry is to devise methods of detecting, characterizing, and monitoring the dynamics of these chemical intermediates. This is an extremely important task for two general reasons. Often these chemical intermediates are simpler and much more reactive "pieces" of larger molecules. Their physical characterization can play a key role in the fundamental understanding of molecular bonding and reaction. In addition, the capability to detect and monitor the transient populations of these inter-

4

Chemistry and Chemical Intermediates in Supersonic Free Jet Expansions

Terry A. Miller

Science 223, 545–553 (10 February 1984)

mediates is important in the understanding of the detailed mechanisms, and hence the control, of many complicated chemical reactions of commercial and environmental importance.

Because of their transient role in chemical reactions, the concentrations of such chemical intermediates are frequently quite low. Often sophisticated spectroscopic techniques are used for their detection and characterization. In this article, I describe some recent work done by our group and by others that combines, for the first time, the power of laser spectroscopic techniques with supersonic free jet expansions for the study of chemical intermediates. The laser techniques give us extremely powerful and sensitive ways in which to produce and detect chemical intermediates. The free jet expansion provides, in many ways, an ideal environment for the study of both the physical characteristics of the intermediates and their reactions.

The Free Jet Expansion

The properties of a free jet expansion, and their potential usefulness, probably were first recognized by Kantrowitz and Grey (3). The correctness of their ideas was almost immediately demonstrated experimentally by Kistiakowsky and Slichter (4), with at least part of the motivation provided by the desire to determine the structure of one of the simplest chemical intermediates, the methyl free radical, CH_3. After that seminal work over 30 years ago, much of the attention to free jet expansions centered on the details of the gas dynamics occurring therein, rather than on their potential use for the study of chemical intermediates.

Before discussing the advances that have made possible the present studies of chemical intermediates I shall summarize some of the relevant properties of free jet expansions themselves. This discussion will be brief and heuristic; there are several fine reviews (5, 6) the reader can consult for additional details. The key feature of the free jet expansion is a small-nozzle orifice that separates a high-pressure gas sample behind the nozzle from a downstream region kept at low pressure by a vacuum pump. The diameter, D, of the nozzle orifice is usually 0.01 to 1 millimeter; the pressure upstream from the nozzle, the stagnation pressure, P_0, ranges from 100 atmospheres to less than 1. Figure 1 depicts the conditions for a typical expansion, one with 10 atmospheres of room-temperature He behind the nozzle. The conditions downstream from the nozzle are indicated for various multiples (X/D) of the nozzle diameter. As indicated in Fig. 1, the pressure 50 nozzle diameters downstream is less than 10^{-7} that of the stagnation pressure. To maintain such a large differential places great demands on the pumping system, and often available pumping speed is the limiting factor in experiments. [To some extent, pumping speed limitations can be overcome by using a pulsed valve on the nozzle, thus limiting the total gas load on the pump— see, for example, the pioneering work of Gentry and his co-workers (7).]

From a molecular point of view, the free jet expansion converts random thermal motion in the gas reservoir to directed motion in front of the nozzle. This is illustrated in Fig. 1 by the conversion of the arrows indicating velocity from random length and direction behind the nozzle to uniform length and direction beyond it. It has been said that the free jet expansion "monochromatizes" the gas velocity. Even though this velocity may be dependent on the details of the expansion, it can be shown to reach a terminal, limiting value. For He, this corresponds to 2×10^5 centimeters per second, which can be compared to 1.3×10^5 centimeters per second, the arithmetical average velocity (6) in a random direction for a static gas of He at 300 K.

Such jets are often referred to as supersonic expansions, which may seem a little surprising since the limiting jet flow velocity is only slightly greater than the average bulb velocity, which indeed is close to the speed of sound (6). However-

er, the local speed of sound is proportional to the square root of the temperature. Thus as the expansion proceeds, the Mach number, M, which is the ratio of the flow velocity to the local speed of sound, increases dramatically. For the portion of the expansion shown in Fig. 1, M has already reached 50. However, M will not increase indefinitely as a function of X/D; the decreasing frequency of collisions causes M to approach a limiting value. Nonetheless for He, M values over 200, which is truly supersonic, have been reported (5).

From our point of view, the most important physical property is not the

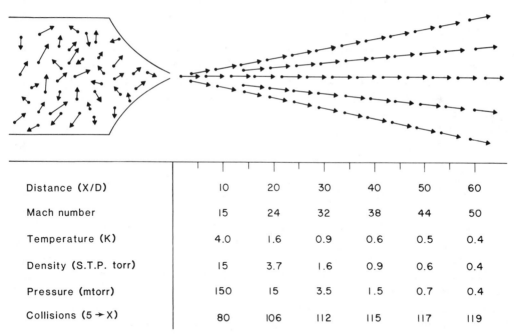

Distance (X/D)	10	20	30	40	50	60
Mach number	15	24	32	38	44	50
Temperature (K)	4.0	1.6	0.9	0.6	0.5	0.4
Density (S.T.P. torr)	15	3.7	1.6	0.9	0.6	0.4
Pressure (mtorr)	150	15	3.5	1.5	0.7	0.4
Collisions (5 → X)	80	106	112	115	117	119

Fig. 1. Schematic diagram of a free jet expansion. Shown below the expansion is a scale measuring distance (X/D) downstream in units of the nozzle diameter (D). Given at various positions downstream are the Mach number, the temperature, the density [in torr at standard temperature and pressure (S.T.P.); 1 S.T.P. torr = 3.54×10^{16} molecules per cubic centimeter], the pressure (in millitorr), and the number of collisions between $X/D = 5$ and the indicated point. It is assumed that the stagnation pressure behind the nozzle is 10 atmospheres of He at 300 K; for the collision calculation, $D = 0.15$ mm, and a collision cross section of 50 square angstroms is assumed.

absolute velocity but the width of the velocity distribution, which constitutes a direct measure of the gas's translational temperature. For the expansion in Fig. 1, this width corresponds to a temperature of less than 10 K at $X/D = 10$ and subsequently drops to less than 1 K; under more extreme conditions the translation temperature can approach 0.01 K. Since the distribution of molecular velocities is narrow, the energies of the collisions that occur in the expansion will be characteristic of these low temperatures. As a result of these collisions, the internal degrees of freedom of the molecules— that is, the vibrational and rotational energy levels—will relax toward these same low temperatures.

As the expansion proceeds, the density of the gas and the probability of collisions decrease. Thus the ultimate distribution of internal excitation in the molecule may not reach precise equilibrium with the translational temperature of the molecule. As a rough rule, rotational-translational equilibrium is usually rapid, and rotational temperatures are rarely greatly different from translational ones in a simple expansion. Vibrational-translational equilibrium is usually less efficient and often dependent on the details of the vibrational structure. Nonetheless, substantial vibrational cooling is also typically observed.

The collision processes in the jet deserve more discussion. The number density drops by a factor of more than 500 in the first ten nozzle diameters, and then, as Fig. 1 shows, drops by a comparably small factor of less than 40 in the next 50 nozzle diameters. Although, it is not possible to draw completely distinct boundaries, it is reasonable to assume that most of the n-body ($n \geq 3$) collisions occur at $X/D \lesssim 5$ for these expansion conditions. Because the initial step in molecular condensation requires n-body ($n \geq 3$) collisions, condensation must initiate mostly for $X/D \lesssim 5$. (The degree of condensation can be reduced by using a mixed gas expansion—for example, an inert gas like He or Ar dilutely seeded with the molecules of interest. In this case, most of the collisions occur among inert gas atoms, between which the intermolecular forces are relatively weak. Thus, in comparison with a pure molecular expansion, condensation is greatly lessened.)

The region very close to the nozzle is difficult to model theoretically and may vary considerably from expansion to expansion because of details in nozzle design. However, the region beyond $X/D \gtrsim 5$, where 2-body collisions predominate, can be rather accurately modeled. Lubman et al. (8) reported very useful results describing the collision frequency in this region. In Fig. 1, the flow velocity was used to convert this collision frequency into the number of 2-body collisions suffered by a typical molecule downstream from the point $X/D = 5$. Most of the collisions have occurred by $X/D = 30$ to 40, but inside this region they are quite frequent.

From our perspective, this region of 2-body collisions may be the most interesting. If "hot" radicals are injected into the jet, these collisions can thermalize them to the low temperature characteristic of the expansion. Similar collisions between reactive species can lead to bimolecular chemical reactions.

Detection of Species in Jets

In many ways the free jet expansion produces an almost ideal source of molecules for study. They are at a very low

50

temperature, so that even complex molecules have only a few populated rotational or vibrational levels. Yet they are isolated in the gas phase, free from any condensed-phase effects. The problem for the chemist is how to sample and interact with these exquisitely prepared molecules. The introduction of most mechanical devices into the jet will cause turbulence, destroying the ordered flow and raising the temperature of the seeded molecules. Great effort has been expended to design carefully shaped "skimmers" to sample physically the cold core of the expansion without introducing undesirable turbulence. Usually these skimmers serve as high-intensity (because of the jet's directed flow), low-temperature sources for molecular beams.

Although considerable work has been done with such molecular beams, probably the most important advance has been the realization that the simplest, least perturbing way to study molecules is in the free jet itself, by using photons that can pass through the expansion without causing any perturbation to the medium.

The first clear demonstration of this fact was a precedent-setting experiment on NO_2. The visible, electronic absorption spectrum of NO_2 under ambient temperature conditions had long been regarded as so complicated and overlapped that a successful analysis was impossible. In the experiment of Smalley et al. (9) a supersonic expansion of NO_2 seeded in Ar was produced. The translational temperature of the beam (where interrogated) was 2 K and the rotational temperature of the NO_2 was 3 K, leaving only two rotational levels with significant populations. The expansion was crossed by a tunable dye laser. The fluorescence generated by the interac-

tion of the laser beam and the cold NO_2 was collected in the third orthogonal direction. Tuning the laser frequency over the NO_2 absorption spectrum resulted in a fluorescence excitation spectrum of NO_2 that for the first time permitted a clear separation, and rotational analysis, of the otherwise badly overlapped vibronic bands.

Since this initial experiment, the use of laser-induced fluorescence for the detection of molecules in free jet expansions has become widespread (5, 6, 10). The low temperatures in the jet have permitted the obtainment of greatly simplified spectra of many complex molecules (11), among them tetracene ($C_{18}H_{12}$), pentacene ($C_{22}H_{14}$), ovalene ($C_{32}H_{14}$), and phthalocyanine ($C_{32}N_8H_{18}$), with molecular weights even exceeding 500. Radiationless transitions have also been studied in simpler molecules. Because of incipient condensation, very weakly bonded molecules—van der Waals molecules—are found to some degree in typical expansions. Because of the low temperatures and collision energies present, these species can survive in the jet. This has opened up a wide area of research (5, 6, 12) on their spectroscopy and photochemistry.

Although not yet as widespread, spectroscopic techniques other than laser-induced fluorescence have been used to study molecules in jets. These include Raman scattering, direct absorption, and multiphoton ionization.

Chemical Intermediates

The combination of optical, particularly laser techniques, with supersonic free jet expansions was initially widely used for the study of chemically stable molecules and their van der Waals adducts. Recently, however, such work has been extended to include highly reactive chemical intermediates. Figure 2 shows a schematic diagram of the apparatus constructed in our laboratory for this purpose. Apparatuses based on similar principles have been developed in other laboratories.

The key complication in studying chemical intermediates is that they cannot simply be entrained in the carrier gas behind the nozzle because of their extreme reactivity. Rather, they must be produced, at some stage, in situ in the expansion either immediately before or after the nozzle. Electron beam irradiation and laser irradiation downstream from the nozzle (Fig. 2) have been the most popular techniques for the production of chemical intermediates in the expansion.

Electron impact production and excitation of ions. In the first experiments producing chemical intermediates, an electron beam was used to produce ions, particularly N_2^+ from N_2. Two features of these experiments (*13*) set them apart from the rest of the work discussed in this article. First, the objective of these experiments was not the elucidation of either the structure or reactions of N_2^+. Rather this work was motivated by a desire on the part of fluid dynamicists to develop a convenient temperature probe for the free jet expansions. The way in which it was hoped that N_2^+ might provide such a probe rested on the second

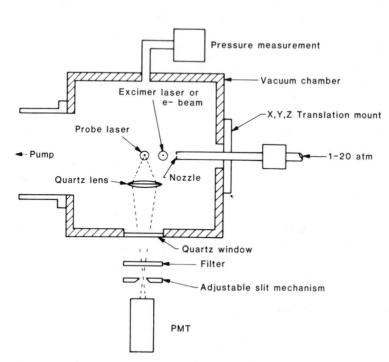

Fig. 2. Experimental apparatus used to study chemistry and chemical intermediates in a free jet expansion. The photolysis laser or electron beam is typically positioned between ~ 5 and 25 (*X/D*). The probe laser will usually be positioned at the photolysis laser position to ~ 80 (*X/D*). For experiments employing only an electron beam and no probe laser, the simple photomultiplier (*PMT*) detection will be replaced by a monochromator and photomultiplier. Similarly, in some cases laser-induced fluorescence may be dispersed by a monochromator.

feature that differentiates this work from the rest. The electron impact process both creates the N_2^+ and provides for its detection in that it creates a fraction of the N_2^+ in an excited electronic state, which decays by the emission of photons to the ground state of the ion. Dispersing this optical emission through a monochromator shows clearly the rotational structure of this spectrum. From an analysis of the relative intensities of the transitions the rotational temperature of the excited B state can be determined.

It was originally hoped that the rotational temperature of the B state would be the same as the translational temperature of the jet at the point sampled. Although this appears to be roughly true, there has been great difficulty establishing the exact relation between these quantities. Recent work (14) shows there is apparently a slight "warming" of the temperature upon electron impact excitation, especially if the energy of the electrons is low.

In the present context, probably the most important aspect of the N_2^+ work is the demonstration that chemically reactive species could be produced within the expansion without grossly distorting its unique properties such as directionality and low temperature. Being a simple diatomic species, N_2^+ had been well characterized in previous, traditional spectroscopic work. However, the electron impact techniques have recently been used by several groups to characterize reactive positive radical ions of triatomic neutrals and larger organic molecules. Species studied (14–19) in this way now number well over a dozen and include such diverse cations as those of CO_2, H_2O, N_2O, hexadiyne (C_6H_6), dichloroacetylene (C_2Cl_2), bromodiacetylene (C_4HBr), trichloroben-

zene ($C_6H_3Cl_3$), and hexafluorobenzene (C_6F_6).

Figure 3 shows the simplification in the emission spectrum that can be achieved in a free jet expansion for a typical, relatively large, organic cation, $C_6F_6^+$. Clearly the resolution provided by the "cold" spectrum of $C_6F_6^+$ can be invaluable in analyzing vibrational structure that can probe the changes in chemical bonding from the neutral to the ion in an aromatic molecule. For systems like the dichloroacetylene cation (17), similarly improved resolution yields the fine structure splitting caused by the spin-orbit interaction. Even for a relatively small ion like CO_2^+, the cold jet spectrum can be useful (18) for probing, for example, perturbed rotational structure.

Laser production and detection of ions. Although the technique of electron impact ionization with simultaneous electronic excitation is a simple means for obtaining spectra of both small and large jet-cooled ions, it is relatively limited with respect to versatility. Because of the very short lifetimes (nanoseconds to microseconds) typically associated with radiating, excited electronic states, production and detection are nearly superimposed in space and time. For many applications, it is desirable to separate these two events.

Figure 2 illustrates the means by which this may be accomplished. Rather than monitoring the emission associated with the ionization process, the detection of the ions is accomplished downstream in the jet by laser-induced fluorescence. This part of the experiment is essentially identical to that described earlier for the observation of cold NO_2. The only substantial difference is that the ionic species being monitored is now being produced in situ in the free jet

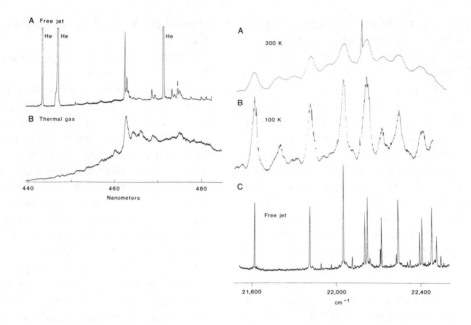

Fig. 3 (left). Experimental traces of the emission spectrum of $C_6F_6^+$ excited by electron impact. In (A), the sample is a free jet expansion of He seeded with C_6F_6, whereas that in trace (B) is a room-temperature sample of pure C_6F_6. The instrumental resolution is similar in the two traces. Lines marked He are due to electronic transitions in the carrier gas.   Fig. 4 (right). Laser excitation spectra of $C_6F_6^+$. In (A) and (B), the $C_6F_6^+$ was produced by Penning ionization in a flow system; (A), the flow was at room temperature, whereas in (B) it was cooled by liquid N_2. For (C), the $C_6F_6^+$ was produced by two-photon photoionization through the use of an ArF laser interacting with C_6F_6 dilutely seeded in a free jet expansion in Ar. The probe laser line width is the same for all three traces; the sharp line in (A) is due to an atomic impurity.

expansion. Other detection schemes could be envisioned, but laser-induced fluorescence has the key advantage of being able to monitor selectively different quantum states of the ion.

Thus far, two methods of ion production have been shown to generate high enough ion densities for successful downstream monitoring of laser-induced fluorescence. One of these is the electron impact ionization (*18, 19*) used in the emission studies. The other is the production of the ions by multiphoton photoionization. A large number of organic ions have been produced in our

laboratory by the absorption of two excimer laser (ArF) photons. At present the two techniques appear to be somewhat complementary. The electron impact method appears to be better suited for light species with high ionization potentials requiring the absorption of many photons. Laser photoionization, however, is a cleaner and less intrusive technique, and it has been successfully applied to a number of larger organic ions.

Figure 4 shows a laser excitation spectrum of $C_6F_6^+$ that was obtained in a jet after production of the ion by two-photon photoionization (*20–22*). For com-

54

parison, the laser excitation spectrum of $C_6F_6^+$ is also shown under conditions in which the ion's temperature was approximately 100 K and 300 K.

Spectra such as those shown in Figs. 3 and 4 have greatly increased our understanding of $C_6F_6^+$ and related organic ions. For instance, it has been established (22) that several substituted benzene ions retaining a degenerate ground electronic state by virtue of either a threefold or sixfold axis of symmetry [for example, $C_6F_6^+$ (sixfold) and sym-trifluoro- or sym-trichlorobenzene cation (threefold)] suffer an interaction between the nuclear and vibrational motion, called the Jahn-Teller effect. The existence of this Jahn-Teller interaction causes the electronic state to be stabilized by 500 to 1000 cm^{-1} in energy and to be geometrically distorted. Experiments have shown that the minimum energy configuration of these benzenoid ions does not correspond to a regular hexagon but rather to one with the internal ring angles distorted from 120 degrees by 1 or 2 degrees, and the C–C bond length alternating in length by approximately 0.02 angstrom. The implications of such geometric distortion for the chemical reactivity of these ions is being explored.

Ionic clusters. The role of ions as preferred sites for nucleation has long been recognized. Indeed, the venerable Wilson cloud chamber is perhaps the example that first comes to mind. In this case a supersaturated vapor is exposed to ionizing radiation such as cosmic rays. The cosmic rays produce ions along their path that serve as nucleation sites for droplet formation, giving rise to the well-known particle track in the chamber.

Ions, of course, constitute preferred nucleation sites because the strongest of the long-range, intermolecular forces are those involving the lowest order multipole, namely, the point charge. The microscopic origin of charged droplets are molecular clusters with one or a few neutral molecules bound to an ion by intermolecular forces. Such ionic clusters have long been thought to have a role in the chemistry of the upper atmosphere, some aspects of combustion, and certain electrical discharges. Indeed the widespread existence of ionic clusters has been demonstrated in numerous mass spectrometric experiments.

However, until recent jet experiments, no other physical identification or characterization of these clusters had been accomplished. Consequently, almost nothing is known about their structure, and much remains to be learned about their growth and role in chemical reactions.

The picture is starting (20, 21, 23) to come into focus with the obtainment of optical spectra for model cluster systems, one or more inert gas atoms bound to an organic cation. For example, distinct spectra have recently been obtained for the addition of one, two, and three He atoms to $C_6F_6^+$. The addition of each of the first two He atoms produces almost equal spectral shifts, implying nearly equivalent bonding sites. However, the spectral shift upon the addition of the third He is only about 15 percent of the value of the first two. It is appealing to suggest that the first two He's are binding equivalently above and below the benzene ring. The third He must then go into a second "solvation sphere," probably binding around the edge of the ring or to another He atom. Similar but somewhat more complicated data are now available for heavier inert gas atoms, and it seems likely that many different kinds of cluster ions can be studied by this technique. Although the picture still is not completely clear, it appears that our knowledge of ionic clusters should increase dramatically in the near future.

Simple neutral free radicals. The discussion to this point has focused almost exclusively on the study of charged chemical intermediates in free jets, but many chemical reactions have, as intermediates, neutral species—usually free radicals. It is not obvious that the techniques applied so successfully to the study of molecular ions in jets can be applied to the study of free radicals. To produce a reactive parent ion from a neutral precursor, one need only remove an electron. As mentioned earlier, removal of an electron will always approximately conserve the rotational and often the vibrational state of the precursor. Thus, given a cold precursor, it is relatively easy to produce a cold parent ion.

In contrast, to produce a neutral free radical (or charged) fragment from a precursor requires the rupture of a chemical bond. Excess energy must be converted into the translational and internal degrees of freedom of the fragments. Thus, in general, a cold precursor will produce vibrationally and rotationally hot, not cold, free radicals (24), as well as imparting a velocity to these species very uncharacteristic of the expansion.

Such a result would at first glance appear to destroy most of the jet's desirable characteristics. There is, however, the possibility that the free radicals, although initially produced hot, can be "thermalized" to the jet's temperature before they are detected or used. Figure 1 indicates that many collisions are possible if the fragments are produced near the nozzle exit. Indeed if radicals are

initially formed translationally hot, they will suffer even more collisions than is shown in Fig. 1, since they will be "out of step" with the rest of the molecules in the jet. However, the extent to which thermalization can be accomplished under normal expansion conditions, is not a priori clear.

Smalley's group at Rice performed the first experiments on free radicals in jets. In the initial experiment (25), they photolyzed ketene, CH_2CO, to produce the methylene free radical, CH_2. They observed that the nascent rotational temperature of CH_2 was moderate but could be seen to cool even further by subsequent collisions in the expansion. Under certain photolysis conditions, considerable vibrational excitation appeared in the CH_2, which was only inefficiently cooled by the expansion.

To try to ensure more efficient cooling, the Smalley group developed (26) a two-stage expansion, which permitted precursor photolysis in a region of moderately high pressure, just before final free expansion. Using this technique, the group observed the methoxy radical, CH_3O, so cold rotationally that only the lowest two or three levels were populated. When the carrier gas was Ar, a van der Waals complex, $Ar \cdot OCH_3$, was also observed.

The pressure in the photolysis region of the two-stage expansion is sufficiently high that newborn radicals can reach roughly a room-temperature equilibrium. Depending on the experiment, this high-pressure section may or may not have desirable effects. It ensures more complete vibrational relaxation than might be expected in the expansion itself. Likewise, the room-temperature equilibrium apparently facilitates van der Waals

56

complex formation in the expansion. On the other hand, the high-pressure region may lead to radical reactions that alter or destroy the species of interest. Moreover, this technique does not permit following the relaxation of the radical from a nascent to cold distribution. (Other techniques involving preexpansion production are discussed below.)

Heaven *et al.* (*27*) performed several experiments in which it was possible to produce hot nascent radicals and follow their relaxation in the jet. Using the apparatus shown in Fig. 2, they studied the photolysis of BrCN with an ArF excimer laser. This technique produces the CN free radical with a very hot rotational and vibrational distribution. Figure 5A shows the laser-induced fluorescence spectrum of CN, with the probe laser nearly superimposed in space and time with the photolysis laser. The observed distribution approaches, if it is not equal to, the nascent distribution. Although the rotational distribution is not exactly a Boltzmann, it can be approximated by a Boltzmann distribution with a temperature of about 4000 K. The fraction of population in the $v = 1$ vibrational level is about 20 percent of that in $v = 0$, which is equivalent to that expected for a vibrational temperature of about 1800 K.

If the probe laser is translated downstream from the photolysis laser, the trace of Fig. 5B is obtained. The CN has now suffered many collisions with the cold Ar carrier gas, causing a massive redistribution of rotational population. The overall distribution is now strongly non-Boltzmann. However, the lowest half-dozen or so levels are in equilibrium with a temperature of about 10 K. The ratio of populations in the $v = 1$ and $v = 0$ levels is essentially unchanged.

If the number of downstream collisions is increased by raising the Ar stagnation pressure toward 20 atmospheres, the only detectable CN radicals are in the lowest J levels characterized by a temperature of about 7 K. If He replaces Ar as the carrier gas, then rotational relaxation is not as complete, but all of the $v = 1$ population is efficiently relaxed.

Radical-radical reactions. For CN, the collisions relaxed the radical's internal energy but were chemically nonreactive. However, it is also possible to study reactive collisions with similar techniques. An example of such a chemical reaction is the one (*28*) forming the S_2 molecule. In this experiment, H_2S is photolyzed by an excimer laser to form predominantly SH+H. Laser-induced detection of SH, immediately after the photolysis pulse, is shown in Fig. 6A. The rotational distribution is nearly nascent and characterized by a rotational temperature of 300 to 400 K.

Figure 6B shows the same spectral region but the probe laser is now displaced downstream by about 5 millimeters in distance and delayed by about 8 microseconds in time. The SH spectrum is now much colder, only the lowest two or three rotational levels having observable populations. However, by far the most interesting feature in Fig. 6B is the newer, much denser spectrum now present, but absent in Fig. 6A. If the probe laser is narrowed by insertion of an etalon, this denser spectrum can be completely resolved and its carrier can be unambiguously identified as the S_2 molecule. Analysis of the spectrum shows that the S_2 present has comparable populations in the lowest five or so vibrational

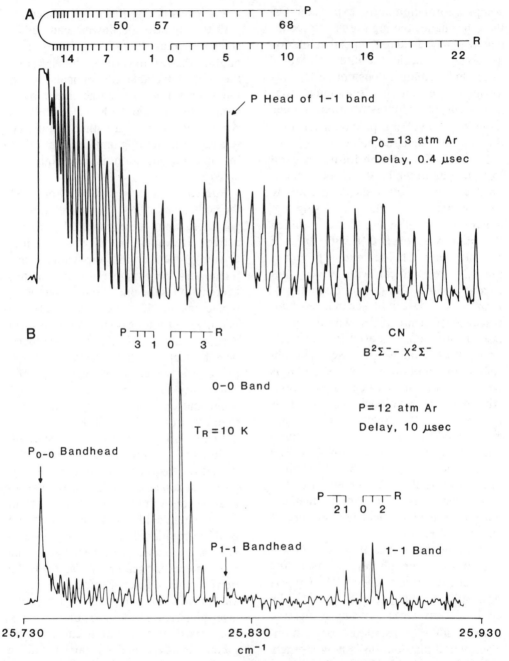

A

P
50   57   68

R
14   7   1   0   5   10   16   22

P Head of 1-1 band

$P_0 = 13$ atm Ar
Delay, 0.4 μsec

B

P   R
3 1 0   3

0-0 Band

$T_R = 10$ K

CN
$B^2\Sigma^- - X^2\Sigma^-$

$P = 12$ atm Ar
Delay, 10 μsec

P_{0-0} Bandhead

P   R
21 0 2

P_{1-1} Bandhead

1-1 Band

25,730   25,830   25,930
cm^{-1}

Fig. 5. Laser-induced fluorescence spectra of the CN free radical produced by the ArF laser photolysis of BrCN in an Ar free jet expansion. In (A), the probe and photolysis laser are nearly overlapped in space and time, and the CN distribution detected is nascent or near nascent. In (B), the probe laser has been displaced downstream by 10 μsec in time and 6 mm in distance. Large redistributions in rotational populations are clear. The rotational assignments for the R and P branches are as indicated.

levels, but its rotational and fine structure populations can be described by a Boltzmann distribution with a temperature of about 20 K.

Since S_2 contains two S atoms, its formation requires the participation of at least two H_2S molecules. There seem to be only two ways for this to occur. Either the initial species to absorb a photolysis photon is a van der Waals cluster, $(H_2S)_n$ with $n \geq 2$, or subsequent radical-radical reactions occur between the radicals created in the initial photolysis of H_2S.

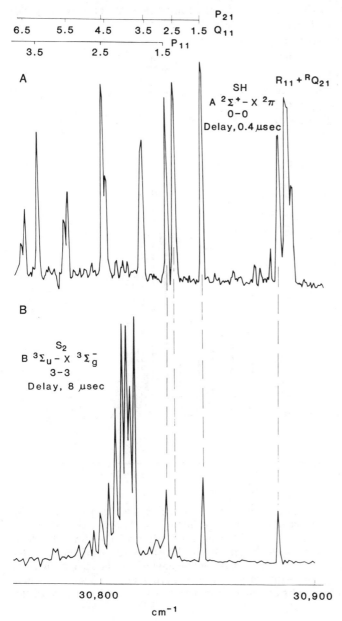

Fig. 6. Laser-induced fluorescence spectrum of the products of an ArF laser photolysis of H_2S seeded in an Ar free jet expansion. In (A), the probe and photolysis lasers are nearly overlapped. The spectrum observed is for the SH free radical with a nearly nascent population distribution. The rotational assignments of the lines are as indicated. In (B), the probe laser is delayed about 8 μsec in time and moved 5 mm downstream. The SH is considerably colder than in (A). The new, denser spectrum can be attributed to the S_2 molecule.

Experiments (28) showed that the product ratio S_2/SH was independent, over a moderately wide range, of the fraction of H_2S in the expansion. This would argue against the cluster explanation for S_2 formation, because species like $(H_2S)_n$ would be expected to increase at least quadratically with H_2S concentration while SH should have only a linear dependence. Similarly, the S_2/SH ratios were measured as a function of downstream position of the photolysis beam. With increasing distance between the photolysis laser and the nozzle, the S_2 concentration fell dramat-

ically. This result is consistent with a radical-radical reaction because, as the photolysis is moved downstream to a lower pressure region of the jet, the probability of reactive collisions rapidly decreases.

The observations are all consistent with the following reaction scheme yielding S_2.

$$H_2S \xrightarrow{h\nu} SH + H$$

$$H + SH \rightarrow H_2 + S$$

$$S + SH \rightarrow S_2(X^3\Sigma_g^-, v'') + H$$

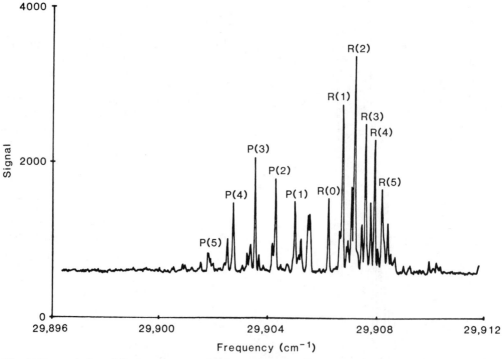

Fig. 7. Laser-induced fluorescence excitation spectrum of the vinoxy free radical (CH_2CHO). The vinoxy was produced by excimer laser photolysis of methyl vinyl ether (CH_2CHOCH_3). The rotationally resolved spectrum of the 010 vibrational band of the $\tilde{B}^2A''-\tilde{X}^2A''$ electronic transition is shown with the probe laser approximately 40 nozzle diameters downstream and the photolysis laser 10 nozzle diameters downstream. A rotational analysis of the spectrum shows the CH_2CHO to have a rotational temperature of ~ 10 K at the probed point. Vinoxy is a near prolate symmetric top and the indicated rotational assignments are in conformance with this approximation.

The rate constants for the various steps in the reaction are known from previous work. Use of these constants and a crude model of the expansion embodying the ideas in Fig. 1 shows (28) that the observed variation of S_2/SH with photolysis position, stagnation pressure, and so on can be semiquantitatively predicted.

In general, the marked change from energetic, ambient temperature collisions to low-energy, ultracold collisions might be expected to alter profoundly many reaction mechanisms. On a macroscopic scale, the different temperature dependencies of rate constants could be expected to change dramatically the balance between competing chemical reactions. In this case, the final product, S_2, found in the free jet expansion is the species that would be expected from a similar photolysis in an ordinary "bulb" experiment. However, in terms of internal energy content, the S_2 formed in the free jet reaction is very different from the S_2 from any other known source. It is vibrationally quite hot, but its rotational and fine structure distributions are quite cold. This unusual combination allows spectroscopic analysis of S_2 and consequently an understanding of its electronic structure, which had not been previously possible.

Large free radicals. While such unusual distributions may be of specialized interest for simple radicals like diatomic S_2, they have considerably wider utility for larger free radicals. Figure 7 shows (29) the spectrum of the vinoxy free radical, CH_2CHO, obtained in a free jet expansion. This radical is widely thought to be an important intermediate in flame chemistry and was earlier observed by laser-induced fluorescence spectroscopy in a room-temperature flow system. In those experiments, no rotational or fine structure resolution was obtained, largely because of the congested nature of the spectrum at room temperature. As Fig. 7 shows, the free jet photolysis of methyl vinyl ether produces the vinoxy radical with an internal temperature of about 10 K where interrogated. A thorough analysis of this spectrum is possible and has yielded detailed information (30) about the geometric and electronic structure of this radical. Recently, similar cold free jet spectra have been reported (29) for other large radicals, including the cyclopentadienyl (C_5H_5) and benzyl ($C_6H_5CH_2$) radicals.

Radical generation by other techniques. In the work described thus far, the chemical intermediates have been produced either by electron bombardment or laser photolysis. Until a few months ago, all production of radicals and ions in free jet expansions had been by one of these two methods. However, several experiments have now been performed that indicate that other techniques may be used to form reactive intermediates in jets.

Obi *et al.* (31) showed that a laser is not necessary for the photochemical production of free radicals in a jet. In their experiment, a simple Hg lamp irradiated a mixture of H_2, NO, Ar, and Hg upstream from the nozzle. The free radical HNO was then formed by the complicated reaction series

$$Hg + h\nu \text{ (Hg lamp)} \rightarrow Hg^*$$

$$Hg^* + H_2 \rightarrow HgH + H$$

$$H + NO + Ar \rightarrow HNO + Ar$$

or

$$HgH + NO \rightarrow HNO + Hg$$

Laser-induced fluorescence spectra well downstream from the nozzle detected

HNO cooled to a rotational temperature of approximately 16 K.

In a recent experiment by Droege and Engelking (32), a corona discharge was established upstream from the nozzle in a mixture of H_2O and He. In this work, optical emission from the discharge afterglow was observed downstream from the nozzle. The emission showed the presence of OH radicals in their excited A $^2\Sigma^+$ electronic state cooled to about 11 K rotationally, but with a vibrational temperature of approximately 3400 K. Although the mechanism for the production of the OH is not firmly established, it has been suggested that $^2\Sigma^+$ OH is produced in the discharge upstream from the nozzle and rotationally cools before it radiatively decays downstream from the nozzle.

Farthing et al. (33) have produced the reactive species, NH_2, BrF, and IF, in a free jet. Their technique involved the reaction of F atoms, produced in a microwave discharge with NH_3, Br_2, and CH_3I, respectively, to produce the above-mentioned species. The chemical reaction takes place immediately behind the nozzle and the reactive intermediates expand through the orifice. Moderate cooling to approximately 100 K is reported by this technique.

It is rather difficult to assess the impact of each of these very recent experiments on the overall field. A major advantage of each of these techniques is continuous, rather than pulsed, production of intermediates. A clear limitation of the techniques is that they all rely on chemistry behind the nozzle. Overall, these techniques clearly add considerable versatility to our methods for the production of chemical intermediates in free jets.

Conclusions

Observation of chemistry and chemical intermediates under the singular conditions of a supersonic free jet expansion has wide-ranging implications. The spectroscopy of large reactive species can be studied at low temperature at which the number of populated quantum states is few. Such simplified spectra usually allow a detailed physical characterization of important chemical intermediates. Several relatively large organic intermediates, both ions and neutral free radicals, have already been studied in this manner. Prospects for future work in this direction are good.

Although the existence of small ionic clusters has been established by mass spectrometric work, their structural characteristics are largely unknown. Their detection by laser-induced fluorescence techniques in free jet expansions opens the door to a much better understanding of the bonding and structure of these important species.

In some cases, chemical intermediates produced upstream from the nozzle survive passage through the nozzle with subsequent cooling in the expansion. Evidence has been presented, in addition, for the occurrence of ion-molecule and radical-radical reactions in the jet itself. The implications of these observations are considerable. For example, the jet may serve as a laboratory prototype capable of simulating the isolation and low temperatures of interstellar chemistry. It also may be possible to develop entirely new chemistry based on the uncommon properties of the jet expansion. At the least, it gives us an opportunity to study old chemical reactions in a new environment, with the likelihood that a more thorough understanding of them will develop.

References and Notes

1. A. Bernthsen, *Textbook of Organic Chemistry* (Vieweg, Braunschweig, ed. 1, 1887; ed. 16, 1924), English edition (Blackie, London, 1922), quoted by G. Herzberg, *The Spectra and Structure of Simple Free Radicals* (Cornell Univ. Press, Ithaca, N.Y., 1971).
2. The aspect of physical stability can be used to distinguish chemical intermediates from transition states in chemical reactions. For the purposes of this article, we assume that true chemical intermediates correspond to local minima in the potential surface along the reaction pathway, whereas transition states correspond to physically unstable energy maxima. If the overall reaction mechanism is altered, the local minimum corresponding to an intermediate may become an absolute minimum.
3. A. Kantrowitz and J. Grey, *Rev. Sci. Instrum.* **22**, 328 (1951).
4. G. B. Kistiakowsky and W. P. Slichter, *ibid.*, p. 333.
5. D. H. Levy, *Annu. Rev. Phys. Chem.* **31**, 197 (1980).
6. R. E. Smalley, L. Wharton, D. H. Levy, *Acc. Chem. Res.* **10**, 139 (1977); D. H. Levy, *Science* **214**, 263 (1981); L. Wharton, D. Auerback, D. H. Levy, R. Smalley, *Advances in Laser Chemistry*, A. H. Zewail, Ed. (Springer-Verlag, New York, 1978), p. 408; D. H. Levy, L. Wharton, R. E. Smalley, *Chemical and Biochemical Applications of Lasers*, C. B. Moore, Ed. (Academic Press, New York, 1977), vol. 2, p. 1; H. W. Liepmann and A. Roshko, *Elements of Gas Dynamics* (Wiley, New York, 1957), p. 40f; S. Dushman, *Scientific Foundations of Vacuum Technique*, J. M. Lafferty, Ed. (Wiley, New York, 1962).
7. C. F. Giese and W. R. Gentry, *Rev. Sci. Instrum.* **49**, 595 (1978); K. Bier and O. Hagena, *Rarefied Gas Dynamics, 4th Symposium*, J. H. deLeeuw, Ed. (Academic Press, New York, 1966), p. 260.
8. D. M. Lubman, C. T. Rettner, R. N. Zare, *J. Phys. Chem.* **86**, 1129 (1982).
9. R. E. Smalley, B. L. Ramakrishna, D. H. Levy, L. Wharton, *J. Chem. Phys.* **61**, 4363 (1974); R. E. Smalley, L. Wharton, D. H. Levy, *ibid.* **63**, 4977 (1975).
10. Review articles in this area include R. E. Smalley, *J. Phys. Chem.* **86**, 3504 (1982); D. H. Levy, *Photoselective Chemistry*, J. Jortner, Ed. (Wiley, New York, 1981), part 1, p. 323; D. H. Levy, C. A. Haynam, D. V. Brumbaugh, *Faraday Discuss. Chem. Soc.* **73**, 137 (1982).
11. A. Amirav, U. Even, J. Jortner, *Opt. Commun.* **32**, 266 (1980); *J. Phys. Chem.* **86**, 3345 (1982); *Chem. Phys. Lett.* **69**, 14 (1980); L. Wharton and D. H. Levy, *J. Chem. Phys.* **69**, 3424 (1978); P. S. H. Fitch, C. A. Haynam, D. H. Levy, *ibid.* **73**, 1064 (1980).
12. W. Klemperer, *J. Mol. Struct.* **59**, 161 (1980).
13. See E. P. Muntz, *Phys. Fluids* **5**, 80 (1962); P. V. Marrone, *ibid.* **10**, 521 (1967); F. Robben and L. Talbot, *ibid.* **9**, 644 (1966); H. Ashkenas, *ibid.* **10**, 2059 (1967); A. E. Kassem and R. S. Hickman, *ibid.* **17**, 1976 (1974); D. Coe, F. Robben, L. Talbot, R. Cattolica, *ibid.* **23**, 7061 (1980).
14. B. M. De Koven, D. H. Levy, H. H. Harris, B. R. Zegarski, T. A. Miller, *J. Chem. Phys.* **74**, 5659 (1981). See also S. P. Hernandez, P. J. Dagdigian, J. P. Doering, *Chem. Phys. Lett.* **91**, 409 (1982); *J. Chem. Phys.* **77**, 6021 (1982).
15. A. Carrington and R. P. Tuckett, *Chem. Phys. Lett.* **74**, 19 (1980); R. P. Tuckett, *Chem. Phys.* **58**, 151 (1981).
16. T. A. Miller, B. R. Zegarski, T. J. Sears, V. E. Bondybey, *J. Phys. Chem.* **84**, 3154 (1980).
17. D. Klapstein, S. Leutwyler, J. P. Maier, *Chem. Phys. Lett.* **84**, 534 (1981); D. Klapstein, J. P. Maier, L. Misev, W. Zambach, *Chem. Phys.* **72**, 101 (1982); D. Klapstein, J. P. Maier, L. Misev, F. Thommen, W. Zambach, *J. Chem. Soc. Faraday Trans. 2*, **78**, 1765 (1982).
18. M. A. Johnson, J. Rostas, R. N. Zare, *Chem. Phys. Lett.* **92**, 225 (1982).
19. M. I. Lester, B. R. Zegarski, T. A. Miller, *J. Phys. Chem.* **87**, 5228 (1983).
20. M. Heaven, T. A. Miller, V. E. Bondybey, *J. Chem. Phys.* **76**, 3831 (1982).
21. T. A. Miller and V. E. Bondybey, *Philos. Trans. R. Soc. London Ser. A* **307**, 617 (1982).
22. _____, in *Molecular Ions: Spectroscopy, Structure, and Chemistry*, T. A. Miller and V. E. Bondybey, Eds. (North-Holland, Amsterdam, 1983), p. 201.
23. L. Di Mauro, M. Heaven, T. A. Miller, *Chem. Phys. Lett.* **104**, 526 (1984).
24. We use the terms "hot" and "cold" in a loose sense in this article. By "cold" we mean any species that is approximately in thermal equilibrium with the low temperatures characteristic of the expansion. "Hot" refers to all species that do not satisfy this criterion irrespective of whether their populations can be characterized by a temperature.
25. D. L. Monts, T. G. Dietz, M. A. Duncan, R. E. Smalley, *Chem. Phys.* **45**, 133 (1980).
26. D. E. Powers, J. B. Hopkins, R. E. Smalley, *J. Phys. Chem.* **85**, 2711 (1981).
27. M. Heaven, T. A. Miller, V. E. Bondybey, *Chem. Phys. Lett.* **84**, 1 (1981).
28. M. Heaven, T. A. Miller, V. E. Bondybey, *J. Chem. Phys.* **80**, 51 (1984).
29. M. Heaven, L. Di Mauro, T. A. Miller, *Chem. Phys. Lett.* **95**, 347 (1983).
30. L. F. Di Mauro, M. Heaven, and T. A. Miller, *J. Chem. Phys.* **81**, 2339 (1984).
31. K. Obi, Y. Matsumi, Y. Takeda, S. Mayama, H. Watanabe, S. Tsuchiya, *Chem. Phys. Lett.* **95**, 520 (1983).
32. A. T. Droege and P. C. Engelking, *ibid.* **96**, 316 (1983).
33. J. W. Farthing, I. W. Fletcher, J. C. Whitehead, *J. Phys. Chem.* **87**, 1663 (1983).
34. I thank my colleagues, particularly V. E. Bondybey, L. Di Mauro, M. Heaven, M. I. Lester, and B. R. Zegarski, for their important contributions to this work.

The study of the chemistry of ions in the gas phase has advanced rapidly in the last decade due to developments in ion trapping technology, light source technology, and ion detection methods. As is the case in many areas of chemistry, new technology has opened the door to new experiments and insights.

Perhaps the single most important motivation for studying ions in the gas phase is that the data obtained from such species will reflect the intrinsic properties of the ion. That is, the structure and reactivity are unperturbed by neighboring molecules or ions in the solid phase or solvent molecules and counterions in the liquid phase. The intrinsic properties of a molecular ion can be analyzed and interpreted in a straightforward manner and perhaps be related to a fundamental property of the ion itself rather than to its surrounding environment. The critical effects of solvation can thus be clarified by an understanding of intrinsic molecular properties. In addition, verification and elucidation of solvation effects has become possible with the ability to generate and study large ionic clusters that are composed of many neutral molecules loosely bound to an ion.

Among the challenging experimental aspects of gas-phase ion chemistry today are the quest for an understanding of geometrical structures and bonding and the quest for a detailed understanding of the intrinsic reactivities and potential energy surfaces for ion-molecule reactions with a view towards evaluating theories of the dynamics of chemical reactions.

Reaction Dynamics

Most early studies of ionic reactions in the gas phase focused on unimolecular

5

Gas-Phase Ion Chemistry

Paul B. Comita
and John I. Brauman

Science 227, 863–869 (22 February 1985)

fragmentations and rearrangements recorded in electron-impact mass spectra. Through the use of new technology and techniques, particularly ion cyclotron resonance (ICR), flowing afterglow, ion beams, and high-pressure mass spectrometry, significant progress has been made in understanding the dynamics and mechanisms of ionic reactions, not only in unimolecular fragmentations but also in reactions of higher order. With the development of new ion trapping technology and detection devices, such as pulsed (and now Fourier transform) ICR and selected ion flow tubes, it has become possible to examine both negative and positive ions with a wide variety of structural features. The dependence of ionic reactions on both temperature and translational energy has been investigated, and with the use of laser sources, the vibrational energy of the ion can be specifically varied to measure its effect on reaction dynamics.

Unimolecular reactions. Unimolecular fragmentation of gas-phase ions can be induced with a number of techniques. Electron impact on a neutral molecule can generate ions with excess internal energy, although the internal energy distribution of the ion population is not well defined. Collisional activation is a widely used technique in which ions are accelerated through a neutral gas. Energetic collisions with the neutral molecules result in electronic, vibrational, and rotational energy transfer to the ions, and those ions excited above a threshold for reaction can then undergo dissociation (*1*).

Unimolecular reactions of ions can in some cases be induced by ultraviolet, visible, or infrared radiation. Ion photodissociation with visible light has been well studied and has become a useful spectroscopic technique that will be discussed further in the section on determination of ion structure.

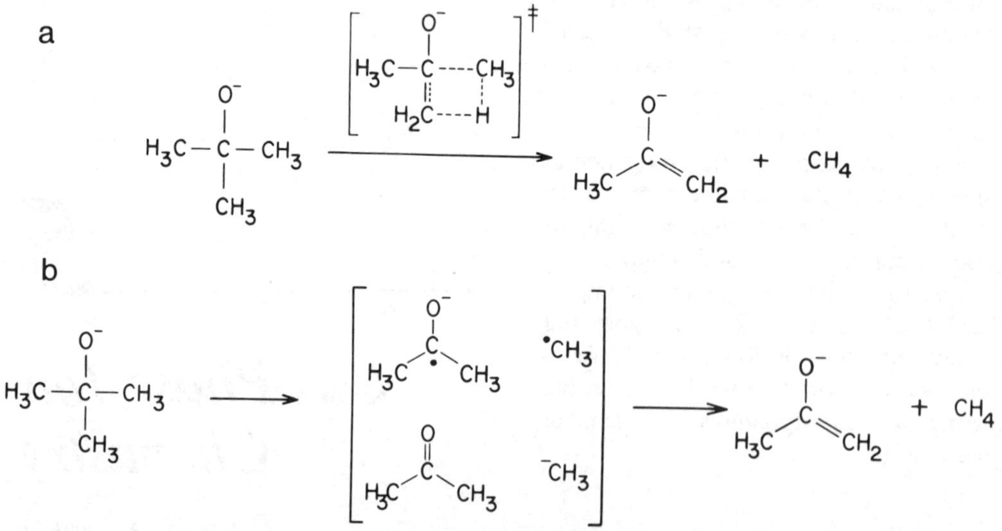

Fig. 1. Two pathways for the IRMP-induced dissociation of *t*-butoxide anion, involving a concerted, four-center elimination (a) and a two-step mechanism with the formation of an ion-molecule complex (b).

Infrared multiple-photon (IRMP) activation using a CO_2 laser is an attractive method for activating ions at low pressure, and it has many features that are advantageous for some experiments (2). For example, collisional effects on the activation and reaction process are avoided since the low pressures ($\sim 10^{-6}$ torr) used result in a time between collisions that is long compared with the time for multiple photon absorption and dissociation. Also, photoproducts can be detected after a single, moderate intensity (unfocused) laser pulse, so primary reactions can be studied unambiguously.

Since the energy required to break covalent bonds is much greater than the energy of infrared photons from a CO_2 laser (approximately 3 kcal per mole), many photons must be absorbed before dissociation can occur. Much theoretical and experimental work has concentrated on the infrared multiple-photon absorption process (3). In general, the absorption of energy into the molecules appears to be a largely incoherent process, and unimolecular reactions induced by this method can be described by statistical reaction-rate theory.

A number of IRMP-induced reactions of negative ions have been investigated with pulsed ICR techniques. One deceptively simple elimination reaction illustrates some of the considerations involved in an investigation of a unimolecular ion reaction. The dissociation of the t-butoxide anion upon irradiation with a CO_2 laser gives rise to the acetone enolate ion by elimination of methane (Fig. 1). Two of the simplest pathways by which this fragmentation could occur are a concerted, four-center elimination of methane and a two-step mechanism involving the formation of an intermediate ion-molecule complex. The operative mechanism was revealed by a study of intramolecular kinetic isotope effects (4). Measurement of both primary and secondary isotope effects was made possible by dissociating specifically deuterated t-butoxide anions and detecting partially deuterated enolate ions. The small, intensity-independent primary effect and the large, intensity-dependent secondary effect were consistent only with a stepwise mechanism in which a rate determining cleavage precedes the proton (or hydrogen atom) transfer.

The nature of the initial cleavage was addressed by further studies of IRMP-induced dissociations of alkoxide anions with various alkyl groups. Specifically, experiments were designed to test whether the first step involved homolytic or heterolytic bond cleavage. Alkoxide (1), with R = CF_3, C_6H_5, and H,

$$ R \!-\!\! \overset{\displaystyle CH_3}{\underset{\displaystyle CH_3}{|}}\!\!-\! O^- \xrightarrow{\ nh\nu\ } RH \ + \ \overset{\displaystyle O^-}{\underset{\displaystyle CH_3 \quad CH_2}{C}} \qquad (1) $$

1

eliminates RH exclusively to form acetone enolate and CF_3H, benzene, and hydrogen, respectively (5). These results are consistent with initial heterolytic cleavage to form the more stable anion R^- in preference to CH_3^-. For R = t-butyl, ethyl, and i-propyl, the leaving group ability was found to be t-butyl- > methyl > i-propyl > ethyl. This order of leaving group ability appears inconsistent with a heterolytic bond cleavage, in contrast to the previous set. It has been explained by a stepwise mechanism involving an intermediate anionic cluster in which an electron is bound nonspecifically by the radical-molecule complex (5). An identical order of leaving groups was also discovered in a completely different ion-molecule reaction but has been attributed by DePuy *et al.* to the relative stabilities of alkyl anions (6).

For negative ions, loss of an electron is generally the lowest energy unimolecular reaction. Electron detachment can be effected with both ultraviolet and visible light as well as infrared radiation. IRMP studies of electron detachment have provided information concerning the intramolecular transfer of vibrational energy to electronic energy in molecules (7). Because electron detachment is slow, bond cleavage frequently is competitive with it, even if the threshold energy of the cleavage is higher.

Generally, IRMP-induced reactions result in the formation of a single product from the lowest energy reaction. In some cases, however, it is possible to activate ions to more than one reaction threshold, and then multiple products are observed. In these multiple-product reactions, the branching ratio depends on the light intensity, offering the possibility of controlling the chemical reaction by varying the laser parameters, including fluence, intensity, and wavelength.

Multiple products in IRMP-induced reactions arise not only from the reactant acquiring internal energy in excess of more than one reaction threshold (true photochemical branching) but also from IRMP excitation of a primary product. In addition, an IRMP-induced reaction may give rise to an intermediate for which several possible reaction channels exist.

In the case of photochemical branching, products resulting from activation above several thresholds can be expected when the thresholds are very close in energy. The dissociation of isotopically labeled alkoxides constitutes an example of such branching. An example of an IRMP-induced reaction that gives rise to an intermediate that subsequently

branches is the dissociation of 1,1,1-trifluoroacetone enolate (8). The reaction products are deprotonated ketene (HC_2O^-) and trifluoromethyl anion (CF_3^-), and the ratio of these products depends on the light intensity. True photochemical branching would result from competition between a four-center concerted elimination of CF_3H and a heterolytic cleavage resulting in CF_3^-. Branching from an intermediate ion-molecule complex that is photochemically generated appears to be reasonable since the activation barriers for four-center elimination are large. Branching would then arise from competition between separation of the complex, giving rise to CF_3^-, and proton transfer, yielding ketene enolate.

Ion-molecule reactions. Ions and neutral molecules have long-range attractive interactions arising from ion-dipole and ion–induced dipole forces. Potential surfaces for ion-molecule reactions have energy minima due to these interactions. Two examples of such potential surfaces are depicted in Fig. 2, one with a single minimum in free energy and the other with a double minimum. At the low pressures typical of ion-molecule experiments (10^{-7} to 1 torr), the intermediate complexes frequently cannot be stabilized by collision prior to dissociation and, therefore, still contain the energy of the attractive stabilization. These intermediates (for example, $[A^- \cdots B]^*$ in Fig. 2) are vibrationally excited. The intermediate dissociates either back to the reactants or to the products, through the entrance or exit channel, at a rate governed by this excitation energy; and, as is the case for many unimolecular reactions, this rate can be modeled accurately by statistical reaction-rate theory.

From such a theoretical modeling of

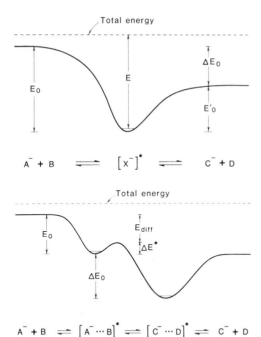

$$A^- + B \rightleftharpoons [X^-]^* \rightleftharpoons C^- + D$$

$$A^- + B \rightleftharpoons [A^- \cdots B]^* \rightleftharpoons [C^- \cdots D]^* \rightleftharpoons C^- + D$$

Fig. 2. (Top) Single-minimum potential energy surface for an ion-molecule reaction. (Bottom) Double-minimum potential energy surface for an ion-molecule reaction.

ion-molecule reactions with single-minimum surfaces, an exothermic reaction of A^- and B to form products would be expected from each collision of A^- and B. In other words, the efficiency of the reaction, the ratio of reaction rate constant to the collision rate constant ($k_{observed}/k_{collision}$), would be equal to one.

Not all ion-molecule reactions have unit efficiencies, and the double-minimum potential surface has been suggested to account for some reactions with low efficiencies (9). For these potential surfaces, the reaction is slow because passage over the central barrier has an unfavorable entropy relative to dissociation back to reactants. The entropy is lower for the forward reaction because its transition state involves the formation

of chemical bonds in which rotations have been converted to vibrations with a consequent decrease in the available number of quantum states. The relative energies of the entrance channel and the central potential energy maximum, E_{diff}, in this model determines the efficiency of the reaction. This model also predicts that reactions that are slow for this reason will become slower as the energy (or temperature) increases (10).

Gas-phase proton transfers are a set of ion-molecule reactions that have been extensively studied to probe their potential energy surfaces. Proton transfers between ions with localized charge (that is, with the charge centered primarily on one atom in the ion) have been shown to have an efficiency ratio equal to one (11). Proton transfers involving ions with delocalized charge, such as enolates, phenoxides, and the benzyl anion, do not have unit efficiencies and can be modeled with a double-minimum potential energy surface with a central barrier (9). These reactions may be slow due to small stabilization energies, presumably from poor hydrogen bonding in the ion-molecule complex. The central barriers may be somewhat higher than those for ions with localized charge due to the structural reorganization or charge localization that are required for the reactions to occur. The dynamics of endothermic proton transfers have been studied using a hydrogen-deuterium exchange technique (12).

The height of the central barrier, also referred to as the activation energy ΔE^*, for ion-molecule reactions can be determined by a kinetic analysis involving the use of RRKM (Rice-Ramsperger-Kassel-Marcus) theory, a statistical reaction rate theory. The difference in energy (E_{diff}) between the entrance channel transition state and the central barrier in this model is adjusted to give agreement with the experimentally observed efficiency. The activation energy ΔE^* can then be determined if the value of the well depth (the energy minimum between the barriers) is known or can be reliably estimated.

A general class of reactions that can be modeled with a double-minimum potential are methyl transfer reactions, nucleophilic displacement (S_N2) reactions on methyl groups (13). Many of these reactions have efficiency ratios of less than one and the central barrier heights can be determined by the above analysis (14). The barrier heights calculated by this method can be correlated with the changes in enthalphy of the reaction by means of a rate-equilibrium relationship such as the one embodied in the Marcus theory.

The Marcus theory separates the activation barrier height into two components, one an intrinsic component and the other arising solely from the overall reaction energetics. That is, the height of the energy barrier for an elementary reaction is related to not only the standard energy change for the reaction but also to an intrinsic barrier, the barrier of a hypothetical reaction for which $\Delta E_0 = 0$. For the methyl transfer reactions, the intrinsic activation energies ΔE^*_0 have been compiled for nucleophile-leaving group pairs (14). The intrinsic barrier for a cross reaction

$$X^- + CH_3Y \rightleftharpoons CH_3X + Y^- \qquad (2)$$

is the same in the forward or reverse direction, and ΔE^*_0 for the identity reaction (X = X')

$$X'^- + CH_3X \rightleftharpoons CH_3X' + X^- \qquad (3)$$

is obtained directly or by assuming that

the ΔE^*_0 for the cross reaction is the mean of the two corresponding identity reactions. Values of ΔE^*_0 (X,X) are listed in Table 1.

Nucleophiles in general appear to have large intrinsic barriers for methyl transfer, especially first row atoms. Strong nucleophiles such as Cl^- and Br^- react rapidly in identity reactions both in the gas phase and in solution. Most of the activation energy for the Cl^- identity reaction in solution is a consequence of solvation effects; F^- and RO^- have high intrinsic barriers in the gas phase as well as additional activation barriers in solution due to solvation. Recent quantum mechanical calculations for the Cl^- identity reactions in the gas phase coupled with statistical mechanical simulations of an aqueous phase potential surface clearly show the influence of solvation: a flattening of the free-energy minima for the ion-molecule complex and a large increase in the free energy of activation (15). Albery, Kreevoy, Lewis, and Kukes have applied Marcus theory to solution phase reactions and have found reasonable and consistent intrinsic barriers for methyl transfer (16). Further studies of gas-phase methyl transfer reactions have shown that the experimental change in ΔE^* with the ΔE_0 of the

Table 1. Intrinsic nucleophilicities of selected anions.

X^-	ΔE^*_0 ($X^- + CH_3X$) (kcal/mol)
Cl^-	10.2
Br^-	11.2
F^-	26.2
CH_3O^-	26.6
t-BuO^-	28.8

reaction (given by the Bronsted coefficient) agrees with that calculated by Marcus theory (*17*).

Ion-molecule complexes. The techniques described in the previous section dealt with understanding the nature of the barrier in a double-minimum potential energy surface of ion-molecule reactions. The nature and magnitude of the minima are also important, because they are the most energetically stable configurations of the neutral and ionic fragment. Ion-molecule complexes such as these are bound by long-range attractive forces, and they are intermediates in gas-phase ion-molecule reactions such as the proton or methyl transfer reactions. The binding energy of the complex, along with the central barrier height and the overall thermochemistry, completely characterize the double-minimum potential energy surface of a reaction.

The values of the binding energies of ion-molecule complexes are thus of great interest. They have been determined by several experimental techniques. High-pressure mass spectrometry has been used by Kebarle to study the temperature dependence of ion-molecule equilibria, leading directly to binding energies (*18*). Both Bartmess and McMahon have determined energy-well depths using pulsed ICR techniques analogous to those developed for determining gas-phase acidities (*19*). Relative binding energies of ion-molecule complexes are obtained by exchange of the ion between two neutral molecules:

$$R_1H\cdots Cl^- + R_2H \rightleftharpoons$$
$$R_2H\cdots Cl^- + R_1H \qquad (4)$$

These experiments have elucidated the factors involved in the binding of alkoxide-alcohol complexes, halide-alcohol complexes, and halide-acid complexes. Single molecule binding energies have been shown to be influenced by the size of the ion and the steric requirements of the neutral species.

The binding of gas-phase ions to molecules provides a means of understanding solvation phenomena and also provides a starting point for the investigation of the transition from gas phase to solution phase. The behavior of large clusters of neutral molecules with an ion may approach the behavior of ions in a condensed phase. A key question is at what point do cluster ions have essentially the properties of an ion in solution?

A number of new techniques have been developed specifically for generating ion clusters in the gas phase and studying their stability. Secondary-ion mass spectrometry has been used to generate positively charged ion clusters of nitrogen and nitrogen oxides (*20*). Collision-induced dissociation has given some information on the internal structure of these clusters, which appear to behave, qualitatively, as a central ion with loosely bound solvating molecules. Negatively charged ion clusters have been generated by collisional electron transfer from alkali atoms to neutral clusters formed by expanding a supersonic molecular beam (*21*). Negatively charged ion clusters of hydrogen halides, $(Cl_2)_n$, $(SO_2)_n$, $(NO_2)_n$, $SO_2(CO_2)_n$, and $(H_2S)_n$ have been formed, and the technique appears to be useful for obtaining electron affinities of clusters only as large as dimers. The influence of the size of the cluster ion on the rate of reaction has been determined for the gas-phase nucleophilic substitution reactions of hydroxide ion in water clusters, but the interpretation of the data is controversial at present (*22*).

Ion Energetics

An understanding of ion energetics is a prerequisite for the formulation and evaluation of dynamical theories of ion-molecule reactions. The thermodynamic parameters that have been shown to be of fundamental importance in this regard are the electron affinities and ionization potentials of molecules and radicals, and the bond strengths of ions and neutral species. Thermochemical cycles can be constructed that connect these quantities to the acidities and basicities of species in the gas phase, properties that have been shown to be substantially different from the corresponding properties of species in condensed phases. New experimental techniques have resulted in extensive tabulations of these thermochemical data, and the understanding of these values has been put on a more solid foundation (23).

Systematic evaluation of the gas-phase acidities of neutral molecules and the proton affinities of negative ions continue to provide the basis for any discussion of ionic stabilities (23). Comparisons of these data with solution-phase values provide important insights into solvation effects, especially those that are anomalous and specific to certain classes of ions (24).

Methods for determining ion-molecule interaction energies and ionic bond strengths as well as the electron affinities of negatively charged ion clusters are among the recent developments in the acquisition of thermodynamic data for gas-phase ions. The measurement of intrinsic ion properties such as these has been useful in gaining an understanding of structure-reactivity relationships for organic molecules. Analogous data for the bond strengths of transition metal ions and their clusters have not been available until quite recently. Metal-hydrogen and metal-carbon bond strengths of many transition metal cations have been obtained and analyzed by Beauchamp, Armentrout, and co-workers, using ion-beam mass spectrometry (25). By adjusting the relative kinetic energies of the transition metal ions, the threshold energies for the formation of a metal-ligand bond can be obtained. Periodic trends in the bond strengths have been analyzed and modeled by simple covalent and electrostatic bonding models. The reactivity and bond energies of transition metal cluster ions, such as Mn_2^+, have been determined (25). Both metal-metal bond energies and ionization potentials have been obtained.

Determination of Ion Structure

Spectroscopy of both neutral and ionic molecules in the gas phase encompasses transitions among rotational, vibrational, and electronic energy levels and can give rise to information leading to a specification of molecular structure. The principle spectroscopic methods used to determine ion structure can be grouped into two categories: those which measure light absorption or emission directly (absorption or emission spectroscopy) and those which measure an event, such as photodissociation or electron photodetachment, that follows absorption of light (action spectroscopy). Both types of spectroscopy give rise to detailed structural information about gas-phase ions.

Conventional absorption spectroscopy of gas-phase ions has developed at a much slower pace than that of neutral, closed-shell molecules. This is due to the difficulty in containing ions at such densities that light absorption can be reliably measured and also in devising detection schemes that minimize high background signals from a neutral precursor. In the past 10 years, with the widespread application of laser technology and much clever experimentation and development of new detection methods, considerable progress has been made.

Absorption spectroscopy. The detection of transitions between rotational states by microwave spectroscopy was achieved initially with two- and three-atom molecular ions (26). Microwave spectra of HCO^+, NNH^+, and CO^+ were measured with a cell that sustained a direct-current glow discharge. Modulation of the source or magnetic field allowed for highly sensitive detection of microwave absorption. Recent work with improved detector sensitivity has permitted the characterization of HOC^+ and HCS^+.

Transitions between rotational states in simple paramagnetic ions have also been observed by laser magnetic resonance, a technique that makes use of a magnetic field to lift the degeneracy of magnetic sublevels of the ion (27). By varying the magnetic field strength, transitions between the sublevels of rotational states can be brought into resonance with an available laser frequency. A number of small ions, such as the positive ions of the hydrogen halides, have been studied with this technique (28). The data obtained have been used to derive structural parameters for the ions, such as rotational constants and centrifugal distortion constants, as well as details of electronic structure.

With vibrational spectroscopy, several absorption experiments have produced results for simple positive ions. These include a difference-frequency laser absorption experiment for H_3^+ (29) and for HeH^+, NeH^+, and H_3O^+ (30) and the measurement of high-resolution spectra of protonated rare gases using Fourier-transform detection methods (31).

More recently, a new technique for detecting infrared absorption of ions has been developed (32). Ion velocities are modulated by alternating the polarity of a glow discharge, thus shifting the ion absorption frequencies according to the Doppler effect. Phase sensitive detection yields the absorption spectrum of the ion without interference from the spectrum of the neutral precursor, which is in much greater abundance. Vibration-rotation transitions have been observed with this technique for HCO^+ (33), HNN^+ (34), H_3O^+ (35), and NH_4^+ (36) as well as other species.

Electronic spectra can be used to study the structures of both the ground and excited states of ions in which the excited states can be reached either by bombarding the corresponding neutral species with electrons or by direct optical excitation. The fluorescence emission spectra consist of vibronic bands that give rise to the determination of vibrational frequencies for ground and excited states and also rotational constants for small ions. Highly precise data have been obtained for moderate size cations by lowering the temperature of the ions. The cooling is accomplished by expanding the sample gas in a supersonic free jet. Recent results from this technique have been reviewed (37, 38). In general, open-shell organic cations that are unsaturated, such as substituted acetylene or benzene ions, have been studied by this method because the energy of the emitted photons lies in the visible region of the spectrum, a favorable region for fluorescence detection. Recent work on laser-induced fluorescence of ionic clusters of fluorinated benzenes and inert gases appears to provide information concerning the binding sites of the inert atoms in the ion (39).

Action spectroscopy. Three general areas of ion spectroscopy can be grouped into the category of action spectroscopy: photodissociation, electron photodetachment, and photoelectron spectroscopy. These methods are characterized by the detection of an event following the absorption of light by the ion.

The technique of using a photodissociation event to study ion structure is made possible by producing an excited state of an ion above a dissociation threshold. The excited state either predissociates by mixing with a repulsive excited state or internally converts to a hot ground state and then dissociates. Spectra are inferred from the wavelength dependence of the dissociation, assuming a constant quantum yield for the process. For example, if all absorption events result in dissociation, then the action spectrum and the absorption spectrum are identical.

Photodissociation experiments have been carried out in ICR spectrometers and have been reviewed (40). These experiments generally have low spectral resolution and give rise to electronic spectra that, in some cases, have vibronic structure (41). Ions are generated and trapped in an ICR cell, and the photodissociation cross sections are calculated by measuring the fractional decrease in the signal of the ion that is being photodissociated.

Ion beams have been employed by many different research groups to study the photodissociation of small ions (42). The technique of photofragment spectroscopy has been used to probe small ions such as O_2^+, Cs_2^+, and SO^+ (43). In general, an ion beam interacts coaxially with continuous-wave laser light producing fragment ions that, in turn, are separated according to their mass by an electrostatic quadrupole, selected according to their energy content, and detected. Photofragment spectroscopy experiments in the ultraviolet region are being performed with an ArF excimer laser to study a variety of small ions (44) and, in the near-ultraviolet and ultraviolet regions, to study the high-resolution spectroscopy of N_2O^+ (45). In the infrared region, photofragment spectroscopy with a continuous-wave CO_2 laser has been used to study vibration-rotation transitions of H_3^+ and HeH^+ ions (46).

A number of advances have been made recently in the area of electron photodetachment of negative ions. In this method, ions are irradiated in an ICR or ion beam; and, depending upon the technique, ion depletion, neutral products, or electron production is measured as a function of light frequency (47). For negative ions, photodetachment produces a neutral species in either its ground state or an excited state. If the geometries of the neutral species and the anion are substantially different, then transitions in which vibrational quantum numbers change become important, and vibrational transitions can be observed. These vibrational transitions manifest themselves as onsets in photodetachment cross sections and have provided information on vibronic transitions, rotational fine structure, and spin-orbit splittings in alkoxides, thioalkoxides, and hydrogen sulfide (48).

Transitions to electronically excited states of anions can be observed as resonances in photodetachment cross sections. The energy of these excited anion

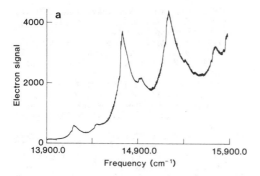

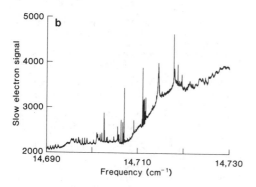

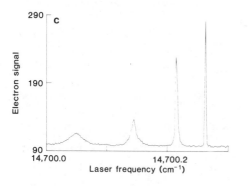

Fig. 3. Electron photodetachment spectra of acetaldehyde enolate: (a) 1 cm^{-1} resolution; (b) data near the 0-0 vibrational transition at 0.03 cm^{-1} resolution; (c) individual rotational transitions fully resolved at a Doppler-limited resolution of 0.0007 cm^{-1}. [Reprinted with permission of R. D. Mead, K. R. Lykke, W. C. Lineberger, J. Marks, J. I. Brauman, *Journal of Chemical Physics* **81**, 4883 (1984). © 1984 American Institute of Physics.]

states usually exceeds the thermodynamic threshold for electron detachment. Excitation to the electronically excited state followed by autodetachment provides a pathway for electron loss in addition to direct photodetachment. Broad unstructured resonances have been observed for large delocalized anions that correspond to electronic transitions to excited anion states (*49*). In the threshold region for photodetachment narrow resonances have been observed that have been assigned as dipole-bound states, diffuse excited states of anions that are analogous to Rydberg states of neutral molecules (*50*). Ultra-high-resolution photodetachment of acetaldehyde enolate and other anions has been carried out using an ion beam and a ring dye-laser operating single-mode (*51, 52*). Spectra for the electron photodetachment of acetaldehyde enolate are shown in Fig. 3. The data display the remarkable difference in the features of the spectrum with progressively higher resolution. High-resolution spectra show that for high rotational energy levels the line widths become immeasurably broad. These energy levels lie above the threshold for direct photodetachment and have very short lifetimes. The lowest rotational levels were not observed because they lie below the threshold of photodetachment and thus are bound. A complete analysis of the transitions between rotational states of the anion ground state and the excited dipole-bound state has yielded spectroscopic constants and geometries for both states.

Conclusion

The impact of gas-phase ion chemistry on the thinking of chemists has been considerable. Intrinsic thermodynamic and kinetic properties have been determined and evaluated with this technique, leading to a clearer understanding of the relation between structure and reactivity in ionic reactions as well as clarifying the role of solvation in the condensed phase. Measurement of thermodynamic and structural parameters by new spectroscopic methods offers the promise of a better understanding of the properties underlying the structures of ionic molecules.

References and Notes

1. R. G. Cooks, in *Collision Spectroscopy*, R. G. Cooks, Ed. (Plenum, New York, 1978), p. 357.
2. W. Tumas, R. F. Foster, J. I. Brauman, *Isr. J. Chem.* **24**, 223 (1984).
3. R. V. Ambartzumian and V. S. Letokhov, in *Chemical and Biochemical Applications of Lasers*, C. B. Moore, Ed. (Academic Press, London, 1977), vol. 3, p. 167; D. M. Golden, M. J. Rossi, A. C. Baldwin, J. R. Booker, *Acc. Chem. Res.* **14**, 56 (1981).
4. W. Tumas, R. F. Foster, M. J. Pellerite, J. I. Brauman, *J. Am. Chem. Soc.* **105**, 7464 (1983).
5. W. Tumas, R. F. Foster, J. I. Brauman, *ibid.* **106**, 4053 (1984).
6. C. H. DePuy, V. M. Bierbaum, R. Damrauer, *ibid.*, p. 4051.
7. R. F. Foster, W. Tumas, J. I. Brauman, *J. Chem. Phys.* **79**, 4644 (1983).
8. C. R. Moylan, J. M. Jasinski, J. I. Brauman, *Chem. Phys. Lett.* **98**, 1 (1983).
9. W. E. Farneth and J. I. Brauman, *J. Am. Chem. Soc.* **98**, 7891 (1976); W. N. Olmstead and J. I. Brauman, *ibid.* **99**, 7203 (1977); O. I. Asubiojo and J. I. Brauman, *ibid.* **101**, 3715 (1979).
10. J. I. Brauman, in *Kinetics of Ion-Molecule Reactions*, P. Ausloos, Ed. (Plenum, New York, 1979), pp. 153–164.
11. D. K. Bohme, in *Interaction Between Ions and Molecules*, P. Ausloos, Ed. (Plenum, New York, 1974), pp. 489–504.
12. J. J. Grabowski, C. H. DePuy, V. M. Bierbaum, *J. Am. Chem. Soc.* **105**, 2565 (1983); R. R. Squires, V. M. Bierbaum, J. J. Grabowski, C. M. DePuy, *ibid.*, p. 5185.
13. D. K. Bohme and L. B. Young, *ibid.* **92**, 7354 (1970); D. K. Bohme, G. I. Mackay, J. D. Payzant, *ibid.* **96**, 4027 (1974); K. Tanaka, G. I. Mackay, D. K. Bohme, *Can. J. Chem.* **54**, 1643 (1976).
14. M. J. Pellerite and J. I. Brauman, *J. Am. Chem. Soc.* **102**, 5993 (1980); M. J. Pellerite and J. I. Brauman, *ibid.* **105**, 2672 (1983).
15. J. Chandrasekhar, S. F. Smith, W. L. Jorgensen, *ibid.* **106**, 3049 (1984).
16. W. J. Albery, *Annu. Rev. Phys. Chem.* **31**, 227 (1980); W. J. Albery and M. M. Kreevoy, *Adv. Phys. Org. Chem.* **16**, 87 (1978); E. S. Lewis and S. Kukes, *J. Am. Chem. Soc.* **101**, 417 (1979).
17. J. A. Dodd and J. I. Brauman, *J. Am. Chem. Soc.* **106**, 5356 (1984).
18. P. Kebarle, *Annu. Rev. Phys. Chem.* **28**, 445 (1977); G. Caldwell and P. Kebarle, *J. Am. Chem. Soc.* **106**, 967 (1984).
19. G. Caldwell, M. D. Rozeboom, J. P. Kiplinger, J. E. Bartmess, *J. Am. Chem. Soc.* **106**, 4660 (1984); J. W. Larson and T. B. McMahon, *ibid.*, p. 517.
20. T. F. Magnera, D. E. David, R. Tian, D. Stulik, J. Michl, *ibid.* **106**, 5040 (1984); D. Stulik, R. G. Orth, H. T. Jonkman, J. Michl, *Int. J. Mass Spectrom. Ion Phys.* **53**, 341 (1983); J. Michl, *ibid.*, p. 255.
21. K. H. Bowen, G. W. Liesegang, R. A. Sanders, D. R. Herschbach, *J. Phys. Chem.* **87**, 557 (1983); E. L. Quitevis, K. H. Bowen, G. W. Liesegang, D. R. Herschbach, *ibid.*, p. 2076.
22. D. K. Bohme and G. I. Mackay, *J. Am. Chem. Soc.* **103**, 978 (1981); D. K. Bohme, A. B. Rakshit, G. I. Mackay, *ibid.* **104**, 1100 (1982); M. Henchman, J. F. Paulson, P. M. Hierl, *ibid.* **105**, 5510 (1983).
23. J. E. Bartmess and R. T. McIver, Jr., in *Gas Phase Ion Chemistry*, M. T. Bowers, Ed. (Academic Press, New York, 1979), vol. 2, p. 87; R. W. Taft, *Prog. Phys. Org. Chem.* **14**, 247 (1983); C. R. Moylan and J. I. Brauman, *Annu. Rev. Phys. Chem.* **34**, 187 (1983).
24. M. Mishima, R. T. McIver, Jr., R. W. Taft, F. G. Bordwell, W. N. Olmstead, *J. Am. Chem. Soc.* **106**, 2717 (1984).
25. M. L. Mandich, L. F. Halle, J. L. Beauchamp, *ibid.*, p. 4403; P. B. Armentrout, S. K. Loh, R. M. Ervin, *ibid.*, p. 1161; K. Ervin, S. K. Loh, N. Aristov, P. B. Armentrout, *J. Phys. Chem.* **87**, 3593 (1983).
26. R. C. Woods, in *Molecular Ions, J:* Berkowitz and K. O. Groeneveld, Eds. (Plenum, New York, 1983), p. 11; R. J. Saykally and R. C. Woods, *Annu. Rev. Phys. Chem.* **32**, 403 (1981).
27. R. J. Saykally, K. G. Lubic, K. M. Evenson, in *Molecular Ions, J.* Berkowitz and K. O. Groeneveld, Eds. (Plenum, New York, 1983), p. 33.
28. R. J. Saykally and K. M. Evenson, *Phys. Rev. Lett.* **43**, 515 (1979); D. Ray, K. G. Lubic, R. J. Saykally, *Mol. Phys.* **46**, 217 (1982); D. C. Houde, E. Schäfer, S. E. Strahan, C. A. Ferrari, D. Ray, K. G. Lubic, R. J. Saykally, *ibid.* **52**, 245 (1984).
29. T. Oka, *Phys. Rev. Lett.* **45**, 531 (1980).
30. P. Bernath and T. Amano, *ibid.* **48**, 20 (1982); M. Wong, P. Bernath, T. Amano, *J. Chem. Phys.* **77**, 693 (1982); P. R. Bunker, T. Amano, V. Spirko, *J. Mol. Spectrosc.* **107**, 208 (1984).
31. J. W. Brault and S. P. Davies, *Phys. Scr.* **25**, 268 (1982); J. W. C. Johns, *J. Mol. Spectrosc.* **106**, 124 (1984).
32. C. S. Gudeman and R. J. Saykally, *Annu. Rev. Phys. Chem.* **35**, 387 (1984).
33. C. S. Gudeman, M. H. Begemann, J. Pfaff, R. J. Saykally, *Phys. Rev. Lett.* **50**, 727 (1983); T. Amano, *J. Chem. Phys.* **79**, 3595 (1983); S. C. Foster, A. R. W. McKellar, T. J. Sears, *ibid.* **81**, 578 (1984).
34. C. S. Gudeman, M. H. Begemann, J. Pfaff, R. J. Saykally, *J. Chem. Phys.* **78**, 5837 (1983).
35. M. H. Begemann, C. S. Gudeman, J. Pfaff, R. J. Saykally, *Phys. Rev. Lett.* **50**, 727 (1983); N. N. Haese and T. Oka, *J. Chem. Phys.* **80**, 572 (1984).
36. M. W. Crofton and T. Oka, *J. Chem. Phys.* **79**, 3157 (1983); E. Schäfer, M. H. Begemann, C. S.

Gudeman, R. J. Saykally, *ibid.*, p. 3159; E. Schäfer, R. J. Saykally, A. G. Robinette, *ibid.* **80**, 3969 (1984).

37. D. Klapstein, J. P. Maier, L. Misev, in *Molecular Ions: Spectroscopy, Structure, and Chemistry*, T. A. Miller and V. E. Bondybey, Eds. (Elsevier, New York, 1983) p. 175.

38. M. I. Lester, B. R. Zegarski, T. A. Miller, *J. Phys. Chem.* **87**, 5228 (1983); T. A. Miller, *Science* **223**, 545 (1984).

39. L. F. Dimanro, M. Heaven, T. A. Miller, *Chem. Phys. Lett.* **104**, 526 (1984).

40. R. C. Dunbar, in *Gas Phase Ion Chemistry*, M. T. Bowers, Ed. (Academic Press, New York, 1984), vol. 3, p. 130; R. Dunbar, in *Molecular Ions: Spectroscopy, Structure, and Chemistry*, T. A. Miller and V. E. Bondybey, Eds. (Elsevier, New York, 1983), p. 231.

41. R. G. Orth and R. C. Dunbar, *J. Am. Chem. Soc.* **104**, 5617 (1982); C. M. Rynard and J. I. Brauman, *Inorg. Chem.* **19**, 3544 (1980).

42. J. T. Moseley and J. Durup, *Annu. Rev. Phys. Chem.* **32**, 53 (1981).

43. J. C. Hansen, J. T. Moseley, P. C. Cosby, *J. Mol. Spectrosc.* **98**, 48 (1983); H. Helm, P. C. Cosby, D. L. Huestis, *J. Chem. Phys.* **78**, 6451 (1983); P. C. Cosby, *ibid.* **81**, 1102 (1984).

44. R. E. Kestiza, A. K. Edwards, R. S. Pandolfi, J. Berkowitz, *J. Chem. Phys.* **80**, 4112 (1984).

45. S. Abed, M. Broyer, M. Carré, M. L. Gaillard, M. Larzillière, *Chem. Phys.* **74**, 97 (1983); R. Frey, R. Kakoschke, E. W. Schlag, *Chem. Phys. Lett.* **93**, 227 (1982).

46. A. Carrington, R. A. Kennedy, T. P. Softley, P. G. Fournier, E. G. Richard, *Chem. Phys.* **81**, 251 (1983); A. Carrington and R. A. Kennedy, *J. Chem. Phys.* **81**, 91 (1984).

47. P. S. Drzaic, J. Marks, J. I. Brauman, in *Gas Phase Ion Chemistry*, M. T. Bowers, Ed. (Academic Press, New York, 1984), vol. 3, p. 167; R. D. Mead, A. E. Stevens, W. C. Lineberger, *ibid.*, p. 214.

48. B. K. Janousek, A. H. Zimmerman, K. J. Reed, J. I. Brauman, *J. Am. Chem. Soc.* **100**, 6142 (1978); B. K. Janousek and J. I. Brauman, *J. Chem. Phys.* **72**, 694 (1980); *Phys. Rev. A* **23**, 1673 (1981).

49. A. H. Zimmerman, R. Gygax, J. I. Brauman, *J. Am. Chem. Soc.* **100**, 5595 (1978); R. Gygax, H. L. McPeters, J. I. Brauman, *ibid.* **101**, 2567 (1979).

50. A. H. Zimmerman and J. I. Brauman, *J. Chem. Phys.* **66**, 5823 (1977); R. L. Jackson, P. C. Hiberty, J. I. Brauman, *ibid.* **74**, 3705 (1981).

51. K. R. Lykke, R. D. Mead, W. C. Lineberger, *Phys. Rev. Lett.* **52**, 2221 (1984).

52. R. D. Mead, K. R. Lykke, W. C. Lineberger, J. Marks, J. I. Brauman, *J. Chem. Phys.* **81**, 4883 (1984).

53. We are grateful to the National Science Foundation and the Petroleum Research Fund, administered by the American Chemical Society, for support of this work. We also thank A. T. Barfknecht and W. Tumas for their many suggestions and for their critical reading of the manuscript.

Dramatic advances have taken place in the generation and application of ultrashort optical pulses. Laser measurement techniques have progressed to the point where investigations of ultrafast phenomena in condensed matter can now be accomplished on the femtosecond (10^{-15} second) time scale. Optical pulses as short as 30 fsec have been generated (*1*) by pulse compression techniques. The application of these advanced techniques is expected to produce new insights into dynamic processes in condensed matter and to have an impact on physics, chemistry, and biology. In this article I describe the basic concepts and new techniques that permit experimental investigations in the femtosecond regime and review the implications and potential areas of application of these measurement tools.

As we move into the femtosecond realm, a whole new range of problems becomes accessible for investigation. If we consider that a 30-fsec pulse corresponds to 1000 cm^{-1}, it is clear that we can study the properties of liquids and solids in a time less than the period of many important vibrations and well under a typical collision time in a liquid. In essence, with these ultrashort optical pulses it should be possible to coherently excite a liquid or solid and resolve the fundamental dephasing mechanisms. On a sufficiently short time scale we should be able to observe the evolution of non-Markovian processes to the statistical limit.

Although picosecond optical pulses have been very important for studying many dynamic processes, increased resolution can make the difference between qualitative and quantitative observations. A good example is the study of the nonradiative relaxation of the molecule

6

Measurement of Ultrafast Phenomena in the Femtosecond Time Domain

C.V. Shank

Science 219, 1027–1031 (4 March 1983)

azulene (2). Early picosecond work could only set limits for the process (3), but with the advent of higher resolution techniques detailed quantitative studies were accomplished (4). But while increased resolution is often valuable for refining our understanding, I believe that the most important effect of femtosecond spectroscopy will be to open up doors to the discovery of new phenomena.

Ultrashort Optical Pulses

During the decade and a half since the first generation of ultrashort laser pulses, optical pulse generation techniques have continued to improve. In Fig. 1 the shortest reported optical pulse width is plotted against year. The first publication on optical pulses in the picosecond range was in 1966 (5). By 1972 pulses on the order of a picosecond were generated with the continuously mode-locked dye laser (6). Improvements of this laser led to the generation of optical pulses shorter than 1 psec in 1974 (7). Within the past 2 years, a new type of passively mode-locked dye laser has extended the reach of attainable pulse widths to less than 100 fsec (8). Most recently, optical pulse compression techniques have led to the shortest reported optical pulse width yet attained, 30 fsec (1). Even shorter optical pulses appear on the horizon. However, since an optical pulse 30 fsec in duration corresponds to only 14 cycles of light, it is clear that we are fast approaching fundamental limits.

Femtosecond Pulse Generation

Techniques

Passive mode locking of the continuous wave (CW) dye laser is now a well-established technique for producing ultrashort optical pulses (9). Figure 2 shows a new kind of dye laser pulse generator called a colliding pulse mode-locked laser (8). This configuration consists of a series of mirrors forming a ring cavity that contains only two essential elements: an optically pumped saturable gain dye (rhodamine 6G) and a saturable absorber dye (diethyloxacarbocyanine iodide) at the two focal points in the cavity. An argon laser operating at 514.5 nanometers is used as an optical pump. The lack of complexity of this configuration accounts in part for its ability to generate femtosecond optical pulses. By minimizing the amount of material in the cavity, the effects of group velocity dispersion are reduced, allowing the cavity to sustain a broad bandwidth of oscillating frequencies necessary to form a short optical pulse.

The mechanism for pulse shortening (10) in this laser configuration is very similar to that in the passively mode-

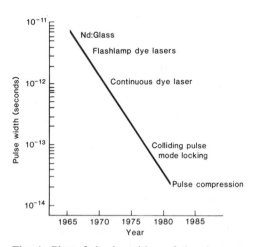

Fig. 1. Plot of the logarithm of the shortest reported optical pulse width versus year. Note that each reduction in pulse width has been accompanied by an advance in pulse generation technology.

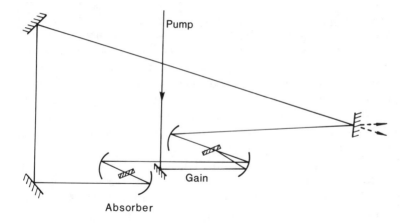

Fig. 2. Laser cavity configuration for a colliding pulse mode-locked dye laser.

Pump

Gain

Absorber

locked dye laser, but with some additions. If we consider an optical pulse traveling around the ring, the pulse is shaped each time it is amplified by the gain dye and absorbed by the saturable absorber. The leading edge of the pulse is preferentially clipped by the saturable absorber. Preferential amplification of the leading edge of the pulse, combined with cavity loss, effectively clips the rear edge of the pulse with each pass around the cavity until a limiting pulse width is achieved. An additional mechanism operative in the colliding pulse ring configuration results from the fact that there are two equally stable but oppositely directed pulses which "collide" as they meet each other traveling around the ring. The energetically most favorable place for the two pulses to meet is in the saturable absorber. Since the pulses are coherent, they can interfere and set up a standing wave pattern in the absorber. The standing wave pattern minimizes the energy lost, because the field is most intense where the absorption is saturated and is weakest in the field minima where the absorption is not saturated. The shortest optical pulses are produced with a thin absorbing region that confines the standing wave field. Figure 3 shows the

experimentally measured autocorrelation function obtained with second-harmonic generation in a crystal of potassium dihydrogen phosphate (KDP). The full width at half-maximum (FWHM) of the optical pulse determined from the autocorrelation function is 65 fsec.

With optical pulse compression techniques even shorter optical pulses can be produced. More than a decade ago, Gires and Tournois (11) and Giordmaine

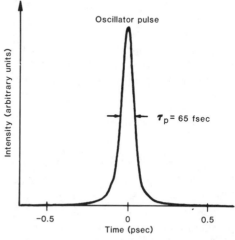

Fig. 3. Autocorrelation function of a 65-fsec optical pulse from a colliding pulse mode-locked dye laser. This autocorrelation measurement was obtained with second-harmonic generation in a KDP crystal.

et al. (*12*) proposed that optical pulses be shortened by adapting microwave pulse compression techniques to the visible spectrum. Optical pulse compression is accomplished in two steps. In the first step, a "chirp" or frequency sweep is impressed on the pulse. Then the pulse is compressed by using a dispersive delay line. A chirp can be impressed on an intense optical pulse simply by passing the pulse through an optical Kerr medium. When an intense optical pulse is passed through a nonlinear medium, the refractive index, n, is modified by the electric field, E,

$$n = n_0 + n_2 <E^2> + \ldots$$

A phase change, $\delta\phi$, is impressed on the pulse

$$\delta\phi = n_2 <E^2> \frac{\omega z}{c}$$

where ω is the frequency, z is the distance traveled in the Kerr medium, and c is the velocity of light. As the intensity of the leading edge of the optical pulse rises rapidly, a time-varying phase or frequency sweep is impressed on the pulse carrier. Similarly, a frequency sweep in the opposite direction occurs as the intensity of the pulse falls on the trailing edge. The amount of frequency sweep is given approximately by

$$\delta\omega = \left(\frac{\omega z n_2}{c}\right)\frac{d}{dt} <E^2(\tau)>$$

A more rigorous approach to this problem requires a solution of the wave equation with the addition of a nonlinear term to account for the Kerr nonlinearity. The problem can be reduced to the solution of the nonlinear Schroedinger equation. Solutions of these equations by numerical techniques have been reported (*13–15*).

Thus far I have only discussed a plane wave solution to the nonlinear frequency problem. In addition, I have neglected self-focusing effects, which are a natural consequence of perturbing the index of refraction of the medium. These problems can be overcome by using a single-mode optical fiber as the Kerr medium (*14, 16, 17*).

Once a chirp has been applied to a pulse, the pulse is passed through a dispersive delay to reassemble all its frequency components in order to achieve compression. A nearly ideal pulse compression device is a pair of parallel gratings (*18*). Each wavelength passing through the grating pair is diffracted at a different angle and follows a different path, giving rise to a wavelength-dependent optical path delay. By properly adjusting the grating spacing, we can provide the right amount of group delay to form the compressed pulse. Figure 4 shows the experimental arrangement for compressing a 90-fsec optical pulse. The optical pulse is focused into a 15-centimeter piece of single-mode polarization preserving optical fiber. The pulse is recollimated with a lens, passed through a grating pair, and sent to the pulse measuring apparatus. The measured autocorrelation function for the compressed pulse is plotted in Fig. 5. It is interesting to note that a 30-fsec optical pulse is significantly broadened when it passes through all the optical elements used for recollimating and directing pulses into the measuring apparatus. Fortunately, the grating pair compresses the chirped pulse and provides a means of compensating the other dispersive elements in the beam as well. Care must be taken when such short pulses are used to make measurements to eliminate possible artifacts resulting from dispersion.

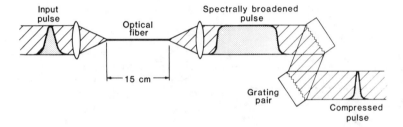

Fig. 4. Experimental arrangement for compressing femtosecond optical pulses.

Pulse compression has a significant advantage over direct pulse generation in a laser cavity for the generation of optical pulses in the femtosecond time regime. As shorter pulses are generated, increasing spectral width is required. In a laser, the gain bandwidth of the lasing medium and the optical cavity determine the lasing bandwidth. In contrast, the Kerr effect is operative from the ultraviolet to the infrared. It appears feasible to compress optical pulses to a few femtoseconds, which is nearly a single optical cycle.

Often it is desirable to amplify short optical pulses and to use nonlinear optical techniques to generate pulses at new frequencies. The peak pulse power coming directly from the dye laser oscillator is in the kilowatt range. An amplifier design (*19*) that can amplify optical pulses to gigawatt power levels and yet preserve the femtosecond pulse width is shown in Fig. 6. A frequency-doubled Nd:YAG (neodymium:yttrium aluminum garnet) laser at 530 nm is synchronized to a mode-locked dye laser and used to pump a four-stage dye amplifier. A saturable absorber dye isolates each stage. A grating pair is used to compensate the dispersion in the dye amplifiers and optical components. Figure 7 shows the autocorrelation function of an amplified pulse with gigawatt peak power and a pulse width of 70 fsec.

Shortly after the first generation of

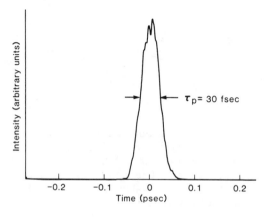

Fig. 5. Measured pulse autocorrelation function of the compressed 30-fsec optical pulse.

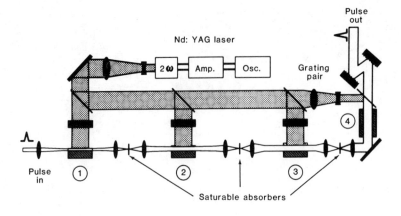

Fig. 6. Diagram of a four-stage dye amplifier for femtosecond optical pulses. With this amplifier a 70-fsec optical pulse can be amplified to gigawatt power levels.

ultrashort optical pulses, Alfano and Shapiro (20) discovered that by focusing an intense optical pulse into several centimeters of almost any clear liquid, a white light continuum pulse could be generated. The precise mechanisms of this generation process are still the subject of some controversy (21).

One of the most severe problems with extending the white light continuum into the femtosecond regime is a sweep in time of the various frequency components brought about by group velocity dispersion. Some time sweep is inevitable because of the broad frequency range over which the continuum extends. We

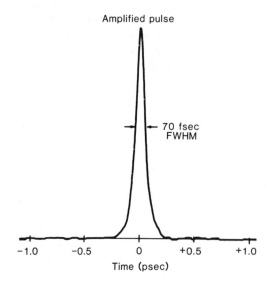

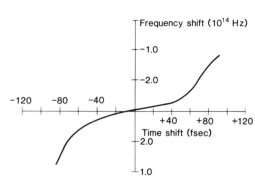

Fig. 7 (left). Pulse autocorrelation function of a 70-fsec amplified optical pulse. Fig. 8 (right). Measured time shift as a function of frequency for a white light femtosecond continuum pulse generated in a 400-μm-thick ethylene glycol stream.

can greatly reduce the sweep by generating the continuum in the shortest possible path length and eliminating lenses by use of reflective optics. We have succeeded in generating an 80-fsec white light pulse by focusing an intense gigawatt 70-fsec optical pulse into a 400-micrometer-thick free-flowing stream of ethylene glycol (22). With an excitation wavelength of 620 nm the generated continuum pulse extended from 0.19 to 1.6 μm.

The measured frequency sweep in the continuum is shown in Fig. 8. Note that over the entire range of the measurement, the time sweep is less than or comparable to the pulse width. We think that most of the observed frequency sweep is a result of the generation process. The shift to lower frequencies at early times, and to higher frequencies at later times, suggests that the self-phase-modulation process described in the section on pulse compression may play an important role in generating white light pulses as well. The spectral width of the white light pulse is too large to be completely described by self phase modulation. At the intensities used to generate this continuum pulse (10^{11} watts per square centimeter) it is not surprising that other nonlinear effects would contribute to the generation process (21).

Femtosecond Spectroscopy

The discussion in the foregoing section was limited to the pulse generation process itself. Now we turn to the application of these short pulses to measurement. A number of novel and clever techniques have been developed in the past decade to measure dynamic processes with short optical pulses (23). Many of these techniques can be directly adapted to measurements in the femtosecond time regime.

An experimental arrangement for measuring time-resolved spectra with femtosecond optical pulses is shown in Fig. 9. Optical pulses of wavelength 620 nm are generated with a colliding pulse mode-locked dye laser. These pulses are then amplified to gigawatt powers with the four-stage amplifier described previously. The amplified pulses are divided into two parts, one for exciting and the other for probing a sample. The excitation pulses are passed through a nonlinear frequency-shifting medium to generate the desired frequency. The frequency shifting can be achieved by using the stimulated Raman effect or some other nonlinear process. The probing pulses pass through a variable path or time delay controlled by a stepper motor and are then focused into an ethylene glycol stream to generate a white light continuum pulse. The white light continuum pulse is further divided into a measuring pulse and a reference pulse, which are directed into a spectrometer with a vidicon or optical multichannel analyzer on the output of the spectrometer. In this way, measurements of optically induced changes in absorption can be performed over a broad spectral range at different time delays following excitation as determined by the path delay controlled by the stepper motor. The resolution is determined by the convolution of the pumping and probing pulse widths. Care must be taken to limit artifacts caused by group delay in the various parts of the white light continuum spectrum.

The application of these measurement techniques is illustrated by some recent work on the dynamics of absorption and

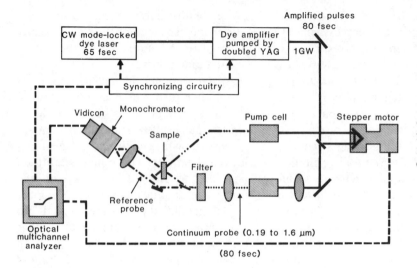

Fig. 9. Schematic of apparatus for femtosecond time-resolved spectroscopy.

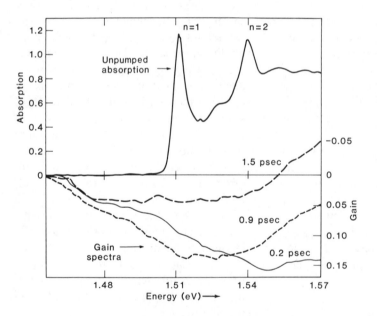

Fig. 10. Time-resolved absorption and gain spectra of an optically excited GaAs-GaAlAs multi-quantum well structure.

gain in a highly excited semiconductor multiquantum well structure made of thin, 200-angstrom layers of GaAs and GaAlAs (24). Time-resolved absorption measurements provide a unique means of observing the influence of hot electrons in optically excited semiconductors (25, 26). In this experiment, we sought to determine the distribution of hot carriers in a two-dimensional semiconductor multiquantum well within the first few hundred femtoseconds after optical excitation. In Fig. 10 time-resolved absorption spectra are plotted at various times after excitation. Note that the scale has been expanded for "gain" or negative absorption spectra. The noise level on these curves is ± 0.005 change in optical density. The solid curve is the optical absorption spectrum before excitation.

The sharp peaks in the absorption spectra are due to free excitons associated with each quantum sublevel. These sharp peaks disappear after intense optical excitation because Debye screening from the dense (10^{13} cm^{-2}) electron-hole plasma screens the electron-hole correlation that gives rise to the bound excitons.

The observation of gain within 200 fsec reveals that a population inversion has been produced on this time scale. At the very earliest times, gain is observed well above the fundamental band edge (1.51 electron volts). Typically, semiconductor lasers emit just below the band edge. With short optical excitation pulses a transient population inversion is produced. As time progresses the carriers cool and recombine, causing the band edge gain to begin return to absorption within the first 1.5 psec. As the carrier population cools the gain moves closer to the band edge. Spectra of this type can be used to determine the temperature of the electron-hole plasma as a function of time (27).

Another example which illustrates the resolving power of our femtosecond measurement system is a recent measurement of the absorption dynamics of optically excited polyacetylene. This material is of interest because it is a prototype one-dimensional semiconductor, and there has been a great deal of work on its optical and electronic properties (28, 29).

Optical excitation of polyacetylene can induce symmetry-lowering distortions in this simple polymer chain that can give rise to induced optical absorptions. Calculations have shown that the time scale for the formation of these induced states can be on the order of 10^{-13} second (30). Figure 11 shows the optical absorption induced at 1.55 μm in polyacetylene following excitation with a 70-fsec optical pulse at 620 nm (31). Note the exponential decay with a time constant of 160 fsec—the shortest time constant yet measured by optical techniques. We consider the rapid recovery of this induced absorption to be the

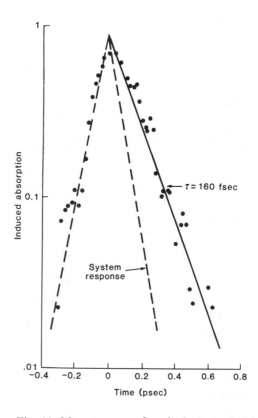

Fig. 11. Measurement of excited state relaxation in polyacetylene. The points are experimental and yield a 160-fsec exponential relaxation time. The dashed curve is the instrument response.

result of a transient excited state absorption produced as the optically excited polymer chain undergoes a dynamic distortion.

Limits

The question inevitably arises of just how short an optical pulse can be generated. Fundamentally, there is no limit other than that imposed by the uncertainty principle, $\Delta\omega\Delta t \sim 1$, where $\Delta\omega$ is the frequency bandwidth and Δt is the pulse width. This requires that we establish coherence over a broad range of frequencies. Typically, for a dye laser the gain bandwidth is on the order of 100 nm, which corresponds to a minimum pulse of a few femtoseconds. Maintaining a fixed phase relationship over such a broad range of frequencies is difficult. Material and optical cavity dispersion tend to broaden the optical pulse. In fact, using such a short pulse to make measurements requires special consideration. Linear dispersion alone will significantly broaden a pulse of a few femtoseconds after it traverses 100 μm in almost any material. In principle, it is possible to compensate for linear dispersion by the pulse compression techniques described earlier, but these techniques become more difficult to apply as the pulses get shorter. New techniques and methods will have to be devised to account for all these problems in order to make meaningful measurements with optical pulses only a few femtoseconds in duration.

Conclusion

I have outlined the latest progress in generating ultrashort optical pulses in the femtosecond time domain and have shown how these advances can be used to make measurements of dynamic processes in solids on this extremely short time scale. These new measurement tools give us a chance to peer into the as yet unexplored world of processes that take place in femtoseconds. Discoveries in a broad range of fields including physics, chemistry, and biology lie ahead.

References and Notes

1. C. V. Shank, R. L. Fork, R. Yen, R. H. Stolen, W. J. Tomlinson, *Appl. Phys. Lett.* **40**, 9 (1982).
2. E. P. Ippen, C. V. Shank, R. L. Woerner, *Chem. Phys. Lett.* **46**, 20 (1977).
3. P. M. Rentzepis, *ibid.* **2**, 117 (1968).
4. C. V. Shank, E. P. Ippen, O. Teschke, R. L. Fork, *ibid.* **57**, 433 (1978).
5. A. J. DeMaria, D. A. Stetser, H. Heynau, *Appl. Phys. Lett.* **8**, 22 (1966).
6. E. P. Ippen, C. V. Shank, A. Dienes, *ibid.* **21**, 348 (1972).
7. C. V. Shank and E. P. Ippen, *ibid.* **24**, 373 (1974).
8. R. L. Fork, B. I. Greene, C. V. Shank, *ibid.* **38**, 671 (1981).
9. C. V. Shank and E. P. Ippen, in *Dye Lasers*, F. P. Schäfer, Ed. (Springer-Verlag, New York, 1972), p. 121.
10. G. H. C. New, *IEEE J. Quantum Electron.* **QE-10**, 115 (1974).
11. F. Gires and P. Tournois, *C.R. Acad. Sci. Paris* **258**, 6112 (1964).
12. J. A. Giordmaine, M. A. Duguay, J. W. Hansen, *IEEE J. Quantum Electron.* **QE-4**, 252 (1968).
13. R. A. Fischer, P. L. Kelly, T. K. Gustafson, *Appl. Phys. Lett.* **14**, 140 (1969).
14. H. Nakatsuka, D. Grischkowsky, A. C. Balant, *Phys. Rev. Lett.* **47**, 1910 (1981).
15. L. F. Mollenauer, R. H. Stolen, J. P. Gordon, *ibid.* **45**, 1095 (1980).
16. R. H. Stolen and C. H. Lin, *Phys. Rev. A* **17**, 1448 (1978).
17. L. F. Mollenauer, R. H. Stolen, J. P. Gordon, *Phys. Rev. Lett.* **45**, 1095 (1980).
18. E. B. Treacy, *IEEE J. Quantum Electron.* **QE-5**, 454 (1969).
19. R. L. Fork, C. V. Shank, R. T. Yen, *Appl. Phys. Lett.* **45**, 223 (1982).
20. R. R. Alfano and S. L. Shapiro, *Phys. Rev. Lett.* **24**, 1980 (1970).
21. D. H. Auston, in *Ultrashort Light Pulses*, S. L. Shapiro, Ed. (Springer-Verlag, New York, 1977).
22. R. L. Fork, C. V. Shank, C. Hirlimann, R. T. Yen, W. J. Tomlinson, *Opt. Lett.*, in press.
23. S. L. Shapiro, *Ultrashort Light Pulses* (Springer-Verlag, New York, 1977).
24. R. Dingle, *Adv. Solid State Phys.* **15**, 671 (1975).
25. C. V. Shank, R. L. Fork, R. F. Leheny, J. Shah, *Phys. Rev. Lett.* **42**, 112 (1979).

26. C. V. Shank, R. L. Fork, B. I. Greene, C. Weisbuch, A. C. Gossard, *Surf. Sci.* **113**, 108 (1982).
27. C. V. Shank, R. L. Fork, R. Yen, C. Weisbuch, J. Shah, *Solid State Commun.*, in press.
28. J. A. Heeger, *Comments Solid State Phys.* **10**, 53 (1981).
29. J. Orenstein and G. L. Baker, *Phys. Rev. Lett.* **49**, 1043 (1982).
30. W. P. Su and J. R. Schrieffer, *Proc. Natl. Acad. Sci. U.S.A.* **77**, 5626 (1980).
31. C. V. Shank, R. Yen, C. Hirlimann, J. Orenstein, G. L. Baker, *Phys. Rev. Lett.* **49**, 1660 (1982).
32. I wish to acknowledge the collaboration of many colleagues at Bell Laboratories in the course of this work.

Part II

Chemical Analysis

Sophisticated instrumentation integrated with powerful computers has created an analytical revolution of broad scope (see box) in the last two decades (*1, 2*). Applications of these systems in such varied areas as basic research, industrial control, environmental monitoring, and health continue to expand faster than anticipated by all but the most optimistic predictions. Analytical instruments, broadly defined, are now essential tools for most chemists ("Chemistry Without Test Tubes;" see cover) (*3*) and for a broad spectrum of physical and biological scientists. Appropriately, fundamental discoveries across these areas have provided the bases for these new methods, as well as a growing challenge to the breadth and flexibility of analytical chemists. Detailed examples of instrumentation leading this revolution are provided by articles in this issue.

Analysis of an atomic or molecular species involves deriving measurable signals which are characteristic of the identity and/or amount of the species. The "wet" analytical chemistry of yesteryear derived these signals by condensed-phase reactions such as acid-base neutralizations for volumetric analysis and precipitations for gravimetric analysis. An analysis requiring many minutes by a trained analyst produced a single value such as milliliters of titrant or grams of precipitate. Thus specificity was dependent on the reaction chemistry, sensitivity on this and the volumetric or gravimetric measurement, and speed on the skill of the analyst. In contrast, computerized instrumental methods can provide much more and better data in much less time, in the most favorable cases even showing a selective "real-time" response for a single atom or molecule and minimizing the require-

7

Trends in Analytical Instrumentation

Fred W. McLafferty

Science 226, 251–253 (19 October 1984)

ments of a trained analyst for operation and interpretation.

The capabilities required of the analytical chemist have increased at least as dramatically. Chemistry was critical to the success of wet analytical methods, so these were largely developed by chemists; fortunately, chemists needing a method in their work were often those best qualified in the requisite chemistry. Modern instrumental methods, however, have originated from a wide variety of disciplines, often in combination, including spectroscopy, nuclear and ion physics, electronics, computer science, and biology. Development of such analytical instrumentation thus usually requires competence in other relevant disciplines without diminishing the requirements in chemistry. Similarly, a key factor in the best modern research in chemistry and other sciences is often the capability to develop and/or utilize such instrumentation.

Analytical chemistry comprises both characterization and measurement of chemical systems—qualitative and quantitative analysis. The accuracy and overall value of a method to its user depend on its sensitivity, specificity, and speed, as well as its simplicity and cost (see box) (4). Sampling can also be of critical importance to the applicability of a method. Each of these aspects will be examined here to define and illustrate the magnitude of this analytical revolution.

Sensitivity

A part per million was considered an impressive sensitivity a generation ago. Now a part per trillion ($1/10^{12}$) sensitivity is achieved for some routine analyses such as that of the notorious dioxin (5), while a special tandem-accelerator mass spectrometer can detect three atoms of ^{14}C in the presence of 10^{16} atoms of ^{12}C (6), corresponding to a radiocarbon age of 70,000 years. A pinhead would occupy a part per trillion of the area of a road from New York to California (2). As infinitesimal as this appears, 10^{12} molecules of molecular weight 600 weigh only 10^{-9} g.

Sensitivity depends on the efficiency of converting the analyte to a measurable species, and on the ability to measure this species. In addition, the actual detection limit (signal-to-noise ratio) is also affected by the response of interfering species present (that is, specificity, discussed below). These measurable species include the photons whose absorption or emission is the basis of a wide variety of spectroscopic methods (x-ray, ultraviolet, infrared, nuclear magnetic resonance, Raman, Mössbauer, and so on), and the charged species which are the bases of electron and mass spectrometry and of detectors (electrochemical, flame ionization, and so on) for chromatography. The efficiency of converting the analyte to photons (and even electrons and ions) has been dramatically improved with intense energy sources such as lasers, synchrotron radiation, and plasmas. Efficiencies approaching 100 percent have been reported for resonance-enhanced multiphoton ionization of atomic (7) and molecular (8) species. Multiplier detectors can respond to the arrival of a single photon or ion, so that these methods can actually detect a single cesium atom (7) or naphthalene molecule (8). The sensitivity obtainable from multidimensional techniques is also dependent on their specificity and speed, as discussed below.

Specificity

The amount and utility of the information from an analytical method are key determinants of the resulting selectivity for the atom or molecule sought; a completely "universal" detector would have to produce an infinite amount of information. The total information obtained depends on the number of dimensions and on the amount of data per dimension. Measurement of refractive index or optical rotation, for example, provides only one value indicating displacement in one dimension. Most instrumental methods are multidimensional, resulting in an exponential increase in information. Common spectroscopic techniques have two dimensions, for example, plotting absorbance against wavelength for infrared spectroscopy or ion abundance against mass-to-charge ratio for mass spectrometry.

The amount of data possible in the X dimension of a spectrum depends on the number of peaks and the number of distinguishable locations (resolution increments) possible for each peak. Reducing the number of peaks in a spectrum to one essentially makes it unidimensional; here specificity is determined by the peak width relative to the response position of other components. Specificity can still be high, such as for chemical reactions of wet methods; with modern autoanalyzers these account for hundreds of millions of routine analyses yearly. Radioimmunoassay (9) measures the amount of unknown antigen by the response to a specific antibody, of which there could be 10^7 to 10^9 giving distinguishable responses. Iron-57 Mössbauer spectra (10) cover an energy range of only $\sim 2 \times 10^{11}$ of the source energy, but absorption values can be measured at

```
Analytical Critera

   Specificity

   Sensitivity

   Speed

   Sampling

   Simplicity

   $
```

~ 20 distinguishable energy values because of the incredible resolution of $\sim 10^{12}$ (a peak width of 5×10^{-8} eV compared to a source energy of 1.4×10^4 eV). Dramatic increases in resolution, and thus information content, have been achieved in many techniques. In the ultraviolet region the molecular absorption spectrum of a compound in solution is a "band" electronic spectrum, with possibly a dozen resolution increments. An orders-of-magnitude increase can result from gas-phase cooling by supersonic expansion to resolve the vibrational and even rotational "line" spectra for measurement by a tunable laser. Gas chromatography with capillary columns (11) makes possible $\sim 10^5$ resolution increments, a substantial improvement over that with packed columns. The potential utility of high-resolution ($\sim 10^5$) mass spectrometry of high mass (<10,000 daltons) compounds (12) is reflected in the $\sim 10^6$ possible resolution increments (13). The number of possible resolution increments for the single value in a spectrum's Y axis (for instance, absorbance or ion abundance)

95

depends on the measurement accuracy as well as the dynamic range; 10 to 10^3 increments are common for spectroscopic methods.

Techniques of even higher information dimensionality are among the most exciting new ways to obtain "needle-in-a-haystack" specificity (4). Such combined techniques can be classified as "cloned" or "hybrid." A prime example of the former is two-dimensional nuclear magnetic resonance (14), which provides information on the spatial relationship of the ^1H, ^{13}C, . . . atoms of a molecule in addition to the information on the type and number of atoms provided by the normal nuclear magnetic resonance spectrum. Tandem mass spectrometry (4, 15) (see cover), which has shown amazing recent growth, can similarly provide an extra dimension of structure information for pure samples, and can also serve as a fast ($<10^{-3}$ second) separator interfaced to a highly specific detector or identifier.

The hybrid multidimensional technique combining gas chromatography with mass spectrometry (GC-MS) is a classic example of analytical instrumentation development, having grown in 20 years from a research curiosity to worldwide sales exceeding those of most types of spectroscopic (noncombined) instruments. GC-MS provides powerful and convenient abilities for separation, identification, and quantitation of complex mixture components which have proved ideal for insect pheromones, environmental pollutants, forensic problems, process control, and so on. The counterpart GC-infrared (GC-IR) (16) and liquid chromatograph-MS systems (17) have advantages in terms of isomer specificity and have applicability to less volatile mixtures.

Multidimensional spatial analysis is important for inhomogeneous samples. Special "microprobe" and "microscope" techniques employ electrons, ions (18), or laser photons as both probes and sensors to map the concentration of a selected element or molecule in two (or even three) dimensions of the sample. Such improved methods for surface analysis have been a key factor in elucidating basic mechanisms of heterogeneous catalysis (1), while nuclear magnetic resonance imaging (14) is revolutionizing medical diagnosis.

Speed, Simplicity

Many analytical problems in research, process control, and toxicant monitoring have critical time requirements. Further, the rate and ease of producing useful information are often key factors in justifying the high purchase and maintenance expenses of modern analytical instruments; the "cost per analysis" must be minimized. Progress in increasing speed has been particularly impressive recently. Although in selected cases this is due to specific instrumentation, such as picosecond lasers (8), the nearly universal application of the dedicated computer to analytical instrumentation in the last decade has given dramatic improvements in speed as well as in accuracy, control, and dependability. The information rate for multidimensional instruments such as GC-MS and two-dimensional nuclear magnetic resonance is so great in most applications that the computer is required for data acquisition and reduction.

Fourier transform techniques have greatly increased both speed and sensitivity (the latter by orders of magnitude) for infrared (*16*), nuclear magnetic resonance (*14*), and mass (*19*) spectroscopy. With these techniques, in contrast to a scanning spectrometer with slits, the collective frequencies representing a spectrum are simultaneously recorded as a function of time, from which the Fourier transform produces the frequency-domain spectrum. Dedicated, reasonably priced array processors can perform several transforms per second. This is particularly valuable for techniques like GC-IR (*16*), allowing several infrared spectra to be recorded across each GC peak.

Relatively complex calculations represent key steps in other methods such as pyrolysis-MS (*20*) and process control (*21*). Although a dedicated computer improves speed, the improvements in simplicity and reliability can be equally important. Computer interpretation of results, such as GC-MS on-line matching of the spectra of eluted components (*22*), increases the productivity of professional personnel.

Sampling

Many analytical problems require special arrangements to make the sample available to the instrument, or vice versa. Elegant examples of the latter were the orbiter and lander instruments on the Viking mission to Mars (*23*). Remote sensing, particularly of thermal or laser-induced emission spectra (*8*), is used routinely for analysis of smokestack effluents or stratospheric components. "Chemical profiling" of a large area for toxic or other critical compounds can be done expeditiously (at 35 miles per hour) by using a van-mounted MS-MS instrument with atmospheric pressure ionization and high-speed cryogenic pumping (*24*).

The Education of Chemists and the Future of Analytical Chemistry

This revolution in analytical instrumentation promises high future productivity in chemical research and manufacturing. The rapid progress in so many areas undergirding these instrumental advances brings new challenges, however, for proper education and training of both student and experienced chemists. In many schools the undergraduate curriculum is undergoing constant updating for this, including experience with relatively sophisticated analytical instrumentation as well as electronics, computers, vacuum systems, and so forth. The continuing education of experienced chemists would appear to be a greater problem. This is much like the computer familiarity of grade school and high school students compared to that of their parents.

As an analytical chemist, I see even more serious problems for us in defining our role in the development and application of such analytical instrumentation. Its promise for all areas of chemistry is much too great for other chemists to wait for us in this matter; note that fewer than half of the articles in this issue are authored by card-carrying analytical chemists. Understanding the chemistry is still central to the success of an analytical method, such as the interplay of species in the solution analyzed, the interaction of molecules with an electrode surface or a chromatographic support, the unimolecular dissociation of organic ions in mass spectrometry, or the folding of polypeptide chains in immunoassay. To the extent that we attain real leadership as chemists in research and applications of such computerized instrumentation, analytical chemistry will be even more exciting and deserving of the respect of other chemists and scientists.

References and Notes

1. Research Briefing Panel on Selected Opportunities in Chemistry, G. C. Pimentel, chairman, in *Research Briefings, 1983* (National Academy Press, Washington, D.C., 1983), pp. 67–92.
2. A. J. Bard, "New horizons in analytical chemistry," paper presented at the AAAS Annual Meeting, New York City, 25 May 1984.
3. P. H. Abelson, *Science* 224, 1297 (1984).
4. F. W. McLafferty, *ibid.* 214, 280 (1981).
5. "2,3,7,8-Tetrachlorodibenzo-*p*-dioxin levels in adipose tissue of Vietnam veterans," M. L. Gross *et al.*, *Environ. Res.* 33, 261 (1984).
6. C. L. Bennett, *Am. Sci.* 67, 450 (1979).
7. G. S. Hurst, *Anal. Chem.* 53, 1448A (1981).
8. D. M. Lubman, R. Naaman, R. N. Zare, *J. Chem. Phys.* 72, 3034 (1980); R. N. Zare, *Science* 226, 298 (1984).
9. R. S. Yalow, *Science* 200, 1236 (1978).
10. J. G. Stevens and L. H. Bowen, *Anal. Chem.* 56, 199R (1984).
11. R. Dandneau, P. F. Bente, III, T. Ronney, R. Hiskies, *Am. Lab. (Fairfield, Conn.)* (September 1979), p. 61; K. Grob, Jr., and K. Grob, *J. Chromatogr.* 207, 291 (1981).
12. H. R. Morris, *Soft Ionization Mass Spectrometry* (Heyden, London, 1981); C. Fenselau, *Anal. Chem.* 54, 105A (1982); I. J. Amster, M. A. Baldwin, M. T. Cheng, C. J. Proctor, F. W. McLafferty, *J. Am. Chem. Soc.* 105, 1916 (1983).
13. This value is dependent on the range of elemental compositions possible for the ions produced from the sample.
14. G. C. Levy, Ed., *NMR Spectroscopy: New Methods and Applications* (ACS Symposium Series, American Chemical Society, Washington, D.C., 1982).
15. F. W. McLafferty, Ed., *Tandem Mass Spectrometry* (Wiley, New York, 1983).
16. P. R. Griffiths, J. A. De Haseth, L. V. Azarraga, *Anal. Chem.* 55, 1361A (1983).
17. M. L. Vestal, *Science* 226, 275 (1984).
18. S. Chandra, D. F. Chabot, G. H. Morrison, A. C. Leopold, *ibid.* 216, 1221 (1982).
19. M. L. Gross and D. L. Rempel, *ibid.* 226, 261 (1984).
20. H. L. C. Meuzelaar, W. Windig, A. M. Harper, S. M. Huff, W. H. McClennen, J. M. Richards, *ibid.*, p. 268.
21. T. Hirschfeld, J. B. Callis, B. R. Kowalski, *ibid.*, p. 312.
22. I. K. Mun, D. R. Bartholomew, D. B. Stauffer, F. W. McLafferty, *Anal. Chem.* 53, 1938 (1981).
23. K. Biemann *et al.*, *J. Geophys. Res.* 82, 4641 (1977).
24. J. B. French, W. R. Davidson, N. M. Reid, J. A. Buckley, in (*15*), chapter 18.
25. I am indebted to the other members of the NRC Committee on Selected Opportunities in Chemistry, especially G. C. Pimentel, chairman, and A. J. Bard; to many contributors to the committee report; and to my colleagues H. D. Abruna and G. H. Morrison for valuable material and discussions. My research involving analytical instrumentation is supported by the National Science Foundation, tne National Institutes of Health, and the Army Research Office.

Mass spectrometry is rich in history and in potential. It has drawn from and influenced many fields, including atomic physics, reaction kinetics, geochronology, chemical analysis, and, most recently, biomedicine. Advances in reaction kinematics and dynamics, data acquisition and reduction, isotope separation, and stable isotope labeling have depended on this instrumental method. Practitioners of mass spectrometry have often been specialists, immersed in their subject and almost insulated by its momentum. This group and biologists, among others, are now in the process of mutual discovery; a valuable outcome calls for a familiarity with the capabilities of each. In this article we seek to provide information on recent developments in mass spectrometry. We also seek to emphasize the potential of a subject which we see as a blunt tool, available and needing only to be sharpened for particular tasks.

Three new capabilities epitomize the present state of mass spectrometry and point up its potential for far-reaching contributions in biology and medicine. The mass range of mass spectrometers has been extended by approximately an order of magnitude in the past decade. While other types of mass analyzers have been used to reach considerably higher masses [for instance, 150,000 in some quadrupole experiments (1)], progress in the well-established sector magnet technology best illustrates this trend. Commercial instruments are now available with mass ranges, at full accelerating voltage, of 7500 compared with 1000 just a few years ago (2). A second capability is desorption ionization, a family of procedures [including fast atom bombardment (FAB), secondary ion mass spectrometry (SIMS), and fission fragment methods] which allows ionic, non-

8

Mass Spectrometry: Analytical Capabilities and Potentials

R.G. Cooks, K.L. Busch, and G.L. Glish

Science 222, 273–291 (21 October 1983)

volatile compounds to be examined by mass spectrometry. These capabilities were reviewed last year in this journal (*3, 4*). The third notable capability is the successful integration of separation and analysis techniques represented by tandem mass spectrometry (MS-MS) (*5, 6*) and the much improved liquid chromatography–mass spectrometry (LC-MS) interfaces (*7*). The latter capability should have an impact comparable to that of gas chromatography–mass spectrometry (GC-MS), but without the limitation to volatile, low-mass compounds. MS-MS is a two-dimensional form of spectrometry which often improves signal-to-noise ratios as well as providing entirely new capabilities, such as that of scanning a mixture for all constituents having particular structural subunits.

Comparable in importance to these particular new capabilities are the texture of the subject and the attitudes of its practitioners. A mass spectrum is a chemical product distribution, not a representation of electromagnetic transitions. A mass spectrometer is appropriately considered as a chemical reactor, a type of chromatograph, or a spectrometer. These roles, together with the fact that mass analysis can be implemented by use of several different physical principles, distinguish mass spectrometry from other forms of spectroscopy. This latitude in methodology, and characteristics of the hardware which demand interaction with the equipment, have made mass spectroscopists likely to modify instrumentation or to develop entirely new instruments. These people constitute a reservoir of talent and experience which can be directed at significant problems in collaboration with the informed biologist. Absence of such interactions in the past is surely responsible for the lack of some desirable capabilities. For example, microscopy is largely a morphological technique, with supporting elemental analysis capabilities. One might have expected that information on molecular distributions would have been a high priority, especially in biological specimens. Development of a capability for molecular microscopy, based on desorption ionization techniques, appears to be feasible.

This paper has three subsequent parts. The first is a summary of some notable recent achievements of mass spectrometry, covering a range of applications. The second seeks to identify and illustrate the forces driving various areas of activity in mass spectrometry. Finally, some views on future directions are given.

Typical Recent Achievements

This section is designed to provide information on the range and significance of work being done with mass spectrometry through selected examples. Other work of equal merit could have been chosen to make similar points.

The GC-MS combination has had a considerable impact on biological and environmental research, providing a specific means of characterizing constituents of mixtures and having the necessary sensitivity and quantitative accuracy to be applicable to trace constituents. This established procedure continues to yield valuable results, as illustrated by the recent discovery of the presence and uneven distribution of the neural excitotoxin, quinolinic acid, in the mammalian brain at the level of nanomoles per gram (*8*). The selective excitatory activity of this molecule suggests that it may be a neurotransmitter. Detection of the acid was achieved by GC-MS of the volatile hexafluoroisopropanol diester derivative with electron impact ionization. Quantitation was based on standard addition and single-ion monitoring. Concentrations of both 5-hydroxytriptamine and quinolinic acid increase after tryptophan administration, although the distributions of the two metabolites are quite different. These important results were obtained by using classical ionization (electron impact) and separation (GC) procedures.

Mass spectrometry has many qualities that make it suitable for metabolic profiling, for example, in measuring the distribution of urinary acids or steroids in individuals (9). Such measurements can have immediate diagnostic value, as in the detection of particular genetic diseases, or they may be aimed at chemical diagnosis of predisease states by comparing profiles with those of normal individuals. A large number of samples must be measured, so high throughput is important, and dozens of compounds must be monitored over a wide range of concentrations, so the method must be widely applicable and have a large dynamic range. GC-MS used with electron ionization has proved equal to this task and has contributed significantly to clinical successes in neuroblastoma cases (10).

Challenging structural problems, including protein structure determinations, can sometimes be solved most efficiently by using a combination of techniques. In several recent examples, Edman sequencing has been combined with mass spectrometry to increase the speed of structure determination and to avoid possible misassignments arising from weaknesses in the individual methods (11–13). Two mass spectrometric methods were used to deduce the structure of the 112-amino-acid antitumor protein macromomycin, derived from a *Streptomyces* culture. Partial acid digestion gave a mixture of small (two- to six-amino-acid) peptides, which was derivatized and analyzed by GC-MS with electron ionization. This information was interpreted in conjunction with Edman degradation data and FAB spectra, which provide molecular weights for large peptides even when they are present in a mixture and examined without derivatization. Another rapid method of protein structure determination is that in which amino acid sequences are arrived at by determining the base sequence of the gene coding for the protein. Here, too, a single method is usefully supplemented by other data, and it has been shown how frameshift errors in the translation method can be avoided by obtaining complementary mass spectrometric data (12). Successes with proteins such as glutamine–transfer RNA synthetase (550 amino acids) have depended on this combination of methods.

Tandem mass spectrometry has been used to discover and establish the structure of metabolites of the drugs primidone, cinromide, and phenytoin in plasma and urine extracts; analyses were performed in less than an hour, using concentrations of 1 to 50 μg/ml (14). The study is predicated on the speed and flexibility of MS-MS scans made with a triple quadrupole instrument, and on the realization that metabolites generally retain a large portion of the parent drug structure. Since no other assumptions about metabolite structure are made, no constraints are introduced, and unknown metabolites are quickly identified within a complex mixture.

Measurements of stable isotope abundances have been the province of mass spectrometry since the discovery of the isotopes of neon and the early abundance measurements used to establish chemical atomic weights. Heavy atom isotope effects have become a valuable tool in enzymology in the past decade, and even whole plant isotope effects are now under study and are providing insights into photosynthesis (15). Kinetic isotope effects are best measured with high-precision isotope ratio spectrometers after chemical conversion of the substrate to CO_2, CO, or N_2. The limitations this imposes in terms of accessibility of the label have been lifted by a technique in which a mixture of compounds is labeled in a remote position and doubly labeled in both the remote position and a position accessible to degradation (16). The measured isotope effect is then the product of that at each of the two positions.

Carbon dating is another isotope ratio measurement that has been dramatically improved. In this measurement, and related work on other decay processes, a tandem accelerator is used as an extremely sensitive mass spectrometer. The ability of mass spectrometers to count single ions of ^{14}C, arriving at a rate of one per second or even less, represents an advantage over radioactive decay methods, which can observe only atoms that actually decay during the analysis period. By accelerating the beam to high energies, interfering polyatomic ions can be completely removed by collision-induced dissociation, $^{12}C/^{14}C$ ratios approaching 10^{15} to 1, corresponding to ages of about 40,000 years, can be measured with a precision of a few percent (17). Measurement of isotope ratios with high precision has also figured in the interpretation of an iridium anomaly found in geological sediments (18).

We conclude this section with two examples of contributions of mass spectrometry outside the area of chemical analysis. Electron impact and photoionization measurements have long provided valuable thermochemical determinations. The quality of mass spectrometric data has been greatly enhanced since the development of equilibrium methods used with high-pressure (19) or long residence time (20) instruments. Measurements of the enthalpy and free energy of naked and partly solvated ions derived from this work have been valuable in understanding solvation. The kinetics of unimolecular decomposition of disolvated ions has been studied by a new procedure to determine gas-phase acidities and basicities (21). The method is rapid and is applicable to impure samples; data on aliphatic alcohols show high precision (± 0.1 kcal mole^{-1}) and provide new insights into factors such as through-space dipolar interactions which affect gas-phase acidities (22).

The second example is also from an active area of research: comparisons of organic reaction mechanisms in solution with those in the gas phase. The field embracing ionic structure (23, 24), thermochemistry (25), and gas-phase reactivity (26–29) is important for theoretical chemistry and, because of the overwhelming importance of ionic intermediates in solution reactions, is a natural focus for inquiries into chemical reactivity. Chemical ionization can be used to generate the anionic and cationic species which are the postulated intermediates in solution reactions, and collision-induced dissociation can then be employed to characterize these ions and their gas-phase reaction products. Parallels have been established between gas-phase reactions and the classical Dieckmann ester condensation (*30*), the Beckmann rearrangement (*31*), the Wagner-Meerwein rearrangement (*32*), and others (*24*). [Agreement in mechanism includes *anti* rather than *syn* rearrangement in the Beckmann reaction (*31*).] Reaction channels not observed in solution can compete in the gas phase, where the absence of solvent has a considerable effect on the relative stability of particular ionic species. Internal solvation is a key to gas-phase reactivity and is the underlying cause of divergence from reactivity in solution, where external solvation is possible. For example, the acid-catalyzed intramolecular cyclization of **1** proceeds via the oxonium ion intermediate **2** which displays quite different behavior in the two phases, yielding CH_3^+ in solution but $CH_3^{\cdot}$ in the gas phase (*24*).

Trends in Mass Spectrometry

In this section we identify and illustrate the forces that are driving developments in mass spectrometry. Four factors are discussed: (i) multidimensional experiments, (ii) ionization phenomena, (iii) direct mixture analysis, and (iv) instrument elaboration.

Multidimensional experiments. The trend toward multidimensional forms of spectroscopy is evident in many areas of chemical analysis. In some ways, multidimensional experiments are an alternative to high-resolution ones—detail is sought through examination of a more extensive data array rather than ever-finer exploration in a single dimension. Advances in data acquisition and processing have prepared the way for these experiments. They also make it possible to optimize the sensitivity-specificity relation by examining only the regions of data space that contain the most information.

Resonance-enhanced multiphoton ionization (*33, 34*) is an extremely efficient method for ionizing gaseous molecules, whose ionization cross sections are strongly wavelength-dependent. Samples can be characterized by intensity distributions in a two-dimensional (mass, wavelength) matrix. In addition to its efficiency, the orthogonal nature of the mass and photoionization information makes this a potentially powerful technique, both for pure compounds and for mixtures. Early examples of applications to polynuclear aromatic compounds appear to be fulfilling this promise (*35*).

A related two-dimensional experiment is that in which photodissociation of a mass-selected ion is followed as a function of the photon energy. In effect, the ion is characterized by its breakdown curve—that is, by the internal energy dependence of its mass spectrum—and the isomer specificity reported is better than that achieved with simpler experiments (36). Rapid, more approximate methods of obtaining the internal energy dependence of mass spectra are also coming into use. In these methods fragmentation patterns are measured as a function of the angle of deflection for collision-induced dissociation at kilo-electron-volt energies (angle-resolved mass spectrometry) (37) or as a function of the collision energy in the electron-volt range (energy-resolved mass spectrometry) (38). Breakdown curves calculated from unimolecular kinetics or measured by the precise but tedious photoion-photoelectron coincidence experiment (39) agree with those obtained by these newer, faster methods. Breakdown curves established by energy-resolved mass spectrometry have been used to optimize experimental conditions, for example, in quantitation of the drug de-acetylmetipranolol in urine (40).

The MS-MS experiment, in which one analyzer transmits a reactant ion and the second a product ion, is a unique form of two-dimensional spectroscopy (5, 6). The analyzed species are strongly coupled through their reactant-product relation (in contrast with the situation in the photoionization experiments already discussed and in the "hyphenated" chromatography-spectroscopy methods). As a result, several informative scan types are accessible: if one fixes the product analyzer, a scan of the analyzer that selects the reactant records all reagents that yield the chosen product. Such parent scans provide information on all constituents of the sample which yield the selected product—that is, all compounds of a particular chemical type. Scans in which the reagent is chosen and the products recorded are termed daughter scans, and they characterize the individual molecular species selected for examination. One can go further: the two mass parameters being selected, say m_1 and m_2, can be surveyed according to a relation of the type $m_1 = m_2 + c$, where c is a constant. Such an expression has meaning only in a homogeneous multidimensional experiment where the same type of property is being examined. The scan just mentioned is useful; it is termed a (constant) neutral loss scan and records all sample constituents that can react by loss of a fragment of mass c. For example, surveys of coal liquids for all phenols (which lose 18 mass units after protonation) can be made with this type of scan, while partially hydrogenated azaaromatics (important in fuel processing) are among several series of compounds easily recognized in such a screen (41).

Ionization. The creation of gas-phase ions is emerging from a period of intense activity during which signal advances were made (42). New procedures, such as fast atom bombardment (43), have been introduced, and advances in other techniques, such as laser desorption (44, 45) and electrohydrodynamic ionization (46), should increase their usefulness.

While efforts to detail the events that occur in the several desorption ionization methods and to optimize these methods for individual problems will continue, several major lessons are now evident. First, the controlling factors in the newer ionization methods are chemical rather than physical. This point was initially made for field desorption (47), but it is also true for other desorption procedures. A variety of physical processes, including different methods of energy deposition, give strikingly similar mass spectra. On the other hand, changes in sample composition on pretreatment with simple reagents can cause profound changes in the nature and abundance of ions recorded in the spectrum. Second, the newer forms of mass spectrometry are more complex than traditional procedures because they are influenced, not only by the unimolecular fragmentations of gas-phase ions, but by ion-molecule reactions in regions of varying pressure (48) and by radical reactions associated with energized condensed-phase material (49). Third, it is striking that biomolecules are rugged enough to survive the vigorous conditions (such as bombardment with ions at millions of electron volts) used to ionize them. Abundant molecular ions are the rule and structure-spectrum correlations are rapidly being developed. These generalizations do not hold with the same force for organometallics or inorganic complexes, where reactions such as clustering, transmetallation, or ligand interchange can often occur (50, 51).

There are corollaries to each of these observations. First, to examine some biomolecules by desorption ionization mass spectrometry, it is useful to derivatize them to convert them into an ionic form (52). Formation of gas-phase ions then requires a simple phase transfer, which is often very efficient. This approach is the reverse of that traditionally used in mass spectrometry, where volatile and nonionic derivatives have been the objective. It is also much easier to effect; simple acid or base treatments often suffice. The second generalization, regarding the complex origin of ions and often poor signal-to-noise ratios in desorption ionization, represents a problem to which MS-MS is a readily available solution. Combinations of MS-MS, particularly with SIMS, FAB, and laser desorption, have resulted in solutions to problems in alkaloid (53), macrocyclic antibiotic (54), and macrocyclic peptide characterization, including corrections of earlier proposed structures (55). Not only is signal-to-noise ratio improved, but structurally diagnostic fragmentations are recorded. The third generalization, that covering analyte reactions during ionization, has led to the use of glycerol and other liquid matrices to reduce such interactions. The problem is evident in exaggerated form in work on frozen nitrogen oxides, where, even under low-flux conditions, cluster ions are generated with stoichiometries different from that of the analyte (56). An alternative solution, which has been used in examining organometallics and inorganic complexes, is the use of solid-salt matrices (57).

Research activity in ionization continues and the following particulars are worth mention. The bombarding species in desorption ionization can be a metal ion such as Hg^+ (58), an organic ion (59), or even a dust particle in the 10^9- to 10^{15}-dalton range (60, 61). Field emission metal ion sources are also useful (62), especially because they do not contribute gas loads to the source and because their intense beams can be finely focused. Many workers have explored liquid matrices in FAB (63), including the use of added noble metal salts to act as cationizing agents (64). This approach, especially with polyisotopic metals, enhances the ability to identify ions of the analyte.

Plasma desorption (PD), originally achieved with ^{252}Cf fission fragments (65), is now often done with beams extracted from nuclear accelerators. The successes of PD are duplicated by particle bombardment sources (SIMS and FAB), which operate in the keV range and are easily retrofitted to commercial mass spectrometers. The time-of-flight mass analyzer used in PD has the advantages of simplicity and a large mass range, and it serves to integrate the low signals available in this experiment. However, it has very low resolution and data acquisition times are usually several hours (*66, 67*). Nevertheless, PD continues to produce molecular weight data on more massive biological compounds than have been measured by other techniques. For example, the molecular weight of a tiger snake venom peptide was measured as 13,284 ± 25 daltons (Fig. 1) (*68*). Fragment ions formed in PD also provide valuable information, as in sequence studies on protected oligonucleotides (*69*) and polypeptide antibiotics (*70*). Intercalation of DNA by polycyclic aromatic compounds may be another area in which this method can make significant contributions (*71*).

The other desorption ionization methods usually give much higher quality mass spectra. FAB has been used for real-time monitoring of protein digestion (*72, 73*). The enzyme and protein are mixed in a glycerol matrix, and intermittent exposure to the atom beam allows peptide production to be monitored for up to 20 minutes inside the mass spectrometer. The observation of the molecular ion from bovine insulin by plasma desorption (*74*) was followed shortly by the same accomplishment at higher resolution with atom bombardment (*75*), and the spectrum of human proinsulin with the intact molecule at 9390 has now been reported (*76*) (Fig. 2). This type of accomplishment shows the promise of mass spectrometry as a tool in genetic engineering.

Field desorption continues to be important in the analysis of biological molecules. This area has long been its forte, although it has also been useful for trace metal, inorganic, organometallic, and isotopic analyses (*77*). Although the

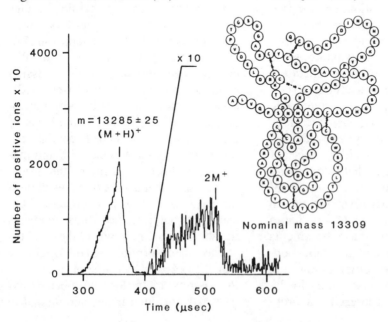

Fig. 1. Molecular weight determination of a tiger snake venom polypeptide. [I. Kamensky *et al.*, *FEBS Letters*, in press, reprinted with permission.]

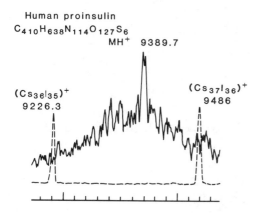

Human proinsulin
$C_{410}H_{638}N_{114}O_{127}S_6$
MH$^+$ 9389.7

$(Cs_{36}I_{35})^+$
9226.3

$(Cs_{37}I_{36})^+$
9486

Fig. 2. Molecular ion region of human proinsulin with inorganic cluster ions used to establish mass scale. [M. Barber, R. S. Bordoli, G. J. Elliott, N. J. Horoch, B. N. Green, *Biochemical and Biophysical Research Communications* **110**, 753 (1983), reprinted with permission.]

technique has a reputation for irreproducibility, this is countered by the excellent results consistently produced by laboratories that pursue the technique.

A recent application of field desorption mass spectrometry (FDMS) has been in the structural analysis of human hemoglobin variants (*78, 79*). FDMS pinpoints the single amino acid substitution which often creates an abnormality but is not detected by electrophoresis or liquid chromatography. The experiment uses a few micrograms of tryptic hydrolysates of the purified hemoglobins. Molecular weights of peptide residues are assigned from the protonated molecular ions; for peptides of mass higher than about 2100 daltons, doubly charged ions (M + 2H)$^{2+}$ are used. Shifts in the amino acid composition of a peptide are identified by the mass shifts between the expected protonated ion and its replacement. For example, the hemoglobin variant that causes sickle cell anemia involves substitution of valine for glutamic acid in a

residue and is easily pinpointed (Fig. 3). FAB has also been used (*79*) for the analysis of hemoglobin variants. The results were roughly comparable with the following exceptions: (i) the FAB spectra contained interfering chemical noise below mass-to-charge ratio (*m/z*) 500, and (ii) the greater intensity of doubly charged ions in the FD spectra increased the accessible mass range and provided a second confirmation of peptide molecular weights. Balanced against this is the relatively short period (about 20 seconds) of stable ion emission in FD as compared to the several minutes of ion emission in FAB.

Direct mixture analysis. In many laboratories, much effort is spent on preparing samples for measurement through extraction, centrifugation, and chromatography. Procedures that reduce this effort deserve attention. MS-MS and LC-MS offer this capability and both are undergoing rapid development.

The mass spectrometer is usually thought of as an analytical device rather than a separator, but the two functions are intimately connected. Large mass spectrometers, known as calutrons, have been used for 40 years to separate and isolate macroscopic amounts of particular elements (*80*). By linking two mass spectrometers in tandem it is possible to employ the first as a separator and the second as an analyzer, and so to perform direct analysis of mixtures. The two main advantages of the method can be illustrated (*81*) by considering a complex coal liquid mixture, the mass spectrum of which is shown in Fig. 4a. The signal due to a dioxin spike is lost in the chemical noise from the other constituents and single-stage mass spectrometry is not capable of analyzing for it. MS-MS filters against chemical noise and allows a

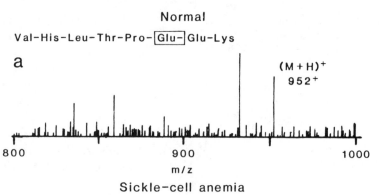

Normal

Val-His-Leu-Thr-Pro-[Glu-]Glu-Lys

a

$(M+H)^+$
952^+

800   900   1000

m/z

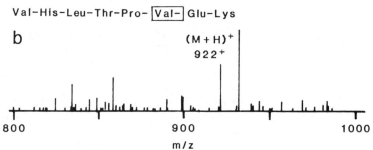

Sickle-cell anemia

Val-His-Leu-Thr-Pro-[Val-]Glu-Lys

b

$(M+H)^+$
922^+

800   900   1000

m/z

Fig. 3. Comparison of field desorption mass spectra, showing amino acid substitution in hemoglobin. [Y. Wada, A. Hayashi, T. Fujita, T. Matsuo, I. Katakuse, H. Matsuda, *Biochemica and Biophysica Acta* **667**, 223 (1981), reprinted with permission.]

high-quality spectrum of the dioxin to be recorded (Fig. 4b). In addition to improving detection limits in this way, tandem mass spectrometry provides alternative scan modes which can be employed to efficiently search the data domain for particular information. For example, chlorinated dioxins are characterized by the loss of COCl, so a scan for this reaction reveals all dioxins present in the mixture (Fig. 4c).

The enhanced detection limits that can be achieved by MS-MS over single-stage mass spectrometry are further illustrated by data on nucleosides. In one case (Fig. 5), 36 pmole of a nucleoside gave a spectrum in which the ion due to the analyte was lost in the chemical noise; however, the same sample gives an MS-MS spectrum that has an excellent sig-

nal-to-noise ratio and is a good match to that of the authentic compound (*82*).

These types of capabilities are beginning to have an impact on pharmacokinetics, where GC-MS-MS can decrease GC-MS detection limits by an order of magnitude. Isosorbide-5-mononitrate, a coronary vasodilator, is metabolized to the glucuronide, which can be determined in urine by a simple MS-MS procedure to 0.1 ng/ml (*40*). The improved detection limit in the MS-MS experiment is the direct result of minimizing interferences; one follows a reaction and not simply the product of a reaction.

The search for new compounds by MS-MS, as opposed to analysis of targeted compounds, is a new and rapidly evolving field. In one approach, illustrated by work on alkaloids, daughter MS-

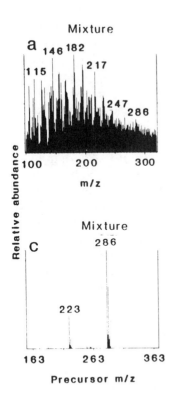

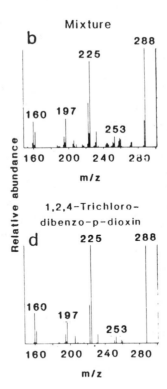

Fig. 4. 1,2,4-Trichlorodibenzo-dioxin spike in a coal liquid is lost in the chemical noise in the mass spectrum (a) but clearly revealed by the MS-MS daughter ion spectrum of the parent ion 288⁺ in the mixture (b), which matches that of the authentic compound (d). A survey of the mixture for chlorinated dioxins is achieved by a neutral loss scan (c). [Reprinted with permission of K. E. Singleton, R. G. Cooks, K. V. Wood, *Analytical Chemistry* **55**, 762 (1983). © 1983 American Chemical Society.]

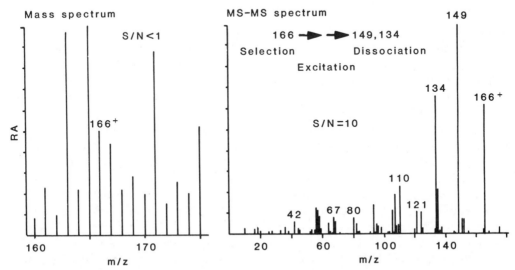

Fig. 5. Improved detection limits in MS-MS over single-stage mass spectrometry for 10 ng of the alkylated nucleoside O⁶-methyldeoxyguanosine. [From (82)]

MS spectra are taken on all major ions in the chemical ionization (*83*) or laser desorption (*53*) mass spectra. By using only a few grams of plant material it has proved possible to discover new alkaloids in several plant species by careful interpretation of the MS-MS spectra. An alternative approach is to use parent scans to search for structural units expected to be present in the compounds of interest.

Both the sensitivity and the speed of analysis available with MS-MS are noteworthy. Tetrahydrocannabinol administered in doses of 0.1 mg/kg can be followed for 8 days down to 10^{-11} g/ml by using a combination of GC-MS with simple MS-MS to avoid extensive sample cleanup (*84*). Sensitivities in the low parts per trillion have been reported in MS-MS studies on animal tissue (*85*). High-resolution mass spectrometry and MS-MS have been used to achieve absolute detection limits of less than 1 pg for tetrachlorodibenzodioxin (*86*), and a GC-MS-MS combination has achieved good signal-to-noise ratios with 250-fg samples (< 20 parts per trillion) analyzed in a mobile laboratory at the rate of 30 samples per day (*87*). In terms of sample throughput, the determination of trichlorophenol in serum at concentrations as low as 1 ppb and a rate of 90 samples per hour is noteworthy (*88*), as is characterization of the foodstuff contaminant (and chemical warfare analog) vomitoxin at 25 pg in wheat at a rate of 10 minutes per sample (*89*). For the latter analysis, gas chromatography with electrochemical detection has comparable sensitivity but is slower and requires prior sample cleanup.

A different approach to the characterization of mixtures of nonvolatile compounds is LC-MS. The first practical LC-MS interface was based on complete removal of the solvent, and temporary storage of the solute during transport by a moving belt or wire into the ion source (*90*). In the source the sample is either thermally desorbed and ionized by electron or chemical ionization, or the belt is bombarded by an energetic beam to create secondary ions (*91, 92*). Improved techniques are based on direct introduction of some or all of the eluant into the source. The large pumping capacities of chemical and atmospheric ionization sources makes it possible to work at flow rates consistent with normal column operation, for example, 2 ml of aqueous mobile phase per minute (*93–97*). The solvent itself acts as the reagent gas in these experiments. An alternative, the thermospray procedure (Fig. 6), does not employ any external ionization technique (*94*). An aerosol generated in the interface is evaporated, and separation of charges present in the nominally neutral solution allows positive- and negative-ion mass spectra to be recorded. Performance, which is still being improved, is illustrated by detection limits of 10 pg (selected ion monitoring) or 1 ng (full spectrum) for β-hydroxyethyltheophylline and by the observation of the protonated molecule of an underivatized decapeptide. The method involving direct liquid introduction and chemical ionization gives comparable data; for example, 50 ng of vitamin B_{12} gives a negative-ion spectrum of high quality (*98*).

Even in the area of elemental analysis, the tendency to minimize separations is being strongly felt. Resonance ionization mass spectrometry (RIMS), one of the most exciting advances in inorganic mass spectrometry of the past decade, employs multiphoton techniques to se-

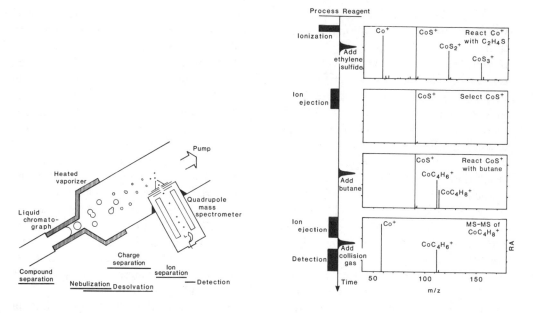

Fig. 6 (left). Liquid chromatograph–mass spectrometer thermospray interface, allowing biomolecules to be examined without an external ion source. [C. R. Blakley and M. L. Vestal, *Analytical Chemistry* **55**, 750 (1983), reprinted with permission. © 1983 American Chemical Society.] Fig. 7 (right). Sequences of steps used to prepare ions, react them, and characterize their products in a computer-controlled FTMS experiment. [From T. Carlin and B. S. Freiser]

lectively ionize particular elements (*99*). In one application of the technique, neodymium, a nuclear fuel by-product used in evaluating reactors, is measured in the presence of samarium. Chemical separation of these elements is difficult and several isotopes interfere isobarically, thus precluding thermal ionization from giving accurate isotope ratios for neodymium (*100*).

Instrumental elaboration. Instrumental developments do not only occur in response to perceived needs; capabilities are sometimes in place before appropriate problems are conceived. The past 12 months have seen the introduction or delivery of the following new commercial instruments: (i) an ion trap, a sophisticated but inexpensive three-dimension-

al quadrupole which forms the basis for a GC-MS instrument, (ii) an inductively coupled plasma (ICP) mass spectrometer, which shows considerable improvements in performance over previous (spark source) mass spectrometric methods of trace metal analysis, (iii) Fourier transform mass spectrometers, which provide high resolution and mass range, MS-MS capabilities, and unrivaled abilities to explore ion-molecule reaction chemistry, (iv) hybrid mass spectrometers, in which a multiple quadrupole section is added to a high-resolution sector instrument to provide both exact mass measurement and standard mass spectrometric capabilities in a single versatile instrument.

Fourier transform mass spectrometry

111

(FTMS) (*101*) illustrates the speed with which instrumental developments are transforming mass spectrometry. The method has arisen from ion cyclotron resonance experiments, which, in turn, derive from the cyclotron principle of mass-to-charge ratio analysis. The high resolution of FTMS is probably its most discussed property. Impressive performance data have been reported, such as a resolution of 1.4 million for m/z 166 from tetrachloroethane (*102*) and $> 10^8$ for m/z 18 from water (*103*). FTMS instruments are also capable of performing MS-MS experiments (*104–106*). Unlike a conventional MS-MS experiment, where the different stages of analysis are separated in space, the separation here is achieved in time. This allows the extension of the experiment to three (MS-MS-MS) or more stages (*107, 108*). Choice of ionization procedures in FTMS is constrained by the powerful magnetic field; laser desorption is more readily implemented than the particle desorption methods (*109, 110*). FTMS instruments require very low pressures for optimum performance. This has made interfacing with chromatography difficult, although a GC-MS combination has been reported (*111*). A possible solution to the pressure limitations is the use of a quadrupole mass filter as a device for injecting ions into the FTMS (*112*). All the "dirty" work can then be done in an environment removed from the FTMS.

The main successes of FTMS have been in the area of ion-molecule reaction chemistry, including gas-phase metal ion chemistry (*113, 114*). Ions may be stored, reacted, and characterized in the cell, using computer-selected reaction times and controlled translational energies. The reaction of CoS^+ with chemistry illustrates some of these capabilities.

Figure 7a shows a series of peaks corresponding to CoS_n^+ ($n = 0$ to 3) generated sequentially by sulfidation of the metal. Ethylene sulfide is added at a peak pressure of 10^{-5} torr and permitted to interact with Co^+ for about 250 msec. The desired product, CoS^+, is then specifically selected by removing all of the other ions from the cell. The CoS^+ is permitted to interact for approximately 1 second with butane at 10^{-7} torr and the appearance of two products is observed (Fig. 7c). Finally, in Fig. 7d the structure of the $CoC_4H_8^+$ product is examined by collision-induced dissociation. The products observed are indicative of a linear Co^+-butene structure resulting from C–H bond activation.

Instruments that go by the name of mass spectrometer are appearing in ever-increasing variety with an astonishing range of applications. Portable instruments of fairly conventional design have long been used for atmospheric and planetary exploration (*115*), and the Viking mission's life probe included a GC-MS which provided the hard numbers (and bad news) regarding life on Mars (*116*). Environmental monitoring in real time with mobile mass spectrometers is an established technology (atmospheric pressure MS-MS) (*117*) and MS-MS procedures used to automate and greatly speed searches for priority and other pollutants have been reported (*118*). On-line instruments are increasingly used in quality control in industrial plants. A method is no sooner established than entirely new procedures are developed to effect it. This is evident in the use of linked scans of sector analyzers (*119, 120*) and the measurement of flight time during passage through a single analyzer (*121*) to effect MS-MS. Naturally, attempts have been made to maximize, in single high-performance instruments, all the capabilities important in organic analysis. The hybrid instruments that combine sectors and quadrupoles typify this attempt. Commercial versions of these instruments offer exact mass measurements, MS-MS spectra, GC-MS, and various ionization procedures, and they facilitate studies on such fundamental topics as kinetic energy releases and translational energy dependence of ion-molecule reactions.

Prospects

It can be predicted with some confidence that mass spectrometry will soon come to be much more widely used, that many experiments will be done on types of instruments now being introduced, and that the data obtained from mass spectrometers will increasingly be other than conventional mass spectra. These conclusions are based on the following: (i) Low-cost mass spectrometers of advanced design have recently become available. These systems have limited mass ranges (650 daltons in one commercial instrument) and are restricted to electron impact on GC eluants; however, they have data acquisition and reduction capabilities that give them considerable power and guarantee them wide distribution. (ii) Simultaneously, high-performance mass spectrometers of a number of different types are being introduced. Multiple sector, multiple quadrupole, hybrid sector-quadrupole and sector-time of flight, Fourier transform, and other instruments are available. Many are capable of exact mass measurement and have other special capabilities such as variable scattering angles, collision energies, and reaction times. (iii) Much

GC-MS work is already being done with selected ion monitoring, rather than by scanning complete mass spectra. The increased variety of experiments possible in multianalyzer mass spectrometers increases the tendency to employ scans or data acquisition schemes optimized to the problem in hand—most notably scans that are selective for the particular compound or group of compounds of interest.

Much of the recent excitement about mass spectrometry has centered on new ionization techniques and new instrumental configurations. Looking ahead, one can see the possibility of new types of experiments in which mass spectrometry is used. One is the use of molecular ion beams to prepare macroscopic amounts of material, particularly in tailoring surfaces to achieve chemical properties desirable in catalysis (122). The variety of chemical species and the control of translational energy and isotopic composition thus available make such synthetic experiments particularly attractive. Ion implantation and metal atom vapor experiments achieve similar objectives, but the even more promising chemical reactions of polyatomic ion beams are essentially unexplored. A second new type of experiment is the imaging of specimens for organic constituents at the micrometer level, using finely focused probe beams in a desorption ionization experiment. Larger scale versions of this experiment have proved successful, including those involving direct examination of the surfaces of paper and thin-layer chromatograms, and of electrophoretograms (123). Elemental analysis by analogous procedures is a well-developed procedure (124).

The new experiments indicated above exemplify the range of mass spectrometry. We noted earlier that the discipline has been inward-looking, so that some of the more remarkable capabilities of mass spectrometry are little known outside the field. An example is the determination of energy transfer from vibrational to translational modes associated with unimolecular dissociation. These quantities can be measured with extraordinary sensitivity by mass spectrometry; energies less than 10^{-4} eV (about 2 calories per mole) are accessible. For example, α-haloacetophenone molecular ions yield the benzoyl cation in a process releasing 5 to 10 calories per mole (125), while charge separation reactions of hydrocarbon dications release up to 3 eV (126).

In the biological sciences, applications of mass spectrometry now lag instrumental capabilities, particularly in regard to mass range and ionization. Many classes of compounds for which mass spectrometry has either been awkward or had limited success—steroid conjugates, glycopeptides, phospholipids, and prostaglandins—now fall squarely within its scope. The high sensitivity of the method will continue to make it an important tool for elucidating the structures of new compounds available in trace amounts, as was the case with the endorphins (127) and the leukotriene slow-reacting substances (128). Clinical applications, as in diagnosing comatose drug overdose patients (129) and treating endocrine dysfunctions (130), could become much more widespread as instrument costs continue to decline. The capabilities of mass spectrometry in mixture analysis should lead to expanded use in metabolic profiling and pharmacokinetics, and its high molecular specificity, sensitivity, and quantitative accuracy make it particularly useful in such work as the examination of methylated DNA, where the site and degree of alkylation are of concern (82), or the direct analysis of antitumor agents in cell cultures (131).

Analytical chemistry has a distinct persona, a mode of approaching its science, which is difficult to define but well illustrated through current developments in mass spectrometry. Creativity is expressed in the architecture of new types of instruments, in new scan and data manipulation modes, in optimizing the examination of multidimensional data domains. The biological scientist should seek familiarity with the underlying framework, as well as utilize the techniques and capabilities which are its most recent products. An active partnership is necessary if we are to continue to adapt the mass spectrometer to new ends, if we are to march, rather than stumble forwards.

References and Notes

1. R. J. Beuhler and L. Friedman, paper presented at the 30th Conference on Mass Spectrometry and Allied Topics, Honolulu, 6 to 11 June 1982; *J. Phys. Chem.* **77**, 2549 (1982).
2. Kratos Analytical Instruments, Ramsey, N.J.
3. K. L. Busch and R. G. Cooks, *Science* **218**, 247 (1982).
4. K. L. Rinehart, Jr., *ibid.*, p. 254.
5. R. G. Cooks and G. L. Glish, *Chem. Eng. News* **59**, 40 (30 November 1981).
6. F. W. McLafferty, *Science* **214**, 280 (1981).
7. P. J. Arpino and G. Guiochon, *J. Chromatogr. Chromatogr. Rev.* **251**, 153 (1982).
8. F. Moroni, G. Lombardi, V. Carla, G. Moneti, in preparation.
9. S. C. Gates and C. C. Sweeley, *Clin. Chem.* **24**, 1663 (1978).
10. S. C. Gates, C. C. Sweeley, W. Krivit, D. DeWitt, B. E. Blaisdell, *ibid.*, p. 1680.
11. T. S. Samy *et al.*, *J. Biol. Chem.* **258**, 183 (1983).
12. K. Biemann, *Int. J. Mass Spectrom. Ion Phys.* **45**, 183 (1982).
13. C. V. Bradley, D. H. Williams, M. R. Hanley, *Biochem. Biophys. Res. Commun.* **104**, 1223 (1982).
14. R. J. Perchalski, R. A. Yost, B. J. Wilder, *Anal. Chem.* **54**, 1466 (1982).
15. M. H. O'Leary, *Phytochemistry* **20**, 553 (1981).

16. _____ and J. F. Marlier, *J. Am. Chem. Soc.* **101**, 3300 (1979).
17. A. E. Litherland, R. P. Benkens, L. R. Kilius, J. C. Rucklidge, H. E. Gore, D. Elmore, K. H. Purser, *Nucl. Instrum. Methods* **186**, 463 (1981).
18. L. W. Alvarez, W. Alvarez, F. Asaro, H. V. Michel, *Science* **208**, 1095 (1980).
19. P. Kebarle, *Annu. Rev. Phys. Chem.* **28**, 445 (1977).
20. C. L. Wilkins and M. L. Gross, *Anal. Chem.* **53**, 1661A (1981).
21. S. A. McLuckey, D. Cameron, R. G. Cooks, *J. Am. Chem. Soc.* **103**, 1313 (1981).
22. G. Boand, R. Houriet, T. Gäumann, *ibid.* **105**, 2203 (1983).
23. F. W. McLafferty, *Acc. Chem. Res.* **13**, 33 (1980).
24. K. Levsen and H. Schwarz, *Mass Spectrom. Rev.* **2**, 77 (1983).
25. M. T. Bowers, Ed., *Gas Phase Ion Chemistry*, (Academic Press, New York, 1979).
26. T. A. Lehman and M. M. Bursey, *Ion Cyclotron Resonance Spectrometry* (Wiley, New York, 1976).
27. R. D. Bowen, D. H. Williams, H. Schwarz, *Angew. Chem. Int. Ed. Engl.* **18**, 451 (1979).
28. C. H. DePuy, J. J. Grabowski, V. M. Bierbaum, *Science* **218**, 955 (1982).
29. J. Allison, R. C. Freas, D. P. Ridge, *J. Am. Chem. Soc.* **101**, 1332 (1979).
30. D. J. Burinsky and R. G. Cooks, *J. Org. Chem.* **47**, 4864 (1982).
31. A. Maquestiau *et al.*, *Nouv. J. Chim.* **3**, 517 (1979).
32. R. Wolfschütz, H. Schwarz, W. Blum, W. J. Richter, *Org. Mass Spectrom.* **14**, 462 (1979).
33. D. M. Lubman, R. Naaman, R. N. Zare, *J. Chem. Phys.* **72**, 3034 (1980).
34. D. M. Lubman and M. N. Kronick, *Anal. Chem.* **54**, 660 (1982).
35. L. Zandee and R. B. Bernstein, *J. Chem. Phys.* **70**, 2574 (1979).
36. I. W. Griffiths, F. M. Harris, E. S. Mukhtar, J. H. Beynon, *Int. J. Mass Spectrom. Ion Phys.* **41**, 83 (1981).
37. J. A. Laramee, J. J. Carmody, R. G. Cooks, *ibid.* **31**, 333 (1979).
38. S. A. McLuckey, G. L. Glish, R. G. Cooks, *ibid.* **39**, 219 (1981).
39. I. Powis, P. I. Mansell, C. J. Danby, *ibid.* **32**, 15 (1979).
40. M. Senn and R. Endele, paper presented at the 30th Conference on Mass Spectrometry and Allied Topics, Honolulu, 6 to 11 June 1982.
41. J. D. Ciupek, R. G. Cooks, K. V. Wood, C. R. Ferguson, *Fuel* **62**, 829 (1983). Compare: K. V. Wood, C. E. Schmidt, R. G. Cooks, B. D. Batts, in preparation.
42. A. Benninghoven, Ed., *Ion Formation from Organic Solids* (Springer-Verlag, New York, 1983).
43. M. Barber, R. S. Bordoli, G. J. Elliott, R. D. Sedgwick, A. N. Tyler, *Anal. Chem.* **54**, 645A (1982).
44. R. B. Van Breemen, M. Snow, R. J. Cotter, *Int. J. Mass Spectrom. Ion Phys.* **49**, 35 (1983).
45. E. Genoyer, R. Van Gricken, F. Adams, D. F. S. Natusch, *Anal. Chem.* **54**, 26A (1982).
46. S. T. F. Lai, K. W. Chan, K. D. Cook, *Macromolecules* **13**, 953 (1980).
47. J. F. Holland, B. Soltmann, C. C. Sweeley, *Biomed. Mass Spectrom.* **3**, 340 (1976).
48. G. M. Lancaster, F. Honda, Y. Fukuda, J. W. Rabalais, *J. Am. Chem. Soc.* **101**, 1951 (1979).
49. L. Kurlansik *et al.*, *Biochem. Biophys. Res. Commun.* **111**, 478 (1983).
50. J. Pierce, K. L. Busch, R. A. Walton, R. G. Cooks, *J. Am. Chem. Soc.* **103**, 2583 (1981).
51. R. Davis, I. F. Groves, J. L. A. Durrant, P. Brooks, I. Lewis, *J. Organomet. Chem.* **241**, C27 (1983).
52. K. L. Busch, S. E. Unger, A. Vincze, R. G. Cooks, T. Keough, *J. Am. Chem. Soc.* **104**, 1507 (1982).
53. D. V. Davis, R. G. Cooks, B. N. Meyer, J. L. McLaughlin, *Anal. Chem.* **55**, 1302 (1983).
54. H. Kambara, in (*42*), p. 101.
55. M. L. Gross *et al.*, *Tetrahedron Lett.* **23**, 5381 (1982).
56. R. G. Orth, H. T. Jonkman, J. Michl, *J. Am. Chem. Soc.* **104**, 1834 (1982).
57. J. L. Pierce, D. W. Wigley, R. A. Walton, *Organometallics* **1**, 1328 (1982).
58. R. Stoll, U. Schade, F. W. Röllgen, U. Giessmann, D. F. Barofsky, *Int. J. Mass Spectrom. Ion Phys.* **43**, 227 (1982).
59. S. S. Wong, R. Stoll, F. W. Röllgen, *Z. Naturforsch, Teil A* **37a**, 718 (1982).
60. W. Knabe and F. R. Krueger, *ibid.*, p. 1335.
61. F. R. Krueger and W. Knabe, *Org. Mass Spectrom.* **18**, 83 (1983).
62. D. F. Barofsky, U. Giessmann, L. W. Swanson, A. E. Bell, *Int. J. Mass Spectrom. Ion Phys.* **46**, 495 (1983).
63. J. Meili and J. Seibl, *ibid.*, p. 367.
64. G. R. Petit, C. W. Holzapfel, G. M. Cragg, C. L. Herald, P. Williams, *J. Am. Chem. Soc.*, in press.
65. R. D. Macfarlane and D. F. Torgerson, *Science* **191**, 920 (1976).
66. R. D. Macfarlane, D. Uemura, K. Ueda, Y. Hirata, *J. Am. Chem. Soc.* **102**, 875 (1980).
67. P. Hakansson *et al.*, *ibid.* **104**, 2948 (1982).
68. I. Kamensky *et al.*, *FEBS Lett.*, in press.
69. C. J. McNeal, K. K. Ogilvie, N. Y. Theriault, M. J. Nesser, *J. Am. Chem. Soc.* **104**, 972 (1982).
70. B. T. Chait, B. F. Gisin, F. H. Field, *ibid.*, p. 5157.
71. S. Della Negra, Y. M. Ginot, Y. Le Beyec, M. Spiro, P. Vigny, *Nucl. Instrum. Methods* **198**, 159 (1982).
72. R. M. Caprioli, L. A. Smith, C. F. Beckner, *Int. J. Mass Spectrom. Ion Phys.* **46**, 419 (1983).
73. L. A. Smith and R. M. Caprioli, *Biomed. Mass Spectrom.* **10**, 98 (1983).
74. P. Hakansson *et al.*, *J. Am. Chem. Soc.* **104**, 2948 (1982).
75. M. Barber *et al.*, *J. Chem. Soc. Chem. Commun.* (1982), p. 936.
76. M. Barber, R. S. Bordoli, G. J. Elliott, N. J. Horoch, B. N. Green, *Biochem. Biophys. Res. Commun.* **110**, 753 (1983).
77. W. D. Lehmann, U. Bahr, H.-R. Schulten, *Biomed. Mass Spectrom.* **5**, 536 (1978).
78. Y. Wada, A. Hayashi, T. Fujita, T. Matsuo, I. Katakuse, H. Matsuda, *Biochim. Biophys. Acta* **667**, 233 (1981).

79. _____, *Int. J. Mass Spectrom. Ion Phys.* **48**, 209 (1983).
80. A. E. Cameron, in *Physical Methods in Chemical Analysis*, W. G. Berl, Ed. (Academic Press, New York, 1961), vol. 4, p. 119.
81. K. E. Singleton, R. G. Cooks, K. V. Wood, *Anal. Chem.* **55**, 762 (1983).
82. D. J. Ashworth *et al.*, in preparation.
83. N. Ferrigni, J. L. McLaughlin, K. E. Singleton, R. G. Cooks, in preparation.
84. D. J. Harvey, J. T. A. Leuschner, W. D. M. Paton, paper presented at the 30th Conference on Mass Spectrometry and Allied Topics, Honolulu, 6 to 11 June 1982.
85. W. R. Davidson, B. A. Thomson, B. I. Shushan, J. E. Fulford, paper presented at the 30th Conference on Mass Spectrometry and Allied Topics, Honolulu, 6 to 11 June 1982.
86. D. J. Harvan, J. R. Hass, J. L. Schroeder, B. J. Corbett, *Anal. Chem.* **53**, 1755 (1981).
87. W. R. Davidson, T. Sakuma, N. Gurprasad, paper presented at the 186th National Meeting, Washington, D.C., 29 August 1983.
88. D. Fetteroff and R. Yost, *Anal. Chem.*, in press.
89. R. D. Plattner, G. A. Bennett, R. D. Stubblefield, paper presented at the 30th Conference on Mass Spectrometry and Allied Topics, Honolulu, 6 to 11 June 1982; *J. Agric. Food Chem.* **31**, 785 (1983).
90. W. H. McFadden, H. L. Schwartz, S. Evans, *J. Chromatogr.* **122**, 389 (1976).
91. R. D. Smith, J. E. Burger, A. L. Johnson, *Anal. Chem.* **53**, 1603 (1981).
92. P. Dobberstein, E. Korte, G. Meyerhoff, R. Pesch, *Int. J. Mass Spectrom. Ion Phys.* **46**, 185 (1983).
93. P. J. Arpino, J. P. Bounine, M. Dedieu, G. Guichon, *J. Chromatogr. Chromatogr. Rev.* **271**, 43 (1983).
94. C. R. Blakley and M. L. Vestal, *Anal. Chem.* **55**, 750 (1983).
95. H. Yoshida *et al.*, *Fres. Z. Anal. Chem.* **311**, 674 (1982).
96. J. D. Henion, *Anal. Chem.* **50**, 1687 (1978).
97. E. C. Horning *et al.*, *J. Chromatogr. Sci.* **112**, 725 (1974).
98. M. Dedieu, G. Devant, C. Juin, M. Hardy, J. P. Bounine, P. J. Arpino, paper presented at the 30th Conference on Mass Spectrometry and Allied Topics, Honolulu, 6 to 11 June 1982.
99. G. S. Hurst, M. G. Payne, S. D. Kramer, J. P. Young, *Rev. Mod. Phys.* **51**, 767 (1979).
100. J. P. Young and D. L. Donohue, *Anal. Chem.* **55**, 88 (1983).
101. M. B. Comisarow and A. G. Marshall, *Chem. Phys. Lett.* **25**, 282 (1974).
102. M. Allemann, H. P. Kellerhals, K. P. Wanczek, *ibid.* **75**, 328 (1980).
103. _____, *Int. J. Mass Spectrom. Ion Phys.* **46**, 139 (1983).
104. R. B. Cody and B. S. Freiser, *ibid.* **41**, 199 (1982).
105. M. Comisarow and A. G. Marshall, paper presented at the 23rd Conference on Mass Spectrometry and Allied Topics, Houston, 1975.
106. G. S. Groenewold and M. L. Gross, *Org. Mass Spectrom.* **17**, 269 (1982).
107. R. B. Cody, R. C. Burnier, C. J. Cassady, B. S. Freiser, *Anal. Chem.* **54**, 2225 (1982).
108. D. L. Miller and M. L. Gross, *J. Am. Chem. Soc.* **105**, 3783 (1983).
109. D. A. McCrery, E. B. Ledford, Jr., M. L. Gross, *ibid.*, p. 1435.
110. E. C. Oryiriuka, R. L. White, D. A. McCrery, M. L. Gross, C. L. Wilkins, *Int. J. Mass Spectrom. Ion Phys.* **46**, 135 (1983).
111. C. L. Wilkins, G. W. Giss, R. L. White, G. M. Brissey, E. C. Onyiriuka, *Anal. Chem.* **54**, 2260 (1982).
112. R. T. McIver, Jr., R. L. Hunter, M. Story, J. Syka, M. Labunsky, paper presented at the 31st Conference on Mass Spectrometry and Allied Topics, Boston, 9 to 13 May 1983.
113. G. D. Byrd, R. C. Burnier, B. S. Freiser, *J. Am. Chem. Soc.* **104**, 3565 (1982).
114. R. W. Jones and R. H. Staley, *ibid.*, p. 2296.
115. J. H. Hoffman, R. R. Hodges, K. D. Duerksen, *J. Vac. Sci. Technol.* **16**, 692 (1979).
116. D. R. Rushneck *et al.*, *Rev. Sci. Instrum.* **49**, 817 (1978).
117. D. A. Lane, B. A. Thomson, A. M. Lovett, N. M. Reid, *Adv. Mass Spectrom.* **8B**, 1480 (1980).
118. D. F. Hunt, J. Shabanowitz, T. M. Harvey, M. L. Coates, *J. Chromatogr. Chromatogr. Rev.* **271**, 93 (1983).
119. B. Shushan and R. K. Boyd, *Anal. Chem.* **53**, 421 (1981).
120. A. F. Weston, K. R. Jennings, S. Evans, R. M. Elliott, *Int. J. Mass Spectrom. Ion Phys.* **20**, 317 (1976).
121. J. T. Stults, C. G. Enke, J. F. Holland, *Anal. Chem.* **55**, 1323 (1983).
122. M. A. LaPack, S. J. Pachuta, K. L. Busch, R. G. Cooks, *Int. J. Mass Spectrom. Ion Phys.*, in press.
123. S. E. Unger, A. Vincze, R. G. Cooks, R. Chrisman, L. D. Rothman, *Anal. Chem.* **53**, 976 (1981). Compare A. Mathey, *Fres. Z. Anal. Chem.* **308**, 249 (1981).
124. D. M. Drummer and G. H. Morrison, *Anal. Chem.* **52**, 591 (1983).
125. R. G. Cooks, K. C. Kim, J. H. Beynon, *Chem. Phys. Lett.* **26**, 131 (1974).
126. T. Ast, J. H. Beynon, R. G. Cooks, *Org. Mass Spectrom.* **6**, 749 (1972).
127. J. Hughes *et al.*, *Nature (London)* **258**, 577 (1975).
128. R. C. Murphy, S. Hammarström, B. Samuelsson, *Proc. Natl. Acad. Sci. U.S.A.* **76**, 4275 (1979).
129. C. E. Costello, H. S. Hertz, T. Sakai, K. Biemann, *Clin. Chem.* **20**, 255 (1974).
130. A. W. Pike, J. Klein, P. V. Fennessey, K. Horwitz, paper presented at the 31st Conference on Mass Spectrometry and Allied Topics, Boston, 9 to 13 May 1983.
131. Y. Tondeur, M. Shorter, M. E. Gustafson, R. C. Pandey, paper presented at the 31st Conference on Mass Spectrometry and Allied Topics, Boston, 9 to 13 May 1983.
132. Support from the National Science Foundation and the Department of Energy and comments from J. Amy, K. Wood, B. Freiser, and N. Delgass are acknowledged.

116

The intimate association of chemistry and light ranges from fireworks displays and the color of solutions and precipitates to the spectroscopic analysis of new and unknown substances. The advent of the laser has only served to strengthen this natural bond, so that no major chemical research laboratory is without lasers today. Because of its high power, directionality, purity of color, and temporal coherence, the laser has become a highly versatile tool, first applied to the study of how chemical reactions occur, then to initiate chemical reactions upon irradiation, and finally as an extremely sensitive and selective means to analyze for the presence of chemical substances of interest. It is this last topic which is the subject of this brief selective review in which outstanding examples of recent advances in chemical analysis based on laser techniques are presented. Laser methodologies promise to improve dramatically the detection of trace substances embedded in "real" matrices, giving the analyst a most powerful means for determining the composition of materials.

Multiphoton Ionization

One of the most promising laser techniques is multiphoton ionization (MPI), in which an atom or molecule absorbs more than one photon to cause ejection of an electron (1, 2). This nonlinear process is made possible by the high intensity of laser light sources. We first review how the MPI process works, then describe some recent applications.

Figure 1 compares multiphoton ionization to single-photon ionization. If the energy of the photon exceeds the ionization energy of the target atom or molecule, the target species will be ionized

9

Laser Chemical Analysis

Richard N. Zare

Science 226, 298–303 (19 October 1984)

117

and can be detected by measuring the subsequent positively charged ion or negatively charged photoelectron (Fig. 1a). If, however, the energy of one photon lies below the ionization threshold, ionization can occur only by the simultaneous absorption of several photons whose energy sum exceeds the ionization potential (Fig. 1, b and c). Multiphoton processes may involve virtual levels (Fig. 1b) which are not eigenstates (real levels) of the isolated atom or molecule. The lifetime for such virtual levels is often on the order of 10^{-15} second. The MPI process is said to be resonant when the energy of an integral number, n, of photons approaches closely the energy of an n-photon–allowed transition. Because real levels have lifetimes typically of 10^{-9} to 10^{-6} second, the probability for absorbing subsequent photons is greatly increased (six orders of magnitude or more). Consequently, nonresonant multiphoton ionization usually requires laser powers of about 1 GW/cm^2, which can be achieved only by tightly focusing powerful pulsed lasers. In contrast, resonant enhanced multiphoton ionization (REMPI), also called resonance ionization spectroscopy (RIS), can be carried out with pulsed lasers of fairly modest intensity (1 MW/cm^2) and under favorable conditions even with continuous-wave laser sources. The choice of the excitation wavelength gives REMPI its selectivity, and it is the availability of powerful tunable lasers that makes REMPI a nearly universal detector. One or more laser photons with the same or different frequencies may be employed, and variations include field ionization of high-lying Rydberg levels (3–5).

Because the laser intensity required for MPI is high, energy levels are often

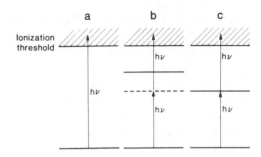

Fig. 1. Schematic energy diagram for (a) one-photon photoionization, (b) nonresonant multiphoton ionization, and (c) resonant multiphoton ionization. The solid horizontal lines represent real levels, the dashed horizontal lines virtual levels. In the multiphoton ionization process the photon frequencies may be the same as drawn (one-color experiment) or different (multicolor experiment).

Stark-shifted. In practice, the laser intensity in the (focal) detection volume is uniform in neither space nor time, causing the target species to be subjected to a wide range of different intensities. This results in broadened line profiles of the resonant MPI process since different shifts occur in the interaction volume. Consequently, REMPI of atomic systems has poor selectivity with regard to different isotopes, although for molecular systems with much larger spectral isotope shifts REMPI can easily be made to be isotopically specific (6). In either case, isotopic analysis is usually realized by combining REMPI with mass analysis, using, for example, time of flight (TOF) (7–9), reflectron (10), quadrupole mass filter (11, 12), or ion cyclotron resonance (13, 14).

Bound-to-continuum transitions having cross sections typically less than 10^{-17} cm^2 are usually the rate-limiting step in REMPI. Whereas the behavior of the MPI rate near resonance may be rather intuitive, the MPI rate between resonances need not show a flat, monotonic behavior. Destructive interference effects, called anti-resonances, are possible and may lead to almost complete cancellation in the MPI rate (2).

At low pressures (below 10^{-5} torr) each target species interacts essentially independently with the radiation field. In this case, the MPI signal is always proportional to the (partial) pressure of the species being detected. At high pressures (1 to 10 torr) collective (coherent) behavior becomes important, leading to such nonlinear processes as third harmonic generation. For intermediate pressures both types of nonlinear processes compete, and above 10^{-3} torr it appears that the MPI signal is often severely reduced because of third harmonic generation and the like (15). At high pressures there are the added problems of space charge, ion-electron recombination, ion cluster formation, and inability to use high-gain multipliers for the detection of charged particles. These considerations set a practical limit on the use of MPI as an analytical tool for gas samples and explain why MPI is usually used as a detector of rarefied gas samples, such as those found in evaporation from hot surfaces or filaments or in atomic or molecular beams.

Most analytical applications of REMPI involve elemental analysis or isotope ratio measurements, or both. Atoms of nearly every element can be detected by REMPI with pulsed tunable dye lasers (16–18). However, this procedure often has two drawbacks. First, the effective volume for detection is about 10^{-3} cm^3 or less because of the need to focus the laser output to obtain sufficiently high intensities to cause efficient ionization. Second, the pulsed nature of the high-intensity laser source implies a

low duty cycle, often of only 5×10^{-6} or less. Both of these factors may cause the rate of material processing to be severely restricted.

Despite these limitations, some spectacular advances have already been achieved in chemical analysis with resonant and nonresonant MPI. One such example is surface analysis by multiphoton ionization, in which a minute fraction of a monolayer is desorbed, ablated, or sputtered from a surface exposed in a high-vacuum environment to a probe beam (ion beam, electron beam, or laser beam), and the resulting neutral component is ionized above the surface. Several groups (19–28) are pursuing the development of this method, which promises to permit trace analysis for some materials at concentrations below parts per billion (ppb) and in special cases parts per trillion (ppt). Immediate applications are in the analysis of semiconductor crystals, whose electrical properties are markedly altered by minute amounts of lattice impurities.

It is useful to make some comparisons with a well-established tool for surface analysis, secondary ion mass spectrometry (SIMS), which involves the analysis of the charged particles ejected from ion-bombarded surfaces (29, 30). The ~10 ppb detection limit in SIMS arises from three sources: (i) the fraction of ejected particles that are charged is often 10^{-3} or less; (ii) the sorting of the charged particles by mass using quadrupole or magnetic sectors involves ion transmissions in the range of 10^{-3} to 10^{-4}; and (iii) secondary ion formation is strongly influenced by the sample matrix, making quantification of measurements extremely difficult.

Surface analysis by MPI detection of the desorbed neutral components appears to be an improvement over the sensitivity and reliability of SIMS and at the same time can be applied to a wide range of surface materials, overlayers, and adsorbates. An important aspect of this laser surface analysis technique is that the desorption step is separated spatially and temporally from the ionization step, unlike the process in SIMS. This separation permits detection of the majority neutral fraction, greater control in the choice and type of the probe and ionizing beams with a resultant reduction in surface damage, a gain in quantitation of the ionization step, and avoidance of the large matrix effects seen when the ionization probability of the analyte varies strongly with the chemical environment. The laser ionization step may be resonantly enhanced or nonresonant. In the former case, as in recent studies of the sputtering of indium from an indium metal surface (19–21), mass analysis may be unnecessary, and there is a corresponding gain in sensitivity. Recent examples of this are the detection of as few as 10^{11} sodium atoms per cubic centimeter in laser ablation of single-crystal silicon (22) and the detection of iron at 50 parts per million (ppm) in iron-doped silicon targets (25, 26).

In the nonresonant case, it may be possible to carry out a simultaneous analysis of all of the neutral components evolving from the surface by using some form of TOF mass spectrometry, which permits simultaneous recording of all the ions. Quantitative analysis of relative amounts of desorbed species then requires knowledge of the relative nonresonant MPI efficiencies at the laser frequency employed. Figure 2 presents a portion of a TOF mass spectrum taken of the standard reference material NBS copper C1252 (27, 28). The room tem-

perature sample was bombarded by a pulsed 1-μA, 2.7-keV argon ion beam, and the sputtered species were ionized by using pulsed (10-Hz) krypton fluoride (KrF) excimer laser radiation at 248 nm with a focused intensity of 7 GW/cm^2. The data collection time was 1 hour. For the mass range displayed in Fig. 2, the elements whose bulk atomic composition values were given by the National Bureau of Standards include ^{107}Ag (with a concentration of 51 ppm), ^{123}Sb (9 ppm), ^{197}Au (11 ppm), ^{206}Pb (4 ppm), and ^{209}Bi (6 ppm). The three largest peaks originate from Cu$_2$, and peaks at 170, 172, and 174 arise from AgCu dimers. Small amounts of other species (Cd, Sn, Ba, Ce, Hg, Pt, Ta, and W, the latter three as contaminants from the sample holder and the Ar$^+$ sputtering source) are also observed. It is estimated that a sensitivity of 1 ppm is obtained with only about 10^{-10} g of the sample removed (27, 28). This method appears to be well suited for studying the kinetics of surface segregation of impurities and differential evaporation.

Another striking application of laser multiphoton ionization to chemical analysis is the demonstration that individual atoms of nearly every element can be counted, one by one, in a gas sample or flow (16, 31, 32). All that is required to achieve this feat is to use the focused output of a pulsed laser having enough intensity that nearly every atom in the focal volume is ionized, that is, to saturate the REMPI process. While such capabilities far exceed the normal needs of elemental analysis, single-atom detection presents the opportunity to search for rare events of great interest. Examples include various nuclear reactions, such as neutrino detection by transmutation of elements, double beta decay

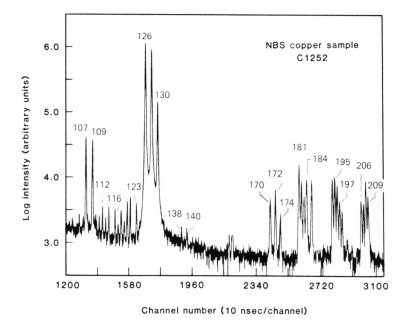

Fig. 2. Segment of a reflection time-of-flight mass spectrum obtained by Ar^+ bombardment of NBS reference material copper C1252 followed by nonresonant multiphoton ionization of the ejected neutral species near the surface with a pulsed KrF excimer laser. The number above a peak is its nominal mass; see the text for identification. This drawing is adapted from Becker and Gillen (27).

events, and the production of short-lived superheavy elements or nuclear isomers. Moreover, the combination of resonant multiphoton ionization of atoms with nearly 100 percent efficiency and high-resolution mass analysis offers the possibility of dramatically improving on isotopic selectivity (33).

An example of the power of this combination is the selective ionization of Lu in the presence of Yb (34, 35). These rare earth elements are notoriously difficult to separate chemically or by simple physical means. Of particular interest is the $^{173}Lu/^{175}Lu$ ratio in mixed lutetium/ytterbium samples in which isobaric interference from ^{173}Yb is severe. This is exacerbated by the fact that ytterbium is an order of magnitude more volatile than lutetium. By using REMPI with a continuous-wave laser scheme, discrimination factors approaching 10^6 have been measured against ^{173}Yb, a selectivity that is

limited not by the REMPI process but by the resolution of the mass spectrometer, which allows tailing of the ^{175}Lu peak to lower mass. There are many other examples of the use of resonant multiphoton ionization to separate isobaric elements (17, 18). The precise measurement of the ratios of isotopes of a single element has many applications, including geochemical dating and the determination of patterns of air, water, and soil circulation.

Perhaps the most advanced application of REMPI for single-atom detection and isotopic ratio analysis is the very recent demonstration that 10^3 atoms of ^{81}Kr can be sorted out and counted in a sample along with 10^7 atoms of ^{80}Kr or ^{82}Kr plus 10^{12} other atoms or molecules, all in about an hour starting from a several liter gas sample (36, 37). Figure 3 is a diagram of the experimental setup, in which (i) a flash lamp–pumped dye laser is fired at krypton atoms condensed on a liquid helium–cooled finger to "bunch" them together in the ionization region (*38*); (ii) after a 7-μsec delay, the krypton atoms are resonantly ionized with a three-photon excitation scheme; (iii) the krypton ions are isotopically selected with a quadrupole mass filter of modest resolution; and (iv) the ions are accelerated to 10 kV onto a Be-Cu target, where they are implanted, while at the same time an electron multiplier with a gain of more than 10^6 counts each implanted ion by detecting the secondary electrons emitted by the target. Since the implanted krypton atoms remain "buried" in the Be-Cu target nearly indefinitely at room temperature, every istopically selected krypton atom will be counted once and only once (provided the sticking coefficient is unity). On the other hand, should a recount be desired, it is possible to expel all the inert gas atoms by heating an appropriately chosen target material under vacuum to an elevated temperature (*39*). This recycling technique may permit isotopic enrichment before final counting, if desired.

While past analytical applications of multiphoton ionization have concentrated almost exclusively on elemental analysis, there is no reason why the same principles will not permit single molecule detection. Already impressive advances have been made in detecting molecular hydrogen and its isotopic analogs (*40–42*), not only in identifying this species but also in determining the relative populations of individual quantum states. With regard to surface analysis, experiments have been reported on the REMPI detection of 10^{-8} to 10^{-10} of a monolayer of anthracene and naphthalene adsorbed on graphite (*23*). Once again it is possible to increase the sensitivity of the MPI detection scheme by accumulation of the

molecules on a cooled substrate followed by pulsed desorption and photoionization above the surface (*43*). Extensive fragmentation of polyatomic ions is common in MPI (*1*), but often can be avoided by the choice of wavelengths and power levels. In either case, there is a great need to develop the necessary data base to allow these methods to become useful analytical tools. Perhaps it will also be possible to exploit the kinetic energy distribution of the photoelectrons, which carries a characteristic signature of the neutral species whose detection is sought.

Laser Fluorimetry

Multiphoton ionization rates are usually limited by the small cross sections (10^{-17} to 10^{-19} cm^2) for bound-to-continuum transitions. On the other hand, typical cross sections for strongly allowed transitions are 10^{-11} to 10^{-14} cm^2. Thus, in cases where laser fluorescence is applicable, the effective detector volume often exceeds 10^3 times that for MPI, which may tip the balance in favor of laser fluorimetry for sensitive detection problems (*33*).

As in the case of MPI, maximum sensitivity is achieved by saturating the fluorescence transition. Such saturation effects may be particularly well suited for the detection of atomic species whose upper level reradiates nearly exclusively to the lower level being pumped. This may result in a "burst" of fluorescence photons when the atom passes through the laser beam (provided the radiative lifetime is short compared to the passage time). By recording such correlated fluorescence bursts a significant discrimination against background interference is achieved, allowing single-atom detection, particularly for single ions confined in a trap (44–48). However, saturated fluorescence may be less useful for the detection and quantitation of molecular species in which the upper level reradiates typically to many lower levels (49).

The detection of atoms by laser-induced fluorescence has been used to advantage in measuring atomic constituents, velocity distributions, and relative populations of excited state and fine structure levels of atoms sputtered from surface samples (50). Important applications of this work involve diagnostics for plasma devices and the determination of wall integrity of confined plasma sources.

Laser fluorimetry is also well suited for trace analysis of complex organic mixtures in solution (51), where a sensitivity limit approaching that of single-molecule detection may be possible (52, 53). For volatile organics, gas chromatography in combination with such detectors as mass spectrometry is often the analytical method of choice, but for nonvolatile organics, a standard method of separation is high-performance liquid chromatography (HPLC) (54). Unfortunately, HPLC generally has poorer reso-

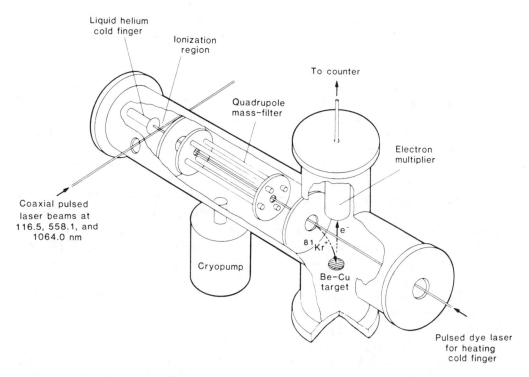

Fig. 3. Schematic diagram for the sorting and counting of ^{81}Kr atoms using the three-photon resonant multiphoton ionization scheme: Kr $4p^6$ + hν (116.5 nm) $\rightarrow 4p^5 5s$ $[\frac{1}{2}]_1$ + hν (558.1 nm) $\rightarrow 4p^5 6p$ $[\frac{1}{2}]_0$ + hν (1064.0 nm) $\rightarrow$ Kr$^+$ + e$^-$. [Reprinted with permission of C. H. Chen, S. D. Kramer, S. L. Allman, G. S. Hurst, *Applied Physics Letters* **44**, 640 (1984). © 1984 American Institute of Physics.]

lution than gas chromatography, and so far it has not been easy to combine it with powerful detectors such as mass spectrometers, although laser desorption mass spectrometry in combination with thermospray sample deposition appears to hold much promise for that purpose (55). The separation power of HPLC can be significantly increased by the use of microcolumns, which are generally of three types: narrow-bore packed columns, packed capillary columns, and open tubular capillaries (56). However, each of these requires analysis of minute quantities of effluent from the microcolumn. Indeed, the burden is placed on detector methods to compensate for the lack of chromatographic resolution of HPLC, and the development of sensitive and selective detectors for this purpose may be the gating factor in the use of microcolumn HPLC. It is in this area that laser methods have the potential for making an important advance (57).

Because of their coherence properties, lasers can deliver to a far-field target a highly collimated beam of light with power levels in excess of 10 mW and in some cases in excess of 10 W. In contrast, a sample exposed to the 254-nm output of a 100-W mercury-arc source receives only ~5 mW, and possibly even less if this radiation cannot be efficiently coupled into a small detection volume. However, laser power alone does not necessarily improve the limits of detection in laser fluorimetry, because these limits are usually set by background interference from the sample on which the fluorescence signal of interest is superimposed. This background interference arises from scattering of the laser light from cell walls (58), elastic (Rayleigh) scattering of the laser light, inelastic (Ra-

man) scattering of the laser light, and fluorescence from other interfering species. A number of groups are actively seeking ways to overcome these limitations. Clearly, one of the easiest improvements is to employ (nonfluorescent) filters that transmit the fluorescence but reject scattered light at other wavelengths. Conventional fluorimetric systems have an excitation bandwidth on the order of 10 nm, so that the Rayleigh and Raman components of the scattered light are at least of comparable width. On the other hand, laser sources are typically narrower than 0.1 nm, and the possible reduction of the width of the Rayleigh and Raman components may aid in their rejection. Further discrimination may also be achieved by placing a polarization analyzer in front of the fluorescence detector, since Rayleigh scattering and the stronger Raman features are generally polarized. Beyond these improvements, a number of other options are available.

One approach is to eliminate interference arising from cell walls by using a "windowless" fluorescence cell in which the flowing sample forms a suspended droplet or a free-falling jet at the exit of the column (59–61). Vibrations and multiple reflection at the droplet-air or jet-air interface may adversely affect the performance of this system, but it has been possible to detect as little as 20 fg (2×10^{-14} g) of fluoranthene in 10 μl of hexane eluting from a microcolumn (61).

Another alternative is to couple the capillary tube to an optical fiber, which can be placed very close to the focused laser beam to achieve excellent collection efficiency (62, 63). Because optical fibers have a critical cone of acceptance, it is thus possible to reject scattering and fluorescence originating from the cell walls.

Yet another possibility is to use the hydrodynamic focusing technique employed in flow cytometry. Here the eluent is confined to the flow along the center of an ensheathing solvent stream under laminar flow conditions (52, 53, 64, 65). The laser excitation source is focused on the sample stream to provide a very small detection volume far removed from the cell walls. Because the sample stream is never in contact with the cell walls, molecular adsorption onto the walls is eliminated. By such means, as few as 22,000 dye molecules of aqueous rhodamine 6G have been detected in a probe volume of 11 pl during the 1-second time constant necessary to reduce the noise in the scattered light by signal averaging (53). At this detection limit, the probability of a single rhodamine 6G molecule being present in the probe volume is 0.6. It seems, then, that with further improvements, true single-molecule detection in condensed media is certainly within grasp.

Each of these flow cell designs has different advantages and drawbacks, and the actual detection limit depends, of course, on the species of interest, especially the relative and absolute spectral locations of the absorption and fluorescence "bands" and the nature of the background interference.

Two other laser schemes to overcome background interference may not be generally applicable, but where it has been possible to employ them the results have been extremely encouraging. One such scheme is two-photon fluorescence excitation, again made possible by the high powers available from pulsed lasers or tightly focused continuous-wave lasers (66, 67). This method provides additional spectral selectivity in HPLC detection based on the selection rules for two-photon transitions. In addition, excitation typically involves the use of visible lasers while fluorescence is observed in the ultraviolet, making the rejection of scattered light a very easy task. Another scheme is to employ time-resolved fluorescence detection, that is, to delay opening the detector until most of the scattered Rayleigh, Raman, and short-lived background fluorescence has passed by it (68, 69). Of course, this requires that (i) the analyte of interest have an intrinsically long-lived fluorescence or can be labeled with a long-lived fluorescent tag, and (ii) the excitation pulse be much shorter than the fluorescence lifetime, τ_f. Considerable success has been achieved. A detection limit of $1.8 \times 10^{-13} M$ for rubrene ($\tau_f = 17$ nsec) has been obtained with a picosecond excitation source (70), and a detection limit of $2 \times 10^{-15} M$ was reported for the complex of europium (Eu^{3+}) with 1,1,1-trifluoro-4-(2-thienyl)-2,4-butanedione ($\tau_f = 420$ μsec) with a 10-nsec pulsed nitrogen laser excitation source (71). Recently, time-resolved photon counting in conjunction with competitive binding fluorescence immunoassay has allowed human immunoglobulin G to be analyzed directly in serum-containing samples through the use of a long-lived fluorescence label (Tb^{3+}) attached to immunoglobulin G by a bifunctional chelating agent (72).

Application of laser fluorimetry to microbore HPLC is still far from routine but appears to hold much promise. This is illustrated in Fig. 4, in which 16 polynuclear aromatic hydrocarbons of low

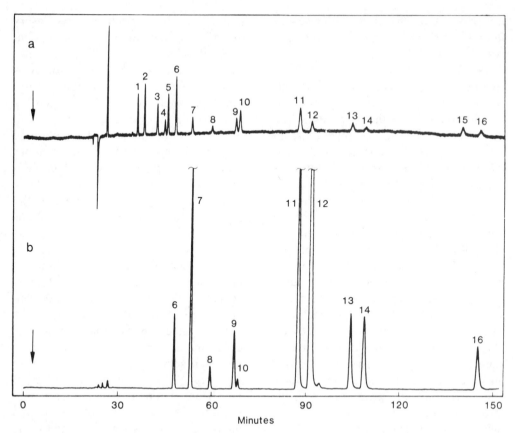

Fig. 4. Chromatograph (73) of polynuclear aromatic hydrocarbon standards obtained with a narrow-bore fused silica capillary (0.2 mm inner diameter, 1.33 m length) packed with Micropak-SP C18 (3 μm): (a) ultraviolet absorbance, and (b) laser fluorescence, 5 nl optical volume, excitation wavelength = 325 nm, fluorescence wavelength = 430 nm. The mobile phase is 92.5 percent aqueous acetonitrile at 1.2 μl/min. The solutes are: 1, naphthalene, 2.5 ng; 2, acenaphthylene, 5.5 ng; 3, acenaphthene, 2.5 ng; 4, fluorene, 0.5 ng; 5, phenanthrene, 0.2 ng; 6, anthracene, 0.2 ng; 7, fluoranthene, 0.5 ng; 8, pyrene, 0.2 ng; 9, benzo[a]anthracene, 0.2 ng; 10, chrysene, 0.2 ng; 11, benzo[b]fluoranthene, 0.5 ng; 12, benzo[k]fluoranthene, 0.2 ng; 13, benzo[a]pyrene, 0.2 ng; 14, dibenz[a,h]anthracene, 0.5 ng; 15, benzo[ghi]perylene, 0.5 ng; and 16, indeno(1,2,3-cd)pyrene, 0.2 ng.

molecular weight are separated on a packed narrow-bore microcolumn (73). The upper trace was obtained with a 254-nm absorption detector, while the lower trace results from fluorescence excitation of the 5-nl volume with a 325-nm, 3-mW helium-cadmium laser viewed by a photomultiplier through a 430-nm inter-ference filter (10 nm full width at half-maximum).

This chromatogram illustrates several features of laser fluorimetry. First, not all compounds are naturally fluorescent. This allows selective detection but may require that the analyte of interest be derivatized to incorporate a fluorescent tag. Second, when laser fluorescence analysis is applicable, the signal-to-noise ratio is far superior to that in ultraviolet absorbance, and readily permits picogram detection levels. Third, the laser system and detection optics required for laser fluorimetric detection with microcolumn HPLC are rather simple and inexpensive—far from the state of the art—so that such instrumentation should be widely applicable. For example, this technique is already being applied to a variety of biomedical problems, including dansyl amino acids (73), derivatized fatty acids (74), bile acids (75), and other lipids (74, 75).

The examples above give only a glimpse into the exciting possibilities laser methods are opening to the analyst. Many other laser-related techniques (76) have gone unmentioned here because of space constraints. Nevertheless, it would seem that as the separation and analysis of mixtures of ever increasing complexity are demanded, laser techniques will continue to offer unique possibilities for achieving trace analysis with unprecedented sensitivity and selectivity. Moreover, laser techniques may permit ultratrace analysis to be carried out in practical situations in which matrix interference often proves to be the real limiting factor.

References and Notes

1. P. M. Johnson and C. E. Otis, *Annu. Rev. Phys. Chem.* **32**, 139 (1981).
2. J. Morellec, D. Normand, G. Petite, *Adv. At. Mol. Phys.* **18**, 97 (1982).
3. G. I. Bekov, V. S. Letokhov, V. I. Mishin, *Pis'ma Zh. Eksp. Teor. Fiz.* **73**, 157 (1977) [*Sov. Phys. JETP* **46**, 81 (1977)].
4. _____, *Pis'ma Zh. Eksp. Teor. Fiz.* **27**, 52 (1978) [*Sov. Phys. JETP Lett.* **27**, 47 (1978)].
5. G. I. Bekov, V. S. Letokhov, O. I. Matveev, V. I. Mishin, *Opt. Lett.* **3**, 159 (1978).
6. D. M. Lubman and R. N. Zare, *Anal. Chem.* **54**, 2117 (1982).
7. D. A. Lichtin, S. Datta-Ghosh, K. R. Newton, R. B. Bernstein, *Chem. Phys. Lett.* **75**, 214 (1980).
8. U. Boesl, H. J. Neusser, E. W. Schlag, *J. Chem. Phys.* **72**, 4327 (1980).
9. J. P. Reilly and K. L. Kompa, *ibid.* **73**, 5468 (1980).
10. U. Boesl, H. J. Neusser, R. Weinkauf, E. W. Schlag, *J. Phys. Chem.* **86**, 4857 (1982).
11. L. Zandee and R. B. Bernstein, *J. Chem. Phys.* **70**, 2574 (1979).
12. D. M. Lubman, R. Naaman, R. N. Zare, *ibid.* **72**, 3034 (1980).
13. M. P. Irion, W. D. Bowers, R. L. Hunter, F. S. Rowland, R. T. McIver, Jr., *Chem. Phys. Lett.* **93**, 375 (1982).
14. T. J. Carlin and B. S. Freiser, *Anal. Chem.* **55**, 955 (1983).
15. D. Normand, J. Reif, J. Morellec, in *Electronic and Atomic Collisions*, J. Eichler, I. V. Hertel, N. Stolterfoht, Eds. (North-Holland, Amsterdam, 1984), pp. 471–486.
16. G. S. Hurst, M. G. Payne, S. D. Kramer, J. P. Young, *Rev. Mod. Phys.* **51**, 767 (1979).
17. D. L. Donohue, J. P. Young, D. H. Smith, *Int. J. Mass Spectrom. Ion Phys.* **43**, 293 (1982).
18. J. D. Fassett, J. C. Travis, L. J. Moore, F. E. Lytle, *Anal. Chem.* **55**, 765 (1983).
19. N. Winograd, J. P. Baxter, F. M. Kimock, *Chem. Phys. Lett.* **88**, 581 (1982).
20. F. M. Kimock, J. P. Baxter, N. Winograd, *Surf. Sci. Lett.* **124**, L41 (1983).
21. _____, *Nucl. Instrum. Methods Phys. Res.* **218**, 287 (1983).
22. S. Mayo, T. B. Lucatorto, G. G. Luther, *Anal. Chem.* **54**, 553 (1982).
23. V. S. Antonov, S. E. Egorov, V. S. Letokhov, A. N. Shibanov, *Pis'ma Zh. Eksp. Teor. Fiz.* **38**, 185 (1983) [*JETP Lett.* **38**, 217 (1983)].
24. J. E. Parks, H. W. Schmitt, G. S. Hurst, W. M. Fairbanks, Jr., *Thin Solid Films* **108**, 69 (1983).
25. D. M. Gruen, M. J. Pellin, C. E. Young, W. F. Calaway, *Res. & Dev.* (March 1984), pp. 153–160.
26. M. J. Pellin, C. E. Young, W. F. Calaway, D. M. Gruen, *Surf. Sci.* **144**, 619 (1984).
27. C. H. Becker and K. T. Gillen, *J. Opt. Soc. Am. B* **2**, 1438 (1985).
28. _____, *Anal. Chem.* **56**, 1671 (1984).
29. R. J. Colton, *J. Vac. Sci. Technol.* **18**, 737 (1981).
30. P. Williams, in *Applied Atomic Collision Physics*, vol. 4, *Condensed Matter*, S. Datz, Ed. (Academic Press, Orlando, Fla., 1983), pp. 327–377.
31. V. S. Letokhov, in *Chemical and Biochemical Applications of Lasers*, C. B. Moore, Ed. (Academic Press, New York, 1980), vol. 5, pp. 1–38.
32. C. Th. J. Alkemade, *Appl. Spectrosc.* **35**, 1 (1981).
33. R. A. Keller, D. S. Bomse, D. A. Cremers, *Laser Focus* (October 1981), pp. 75–80.
34. C. M. Miller and N. S. Nogar, *Anal. Chem.* **54**, 1606 (1983).
35. N. S. Nogar, S. W. Downey, C. M. Miller, "Analytical Capabilities of RIMS: Absolute Sensitivity and Isotopic Analysis," *Resonance Ionization Spectroscopy-Conference Series No. 71*, G. S. Hurst and M. G. Payne, Eds. (Institute of Physics, Boston, 1984), pp. 91–95.
36. S. D. Kramer, C. H. Chen, M. G. Payne, G. S. Hurst, B. E. Lehmann, *Appl. Opt.* **22**, 3271 (1983).

37. C. H. Chen, S. D. Kramer, S. L. Allman, G. S. Hurst, *Appl. Phys. Lett.* **44**, 640 (1984).
38. G. S. Hurst, M. G. Payne, R. C. Phillips, J. W. T. Dabbs, B. E. Lehmann, *J. Appl. Phys.* **55**, 1278 (1984).
39. C. H. Chen, G. S. Hurst, M. G. Payne, *Chem. Phys. Lett.* **75**, 473 (1980).
40. E. E. Marinero, C. T. Rettner, R. N. Zare, *Phys. Rev. Lett.* **48**, 1323 (1982).
41. H. Rottke and K. H. Welge, *Chem. Phys. Lett.* **99**, 456 (1983).
42. A. H. Kung, N. A. Gershenfeld, C. T. Rettner, D. S. Bethune, E. E. Marinero, R. N. Zare, *Laser Techniques in the Extreme Ultraviolet*, S. E. Harris and T. B. Lucetoro, Eds. (American Institute of Physics, Boulder, Colo., 1984), pp. 10–22.
43. S. E. Egorov, V. S. Letokhov, A. N. Shibanov, *Chem. Phys.* **85**, 349 (1984).
44. G. W. Greenlees, D. L. Clark, S. L. Kaufman, D. A. Lewis, J. F. Tonn, J. H. Broadhurst, *Opt. Commun.* **23**, 236 (1977).
45. V. I. Balykin, V. S. Letokhov, V. I. Mishin, V. A. Semchishen, *Pis'ma Zh. Eksp. Teor. Fiz.* **26**, 492 (1977) [*Sov. Phys. JETP Lett.* **26**, 357 (1977)].
46. W. Neuhauser, M. Hohenstatt, P. E. Toschek, H. Dehmelt, *Phys. Rev. A* **22**, 1137 (1980).
47. C. L. Pan, J. V. Prodan, W. M. Fairbank, Jr., C. Y. She, *Opt. Lett.* **5**, 459 (1980).
48. D. J. Wineland and W. M. Itano, *Phys. Lett. A* **82**, 75 (1981).
49. R. Altkorn and R. N. Zare, *Annu. Rev. Phys. Chem.* **35**, 265 (1984).
50. D. M. Gruen, M. J. Pellin, C. E. Young, M. H. Mendelsohn, A. B. DeWald, *Phys. Scr.* **T6**, 42 (1983).
51. A. B. Bradley and R. N. Zare, *J. Am. Chem. Soc.* **98**, 620 (1976).
52. N. J. Dovichi, J. C. Martin, J. H. Jett, R. A. Keller, *Science* **219**, 845 (1983).
53. N. J. Dovichi, J. C. Martin, J. H. Jett, M. Trkula, R. A. Keller, *Anal. Chem.* **56**, 348 (1984).
54. R. E. Majors, H. G. Barth, C. H. Lochmüller, *ibid.*, p. 300R.
55. E. D. Hardin *et al.*, *ibid.*, p. 2.
56. M. Novotny, *ibid.* **53**, 1294A (1981).
57. R. B. Green, *ibid.* **55**, 20A (1983).
58. J. W. Lyons and L. R. Faulkner, *ibid.* **54**, 1960 (1982).
59. G. J. Diebold and R. N. Zare, *Science* **196**, 1439 (1977).
60. E. Voigtman, A. Jurgensen, J. D. Winefordner,

61. S. Folestad, L. Johnson, B. Josefsson, *ibid.* **54**, 925 (1982).
62. M. J. Sepaniak and E. S. Yeung, *J. Chromatogr.* **190**, 377 (1980).
63. E. S. Yeung and M. J. Sepaniak, *Anal. Chem.* **52**, 1465A (1980).
64. L. W. Hershberger, J. B. Callis, G. D. Christian, *ibid.* **51**, 1444 (1979).
65. T. A. Kelly and G. D. Christian, *ibid.* **53**, 2110 (1981).
66. M. J. Sepaniak and E. S. Yeung, *ibid.* **49**, 1554 (1977).
67. _____, *J. Chromatogr.* **211**, 95 (1981).
68. I. Wieder, in *Immunofluorescence and Related Staining Techniques*, W. Knapp, K. Holubar, G. Wick, Eds. (Elsevier/North-Holland, New York, 1978), pp. 67–80.
69. E. Soini and I. Hemmila, *Clin. Chem.* **25**, 353 (1979).
70. G. R. Haugen and F. E. Lytle, *Anal. Chem.* **53**, 1554 (1981).
71. S. Yamada, F. Miyoshi, K. Kano, T. Ogawa, *Anal. Chim. Acta* **127**, 195 (1981).
72. J. E. Kuo, K. H. Milby, W. D. Hinsberg III, P. R. Poole, V. L. McGuffin, R. N. Zare, *Clin. Chem.* **31**, 50 (1985).
73. V. L. McGuffin and R. N. Zare, "Applications of Laser Fluorimetry to Microcolumn Liquid Chromatography," *ACS Symposium Series of Chromatography and Separation Chemistry*, S. Ahuja, Ed. (American Chemical Society, Washington, D.C., 1985).
74. V. L. McGuffin and R. N. Zare, *Applied Spectroscopy* **39**, 847 (1985).
75. J. Gluckman, D. Shelly, M. Novotny, *J. Chromatogr.* **317**, 443 (1984).
76. For recent reviews see D. S. Kliger, Ed., *Ultrasensitive Laser Spectroscopy* (Academic Press, New York, 1983); R. A. Keller, Ed., "Laser-based ultrasensitive spectroscopy and detection V," *Proc. Soc. Photo-Opt. Instrum. Eng.* **426** (1983); G. M. Hieftje, J. C. Travis, F. E. Lytle, Eds., *Lasers in Chemical Analysis* (Humana, Clifton, N.J., 1981).
77. I gratefully acknowledge the useful comments and suggestions I received on this manuscript from a number of workers in the field including C. H. Becker, R. B. Bernstein, K. T. Gillen, G. S. Hurst, R. A. Keller, C. H. Lochmüller, D. M. Lubman, F. E. Lytle, V. L. McGuffin, C. M. Miller, M. Novotny, J. E. Parks, M. G. Payne, C. E. Young, J. D. Winefordner, and E. S. Yeung. This work was supported by NIH grant 9R01 GM 29276.

Anal. Chem. **53**, 1921 (1981).

Part III

Theoretical Chemistry

With the invention of quantum mechanics in the 1920's, the foundation was laid for a first-principles explanation for all of chemistry. Starting with the Schrödinger equation,

$$H_e \ \Psi_i(r_e,R_n) = E_i(R_n) \ \Psi_i(r_e,R_n) \qquad (1)$$

where e stands for electron; n, nucleus; and i, the ith excited state, one could solve for the wave function Ψ_i, which is the basis for all physical and chemical properties of the system. The ability to solve Eq. 1 did not, however, lead to the demise of experimental chemistry because the solutions are straightforward only for simple systems such as H_2. Even so, approximate solutions of Eq. 1 led pioneers such as Pauling (*1*), Mulliken (*2*), and others to a conceptual description of bonding that shaped chemistry into its modern form. Until the 1970's, quantitative results from Eq. 1 often disagreed with experimental results. In that era, one could not trust theory (except when it was applied to small molecules such as H_2) unless it was confirmed by experiment, and theorists would generally have been foolhardy to suggest that theory was correct when there was disagreement with experiment. The remarkable change in that situation over the last 15 years will be illustrated with several case histories.

The Coming of Age for

Quantitative Quantum Chemistry

Bent versus linear CH₂. The chemically important but very reactive species CH_2 has been a particular challenge for both experiment and theory. In one of the first ab initio (first principles) calculations on a triatomic molecule, Foster and Boys (*3*) in 1960 predicted a bent

10

Theoretical Chemistry Comes Alive: Full Partner with Experiment

William A. Goddard III

Science 227, 917–923 (22 February 1985)

geometry (128°), but simultaneous spectroscopic studies (4) concluded that the molecule is linear. The calculations were considered too crude to be relevant. Everyone believed the molecule to be linear until the increasingly accurate theoretical studies in the late 1960's finally led Bender and Schaefer to insist in 1970 that CH_2 is bent by 135° (5). Indirect experimental evidence for such highly bent CH_2 came quickly (6), followed by a reinterpretation of the spectroscopic studies (7) to confirm the bent geometry predicted by theory. The currently accepted experimental value for the bond angle is 133.84° (8).

The singlet-triplet gap in CH_2. Carbon has four valence electrons and four bonding orbitals. Bonding carbon to two hydrogens uses two carbon electrons and two carbon orbitals for the C–H bonds, leaving carbon with two nonbonding electrons and two nonbonding orbitals (denoted σ and π). This results in two low-lying electronic states having very different chemistries, the triplet and singlet states (Fig. 1).

The triplet state is lower in energy, but by how much? Indirect photochemical experiments performed from 1967 to 1971 yielded values for the difference in energy between the singlet and triplet states), ΔE_{ST}, of 2.5 kcal (9), 1 to 2 kcal (10), and 10 ± 3 kcal (11). Early theoretical estimates were usually greater than 20 kcal, but by 1972 a group at Caltech (12) obtained ΔE_{ST} = 11.5 kcal using generalized valence bond theory, and a group at University of California, Berkeley (13), obtained 13 kcal with an alternative theory, configuration interaction. Although it is difficult to assign error bars to such calculations, the theorists believed that their results were good to ±3 kcal, which was in serious disagreement with some experiments. However, all these experiments were indirect and had serious interpretational uncertainties. Over the next few years, interpretation of the indirect experiments converged on ΔE_{ST} = 8 ± 1 kcal (14), in reasonable agreement with the accepted theoretical value, ΔE_{ST} = 11 ± 3.

In 1976 this pleasant situation was torn asunder by a "bombshell" from Colorado. Lineberger and his co-workers (15) conducted an elegant experiment in which they formed a beam of CH_2^- and used a laser to remove the extra electron. Under these conditions, either the π electron leaves to form singlet CH_2 or the σ electron leaves to form triplet CH_2, so both the singlet and triplet states were observed directly. The experimental results are sketched in Fig. 2. The value for ΔE_{ST} obtained directly from the spectrum, 19.4 kcal, was well outside the range of expected error for either theory or previous (indirect) experiments.

This startling result shattered the complacency of theoretical and experimental chemical physicists. How could everyone be so far wrong? A flurry of experimental and theoretical activity ensued. Theoretical methods had advanced considerably over the years since 1972, and in 1977 a number of extensive calculations were reported (16), all indicating ΔE_{ST} to be about 10.4 kcal. New indirect experiments continued to give values around 8 ± 1 kcal (17). Consternation and concern reigned for a year or two as these results were appearing. What could be causing the discrepancy?

Finally, in 1978 the Caltech theorists published a paper (18) in which they used theoretical results to calculate the energy of vibrational levels, which they then used to reinterpret the direct experiments. The result is shown in Fig. 3. The

Triplet CH₂
Bond angle 134°
σ orbital: one electron
π orbital: one electron

Singlet CH₂
Bond angle 103°
σ orbital: two electrons
π orbital: zero electrons

Doublet CH_2^-
Bond angle 103°
σ orbital: two electrons
π orbital: one electron

Fig. 1. States of CH_2 and CH_2^-

Caltech group claimed that three observed bands (A, B, and C) were hot bands (from vibrationally excited CH_2^- molecules) and that these bands would disappear if CH_2^- were in its true vibrational ground state. This reinterpretation led to an experimental value, $\Delta E_{ST} = 9.0$ kcal (18), in reasonable agreement with the best theoretical value (10 kcal) and the best indirect experimental value (8 kcal). The experimentalists were not convinced, but the theorists were, which

led the Caltech and Colorado researchers to bet several bottles of expensive French champagne on whether the theorists' interpretation of the experiment was correct. For 3 years the Colorado experimentalists tested this reinterpretation, attempting to find conditions under which the intensities of bands A, B, and C would change. No such evidence was found (19) and the bets stood unpaid. Finally, on 29 March 1984, the Colorado group completed the construction of an

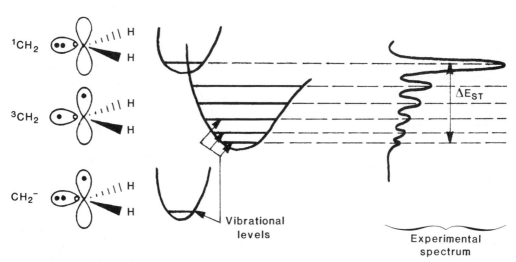

Fig. 2. Photoionization from CH_2^- ($\sigma^2\pi^1$) to yield a singlet state CH_2 ($\sigma^2\pi^0$) with the same geometry results in a large peak in the spectrum. Photoionization to yield a triplet CH_2 ($\sigma^1\pi^1$) leads to a different geometry and hence a number of small peaks corresponding to various vibrational levels of triplet CH_2. The difference between the large peak (singlet state) and the lowest triplet peak yields a direct measure of ΔE_{ST}.

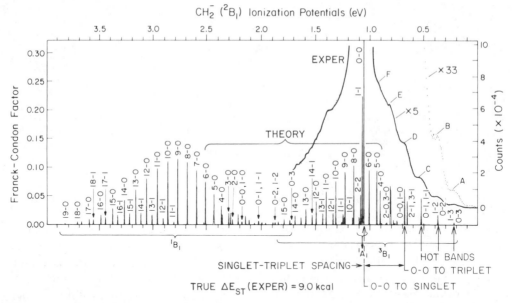

Fig. 3. Predicted photoelectron spectrum (from theory, 1978) and experimental spectrum (1976). From the theory, peaks A, B, and C were assigned as hot bands, with peak D as the 0-0 transition to the triplet state.

apparatus that would eliminate the possibility of hot bands, and the first experiment was run with CH_2. Bands A, B, and C disappeared (20, 21). The theorists were right! The new direct experiments yielded ΔE_{ST} = 9.0 kcal, and the French champagne began its journey from the University of Colorado to Caltech (22).

Such dramas are continuously being played out as science lurches forward in its efforts to elucidate various phenomena. The new feature here for chemistry was that the theorists were so confident of the accuracy of their results that they hung tough in the face of disagreements with experiment, and then with all-consuming gall they reinterpreted the experiments.

What lies ahead? In the early 1980's, systematic approaches to accurate methods with direct calculations of the first derivatives of the total molecular energy

(the forces on the atoms) and the second derivatives of the energy have led to automatic programs for accurate calculation of the geometries and energies of molecules and radicals that can routinely handle systems with up to three or four heavy atoms (for example, carbon plus hydrogens sprinkled hither and yon (23, 24). Indications are that with slight corrections to the ab initio results (24), accuracies of ±3 kcal are obtained even for these highly strained species.

As we approach the mid-1980's, it appears that through use of effective potentials (25) to represent the core electrons, comparably accurate results will soon be routine for species containing several heavy atoms such as silicon, germanium, tin, and their neighbors in the periodic table.

Quantitative quantum chemistry has arrived. The theory is now sufficiently accurate that theorists and experimentalists are becoming equal partners in chemical research.

Recent Progress in

Qualitative Quantum Chemistry

The above applications illustrate how it is possible to gain quantitatively accurate data from theoretical studies of systems that are hard to study experimentally (reactive intermediates and excited states). This is most valuable but not sufficient. If theory could provide only exact results for any desired property of any system, it would not be any better than a collection of good experiments. What theory uniquely provides is the qualitative principles responsible for the results from a particular experiment or calculation. With proper understanding of the principles, one can predict how new systems will act in advance of either experiment or quantitative calculations. The following case studies illustrate the process.

Olefin metathesis. Several important chemical processes involve olefin metathesis, which is the breaking of carbon-carbon double bonds and the piecing together of the fragments from different molecules to make two new olefins, distinct from the starting materials. For example, the Phillips Triolefin Process (26, 27)

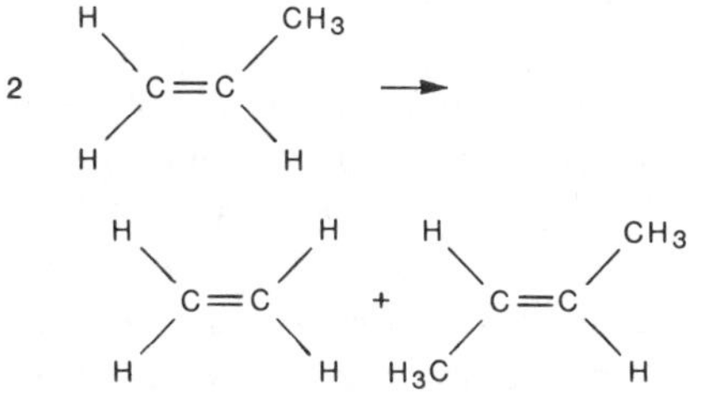

$$2 \quad H_2C{=}C(CH_3)H \longrightarrow$$

$$H_2C{=}CH_2 \; + \; H(CH_3)C{=}C(CH_3)H \qquad (2)$$

was used until 1973 to convert (then cheap) propylene into butene, which in turn was used to produce butadiene, an important starting material for many chemicals. In a Goodyear process (26, 27) cyclic olefins are metathesized to form special polymers.

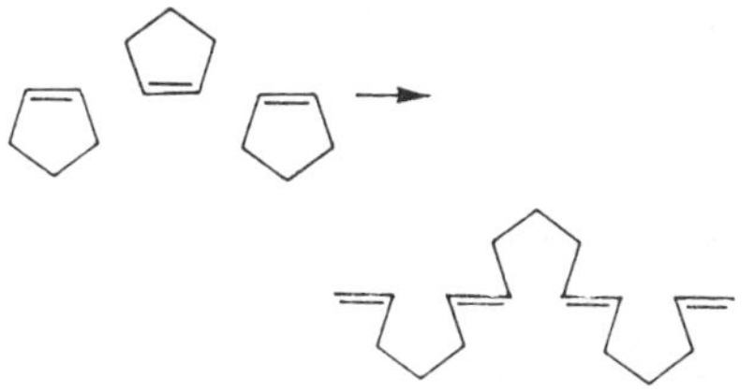

$$(3)$$

These metathesis reactions do not proceed by themselves; they need a catalyst. Generally, a transition metal compound such as WCl_6 or $MoCl_6$ is used together with a number of additives such as organic aluminum compounds or organic phosphorus compounds. In the early days (up to about 1972), it was thought that the metal served to coordinate two olefins and that this somehow allowed a reaction mixing up the carbons.

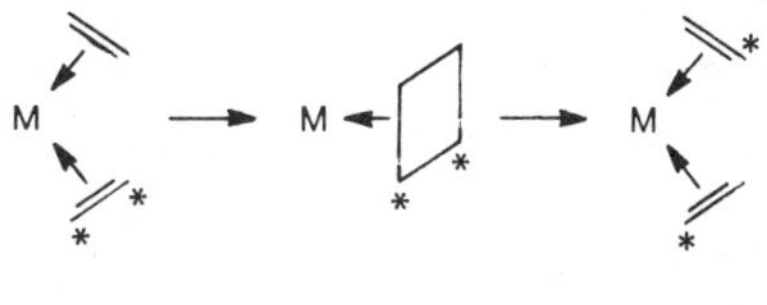

$$(4)$$

However, a series of ingenious experimental studies (28), all indirect, established that the catalytic reactions involve active species having a metal-carbon double bond or a four-membered ring with a metal at one of the corners

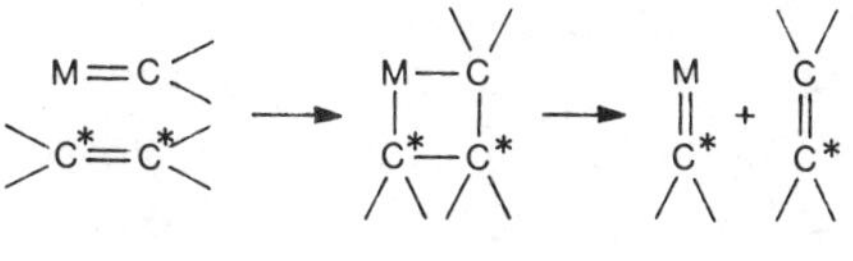

$$(5)$$

135

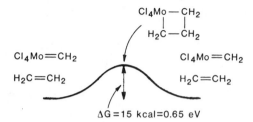

Fig. 4. Free energy along the reaction path (schematic) for metathesis.

where M = Mo or W. Back in 1970, when the earliest of these studies was published, postulating such species was a bold step since no such species had yet been observed directly. However, researchers later synthesized such species and showed that they are tenable.

In the late 1970's, a group at Caltech initiated a series of studies (29) to examine the mechanism of the reaction in Eq. 5. The key intermediate had not yet been characterized experimentally, and there were a number of questions concerning the nature of the metal-carbon bonds in these systems. In addition, no experimental estimates of the energetics were available, so the quantitative results of theory would be valuable. Because good metathesis catalysts are based on tungsten or molybdenum, the group at Caltech examined the process in Eq. 5 and found the results shown in Fig. 4 (29). Since metathesis proceeds rapidly at −50°C, they felt that the 15-kcal energy barrier for Eq. 5 was too high to permit that reaction to be in the catalytic sequence. Because this was the first time that first-principles calculations had been performed for such large systems, the theorists at first worried that there might be bad approximations in the calculations. However, several months of tests confirmed their results. Their next thought was that maybe the active catalyst was some other Mo=CH$_2$ species. Unfortunately, under catalytic conditions there are a large number of additives present (typically, species such as Sn(CH$_3$)$_4$ or Cl$_2$AlCH$_3$ and PR$_3$ together with small amounts of alcohol), introducing an enormous number of possibilities, certainly too many to permit calculations for each. What was needed was more insight into how the energetics in Fig. 4 might be controlled by other ligands.

The Caltech group then did what was natural under these circumstances: they dropped the project. Next they started examining how olefins react with metal oxo bonds,

$$Cl_4Mo{=}O \; + \; H_2C{=}CH_2 \; \longrightarrow$$
$$\mathbf{1}$$

$$\begin{array}{c} Cl_4Mo{-}O \\ | \qquad | \\ H_2C{-}CH_2 \end{array} \quad (6)$$
$$\mathbf{2}$$

$$Cl_2Mo{\overset{O}{\underset{O}{\lessgtr}}} \; + \; H_2C{=}CH_2 \; \longrightarrow$$
$$\mathbf{3}$$

$$\begin{array}{c} O \\ \| \\ Cl_2Mo{-}O \\ | \qquad | \\ H_2C{-}CH_2 \end{array} \quad (7)$$
$$\mathbf{4}$$

which led to quite unexpected results. Although the Mo=O bonds in **1** and **3** would be expected to be very similar, the Caltech group found

$$\Delta G_{(6)} = +44 \text{ kcal} = +1.9 \text{ eV}$$
$$\Delta G_{(7)} = -21 \text{ kcal} = -0.9 \text{ eV}$$

where $\Delta G_{\text{(Eq. No.)}}$ is the change in free energy at room temperature of the corresponding reaction. That is a difference of 65 kcal (2.8 eV) for what would have been expected to be very similar reactions. Obviously something profound must be happening in these systems, and

136

an examination of the wave functions quickly supplied the answer.

The molybdenum-oxygen bonds in the dioxo species **3** are quite covalent, each consisting of a Mo–O σ bond and a M–O π bond, analogous to the C=O bond in formaldehyde ($H_2C=O$). Of the six valence electrons on molybdenum, two are involved in the two ionic Mo–Cl bonds, and four participate in the two double bonds to the two oxo groups. Similarly, in the monoxo species **1**, four of the six valence electrons in the molybdenum participate in ionic Mo–Cl bonds, leaving two to bond to the single oxo group. However, there is a surprise with this system. Both of those molybdenum electrons get into $d\pi$ orbitals and make two π bonds to the oxygen. (If the bond axis is z, then one electron is in the xz plane, the other in the yz plane.) Making these two π bonds requires two singly occupied $p\pi$ orbitals on oxygen (p_x and p_y). That leaves two electrons in the oxygen p_z orbital, which overlaps the empty molybdenum d_{z^2} orbital to make a partial σ bond. The end result is a partial triple bond in **1**, much like the bond in carbon monoxide, C≡O; and the M=O bond in **1** is about 30 kcal stronger than those in **3**.

If the molybdenum-oxygen triple bond is so good, why does it not form in the dioxo species **3**? The problem is that if one oxo group in **3** were to make a triple bond, it would utilize both $d\pi$ electrons, leaving nothing for the other oxo group. The compromise is for each to make double bonds. This is analogous to the comparison of O=C=O and C≡O; for CO_2 each oxygen can only make a double bond, whereas in CO, the oxygen can make a partial triple bond.

How do these interpretations explain the 65 kcal difference in energetics for Eqs. 6 and 7? First, in Eq. 6, we must break a stronger Mo=O bond than in Eq. 7, accounting for about 30 kcal of the difference. But what is the origin of the other 30 kcal? To see this, examine product **4** in Eq. 7 more carefully. Of the six valence electrons on the molybdenum, four are used in σ bonds (two to chlorine, one to oxygen, and one to carbon), leaving two electrons to bond to the spectator oxygen (the oxo group that is not changed by the reaction). Thus, in **4** no other ligands except the spectator oxygen can use molybdenum π orbitals; hence, molybdenum can use both $d\pi$ orbitals to make a partial triple bond to the oxygen. It could not do so in the reactant **3**, since the second oxygen also needs to bond to the molybdenum $d\pi$ orbital. Thus, the spectator oxo group changes from having a double bond to having a partial triple bond at the same time that the olefin reacts with the other oxygen. This stabilizes the reaction intermediate by an extra 30 kcal, explaining the observations. These studies led to the following principle: a spectator oxygen adjacent to a ligand X that has a double bond to the molybdenum

promotes reaction at X by stabilizing the resulting intermediate by ~30 kcal.

Immediately, the Caltech group believed that they had the salient clue to the metathesis reaction. Maybe the catalyst was not $Cl_4Mo=CH_2$ but rather the oxo-methylidene (**5**) with its spectator oxo group. With this spectator oxo group they expected the reaction intermediate (**6**) to be stabilized by ~30 kcal and hence for the process in Eq. 8 to be

$$Cl_2Mo \overset{O}{\underset{CH_2}{\lessgtr}} \quad + \quad \overset{CH_2}{\underset{CH_2}{||}} \quad \longrightarrow \quad Cl_2Mo \overset{O}{\underset{H_2C-CH_2}{\overset{|}{\underset{|}{\bigg|}}}} \quad (8)$$

5 **6**

dative dehydrogenation of methanol to formaldehyde (*34*).

$$H_3COH \ + \ O_2 \ \xrightarrow[\substack{MoO_3 \\ 200-300^\circ C}]{} \ H_2C{=}O \ + \ H_2O \quad (9)$$

exothermic by ~15 kcal. They checked with their computer and found, sure enough, that $\Delta G_{(8)} = -24$ kcal and that, indeed, the spectator oxo had a double bond in **5** and a partial triple bond in **6**.

But could this species have been formed in the experimental solutions where catalysis was observed? Indeed, all metathesis experiments involved reaction mixtures containing some source of oxygen (*30, 31*). In fact, Muetterties had shown that rigorous exclusion of oxygen killed the catalyst (*31*).

In a good catalyst, all reaction steps are thermoneutral ($\Delta G \sim 0$). The Caltech group showed that the presence of Lewis acids ($AlCl_3$) or Lewis bases (PR_3), both of which were present in most experimental catalytic systems, would tend to make ΔG for Eq. 8 near zero, leading to the catalytic sequence in Fig. 5. With that, they submitted their papers. Meanwhile, various experimental groups were exploring the chemistry of molybdenum-oxo systems; and simultaneous with publication of the theory, Schrock and his co-workers reported the synthesis of a species like **5** and showed that it undergoes metathesis (*32*). Sometime later, Muetterties and Band carried out a detailed analysis of their metathesis system and provided strong, although indirect, evidence that species **5** is formed and serves as the active catalyst in that system (*33*).

Heterogeneous oxidations. The Caltech group later became interested in some commercially important heterogeneous catalysts responsible for the oxi-

Numerous experimental studies of this system had been made, yet there had been no evidence for which sort of surface sites were involved. Since bulk molybdates generally contain MoO_6 (a distorted octahedron) or MoO_4 (a tetrahedron), the theorists believed that the stable surface sites in an oxidative atmosphere would be either **7** or **8**,

7

8

and they proceeded to examine various reaction steps with each. They found, for Eq. 10, $\Delta H_{(10)} = +22$ kcal/mol and $\Delta G_{(10)}$ (25°C) $= +33$ kcal/mol and, for Eq. 11, $\Delta H_{(11)} = -9$ kcal/mol and $\Delta G_{(11)}$ (25°C) $= +2$ kcal/mol (*33*); so a surface dioxo unit is required for chemisorption of methanol. (The spectator oxo group is crucial to this stabilization.) To complete the reaction, one C–H bond must be broken. The Caltech group concluded that this requires a second dioxo unit and is favorable only because it is promoted by a spectator oxo group, as shown in Eq. 12 in which $\Delta H_{(12)} = +6$ kcal/mol and $\Delta G_{(12)}$ (300°C) $= +6$ kcal/mol.

138

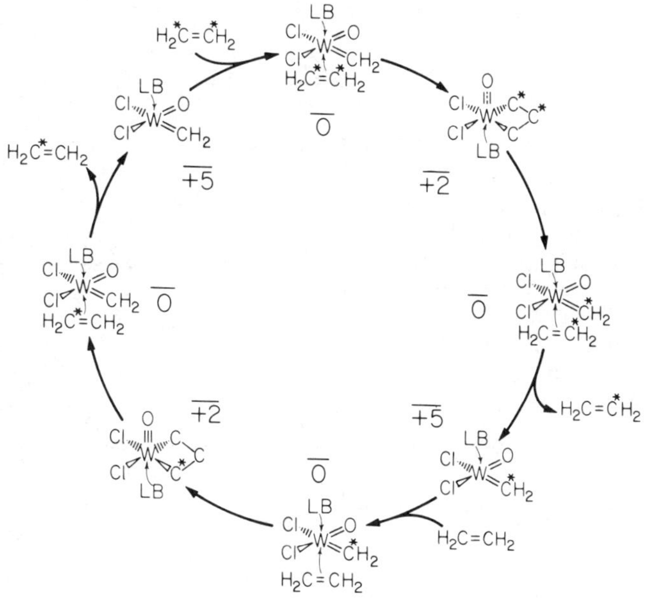

Fig. 5. Catalytic mechanism for metathesis.

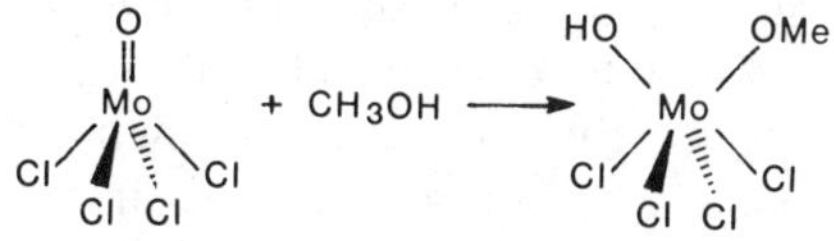

$$\text{(10)}$$

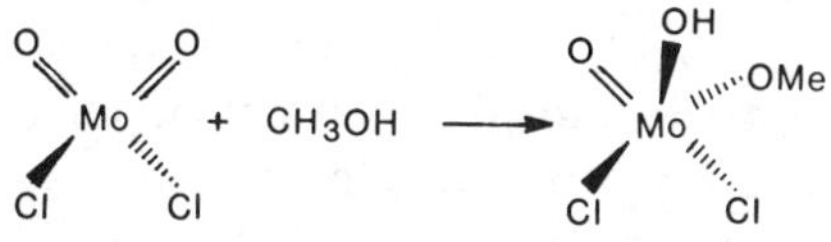

$$\text{(11)}$$

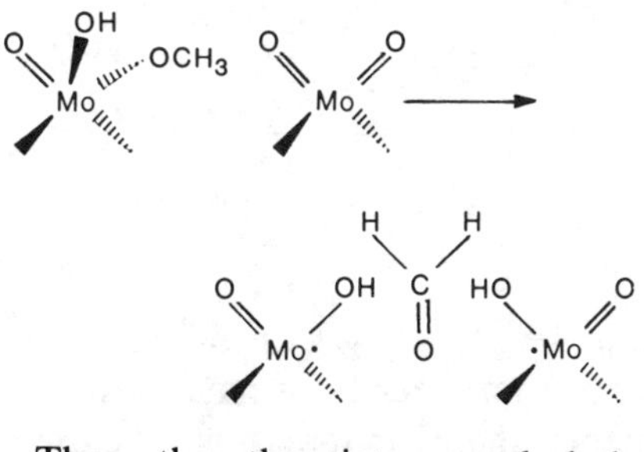

$$\text{(12)}$$

Thus the theorists concluded that selective oxidative dehydrogenation of CH_3OH to CH_2O requires a dual set of adjacent dioxo units. For each dioxo unit, one of the two oxygens can extract a hydrogen while the other oxygen can provide the spectator oxo stabilization.

The spectator-oxo effects are crucial in making the chemisorption, Eq. 11, exothermic and keeping the C–H cleavage, Eq. 12, only slightly endothermic. In this description one would expect Eq. 12 to be the rate-determining step of the process, which has been proved by recent experiments (34, 36).

Now the question is, does MoO_3 have a surface with the requisite configuration of adjacent dioxo sites? Indeed, as indicated in Fig. 6, the crystal structure (37) of MoO_3 shows that the (010) surface has exactly the configuration needed for the reactions shown in Eqs. 11 and 12, whereas the other low-index faces of MoO_3 do not have the requisite combination of dioxo units. This conclusion that adjacent dioxo units are essential for the catalytic dehydrogenation of CH_3OH is strongly supported by recent experimental studies (38). In the presence of O_2, the MoO_3 (010) surface is highly selective for formation of H_2CO and is

139

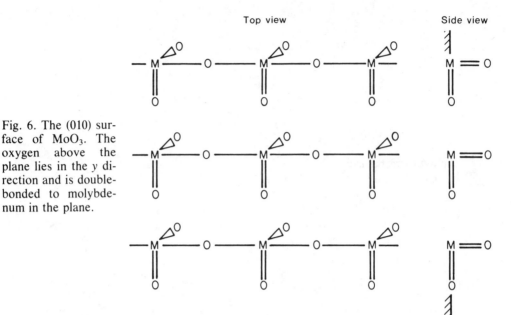

Top view   Side view

Fig. 6. The (010) surface of MoO₃. The oxygen above the plane lies in the y direction and is double-bonded to molybdenum in the plane.

responsible for nearly all production of H₂CO; the other surfaces give rise to CH₂(OCH₃)₂, ether, and very little H₂CO.

The general wisdom had been that the (010) surface of MoO₃ would not be reactive compared with other surfaces because it has no broken chemical bonds (39). The theorists' mechanism suggested that this simple reasoning was inadequate for MoO₃ (010), and indeed experiments (38) have provided strong evidence that MoO₃ (010) is the important surface for formation of H₂CO. It is interesting to note here that exposure of MoO₃ (010) to CH₃OH without O₂ did not lead to reaction (38, 40). Presumably, without O₂ the catalyst loses some of its surface oxygen and hence loses the dioxo units required for the chemistry.

The specific mechanism proposed here, involving a dual dioxo catalytic site, is susceptible to many experimental tests. Such interplay between theory and

experiment will promote the development of a much more detailed understanding of the fundamental chemical mechanisms of heterogeneous catalytic reactions. Catalytic sites composed of collections of surface dioxo units are also expected to be important in selective oxidation and ammoxidation reactions, Eq. 13 (41).

$$\diagup\!\!= \; + \; O_2 \; \longrightarrow \; \diagdown\!\!=\!\diagup^{C=O} \qquad (13)$$
$$\underset{NH_3}{\searrow} \;\; \diagdown\!\!=\!\diagup^{C\equiv N}$$

Summary. Starting with the ideas abstracted from a series of metathesis experiments, theorists found that the theoretical results did not fit the simplest current mechanism and were sidetracked into some parallel studies to develop a better understanding of the basic processes. The result was a new principle, that of spectator oxygen stabilization, which not only provided the missing component in the understanding of metathesis but also provided a framework for understanding a number of other catalytic processes. Indeed, this principle provides a tool that could be useful in designing new catalysts.

One indication of the present state of modern theory is that, when faced with disagreement between theory and experiment, the theorists were sufficiently confident of their results that they continued to examine possible reinterpretations of the experiments until they stumbled onto the key idea. In the mid-1970's, one might have attributed a 20 kcal discrepancy to problems with the theory and gone on to other endeavors, unaware of the existence of any experimental problems. It is also well to emphasize the distinctions between the two aspects of theory being discussed here. First, recent advances in theoretical methods and in the hardware theorists use (high-speed computers cheap enough to belong to individual departments or individual investigators) permit researchers to calculate accurate properties for species that are almost impossible to study experimentally. Although very important, this aspect of theory need not provide new insight. Critical for real advances in chemistry is the second aspect of theory, in which researchers abstract from calculations and experiments the principles and concepts responsible for a particular set of results. With these concepts scientists unfamiliar with the intricacies and pedantics of theory can design materials or catalysts by predicting how the chemistry would be modified by various changes in the metals, ligands, structural environment, and so on. Such concepts allow the scientist to circumvent the numerous tests and experiments traditional to such endeavors.

Onward to Simulation

Is this the end of the story, with theory becoming increasingly able to predict energy surfaces and the detailed sequence of steps in a reaction? No. Even if the energetics for all the possible reaction steps of a catalytic reaction were known, we would still not be satisfied. During the operating lifetime of a real catalyst, its surface is exposed to reactants, reaction intermediates, products, poisons, promoters, and a variety of temperatures and pressures. The ultimate theoretical description of a real system would be to simulate it on a computer. The computer would calculate the motions of molecules and surface atoms as they react, rearrange, migrate, desorb, and so forth, and it would display these motions on a high-speed graphics terminal so that the scientist could change the conditions (temperature, pressure, reactants, surface plane) and visually observe the consequences of those changes. There are a number of formidable problems to solve before this idyllic situation will prevail. Even so, the difficulties seem analogous to those facing quantum chemists 25 years ago.

Some progress in making such simulations has already been made for biological systems, as illustrated by the following example. Thermolysin is a thermophilic (resistant to high temperatures) protease selective for cleaving peptide bonds with hydrophobic residues such as phenylalanine or leucine (42). It is also a good model for angiotensin-converting enzyme (ACE) in the sense that good inhibitors for one are good inhibitors for the other (42, 43). Indeed, a long-term research program at Merck Sharp & Dohme for the design of inhibitors for thermolysin and ACE recently culminated in a new product that is quite effective in hypertension crises (43).

Thermolysin has 2895 atoms (including 443 hydrogens), four calciums, and a zinc). Describing the motions of thermolysin plus a sheath of 170 water molecules by using empirical theoretical force fields and calculating the structure from first principles (44) leads to a structure in excellent agreement with experiment (42); in fact, the agreement is so good that the differences are not visible in a diagram of the entire molecule suitable for publication. Of course, calculations can also be used to model the enzyme in solution, the biologically relevant situation. The photograph in the front of the book shows the surface of the enzyme near the active site and the predicted structure of a new Merck inhibitor (CLT) bound to this site (45). In the calculations, the structure of the complete enzyme-inhibitor complex is allowed to relax so that the optimum structure can be calculated. An interesting aspect of this CLT-thermolysin study is that the structure of the complex was predicted without knowledge of experimental structural data, which were being determined simultaneously (46). Subsequent comparison of the predicted and experimentally determined structures shows an excellent fit. Theory can, of course, provide much more information than merely the structure of enzyme-inhibitor complexes in solution. The energetics of interaction can be partitioned into components that can be used in designing better inhibitors. In addition, such theoretical studies can be carried out as a function of temperature, providing a model of the dynamics of the system.

Summarizing then, the first point is that theoretical chemistry is coming of age. Theorists are now in a position to tackle many important chemical, biological, and materials problems on an equal footing with experimentalists. The second, and perhaps most important, point is that theoretical chemistry is becoming more and more fun, as the role of theory expands toward the mainstream of nearly all physical, chemical, and biological phenomena relevant to society. The ability of theory, at last, to contribute substantially to the elucidation of interesting catalytic processes bodes well for the future. A new age is approaching in which theory and experiment working together will sort out the most intimate details of catalytic processes and translate the data into a conceptual form that other chemists and engineers can use to design new processes.

References and Notes

1. L. Pauling, *The Nature of the Chemical Bond* (Cornell University Press, Ithaca, N.Y., 1960).
2. R. S. Mulliken, *Rev. Mod. Phys.* **4**, 48 (1932).
3. J. M. Foster and S. F. Boys, *ibid.* **32**, 305 (1960); A. Padgett and M. Kraus, *J. Chem. Phys.* **32**, 189 (1960).
4. G. Herzberg and J. Shoosmith, *Nature (London)* **183**, 1801 (1959); G. Herzberg, *Proc. R. Soc. London Ser. A* **262**, 291 (1961); _____ and J. W. C. Johns, *ibid.* **295**, 107 (1966).
5. C. F. Bender and H. F. Schaefer III, *J. Am. Chem. Soc.* **92**, 4984 (1970); J. F. Harrison and L. C. Allen, *ibid.* **91**, 807 (1969).
6. E. Wasserman, W. A. Yager, V. J. Kuck, *Chem. Phys. Lett.* **7**, 409 (1970); E. Wasserman, V. J. Kuck, R. S. Hutton, W. A. Yager, *J. Am. Chem. Soc.* **92**, 7491 (1970); R. A. Bernheim, H. W. Bernard, P. S. Wang, L. S. Wood, P. S. Skell, *J. Chem. Phys.* **53**, 1280 (1970).
7. G. Herzberg and J. W. C. Johns, *J. Chem. Phys.* **54**, 2276 (1971).
8. P. R. Bunker and P. Jensen, *ibid.* **79**, 1224 (1983).
9. M. L. Halberstadt and J. R. McNesby, *J. Am. Chem. Soc.* **89**, 3417 (1967).
10. R. W. Carr, Jr., T. W. Eder, M. G. Topor, *J. Chem. Phys.* **53**, 4716 (1970).
11. W. L. Hase, R. J. Phillips, J. W. Simons, *Chem. Phys. Lett.* **12**, 161 (1971).

12. P. J. Hay, W. J. Hunt, W. A. Goddard III, *ibid.* **13**, 30 (1972).
13. C. F. Bender, H. F. Schaefer III, D. R. Franceschetti, L. C. Allen, *J. Am. Chem. Soc.* **94**, 6888 (1972).
14. H. M. Frey and G. J. Kennedy, *J. Chem. Soc. Chem. Commun.* (1975), p. 233; J. W. Simons and R. Curry, *Chem. Phys. Lett.* **38**, 171 (1976); F. Lahfani, *J. Phys. Chem.* **80**, 2623 (1976).
15. P. F. Zittel *et al.*, *J. Am. Chem. Soc.* **98**, 373 (1976).
16. L. B. Harding and W. A. Goddard III, *J. Chem. Phys.* **67**, 1777 (1977); B. O. Roos and P. M. Siegbahn, *J. Am. Chem. Soc.* **99**, 7716 (1977); R. R. Lucchese and H. F. Schaefer III, *ibid.*, p. 6766; C. W. Bauschlicher and I. Shavitt, *ibid.* **100**, 739 (1978).
17. R. K. Lengel and R. N. Zare, *J. Am. Chem. Soc.* **100**, 7495 (1978).
18. L. B. Harding and W. A. Goddard III, *Chem. Phys. Lett.* **55**, 217 (1978).
19. P. C. Engelking *et al.*, *J. Chem. Phys.* **74**, 5460 (1981).
20. D. G. Leopold, K. K. Murray, W. C. Lineberger, *ibid.* **81**, 1048 (1984).
21. In the meantime, other direct experiments had led to $\Delta E_{ST} = 8.5 \pm 0.7$ [C. C. Hayden, D. M. Neumark, K. Shobatake, R. K. Sparks, Y. T. Lee, *ibid.* **76**, 3607 (1982)] and $\Delta E_{ST} = 9.05 \pm 0.06$ [A. R. W. McKellar *et al.*, *ibid.* **79**, 5251 (1983)].
22. W. C. Lineberger and G. B. Ellison, private communication.
23. J. A. Pople *et al.*, *Int. J. Quantum Chem. Symp.* **13**, 225 (1979); *ibid.* **15**, 269 (1981); P. Saxe, Y. Yamaguchi, H. F. Schaefer III, *J. Chem. Phys.* **77**, 5647 (1982).
24. C. F. Melius and J. S. Binkley, *ACS Symp. Ser.* **249**, 103 (1984).
25. C. F. Melius and W. A. Goddard III, *Phys. Rev. A* **13**, 1528 (1974); A. Redondo, W. A. Goddard III, T. C. McGill, *Phys. Rev. B* **15**, 5038 (1977); P. J. Hay, W. R. Wadt, L. R. Kahn, F. W. Bobrowicz, *J. Chem. Phys.* **69**, 984 (1978); W. C. Lee, Y. S. Ermler, K. S. Pitzer, A. D. McLean, *ibid.* **70**, 288 (1979).
26. G. W. Parshall, *Homogeneous Catalysis* (Wiley, New York, 1980), pp. 174–178.
27. R. H. Grubbs, *Prog. Inorg. Chem.* **24**, 1 (1978); N. Calderon, J. P. Lawrence, E. A. Ofstead, *Adv. Organomet. Chem.* **17**, 449 (1979).
28. J. L. Herisson and Y. Chauvin, *Makromol. Chem.* **141**, 161 (1970); R. H. Grubbs, P. L. Burk, D. D. Carr, *J. Am. Chem. Soc.* **97**, 3265 (1975); T. J. Katz and J. McGinnis, *ibid.* **97**, 1592 (1975); R. R. Schrock, *ibid.* **96**, 6796 (1974); M. Ephritikhine, M. L. H. Greene, R. E. MacKenzie, *J. Chem. Soc. Chem. Commun.* (1976), p. 619; P. Foley and G. M. Whitesides, *J. Am. Chem. Soc.* **101**, 2732 (1979); J. Rajaram and J. A. Ibers, *ibid.* **100**, 829 (1978).
29. A. K. Rappé and W. A. Goddard III, *Nature (London)* **285**, 311 (1980); *J. Am. Chem. Soc.* **102**, 5114 (1980); *ibid.* **104**, 448 (1982).
30. J. M. Basset, G. Coudurier, R. Mutin, H. Proliaud, Y. Trambouze, *J. Catal.* **34**, 196 (1974).
31. M. T. Mocella, R. Rovner, E. L. Muetterties, *J. Am. Chem. Soc.* **98**, 1689 (1976).
32. J. H. Wengrovious, R. R. Schrock, M. R. Churchill, J. R. Missert, W. J. Youngs, *ibid.* **102**, 4515 (1980).
33. E. L. Muetterties and E. Band, *ibid.*, p. 6574.
34. C. J. Machiels and A. W. Sleight, *Proceedings of the Fourth International Conference on Chemistry and Uses of Molybdenum*, H. F. Barry and P. C. H. Mitchell, Eds. (Climax Molybdenum Company, Ann Arbor, Mich., 1982), p. 411.
35. J. N. Allison and W. A. Goddard III, *J. Catal.*, in press.
36. F. Trifiro, S. Notarbartolo, I. Pasquon, *ibid.* **22**, 324 (1971).
37. L. Kihlborg, *Ark. Kemi* **21**, 155 (1983).
38. J. M. Tatibouët and J. E. Germain, *J. Catal.* **72**, 375 (1981); J. M. Tatibouët, J. E. Germain, J. L. Volta, *ibid.* **82**, 240 (1983).
39. L. E. Firment and A. Ferretti, *Surf. Sci.* **129**, 155 (1983).
40. F. Ohuchi, L. E. Firment, U. Chowdry, A. Ferretti, *J. Vac. Sci. Technol. A* **2**, 1022 (1984).
41. J. D. Burrington, C. T. Kartisek, R. K. Grasselli, *J. Catal.* **81**, 489 (1983).
42. K. Morihara and H. Tsuzuki, *Eur. J. Biochem.* **15**, 374 (1970); W. R. Kester and B. W. Matthews, *Biochemistry* **16**, 2506 (1977).
43. D. G. Hangauer, A. F. Monzingo, B. W. Matthews, *Biochemistry*, in press.
44. B. D. Olafson and W. A. Goddard III, in preparation.
45. Prepared by B. D. Olafson and S. L. Mayo using the BIOGRAF/I software on the Evans & Sutherland PS-300 Graphics Terminal attached to a DEC VAX 11/780 host computer.
46. B. W. Matthews, private communication.
47. Contribution No. 7139 from the Arthur Amos Noyes Laboratory of Chemical Physics. Parts of the work reported here were supported by grants from the National Science Foundation (CHE83-18041 and DMR82-15650), the Donors of the Petroleum Research Fund of the American Chemical Society (13110-AC5,6), and by contracts with the Department of Energy (Jet Propulsion Laboratory, Energy Conversion and Utilization Technologies Program) and Shell Development Company. It is also a pleasure to acknowledge the excitement and fun of interacting with the excellent graduate students at Caltech. I particularly thank Dr. Larry Harding, Argonne National Laboratory, for his work on the CH_2 molecule; Professor Tony Rappé, Colorado State University, for his work on metathesis; Dr. Janet Allison for her work on the reactions on MoO_3; and Dr. Barry Olafson for his work on thermolysin.

143

A remarkable revival of the van der Waals picture of liquids occurred during the past two decades. This renaissance was spurred by the discovery (1) from computer simulations that a system of hard spheres (impenetrable "billiard balls") has a first-order fluid-solid transition that is intimately related to the freezing and melting transitions of real materials (2). The van der Waals picture stresses the dominant role of the short-ranged harshly repulsive intermolecular forces (which are nearly hard core interactions) in determining the structural arrangements of molecules in a liquid, while neglecting the influence of the longer ranged attractive interactions on the structure. Though originally developed to describe the liquid-gas critical point (3), this approach is now known to be most useful and accurate at the high densities that characterize a liquid away from the critical point, since at those densities the nearly incompressible nature of the fluid tends to inhibit the fluctuations that would invalidate a van der Waals theory.

The successful exploitation of this picture is found in textbook (4) descriptions of perturbation theories of simple atomic liquids. But the range of utility of the van der Waals picture is far broader than this limited application might suggest. Its validity and usefulness have been documented in numerous studies extending from computer simulations of condensed materials, to analytical equilibrium theories of polyatomic organic liquids and mixtures, to models of transport, and even to vibrational relaxation in liquids. In this article we review this powerful perspective and some recent developments.

11

Van der Waals Picture of Liquids, Solids, and Phase Transformations

David Chandler,
John D. Weeks,
and Hans C. Andersen

Science 220, 787–794 (20 May 1983)

The Basic Idea

According to the van der Waals picture, the average relative arrangements and motions of molecules in a liquid (that is, the intermolecular structure and correlations) are determined primarily by the local packing and steric effects produced by the short-ranged repulsive intermolecular forces. Attractive forces, dipole-dipole interactions, and other slowly varying interactions all play a minor role in the structure, and in the simplest approximation their effect can be treated in terms of a mean field—a spatially uniform background potential—which exerts no intermolecular force and hence has no effect on the structure or dynamics but merely provides the cohesive energy that makes the system stable at a particular density and pressure.

Thus an atom in a monatomic fluid is like a billiard ball and molecules are much like the familiar space-filling models. If one imagines a collection of such objects moving about within a certain volume, colliding elastically with each other and oblivious to any attractive or long-ranged forces between them, then one has a physical picture that captures many essential features of liquid structure and dynamics on a molecular level.

In the modern literature, perhaps the first explicit statements of this picture for monatomic liquids were given by Reiss (5) and by Longuet-Higgins and Widom (2). The idea is much older, however, since it is the primary (though not sole) physical content of the van der Waals equation of state (3). For that reason we attribute the concept to van der Waals, but this historical expedient is not meant to detract from the significance of contributions made by many recent workers to our current understanding of the idea.

The attractive intermolecular interactions rigorously have no effect on the structure only in the hypothetical and unrealistic limit in which the attractive interactions are both infinitely weak and infinitely long-ranged (3, 6, 7). In that case, each particle in the system feels attractive interactions from all the other particles. These interactions exert no net vector force, while the resulting potential energy is accurately described in terms of a spatially uniform mean field. What is significant, however, is that the attractions often continue to have little influence on the structure of real condensed materials, where every atom has several nearest neighbors and packing suppresses large density fluctuations. The spatial variation of real attractions in the intermolecular structures allowed by the repulsive forces is weak enough that the van der Waals uniform mean field treatment remains accurate and provides useful predictive power.

Part of the explanation for this fact rests on the differing physical bases for repulsive and attractive intermolecular forces. In cases where the van der Waals picture is useful, the attractions usually arise from electrostatic effects such as fluctuating dipole-induced dipole interactions (the so-called dispersion interactions). These forces are not associated with significant distortions of intramolecular charge distributions, and hence their magnitudes usually are not large. Typically, the spatial variations of attractive interactions are $\sim k_B T_t$ per one molecular diameter, where k_B is Boltzmann's constant and T_t is the triple point temperature of the material.

In contrast, the short-ranged repulsions arise from the Pauli exclusion principle, which does not allow electrons on different molecules to be in the same part of space at the same time. Any attempt to push two molecules together too closely will cause the electron clouds on each molecule to distort in such a way as to resist such overlap. The forces generated in this way are very strong and rapidly varying functions of the molecular positions and orientations. Indeed, the typical spatial variation for these short-range repulsions is ~ $k_b T_t$ per one-tenth of an atomic or molecular diameter. These forces are responsible for the nearly incompressible nature of many dense fluids, since neighboring particles are close enough together that an attempt to decrease the volume will be strongly resisted by the repulsive forces.

In many cases the repulsive forces are sufficiently harsh that one can approximate their effect by hard core interactions, though notable exceptions exist, as discussed below. We define a van der Waals material as a system composed of hard core molecules, the sizes and shapes of which are chosen to mimic the repulsive branches of the intermolecular potentials, and held at a particular density by a uniform mean field potential. The thesis of this article is that the properties of many real condensed matter systems can be accurately described by using the appropriately chosen van der Waals material.

Simple Atomic Liquids

To establish the quantitative validity of these statements, it is useful to consider, as an example, the interparticle correlations in a monatomic fluid such as liquid argon. The radial distribution function, $g(r)$, provides a simple mathematical description of the structure of an atomic liquid (4). It is defined by

$$\rho g(r) = \text{average density of atoms at } r \text{ given that another atom is located at the origin} \quad (1)$$

where ρ is the average number of atoms per unit volume, N/V. The radial distribution function gives information about the average relative arrangements of pairs of atoms. More complex descriptions of the structure will involve triples, quadruples, and so on, but $g(r)$ suffices for many purposes. It is the pair function, $g(r)$, that is measured by diffraction experiments, since the scattering of neutrons or x-ray radiation from a liquid is dominated by the interference from the distributed pairs of scattering centers (4).

The total potential energy of an atomic fluid is often represented by a sum of radially symmetric pair potentials (8), $w(r)$, like that pictured in Fig. 1. The extensively studied Lennard-Jones fluid has the potential

$$w(r) = 4\epsilon[(\sigma/r)^{12} - (\sigma/r)^6] \quad (2)$$

The properties of this fluid are known from the results of computer simulations (9–11), and, with appropriately chosen values of the energy and length scale parameters ϵ and σ, it serves as an accurate model for real atomic liquids such as argon (12).

The intermolecular force is given by the gradient of the potential $-dw(r)/dr$; hence particles repel each other at separations less than the potential minimum at $r_0 = 2^{1/6}\sigma$. The repulsive branch at $r < r_0$ should be carefully distinguished from the merely positive portion of the potential at $r < \sigma$, or from the first term in the arbitrary algebraic combination in Eq. 2, since only the sum has physical significance.

As first pointed out by the authors (13, 14), one can directly test the differing roles of attractions and repulsions as

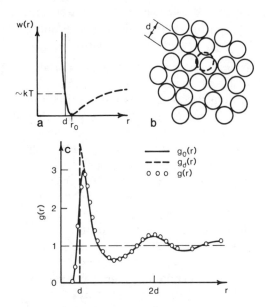

Fig. 1. Structure and pair interactions of a simple liquid. A schematic view of a region of a liquid composed of spherical particles interacting via a pair potential $w(r)$ is shown in (b). The repulsive branch of the potential at $r < r_0$ is indicated by a solid curve. For the Lennard-Jones system, the repulsive force reference system potential, $u_0(r)$, is $w(r) + \epsilon$ for $r < r_0$ and zero otherwise, as given in Eq. 3. A hard sphere potential, $u_d(r)$, which is infinite for $r < d$ and zero otherwise, is associated by Eq. 10 with $u_0(r)$ as shown schematically in (a). The radial distribution function, $g(r)$, for the Lennard-Jones liquid at a state near the triple point with $\rho\sigma^3 = 0.85$ and $k_B T/\epsilon = 0.88$ is plotted in (c). It is compared with $g_0(r)$ and $g_d(r)$, the radial distribution functions of the repulsive force systems with pair potentials $u_0(r)$ and $u_d(r)$, at the same temperature and density. [Reproduced, with permission, from the *Annual Review of Physical Chemistry*, vol. 29. © 1978 by Annual Reviews Inc.]

assumed in the van der Waals picture for such a model of an atomic liquid by comparing its $g(r)$ to $g_0(r)$, the radial distribution function at the same temperature and density produced solely by the repulsive forces. (Throughout this article, the subscript zero refers to the repulsive forces.) The latter is determined by studying the properties of a system with the repulsive pair potential

$$u_0(r) = w(r) + \epsilon \qquad r \leq r_0$$
$$= 0 \qquad r > r_0 \qquad (3)$$

The properties of the hypothetical fluid for which the total potential energy is the sum of repulsive pair potentials, $u_0(r)$, have been studied by computer simulation and by analytical theory (see below).

In Fig. 1, we compare the repulsive force fluid $g_0(r)$ with that for the full Lennard-Jones liquid at a typical high-density and low-temperature thermodynamic state near the triple point. We see from Fig. 1 that

$$g(r) \approx g_0(r) \qquad (4)$$

is an excellent approximation. This striking correspondence (*13, 14*) provides direct verification of the van der Waals picture, and occurs because the density is sufficiently high that neighboring particles are extremely close to one another, as is evident from the position of the first peak in $g(r)$. The change in energy associated with any local (that is, short wavelength) displacement, such as the one depicted in Fig. 1 with the dashed circle, will clearly be dominated by the interparticle repulsive forces. The attractive forces, on the other hand, are much weaker and tend to cancel one another, leaving only an averaged uniform background energy.

Hard Sphere Model

Perhaps more remarkable is the fact, also illustrated in Fig. 1, that there exists a hard sphere system for which the radial distribution function $g_d(r)$ is closely related, indeed (except for r less than r_0) nearly identical to $g_0(r)$. The hard sphere fluid is characterized by the magnitude of the sphere diameter, d, appearing in the hard sphere potential

$$u_d(r) = \infty \qquad r \leq d$$
$$= 0 \qquad r > d \qquad (5)$$

(Throughout, we use the subscript d to refer to hard spheres of diameter d.) The similarity between the structure of the continuous repulsive force system and the appropriately chosen hard sphere fluid is apparent despite obvious differences at short times (more precisely, at high frequencies) in the impulsive hard sphere dynamics and the dynamics arising from a continuous interaction. The agreement is evidently a consequence of the fact that the length scale over which the pair potentials for the two differ significantly, roughly $r_0 - d$, is relatively small compared to d, the single length scale of the hard sphere fluid. The softness (non–hard core nature) of a realistic repulsive potential should therefore play a minor role in the structure except at short wavelengths.

It is for this reason that the similarity between a monatomic liquid and the hard sphere fluid appears especially striking when pair correlations are described with the structure factor

$$S(k) = 1 + \rho \int e^{i\mathbf{k}\cdot\mathbf{r}}[g(r) - 1] \, d^3\mathbf{r} \qquad (6)$$

While short wavelength differences are noticeable in the main peaks of $g_0(r)$ and $g_d(r)$, the differences between $S_0(k)$ and $S_d(k)$ are not substantial until one considers relatively large values of the wave vector k, where $S(k)$ is already close to its asymptotic value of unity.

The Fourier-transformed representation of the radial distribution function is directly determined by scattering experiments. The close correspondence between the structure factors of real liquids and that of the hard sphere fluid was established empirically in the 1960's by Ashcroft and Lekner (15) with their hard sphere model of liquid metals and by Verlet (10) in his computer simulation study of the Lennard-Jones fluid. Verlet showed that at a given density ρ and temperature T, a hard sphere diameter $d(T,\rho)$ could be chosen such that the liquid structure factor $S(k)$ was accurately fit by the hard sphere fluid structure factor, $S_d(k)$. In effect, this procedure determined the "size" of the particles, and its success clearly pointed toward the qualitative validity of the van der Waals picture of liquids.

WCA Theory

Motivated by this discovery and by Longuet-Higgins and Widom's compelling physical arguments (2) in favor of the van der Waals concept, we developed a quantitative theoretical explanation of these observations in what has become known as the WCA theory of liquids (13, 14, 16), the principal results of which we have already discussed in connection with Fig. 1. The WCA theory exploits the van der Waals picture. As a result, the first and crucial step in the development is the division of the intermolecular potential into the short-ranged repulsive portion and the longer ranged more slowly varying part. For monatomic liquids, the division we introduced is unambiguous as described above. The repulsive branch is given uniquely by $u_0(r)$ in Eq. 3, and the remainder

149

$$u(r) = w(r) - u_0(r) \qquad (7)$$

contains all the attractions and no other forces (see Fig. 2). According to the van der Waals picture, we may neglect the role of $u(r)$ in determining $g(r)$, and the problem of explaining the validity of the hard sphere model reduces to the calculation of $g_0(r)$ and showing how this function is related to the properties of the hard sphere fluid.

The radial distribution function due to the repulsive forces depends upon the potential $u_0(r)$ through the Boltzmann factor, $\exp[-u_0(r)/k_B T]$, and complicated integrals involving this factor. This function rises from zero to one over a small range of r values, while the Boltzmann factor for the hard sphere potential, $\exp[-u_d(r)/k_B T]$, is a step function that changes from zero to one at precisely $r = d$. This difference, illustrated in Fig. 3, gives rise to the differences between $g_0(r)$ and $g_d(r)$ shown in Fig. 1. These can be understood by introducing the indirect (or cavity) distribution function, $y_0(r)$, defined by

$$g_0(r) = \exp[-u_0(r)/k_B T]y_0(r) \qquad (8)$$

The Boltzmann factor describes the effect of the direct interaction between a pair of particles separated by a distance r (such as particles 1 and 2 in Fig. 3). That factor would be the full $g_0(r)$ in the dilute gas phase. In a liquid, however, the correlations between that pair of particles are affected by all the surrounding particles, and this effect is described by $y_0(r)$. [In the limit $\rho \to 0$, there is no surrounding environment and $y_0(r)$ tends to unity.] Since we envision a situation in which $r_0 - d$ is much smaller than d, it is reasonable to approximate the indirect effects by those of a hard sphere fluid at the same density with an appropriately chosen diameter d and therefore write

$$y_0(r) \approx y_d(r) \qquad (9)$$

This approximation is the first term in a systematic functional expansion of $y_0(r)$ about $y_d(r)$ (16, 17). Its physical meaning is illustrated schematically in Fig. 3b.

The systematic expansion also provides a simple way of choosing the appropriate value of the hard sphere diameter for any density and temperature. This criterion, first proposed in a somewhat different context by Percus and Yevick (18), is

$$\int d^3 \mathbf{r} y_d(r) \, \Delta f(r) = 0 \qquad (10)$$

Fig. 2. The unique separation of the Lennard-Jones potential $w(r)$ into a part $u_0(r)$ which gives repulsive forces identical to those found in $w(r)$ (and no attractive forces) and a part $u(r)$ which contains all the attractive forces (and no repulsive forces). Since the repulsive forces are equal, the potentials $w(r)$ and $u_0(r)$ can differ by only a constant for

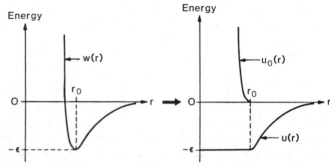

$r < r_0$. The value ϵ (see Eq. 3) of that constant is determined by requiring that $u_0(r)$ vanish at r_0 where the repulsive force vanishes. Note that the remainder, $u(r)$, is smooth and relatively slowly varying with this separation.

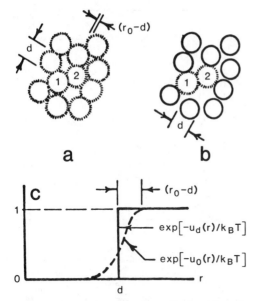

a   b

c

(r_0-d)

$\exp[-u_d(r)/k_BT]$

$\exp[-u_0(r)/k_BT]$

d   r

Fig. 3. Relation between a fluid with soft repulsive force (schematically indicated by fuzzy lines) and the hard sphere fluid (shown by solid lines). (a) All particles interact with continuous repulsive forces and $r_0 - d$ gives a measure of the softness of the repulsive forces. (b) Illustration of Eq. 9, where the effects of the environment on the correlations between particles 1 and 2 are approximated by those of a hard sphere fluid. (c) Boltzmann factors for the soft and hard sphere fluids.

Here $\Delta f(r)$ is the difference between the Boltzmann factors for $u_d(r)$ and $u_0(r)$, as shown in Fig. 3c. Thus the diameter is chosen so that the difference $\Delta f(r)$ vanishes when integrated over the volume while taking account of the environment through the factor $y_d(r)$. The resulting $d(T,\rho)$ is a weakly decreasing function of temperature and/or density, in just the way Verlet's empirically determined diameter (10) was. Indeed, the theory reproduces Verlet's diameters to better than 1/2 percent (14, 16). By combining the computed diameter with Eqs. 4 and 9, one obtains the formula

$$g(r) \approx g_0(r) \approx \exp[-u_0(r)/k_BT]y_d(r)$$
$$(11)$$

which is indistinguishable on the scale of the graph from the solid line in Fig. 1. Equation 11 thus provides a complete theory for the pair structure of a fluid in terms of the pair correlations of a hard sphere fluid. The latter are known in analytical form from the results of approximate theory (19) and "exact" computer simulations (20). Further, it is not difficult to show that, except for large wave vectors, Eqs. 6, 10, and 11 imply $S(k) \approx S_d(k)$, thus justifying the hard sphere model for the structure factor.

Thermodynamic Perturbation Theory

An immediate consequence of the structural ideas underlying the WCA theory is a simple theory for the thermodynamic properties. In particular, when $g(r) \approx g_0(r)$, thermodynamic properties can be obtained from first-order perturbation theory, which estimates the role of attractions by performing a reference system (that is, repulsive force system) average of the energy. This is analogous to the familiar perturbation result of Schrödinger quantum mechanics in which one computes the expectation value of the energy by employing the wave function of the unperturbed Hamiltonian. The standard first-order result (4) for the Helmholtz free energy per unit volume is

$$A/V \approx A_0/V + (\rho^2/2) \int g_0(r)u(r)d^3\mathbf{r}$$
$$(12)$$

where A_0 is the free energy of the reference fluid, which can be related to the free energy of the hard sphere fluid by the same procedure used to derive Eq. 9. Corrections to this formula are obtained from corrections to $g(r) \approx g_0(r)$, and if the attractions truly had no effect on the structure, Eq. 12 would be exact. When ordered in powers of $u(r)/k_B T$, the correction terms to Eq. 12 generate the thermodynamic perturbation series first discussed in this context by Zwanzig (*21*) and by Buff and Schindler (*22*). Fortunately, when the van der Waals picture is accurate, these complicated correction terms are negligibly small (*14, 20, 23*) and calculations are simple enough to be performed on a desk calculator (*20, 24*).

Other thermodynamic perturbation theories, using different potential separations, have relied on the corrections to Eq. 12 rather than the accuracy of the van der Waals picture. The best known of these alternatives is the Barker-Henderson (BH) theory (*25*), which was widely influential and stimulated much interest in the theory of liquids. In the BH theory the potential, $w(r)$, is divided into its positive ($r < \sigma$) and negative ($r > \sigma$) parts. Such a separation seems reasonable for a discussion of a dilute gas, where particles are often far apart and the natural zero of energy is the large separation value of the potential. The close proximity of neighbors in a dense liquid, however, makes it profitable to focus on the change in energy for small displacements, that is, the force between particles, and allow the uniform background to rescale the zero of energy to the potential minimum at r_0, where the force changes from repulsive to attractive.

Since the BH reference potential leaves out the repulsive forces between σ and r_0, it underestimates the "size" of the particles, the most important structural parameter in the van der Waals picture. The degree to which this issue is important is illustrated in Fig. 4. With

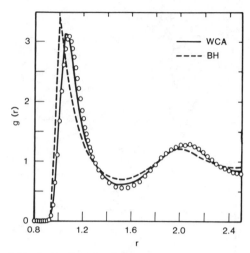

Fig. 4. Radial distribution functions for the repulsive force fluid $g_0(r)$ (solid line), the Barker-Henderson reference fluid $g_{BH}(r)$ (dashed line), and the Lennard-Jones liquid (circles) at a state near the triple point with $\rho\sigma^3 = 0.87$ and $k_B T/\epsilon = 0.75$ as determined by molecular dynamics simulations (*75*). Here r is measured in units of σ.

the BH reference system, $g(r) \approx g_0(r)$ is not a good approximation and the complicated second-order correction term to Eq. 12 is required to achieve accuracy comparable to that of the first-order WCA theory. While such calculations are possible for atomic fluids, they are generally impractical for polyatomic systems. We believe that the van der Waals picture, with its emphasis on optimizing the simple first-order term, offers a more useful starting point, both conceptually and for practical calculations.

Lesson

The preceding discussion of simple monatomic liquids has emphasized that the accuracy of the van der Waals picture depends on a clear separation of the intermolecular potential into a short-ranged harshly repulsive portion and a longer ranged (usually attractive) part which is a relatively slowly varying function of atomic coordinates. When such a separation exists, the intermolecular structure of the dense fluid can be accurately described by the van der Waals picture. Following standard chemical practice, we call such systems nonassociated liquids.

Often the repulsive forces are harsh enough that these systems can be modeled as hard core van der Waals materials. This provides a significant reduction of the complexities that would follow from realistic representations of intermolecular interactions, and it permits a simple phenomenological description of systems whose potentials are not accurately known by taking the size and shape parameters associated with the hard core model as adjustable parameters.

The problems associated with understanding the van der Waals material are those of determining the entropic effects of packing. The van der Waals picture itself gives no information on how these packing effects should be calculated. Thus it does not provide a self-contained theory, but rather a physical picture that focuses attention on the dominant role of correlations produced by hard cores. Fortunately, the statistical mechanics of hard core systems, while nontrivial, can often be successfully analyzed in terms of accurate approximations, and the solutions of these problems can usually be easily visualized.

For example, when considering the hard sphere model of liquid argon, the properties of this van der Waals material can be well estimated from the analytical solutions of the Percus-Yevick equation (19) or even more accurate approximations (26) for the hard sphere fluid. The

pictures drawn from these calculations resemble both the qualitative observations of a child playing with a box of marbles and those of more detailed statistical analyses of randomly packed spheres as performed by Bernal (27) and co-workers. Similarly, the arrangements of molecules in liquid benzene (28, 29) are similar to the average arrangements of neighboring Cheerios in a bowl of the breakfast cereal, and a solution of argon in benzene should be similar to the structure achieved when blueberries are mixed in the Cheerios.

Exceptions and Qualifications

Just as nature is abundant with examples of condensed phases that can be successfully modeled as van der Waals materials, it is also not difficult to identify those which cannot be. The hydrogen bonds, which give rise to the local tetrahedral ordering of molecules in aqueous systems (30), and the ionic interactions, which produce charge layering in fused salts (31), are examples of attractive forces that are so strong that they are competitive with typical repulsions. For these cases the van der Waals picture breaks down.

In the absence of detailed knowledge of intermolecular forces, one may never be sure if this simplified characterization of a particular liquid is valid. However, if there are no intermolecular hydrogen bonds, directional intermolecular covalent bonds, or ionic forces, then one may be fairly confident that the fluid is nonassociated and that at high densities it can be modeled as a van der Waals material. Liquids composed of CCl_4, C_6H_6, and even CH_3CN (a molecule with dipole moment of 4 debyes) are all examples of nonassociated liquids and, as discussed below, all have been successfully analyzed with the van der Waals picture.

Of course, even for these cases where there are no strong associative forces, the picture will break down at low densities where the compressibility is sufficiently high to allow for relatively long wavelength fluctuations [that is, at lower densities, the repulsive cores are not nearly as effective in screening (16) the interparticle correlations caused by the attractions]. For example, Eq. 12 predicts incorrect (classical) behavior at the critical point.

Still another source of inaccuracies in the van der Waals picture exists in liquid mixtures. While the high density of the fluid will hinder fluctuations in the total density, concentration fluctuations are not necessarily suppressed. The latter can lead to important structural effects which must be associated with the longer ranged interactions. This point is illustrated by a model system invented and studied by Alder and co-workers (32) which is composed of hard spheres of diameter σ mixed with a square-well species with the same hard core diameter and an attractive well of range 1.5σ. The attractions between the square wells tend to make these particles cluster, and this clustering can occur without changing and competing with the excluded volume correlations produced by the hard cores.

The use of hard cores to mimic the effects of realistic repulsions provides an enormous practical simplification when applying the van der Waals picture. Unfortunately, it can also be a noticeable source of error. At extremely high densities or for soft enough potentials, the differences from the hard core model must become more significant than the short wavelength differences seen in liquid argon and easily understood with WCA theory. Examples are found in the comparison of the hard sphere structure factor with $S(k)$ for liquid metals (33) (where the effective repulsive cores between atoms are relatively soft), and when the structure of the hard sphere solid is compared with that of the Lennard-Jones solid (34) (where the density is very high). These are practical quantitative issues, however, and they do not contradict the qualitative validity of the hard sphere model or the general van der Waals picture. Indeed, in both examples mentioned the attractive interactions have little effect on the structure.

Finally, in all these remarks it is assumed that it is possible to make a physically meaningful separation of the interparticle potential into a short-ranged harshly repulsive portion and a relatively slowly varying part. This is an essential step in developing an accurate theory based on the van der Waals picture. If an incorrect separation is made, fluids as simple as liquid argon or methane would appear to be exceptions to the van der Waals picture. Identifying this separation for molecular fluids where the intermolecular potential depends on many variables can be difficult, and a useful separation may not always be possible.

Several Examples

We now list several successful examples of condensed phase theories based on the van der Waals perspective.

Molecular liquids, structure. The intermolecular pair correlations of many polyatomic fluids have been interpreted with the reference interaction site model (RISM) theory (35). In this theory, molecules are assumed to be composed of overlapping rigid spheres which are space-filling representations of the atoms. A picture of an acetonitrile molecule modeled in this way is shown in Fig. 5. The resulting total molecular shape is therefore nonspherical, giving rise to significant orientational correlations, though these correlations have little to do with those induced by molecular dipoles. The theory then employs an integral equation (called the RISM equation) which yields an approximate though accurate treatment of the pair correlations

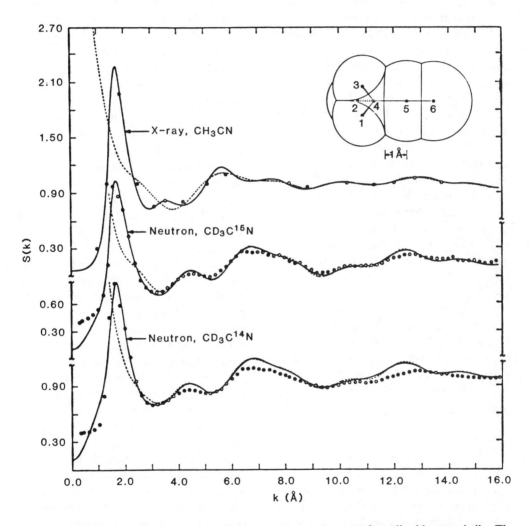

Fig. 5. Structure factors, $S(\lambda)$, for scattering neutrons and x-rays from liquid acetonitrile. The circles represent experimental points. The solid lines are the predictions of the RISM theory, using the hard core model drawn at the upper right (atoms 1, 2, and 3 are hydrogens; atoms 4 and 5 are carbons; and atom 6 is nitrogen). The dashed line shows the scattering cross section associated with only single independent molecules as opposed to the full liquid intermolecular correlations. [Adapted with permission of D. Chandler, *Molecular Physics* **36**, 215 (1978).]

associated with the packing of molecules in such a system (36). The theory has succeeded in interpreting the results of scattering experiments performed on various liquids including CCl_4 (28), C_6H_6 (28, 29), CS_2 (28, 37), $CHCl_3$ (38), and CH_3CN (39). Figure 5 shows a representative comparison between theory and experiment for CH_3CN.

Molecular liquids, diffusion. The hard sphere model for self-diffusion in liquids dates back to the time of Enskog (40). Quantitative calculations, however, began with the computer simulation studies of Alder and co-workers (41). With the simulation results in hand, Levesque and Verlet (11) showed that the self-diffusion constant of an atomic liquid can be well approximated by that of the hard sphere fluid. For molecular fluids, however, the nonspherical shape of particles introduces the additional feature of rotation-translation coupling. This coupling tends to lower the diffusion constant from what one would find in its absence since the coupling introduces additional channels for dissipating velocity correlations.

Although a quantitative theory directly treating diffusion in a fluid of hard nonspherical objects has not yet been developed, it can be argued (42) that a reasonable account of this phenomenon is given by

$$D \approx AD_d, \qquad 0 < A \leqslant 1, \quad (13)$$

where D is the self-diffusion constant of the molecular liquid, D_d is the diffusion constant of the hard sphere fluid at the same packing fraction (that is, ρ times the space filling volume of a molecule or particle in the fluid), and the parameter A differs from unity according to the extent of the translation-rotation coupling. For liquid CCl_4 (42), $A \approx 1/2$. Equation 13 and its extension to the mutual diffusion

constants of mixtures (43) have been successfully applied to numerous liquids and have provided a law of corresponding states for diffusion in nonassociated liquids (44). The model also provides a basis for understanding Hildebrand's concept of fluidity, which says that the inverse of viscosity is proportional to the free volume available to molecules in a liquid (45). Further, the extension of this model to rotational diffusion (46) has provided a means for interpreting the density and temperature dependence of rotational relaxation times measured by nuclear magnetic resonance in terms of the average collision frequencies in a hard sphere fluid (47).

Liquid metals. While properties associated with the conduction electrons are beyond the scope of a simple van der Waals theory, the hard sphere model has proved successful in several studies of the structure (15, 48, 49), diffusion (50–52), and viscosity (53) of liquid metals. The various workers differ in the procedures used to estimate the temperature-dependent hard sphere diameters, but they all find reasonable values for those parameters. The model succeeds at predicting both the magnitude and the temperature dependence of the transport coefficients. For the structure factor, $S(k)$, precise agreement with the hard sphere model is not to be expected because of the effects of the softness of the interatomic potentials. Indeed, it has been shown how information concerning the softness of the potential can be extracted from the deviations of the experimental data from the hard sphere structure factor, $S_d(k)$ (33, 49).

Waseda (48) reviewed data for 42 liquid metals and concluded that the hard sphere fluid provides a suitable model for understanding all of them. Furthermore, he classified them into three categories according to the extent of the similarity between the measured structure and the hard sphere structure. Thirty-three of the 42, including all the transition metals studied, are in the category that agrees most closely with the hard sphere fluid. Waseda also studied metal-metal alloys and concluded that, while many of them had structures consistent with the structure of a fluid of a mixture of hard spheres, some metal-metal alloys showed evidence of compound formation, indicating that they are not well described as van der Waals materials.

Solutions. No discussion of the van der Waals picture would be complete without mentioning Hildebrand's regular solution theory (*54*). It is a remarkable observation (*55*) that when hard spheres of various different sizes appropriate for solutions of real molecules are mixed together in such a way that the total volume is unchanged, the entropy of mixing is well approximated by the ideal solution formula. While the fundamental reasons for this behavior are not entirely understood, the implication of this observation is clear. According to the van der Waals picture, entropic properties are determined by packing effects, and therefore the entropy of mixing of solutions of approximately spherical molecules at constant total volume should be the same as that for an ideal solution. Indeed, this behavior is the premise of regular solution theory, which has been developed into a very successful way of estimating the thermodynamic properties of solutions, and the many liquid mixtures which act in this way are called regular solutions (*54*). In more quantitative work, Snider and Herrington (*56*) applied the theory of Longuet-Higgins and Widom (*2*) to a number of binary mixtures and found generally good agreement with experiment.

Vibrational dephasing. The relaxation of intramolecular vibrational modes of molecules in condensed phases has been the focus of much recent research (*57, 58*). The dephasing processes are strongly affected by the fluctuating force fields associated with molecules that move about in the neighborhood of the tagged molecule. Van der Waals theories for these fluctuating forces were developed by Fischer and Laubereau (*59*) and by Oxtoby (*60*). In these theories, fluctuations are assumed to scale with the collision frequency (or viscosity) and the dephasing rate is therefore predicted to be proportional to the collision frequency (or viscosity) in a hard sphere fluid. However, experiments by Jonas and co-workers (*61*) show that dephasing rates increase with pressure or density to a much smaller extent than predicted by the collision theories. In fact, in certain cases, the dephasing rate decreases with increasing liquid density. An explanation of this curious behavior was recently developed (*62*) by considering the different roles of repulsions and attractions. While the fluctuations in the former are short wavelength and do increase with pressure as predicted by hard sphere collision frequencies, the fluctuations in the latter are relatively long wavelength and as such tend to decrease with increasing density or pressure. Since vibrational modes of molecules will couple to the environment with both types of interactions, two competing effects are present. A van der Waals treatment with this perspective succeeded in explaining the dephasing data in a variety of systems (*62*).

Liquid crystals. Nematic liquid crystals are made up of long molecules whose axes are oriented in roughly the same direction. Onsager (*63*) first used the van der Waals picture to study such systems, but his treatment of the packing effects is valid only in an unphysical limit. Modern approaches, such as the scaled particle theory (*64*), have improved the situation somewhat, but to date no fully satisfactory theory for the statistical mechanics of long hard objects has been applied. As a recent simulation on an idealized system of hard platelets has indicated (*65*), such a theory should capture many qualitative features of the nematic phase and the transition from the isotropic liquid, although the realistic shape and/or flexibility of the long molecules will undoubtedly have to be taken into account in comparisons with real experiments.

Freezing and melting. Since packing considerations induced by the repulsive forces dominate the structure of dense nonassociated liquids, it is natural to suppose they will play an equally important role at the still higher densities of the crystalline or amorphous solid states and in the fluid-solid transition. The good agreement between the distribution functions in Fig. 6 shows the validity of this idea for the Lennard-Jones solid near the triple point. Both the repulsive force and Lennard-Jones systems have a face-centered cubic (fcc) lattice structure, consistent with efficient packing of the repulsive cores. The peaks in the (angularly averaged) $g(r)$ are broadened by thermal vibrations but clearly show the successive neighbor shells in the fcc lattice. As first recognized by Longuet-Higgins and Widom (*2*), the attractive interactions in the van der Waals picture play an essentially passive role in the fluid-solid transition, widening the region of two-phase coexistence over that

found in the repulsive force system alone (*66*), but introducing no new structural correlations (*67*).

An exception to this picture, related to that discussed earlier for liquid mixtures, arises when there are different solid structures for the repulsive force system with very nearly the same free energy (*68*). For example, the fcc structure is only very slightly favored by entropy over the hexagonal close-packed structure for hard spheres and both structures have the same limiting close-packed density (*69*). In such a case, the detailed form of the attractive interactions can favor one nearly degenerate structure over the other.

It is interesting to compare the solid $g(r)$ to that of the coexisting liquid, also shown in Fig. 6. The higher density of the solid has been accommodated by a

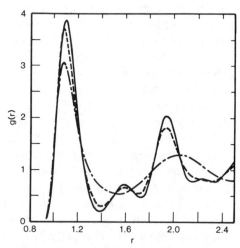

Fig. 6. Radial distribution functions for the Lennard-Jones solid (solid line) and the repulsive force solid (dashed line) at a state near the triple point with $\rho\sigma^3 = 0.98$ and $k_B T/\epsilon = 0.75$. For comparison, $g(r)$ for the coexisting Lennard-Jones liquid (chain-dotted line) at $\rho\sigma^3 = 0.87$ (shown in Fig. 4) is also included. Here r is measured in units of σ.

transition to a topologically different structure—an ordered fcc lattice where molecules in one layer fit into the interstitial "holes" in the layers above and below. Although the high density and interlocking structure of the solid prevent significant diffusion or particle exchange, there is still freedom of motion for a molecule within the "cage" formed by its nearest neighbors as well as long-wavelength distortions of the cage structure itself. This configurational freedom provides a source of entropy which stabilizes the ordered solid, relative to a disordered structure, at high densities. A study of the repulsive force system, and particularly the limiting case of the hard sphere system, is instructive in bringing out the importance of density, packing efficiency, and entropy in the fluid-solid transition. Such features are often slighted in conventional treatments of the solid based on the harmonic approximation.

Amorphous and glassy materials. The hard sphere model also gives us insight into nonequilibrium amorphous and glassy solid structures. Computer simulations have shown that for densities $\rho > \rho_m$, where ρ_m is 73 percent of the close-packed density ρ_{cp}, the ordered fcc solid structure is the thermodynamically stable phase (69). However, it is possible to arrange hard spheres in a disordered structure at densities greater than ρ_m. Following Bernal (70), Scott (71), and Finney (72), one can experimentally produce such packings by jamming an array of ball bearings into an irregularly shaped container so that nearest neighbor spheres are touching one another. If the walls of the container are corrugated so that the "nucleation" of a close-packed solid layer is inhibited, usually one ends up with a disordered arrangement which is called a random close-

packed structure. The maximum packing density that can be so achieved is only about 85 percent of ρ_{cp}.

Bernal (70) suggested that the random close-packed structure be taken as a model for the equilibrium structure of a dense fluid. This suggestion was historically very important, since it emphasized the role of repulsive forces and packing in determining the structure of a fluid and pointed out the topologically disordered nature of the fluid state at a time when lattice models and hole theories of liquids were very popular. However, in detail the model is unsatisfactory. Particles are jammed in one configuration so that very little diffusion or particle exchange is possible. Yet the ability to change configurations—that is, to flow—is one of the characteristic differences between a fluid and a solid and a major source of the increase in entropy on melting.

Rather, it has been recognized that the random close-packed models are more appropriate to describe the properties of supercooled liquids and, in particular, liquid metal mixtures below the "glass transition." By very rapidly quenching a liquid, it is sometimes possible to avoid crystallization and achieve a metastable glassy state in which one particular configuration of the disordered liquid is "frozen in." Diffusion and structural relaxation occur at rates many orders of magnitude lower than that observed in the stable fluid state. The fact that the effective hard sphere diameter increases as the temperature is decreased allows a qualitative understanding of one important source of the reduced freedom of motion in the supercooled glassy state. For many purposes it is useful to picture particles in the glass as locked into place by the repulsive cores of their neighbors, as suggested by the random close-packed model.

Of course, any static model cannot describe the detailed properties of a non-equilibrium system, which depend on the past history, the rate of cooling, the amount of structural relaxation, and other features which only a dynamical theory could hope to explain. Nonetheless, such is the dominance of packing considerations that a model which correctly takes them into account will correctly describe many gross structural features. Thus the random close-packed models serve as the starting point for more sophisticated (and complicated) treatments (73).

As we apply the van der Waals ideas to more and more complex systems we will undoubtedly find that a more detailed analysis is needed for quantitative agreement with experiment. The case of simple liquids is somewhat anomalous, in that the van der Waals picture provides so complete a description. However, in more complex systems, these ideas may prove even more important, not as a quantitative theory, but as a conceptual guide in emphasizing the important packing constraints which more detailed theories must satisfy. Even today, a century after the time of van der Waals, these ideas continue to play a central role in understanding the properties of condensed matter.

References and Notes

1. B. J. Alder and T. E. Wainwright, *J. Chem. Phys.* **27**, 1208 (1957); *ibid.* **33**, 1439 (1960); W. W. Wood and J. D. Jacobson, *ibid.* **27**, 1207 (1957).
2. H. C. Longuet-Higgins and B. Widom, *Mol. Phys.* **8**, 549 (1964); B. Widom, *Science* **157**, 375 (1967).
3. Both technical and historical articles are found in the volume *Physica* **73** (1973), which celebrates the centennial of the van der Waals equation. A more recent review of the subject is given by J. L. Lebowitz and E. M. Waisman [in *The Liquid State of Matter: Fluids, Simple and Complex*, E. W. Montroll and J. L. Lebowitz, Eds. (North-Holland, Amsterdam, 1982), p.1]. See also S. G. Brush, *The Kind of Motion We Call Heat* (North-Holland, New York, 1976).
4. J. P. Hansen and I. R. McDonald, *Theory of Simple Liquids* (Academic Press, New York, 1976).
5. H. Reiss, *Adv. Chem. Phys.* **9**, 1 (1965).
6. J. L. Lebowitz and O. Penrose, *J. Math. Phys.* **7**, 98 (1966).
7. Boltzmann was aware that very long ranged attractive interactions are needed to give a rigorous derivation of the van der Waals equation of state. On seeing Boltzmann's results, van der Waals commented that he never made this assumption and thought such a force law improbable. See L. Boltzmann, *Lectures on Gas Theory*, translated by S. G. Brush (Univ. of California Press, Berkeley, 1964), p. 375.
8. While pairwise additive models of this sort are nontrivial, they do neglect complications that can be of quantitative interest. For example, a detailed description of the thermodynamics of fluids may sometimes require the inclusion of electrostatic three-body interactions such as the Axelrod-Teller potential. However, unless artificially extrapolated to short distances, these interactions are more slowly varying than those already contained in $w(r)$. Hence the validity of the van der Waals picture should remain unaffected by these complications.
9. L. Verlet, *Phys. Rev.* **159**, 98 (1967).
10. _____, *ibid.* **165**, 201 (1968).
11. D. Levesque and L. Verlet, *Phys. Rev. A* **2**, 2514 (1970); _____, *J. Kürkijarvi, ibid.* **7**, 1690 (1973).
12. J. L. Yarnell, M. J. Katz, R. G. Wenzel, S. H. Koenig, *ibid.* **7**, 2130 (1973).
13. D. Chandler and J. D. Weeks, *Phys. Rev. Lett.* **25**, 149 (1970).
14. J. D. Weeks, D. Chandler, H. C. Andersen, *J. Chem. Phys.* **54**, 5237 (1971).
15. N. W. Ashcroft and J. Lekner, *Phys. Rev.* **145**, 83 (1966).
16. H. C. Andersen, D. Chandler, J. D. Weeks, *Adv. Chem. Phys.* **34**, 105 (1976).
17. H. C. Andersen, J. D. Weeks, D. Chandler, *Phys. Rev. A* **4**, 1597 (1971). See also R. Jacobs and H. C. Andersen (49).
18. J. K. Percus and G. J. Yevick, *Phys. Rev. B* **136**, 290 (1964).
19. _____, *Phys. Rev.* **110**, 1 (1958); E. Thiele, *J. Chem. Phys.* **39**, 474 (1963); M. S. Wertheim, *Phys. Rev. Lett.* **10**, 321 (1963).
20. L. Verlet and J. J. Weis, *Phys. Rev. A* **5**, 939 (1972).
21. R. Zwanzig, *J. Chem. Phys.* **22**, 1420 (1954).
22. F. P. Buff and F. M. Schindler, *ibid.* **29**, 1075 (1958).
23. J. D. Weeks, D. Chandler, H. C. Andersen, *ibid.* **55**, 5422 (1971).
24. When $u(r)$ is slowly varying, the integral in Eq. 12 is a weak function of the temperature and density and for many qualitative purposes can be replaced by some constant average value, say a. Equation 12 then leads to a modern version of the van der Waals equation of state in which the inaccurate treatment of the repulsive forces

found in the original version is replaced by the correct three-dimensional description. This equation served as the basis for the Longuet-Higgins and Widom theory of freezing.

25. J. A. Barker and D. Henderson, *J. Chem. Phys.* **47**, 4714 (1967); *Rev. Mod. Phys.* **48**, 587 (1976).
26. E. M. Waisman, *Mol. Phys.* **25**, 45 (1973).
27. J. D. Bernal, *Proc. R. Soc. London Ser. A* **280**, 299 (1964).
28. L. J. Lowden and D. Chandler, *J. Chem. Phys.* **61**, 5228 (1974).
29. A. H. Narten, *ibid.* **67**, 2102 (1977).
30. F. H. Stillinger, *Adv. Chem. Phys.* **31**, 1 (1975).
31. J. P. Hansen and I. R. McDonald, *Phys. Rev. A* **11**, 2111 (1975).
32. B. J. Alder, W. E. Alley, M. Rigby, *Physica (Utrecht)* **73**, 143 (1974); S. H. Sung, D. Chandler, B. J. Alder, *J. Chem. Phys.* **61**, 932 (1974).
33. J. P. Hansen and D. Schiff, *Mol. Phys.* **25**, 1281 (1973).
34. J. J. Weis, *ibid.* **28**, 187 (1974).
35. D. Chandler and H. C. Andersen, *J. Chem. Phys.* **57**, 1930 (1972); D. Chandler, *Mol. Phys.* **31**, 1213 (1976); in *The Liquid State of Matter: Fluids, Simple and Complex*, E. W. Montroll and J. L. Lebowitz, Eds. (North-Holland, Amsterdam, 1982), p. 275.
36. D. Chandler, C. S. Hsu, W. B. Street, *J. Chem. Phys.* **66**, 5231 (1977); C. S. Hsu, D. Chandler, L. J. Lowden, *Chem. Phys.* **14**, 213 (1976).
37. S. I. Sandler and A. H. Narten, *Mol. Phys.* **32**, 1543 (1976).
38. C. S. Hsu and D. Chandler, *ibid.* **37**, 299 (1979).
39. _____, *ibid.* **36**, 215 (1978).
40. D. Enskog, *K. Sven. Vetenskapsakad. Handl.* **63** (No. 4) (1921); S. Chapman and T. G. Cowling, *The Mathematical Theory of Non-Uniform Gases* (Cambridge Univ. Press, Cambridge, ed. 3, 1970), chapter 16.
41. B. J. Alder and T. E. Wainwright, *Phys. Rev. Lett.* **18**, 988 (1967); *J. Phys. Soc. Jpn.* **26**, 267 (1968); *Phys. Rev. A* **1**, 18 (1970); B. J. Alder, D. M. Gass, T. E. Wainwright, *J. Chem. Phys.* **53**, 3813 (1970).
42. D. Chandler, *J. Chem. Phys.* **62**, 1358 (1975).
43. K. J. Czworniak, H. C. Andersen, R. Pecora, *Chem. Phys.* **11**, 451 (1975); S. Bertucci and W. H. Flygare, *J. Chem. Phys.* **63**, 1 (1975).
44. H. J. Parkurst and J. Jonas, *J. Chem. Phys.* **63**, 2698 (1975); R. J. Finney, M. Fury, J. Jonas, *ibid.* **66**, 760 (1977); L. A. Woolf, *J. Chem. Soc. Faraday Trans.* **78**, 593 (1978); K. R. Harris, *Physica (Utrecht)* **93A**, 593 (1978).
45. J. H. Hildebrand, *Science* **174**, 490 (1971).
46. D. Chandler, *J. Chem. Phys.* **60**, 3508 (1974).
47. J. De Zwann and J. Jonas, *ibid.* **62**, 4036 (1975); *ibid.* **63**, 4606 (1975).
48. Y. Waseda, *The Structure of Non-Crystalline Materials* (McGraw-Hill, New York, 1980).
49. R. Jacobs and H. C. Andersen, *Chem. Phys.* **10**, 73 (1975).
50. P. Ascarelli and A. Paskin, *Phys. Rev.* **165**, 222 (1968).
51. C. J. Vadovic and C. P. Colver, *Phys. Rev. B* **1**, 4850 (1970).
52. P. Protopapas, H. C. Andersen, N. A. D. Parlee, *J. Chem. Phys.* **59**, 15 (1973).
53. _____, *Chem. Phys.* **8**, 17 (1975).
54. J. H. Hildebrand, J. M. Prausnitz, R. L. Scott, *Regular and Related Solutions* (Van Nostrand Reinhold, New York, 1970).
55. B. J. Alder, *J. Chem. Phys.* **40**, 2464 (1964).
56. N. S. Snider and T. M. Herrington, *ibid.* **47**, 2248 (1967).
57. D. W. Oxtoby, *Adv. Chem. Phys.* **40**, 1 (1979).
58. A. Laubereau and W. Kaiser, *Rev. Mod. Phys.* **50**, 607 (1978).
59. S. F. Fischer and A. Laubereau, *Chem. Phys. Lett.* **35**, 6 (1975).
60. D. W. Oxtoby, *J. Chem. Phys.* **70**, 2605 (1979).
61. W. Schindler and J. Jonas, *ibid.* **72**, 5019 (1980); W. Schindler, P. T. Sharko, J. Jonas, *ibid.* **76**, 3493 (1982).
62. K. S. Schweizer and D. Chandler, *ibid.* **76**, 2296 (1982).
63. L. Onsager, *Ann. N.Y. Acad. Sci.* **51**, 627 (1949).
64. See, for example, M. A. Cotter, *J. Chem. Phys.* **66**, 1098 (1977).
65. D. Frenkel and R. Eppenga, *Phys. Rev. Lett.* **49**, 1089 (1982).
66. Note, however, that the softness of the repulsive cores tends to reduce the width from that found in the hard sphere system. [See W. G. Hoover, S. G. Grey, K. W. Johnson, *J. Chem. Phys.* **55**, 1128 (1971).] Thus both the effect of the attractions and the finite softness of the repulsions must be considered for detailed comparison with experiment.
67. Recently, it was suggested that some two-dimensional systems may have a higher order melting transition into a "hexatic phase" with positional correlations of a fluid but a remnant of the long-ranged angular order of the solid. [See B. I. Halperin and D. R. Nelson, *Phys. Rev. Lett.* **41**, 121 (1978).] This possibility is consistent with the van der Waals picture provided the repulsive force system itself has a higher order transition. However, most simulations for two-dimensional systems with harshly repulsive forces have thus far indicated only a conventional first-order transition. See Weeks and Broughton (*75*) for further discussion of this point.
68. In general, at a fixed density and temperature the equilibrium state is one which minimizes the Helmholtz free energy $A = E - TS$ where E is the energy and S the entropy.
69. B. J. Alder, W. G. Hoover, D. A. Young, *J. Chem. Phys.* **49**, 3688 (1968).
70. J. D. Bernal, *Trans. Faraday Soc.* **33**, 27 (1937); *Nature (London)* **183**, 141 (1959).
71. G. D. Scott, *Nature (London)* **188**, 908 (1960).
72. J. L. Finney, *Proc. R. Soc. London Ser. A* **319**, 479 and 495 (1970).
73. See, for example, G. S. Cargill, *Solid State Phys.* **30**, 227 (1975); T. H. Gaskell, *J. Phys. C* **12**, 4337 (1979).
74. D. Chandler, *Annu. Rev. Phys. Chem.* **29**, 441 (1978).
75. J. D. Weeks and J. Q. Broughton, *J. Chem. Phys.* **78**, 4197 (1983).
76. We are grateful to P. C. Hohenberg, W. van Saarloos, and F. H. Stillinger for helpful comments. This work was supported in part by grants CHE82-00688 and CHE81-07165 from the National Science Foundation.

Heterogeneous catalysis is crucial in the production of most industrial chemicals and therefore has become a focus of intense efforts in both industry and academia. Although recent developments in catalysis are impressive, they fall short of chemists' expectations. Catalyst development is still a matter of trial-and-error efforts, intuitive assessments, and, of course, a great deal of luck. The reason for this is that the course of a catalytic reaction is determined by many factors, both thermodynamic and kinetic, and the composition and structure of a catalyst may be critical. An important point is that all heterogeneous processes, whether simple or complex, have chemisorption as a necessary first step. Thus, the dissection of catalytic processes into primary events amenable to scientific inquiry must begin with an understanding of chemisorption.

Tremendous advances in describing chemisorption phenomena have led to the acquisition of an enormous amount of diverse information (1–5), but understanding has not kept pace with the accumulation of facts. Having encountered some troublesome examples in our practical work and failing to find coherent explanations within the current theoretical models, we searched for a better explanation of the paradoxical chemisorption phenomena. In this article, we describe some of our theoretical results, which concern the most fundamental aspects of chemisorption. Specifically, we describe relations among seemingly disparate aspects of chemisorption, such as adsorbate bond activation, the heat of chemisorption, adsorbate registry and stereochemistry, barriers for adsorbate surface migration and for adsorbate (molecular) dissociation, work-function changes, and core binding–energy shifts.

12

Toward a Coherent Theory of Chemisorption

Evgeny Shustorovich and Roger C. Baetzold

Science 227, 876–881 (22 February 1985)

Our approach combines, in a complementary fashion, both analytical and computational facets.

New Theoretical Developments

Periodic regularities of the heat of chemisorption. The heat released on chemisorption (Q) determines the course of surface reactions, so that knowledge of the variations of Q—especially its periodic behavior—is a primary theoretical target. For atomic radicals such as H, O, and N, the values of Q monotonically decrease from left to right along a particular series of the periodic table and typically down a column. This decrease is not significant for monovalent H but becomes quite apparent for divalent O and especially trivalent N, although the values are not simply proportional to the number of unpaired valence electrons. The periodic regularities for molecules are less pronounced and more complex. In particular, for strong acceptors such as CO and NO, Q changes only slightly and rather nonmonotonically, showing even a reverse trend compared with adatoms. Some representative data are given in Table 1. The current theoretical models are able to describe the monotonic atomic patterns (*6–8*) but fail to reproduce the nonmonotonic molecular behavior (*7*).

In an attempt to achieve a more coherent understanding, we developed a simple model of Q (*5, 9, 10*). This model is based on perturbation theory and uses the constant d-density approximation, which is effective in treating the cohesion energy regularities for transition metals and their alloys (*11*). The notations used are given in Fig. 1.

The major feature of the metal band structure is the presence of a huge reser-

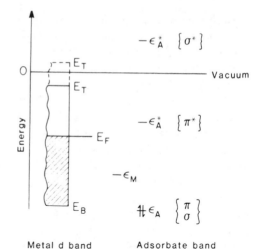

Fig. 1. Energy diagram of the metal-adsorbate band-structure interactions. The metal d band is spread out (E_B is the bottom energy, E_T is the top energy, which may be either below or above vacuum; E_F is the Fermi energy). The adsorbate bands at least for low coverage are very narrow (almost degenerate). Shown are typical positions of the adsorbate levels, occupied ε_A (σ or π) and vacant ε_A^* (π^* below vacuum but σ^* above vacuum), as well as a position of an atomic metal d orbital ε_M.

voir of electrons and electronic states and the presence of the Fermi energy E_F separating the occupied part of the d band of width W^{occ} from the vacant part of width W^{vac} ($W = W^{occ} + W^{vac}$). The Fermi energy does not change significantly under chemisorption. The value of Q includes a contribution from the direct metal-adsorbate interaction, similar to that in coordination compounds, and a contribution from the redistribution of electron density required to maintain the constant E_F specific for chemisorption bonding. As a result, the structure of Q appears to differ for donor, acceptor, and radical adsorbates, as do their periodic variations.

Within the Hückel-type approximation, the first-order perturbation results are as follows:

1) For a lone-pair donor adsorbate, Q^D is roughly proportional to the metal d-hole count $N_h = 10\, W^{vac}/W$ and inversely proportional to $E_F - \epsilon_A$, namely

$$Q^D \propto \frac{\beta^2 N_h}{E_F - \epsilon_A} \qquad (1)$$

where β is the Hückel resonance integral and ϵ_A is the lone-pair energy. Because $|E_F|$ increases only slightly from left to right along the transition series but N_h decreases significantly, one can expect Q to decrease monotonically in this direction and to be especially small for the late transition metals (N_h is very small, so that the bonding situation is similar to that in the He dimer).

2) For a vacant-orbital acceptor adsorbate, Q^A may depend on how close the orbital energy ϵ_A^* is to E_F and the value of the resonance integral β^*. In particular, for moderate acceptors where $\beta^*/(\epsilon_A^* - E_F) \le 1$, we have

$$Q^A \propto \frac{\beta^{*2} N_d}{\epsilon_A^* - E_F} \qquad (2)$$

that is, Q^A is proportional to the d occupancy $N_d = 10\, W^{occ}/W$, which is just opposite to the dependence of Q^D on N_d in Eq. 1 because $N_h = 10 - N_d$. For strong acceptors with $\beta^*/(\epsilon_A^* - E_F) >> 1$, the dependence of Q^A on N_d and N_h appears to be nonmonotonic and may show parabolic character analogous to classic cohesive energy behavior (11), namely

$$Q^A \propto \beta^* N_d N_h \qquad (3)$$

Such a behavior is typical for CO (Table 1).

3) For a radical adsorbate, in which a singly occupied orbital lies below the Fermi level ($\epsilon_A < E_F$), there can be significant charge transfer, so that some self-consistent adjustment of the initial energy ϵ_A and of the resulting occupancy of the chemisorption levels must be made. The resulting heat of chemisorption Q^R for a k-valent radical is

$$Q^R = (E_F - \epsilon_A)^2/2U_A +$$
$$kn\beta^2[\ln(W^{vac}W^{occ}/\beta^2) + 3/2]/W \qquad (4)$$

where U_A and n are scaled parameters (10). Quantitative estimates made with Eq. 4 are highly accurate, errors being typically less than 5 to 10 percent (10).

Table 1. Heats of chemisorption for some atomic and molecular adsorbates. Examples are given for ordered transition metal surfaces of high atomic density with the stated Miller indices. The first three metal entries compare changes within a column, and the latter three compare changes across the 5d series (3–5, 10).

Metal surface	Heat of chemisorption (kcal/mol) for				
	H	O	N	CO	NO
fcc Ni(111)	63	115–130	135	27	25
fcc Pd(111)	62	87	130	34	31
fcc Pt(111)	60*	85	127	32	27
fcc Ir(111)	63	93	127	34	20
bcc W(110)	68	104–129	155	27	D†

*From R. J. Madix (37).   †Dissociated.

Because $|E_F|$ increases monotonically from left to right within groups VI through VIII of the periodic table, the E_F-dependent term in Eq. 4 decreases in this direction. We can also foresee that Q^R will increase as k increases but at less than the first power, because only the second term in Eq. 4 is k-dependent. Thus, all the major periodic trends in Q mentioned above can be understood in terms of Eqs. 1 to 4 [for computational details, see (5, 12)].

The crucial role of the antibonding adsorbate orbitals in bond activation. Saturated molecules such as H_2 or CH_4 and lone-pair molecules such as NH_3 have very high energy vacant antibonding σ^* orbitals well above the vacuum level. For this reason, the σ^* ligand orbitals in transition-metal complexes are commonly neglected in descriptions of bonding, and saturated or lone-pair molecules are considered to be exclusive donors. For example, recent theoretical analyses (13) of hydrogen addition to or elimination from transition-metal complexes have not explicitly considered the d-σ^* interactions.

But even when these molecular rules of the game are applied, they may fail in the case of chemisorption. First, the Fermi level that separates occupied metal states from vacant states becomes an analog of the molecular frontier orbitals, highest occupied and lowest vacant. The Fermi energies and the frontier orbital energies differ significantly. Typical values of $|E_F| = 4.5$ to 5.5 electron volts, as defined by the transition-metal work function, can be contrasted with the atomic d-orbital ionization potentials ranging from 8 to 12 eV. Thus, metal surfaces are better electron donors than metal complexes or clusters and interact more significantly with σ^* and π^* ad-

sorbate vacant orbitals. Second, the antibonding σ^* and π^* orbitals overlap with the metal d orbitals more strongly than do their bonding σ and π counterparts. The reason, which went unnoticed until recently (14), is that the normalized LCAO-MO coefficients are larger in $\sigma^*(\pi^*)$ than they are in $\sigma(\pi)$ (LCAO-MO is a linear combination of atomic orbitals representing a molecular orbital). For example, the bonding ψ and antibonding ψ^* LCAO-MO for a homonuclear molecule X_2, such as H_2, N_2, or O_2, are

$$\psi = [2(1 + S)]^{-1/2} (\chi_1 + \chi_2)$$

$$\psi^* = [2(1 - S)]^{-1/2} (\chi_1 - \chi_2) \qquad (5)$$

where S in an overlap integral between the interacting atomic orbitals χ_1 and χ_2. Thus, in a linear fragment M-$X_{(1)}$-$X_{(2)}$ we have for the interacting metal orbital $\chi_M(d_{z^2}$ or $d_{xz})$

$$\beta = <\chi_M|H|\psi>$$
$$= [2(1 + S)]^{-1/2}(\beta_1 + \beta_2) \qquad (6)$$

$$\beta^* = <\chi_M|H|\psi^*>$$
$$= [2(1 - S)]^{-1/2}(\beta_1 - \beta_2) \qquad (7)$$

where H is the Hamiltonian. Assuming $|\beta_1| >> |\beta_2|$, since the resonance integral is a strong function of the metal-adsorbate distance, we have

$$\beta^{*2}/\beta^2 \approx (1 + S)/(1 - S) \qquad (8)$$

Thus, for typical values of $S = 0.3$ to 0.6, $(\beta^*)^2/\beta^2 \approx 2$ to 4. Because typically $E_F - \epsilon_A \gg \epsilon_A^* - E_F$, both the numerator and the denominator favor acceptor bonding when we compare Eqs. 1 and 2.

Table 2 illustrates the metal-adsorbate (M–A) charge transfer found by straightforward calculations for H, Cl, CH_4, CO, and NH_3 on an fcc(111) surface of a five-

Table 2. Effective charge q and polarization dipole change $\Delta\mu$ under chemisorption. Results are typical calculations for an fcc(111) surface of a five-layer metal film of a late transition metal (15). See text.

Ad-sorbate	q (atomic units)	$\Delta\mu$ (Debye)
H*	−0.37	−0.05
Cl*	−0.54	−0.14
CH_4*	−0.08	−0.22
CH_4†	−0.17	−0.04
CO*	−0.30	−0.19
NH_3†	−0.03	−0.15

*On-top site.   †Hollow C_{3v} site.

layer metal film (15). Even CH_4 and NH_3 behave as acceptors. The predominant role of σ* in chemisorption was further corroborated in a subsequent comprehensive study (16).

Adsorbate-induced surface polarization. The energy required to remove an electron from a metal to vacuum is the work function φ. It is a common practice to explain changes in the metal work function $\Delta\phi$ induced by chemisorption solely through formation of an M–A electrostatic dipole moment (1). If the work function decreases ($\Delta\phi < 0$), the adsorbate A is deduced to be more electropositive than the metal M, or if the work function increases ($\Delta\phi > 0$), A is considered to be more electronegative. The behavior of $\Delta\phi$ found experimentally for various adsorbates and surfaces is rather bizarre if the explanation described above is universally true. For example, for a given metal M, adatoms such as H, N, S, and Cl often give $\Delta\phi < 0$ on highly dense surfaces but $\Delta\phi > 0$ on low-atomic-density surfaces (17, 18). Moreover, the sign of $\Delta\phi$ for a given surface may depend upon the adsorbate coverage.

For example, $\Delta\phi < 0$ at low coverage of Cl on Pt(111) and $\Delta\phi > 0$ at high coverage (18).

The conventional explanation of the reversal of the sign of $\Delta\phi$ is the reversal of the M–A electrostatic dipole moment, $M^{\delta-}$–$A^{\delta+}$ ($\Delta\phi < 0$) or $M^{\delta+}$–$A^{\delta-}$ ($\Delta\phi > 0$), respectively. But it is hard to comprehend how adsorbate atoms that are more electronegative than the metal surfaces may become electropositive. It appears that the conventional interpretation of the work-function change under chemisorption is strongly deficient, and nonelectrostatic factors must be considered.

An explanation of this behavior comes from the fact that, for a particular metal, its surfaces have different values of φ, which are larger for densely packed surfaces than for loosely packed surfaces. This anisotropy in φ may be as large as 1 eV and often exceeds $\Delta\phi$ caused by chemisorption. The only source of the anisotropy in φ is surface polarization leading to formation of the intrinsic surface dipole moment (19), because clean metal surfaces are electrostatically neutral. We therefore assumed that this surface dipole moment is affected by chemisorption as well. The adsorbate-induced change $\Delta\phi$ consists of two contributions, $\Delta\phi_{ext}$ from the external electrostatic dipole moment and $\Delta\phi_{int}$, the new term due to internal polarization dipole moment

$$\Delta\phi = \Delta\phi_{ext} + \Delta\phi_{int} \qquad (9)$$

We have analyzed this new $\Delta\phi_{int}$ term within the LCAO-MO tight-binding approximation, both analytically by first-order perturbation theory (20) and computationally by straightforward metal-film calculations (15). We found that all atomic and molecular adsorbates cause uniform changes in surface polarization,

167

shown schematically in Fig. 2c. More precisely, all adsorbates decrease the surface d density and induce the out-of-phase d-p orbital rehybridization, leading to formation of the surface dipole moment directed to the bulk and therefore decreasing the metal work function ($\Delta\phi < 0$). This uniform pattern for adsorbates as diverse as H, Cl, NH_3, CO, and CH_4 is illustrated by metal-film calculations (Table 2).

As indicated above, most adsorbates behave as acceptors on metal surfaces. Thus, in Eq. 9 the two terms typically have opposite signs ($\Delta\phi_{ext} > 0$, whereas $\Delta\phi_{int} < 0$) so that the resulting sign of $\Delta\phi$ is determined by a balance of these opposite contributions. Our model predicts that the electrostatic term $\Delta\phi_{ext} > 0$ will be the least positive and the polarization term $\Delta\phi_{int} < 0$ will be the most negative on the most densely packed surface. These surfaces are prone to show a decrease in work function ($\Delta\phi < 0$) under chemisorption. This conclusion, based on our model, makes understandable the seemingly bizarre patterns of $\Delta\phi$ mentioned above. Most recently, our model has gained further support from the ab initio band-structure calculations for sulfur adsorbed on Rh(100) (21), which revealed an unexpected decrease in work function despite the conventional charge transfer $Rh^{\delta+}$–$S^{\delta-}$, and this decrease ($\Delta\phi < 0$) was due to the surface polarization (rehybridization).

The core binding energy of an atom is the energy required to ionize a core electron of the atom. The core binding energy of atoms on the surface differs from that in the interior of transition metals. The peak of the core binding energy on the surface may be shifted by the presence of adsorbates. Our model allows an explanation of these shifts in terms of surface rehybridization rather than charge transfer. It is commonly accepted (22–24) that for late transition metals with more-than-half-occupied d bands, the local d density increases as the number of the metal nearest neighbors decreases; the opposite behavior is found for early transition metals with less-than-half-occupied d bands. For example, it has been shown experimentally (22) and theoretically (23, 24) that, for metals on the right half of the $5d$ transition series such as W, Ir, and Pt, the $4f_{7/2}$ core binding energies decrease from the bulk to surface metal atoms. Again, no charge separation can be invoked to explain this behavior for clean metal surfaces, but rather a d-sp rehybridization affecting the Coulomb potential in the metal core. We saw above that the persistent pattern of chemisorption is a decrease of the surface d density. Thus, our model predicts that all adsorbates will increase core binding energy.

We tested our polarization mechanism by experimentally measuring surface $4f_{7/2}$ core shifts for ordered Pt surfaces chemisorbing diverse adsorbates such as CO, NH_3, and K (25). For each adsorbate, the $4f_{7/2}$ surface core binding energy increases. Final-state relaxation effects, which are often discussed in photoemission (24, 26), cannot explain the direction of the shift. We conclude that chemisorption phenomena such as the surface core shifts and work-function changes are primarily determined by the adsorbate-induced surface polarization or rehybridization but not the adsorbate-surface charge transfer.

Surface migration and dissociation of adsorbates. Surface migration phenomena involving lateral motion of an adsorbate along a metal surface are potentially rate limiting in desorption of dissociated

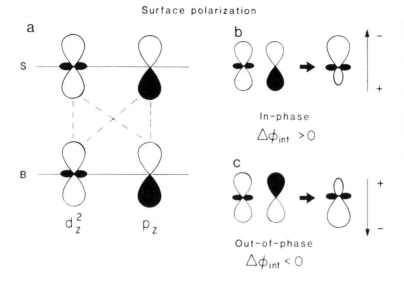

Surface polarization

a

S

B

d_z^2 p_z

b

In-phase

$\triangle\phi_{int} > 0$

c

Out-of-phase

$\triangle\phi_{int} < 0$

Fig. 2. (a) Mutual orientation of d_{z^2} and p_z atomic orbitals for the surface and bulk layers. The dashed lines show how the atomic orbital lobes interact. Formation of the surface dipole moment for the (b) in-phase $S_d + S_p$ and (c) out-of-phase $S_d - S_p$ mixing is also shown. [Reprinted with permission from *Solid State Communications* **38**, E. Shustorovich, p. 493. © 1981, Pergamon Press, Ltd.] The out-of-phase mixing always prevails. See text.

species in many catalytic processes (*2*). Experimental observations show that the migration activation barrier ΔE^* typically equals 10 to 25 percent of the heat of chemisorption Q (*1–3*), but there were no specific theoretical arguments explaining this experimental range.

Migration and dissociation of adsorbates involves changes in the coordination mode of the metal site M_n–A, where n is the coordination number of adsorbate A, and in the M–A distances. Thus, it is crucial to choose a potential for the M–A interaction that can reproduce the equilibrium minima as well as other points on the potential energy curves. We have chosen the simplest potential, namely, the Morse potential (Eq. 10), following the "Occam's razor" philosophy. More specifically, we describe each two-center M–A interaction by Eq. 10, where the total energy $E(r, x)$ relates to the bond order x (Eq. 11), which is an exponential function of the M–A distance r. Here a is a screening parameter, and Q_0 and r_0 are

the bond energy and distance at equilibrium

$$E(x) = Q_0(x^2 - 2x) \qquad (10)$$

$$x = \exp[-(r - r_0)/a] \qquad (11)$$

where $x = 1$, by definition. Some justification of the Morse potential curve comes from its similarity to the universal binding-energy plots recently demonstrated numerically (*27*) and analytically (*28*) for various cases of metallic binding. We have further assumed that the multicenter M_n–A interactions are pairwise additive and that the total M_n–A bond order is normalized to unity and conserved along a migration path up to the dissociation point. One can add that bond-order conservation for various gas-phase three-center interactions is known to be a very accurate criterion (*29*).

In the analytical model we have limited the M_n–A interactions to nearest metal neighbors. In the straightforward computations, we have treated a five-layer metal film for which the number of metal atoms involved exceeded 1000. Also, in the film computations, an energy-minimization criterion along with bond-order conservation was examined. The analytical results are rigorous (29) and fully consistent with the computational analyses for a broad range of the relevant parameters (30). Since the atomic migration patterns might differ greatly from the molecular migration patterns, we describe them separately. Because the model does not explicitly take into account the adsorbate-adsorbate interactions, all of the results, strictly speaking, should be assigned to low (zero) coverage.

Atomic Migration

The M_n–A bonding energy Q_n is a monotonic function of the coordination number n for atomic adsorbates A

$$Q_n = Q_0(2 - 1/n) \qquad (12)$$

where Q_0 is the two-center M–A bond energy in the on-top position. On highly symmetric surfaces with regular unit meshes M_n such as equilateral triangles, $n = 3$ for fcc(111), or squares, $n = 4$ for fcc(100), the hollow positions of the highest coordination should always be preferred, in full agreement with experiment (1, 4). The relevant value of Q_n (Eq. 12) can be identified with the heat of chemisorption Q_A. Migration on such surfaces is always confined to the hollow → bridge → hollow path with the barrier

$$\Delta E^* = k_n Q_n, \qquad k_n = \frac{n - 2}{4n - 2} \qquad (13)$$

where the proportionality coefficients k_n fall well within the experimental range of $k = 0.1$ to 0.3 (1–3). For the first time, theory explained the interrelation between the migration barrier and the heat of chemisorption mentioned earlier. A typical energy profile computed for atomic migration on a fcc(111) surface is shown in Fig. 3.

Molecular Migration

A diatomic molecule AB interacts with a metal surface separately through the A and B ends within the pairwise additive scheme. Thus, the heat of molecular chemisorption Q_{AB} includes two atomic contributions, Q_A and Q_B, which may have the same or opposite signs. The former case ($Q_A > 0$, $Q_B > 0$) corresponds to a donor AB, with the in-phase LCAO-MO, σ or π, being responsible for the M–AB bonding. The latter case ($Q_A > 0$, $Q_B < 0$) corresponds to an acceptor AB, for which the M–AB bonding involves primarily the out-of-phase LCAO-MO, σ^* or π^* (29–31). Similar results were found for the cases of the fixed and varied A–B bond order. Atom-like migration patterns are predicted (29, 31) for strong donor molecules for which ΔQ monotonically increases with n. This leads to a preferred hollow site and a migration activation barrier identified with the hollow-bridge energy difference. A migration energy profile for a strong donor A_2 on an fcc(111) surface is shown in Fig. 4. The similarity to the atomic migration profile in Fig. 3 is remarkable. The analytical model, in contrast, predicts (29, 31) for acceptor admolecules a nonmonotonic ΔQ versus n energy profile. The preferred site will be typically of low coordination number,

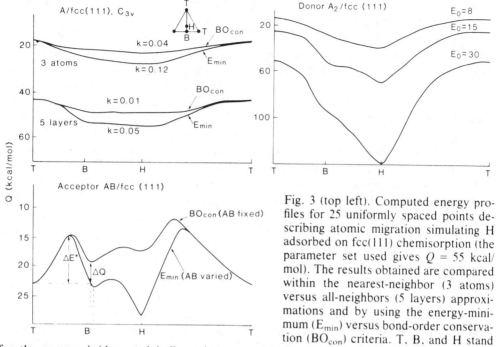

Fig. 3 (top left). Computed energy profiles for 25 uniformly spaced points describing atomic migration simulating H adsorbed on fcc(111) chemisorption (the parameter set used gives $Q = 55$ kcal/mol). The results obtained are compared within the nearest-neighbor (3 atoms) versus all-neighbors (5 layers) approximations and by using the energy-minimum (E_{min}) versus bond-order conservation (BO_{con}) criteria. T, B, and H stand for the on-top, bridge, and hollow sites, respectively. The M_n–A bond energy increases monotonically as the effective coordination number n increases along the series T < B < H for both E_{min} and BO_{con} procedures, the latter giving a shallower curve. The values of $k = \Delta E^*/Q$ are also shown (31). Fig. 4 (top right). Computed energy profiles for 25 uniformly spaced points describing a homonuclear donor molecule A_2 on a five-layer fcc(111) film. The upright geometry is kept throughout the migration path. The results (within the E_{min} procedure) were obtained for fixed values of r_0 (1.9 Å) and a (0.43 Å) with E_0 varied. The patterns are rather atom-like (see Fig. 3) (31). Fig. 5 (bottom left). Computed energy profiles for 25 uniformly spaced points describing a heteronuclear acceptor molecule AB (upright configuration) on a five-layer fcc(111) film, where A and B are different. Both the energy minimum (E_{min}) and bond-order conservation (BO_{con}) procedures were used. The parameters for atom A are $a = 0.271$ Å; $r_0 = 1.9$ Å, and $Q_0 = 80$ kcal/mol. The parameters for atom B are $a = 0.901$ Å, $r_0 = 1.0$ Å, and $Q_0 = 3812$ kcal/mol. Nonmonotonic energy barriers are found for fixed A–B bond length (1.178 Å) or variable A–B bond length (31).

and the activation barrier may be much larger than the energy difference between the total energy minima, which may not necessarily correspond to symmetric sites. This explains why CO usually prefers the on-top or bridge sites but seldom the hollow sites on flat surfaces. This puzzling observation is a reversal of the atomic behavior. Figure 5 shows that

the M_n–CO energy varies nonmonotonically with n on an fcc(111) surface and that a significant barrier exists between the deepest minima. Furthermore, one of the minima corresponds to a nonsymmetric site. This behavior, discussed above as a qualitative possibility, is consistent with experimental data recently obtained for CO migration on Pt(111).

The energy difference between the preferred on-top adsorption site of the upright geometry and bridge site was estimated to be less than 1 kcal/mol (32), but the migration barrier is 7 kcal/mol (33).

Dissociation of Adsorbates

A catalytically important chemisorption process is molecular dissociation on a surface. Thermodynamically, for a diatomic molecule AB to dissociate, the heat of chemisorption of atomic constituents A (Q_A) and B (Q_B) must exceed the gas-phase dissociation energy D_{AB}. Too often, this necessary thermodynamic condition is not sufficient because the dissociation process has a large activation barrier ΔE^*_{AB} that makes the reaction rate too slow at common temperatures. Again, no theoretical model has explicitly related ΔE^*_{AB} with $Q_A(Q_B)$ or D_{AB}, or with other observables.

Within our Morse potential approach based on M_n–AB bond-order conservation, the dissociation activation barrier ΔE^*_{AB} reads as follows (34)

$$\Delta E^*_{AB} = D_{AB} - (Q_A + Q_B) + \frac{Q_A Q_B}{Q_A + Q_B}$$

(14)

where D_{AB} is the gas-phase dissociation energy and $Q_A(Q_B)$ is the heat of atomic chemisorption. In the homonuclear case A_2 (A = B), we have

$$\Delta E^*_{A_2} = D_{A_2} - kQ_A \text{ for } k = 3/2 \quad (15)$$

that is, $\Delta E^*_{A_2}$ is linearly dependent on Q_A, with slope $k = 3/2$. For H_2, O_2, and N_2 on various surfaces of Fe, Ni, Co, W, and Pt, the experimental values of k lie within the range 1.4 to 1.7 (3, 35) close to the theoretical value of 1.5. Unfortunately, almost no experimental data on

heteronuclear ΔE^*_{AB} are available, so that Eq. 14 cannot be directly verified.

Equations 14 and 15 predict atomic heats of chemisorption to be the only variable components of the dissociation barrier. Thus, periodic regularities of ΔE^*_{AB} can be deduced from the periodic regularities of Q_A (Q_B), discussed in detail above. The results correlate well with the experimental observations (1–3, 34, 35).

The value of ΔE^*_{AB} (Eq. 14) does not explicitly depend on the molecular heat of chemisorption Q_{AB}. Although Q_{AB} may somehow affect ΔE^*_{AB} [which may be the reason for the observed 1.4 to 1.7 range for k (3, 35) rather than the constant value of 1.5 (Eq. 15)], the Q_{AB} contribution appears to be minor. We stressed above that molecular heats of chemisorption (unlike atomic ones!) vary slightly from left to right along the transition series. Moreover, they vary non-monotonically and can even increase (Table 1), which shows no correlation with the periodic trends in ΔE^*_{AB}.

Atomic versus Molecular Chemisorption

It is common to talk about the gas-solid phase interactions without specifying the atomic or molecular state of the adsorbate. One tacitly assumes that the general regularities of the metal-adsorbate interactions are qualitatively similar for the atomic and molecular species. As a result, many generalizations about chemisorption phenomena have been made (and are still being made) by extrapolating atomic chemisorption findings to molecular chemisorption, and vice versa. We saw, however, that differences between the atomic and molecular chemisorption patterns are typical, whereas similarities are rather exceptional. Examples we have discussed are periodic changes of the heat of chemisorption, adsorbate registry, the migration energy profile, and the nature of the migration barrier. Although this pattern was well known from experimental work, its importance was underestimated because there was no theoretical explanation for the occurrence of such differences.

In constructing our theoretical model, we tried to account for the differences in atomic and molecular chemisorption patterns. Our general conclusion is that strong donor admolecules may closely mimic adatoms in their relatively simple behavior, but acceptor admolecules will typically show distinct and complicated patterns. Our model resolves many seeming contradictions with the prediction that on metal surfaces (unlike clusters!) most molecules, including even saturated molecules, will behave as effective acceptors.

Concluding Remarks

Although the conclusions based on our model are contrary to some commonly held perceptions, the picture as a whole fits experimental findings well. Our goal was a broad and coherent understanding of a variety of chemisorption phenomena with considerable relevance to heterogeneous catalysis (36). The new developments we have described constitute a first step in this direction. We hope that our findings will stimulate further theoretical and experimental analyses of this area of important industrial applications.

References and Notes

1. See, for example: G. Somorjai, *Chemistry in Two Dimensions: Surfaces* (Cornell Univ. Press, Ithaca, N.Y., 1981); T. N. Rhodin and G. Ertl, Eds., *The Nature of the Surface Chemical Bond* (North-Holland, Amsterdam, 1978).
2. L. D. Schmidt, in *Interactions on Metal Surfaces*, R. Gomer, Ed. (Springer, Berlin, 1975), pp. 70–71.
3. G. Ertl, in *The Nature of the Surface Chemical Bond*, T. N. Rodin and G. Ertl, Eds. (North-Holland, Amsterdam, 1978), chapter 5, p. 315.
4. E. L. Muetterties, T. N. Rhodin, E. Band, C. F. Brucker, W. R. Pretzer, *Chem. Rev.* **79**, 91 (1979).
5. E. Shustorovich, R. D. Baetzold, E. L. Muetterties, *J. Phys. Chem.* **87**, 1100 (1983).
6. C. M. Varma and A. J. Wilson, *Phys. Rev. B* **22**, 3795, 3805 (1980).
7. W. Andreoni and C. M. Varma, *ibid.* **23**, 437 (1981).
8. M. C. Desjonquères and D. Spanjaard, *J. Phys. C* **16**, 3389 (1983); C. Thuault-Cytermann, M. C. Desjonquères, D. Spanjaard, *ibid.*, p. 5689; P. Norlander, S. Holloway, J. K. Nørskov, *Surf. Sci.* **136**, 59 (1984).
9. E. Shustorovich, *Solid State Commun.* **44**, 567 (1982).
10. _____, *J. Phys. Chem.* **88**, 1927, 3490 (1984).
11. See, for example: J. Friedel [in *The Physics of Metals*, J. M Ziman, Ed. (Cambridge Univ. Press, London, 1969)], W. A. Harrison [*Electronic Structure and the Properties of Solids* (Freeman, San Francisco, 1980), chapter 20, p. 476], and D. G. Pettifor [*Phys. Rev. Lett.* **42**, 846 (1979)].
12. R. C. Baetzold, *Solid State Commun.* **44**, 781 (1982).
13. K. Tatsumi, R. Hoffmann, A. Yamamoto, J. K. Stille, *Bull. Chem. Soc. Jpn.* **54**, 1857 (1981); A. C. Balazs, K. H. Johnson, G. M. Whitesides, *Inorg. Chem.* **21**, 2162 (1982).
14. E. Shustorovich, *J. Phys. Chem.* **87**, 14 (1983).
15. R. C. Baetzold, *ibid.*, p. 3858; *J. Am. Chem. Soc.* **105**, 4271 (1983); *Phys. Rev. B* **29**, 4211 (1984).
16. J.-Y. Saillard and R. Hoffmann, *J. Am. Chem. Soc.* **106**, 2006 (1984).
17. D. L. Adams and L. H. Germer, *Surf. Sci.* **27**, 21 (1971); F. Bonczek, T. Engel, E. Bauer, *ibid.* **97**, 595 (1980); K. Christmann and G. Ertl, *ibid.* **60**, 365 (1976); Z. T. Stott and M. P. Huges, *ibid.* **126**, 455 (1983).
18. W. Erley, *ibid.* **94**, 281 (1980).
19. See, for example, R. Smoluchowski [*Phys. Rev.* **60**, 661 (1941)] and A. Modinos [*Surf. Sci.* **75**, 327 (1978)].
20. E. Shustorovich, *Solid State Commun.* **38**, 493 (1981); *J. Phys. Chem.* **86**, 3114 (1982); _____ and R. C. Baetzold, *Appl. Surf. Sci.* **11/12**, 693 (1982).
21. P. J. Feibelman and D. R. Hamann, *Phys. Rev. Lett.* **52**, 61 (1984).
22. P. H. Citrin, G. K. Wertheim, Y. Bayer, *Phys. Rev. Lett.* **41**, 1425 (1978); P. Heimann, J. F. van der Veen, D. E. Eastman, *Solid State Commun.* **38**, 595 (1981); J. F. van der Veen, F. J. Himpsel, D. E. Eastman, *Phys. Rev. Lett.* **44**, 189 (1980); T. M. Duc, G. Guillot, Y. Lassailly, Y. Lecante, Y. Jugnet, J. C. Vedrina, *ibid.* **43**, 789 (1979); G. Treglia, M. C. Desjonquères, D. Spanjaard, Y. Lassailly, C. Guillot, Y. Jagnet, T. M. Duc, J. Lecante, *J. Phys. C* **14**, 3463 (1981).

23. J. A. Appelbaum and D. R. Hamann, *Solid State Commun.* **27**, 881 (1978); P. J. Feibelman and D. R. Hamann, *ibid.* **31**, 413 (1979); M. C. Desjonquères, D. Spanjaard, Y. Lassailly, C. Guillot, *ibid.* **34**, 807 (1980); A. Rosengren and B. Johansson, *Phys. Rev. B* **22**, 3706 (1980).

24. P. H. Citrin and G. K. Wertheim, *Phys. Rev. B* **27**, 3176 (1983).

25. G. Apai, R. C. Baetzold, E. Shustorovich, R. Jaeger, *Surf. Sci. Lett.* **116**, L191 (1982); R. C. Baetzold, G. Apai, E. Shustorovich, R. Jaeger, *Phys. Rev. B* **26**, 4022 (1982); G. Apai, R. C. Baetzold, P. J. Jupiter, A. J. Viescas, I. Lindau, *Surf. Sci.* **134**, 122 (1983); R. C. Baetzold, G. Apai, E. Shustorovich, *Appl. Surf. Sci.* **19**, 135 (1984).

26. J. R. Smith, F. J. Arlinghaus, J. G. Gay, *Phys. Rev. B* **26**, 1071 (1982).

27. J. H. Rose, J. R. Smith, J. Ferrante, *ibid.* **28**, 1835 (1983); J. H. Rose, J. R. Smith, F. Guinea, J. Ferranete, *ibid.* **29**, 2963 (1984).

28. D. Spanjaard and M. C. Desjonquères, *ibid.* **30**, 4822 (1984).

29. E. Shustorovich, *J. Am. Chem. Soc.* **106**, 6479 (1984).

30. R. C. Baetzold, *Surf. Sci.* **150**, 193 (1985).

31. E. L. Muetterties, E. Shustorovich, R. C. Baetzold, in preparation.

32. G. Ertl, M. Neumann, K. M. Streit, *Surf. Sci.* **64**, 393 (1977); P. R. Norton, J. W. Goodale, E. B. Selkirk, *ibid.* **83**, 189 (1979).

33. B. Poelsema, L. K. Verheij, G. Comsa, *Phys. Rev. Lett.* **49**, 1731 (1982).

34. E. Shustorovich, *Surf. Sci.* **150**, L115 (1985).

35. G. Ertl, S. B. Lee, M. Weiss, *ibid.* **114**, 515 (1982); J. L. Gland, B. A. Sexton, G. B. Fisher, *J. Catal.* **95**, 587 (1980); C. T. Campbell, G. Ertl, H. Kuipers, J. Segner, *Surf. Sci.* **107**, 220 (1981); J. Lee, R. J. Madix, J. E. Schlaegel, D. J. Auerbach, *ibid.* **143**, 626 (1984).

36. Most recently, we have extended our bond-order-conservation model to treat various coverage and coadsorption (promoting and positioning) effects, in good agreement with experiment (E. Shustorovich, *Surf. Sci.* **163**, L645 (1985).

37. R. J. Madix, personal communication.

38. We express our deep gratitude to the late Earl L. Muetterties, whose unique intuition often guided our choice of problems.

Part IV
Chemical Catalysis

This will be an account in historical perspective of the development of part of the field of chemistry that I have been active in for most of my professional life, the field that is loosely described by the phrase "electron transfer in chemical reactions." In the short time available to me for the preparation of this paper, I can't hope to provide anything significant in the way of original thought. But I can add some detail to the historical record, especially on just how some of the contributions which my co-workers and I have made came about. This kind of information may have some human interest and may even have scientific interest of a kind which cannot easily be gathered from the scientific journals. For publication there, the course of discovery as it actually took place may be rewritten to invest it with a logic that it did not fully acquire until after the event.

Simple electron transfer is realized only in systems such as $Ne + Ne^+$. The physics already becomes more complicated when we move to $N_2 + N_2^+$, for example; and with the metal ion complexes which I shall deal with, where a typical reagent is $Ru(NH_3)_6^{2+}$, and where charge trapping by the solvent, as well as within the molecule, must be taken into account, the complexity is much greater. Still, a great deal of progress has been made by a productive interplay of experiment, qualitative ideas, and more sophisticated theory, involving many workers. Because of space limitations, I will be unable to trace all the ramifications of the field today, and will emphasize the earlier history of the subject, when some of the ideas basic to the field were being formulated. This choice of emphasis is justified because, by an accident of history, I was a graduate student at the University of

13

Electron Transfer Between Metal Complexes: Retrospective

Henry Taube

Science 226, 1028–1036 (30 November 1984). Copyright © 1984 by the Nobel Foundation. This chapter is the basis of the lecture delivered by Henry Taube in Stockholm on 8 December 1983 when he received the Nobel Prize in Chemistry.

California, Berkeley, about the time the first natal stirrings of the subject of this article occurred and at a place where these stirrings were most active. As a result, I may be in a unique position to deal knowledgeably and fairly with the early history of the subject. The emphasis on the early history is all the more justified because most of the topics touched on in this article, and also closely related topics, are brought up to date in a very recent volume of the series *Progress in Inorganic Chemistry* (1).

Chemical reactions are commonly classified into two categories: substitution or oxidation-reduction. The latter can always be viewed as involving electron transfer, though it is agreed that when we consider the mechanisms in solution, electron transfer is not as simple as it is in the $Ne + Ne^+$ case. Rearrangement of atoms always attends the changes in electron count at each center, and these must be allowed for. I will, however, simplify the subject by considering only processes of simple chemistry: those in which electron transfer leaves each of the reaction partners in a stable oxidation state. While substitution reactions can be discussed without concern for oxidation-reduction reactions, the reverse is not true. The changes that take place at each center when the electron count is changed are an essential part of the "electron transfer" process, and may be the dominating influence in fixing the rate of the reaction. Moreover, most of the early definitive experiments depended on exploiting the substitution characteristics of the reactants and of the products. Thus, the attention which will be devoted to the substitution properties of the metal ions is not a digression but is an integral part of the subject.

An appropriate place to begin this account is with the advent of artificial radioactivity. This enormously increased the scope of isotopic tracer methods applied to chemistry, and made it possible to measure the rates of a large number of oxidation reduction reactions such as:

$$*Fe^{2+}(aq) + Fe^{3+}(aq) =$$
$$*Fe^{3+}(aq) + Fe^{2+}(aq) \qquad (1)$$

[The first demonstration of a redox exchange was made by von Hevesy and co-workers (2), who used naturally occurring isotopes to follow Pb(IV)/Pb(II) exchange in acetic acid.] Because chemists there were involved in the discovery of many of the new isotopes (3), an early interest in this kind of possibility developed in the chemistry community at the University of California, Berkeley, and was already evident when I was a graduate student there (1937–1940). Mention is made in a review article by Seaborg (3), devoted to artificial radioactivity, of an attempt (4) to measure the rate of the $Fe^{3+/2+}$ exchange in aqueous chloride media, the result of this early attempt being that the exchange was found to be complete by the time the separation of Fe(III) from Fe(II) was made. It was appreciated by many that the separation procedure, in this case extraction of the Fe(III)-Cl^- complex into ether, might have induced exchange. It was also appreciated that Cl^- might have affected the reaction rate, possibly increasing it, and that quite different results might be obtained were the experiment done with Cl^- being replaced by an indifferent anion.

There were several reasons for the interest, among many physical-inorganic chemists, in a reaction such as 1. That the interest in chemical applications of the new isotopes was keen in Berkeley may be traced in part to the involvement of much of the research body in teaching in the introductory chemistry course. We all had a background of qualitative observations on oxidation-reduction reactions of simple chemistry—as an example, on the reaction of Ce(IV)(aq) with Fe^{2+}(aq)—from experience in qualitative or quantitative analysis. Still, to my knowledge, at the time I was a graduate student, not a single measurement had been made of the rate of this kind of reaction. That a field of research, which has since grown enormously, was started by studying "self-exchange reactions" (5) such as 1, rather than net chemical changes (descriptor "cross-reaction") (5), may reflect the intervention of a human factor. Measuring the rate of a virtual process such as 1, today made commonplace for many systems by the introduction of new spectroscopic methods, seemed more glamorous than measuring the rate of oxidation of V^{2+}(aq) by Fe^{3+}(aq), for example. But I also recall from informal discussions that it was felt that driving force would affect the rate of reaction, and thus there would be special interest in determining the rates for reactions for which $\Delta G°$ (except for the entropic contributions to the driving force) is zero.

The interest in the measurement of the rates of self-exchange reactions which I witnessed as a graduate student is not reflected in the literature of the years immediately following. Many of those who might have had plans to do the experiments were engaged in war-related activities. After the war, at least five different studies on the rate of reaction 1, all carried out in noncomplexing media, were reported, with conflicting results, some indicating a half-life for exchange on the order of days at concentration levels of $10^{-2}M$. The discrepancies led to considerable controversy, and in informal discussions, strong opinions were

expressed on just what the true rate of self-exchange might be. The basis for this kind of judgment, exercised in the absence of any body of quantitative measurements, is worth thinking about. I believe it reflected an intuitive feeling that there would be a relation between the rates of self-exchange and of the related cross-reactions, and of course each of us had at least some qualitative information on redox rates for the Fe^{3+}/Fe^{2+} couples. The definitive measurements on the rate of reaction 1 in noncomplexing media were made by Dodson (6). These measurements were soon extended (7) to reveal the effect of $[H^+]$ and of complexing anions on the rate and yielded rate functions such as $[Fe^{3+}][X^-][Fe^{2+}]$ (because substitution is rapid compared to electron transfer, this is kinetically equivalent to $[FeX^{2+}][Fe^{2+}]$ and to $[Fe^{3+}][FeX^+]$), in addition to $[Fe^{3+}][Fe^{2+}]$, none specifying a unique structure for an activated complex. Particularly the terms involving the anions provided scope for speculation about mechanism. The coefficient for the simple second-order function was found (7) to be $4M^{-1}$ sec^{-1} at 25°C, $\mu = 0.5$, and those who had argued for "fast" exchange won out.

Another important experimental advance during the same period, important for several reasons, was the measurement (8) of the rate of self-exchange for $Coen_3^{3+/2+}$ ($k = 5.2 \times 10^{-5}M^{-1}$ sec^{-1} at 25°C, $\mu = 0.98$). This is, I believe, the first quantitative measurement of a rate for a self-exchange reaction and it may also be the first time that the oxidizing capability of a cobalt(III) ammine complex was deliberately exploited. In the article of (8), the rate of the reaction of $Co(NH_3)_6^{3+}$ with $Coen_3^{2+}$ is also reported; this is, I believe, the first deliberate measurement of the rate of an electron transfer cross-reaction. In contrast to the $Fe^{3+/2+}$ system, where both reactants are labile to substitution, $Coen_3^{3+}$ is very slow to undergo substitution and thus an important feature of the structure of the activated complex for the $Coen_3^{3+/2+}$ self-exchange appeared to be settled. In considering the observations, the tacit assumption was made that the coordination sphere of $Coen_3^{3+}$ does not open up on the time scale of electron transfer, so that it was concluded that the activated complex for the reaction does not involve interpenetration of the coordination spheres of the two reactants. We were thus obliged to think about a mechanism for electron transfer through two separate coordination spheres (descriptor "outer-sphere" mechanism) (9).

In 1951, an important symposium on electron transfer processes was held at the University of Notre Dame, and the proceedings are reported in the *Journal of Physical Chemistry*, volume 56 (1952). Though the meeting was organized mainly for the benefit of the chemists, the organizers had the perspicacity to include physicists in the program. Thus, the gamut of interests was covered, ranging from electron transfer in the gas phase in the simplest kind of systems, for example $Ne + Ne^+$, to the kind of system that the chemist ordinarily deals with. Much of the program was devoted to experimental work, the chemistry segment of which included reports on the rates of self-exchange reactions as well as of reactions involving net chemical change—but none on cross-reactions. Two papers devoted to theory merit special mention: that by Holstein (10), whose contributions to basic physics are now being applied in the chemistry community, and the paper by Libby (11), in which he stressed the relevance of the Franck-Condon restriction (12) to the electron transfer process, and applied the principle in a qualitative way to some observations. It is clear from the discussion which several of the papers evoked that many of the participants had already appreciated the point which Libby made in his talk. Thus, in the course of the discussion, the slowness of the self-exchange in the cobaltammines was attributed (13) to the large change in the Co-N distances with change in oxidation state, then believed to be much larger than it actually is. During the meeting, too, the distinction between outer- and inner-sphere activated complexes was drawn, and the suggestion was made that the role of Cl^- in affecting the rate of the $Fe^{3+/2+}$ self-exchange might be that it bridges the two metal centers (14). But, of course, because of the lability of the high-spin Fe(II) and Fe(III) complexes, no unique specification of the geometry of the activated complex can be made on the basis of the rate laws alone. During the discussion of Libby's paper, a third kind of mechanism was proposed (15), involving "hydrogen atom" transfer from reductant to oxidant.

Preparation

My own interest in basic aspects of electron transfer between metal complexes became active only after I came to the University of Chicago in 1946. During my time at Cornell University (1941–1946), I had been engaged in the study of oxidation-reduction reactions, and I was attempting to develop criteria to distinguish between $1e^-$ and $2e^-$ redox changes and, as an outgrowth of this interest, using $1e^-$ reducing agents to generate atomic halogen, X, and studying the ensuing chain reactions of X_2 with organic molecules. My eventual involvement in research on electron transfer between metal complexes owes much to the fact that I knew many of the protagonists personally (A. C. Wahl, C. N. Rice, C. D. Coryell, C. S. Garner; the first two were fellow graduate students at Berkeley), and to the fact that I had W. F. Libby, with whom I had many provocative discussions, and J. Franck, as well as F. H. Westheimer, as colleagues at the University of Chicago. By the time of the meeting at the University of Notre Dame (which I did not participate in), I appreciated in a general way the special advantages which the $Cr^{3+/2+}$ couple offered in the investigation of the mechanism of electron transfer, and outlined my ideas to N. Davidson when he visited me en route to the meeting, but the experimental work which led to the first two papers (16, 17) was not done until 1953. In the interim, I had failed to interest any of my graduate students in the work, because, of course, no one foresaw what it might lead to, and because it seemed less exciting than other work in progress in my laboratories, much of it concerned with isotopic effects, tracer and kinetic. The bulk of the work reported in the two papers just cited was done by my own hands; I shall now outline the background for those early experiments.

My interest in coordination chemistry did not develop until I elected it as a topic for an advanced course given soon after coming to the University of Chicago. Instead of using the standard textbook material, I used as major source the relevant volume of the reference series by Gmelin in which the chemistry of the cobaltammines is described. At this time I already had a good background in the literature devoted to substitution at carbon and understood the issues raised in that context, and I soon became interested in raising the same issues for substitu-

tion at metal centers. Furthermore, it was evident that the complexes based on metal ions which undergo substitution slowly were readily amenable to experimental study. I became curious as well about the reasons underlying the enormous differences in rates of substitution for metal ions of the same charge and (approximately) the same radii. The ideas that resulted were presented in my course the next time it was given, but the extensive literature study that led to the paper published in February 1952 (18) was not done until 1949, when I was on leave from the University of Chicago as a Guggenheim fellow.

In this paper, a correlation with electronic structure was made of observations, mainly qualitative ("labile" complexes arbitrarily defined as those whose reactions appear to be complete on mixing and "inert" as those for which continuing reaction can be observed), for complexes of coordination number 6. To make the correlation, it was necessary to break away from the practice which was common in the United States of classifying complexes as "ionic" and "covalent" according to electronic structure. Thus, for example, the comparison of the affinities of Cr(III) and Fe(III) (the latter high-spin and thus "ionic") convinced me that in the Fe(III) complexes, the bonds to the ligands might actually be somewhat more covalent than in the Cr(III) complexes. Furthermore, it appeared to me that in earlier discussions of relative rates of substitution, where attempts were made to understand the observations in terms of the existing classification, there was a failure to distinguish between thermodynamic stability and inertia, the latter being understood as referring solely to rate. The affinity of Cl^- for Cr^{3+}(aq) is considerably less than it is for Fe^{3+}(aq), yet the aquation rate of $CrCl^{2+}$(aq) is much less than it is for $FeCl^{2+}$(aq). Rates, of course, cannot be accounted for by considering ground-state properties alone, but the stability of the activated complex relative to the ground state must be taken into account. When the effect of electronic structure on the relative stabilities is allowed for, a general correlation of rates with electronic structure emerges. [In the language of ligand field theory, for complexes of coordination number 6, substitution tends to be slow when the metal ions have each of the πd (nonbonding) orbitals, but none of the σd (antibonding) orbitals, occupied. The specific rates of exchange of water between solvent and the hexaaquo ions of V^{3+} (πd^2), Cr^{3+} (πd^3), and Fe^{3+} ($\pi d^3 \sigma d^2$) are 1×10^3, 5×10^{-6}, and $\sim 1 \times 10^3$ sec^{-1}, respectively (19).]

A shortcoming of this early effort is that such rationalizations of the correlation as were offered were given in the language of the valence bond approach to chemical binding because, at the time I wrote the paper, I did not understand the principles of ligand field theory even in a qualitative way. The valence bond approach provides no simple rationale of the difference in rates of substitution for labile complexes, and these have been found to cover a very wide range, thanks to a pioneering study by Bjerrum and Poulsen (20), in which they used methanol as a solvent to make possible measurements at low temperature, and those of Eigen (21), in which he introduced relaxation methods to determine the rates of complex formation for labile systems.

Activated complexes have compositions and structures, and it is necessary to know what these are if rates are to be understood. It is hardly likely that these features can be established for the activated complexes if they are not even known for the reactants, and in 1950 we were not certain of the formula for any aquo cations in water. It seemed to me important, therefore, to try to determine the hydration numbers for aquo cations. Hydration number as I use it here does not mean the average number of water molecules affected by a metal ion as this is manifested in some property such as mobility but has a structural connotation: how many water molecules occupy the first sphere of coordination? Because the rates of substitution for Cr^{3+}(aq) were known to be slow (22), J. P. Hunt and I undertook to determine the formula for Cr^{3+}(aq) in water (23), with some confidence that we would be successful even with the slow method we applied, isotopic dilution using ^{18}O-enriched water. That the formula turned out to be $Cr(H_2O)_6^{3+}$ was no great surprise—although I must confess that at one point in our studies, before we had taken proper account of isotopic fractionation effects, $Cr(H_2O)_7^{3+}$ was indicated and, faced with the apparent necessity, I was quite prepared to give up my preconceived notions. It was also no great surprise that the exchange is slow ($t_{1/2}$ at 25°C, ~40 hours). Even so, the experiments were worth doing. They were the first of their kind, and they attracted the attention also of physical chemists, many of whom were astonished that aquo complexes could be as kinetically stable as our measurements indicated, and were impressed by the enormous difference in the residence time of a water molecule in contact with a cation compared with water molecules just outside. That we dealt with hydration in terms of detailed structure rather than in terms of averaged effects may have encouraged the introduction into the field of other methods, such as NMR, to make the distinction between cation-bound and free water (19). As I will now detail, it also led directly to the experiments described in the papers of (16) and (17).

The Inner-Sphere Activated Complex

R. A. Plane undertook to measure the rate of self-exchange for $Cr(H_2O)_6^{3+}$-$Cr^{2+}(aq)$ by using $Cr^{2+}(aq)$ as a catalyst for the exchange of water between $Cr(H_2O)_6^{3+}$ and solvent. The expected catalysis was found, but owing to our inexperience in handling the air-sensitive catalyst, the data were too irreproducible to lead to a value for the self-exchange rate. Catalysis on electron transfer was expected because the aquo complex of $Cr^{2+}(aq)$ was known to be much more labile than $Cr(H_2O)_6^{3+}$—the lability is now known (24) to decrease by a factor of at least 10^{14} when $Cr^{2+}(aq)$ is oxidized to $Cr^{3+}(aq)$ (note that Cr^{2+}, but not Cr^{3+}, has an antibonding electron). It occurred to me in the course of Plane's work that it would be worthwhile to test the potential of the Cr(III)/Cr(II) couple for diagnosis of mechanism with a nonmetal oxidant. Following up on the idea, I did a simple test tube experiment, adding solid I_2 to a solution of $Cr^{2+}(aq)$ which Plane had prepared for his own experiments. I observed that reaction occurs on mixing, that the product solution is green, and that the green color fades slowly, to produce a color characteristic of $Cr(H_2O)_6^{3+}$. The fading is important because it demonstrates that $(H_2O)_5CrI^{2+}$, which is responsible for the green color, is unstable with respect to $Cr(H_2O)_6^{3+} + I^-$, and thus we could conclude that the chromium-iodine bond is established before Cr(II) is oxidized.

The principle having been demonstrated with a nonmetal oxidant, I turned to the problem of finding a suitable metal complex as oxidant. What was needed was a reducible robust metal complex, having as ligand a potential bridging group, and the idea of using $(NH_3)_5CoCl^{2+}$ surfaced during a discussion of possibilities with another of my then graduate students, R. L. Rich. Because at that time virtually nothing was known about the rates of reduction of Co(III) ammines and because they were not thought about as useful oxidants, I was by no means sanguine about the outcome of the first experiment, which again was done in a test tube. I was delighted by the outcome. Reaction was observed to be rapid [the specific rate has since been measured (25) as 6×10^5 $M^{-1} \sec^{-1}$ at 25°C] and the green color of the product solution indicated that $(H_2O)_5CrCl^{2+}$ is formed. Further work (16, 17) showed that this species is formed quantitatively, and that in being formed it picks up almost no radioactivity when labeled Cl^- is present in the reaction solution, thus demonstrating that transfer is direct, that is, Cl^- bridges the two metal centers, and this occurs before Cr^{2+} is oxidized.

$$(NH_3)_5CoCl^{2+} + Cr(H_2O)_6^{2+} \rightleftarrows$$
$$(NH_3)_5Co^{III}Cl \ldots Cr^{II}(H_2O)_5 \rightarrow$$
$$\text{precursor complex}$$

$$[(NH_3)_5Co \ldots Cl \ldots Cr(H_2O)_5]^{4+} \rightarrow$$
$$\text{activated complex}$$
$$(NH_3)_5Co^{II} \ldots ClCr^{III}(H_2O)_5 \rightarrow$$
$$\text{successor complex}$$

$$Co(H_2O)_6^{2+} + NH_4^+ + (H_2O)_5CrCl^{2+}$$

These early results were presented at a Gordon conference on inorganic chemistry, which was held only a short time

183

after they had been obtained, and they were received with much interest. E. L. King was present at the meeting, and together we planned the experiment in which self-exchange by an "atom" transfer mechanism was first demonstrated (26).

$$(H_2O)_5*CrCl^{2+} + Cr^{2+}(aq) =$$

$$*Cr^{2+}(aq) + (H_2O)_5CrCl^{2+} \quad (2)$$

Note that in a process of this kind, the bridging group will remain bound to a Cr(III) center, but the auxiliary ligands about the original Cr(III) center will be exchanged or replaced because of the high lability of Cr(II).

[Several years later, King and co-workers (27) extended the self-exchange work to include other halides as bridging ligands, and still later (28) encountered the first example of "double bridging."

$$cis\text{-}[(H_2O)_4*Cr(N_3)_2]^+ + Cr^{2+}(aq) =$$

$$*Cr^{2+}(aq) + cis\text{-}[(H_2O)_4Cr(N_3)_2]^+$$

A good case can be made for this also being the first demonstration of "remote attack," that is, inner-sphere electron transfer where more than one atom separates the metal atoms in the activated complex.]

There is a brief hiatus in my work after the early experiments on inner-sphere mechanisms, caused by my taking leave from the University of Chicago in 1956. But before leaving, I hurriedly did some experiments (29) which, though semi-quantitative at best, showed that not only atoms but groups such as N_3^-, NCS^-, and carboxylates transfer to chromium when the corresponding pentaamminecobalt(III) complexes are reduced by $Cr^{2+}(aq)$, and that there are large differences in rate for different dicarboxylate complexes (maleate much more rapid than succinate). Moreover, it

was shown that Cr(II) in being oxidized can incorporate other ligands such as $H_2P_2O_7^{2-}$ which are present in solution. In this paper the possibility of electron transfer through an extended bond system of a bridging group was first raised, but was by no means demonstrated by the results. While I was away, Ogard (30) began his studies on the rates of aquation of $(NH_3)_5CrX^{2+}$ (X = Cl, Br, I) catalyzed by Cr^{2+}. By following the arguments made in connection with reaction 2, it can be seen that if an inner-sphere mechanism operates, $(H_2O)_5CrX^{2+}$ and NH_4^+ (acid solution) will be products. The contrast with uncatalyzed aquation is worth noting, where $(NH_3)_5CrH_2O^{3+}$ and X^- are the expected products.

General Progress

The hiatus referred to above provides me with a suitable opportunity to outline some of the advances that were being or were soon to be made on other fronts. An important one is that the rates of numerous self-exchange reactions were being measured. Here I want especially to mention the contributions from the laboratories of A. C. Wahl and C. S. Garner. Some of the experiments by Wahl and co-workers made use of rapid mixing techniques—see, for example, the measurement (31) of the rate of self-exchange for $MnO_4^{1-/2-}$. An important experimental contribution was made by N. Sutin, who introduced the rapid flow method (32) (and later other rapid reaction techniques) into this field of study, and who helped others, including myself, to get started with the rapid flow technique. A spate of activity on the measurements of the rates of cross-reactions followed, motivated in large part by the

184

desire to test the validity of the cross-reaction correlation (5).

At the quantitative theoretical level, attempts were being made to account for the barrier to reaction attending encounter and separation and that contributed by charge trapping within the metal complex and by the surrounding medium. The papers which most influenced the experimentalists, at least during the formative period in question, are those by Marcus (33) and Hush (34), dealing with adiabatic (35) electron transfer. Other theoretical approaches were being advanced during this period and, in some, attempts were made to account for non-adiabaticity, and the various theories are compared and evaluated in (33). [The current state of theory can be gathered from a recent article by Sutin (36).] Because this very important aspect of electron transfer reactions is not dealt with in this paper, it is essential to mention that the processes as they occur at electrodes were not being overlooked.

The correlation of the rates of cross-reactions with the rates of the component self-exchange reactions (5) has been widely applied, especially to outer-sphere reactions. The limits of its validity were clearly set down by the author: allowance must be made for the work of bringing the reaction partners together and separating the products, electron delocalization must be great enough to ensure adiabatic behavior, but not so great as materially to reduce the activation energy. (The last condition, it should be noted, does not necessarily limit the applicability of the Marcus equation to outer-sphere processes.) Hush's treatment (34) also leads to a correlation of rates of self-exchange and cross-reactions. It also takes into account the contribution by driving force,

and in fact the first calculation of the rate of a cross-reaction, in this instance

$$Pu^{3+}(aq) + PuO_2^{2+}(aq) =$$
$$Pu^{4+}(aq) + PuO_2^{+}(aq)$$

from those of the self-exchange processes and the equilibrium constant appears in (34).

Theory of another kind has profoundly influenced the development of the field, even though it is qualitative. It responds to the question of how the choice of mechanism, and relative rates, are to be understood in terms of the electronic structures of the reactants. As is true also of rates of substitution, the observations are so sensitive to electronic structure that even qualitative ideas are useful in correlating observations and in pointing the way to new experiments. Orgel (37) early applied qualitative ligand field theory in discussing the inner-sphere mechanisms for the reduction of Cr(III) and low-spin Co(III) complexes. When electron transfer takes place through a bridging group, it is important to distinguish a chemical or "hopping" mechanism—here a low-lying orbital of the ligand is populated by the reductant, or a hole is generated by the oxidant in an occupied orbital—from resonance transfer—that is, electron tunneling through the barrier separating the two metal centers. This distinction was drawn rather early by George and Griffith (38), who moreover proposed alternative mechanisms for resonance transfer. Shortly thereafter, Halpern and Orgel (39) gave a more formal treatment of resonance transfer through bridging ligands. Concerns about the relation between electronic structure and the observations on electron transfer strongly influenced my own work, but before tracing this theme I want to report on the progress made,

185

mainly by others, in extending the descriptive chemistry of electron transfer reactions.

The unambiguous demonstration of an inner-sphere mechanism in a sense completed a second dimension in the field of electron transfer mechanisms. That in certain systems reaction perforce took place by an outer-sphere mechanism had long been known, but until the experiments of (16) and (17) were done the inner-sphere mechanism was only conjecture and, as frequently happens in research in chemistry, only after conjecture, however reasonable, is upgraded by proof is it accepted as a base for further development. That the distinction between the two reaction classes is meaningful, not only in terms of chemistry but also in rates, will be illustrated by a single comparison: the specific rates of reaction of $Cr^{2+}(aq)$ with $(NH_3)_5CoCl^{2+}$ is $\sim 10^8$ greater than it is with $Co(NH_3)_6^{3+}$ (40) (the latter can react only by an outer-sphere mechanism). The classification of reaction paths as inner sphere or outer sphere, on the basis of rate comparisons, involving effects (such as those exerted by nonbridging ligands) established with reactions of known mechanism, became the focus of experiment and discussion when direct proof based on product or intermediate identification was lacking.

Early in the 1960's, new metal centers were added to the roster of those proved to react by inner-sphere mechanisms. For $Co(CN)_5^{3-}$ as reducing agent (41), the demonstration of an inner-sphere path again depended on the characterization of product complexes by orthodox means. Sutin and co-workers have been particularly resourceful in using flow techniques to provide direct proof of mechanism for oxidizing centers ordinarily considered as labile to substitution: thus note the proof of inner-sphere mechanism for $FeCl^{2+}(aq) + Cr^{2+}(aq)$ (42), $CoCl^{2+}(aq) + Fe^{2+}(aq)$ (43), even the much studied $Fe^{3+/2+}$ exchange (44). It was early appreciated (17) that atom or group transfer is not a necessary concomitant of an inner-sphere process. Whether the bridging group transfers to reductant, remains with the oxidant, or transfers from reductant to oxidant depends on the substitution labilities of reactants and products. Early qualitative observations (17) on the $Cr^{2+}(aq) + IrCl_6^{2-}$ system had apparently exposed an example of reaction by an inner-sphere mechanism, but leading to no net transfer of the bridging atom. Here $(H_2O)_5CrClIrCl_5$ is formed as an intermediate, but this then aquates to $Cr(H_2O)_6^{3+} + IrCl_6^{3-}$ [later work (45) has shown the inner-sphere path to be minor compared to the outer-sphere, and that $Cr(H_2O)_6^{3+}$ is the lesser product of the former path]. Experiments (46, 47) with $V(H_2O)_6^{2+}$ as reducing agent provided numerous examples of systems in which substitution on the reducing complex is rate-determining for the net redox process [note that $V(H_2O)_6^{2+}$ because of its electronic structure is expected (18) to undergo substitution relatively slowly]. Unstable forms of linkage isomers were prepared by taking advantage of the chemistry of the inner-sphere mechanism: $(H_2O)_5CrSCN^{2+}$ (48) by the reaction of $Cr^{2+}(aq)$ with $FeNCS^{2+}$ (49); $(H_2O)_5CrNC^{2+}$ (50) by the reaction of $Cr^{2+}(aq)$ with $(NH_3)_5CoCN^{2+}$. Oxygen atom transfer was shown to be complete (51) in the reaction of $(NH_3)_5CoOH_2^{3+}$ with $Cr^{2+}(aq)$. [The path involving direct attack on the aquo complex was later (52) shown to be unobservable compared to attack on the hydroxo. Bridging by

H₂O has to date not been demonstrated.] The inner-sphere path was demonstrated (53) also for net $2e^-$ processes in an elegant series of studies involving the reactions of Pt(IV) complexes with those of Pt(II). In an important departure, Anet (54) showed that the "capture" property of Cr^{2+}(aq) in being oxidized can be exploited to produce complexes in which an organic radical is ligated to Cr(III). An entire chapter of the volume of (1) is devoted to the chemistry of similar organochromium complexes (55).

Electronic Structure and Mechanism

A major theme in my own research, on returning from leave, was to try to understand the large differences in rate, noted qualitatively in my own early work (29) and later made more quantitative, for the reactions Cr^{2+}(aq) with carboxylate complexes of $(NH_3)_5Co(III)$. Many of the ligands were dicarboxylic acids, and to explain the observation that when a conjugated bond system connects the two carboxylates, reaction is usually more rapid than it is for the saturated analog, it seemed reasonable then to assume that in the case of conjugated ligands the reducing agent attacks the exo carboxyl (remote attack), the conjugated bond system serving as a "conduit" for electron transfer. In adopting this view the tacit assumption was made that the reactions are nonadiabatic so that the extent of electronic coupling would be reflected in the rate. In retrospect, this assumption is naïve because the effect of conjugation would be exerted even if the reducing agent attacked at the endo carboxyl.

A false start was made in demonstrating remote attack defined as above. Acti-

vation effects accompanying electron transfer were reported (56) which, if true, would have constituted proof of remote attack for these systems. These effects could not be reproduced (57) in later work (I had by now moved from University of Chicago to Stanford University). Remote attack for the large organic ligands was finally demonstrated (58) in the reaction of

$$(NH_3)_5CoN \left(\bigcirc \right) - \underset{\underset{NH_2}{|}}{C} = O^{3+}$$

with Cr^{2+}(aq). This work, which also yielded a measurement of the rate of reaction of Cr^{2+}(aq) with the analogous Cr(III) species, provided the clue to understanding, at least in a qualitative way, the rate differences observed for different conjugated ligands. The astonishing result was that the rate of reduction of the Co(III) complex is only about tenfold greater than that of the Cr(III) complex. When the bridging group is a nonreducible species such as acetate, the ratio is $>10^4$. The insensitivity of rate to the nature of the oxidant suggested that the electron does not transfer directly from Cr(II) to the oxidizing center but that the mechanism rather involves the $1e^-$ reduction (59) of the ligand by the strong reducing agent Cr^{2+}, followed by reduction of the oxidizing center by the organic radical—that is, a "hopping" mechanism obtains. This view provided satisfactory rationalizations of most of the observations of rates made with organic ligands. For example, the fact that the rate is considerably greater for $HO_2C-CH=CH-CO_2^-$ (fumarate) than for $CH_3CO_2^-$ as ligand on Co(III) may have little to do with the opportunity that Cr(II) has to attack the remote carboxyl in the former case, and only reflects the

187

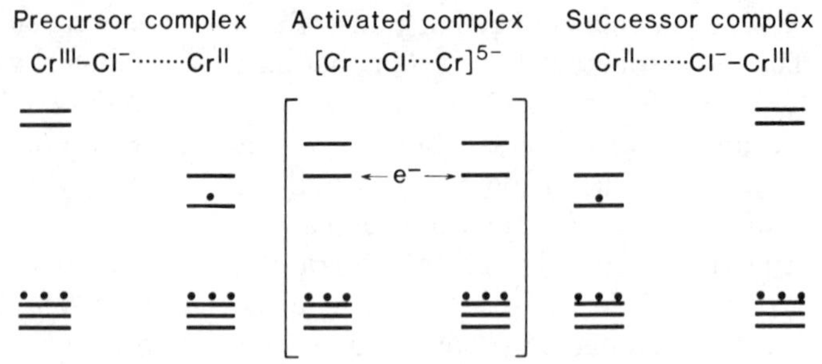

Precursor complex $Cr^{III}-Cl^-\cdots\cdots Cr^{II}$ Activated complex $[Cr\cdots Cl\cdots Cr]^{5-}$ Successor complex $Cr^{II}\cdots\cdots Cl^--Cr^{III}$

Fig. 1. Electronic structure and "atom" transfer in the self-exchange reaction: $(H_2O)_5CrCl^{2+} + Cr(H_2O)_6^{2+}$.

fact that fumarate can be reduced by Cr^{2-}(aq). Moreover, the otherwise puzzling observation that the rate for the fumarate complex is increased (60) by H^- is now easily understood; positive charge added to the ligand makes it easier to reduce. That many reactions of the class under consideration proceed by a stepwise mechanism has been convincingly demonstrated and amply illustrated in subsequent work, most of it done by Gould and co-workers (61).

The rationalization offered for the operation of the stepwise mechanism in the Co(III)-Cr(II) systems is that the carrier orbital on the ligand has π symmetry, while the donor and acceptor orbitals have σ symmetry. This renders as highly improbable an event in which the four conditions: Franck-Condon restrictions at each center, and the symmetry restrictions at each center, are simultaneously met. Whether or not this is the correct explanation, it led me to search for an oxidizing center of the ammine class in which the acceptor orbital has π symmetry.

When the important condition that the complexes undergo substitution slowly was added, only one couple within the entire periodic table, $Ru^{3+/2+}$ ($\pi d^5/\pi d^6$) qualified (62). In principle, the $Os^{3+/2+}$ couple is also a candidate, but unless some strong π acid ligands are present,

the couple is too strongly reducing to be useful. The ruthenium species had the added advantage that much more in the way of preparative work was known (63), and further that the redox potentials are close to those of the much studied cobalt couples. Since the π orbital on Ru(III) can overlap with the π^* orbital on the ligand, we expected that the "hopping" mechanism would no longer obtain. Reaction with Cr^{2+}(aq) is in fact much more rapid (2×10^4) than it is in the case of the Co(III) isonicotinamide complex, and moreover, the rates are now quite sensitive to changes in the redox potential of the oxidizing center (64). The chemistry also differs in an interesting way from the Co(III)-Cr(II) case. The bond between Ru and the ligand is not severed when Ru(III) is reduced to Ru(II), and a kinetically stable binuclear intermediate is formed, as is expected (18), from the electronic structures of the products $-\pi d^6$ for Ru(II) and πd^3 for Cr(III).

Though the main intent of this article is to provide historical background rather than to develop the subject itself in detail, because the reaction properties are so sensitive to electronic structure it may be appropriate in concluding this section to illustrate the connection with a few examples. Effects arising from differences in electronic structure are manifested in several different ways: by affecting the rates of substitution, they can affect the choice of mechanism and, for an inner-sphere reaction path, can determine whether binuclear intermediates are easily observable, and whether there is net transfer of a group from one center to another; even after the precursor complex is assembled, orbital symmetry can affect the mechanism itself, as in the example offered in an earlier paragraph, and can profoundly affect the rate of conversion of the precursor to successor complex.

The Cr(III)/Cr(II) ($\pi d^3/\pi d^3 \sigma d^1$) and Ru(III)/Ru(II) ($\pi d^5/\pi d^6$) couples offer perhaps the greatest contrasts in behavior. It should be noted that the σ electron in Cr(II) is antibonding, thus accounting for the tetragonal distortion in Cr(II) complexes and their enormous lability compared to those of Cr(III). By contrast, the complexes of both oxidation states of ruthenium undergo substitution slowly, with the useful exception of water as a ligand, where the residence time on Ru(II) is a fraction of a second.

The reducing agent Cr(II) shows preference for an inner-sphere mechanism, and this is especially marked if the acceptor orbital has σ symmetry. There is a great economy of motion for electron transfer by an inner-sphere path for the σ donor–σ acceptor cases which arises from the reciprocity at the two centers of the events which are required in overcoming the inner-sphere Franck-Condon barrier. This point is illustrated for the $(H_2O)_5 *CrCl^{2+} + Cr(H_2O)_6{}^{2+}$ self-exchange reaction in Fig. 1, where the electronic levels are shown for the precursor complex, for the activated complex, and for the successor complex. Motion of the bridging Cl^- from Cr(III) to Cr(II) lowers the energy of an accep-

tor orbital on Cr(III), and at the same time raises that of the donor orbital on Cr(II), and although other nuclear motions are also required, there is some correlation of the events required for activation to electron transfer, a correlation which is absent in the case of an outer-sphere mechanism. The high substitution lability of Cr(II) of course means that the precursor complex can be formed rapidly.

The comparison of the rates of self-exchange for $Cr(H_2O)_6{}^{3+}/Cr(H_2O)_6{}^{2+}$ versus $(NH_3)_5RuH_2O^{3+}/(NH_3)_5RuH_2O^{2+}$ and how these respond when the higher oxidation state for each couple is converted to the hydroxo complex is quite instructive. The upper limit for the specific rate of self-exchange for the Cr(III)/Cr(II) couple is $2 \times 10^{-5} M^{-1}$ sec^{-1} (65, 66); although it has not been directly measured for the ruthenium couple, the specific rate can reasonably be taken to be close to that (67) for $Ru(NH_3)_6{}^{3+/2+}$, namely $\sim 1 \times 10^3 M^{-1}$ sec^{-1}. Because the redox change for the ruthenium couple involves a π electron, it causes only a small change in the dimensions of the complex (68). Thus the Franck-Condon barrier to electron transfer arising from inner-sphere reorganization is small, and facile transfer by an outer-sphere mechanism is observed. By contrast, because Cr(III) has no antibonding electrons and the electron added on reduction is antibonding, there is a large change in dimensions and shape attending the reduction, and the slowness of the self-exchange can be attributed in part at least to the inner-sphere barrier (69). Why the water molecule, which as a ligand on the oxidizing agent still has available an electron pair for sharing with the reducing agent, is such a poor bridging group remains to be understood.

On deprotonation of a water molecule in each oxidant the inner-sphere path for the Cr(III)/Cr(II) system opens up, and a marked increase in rate is observed (65) ($k = 0.66M^{-1}$ sec^{-1} at 25°C)—the increase may be as large as a factor (66) of 10^9. The self-exchange rate for $(NH_3)_5RuOH^{2+} + (NH_3)_5RuOH_2^{2+}$ has not been measured, but it can be asserted with confidence that the rate by either an inner-sphere path or an orthodox outer-sphere path will be much less than it is for the aquo couple. The rate by the inner-sphere path will be limited by the rate of bridge formation, and thus will be no greater than $1 \times 10^{-2}M^{-1}$ sec^{-1} [neutral ligands in substituting on $(NH_3)_5RuOH_2^{2+}$ show specific rates (70) of the order of $0.1M^{-1}$ sec^{-1}]. The orthodox outer-sphere path now has a composition barrier, as well as a Franck-Condon barrier to overcome [K_{eq} for the production of $(NH_3)_5RuOH^+$ and $(NH_3)_5RuOH_2^{2+}$ from $(NH_3)_5RuOH^{2+}$ and $(NH_3)_5RuOH_2^{2+}$ is $\sim 10^{-9}$]. Reaction by "hydrogen atom" transfer (15) is a reasonable possibility, that is, electron transfer concomitant with proton transfer from the Ru(II) complex to the Ru(III), and some evidence in support of this kind of mechanism has been advanced to explain observations (71) made in the oxidation by Fe^{3+} and $FeOH^{2+}$ of $(NH_3)_5RuOH_2^{2+}$. Reaction by such a path might be quite facile and a specific rate in excess of $1M^{-1}$ sec^{-1} would be strong evidence in its favor.

Applications of the Ru(III)/Ru(II) Couple

Some unexpected benefits have accrued from introducing Ru(III)/Ru(II) ammine couples into this field of research. Ruthenium(II) engages in back bonding interactions to a degree unprecedented among the dipositive ions of the first transition series. A discovery (72) which forcefully brought this message home is that $(NH_3)_5RuN_2^{2+}$ is formed in aqueous solution by the direct reaction of $(NH_3)_5RuOH_2^{2+}$ with N_2. When the heteroligand in $(NH_3)_5RuL^{2+}$ is pyridine or a derivative, the complexes of both oxidation states are slow to undergo substitution, and by changing the number of π acid ligands, a versatile series of outer-sphere redox couples is made available, spanning a range in redox potential of over 1 volt. When derivatives of the Os(III)/Os(II) ammines are included, the useful range in aqueous solution is extended by approximately 0.5 volt. These reagents are finding wide application in research on redox processes.

Intramolecular electron transfer. It has occurred to many that in trying to arrive at a basic understanding of electron transfer processes, it would be a great advantage if the reactions could be studied in an intramolecular mode rather than, as is commonly done, in the biomolecular (or intermolecular) mode, particularly if the geometric relation between the two metal centers were unambiguously defined. Such systems had been encountered in studies of "induced" electron transfer (73); when a powerful $1e^-$ oxidant acts on

for example, the resulting coordinated organic radical can undergo intramolecular electron transfer, the oxidation of the ligand to the carboxylic acid being completed by Co(III). In some cases, intramolecular electron transfer can be intercepted by reaction of the radical with the external oxidant, but, at best, only relative rates were obtained for these systems. In an elaboration of this kind of approach, in which pulse radiolysis is used to convert the organic ligand to a radical—usually by reduction—intramolecular transfer rates can be measured (74). These results are important in their own right, but they do not substitute for experiments in which metal-to-metal transfer is studied.

A strategy for dealing with the metal-to-metal case was devised (75), which depends on the special properties of the Co(III)/Co(II) and Ru(III)/Ru(II) couples. The principle is the following. When a molecule (76) such as

$$(NH_3)_5Co^{III}N\bigcirc-\bigcirc NRu^{III}(NH_3)_5$$

which has both metal centers in the oxidized state, is treated with an external reducing agent, Ru(III) is reduced more rapidly than Co(III). This is a direct result of the differences in electronic structure, πd^6 and πd^5 for Co(III) and Ru(III), respectively, the former requiring much more in the way of reorganization energy because the incoming electron is antibonding. In a subsequent step, Ru(II) reduces Co(III) by an intramolecular process, at least if the solution of the binuclear complex is sufficiently dilute.

The first method (75) devised to produce the [III,III] molecule is rather ingenious, but it involves many steps, and it has the disadvantage that SO_4^{2-} rather than NH_3 is *trans* to the pyridine on Ru(III). Schäffer (77) has greatly simplified the preparative procedure by taking advantage of chemistry developed by Sargeson and co-workers (78) and has studied intramolecular electron transfer for

$$(NH_3)_5Co^{III}NC-\bigcirc-CNRu^{II}(NH_3)_5$$

and for the related molecules with the *ortho* and *meta* isomers as the bridging ligands. Quite independently of our work, Gaswick and Haim (79) have done experiments similar to those described, but with $Fe(CN)_5H_2O^{3-}$ as the reducing agent. Substitution on $Fe(CN)_5H_2O^{3-}$ takes place readily so that the simple mixing procedure attempted by Roberts (75a) often can be used with this particular kind of reducing agent.

A point of interest in all of these studies is to learn how the rate of reaction responds to changes in the structure of the bridging group. In the bipyridine case, the coupling between the pyridine rings has been modified (76); since the immediate environment about each metal is left unaltered, the driving force for the reaction is but little affected, and changes in rate can then be attributed to changes in electronic coupling. The results obtained in studies of this kind are outlined in a recent review article by Haim (80). Here I will only mention an extension of this kind of strategy to a system of biochemical interest. Gray and co-workers (81) and Isied et al. (82) have succeeded in placing $(NH_3)_5Ru$ on cytochrome c at a position remote from the porphyrin [Ru(III)–Fe(III) separation, 15 Å]. Different pulse methods were used by the two groups to reduce Ru(III) preferentially over Fe(III), and though the results of the two studies differ somewhat (20 ± 5 sec^{-1} and 82 ± 20 sec^{-1}, respectively), it is clear that the general strategy is effective.

Robust mixed valence molecules. The resurgence of interest (83) in the properties of mixed valence compounds can be traced to review articles (84–86) which

191

appeared in 1967, and to the first deliberate synthesis of a robust mixed valence molecule, the species shown below, which is commonly referred to as the Creutz-Taube (87) ion.

$$[(NH_3)_5RuN\bigcirc NRu(NH_3)_5]^{5+} \quad I$$

[Quite independently of our work, Cowan and Kaufman (88) prepared a molecule based on the ferricinium/ferrocene couple.] Peter Ford and I first produced the Creutz-Taube ion in 1967. In undertaking its preparation, we were motivated by simple curiosity rather than by questions which might arise from a deep understanding of the issues raised by the properties of the mixed valence compounds. The fully reduced ([II,II]) state is readily prepared by direct substitution, using pyrazine and $(NH_3)_5RuOH_2^{2+}$, and in undertaking the project we were taking advantage of our knowledge of affinities and rates of substitution for both oxidation states of ruthenium. Complexes with Ru(II) attached to heterocyclic nitrogen show very strong absorption in the visible region of the spectrum ($\pi^* \leftarrow \pi d$) (89) and on observing that the quality of the color was not significantly altered when the [II,II] species is half oxidized, we did not pursue the matter further. Fortunately, Carol Creutz took up the subject again. The electrochemical results which she obtained about June 1968 showed that the mixed valence state is very stable relative to the isovalent, and this suggested to us that electronic coupling in the mixed valence species is strong. By now, the review papers by Hush (85) and Robin and Day (86) had appeared, and taking their content to heart we felt certain that an intervalence band must exist, which Carol Creutz then located in the near infrared region ($\lambda_{max} = 1570$ nm) where it does not affect the color (heretofore this region of the spectrum had been little investigated by chemists). Intervalence absorption corresponds to using the energy of a photon to transfer an electron from the reduced to the oxidized metal center, subject to the Franck-Condon restriction. Intervalence absorption confers on Prussian blue its characteristic blue color. The intervalence absorption is at a longer wavelength for species I than for Prussian blue because the two iron sites in the latter are not substitutionally equivalent. This leads to a ground-state energy difference which is then added to that associated with the Franck-Condon barrier when the process is induced by a quantum of light.

One of my main interests in the field of mixed valence molecules has been to explore and to try to understand the energetics of the systems. I will illustrate by a single example the kind of conclusion which we have reached in pursuing these interests and where we have relied on theory introduced into the field by Hush (85), Mulliken (90), and, for the correlation of extent of electron delocalization with electronic structure, Mayoh and Day (91), and choose for illustration the localized mixed valence molecule (92)

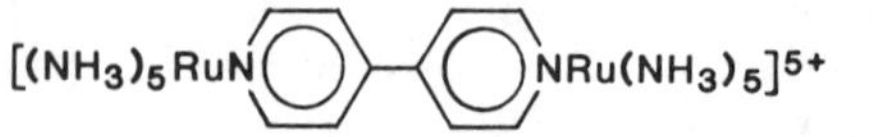

$$[(NH_3)_5RuN\bigcirc-\bigcirc NRu(NH_3)_5]^{5+}$$

The stability conferred on the ground state of the molecule by charge delocalization is only of the order of 50 cal, far below the upper limit of 5×10^2 cal set by the electrochemical results, which measure the total stabilization of the mixed valent compared to the isovalent state. When the nuclear coordinates about each center are adjusted so that the Franck-Condon condition is met, the energy separating the bonding and antibonding states which arise from electron delocalization is calculated as 2.2 kcal (*93*). This is taken to be sufficient to ensure adiabatic transfer (*94*), in agreement with the conclusion reached in the course of studying intramolecular electron transfer in Co(III)-Ru(II) systems with 4,4'-bipyridine and related molecules as bridging ligands (*76*). If electron transfer is assumed to be adiabatic, the specific rate for intramolecular electron transfer is calculated as $3 \times 10^8 \, \text{sec}^{-1}$, in reasonable agreement with an estimate ($1.6 \times 10^8 \, \text{sec}^{-1}$) reached from measurements of intermolecular electron transfer rates for pyrinedinepentaamineruthenium species (*95*).

Mixed valence molecules have been prepared (*96*) which are delocalized even though the bridging group is so large that direct metal-to-metal orbital overlap cannot be responsible for the delocalization. These have remarkably interesting properties in their own right and are the subject of current studies (*97*).

Concluding Remarks

In this article I have focused rather narrowly on electron transfer reactions between metal complexes. The separation of this subclass from other possible ones which can be assembled from the reactant categories: metal complexes, organic molecules (*98*), molecules derived from other nonmetallic elements, any of the above in excited states (*99, 100*), electrodes, is not totally arbitrary, as it might be were it dictated solely by limitations of space. Admittedly, all the possible electron transfer processes are governed by the same principles, at least when these are stated in a general enough way. But as these principles manifest themselves in the different subclasses, the descriptive chemistry can be quite different, and these differences are the fabric of chemistry. The subclass which has been treated has perhaps been the most thoroughly studied, yet, as the article by Sutin shows (*36*), our understanding at a basic level is far from complete.

Even for the much studied Fe^{3+}/Fe^{2+} self-exchange reaction, which served to introduce this subject, the important question of whether the reaction is adiabatic or not has not been settled to everyone's satisfaction. Still, a great deal of progress has been made. The descriptive matter has increased enormously since 1940, and our understanding of it, in both scope and depth, has more than kept pace with observations. A great deal of progress has also been made in many of the other subclasses—for example, in the study of electrode processes and in "atom transfer" reactions (as a specific case, the use of transition metal species to carry the oxidizing capacity of O_2 to a substrate such as an organic molecule). Both are of the greatest importance in industrial applications, and the latter also in reaching an understanding of the chemistry of living cells. Because the subclasses are interrelated, progress in one speeds progress in another.

References and Notes

1. S. J. Lippard, Ed., *Progress in Inorganic Chemistry* (Wiley, New York, 1983), vol. 30.
2. G. von Hevesy and L. Zechmeister, *Ber* **53**, 410 (1920).
3. G. T. Seaborg, *Chem. Rev.* **27**, 199 (1940).
4. J. Kennedy, S. Ruben, G. T. Seaborg, *ibid.*, p. 256.
5. This descriptor and that following were introduced by R. A. Marcus [*Faraday Soc. Discuss.*

29, 21 (1960); *J. Phys. Chem.* **67**, 853 (1963)] in proposing an equation which correlates the specific rate for a "cross reaction" such as

$$Fe^{3+} + V^{2+} = Fe^{2+} + V^{3+}$$

with the specific rates of the component "self-exchange" reactions $Fe^{3+/2+}$ and $V^{3+/2+}$. Allowance is made in the equation for the effect of driving force on the rate.

6. R. W. Dodson, *J. Am. Chem. Soc.* **72**, 3315 (1950).
7. J. Silverman and R. W. Dodson, *J. Phys. Chem.* **56**, 846 (1952).
8. W. B. Lewis, C. D. Coryell, S. W. Irvine, Jr., *J. Chem. Soc.* (1949), p. S386.
9. The descriptor "inner sphere" is used for a reaction in which oxidant and reductant metal centers are linked through primary bonds to a bridging group. In the early work it was assumed that the bridging group would play a special role in the electron transfer reaction. In more recent work on intramolecular transfer (vida infra), examples have been found of systems in which the bridging group seems to serve only to hold the two partners together. "Inner sphere" or "outer sphere"?
10. T. Holstein, *J. Phys. Chem.* **56**, 832 (1956).
11. W. F. Libby, *ibid.*, p. 863.
12. As a historical note, I wish to add that James Franck was very much interested in electron transfer in chemical reactions and fully appreciated the importance of keeping the Franck-Condon restriction in mind in trying to understand the observations.
13. H. C. Brown, discussion of the paper of Libby (*11*, p. 868).
14. W. F. Libby, discussion following his paper (*11*, p. 866). See also H. C. Brown, discussion following paper of Silverman and Dodson (*7*, p. 852).
15. R. W. Dodson and N. Davidson, discussion following the paper of Libby (*11*, p. 866).
16. H. Taube, H. Myers, R. L. Rich, *J. Am. Chem. Soc.* **75**, 4118 (1953).
17. H. Taube and H. Myers, *ibid.* **26**, 2103 (1954).
18. H. Taube, *Chem. Rev.* **50**, 69 (1952).
19. The subject of ionic hydration is brought up to date in an article by J. P. Hunt and H. L. Friedman, in (*1*), p. 359.
20. J. Bjerrum and K. Poulsen, *Nature (London)* **169**, 463 (1952).
21. M. Eigen, *Faraday Soc. Discuss. No. 17* (1954), p. 1.
22. J. Bjerrum, *Z. Phys. Chem.* **59**, 336 (1907); *ibid.*, p. 581; *Z. Anorg. Allg. Chem.* **119**, 179 (1921).
23. J. P. Hunt and H. Taube, *J. Chem. Phys.* **18**, 757 (1950); *ibid.* **19**, 602 (1951).
24. C. W. Meredith, *USAEC Rep. UCRL-11704* (1965), Berkeley, Calif. Work done under supervision of R. E. Connick.
25. J. P. Candlin and J. Halpern, *Inorg. Chem.* **4**, 766 (1965).
26. H. Taube and E. L. King, *J. Am. Chem. Soc.* **76**, 4053 (1954).
27. D. L. Ball and E. L. King, *ibid.* **80**, 1091 (1958).
28. R. Snellgrove and E. L. King, *ibid.* **84**, 8609 (1962).
29. H. Taube, *ibid.* **77**, 4481 (1955).
30. A. E. Ogard and H. Taube, *ibid.* **80**, 1084 (1958).
31. L. Gjertsen and A. C. Wahl, *ibid.* **81**, 1572 (1959); J. C. Sheppard and A. C. Wahl, *ibid.* **79**, 1020 (1957).
32. N. Sutin and B. M. Gordon, *ibid.* **83**, 70 (1961).
33. R. A. Marcus, *Annu. Rev. Phys. Chem.* **15**, 155 (1964); also see earlier papers.
34. N. S. Hush, *Trans. Faraday Soc.* **57**, 557 (1961).
35. Consider the process:

$$A \cdot B^- \to [A \cdot B^-] \to A^- \cdot B$$
$$\quad I \qquad\qquad II$$

State I is the precursor complex; in state II the energy is independent of whether the electron is on atom A or B—that is, the Franck-Condon condition has been met. In adiabatic transfer, electron delocalization is great enough so that whenever state II is reached, electron transfer can take place, and the rate of the chemical reaction is determined solely by the rate at which state II is reached. In nonadiabatic transfer, the system passes through state II a number of times before electron transfer occurs, and both the Franck-Condon barrier and the frequency of electron jump in state II are rate-determining.

36. N. Sutin, in (*1*), p. 441.
37. L. E. Orgel, *Rept. Xe Cons. Inst. Int. Chem. Solvay* (1956), p. 289.
38. P. George and J. Griffith, *Enzymes* **1**, 347 (1959).
39. J. Halpern and L. E. Orgel, *Discuss. Faraday Soc.* **29**, 32 (1960).
40. A. M. Zwickel and H. Taube, *J. Am. Chem. Soc.* **81**, 2915 (1959).
41. J. Halpern and S. Nakamura, *ibid.* **87**, 3002 (1965).
42. G. Dulz and N. Sutin, *ibid.* **86**, 829 (1964).
43. T. J. Connochioli, G. H. Nancolles, N. Sutin, *ibid.*, p. 1453.
44. _____, *ibid.*, p. 459.
45. A. G. Sykes and R. N. V. Thorneley, *J. Chem. Soc.* (1970), p. A232.
46. J. H. Espenson, *J. Am. Chem. Soc.* **89**, 1276 (1967).
47. H. J. Price and H. Taube, *Inorg. Chem.* **7**, 1 (1968).
48. A. Haim and N. Sutin, *J. Am. Chem. Soc.* **87**, 4210 (1965).
49. S. Fronaeus and R. Larsson, *Acta Chem. Scand.* **16**, 1447 (1962).
50. J. H. Espenson and J. P. Birk, *J. Am. Chem. Soc.* **87**, 3280 (1965); *ibid.* **90**, 1153 (1968).
51. W. Kruse and H. Taube, *ibid.* **82**, 526 (1960).
52. D. L. Toppen and R. G. Linck, *Inorg. Chem.* **10**, 2635 (1971).
53. F. Basolo, M. L. Morris, R. G. Pearson, *Discuss. Faraday Soc.* **29**, 80 (1960).
54. F. A. L. Anet and E. Leblanc, *J. Am. Chem. Soc.* **79**, 2649 (1957).
55. J. H. Espenson, in (*1*), p. 189.
56. Last in the series: R. T. M. Fraser and H. Taube, *J. Am. Chem. Soc.* **83**, 2239 (1961).
57. I owe my associates during my first years at Stanford an enormous debt of gratitude for helping to set the record straight. Special thanks are due to E. S. Gould, who first uncovered discrepancies, and to J. K. Hurst, who repeated much of the dubious earlier work.

58. F. R. Nordmeyer and H. Taube, *J. Am. Chem. Soc.* **88**, 4295 (1966); *ibid.* **90**, 1162 (1968).
59. The relation between reducibility of the ligands and their effectiveness in mediating electron transfer was developed in an earlier paper [E. S. Gould and H. Taube, *ibid.* **86**, 1318 (1964)].
60. D. K. Sebera and H. Taube, *ibid.* **83**, 1785 (1961).
61. See, for example, E. S. Gould, *ibid.* **94**, 4360 (1972).
62. J. F. Endicott and H. Taube, *ibid.* **84**, 4989 (1962); *ibid.* **86**, 1686 (1964); *Inorg. Chem.* **4**, 437 (1965).
63. K. Gleu and K. Breuel, *Z. Anorg. Allg. Chem.* **237**, 335 (1938).
64. R. Gaunder and H. Taube, *Inorg. Chem.* **9**, 2627 (1970).
65. A. Anderson and N. A. Bonner, *J. Am. Chem. Soc.* **76**, 3826 (1954).
66. Indirect evidence, which is quite convincing, suggests the outer-sphere self-exchange rate for $Cr(H_2O)_6^{3+}/Cr(H_2O)_6^{2-}$ to be $\sim 5 \times 10^{-10}$ M^{-1} sec^{-1} [W. S. Melvin and A. Haim, *Inorg. Chem.* **16**, 2016 (1977)].
67. T. J. Meyer and H. Taube, *ibid.* **7**, 2369 (1968).
68. H. D. Stynes and J. A. Ibers, *ibid.* **10**, 2304 (1971).
69. Calculations of the barrier associated with inner-sphere electron reorganization leave room for a nonadiabaticity factor of a few orders of magnitude. See J. F. Endicott, K. Krishan, T. Ramasami, F. P. Rotzinger, in (*1*), p. 141.
70. S. S. Isied and H. Taube, *Inorg. Chem.* **15**, 3070 (1976).
71. T. J. Meyer and H. Taube, *ibid.* **7**, 2361 (1968).
72. D. E. Harrison and H. Taube, *J. Am. Chem. Soc.* **89**, 5706 (1967).
73. R. Robson and H. Taube, *ibid.*, p. 6487; J. French and H. Taube, *ibid.* **91**, 6951 (1969); earliest example: P. Saffir, *ibid.* **82**, 13 (1960).
74. M. Z. Hoffman and M. Simic, *ibid.* **94**, 1957 (1972).
75. S. S. Isied, *ibid.* **95**, 8198 (1973).
75a. In an earlier effort, Kirk Roberts tried the simple procedure of mixing the Co(III) complex with $(NH_3)_5RuOH_2^{2+}$. Substitution is too slow relative to intramolecular transfer for the method to work in these systems.
76. H. Fischer, G. M. Tom, H. Taube, *J. Am. Chem. Soc.* **98**, 5512 (1976).
77. L. Schäffer, work in progress.
78. N. E. Dixon, G. A. Lawrence, P. A. Lay, A. M. Sargeson, *Inorg. Chem.* **22**, 846 (1983).
79. D. G. Gaswick and A. Haim, *J. Am. Chem. Soc.* **96**, 7845 (1974).
80. A. Haim, in (*1*), p. 273.
81. J. R. Winkler, D. G. Nocera, K. M. Yocom, E. Bordignon, H. B. Gray, *J. Am. Chem. Soc.* **104**, 5798 (1982).
82. S. S. Isied, G. Worosila, S. J. Atherton, *ibid.*, p. 7659.
83. The level of current activity in the field can be gauged by the recent review of the subject of mixed valence molecules based on $\pi d^5/\pi d^6$ couples by C. Creutz [see (*1*), p. 1].
84. G. C. Allen and N. S. Hush, *Prog. Inorg. Chem.* **8**, 357 (1967).
85. N. S. Hush, *ibid.*, p. 391.
86. M. B. Robin and P. Day, *Adv. Inorg. Chem. Radiochem.* **10**, 247 (1967).
87. C. Creutz and H. Taube, *J. Am. Chem. Soc.* **91**, 3988 (1969); *ibid.* **95**, 1086 (1973).
88. D. O. Cowan and F. Kaufman, *ibid.* **92**, 219 (1970).
89. P. Ford, deF. P. Rudd, R. Gaunder, H. Taube, *ibid.* **90**, 1187 (1968).
90. R. S. Mulliken and W. B. Person, *Molecular Complexes* (Wiley, New York, 1969), chap. 2.
91. B. Mayoh and P. Day, *J. Am. Chem. Soc.* **94**, 2885 (1972); *Inorg. Chem.* **13**, 2273 (1974).
92. G. M. Tom, C. Creutz, H. Taube, *J. Am. Chem. Soc.* **96**, 7828 (1974).
93. J. E. Sutton, P. M. Sutton, H. Taube, *Inorg. Chem.* **18**, 1017 (1979); J. E. Sutton and H. Taube, *ibid.* **20**, 3125 (1981).
94. N. Sutin, in *Inorganic Biochemistry*, G. L. Eichhorn, Ed. (American Elsevier, New York, 1973), vol. 2, p. 611.
95. G. M. Brown, H. J. Krentzien, M. Abe, H. Taube, *Inorg. Chem.* **18**, 3374 (1979).
96. P. A. Lay, R. H. Magnuson, H. Taube, *J. Am. Chem. Soc.* **105**, 2507 (1983).
97. Spectroscopic studies by J. Ferguson and co-workers (Australia National University) are in progress.
98. R. A. Sheldon and J. Kochi, *Metal Catalyzed Oxidations of Organic Compounds* (Academic Press, New York, 1981).
99. P. Ford, D. Wink, J. Dibenedetto, in (*1*), p. 213.
100. T. J. Meyer, in (*1*), p. 389.
101. Many of the co-workers who have contributed to progress in the subject of this article are cited in the references, and this is an implicit acknowledgment of their contributions. Because this account is incomplete, others who have contributed directly to this work have not been cited, nor still others who have had interests that are not reflected in the account I have given. I am grateful to them all for the help they have given, and for what I have learned in the course of working with them. The nature of the contributions made by my co-workers is not evident either from the acknowledgment I have made, or from my expression of gratitude. I need to add that I have always relied on the independence of my co-workers and to a large extent my contribution to the effort has been that of maintaining continuity. I also wish to acknowledge financial support of my research by the agencies of the U.S. government, beginning with the Office of Naval Research in about 1950. Later, I derived partial support from the U.S. Atomic Energy Commission, and still later from the National Science Foundation and the National Institutes of Health (General Medical Sciences). The Petroleum Research Fund of the American Chemical Society has also been a source of research support.

Modern organometallic chemistry is the realm of the possible, in several senses. Out of the collection of molecules in Fig. 1, none was known 30 years ago. Over these years our knowledge of the decisive role of metal ions at the active sites of many enzymes, a function that has evolved over millennia, has increased tremendously. There is some connection, of course, between this natural metal-based chemistry and the multitude of new inorganic compounds. But the latter are, by and large, the handiwork of man. Not that they are necessarily made by the most rational means. Nevertheless, they are creatures of the laboratory, owing their existence to the flask and the dry box and often incapable of withstanding the rigors of an aqueous, oxygen-rich, biological or geochemical environment (1).

Organometallic complexes of the transition metals are also in the realm of the possible because there seems to be no limit to their geometrical and stereochemical beauty, derived here from simplicity and there from complexity. We could not have foretold that a hydrogen atom would find its way smack in the middle of an octahedron of ruthenium atoms (1) a structure utterly simple in its

1    $HRu_6(CO)_{18}^{-}$

Cartesian geometry (2). Nor would we have predicted, yet we cannot help but appreciate, the staggering variety of ways that a metal cluster, $Ru_3(CO)_{12}$, has of binding, dismembering, and reassembling acetylenes (3) (Fig. 2).

14

Theoretical Organometallic Chemistry

Roald Hoffmann

Science 211, 995–1002 (6 March 1981)

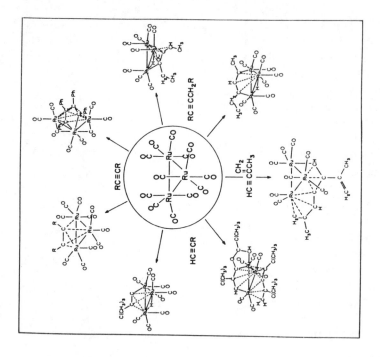

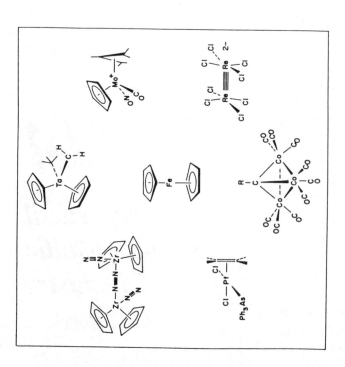

Fig. 1. Some typical organometallic complexes. Purists may carp at the inclusion of $Re_2Cl_8{}^{2-}$, but it is so pretty that, despite the lack of carbon, I like to think of it as an honorary organometallic molecule.   Fig. 2. Those products of the reaction of various acetylenes with $Ru_3(CO)_{12}$ whose structures are established. The structures shown by no means exhaust the total set of complex products of these reactions.

The admission that we lack predictive power in this field is a reaffirmation of the experimental nature of chemistry, and a challenge. The challenge is to use the apparatus of modern quantum mechanics, perforce in approximate form, to achieve some degree of understanding of the electronic structure of organometallic molecules.

Ligands and Orbitals

In the organometallic molecules in Fig. 1 we perceive one or more transition metal atoms and some attached groups. These groups are termed ligands. They span a range of complexity from simple atoms such as Cl^-, through diatomic molecules such as carbon monoxide or carbonyl (CO), nitric oxide, and molecular nitrogen, larger groupings such as phosphines (PR_3) and alkyls (CH_3), to still larger, more complicated, clearly organic moieties endowed with substantial stability on their own such as ethylene ($H_2C=CH_2$) or groupings that kinetically are not very stable at all such as methylene (CH_2) or allyl ($H_2C\cdots CH\cdots CH_2$).

All the ligands must possess an electronic arm by which they attach themselves to the metal. This is their basicity, their donor function, and it consists of one or more pairs of electrons used for bonding to the metal. To display the donor function explicitly, I introduce the convention of displaying the donor's electrons in pairs. A negative charge is thereby imposed on some of these donors: CO is neutral (2), but a methyl group is anionic (3), which makes it anal-

ogous to a phosphine (4). The quantum mechanical description of a donor or base is a doubly occupied orbital (indicated by the broad horizontal bar with two strokes through it, standing for the electron pair), at relatively high energy. The C_2H_4 molecule carries its donor function in a π orbital (5), whereas the ubiquitous cyclopentadienyl (6) has three such orbitals and so is counted as $C_5H_5^-$

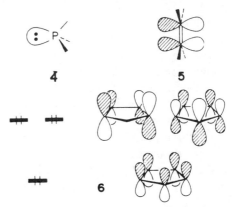

(Cp$^-$) (4). In 5 and 6, I introduce a graphic notation that will be used throughout this article: shading or lining of an orbital lobe implies the positive phase of the wave function, and no shading means the negative phase.

Some ligands, indeed the best ones, are also Lewis acids, carrying an acceptor function as well. In orbital language what makes for an acceptor is a relatively low-energy unfilled orbital. Those of methyl are too high in energy, but CO and C_2H_4 both have π^* orbitals that serve very well (7 and 8):

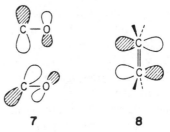

199

The ability to both donate electrons to the metal and accept other electrons back is the hallmark of the strongest bonding ligands. Still another determinant of a good ligand is its size, to be precise the size of its orbitals, relative to a typical metal atom. This is what makes the Cp unit, an aromatic system but nevertheless not very stable on its own, such a common ligand in organometallic compounds. Its diameter, 2 to 3 angstroms, matches very well the span of a typical transition metal d orbital.

The center of all the activity is the metal atom. The three transition metal series (Fig. 3) provide the same set of valence orbitals, albeit differing in principal quantum number. These frontier orbitals are the five nd ($3d$ or $4d$ or $5d$, depending on the transition series), one $(n + 1)s$, and three $(n + 1)p$. With these nine orbitals (9), the metal brings a variable

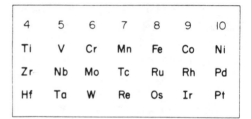

Fig. 3. That portion of the periodic table containing the transition elements. The number heading each column is the number of electrons available for bonding in the neutral form of the atom.

periodic table. Still greater freedom (and control) is provided by the potential of adjusting orbital size by changing the metal, its oxidation state, or switching from the first transition series to the second or third. No wonder that the variety of geometry, bonding strength, and function in organometallic complexes seems to be infinite.

Interaction Diagrams, Fragments, and a Molecular Orbital Theory of Bonding

Let us investigate a specific and typical organometallic molecule and determine what holds it together. The example is $Fe(CO)_4(C_2H_4)$, iron tetracarbonyl ethylene, an iron complex typical in that it contains several CO groups and an olefin ligand. It is a relatively unstable orange-yellow oil, first made around 1963 (5). The geometry of the molecule, known from the crystal structures of the parent molecule and numerous derivatives (6), is shown in **10**.

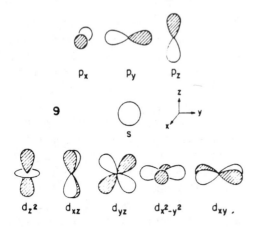

number of electrons, indicated by the numeral above each column in Fig. 3.

The versatility of the transition metals is now fully displayed. There is spatial flexibility—enough orbitals to provide bonding in just about any direction—and there is electronic tuning—the possibility of modifying the electronic count, sometimes drastically, by moving across the

10

It may be described alternatively as a trigonal bipyramid (two CO groups along the axis, two CO groups and the C_2H_4 carbons in the equatorial plane) or as an octahedron, the two C_2H_4 carbons approximately completing this common coordination polyhedron.

The C_2H_4 in this complex has its C–C bond stretched (~1.42 Å) (6), and its hydrogens somewhat pinned back. Yet it is clearly recognizable as the C_2H_4 unit that it was in the free state, and indeed it may be set free under suitable conditions. The fact that this transition metal complex contains a C_2H_4 ligand suggests a partitioning of the molecule into $Fe(CO)_4$ and C_2H_4. This is a theoretical fragmentation. It may not be the way the molecule is made, but we are not barred from constructing the compound in our minds, on paper, or in a computer in this way.

In order to examine how this molecule is made, we need to write down the orbitals of C_2H_4 and $Fe(CO)_4$ and then allow them to interact. The orbitals of C_2H_4 are well known (4): crucial to any bonding scheme are the frontier π and π^* levels of 11

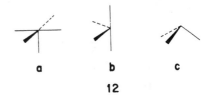

11

What about the orbitals of $Fe(CO)_4$, the inorganic moiety? This fragment, as well as other typical ML_n components of organometallics shown in Fig. 1, is made up of a metal atom and several ligands. The orbitals of such ML_n fragments are

very easy to construct, in either a molecular orbital or a valence bond way (7). Let us do so.

In the geometry it displays in the olefin complex, $Fe(CO)_4$ clearly looks like an octahedron with two *cis* ligands missing (12b). Most other popular fragments, for example, square pyramidal ML_5 (12a) and pyramidal ML_3 (12c), may also be re-

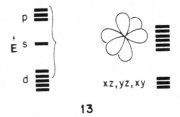

12

lated geometrically to an octahedron. So let us prepare a metal center for octahedral bonding. From its nine valence orbitals (9) one may form six equivalent hybrids pointing toward the vertices of an octahedron. These are d^2sp^3 hybrids, that is, they use up two d functions and all of the s and p orbitals (8). The hybrids are shown schematically in 13, along

$$p \; \underline{\underline{\underline{}}}$$
$$E \; s \; \underline{}$$
$$d \; \underline{\underline{\underline{}}}$$
$$xz, yz, xy \; \underline{\underline{}}$$

13

with the three d orbitals, d_{xy}, d_{xz}, and d_{yz}, that remain unhybridized.

Now suppose we need to know the orbitals of the ML_5 fragment (12a). The unspecified ligand L is so far characterized only by a directed orbital containing two electrons—it could be any of the Lewis bases (2 through 5). Five such ligands coming in along the octahedral directions interact with five of the six hybrids, as shown in Fig. 4.

"Interaction" is an important word in our quantum chemical vocabulary. It has the sense of forming from two initially isolated wave functions, orbitals, two new combinations. One is a bonding, in-phase combination, and the other is an antibonding, out-of-phase, noded mixture. The simplest manifestation of this most familiar phenomenon is the formation of H_2 (**14**), where the bonding combination is σ_g and the antibonding combination is σ_u.

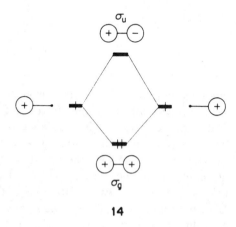

14

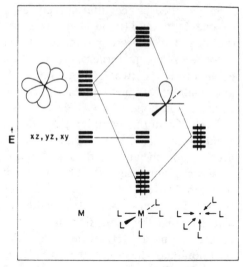

Fig. 4. Five ligands, Lewis bases, are at right. At left are the valence orbitals of a typical transition metal center prepared for octahedral bonding. The five ligands come in along five of the six octahedral directions. The resultant orbitals of ML_5 are shown in the center.

As shown in the center of Fig. 4, five metal-ligand bonding combinations with reduced energies carry the ten electrons that the five bases bring to the metal. Five metal-ligand antibonding combinations have increased energies. One hybrid orbital remains unused, along with the three unhybridized d functions. These four orbitals are the valence or frontier levels of the ML_5 fragment. They carry all the metal d electrons (precisely how many depends on the metal), and they will interact with any other ligands that come near.

The patterns of interaction for ML_4 and ML_3 are no different. Four or three bases come up, interacting with four or three of the octahedral hybrids of structure **13**, leaving two or three hybrid orbitals, respectively, untouched. The primitive picture, not a bad one at that,

of ML_5, ML_4, and ML_3 is then given in **15**. In each ML_n fragment there are three orbitals of lower energy, essentially xz, yz, and xy, and $6-n$ hybrids of higher energy, pointing toward the octahedral sites.

15

One additional bit of preparation is still needed before we proceed with our analysis of the attraction of $Fe(CO)_4$ for

202

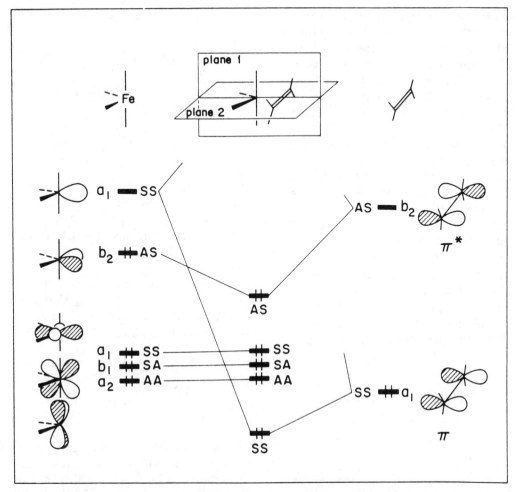

Fig. 5. Interaction diagram for $Fe(CO)_4(C_2H_4)$. The C_2H_4 is at right, the $Fe(CO)_4$ fragment at left. The orbitals are classified in two ways, according to their symmetry properties in the C_{2v} group of the complex or, equivalently, according to their reflection properties in the two indicated planes.

C_2H_4. The two higher-lying hybrids in **15b** are localized. They must be prepared for bonding by the formation of delocalized combinations. In the jargon of the trade, the orbitals need to be symmetry-adapted. In this particular case nothing more complicated than taking a sum and a difference of the localized orbitals is required, as in **16** (9).

Group theory, that marvelous mathematical edifice which combines elegance with labor saving, makes the work of constructing the interaction diagram of Fig. 5 easier (10). The orbitals are classified both in their appropriate C_{2v} symmetry and according to their reflection properties (S = symmetric, A = antisymmetric) in the two mirror planes. The metal fragment b_1 and a_2 orbitals do not have the correct symmetry to mix with either π or π^* of C_2H_4. There are two metal-based a_1 orbitals that are of the same symmetry as the C_2H_4 π. The upper orbital, $2a_1$, a hybrid of d, s, and p on the metal, is better disposed for interaction with the C_2H_4 π than the lower $1a_1$, primarily metal d. Thus one bonding interaction is between the π of C_2H_4 and $2a_1$, another between C_2H_4 π^* and Fe(CO)$_4$ b_2. The lower-energy component of each interaction is sketched in 17 and 18. All that remains is to place ten

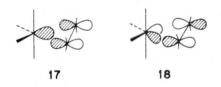

17   18

electrons in the final level scheme, two from the C_2H_4 π and eight from the iron atom. These electrons fill five orbitals, two of which are bonding (17, 18) and three of which are nonbonding. There is a substantial gap between filled and unfilled levels, a sign of kinetic and thermodynamic stability in organic and inorganic molecules.

The nature of the bonding in this organometallic complex differs in no substantial way from the bonding in any molecular aggregate. For instance, the H–Cl bond in hydrogen chloride forms by the interaction of a chlorine 3p orbital with a hydrogen 1s orbital, placing an electron pair in the bonding combination, as in 19.

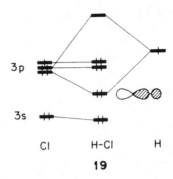

3p

3s

Cl   H–Cl   H

19

Two chlorine lone pairs remain unperturbed by this σ bonding. In Fe(CO)$_4$(C$_2$H$_4$), two bonds form between the metal fragment and the C_2H_4, and three lone pairs remain essentially unaffected.

A certain perceptual shading of the bonding picture of Fe(CO)$_4$(C$_2$H$_4$) is worth pursuing. Let us redraw the bonding combinations 17 and 18 in 20 and 21.

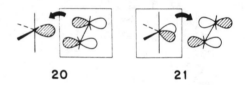

20   21

The box outlines signify the parentage of these orbitals. In our conceptual fragmentation of Fig. 5 the two electrons in 20 initially belonged entirely to the C_2H_4. The resultant bonding molecular orbital (20) still is occupied by two electrons, but now they are delocalized over both fragments. Thus some electron density has been shifted to the Fe(CO)$_4$ in the course of bond formation. Similarly the two electrons in 21 initially were assigned to the Fe(CO)$_4$ fragment; bonding has shifted some electron density in the reverse direction, from the metal to the olefin.

Forward and back donation have been part of the theoretical framework of inorganic chemistry from the early 1950's. The picture I have just described is the

Dewar-Chatt-Duncanson model of olefin bonding (*11*). In the 1950's several intellectual streams combined into a torrent—the renaissance of modern inorganic and organometallic chemistry. The Dewar-Chatt-Duncanson model was one of the streams. My choice of the other rivers includes the metallocenes, the rediscovery of crystal field theory, Ziegler-Natta catalysis, and three-dimensional crystal structure determination.

Geometries and Conformations

One may inquire whether the geometry we assumed for $Fe(CO)_4(C_2H_4)$, **10**, is preferred to one in which the C_2H_4 is twisted by 90° around the metal to the C_2H_4 midpoint axis, as in **22**. This is a

22

conformational question, a query regarding the equilibrium or most stable geometry of the complex. Once one has used the fragment decomposition to analyze the bonding in **10**, the approach to the alternative geometry is evident. One uses the same fragments and reconstructs **22**. There had better be a difference, and Fig. 6 shows what it is.

The C_2H_4 π interacts with $Fe(CO)_4$ a_1 (*10*) approximately to the same extent in the two conformations. A differential is provided by the C_2H_4 acceptor function, the π^* orbital, of b_2 symmetry (*10*) in **10**, b_1 in **22**. Orbital interactions are governed by the usual perturbation theoretical expression (*4, 12*)

$$\Delta E = \frac{|H_{ij}|^2}{E_i - E_j}$$

which says that the extent of interaction between two orbitals i and j is determined by a quotient. The numerator is the matrix element of the interaction squared. This matrix element H_{ij} is a function of the overlap between the two orbitals. If not directly proportional to the overlap, H_{ij} is at least large when the overlap of orbitals i and j is large and small when it is small. The denominator of the perturbation expression makes ΔE inversely proportional to the difference in energy between the levels so that the closer two energy levels are, the more strongly they will interact (*4, 12*). On both counts the C_2H_4 π^* interaction in the initial conformation **10** is better. The $Fe(CO)_4$ orbital that it mixes with is higher in energy, closer to π^* of C_2H_4. And the overlap is better; the b_2 hybrid of $Fe(CO)_4$ is nicely directed away from the metal, toward the C_2H_4, whereas b_1 is an unhybridized, mainly d orbital. The "backbonding" b_2 interaction in the conformation in which C_2H_4 lies in the equatorial plane is better. There is a substantial rotational barrier in these complexes, 10 to 15 kilocalories per mole (*13, 14*).

This brief introduction to one methodology of analyzing preferred geometries of molecules in no way exhausts the geometrical richness of inorganic stereochemistry. In addition to orientational degrees of freedom of coordinated groups, one has in complexes every type of isomerism—positional, geometrical, optical—in detail the equivalent of the same beautiful, inherently chemical phenomena in organic chemistry. But one has an added range of coordination number and geometry. Not just two, three, or

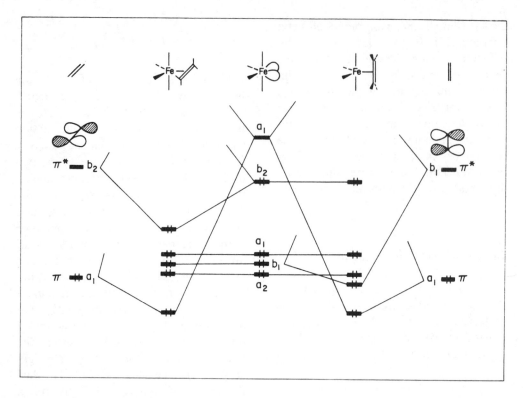

Fig. 6. Interaction diagram for two conformations of Fe(CO)$_4$ (C$_2$H$_4$), differing in the rotational disposition of the C$_2$H$_4$ ligand.

four atoms may be bound to a central one, but five, six, seven, and eight, or even more ligands are commonplace. Each coordination number has associated with it several polyhedral shapes. For instance, in 7-coordination, a typical environment for molybdenum, tungsten, and uranium, one finds geometries as seemingly diverse (but in fact very closely related to each other) as a pentagonal bipyramid (**23a**), a capped octahedron (**23b**), and two kinds of capped trigonal prisms (**23c** and **23d**). The stereochemical diversity of these structural types is fascinating (*15*).

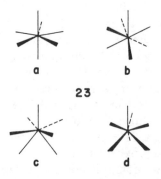

Fragments at Work

The fragment approach is one theoretical tool, one piece of interpretative ma-

206

chinery, that many investigators (16) have found useful in trying to make sense of the immense variety of complex structures that have poured forth from the cornucopia of the synthetic organometallic chemist. By way of example, let me trace the multitudinous uses to which my collaborators and I have put one such fragment, ML_3.

The orbitals of a C_{3v} ML_3 fragment were sketched in **15c**. There are three low-energy orbitals and three hybrids above. This fragment comes in complexes with from zero to ten electrons. Here are some of the things that can be done with it.

1) Once the lower three orbitals are occupied by six electrons, the fragment seeks three bases (six electrons) to interact with the upper three hybrids. These six electrons may come from the ligands, as in the restored octahedron Cr(CO)_6 (**24**) or from a six-electron organic donor, as in (arene)Cr(CO)_3 (**25**), or CpFe(CO)_2CH_3 (**26**) (here we have for-

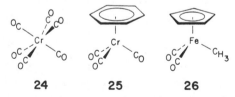

24	**25**	**26**

mally a Cp⁻ and a CH_3⁻, which makes the iron Fe^{2+}, d^6). Or some of the electrons may come from the metal and some from the ligand, as in the complexes **27** through **29**. The last compound is an example of the remarkable capability of transition metal fragments to stabilize unstable organic moieties (17).

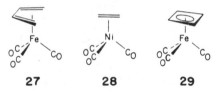

27	**28**	**29**

2) Conformational preferences and rotational barriers spanning a wide energy range may be explicated (18). Trimethylenemethane iron tricarbonyl prefers geometry **30b** over **30a** by nearly 20 kcal/mole (19).

The barrier to internal rotation in benzene chromiumtricarbonyl (**25**) is small. However, when a donor is substituted on the benzene ring, the equilibrium conformation is **31**, but, when an acceptor is substituted, it is **32**. In each case substantial barriers to internal rotation appear (18).

The preferred geometry of polyene-ML_3 complexes is **33**, one arm of the tricarbonyl under the "open" end of the polyene, unless the metal is electron-

33

deficient. Then the preferred conformation changes. Thus a Cr(CO)_3 orientation may be coupled with the norcaradiene-cycloheptatriene equilibrium, as in **34** (20).

3) A proton moving across the face of a cyclopentadienyl ring "sits down" directly above a carbon atom (cyclopentadiene, **35**) and moves about, with substantial activation energy, by a sequence of 1,5 sigmatropic shifts (*21*). So do a methyl and other group IV ER_3 groups (where E = Si, Ge, Sn, or Pb), the activation energy for the process falling as one descends group IV (*22*). In contrast, an $Mn(CO)_3^+$ group places itself directly over the center of the ring (**36**). So do CuL^+, BeR^+, Li^+, Tl^+, and In^+.

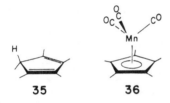

35   **36**

More complicated patterns of equilibrium placement and mobility occur with larger rings. Sometimes the barriers to rearrangement are substantial [22 kcal/mole in **37** (*23*)], whereas in not distantly related molecules they are much smaller [~8 kcal/mole in **38** (*24*)]. A fragment analysis will explain this trend (*25*).

CH₂
Fe(CO)₃

(CO)₃Fe —

37   **38**

Tours of ML_n groups across the face of larger organic ligands may be subject to substantial symmetry-imposed barriers (*26*). For instance, the rearrangement of naphthalene-$Cr(CO)_3$, in which the $Cr(CO)_3$ moves in a least-motion way from one ring to another (dashed line in **39**), should not be as easy as the roundabout excursion marked by the solid lines (*25*).

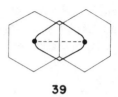

39

4) Two ML_3 groups brought up to each other make M_2L_6. For a d^3 electron count, these are the remarkable d^3 triple-bonded complexes (**40**) of Chisholm and Cotton (*27*). We think that with small ligands these might be eclipsed (*28*), but others do not agree (*29*).

M = Mo, W

L = NR_2, CH_2R, Cl

40

Next one can insert other atoms or molecules in the middle, in the plane marked in **41**. The insertion of three hydrides gives rise to **42**, a reasonably well-known structural type for d^6 and d^7 metal centers (M = Co, Fe, Ir, Re) (*30*). Incorporation of three CO groups produces the d^8 $Fe_2(CO)_9$ (**43**). If three chlorides are inserted, we have the remarkable d^3 series $M_2Cl_9^{3-}$ (M = Cr, Mo, W) (**44**), in which a striking range of metal-metal bonding is traversed as one scans down the group, from M = chromium where there is no metal-metal bond, six unpaired electrons, and an elongated bioctahedron, to M = tungsten, with a triple bond, diamagnetic properties, and a squashed bioctahedron (*30*).

41   **42**

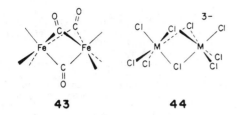

43   **44**

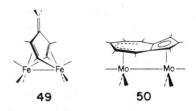

49   **50**

Sandwich molecules, with a metal atom between two organic fragments, exemplified by ferrocene, should appear on any flag of modern inorganic chemistry. It occurred to my colleagues and me that an organic ligand could be inserted in between the $M(CO)_3$ units as well, to give the inverse sandwich (**45**), stable for certain electron counts (*31*). Remarkably, several of these molecules have been made (*32*).

45

5) The M_2L_6 unit, bent back a little (**46**), serves as a building block for an

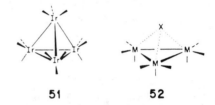

46

enormous variety of M_2L_6(ligand) complexes. The acetylene complex (**47**), the ferrole (**48**), the flyover bridge (**49**), and the azulene complex (**50**) are examples (*33*). Each molecule has intriguing conformational choices available to it.

6) Four $Ir(CO)_3$ fragments may be combined to give an inorganic tetrahedrane (**51**). A general theory of bonding in this and substantially more complicated clusters, developed by Wade [see (*16*)] and Mingos [see (*16*)], is one of the most important conceptual developments in modern inorganic theory. Three $M(CO)_3$ units may be combined to form an $M_3(CO)_9$ fragment (**52**) that can be

51   **52**

capped in many ways, by a chalcogen atom, by CR, by a Cp ring, or by another metal (*34*). An isoelectronic Co_3Cp_3 unit can split an acetylene and keep the two CR pieces thereof as caps (*35*); $Ru_3(CO)_{12}$ can dismember C_2H_4 in at least two distinct ways (**53**) (*36*). It seems

$$Ru_3(CO)_{12} \xrightarrow{C_2H_4}$$

53

likely that we will eventually find out how these extreme disruptions of bonding occur. For the moment, the theoreti-

209

cal analysis is limited to the geometrical preferences of the products (34).

An understanding of bonding and equilibrium structure has been the main subject of this discussion. Only the simplest reaction type, unimolecular rearrangement, has figured in the above examples. Yet chemistry is in large part molecules in motion, it is chemical reactivity. Some progress has been made in the analysis of simple organometallic reactions (37). But the broad outlines of a theory of organometallic reactivity, the kind of framework that would allow us to anticipate approximately the range of products exhibited in Fig. 2, is still lacking. Our relative ignorance in this area is not a cause for despair. On the contrary, it makes theoretical work in this realm of the possible a great deal of fun.

References and Notes

1. For an introduction to the multitude of new organometallic molecules, see F. A. Cotton and G. Wilkinson, *Advanced Inorganic Chemistry* (Interscience, New York, ed. 4, 1980); J. Huheey, *Inorganic Chemistry* (Harper & Row, New York, ed. 2, 1978); K. F. Purcell and J. C. Kotz, *Inorganic Chemistry* (Saunders, Philadelphia, 1977).
2. C. R. Eady, B. F. G. Johnson, J. Lewis, M. C. Malatesta, P. Machin, M. McPartlin, *Chem. Commun.* **1976**, 945 (1976); P. F. Jackson, B. F. G. Johnson, J. Lewis, P. Raithby, M. McPartlin, W. H. Nelson, K. D. Rouse, J. Allibon, S. A. Mason, *ibid.* **1980**, 295 (1980).
3. O. Gambino, E. Sappa, A. M. Manotti Lonfredi, A. Tiripicchio, *Inorg. Chim. Acta* **36**, 190 (1979), and references therein.
4. Two good guides to the orbitals of organic molecules are E. Heilbronner and H. Bock, *The HMO [Hückel Molecular Orbital] Model and Its Application* (Wiley-Interscience, New York, 1976) and W. L. Jorgensen and L. Salem, *The Organic Chemist's Book of Orbitals* (Academic Press, New York, 1973).
5. H. D. Murdoch and E. Weiss, *Helv. Chim. Acta* **46**, 1588 (1963); N. von Kutepow, E. Zahn, K. Eisfeld, German Patent 1,158,511.
6. There are many structures of olefins complexed to a 5-coordinate d^8 center. References may be found in T. A. Albright, R. Hoffmann, J. C. Thibeault, D. L. Thorn, *J. Am. Chem. Soc.* **101**, 3801 (1979). The first crystal structure determination of an $Fe(CO)_4$(alkene) complex was carried out by A. R. Luxmoore and M. R. Truter, *Acta Crystallogr.* **15**, 1117 (1962).

7. M. Elian and R. Hoffmann, *Inorg. Chem.* **14**, 1058 (1975); M. Elian, M. M. L. Chen, D. M. P. Mingos, R. Hoffmann, *ibid.* **15**, 1148 (1976); R. Hoffmann, T. A. Albright, D. L. Thorn, *Pure Appl. Chem.* **50**, 1 (1978).
8. L. Pauling, *The Nature of the Chemical Bond* (Cornell Univ. Press, Ithaca, N.Y., ed. 3, 1960).
9. The reader may wonder why the antisymmetric b_2 combination is at lower energy. This is a consequence of its being mainly nd with some $(n + 1)p$ admixture, whereas the symmetric a_1 combination is mainly $(n + 1)s + (n + 1)p$, with some nd.
10. For a readable introduction to group theory, see F. A. Cotton, *Chemical Applications of Group Theory* (Wiley-Interscience, New York, ed. 2, 1971).
11. M. J. S. Dewar, *Bull. Soc. Chim. Fr.* **18**, C79 (1951); J. Chatt and L. A. Duncanson, *J. Chem. Soc.* **1955**, 2939 (1955).
12. R. Hoffmann, *Acc. Chem. Res.* **4**, 1 (1971).
13. The barrier to rigid rotation of C_2H_4 is probably larger. The observed numbers refer to a process in which Berry pseudo-rotation at the metal is undoubtedly coupled to simple rotation of the olefin [5-coordination in (15)]. For experimental work on this problem, see: P. W. Clark and A. J. Jones, *J. Organomet. Chem.* **122**, C41 (1976); L. Kruczynski, J. L. Martin, J. Takats, *ibid.* **80**, C9 (1974); A. L. Onderdelinden and A. van der Ent, *Inorg. Chim. Acta* **6**, 420 (1972); T. Kaneshima, Y. Yumoto, K. Kawakami, T. Tanaka, *ibid.* **18**, 29 (1976); H. C. Clark and L. E. Manzer, *Inorg. Chem.* **13**, 1996 (1974); L. Kruczynski, L. K. K. LiShingMan, J. Takats, *J. Am. Chem. Soc.* **96**, 4006 (1974); S. T. Wilson, N. J. Coville, J. R. Shapley, J. A. Osborn, *ibid.*, p. 4038; J. A. Segal and B. F. G. Johnson, *J. Chem. Soc. Dalton Trans.* **1975**, 677 (1975); *ibid.*, p. 1990. Theoretical studies may be found in (14).
14. See Albright *et al.* (6); J. Demuynck, A. Strich, A. Veillard, *Nouveau J. Chim.* **1**, 217 (1977).
15. For some leading references on 5-, 6-, 7-, and 8-coordination and the subject of geometrical flexibility within each coordination number, see the following.
   For 5-coordination: E. L. Muetterties and R. A. Shunn, *Q. Rev. Chem. Soc.* **20**, 245 (1966); R. Hoffmann, J. M. Howell, E. L. Muetterties, *J. Am. Chem. Soc.* **94**, 3047 (1972); E. L. Muetterties and L. J. Guggenberger, *ibid.* **96**, 1748 (1974); L. Sacconi, *Pure Appl. Chem.* **17**, 95 (1968); J. S. Wood, *Prog. Inorg. Chem.* **16**, 227 (1972); A. Rossi and R. Hoffmann, *Inorg. Chem.* **14**, 365 (1975).
   For 6-coordination: R. Eisenberg, *Prog. Inorg. Chem.* **12**, 295 (1970); D. L. Kepert, *ibid.* **23**, 1 (1977), and references therein; R. A. D. Wentworth, *Coord. Chem. Rev.* **9**, 171 (1972); R. Hoffmann, J. M. Howell, A. R. Rossi, *J. Am. Chem. Soc.* **98**, 2484 (1976).
   For 7-coordination: M. G. B. Drew, *Prog. Inorg. Chem.* **23**, 67 (1977); D. L. Kepert, *ibid.* **25**, 41 (1979), and references therein; E. L. Muetterties and C. M. Wright, *Q. Rev. Chem. Soc.* **21**, 109 (1967); R. Hoffmann, B. F. Beier, E. L. Muetterties, A. R. Rossi, *Inorg. Chem.* **16**, 511 (1977).
   For 8-coordination: S. J. Lippard, *Prog. Inorg. Chem.* **8**, 109 (1967); *ibid.* **21**, 91 (1976); D. L. Kepert, *ibid.* **24**, 179 (1978); R. V. Parish,

Coord. Chem. Rev. **1**, 439 (1966); M. G. B. Drew, *ibid.* **24**, 179 (1977); J. K. Burdett, R. Hoffmann, R. C. Fay, *Inorg. Chem.* **17**, 2553 (1978), and references therein.

16. The fragment approach is in no way our creation but has been used extensively by others, for example: S. F. A. Kettle, *J. Chem. Soc. A.* **1966**, 420 (1966); *Inorg. Chem.* **4**, 1661 (1965); M. L. H. Green, *Organometallic Compounds* (Methuen, London, 1968), vol. 2, p. 115; K. Wade, *Chem. Commun.* **1971**, 792 (1971); *Inorg. Nucl. Chem. Lett.* **8**, 559 (1972); *ibid.*, p. 563; *Electron Deficient Compounds* (Nelson, London, 1971); P. S. Braterman, *Struct. Bonding (Berlin)* **10**, 5 (1972); F. A. Cotton, W. T. Edwards, F. C. Rauch, M. A. Graham, R. N. Perutz, J. J. Turner, *J. Coord. Chem.* **2**, 247 (1973); D. V. Korolkov and H. Miessner, *Z. Phys. Chem. (Leipzig)* **253**, 25 (1973); D. M. P. Mingos, *Nature (London) Phys. Sci.* **236**, 99 (1972); *J. Chem. Soc. Dalton Trans.* **1977**, 602 (1977); *Adv. Organomet. Chem.* **15**, 1 (1977); J. K. Burdett, *J. Chem. Soc. Faraday Trans.* 2 **70**, 1599 (1974); *Inorg. Chem.* **14**, 375 (1975); T. H. Whitesides, D. L. Lichtenberger, R. A. Budnik, *Inorg. Chem.* **14**, 68 (1975); B. R. Bursten and R. F. Fenske, *ibid.* **18**, 1760 (1979); D. L. Lichtenberger and T. L. Brown, *J. Am. Chem. Soc.* **100**, 366 (1978); D. L. Lichtenberger and R. F. Fenske, *ibid.* **98**, 50 (1976); *J. Organomet. Chem.* **117**, 253 (1976); P. Hofmann, *Angew. Chem.* **89**, 551 (1977); *ibid.* **91**, 591 (1979).

17. See Bursten and Fenske [in (*16*)] for a perceptive and detailed analysis of cyclobutadiene iron tricarbonyl.

18. T. A. Albright, P. Hofmann, R. Hoffmann, *J. Am. Chem. Soc.* **99**, 7546 (1977).

19. E. S. Magyar and C. P. Lillya, *J. Organomet. Chem.* **116**, 99 (1976); R. J. Clark, M. P. Abraham, M. A. Busch, *ibid.* **35**, C33 (1972).

20. T. A. Albright, R. Hoffmann, P. Hofmann, *Chem. Ber.* **111**, 1591 (1978).

21. V. A. Mironov, E. V. Sobolev, A. N. Elizavera, *Dokl. Akad. Nauk SSSR* **143**, 1112 (1962); S. McLean and R. Haynes, *Tetrahedron Lett.* **1964**, 2385 (1964); S. McLean, C. J. Webster, R. J. D. Rutherford, *Can. J. Chem.* **47**, 1555 (1969); W. R. Roth, *Tetrahedron Lett.* **1964**, 1009 (1964); _____ and J. Konig, *Justus Liebigs Ann. Chem.* **699**, 24 (1966); G. I. Avramenko, N. M. Sergeyev, Yu. A. Ustynyuk, *J. Organomet. Chem.* **37**, 89 (1972); A. J. Ashe III, *J. Am. Chem. Soc.* **92**, 1233 (1970); R. B. Woodward and R. Hoffmann, *ibid.* **87**, 2511 (1965); *Angew. Chem.* **81**, 797 (1969).

22. See the references in Nguyen Trong Anh, M. Elian, R. Hoffmann, *J. Am. Chem. Soc.* **100**, 110 (1978).

23. K. J. Karel and M. Brookhart, *ibid.*, p. 1619; B. E. Mann, *J. Organomet. Chem.* **141**, C33 (1977).

24. G. Kreiter, A. Maasbol, F. A. L. Anet, H. D. Kaesz, S. Winstein, *J. Am. Chem. Soc.* **88**, 3444 (1966); F. A. Cotton, A. Davison, J. W. Faller, *ibid.*, p. 4507; C. E. Keller, B. A. Shoulders, R. Petit, *ibid.*, p. 4760; F. A. L. Anet, H. D. Kaesz,

A. Maasbol, S. Winstein, *ibid.* **89**, 2489 (1967); F. A. Cotton and D. L. Hunter, *ibid.* **98**, 1413 (1976).

25. T. A. Albright, in preparation.

26. For an experimental demonstration of this effect in fluorenyl-FeCp or fluorenyl-$Mn(CO)_3$ complexes, see: P. M. Treichel and J. W. Johnson, *Inorg. Chem.* **16**, 749 (1977); J. W. Johnson and P. M. Treichel, *J. Am. Chem. Soc.* **99**, 1427 (1977).

27. M. H. Chisholm and F. A. Cotton, *Acc. Chem. Res.* **11**, 356 (1978), and references therein.

28. T. A. Albright and R. Hoffmann, *J. Am. Chem. Soc.* **100**, 7736 (1978); A. Dedieu, T. A. Albright, R. Hoffmann, *ibid.* **101**, 3141 (1979).

29. B. E. Bursten, P. A. Cotton, J. C. Green, E. A. Seddon, G. G. Stanley, *ibid.* **102**, 4579 (1980); M. B. Hall, *ibid.*, p. 2104.

30. See references in R. H. Summerville and R. Hoffmann, *J. Am. Chem. Soc.* **101**, 3821 (1979).

31. J. Lauher, M. Elian, R. H. Summerville, R. Hoffmann, *ibid.* **98**, 3219 (1976).

32. W. Siebert and K. Kinberger, *Angew. Chem.* **88**, 451 (1976); *Angew Chem. Int. Ed. Engl.* **15**, 434 (1976); G. E. Herberich, J. Hengesbach, U. Kölle, G. Huttner, A. Frank, *Angew. Chem.* **88**, 450 (1976); *Angew Chem. Int. Ed. Engl.* **15**, 433 (1976).

33. See references in D. L. Thorn and R. Hoffmann, *Inorg. Chem.* **17**, 126 (1978).

34. See references in B. E. R. Schilling and R. Hoffmann, *J. Am. Chem. Soc.* **101**, 3456 (1979); *Acta Chem. Scand. Ser. B* **33**, 231 (1979).

35. J. R. Fritch, K. P. C. Vollhardt, M. R. Thompson, V. W. Day, *J. Am. Chem. Soc.* **101**, 2768 (1979); H. Yamazaki, Y. Wakatsuki, K. Aoki, *Chem. Lett.* **1979**, 1041 (1979).

36. A. J. Deeming and M. Underhill, *J. Chem. Soc. Dalton Trans.* **1974**, 1415 (1974); *J. Organomet. Chem.* **42**, C60 (1972); *Chem. Commun.* **1973**, 277 (1973).

37. S. Komiya, T. A. Albright, R. Hoffmann, J. Kochi, *J. Am. Chem. Soc.* **98**, 7255 (1976); D. L. Thorn and R. Hoffmann, *ibid.* **100**, 2079 (1978); H. Berke and R. Hoffmann, *ibid.*, p. 7224.

38. Many people, able young scientists all, have participated in the work of the Cornell group over the past few years. They include James Howell, Angelo Rossi, Jack Thibeault, Notker Rösch, P. Jeffrey Hay, Mihai Elian, Nguyen Trong Anh, Joseph W. Lauher, Maynard M. L. Chen, Birgitte Schilling, David L. Thorn, Richard H. Summerville, Daniel L. Dubois, D. M. P. Mingos, Jeremy Burdett, Thomas A. Albright, Peter Hoffmann, Prem Mehrota, Heinz Berke, Alain Dedieu, Myung-Hwan Whangbo, Armel Stockis, E. D. Jemmis, Sason Shaik, and Allan R. Pinhas. I thank them all. Supported by the National Science Foundation. This article was written in the course of a visit to the University of North Carolina at Chapel Hill, where the hospitality of M. Brookhart and J. Templeton was appreciated.

Over the past 20 years techniques have been developed that permit the investigation, on the molecular level, of surfaces in ultrahigh vaccuum and at solid-gas interfaces (Table 1). With these techniques, the structure and composition of surfaces and the oxidation states of surface atoms can be identified. The bonding of atoms and molecules in surface monolayers has been studied by surface crystallography and spectroscopies of different types (*1, 2*). Molecular-beam and low-pressure studies probe the energy transfer that occurs upon collision of a gaseous atom or molecule with the surface, the interaction potential of that collision, and the elementary steps of surface reactions: adsorption, bond breaking, and desorption (*1, 2*).

The intellectual challenge to understand the properties of surface monolayers made surface science an important topic of academic research in chemistry and chemical engineering departments. This research has influenced many technologies in which surfaces optimize reactivity, chemical change, or charge transport, including catalysis, electrochemistry, photography, and electronic circuitry. Modern surface science can be used to improve existing systems and serves as the foundation for the development of new devices. By converting many standard technologies to high technology, surface science induces rapid development in both technology and long-range basic research, since these two areas are interrelated.

In this review I shall concentrate on my work in the surface science of catalysis and the influence this work has had on catalyst manufacture. As a result of recent research, catalyst design and fabrication has developed from an art to a science. Since catalysts lie at the heart of

15

Surface Science and Catalysis

Gabor A. Somorjai

Science 227, 902–908 (22 February 1985)

Table 1. Frequently used techniques of surface science for studies of catalysts.

Table 1. Frequently used techniques of surface science for studies of catalysts.

Electron scattering
Electron spectroscopies (x-ray photoelectron spectroscopy, high resolution electron energy loss spectroscopy, Auger electron spectroscopy)
Low-energy electron diffraction
Electron microscopy

Photon scattering
(high and low intensities)
Spectroscopies (infrared, Fourier-transform infrared, Raman, solid-state nuclear magnetic resonance, electron spin resonance, extended x-ray absorption fine structure, near-edge x-ray absorption fine structure, laser)
Grazing-angle x-ray diffraction

Molecule and ion scattering
Molecular beam–surface interaction
Secondary ion mass spectroscopy
Ion scattering spectroscopy

Other techniques
Radiotracer labeling
Mössbauer spectroscopy
Thermal desorption

most chemical processes, this development affects the future of chemical and petrochemical technologies and the ability to produce, convert, and transport energy.

Catalytic Versus Stoichiometric Reactions

One of the deficiencies of our chemical education system is the lack of emphasis on catalytic processes. A molecule adsorbs on a catalyst surface; it chemically rearranges while visiting several reaction sites by surface diffusion; and then it desorbs, as the product, to the gas phase. During the lifetime of a good catalyst, the reaction turns over a million times (10^6 product molecules per site) or more. If the reactant forms strong chemical bonds upon adsorption on the surface there is no catalysis, and the result is a stoichiometric reaction with a turnover of one (1 product molecule per site). Chemists are usually taught only about stoichiometric reactions during their formal training; yet many important life-sustaining reactions, including photosynthesis, the biological processes of our bodies, and the synthesis of ammonia, are catalytic.

During the catalytic process, therefore, a reactant cannot be strongly bound since that would poison the catalyst. However, if the bonding were too weak there would be no opportunity for chemical bond breaking. Thus, bonding of intermediate strength between the catalyst and the surface is needed, and surface sites are required where both bond breaking and bond formation are possible within the residence time of the intermediates.

Selected Properties of Surfaces

The structure of solid surfaces. The surfaces of single crystals of metals exhibit a variety of structures, depending on the angle of cut. The surface may be flat and close-packed where each atom is surrounded by a large number of nearest neighbors. It may be "rough" with atom-size openings between surface atoms that expose atoms in the second or even in the third layer to the incoming molecules. It may be stepped with terraces several atoms wide separated by steps the height of one atom, and there may be kinks in the steps. Small particles contain sites with many or all of these surface characteristics. Very small particles may have all of their atoms at the surface (dispersion of unity) in close-packed configurations. As a particle grows, the relative concentrations of terraced, stepped, and kinked sites are altered. If the particle grows slowly, it is bound mostly by close-packed surfaces that are thermodynamically stable. However, rapid and kinetically controlled growth processes can stabilize rough, more open surfaces.

Surface crystallography studies by low-energy electron diffraction (LEED) indicate that the distance between the first and second layer of atoms at the surface can be considerably shorter than interlayer distances in the bulk (3). The more open or rough the surface is, the greater this contraction. The contraction observed at the surface is called relaxation. It is the consequence of the anisotropic surface environment and is due to the reduction of the number of nearest neighbors around atoms at the surface.

Frequently, atoms at the surface will seek new equilibrium positions to optimize their bonding configuration to such an extent that the surface reconstructs. That is, the periodic arrangement of surface atoms becomes different from what is expected from inspection of the bulk unit cell. The surface becomes buckled (Fig. 1a), or there are periodically arranged dimers (Fig. 1b), or there are surface vacancies. The surface structure of clean solids is as fascinating as it is unpredictable at present in the absence of sufficient experimental information.

Surface composition. Frequently, impurities that are present in minute amounts (parts per million) in the bulk of the solid segregate on the surface in such large concentrations that they could completely cover it. This occurs if their chemical bonds to surface atoms are strong and if their presence greatly reduces the positive free energy of the surface as compared to that of the clean surface. Similar thermodynamic characteristics are responsible for the enrichment of one constituent at the surface of binary or multicomponent alloys. There is excess silver at Ag-Au alloy surfaces, copper at Cu-Ni surfaces, and gold at Au-Sn surfaces. Thermodynamic models have been developed to predict the composition of a clean surface for a particular alloy (1, 4), but adsorbed atoms or molecules can alter the surface composition from what is predicted by forming stronger chemical bonds with one of the alloy constituents.

When alloy particles are produced with all or most of the atoms at the surface, their phase diagram is different from that in three dimensions. Often called bimetallic clusters (5), these particles form solid solutions and exhibit complete miscibility when phase separation is observed in the bulk phases. This unique behavior in two dimensions has been treated theoretically.

Adsorbed Atoms and Molecules

The surface chemical bond. Surface crystallography studies by LEED have revealed the formation of a large number of ordered atomic and molecular layers of adsorbed species on surfaces of crystals (1, 6). The location of the adsorbates, their bond lengths, and their bond angles have been determined. On flat surfaces with high symmetry, atoms like to occupy sites of high coordination. Some of the structures that are encountered most frequently are shown in Fig. 2. Sites with tetracoordination and tricoordination are frequently occupied first by atoms such as sulfur, iodine, oxygen, or carbon. An interesting bonding situation is encountered with small atoms, such as hydrogen and nitrogen, that can lie under the first atomic layer instead of on top of it.

Among small molecules, CO has been studied most frequently. The molecule usually adsorbs with its C–O bond axis perpendicular to the surface and binds to metal surfaces through the carbon atom (7). It can occupy a top site (Fig. 3a) or a bridge site (Fig. 3b) with only a small difference in its heat of adsorption at these two locations. However, as the coverage of CO is increased above one half of a monolayer, some of the CO molecules occupy sites of lower symmetry (8) to stay apart as much as possible because of repulsive interactions among the molecules. The average heat of adsorption then declines rapidly with coverage, as shown in Fig. 3C.

a   Two-bridge   Top-center

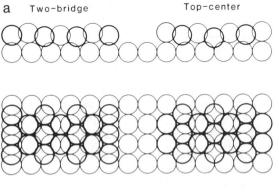

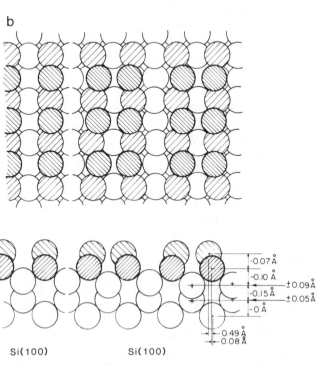

Fig. 1. (a) Structure of the reconstructed Ir(100) crystal face, as determined by surface crystallography. (b) Top and side views of ideal, bulk-like Si(100) at the left and the Si(100) reconstructed real surface at the right. Layer-spacing contractions and intralayer atomic displacement relative to the bulk structure are given; shading differentiates surface layers.

b

Si(100)   Si(100)

-0.07 Å
-0.10 Å
±0.09 Å
-0.15 Å
±0.05 Å
-0 Å

0.49 Å
0.08 Å

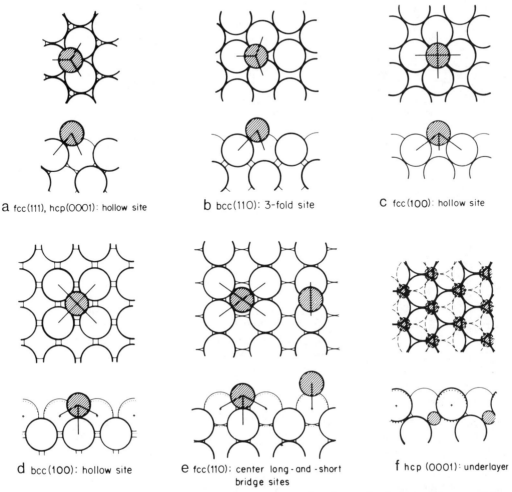

Fig. 2. Top and side views (in top and bottom sketches of each panel) of adsorption geometries on various metal surfaces. Adsorbates are drawn shaded. Dotted lines represent clean surface atomic positions; arrows show atomic displacements due to adsorption. Abbreviations: face-centered cubic, fcc; hexagonal close packing, hcp; and body-centered cubic, bcc.

The structure of ethylene adsorbed on metal surfaces has also been studied in great detail (9). This molecule loses a hydrogen atom and forms a stable ethylidyne C_2H_3 species on the (111) crystal faces of platinum, rhodium, and palladium at room temperature. The C–C bond is perpendicular to the metal surface and elongated to a single bond. The carbon that bonds to the metal is in a tricoordi-

nated site and forms short strong bonds to the metal, as seen in Fig. 4. The structure is similar to the structures of several organometallic clusters, indicating that the surface chemical bond is localized and cluster-like.

One of the striking properties of the surface chemical bonds of adsorbed organic molecules is their temperature-dependent variation (10). Upon heating, a

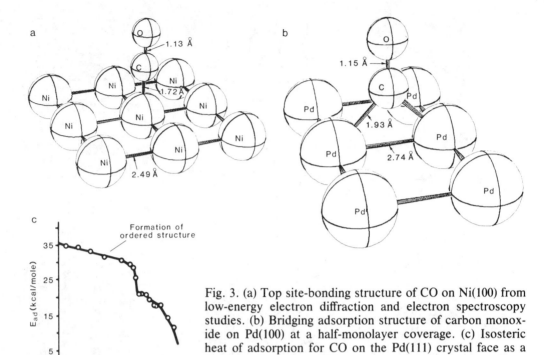

a

1.13 Å

O

C

Ni
Ni
Ni
1.72 Å
Ni
Ni
Ni
Ni
Ni
Ni
2.49 Å

b

O

1.15 Å →

C

Pd
Pd
1.93 Å
Pd
Pd
2.74 Å
Pd
Pd

c

Formation of
ordered structure

35

25

E_{ad} (kcal/mole)

15

5

0   0.2   0.4   0.6   0.8
Coverage
(fraction of a monolayer)

Fig. 3. (a) Top site-bonding structure of CO on Ni(100) from low-energy electron diffraction and electron spectroscopy studies. (b) Bridging adsorption structure of carbon monoxide on Pd(100) at a half-monolayer coverage. (c) Isosteric heat of adsorption for CO on the Pd(111) crystal face as a function of coverage (26).

sequential loss of hydrogen occurs and the molecules form organic fragments, CH, CH_2, C_2H, and so on, with well-defined stoichiometry (Fig. 4). Only at higher temperatures (≥800K) will the fragments lose all their hydrogen and form a graphitic or carbidic overlayer, the most stable structure thermodynamically. The fragments are stable in a well-defined temperature range and the surface remains chemically reactive and catalytically active in their presence. The metal surface is deactivated only when the graphitic overlayer is produced.

On more heterogeneous surfaces molecules usually form stronger chemical bonds at kinks and steps than on the flat portions of the surfaces. Thermal desorption spectroscopy reveals (Fig. 5)

that CO molecules or H atoms adsorbed at kink sites desorb at higher temperatures (10) than those adsorbed at steps, and those adsorbed at flat terraces are

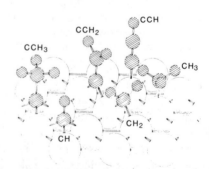

CCH

CCH₂

CCH₃

CH₃

CH₂

CH

Fig. 4. Schematic representation of the various organic fragments present on metal surfaces at high temperature. The presence of CH, C_2, C_2H, CH_2, and CCH_3 has been detected.

218

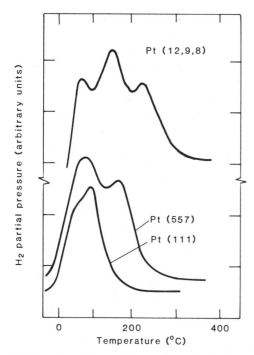

Fig. 5. Thermal desorption spectra for hydrogen chemisorbed on flat Pt(111), stepped Pt(557), and kinked Pt(12,9,8) surfaces.

bound even less strongly. The chemical bond of adsorbates is thus sensitive to surface structure, so it is not surprising that one observes a sequential filling of sites upon exposure of the clean surface to adsorbates, the most strongly binding sites filling up first.

Coadsorption on surfaces. Often, when two different species are adsorbed, they markedly influence each other's bonding. For example, when CO and benzene are adsorbed together on Pt or Rh (111) surfaces, ordered structures form with both CO and benzene in the same unit cell (*11*). There is clearly an attractive interaction between these two molecules in the adsorbed monolayer. When sulfur is adsorbed with other molecules on metal surfaces, it blocks sites thereby reducing the concentration of

the other adsorbed species (for example, sulfur blocks the adsorption of two hydrogen atoms per sulfur or one CO per sulfur on molybdenum single-crystal surfaces) (*12*). By filling up those sites first that would adsorb molecules very strongly, sulfur attenuates the strong chemical interaction between the metal and the coadsorbed molecule. For example, a clean molybdenum metal surface would decompose thiophene, C_4H_5S, an important sulfur-containing molecule. When thiophene is coadsorbed on molybdenum that is partially covered with sulfur, it is adsorbed as a molecule because the sites where it would decompose are blocked by sulfur.

Alkali metals exert pronounced electronic effects when coadsorbed with other molecules (such as CO, N_2, and hydrocarbons) on transition metal surfaces. By donating electrons to the transition metal, potassium, for example, becomes ionized. The excess charge donated to the transition metal finds its way to the molecular orbitals of the coadsorbed molecules if those orbitals overlap with the charge density of the metal. This is the case for CO when it is coadsorbed with potassium: the C–O bond is greatly weakened and the metal-carbon bond strengthened by as much as 12 kcal (*13*). Such modification of bonding leads to CO dissociation on rhodium surfaces in the presence of potassium, whereas CO adsorbs only molecularly in the absence of potassium. The presence of alkali metals strongly modifies the reaction path during CO hydrogenation, altering the product distribution (more dissociative CO adsorption and less hydrogenation lead to the formation of alkanes and alkenes of higher molecular weight) (*13*). The N_2 bond is also weakened by alkali metals, and as a result the presence of alkali metals increases the rate of NH_3 synthesis over iron surfaces. On the other hand, coadsorbed alkali metal atoms decrease the reaction rates of hydrocarbons on transition metal surfaces by strengthening the C–H bonds, which therefore break less readily (14).

Thus, the coadsorption of two species can markedly influence the chemical bonding of adsorbed molecules. An attractive adsorbate-adsorbate interaction may induce ordering or modify bonding. The blocking of chemically active sites by one of the adsorbates can attenuate the chemical interaction of the other adsorbate with the surface. Charge transfer from one of the coadsorbed species to the other through the metal substrate can strongly influence its chemical bond. Finally, certain adsorbates (such as potassium and sulfur), by adsorbing more strongly on one crystal face than another, can induce restructuring of the whole surface.

The Surface Science of Catalytic Reactions

Strongly chemisorbed molecules rarely participate directly in catalytic reactions. Because of their tenacious bonding, their surface-residence times are too long for the reaction to proceed with rapid turnover, as required by catalytic process. They are surface compounds not unlike the stable compounds that form during stoichiometric reactions in the gas phase, in solutions, or in the solid state. There must be other surface sites that adsorb the reactant molecules only weakly and carry out the catalytic process with high turnover rates. Because of the decreasing heat of chemisorption with increasing coverage (Fig. 3c), cata-

lytic processes are likely to occur at high surface coverages.

There are three molecular components of heterogeneous catalysis that control reaction rates and product selectivity. These are: (i) the atomic surface structure, (ii) an active overlayer that is deposited before or during the reaction on the active catalyst surface, and (iii) the oxidation states of the surface atoms. The studies that verified the important roles of these surface components have mostly been of model catalysts, single-crystal surfaces of well-defined structure and composition. The use of these model systems required mating the ultrahigh-vacuum and high-pressure techniques used for characterizing the surface structure and composition and the chemical reactivity of the surface, respectively. An apparatus that can operate in the pressure range of 10^{-10} to 10^5 torr has been developed in our laboratory for this purpose, and it is widely used by surface scientists (15).

Surface-structure sensitivity. Rough surfaces are much better catalysts than smooth surfaces. For example, the (111) crystal face of iron produces ammonia from N_2 and H_2 at about 500 times the rate of the (110) face under industrial conditions, 20 atm and 450°C (Fig. 6) (16). The (11$\bar{2}$1) or (11$\bar{2}$0) crystal faces of Re produce ammonia 2000 times faster than the close-packed (0001) Re surface. These catalytically active metal surfaces contain many atoms with large numbers of nearest neighbors (high coordination). In addition, the top layers have an open structure, so atoms with high coordination in the second layer are also accessible to incoming reactant molecules. Thus, it appears that sites with highly coordinated atoms are key components of catalysts with high turnover rates.

Stepped surfaces and surfaces with large concentrations of kinks in the steps are also active in carrying out catalytic reactions of many types at high rates [for example, the hydrogenolysis of isobutane on kinked Pt surfaces (17) and the exchange of H_2 and D_2 at stepped Pt surfaces (18)]. At these defect sites atoms of both low and high coordination are exposed, the atoms at the bottom of the step having a high coordination and those at the top of the step having low coordination. Thus, such sites provide both high catalytic activity and strong chemical bonding, at the bottom and at the top of the step, respectively.

A recently developed theory (19) of metal catalysis proposes correlations between catalytic activity and local electronic fluctuations of low energy in transition metals. Maximum electronic fluctuations take place at metal sites of high coordination. The most active surfaces for carrying out structure-sensitive catalytic reactions are atomically rough surfaces, those that expose large numbers of nonmagnetic or weakly magnetic neighboring atoms to the reactant molecules in the first or second layer and those that are stepped or kinked.

The electronic fluctuations of importance include charge fluctuations, configuration fluctuations, spin fluctuations, and term and multiplet fluctuations. Configuration fluctuations are the largest for those atoms with a large number of metallic neighbors; thus, they should be more catalytically active. In addition, the catalytic activity of the transition metal atoms is directly related to their unfilled d-shells.

The role and reactivity of strongly adsorbed overlayers. It is important to point out that not all catalytic reactions are structure-sensitive and that struc-

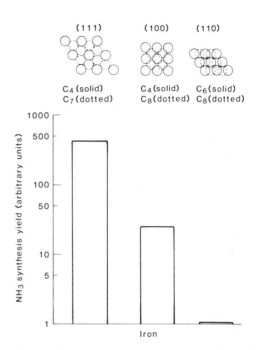

Fig. 6. The sensitivity of the iron-catalyzed ammonia synthesis to the structure of the iron surface.

ture-insensitive processes may be mediated by tenaciously held overlayers of strongly adsorbed molecules that insulate the metal from direct contact with some of the reactants. Indeed, the overlayer itself plays an important role in certain catalytic reactions. One of these reactions is the hydrogenation of ethylene, C_2H_4 (20). When C_2H_4 is introduced along with H_2 in a steady flow over platinum or rhodium surfaces at 300K and 1 atm, hydrogenation to C_2H_6 occurs with very high rates of turnover. An ethylidyne overlayer forms almost instantaneously, however, and ^{14}C labeling and spectroscopic studies indicate that the hydrogenation occurs by the mediation of this overlayer, which permits H_2 adsorption and atomization on the metal surface but blocks the direct adsorption of C_2H_4 on the metal. A mechanism that explains all of the experimental data is shown in Fig. 7. The hydrogen transfer via formation of an ethylidene intermediate is implicated as the key hydrogenation step, and ethylene does not directly contact the metal surface.

Of course this is not the only mechanism for C_2H_4 hydrogenation. At higher temperatures the strongly adsorbed C_2H_3 groups are rehydrogenated at sufficiently high rates to continuously expose the bare metal sites to the incoming C_2H_4 molecules. At electrode surfaces in aqueous solution, C_2H_4 hydrogenation occurs in the presence of an external potential over a metal surface that is covered with such a large concentration of hydrogen atoms that the hydrogenation of approaching ethylene molecules takes place instantaneously without the formation of ethylidyne groups.

For most hydrocarbon conversion reactions (isomerization, dehydrocyclization, and hydrogenolysis) both the catalyst's surface structure and hydrogen transfer by the strongly held carbonaceous deposit play important catalytic roles (21). Thus, the reaction conditions are adjusted to continuously expose some of the bare metal sites to the incoming reactants, which also migrate onto the strongly held deposit to participate in chemical interactions before the final product molecule desorbs. A model

of the platinum surface that is active for the conversion of organic molecules in isomerization, dehydrocyclization, and hydrogenolysis reactions indicates the importance of both the metal surface structure and the carbonaceous deposit that performs hydrogen transfer (Fig. 8).

The oxidation state of surface atoms. Several experimental studies in recent years have indicated the importance of oxidation states other than the zero valent metallic state for catalyzed reactions. The hydrogenation of CO over rhodium was reported to yield predominantly C_2 oxygenated products, acetaldehyde and acetic acid, under certain experimental conditions (22). In our studies of unsupported polycrystalline rhodium foils, we detected mostly methane, and only small amounts of ethylene and propylene, under similar conditions. This latter product distribution was identical to that obtained by others from supported rhodium catalysts as was the value of the activation energy for methanation that we found: 24 kcal per mole (22). It appears that after the dissociation of CO, the organic molecules form by the rehydrogenation of CH_x units in a manner similar to alkane and alkene production from mixtures of CO and H_2 over other transition metal catalysts (iron, ruthenium, and nickel). However, when rhodium oxide (Rh_2O_3) was the catalyst, large concentrations of oxygenated C_2 or C_3 hydrocarbons were produced, includ-

Fig. 7. Proposed mechanism for the rapid, structure-insensitive hydrogenation of ethylene.

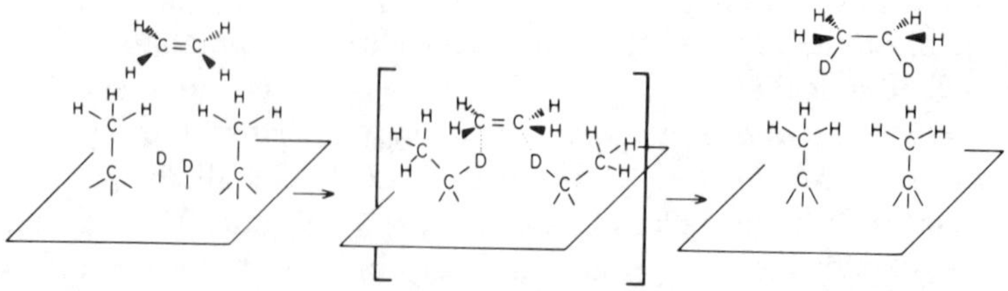

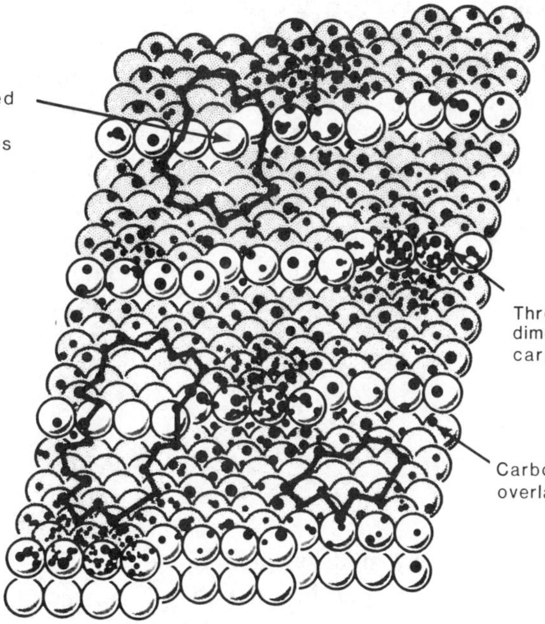

Uncovered
ensemble
of Pt sites

Three-
dimensional
carbon island

Carbonaceous
overlayer

Fig. 8. Model for the working platinum catalyst that was developed from a combination of studies of single-crystal surfaces and studies of the hydrocarbon reaction rates on these same surfaces.

ing ethanol, acetaldehyde, and propionaldehyde (22). Furthermore, the addition of C_2H_4 to the CO and H_2 mixture yielded propionaldehyde, indicating the carbonylation ability of Rh_2O_3. Under similar conditions and over rhodium metal, C_2H_4 was quantitatively hydrogenated to ethane, and carbonylation activity was totally absent. Clearly, rhodium ions of higher oxidation state were necessary to produce the oxygenated molecules. Similar behavior was exhibited by molybdenum. From CO and H_2, molybdenum metal produced methane and alkanes or alkenes that contain two or three carbons, whereas molybdenum compounds that contain molybdenum ions in higher oxidation states produced alcohols.

The marked change of selectivity in reactions of CO and H_2 upon alteration of the oxidation state of the transition metal is due largely to the change in the heats of adsorption of CO and H_2 as the oxidation state of the transition metal ion

is varied (22). The CO adsorption energy is decreased upon oxidation while the heat of adsorption of H_2 is increased. This in turn changes the relative surface concentrations of CO and H_2. In addition, the metal is primarily active for hydrogenation and CO dissociation, whereas the oxide can perform carbonylation and has reduced hydrogenation activity.

One of the difficulties in preparing selective catalysts for hydrocarbon conversion is the poor thermodynamic stability of the higher oxidation states of transition metal ions in a reducing environment. It appears that the strong interaction between the metal and support that permits the incorporation of the high-oxidation-state transition metal ion into the supporting refractory oxide or sulfide crystal lattice often provides for the kinetic stability of the desired oxidation state, so long as the catalytic reaction temperatures are appreciably below the decomposition temperature of the binary oxide so prepared.

Another example of the importance of the changing oxidation state of transition metal ions at the surface is the catalytic cycle leading to the photocatalyzed dissociation of water on strontium titanate, SrTiO$_3$, surfaces (Eq. 1) (23). The oxide

$$-\overset{\overset{\displaystyle OH^-}{|}}{Ti}{}^{4+}-O^{2-}-\overset{\overset{\displaystyle OH^-}{|}}{Ti}{}^{4+}-\ \xrightarrow{2h\nu}\ -\overset{\overset{\displaystyle \cdot OH}{|}}{Ti}{}^{3+}-O^{2-}-\overset{\overset{\displaystyle \cdot OH}{|}}{Ti}{}^{3+}-\ \longrightarrow$$

$$-\overset{|}{\underset{|}{Ti}}{}^{3+}-O^{2-}-\overset{|}{\underset{|}{Ti}}{}^{3+}-\ +\ H_2O_2\ \xrightarrow{H_2O}$$

$$-\overset{|}{Ti}{}^{4+}\overset{\displaystyle O^{2-}}{\diagdown\diagup}O^{2-}-\overset{|}{Ti}{}^{4+}-\ +\ H_2 \qquad (1)$$

surface is completely hydroxylated in the presence of water, and the titanium ions are in the Ti^{4+} oxidation state. When the surface region is irradiated with light of 3.1 eV or more, electron-hole pairs are generated. The electron is utilized to reduce the Ti^{4+} to Ti^{3+}. The electron vacancy induces charge transfer from the hydroxyl group, which produces OH radicals that dimerize to H$_2$O$_2$, and splits off oxygen, which evolves. The reduced surface containing Ti^{3+} can now adsorb another water molecule, which acts as an oxidizing agent to produce Ti^{4+} and a hydroxylated surface again, evolving hydrogen in the process (23). Clearly, changes of oxidation states of transition metal ions are frequently indispensable reaction steps in catalytic processes (23).

We have thus identified several of the molecular components of heterogeneous catalysis. Models that emphasize the importance of the surface structure of catalysts for selectivity, the presence and involvement of organic fragments in HC conversion reactions, and the need for

various oxidation states of surface atoms to obtain desired reaction products are not new to the literature of catalysis. However, the direct relation between these molecular parameters and the catalytic behavior eluded the practitioners of catalysis in the past. Modern surface science provides techniques for determining the atomic structure, composition, and oxidation state of surface atoms and the molecules adsorbed in the monolayer. When these data are combined with studies of the kinetic parameters of catalytic reactions (rates, selectivities, activation energies), the all important relations between the molecular components on the catalyst surface and the high-pressure catalytic reaction behavior can be established.

Direct Study of High-Surface-Area Catalyst Systems

There are many catalyst systems in which the catalyst must be investigated directly rather than by use of models. One of these systems is the zeolites (24), alumina silicates with large surface areas and pores of molecular dimensions. These are the catalysts used by the chemical and petroleum industries for alkylation, cracking, hydrocracking, isomerization, cyclization, and dehydrocyclization. Solid-state nuclear magnetic resonance (NMR) spectroscopy, x-ray photoelectron spectroscopy (XPS), and extended x-ray adsorption fine structure (EXAFS) studies with high-intensity synchrotron radiation, along with a variety of chemical probes, were used to determine the structure and composition of these crystalline materials. Metal-oxide catalyst systems that consist of metals dispersed as small particles on oxide supports with large surface areas (alumina, silica, titanium oxide, zirconia, lanthanum oxide, and magnesium oxide) have also been studied extensively, revealing not only the atomic structure and size of the metal particles and the pore structure of the oxides but also the catalysts' nucleation in solution as clusters and their growth, which control their final structure and composition.

Catalysts have a finite lifetime and thus the causes of their deactivation and the development of methods for their regeneration is a continuous concern to practitioners of catalysis science. The new diagnostic techniques developed by surface science have greatly improved our capabilities to extend catalyst life.

New Generations of Catalysts and New Reactions

The tools of surface science were first used to study, in molecular detail, the working of catalysts that are important in technology. These included platinum, which is used in the production of high-octane gasoline because of its ability to convert saturated alkanes to aromatic, cyclic, and branched organic molecules; silver, for the partial oxidation of ethylene to ethylene oxide; iron, for the ammonia synthesis; and molybdenum, for the hydrodesulfurization and hydrodenitrogenation of organic molecules. From this work, the structure and composition of the catalysts became known on the atomic scale along with the oxidation states of surface atoms, and these observations could be correlated with the reaction rates and product distributions.

The role of additives, which are commonly called promoters because they improve catalytic rates and selectivities to produce the desired molecules, has been unraveled in many cases. Alkali metals, halogens, and transition metals that are used as alloying agents were among the first promoters subjected to detailed surface-science studies. The important role of oxide supports with high surface areas in stabilizing the structure and oxidation states of transition metal catalysts and modifying their chemistry was uncovered. By appropriate variation of the catalyst's composition or structure and of the additives, catalytic behavior could be systematically modified, resulting in even more selective and active catalyst systems. One example of this development is the generation of catalytically superior zeolites with high ratios of silicon to aluminum. The research that gave rise to them, in turn, made possible the discovery of zeolites made of aluminum phosphates. It now appears that crystalline solids of various chemical compositions that contain channels of molecular dimensions similar to those in alumina silicates can be attained. New generations of platinum, silver, and molybdenum catalysts of higher selectivity and longer life have also appeared.

The energy crisis accelerated the development of chemical technologies that produce chemicals and fuels from carbon monoxide and hydrogen. Surface science techniques were used extensively in this research (25). One discovery from this work was that the strong interaction between the metal and the oxide support can be used to control the oxidation state and the surface structure of catalysts. Another was the discovery of surface compounds with unusual thermodynamic and catalytic properties that are formed from two or more metals (often called bimetallic clusters).

Surface science techniques are increasingly being used to develop catalysts for reactions that have not received much attention. These include the photocatalyzed dissociation of water, the use of carbon dioxide as a source of carbon reactant, new methods of nitrogen fixation, and the partial oxidation of methane. The surface reactions of molecules in their excited states are also being studied.

While most catalysis studies have focused on reactions that occur at the solid-gas interface, because of the available techniques, research is now being expanded to include catalysis at solid-liquid interfaces. Reactions of electrode surfaces, of colloid surfaces, and in biological systems, all of which occur at these interfaces, will benefit greatly from scrutiny at the molecular level. Laser spectroscopies and solid-state NMR spectroscopy are among the techniques that appear most promising for these studies.

A number of academic institutions now focus on surface science and catalysis research, and there is strong industrial participation in the newly formed research centers intended to educate scientists and engineers in this subdiscipline of chemistry. As catalysis-based technologies are converted to high technologies, the design and development of new catalyst systems will occur with greater frequency and the development of science and technology will continue at an accelerated pace. The future is indeed bright for surface science and for catalysis science.

References and Notes

1. G. A. Somorjai, *Chemistry in Two Dimensions: Surfaces* (Cornell Univ. Press, Ithaca, N.Y., 1981).
2. B. Davis and W. Hettinger, Eds., *ACS Monogr.* 222 (1983); *Proc. Welch Conf.* 25 (1981); M. Boudart, *Kinetics of Chemical Processes* (Prentice-Hall, Englewood Cliffs, N.J., 1968).
3. G. A. Somorjai and M. A. Van Hove, *Adsorbed Monolayers on Solid Surfaces, Structure and Bonding* (Springer-Verlag, New York, 1979); J. B. Pendry, *Low Energy Electron Diffraction* (Academic Press, New York, 1974); M. A. Van Hove and S. Y. Tong, *Surface Crystallography by LEED* (Springer-Verlag, New York, 1979).
4. S. Overbury, P. Bertrand, G. A. Somorjai, *Chem. Rev.* 75, 550 (1975).
5. J. Sinfelt, *Science* 195, 643 (1977).
6. H. D. Shih, F. Jona, D. W. Jepsen, P. M. Marcus, *Surf. Sci.* 60, 445 (1976); R. L. Strong, B. Firey, F. W. de Wette, J. L. Erskine, *J. Electron Spectrosc. Relat. Phenom.* 29, 187 (1983); R. J. Behm, V. Penka, M.-G. Cattania, K. Christmann, G. Ertl, *J. Chem. Phys.* 78, 7486 (1983).
7. J. T. Yates, T. E. Madey, Jr., J. C. Campuzano, in *Chemical Physics of Solid Surfaces and Heterogeneous Catalysis Series*, vol. 3, *Chemisorption Systems*, D. A. King and D. P. Woodruff, Eds (Elsevier, New York, in press); C. B. Duke, *Appl. Surf. Sci.* 11-12, 1 (1982); R. J. Koestner, M. A. Van Hove, G. A. Somorjai, *Surf. Sci.* 107, 439 (1981); H. J. Behm, K. Christmann, G. Ertl, M. A. Van Hove, *ibid.* 88, L59 (1979).
8. M. A. Van Hove, R. J. Koestner, J. C. Frost, G. A. Somorjai, *Surf. Sci.* 129, 482 (1983).
9. L. L. Kesmodel, L. H. Dubois, G. A. Somorjai, *J. Chem. Phys.* 7, 2180 (1979); R. J. Koestner, M. A. Van Hove, G. A. Somorjai, *Surf. Sci.* 121, 321 (1982); P. Skinner *et al.*, *J. Chem. Soc. Faraday Trans.* 2, 77, 1203 (1981); L. H. Dubois, D. G. Castner, G. A. Somorjai, *J. Chem. Phys.* 72, 5234 (1980); H. Ibach and D. L. Mills, *Electron Energy Loss Spectroscopy and Surface Vibrations* (Academic Press, New York, 1982), p. 326; J. A. Gates and L. L. Kesmodel, *Surf. Sci.* 124, 68 (1983).
10. M. Salmeron and G. A. Somorjai, *J. Phys. Chem.* 86, 341 (1982); C. Minot, M. A. Van Hove, G. A. Somorjai, *Surf. Sci.* 127, 441 (1982).
11. M. Mate and G. A. Somorjai, *Surf. Sci.*, in press.
12. A. J. Gellman, M. H. Farias, M. Salmeron, G. A. Somorjai, *ibid.* 136, 217 (1984); M. H. Farias, A. J. Gellman, R. R. Chianelli, K. S. Liang, G. A. Somorjai, *ibid.* 140, 181 (1984).
13. J. E. Crowell, E. L. Garfunkel, G. A. Somorjai, *ibid.* 121, 303 (1982); E. L. Garfunkel, J. E. Crowell, G. A. Somorjai, *J. Phys. Chem.* 86, 310 (1982); J. E. Crowell, W. T. Tysoe, G. A. Somorjai, in preparation; G. A. Somorjai, Plenary Lecture for International Congress on Catalysis, Berlin, July 1984; G. Ertl, *Proc. Welch Conf.* 25, 179 (1981).

14. F. Zaera and G. A. Somorjai, *J. Catal.* **84**, 375 (1983).
15. A. L. Cabrera, N. D. Spencer, E. Kozak, P. W. Davies, G. A. Somorjai, *Rev. Sci. Instrum.* **53**, 1888 (1982).
16. N. D. Spencer, R. C. Schoonmaker, G. A. Somorjai, *J. Catal.* **74**, 129 (1982); M. Asscher and G. A. Somorjai, *Surf. Sci.* **143**, L389 (1984); M. Asscher, J. Carrazza, M. Khan, K. Lewis, G. A. Somorjai, in preparation.
17. S. M. Davis, F. Zaera, G. A. Somorjai, *J. Am. Chem. Soc.* **104**, 7453 (1982).
18. M. Salmeron, R. J. Gale, G. A. Somorjai, *J. Chem. Phys.* **67**, 5324 (1977); M. Salmeron, R. J. Gale and G. A. Somorjai, *ibid.* **70**, 2807 (1979).
19. L. M. Falicov and G. A. Somorjai, *Proc. Natl. Acad. Sci. U.S.A.*, in press.
20. F. Zaera and G. A. Somorjai, *J. Am. Chem. Soc.* **106**, 2288 (1984); B. E. Koel, B. E. Bent, G. A. Somorjai, in preparation; A. Wieckowski *et al.*, in preparation.
21. S. M. Davis, F. Zaera, G. A. Somorjai, *J. Catal.* **85**, 206 (1984); S. M. Davis, F. Zaera, G. A. Somorjai, *ibid.* **77**, 439 (1982); F. Zaera and G. A. Somorjai, *J. Phys. Chem.* **86**, 3070 (1982).
22. B. A. Sexton and G. A. Somorjai, *J. Catal.* **46**, 167 (1977); P. R. Watson and G. A. Somorjai, *ibid.* **74**, 282 (1982).
23. F. T. Wagner, S. Ferrer, G. A. Somorjai, *ACS Symposium Ser.* **146**, 159 (1981); F. T. Wagner and G. A. Somorjai, *J. Am. Chem. Soc.* **102**, 5494 (1980); Van Damme and K. Hall, *ibid.* **101**, 4373 (1980).
24. J. A. Rabo, Ed., *ACS Monograph. 171* (1976); J. Sinfelt, *Bimetallic Catalysts* (Wiley, New York, 1983).
25. For examples, see the *Journal of Catalysis, Surface Science,* and the *Journal of Physical Chemistry.*
26. H. Conrad, E. Ertl, J. Koch, E. E. Latta, *Surf. Sci.* **43**, 462 (1974).
27. This work was supported by the director, Office of Energy Research, Office of Basic Energy Sciences, Materials and Chemical Sciences Divisions of the U.S. Department of Energy under contract number DE-AC03-76SF00098.

The use of chiral catalysts to obtain high optical yields in the asymmetric hydrogenation of prochiral olefinic substrates represents one of the most impressive achievements to date in catalytic selectivity, rivaling the corresponding stereoselectivity of enzymic catalysts (*1*). Notably high optical yields, approaching 100 percent enantiomeric excess (that is, the excess of one of the mirror-image isomers) have been achieved in the hydrogenation of α-acylaminoacrylic acid derivatives such as **1**

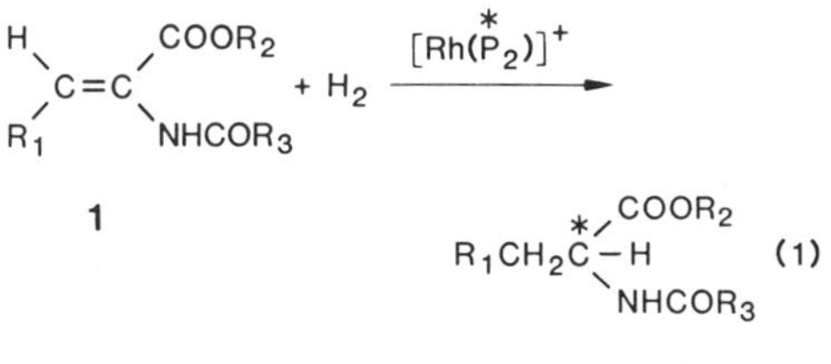

$$\begin{array}{c} \underset{R_1}{\overset{H}{>}}C=C\underset{NHCOR_3}{\overset{COOR_2}{<}} + H_2 \xrightarrow{[Rh(\overset{*}{P_2})]^+} \\ \mathbf{1} \qquad R_1CH_2\overset{*}{C}\underset{NHCOR_3}{\overset{COOR_2}{-}}H \qquad (1) \end{array}$$

to the corresponding amino acid derivatives (reaction 1) with cationic rhodium complexes containing chiral phosphine (especially chelating diphosphine) ligands as catalysts (*1–4*). The commercial synthesis of L-dopa (3,4-dihydroxyphenylalanine) by such a route constitutes an important practical application of this extraordinarily stereoselective catalysis.

Examples of some of the chiral ligands that are effective in such asymmetric catalytic hydrogenation reactions are depicted by structures **2** to **7**.

These catalyst systems are impressive not only for their remarkable stereoselectivities, but also for their very high activities. Extrapolation from low-temperature measurements (*5*) yields turnover frequencies under saturation conditions approaching 10^2 per second at room temperature for reaction 1 catalyzed by cationic rhodium complexes of DIPHOS (**8**) and its chiral derivatives. Even higher catalytic activities are exhibited by rhodium complexes of chelat-

16

Mechanism and Stereoselectivity of Asymmetric Hydrogenation

Jack Halpern

Science 217, 401–407 (30 July 1982)

ing diphosphines that form larger chelate rings, for example, DIOP (**3**) (*6*). Such activities are unusually high for homogeneous hydrogenation catalysts (*7*) and, indeed, lie well up on the scale of activities characteristic of enzymes. Thus, in respect of both selectivity and rate, the behavior of these synthetic catalysts rivals, to an unprecedented degree, that of enzymic catalysts.

This article deals with the kinetic and mechanistic features of these catalyst systems and with the origin of their remarkable stereoselectivities.

Scope of Asymmetric Catalytic Hydrogenation

Asymmetric induction has been achieved with a wide variety of chiral homogeneous hydrogenation catalysts and with a considerable range of substrates (*1–4*). However, only with a relatively limited combination of catalysts

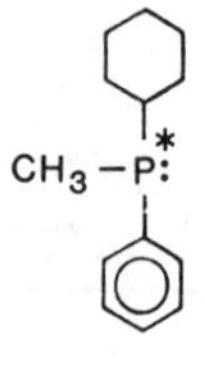

$(CH_3)_3CO-C-N$

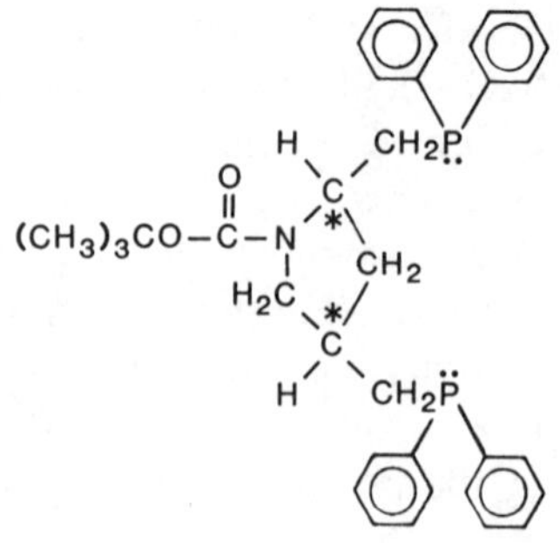

(BPPM)
4

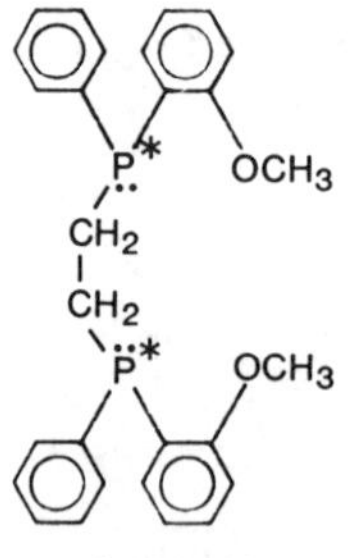

(DIPAMP)
5

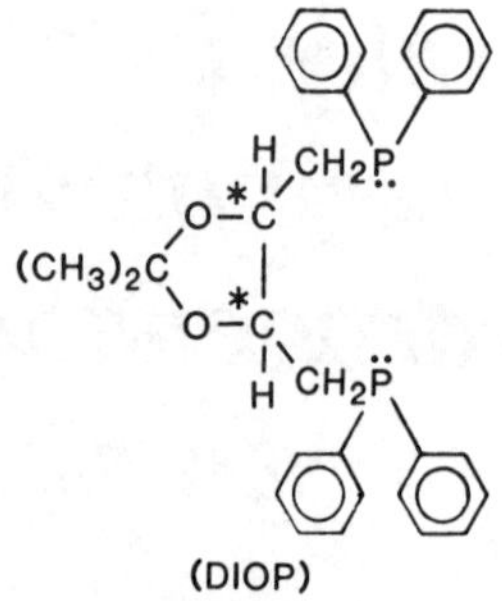

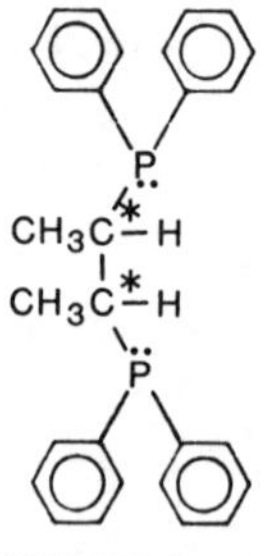

(CHIRAPHOS)
6

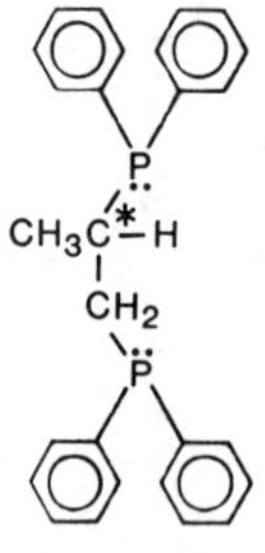

(PROPHOS)
7

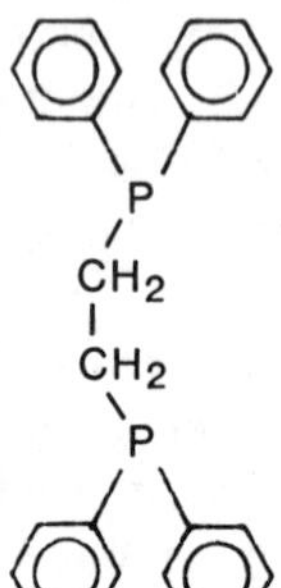

CH_3-P

(PAMP)
2

(DIPHOS)
8

$(CH_3)_2C$

(DIOP)
3

and substrates have very high optical yields ($\gtrsim 95$ percent enantiomeric excess) been obtained—most readily and consistently in the hydrogenation of (*Z*)-α-acylaminoacrylic acids or esters (**1**) catalyzed by cationic rhodium phosphine complexes in polar solvents such as alcohols or acetone. Such catalytic complexes typically contain a chelating diphosphine ligand, although reasonably high optical yields also have been obtained with catalysts containing two monodentate phosphine ligands such as **2** (in both cases Rh:P = 1:2). The range of chiral phosphine ligands, exemplified by **2** to **7**, is extensive and varied (*1, 8*). However, one common feature appears to be the presence on each coordinating phosphorus atom of two ring substituents, such as a phenyl, a substituted phenyl or, occasionally, a cyclohexyl ring. Equally satisfactory results ($\gtrsim 95$ percent enantiomeric excess) have been achieved whether the site of chirality is the coordinating phosphorus atom (**5**) or a substituent or backbone carbon atom (**3, 4, 6,** or **7**). To date only a few other prochiral

$$CH_2=C \begin{array}{c} \overset{O}{\overset{\|}{C}OCH_3} \\ CH_2COCH_3 \\ \overset{\|}{O} \end{array} \qquad CH_2=C \begin{array}{c} \overset{O}{\overset{\|}{C}OC_2H_5} \\ OCCH_3 \\ \overset{\|}{O} \end{array}$$

$$\textbf{9} \qquad\qquad \textbf{10}$$

substrates, for example, certain itaconic acid derivatives (**9**) and certain vinyl acetate derivatives (**10**), have been successfully hydrogenated with high optical yields (*9*).

Mechanism of [Rh(DIPHOS)]$^{+}$–

Catalyzed Hydrogenation

The various studies cited above have served to define the scope of asymmetric catalytic hydrogenation and to delineate the empirical dependence of the rates and stereoselectivities of these reactions on electronic and structural features of the catalysts and substrates. However, a fundamental understanding of these themes, and a rational approach to the

design and modification of such catalysts, must rest ultimately upon a detailed understanding of the catalytic mechanisms. Furthermore, to accommodate the stereochemical features of the reactions, such a mechanistic description must encompass the actual interception and structural characterization of the stereoregulating intermediates as well as the kinetic information necessary to define the reaction pathway and to establish the roles (if any) in the actual catalytic cycle of species that are identified as being present in the reaction system. Kinetic measurements are essential for the elucidation of any catalytic mechanism since catalysis, by definition and significance, is purely a kinetic phenomenon.

Since several of the most effective and widely used chiral ligands for homogeneous catalytic hydrogenation (for example, **5, 6,** and **7**) are simple derivatives of the familiar achiral ligand, 1,2-bis(diphenylphosphino)ethane (DIPHOS, **8**), our initial studies were directed at catalytic systems containing the latter (*10*). As subsequent studies demonstrated, the use of such an achiral "model" as a reference constitutes not only a matter of convenience, but also may be important for the recognition, through appropriate comparisons, of those features of the chiral systems that are distinctively associated with their chiral properties (for example, rates).

The catalyst precursors used in such hydrogenation reactions typically are diene adducts such as [Rh(DIPHOS)-(NOR)]$^{+}$ (**11**) (NOR, norbornadiene). We found that H$_2$ reacts rapidly with such complexes in methanol (S') and related solvents according to reaction 2 and forms the solvated complex [Rh(DIPHOS)S$_2$']$^{+}$ (**12**; abbreviated [Rh(DI-

11

$+$ norbornane   (2)

12

PHOS)]$^+$). The latter, accordingly, is the starting point of the catalytic hydrogenation cycle (*10*).

The catalytic mechanism, as deduced from studies encompassing kinetic measurements as well as the characterization of several intermediates by spectroscopic [notably nuclear magnetic resonance (NMR)] and structural methods, is depicted in Fig. 1 (*5, 10*) for the prototype substrate methyl-(*Z*)-α-acetamidocinnamate (MAC, **13a**). The kinetic parameters are summarized in Table 1 (*5, 6*).

13a (MAC) R = CH$_3$

13b (EAC) R = C$_2$H$_5$

The formation of the [Rh(DIPHOS)-(MAC)]$^+$ adduct (**15**) is rapid and essentially complete ($K_1 = k_1/k_{-1} = $ [**15**]/[**14**][MAC] $\approx 2 \times 10^4 M^{-1}$ at 25°C, where K_1 is the equilibrium constant, and k_1 and k_{-1} are rate constants) even at moderate ($\gtrsim 0.1M$) MAC concentrations. The structure of **15** (Fig. 1) was established by NMR (^{31}P, ^{13}C, and ^{1}H) spectroscopy and by single crystal x-ray analysis of the BF$_4$ salt, revealing chelation of the MAC substrate to the Rh

atom through the carbonyl oxygen of the amide group as well as through normal symmetrical (η^2) coordination of the C=C bond (*11*).

At room temperature, the second step of the catalytic cycle (corresponding to k_2), the reaction of [Rh(DIPHOS)-(MAC)]$^+$ with H$_2$, was found to be rate-determining for the overall catalytic hydrogenation reaction. However, the final product-forming step, corresponding to the rate law $-d[\textbf{17}]/dt = k_4[\textbf{17}]$ exhibited a sufficiently higher activation enthalpy compared with k_2 (17.0 as compared with 6.3 kilocalories per mole) that this step became rate-limiting below $-40°C$, permitting the intermediate **17** to be intercepted and characterized. Figure 1 depicts the structure of **17** as deduced from ^{1}H, ^{31}P, and ^{13}C NMR spectral measurements (*5*). These measurements also served to establish that H-transfer from Rh during the migratory insertion step (**16** → **17**) occurs to the β-carbon atom of the C=C bond while the α-carbon atom becomes bonded to Rh. Although such hydridoalkyl complexes frequently have been postulated as intermediates in homogeneous catalytic hydrogenation reactions (*7, 12*), this is the first time that such an intermediate has actually been intercepted and characterized, and the product C–H bond–forming reductive elimination step has been directly observed.

The only intermediate in the proposed catalytic cycle that has not thus far been intercepted and directly characterized is **16**. The formation of such an intermediate seems highly likely by analogy with the well-recognized oxidative addition reactions of H$_2$ with other Rh^I and related d^8 complexes, particularly in view of the similarity of the activation parameters of k_2 to those that have been deter-

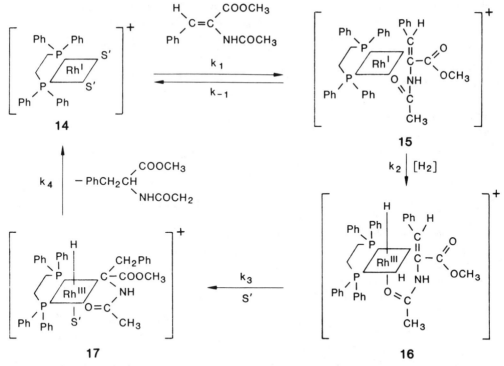

Fig. 1. Mechanism of the [Rh(DIPHOS)]⁺–catalyzed hydrogenation of methyl-(Z)-α-acetamido-cinnamate (MAC).

mined directly for such oxidative addition reactions (7, 13). Failure to intercept **16** apparently reflects the very rapid transformation of **16** to **17**. Indeed, it is the high rate of this step which accounts for the unusually high activity of these catalyst systems and which made possible the interception of the hydridoalkyl intermediate **17**.

The mechanism depicted by Fig. 1 extends also to the [Rh(DIPHOS)]⁺–catalyzed hydrogenation of other olefinic substrates including simple olefins such as 1-hexene (10). However, while such olefins also form intermediate adducts analogous to **15**, the formation constants (corresponding to k_1/k_{-1}) typically are much lower than for (Z)-α-acylamino-acrylic acid derivatives; for example,

$2M^{-1}$ for 1-hexene and $3M^{-1}$ for methyl acrylate, compared to approximately $2 \times 10^4 M^{-1}$ for MAC.

Table 1. Kinetic parameters (rate constant; $\Delta H\ddagger$, the enthalpy of activation; and $\Delta S\ddagger$, the entropy of activation) for the [Rh(DI-PHOS)]⁺–catalyzed hydrogenation of MAC in methanol according to Fig. 1.

Rate constant (units)	k (25°C)	$\Delta H\ddagger$ (kcal/mole)	$\Delta S\ddagger$ (cal/mole °K)
k_1 (M^{-1} sec^{-1})	1.4×10^4		
k_{-1} (sec^{-1})	5.2×10^{-1}	18.3	+2
k_2 (M^{-1} sec^{-1})	1.0×10^2	6.3	−28
k_3 (sec^{-1})	> 1		
k_4 (sec^{-1})	23	17.0	+6

233

Origin of Enantioselection

With the exception of the features identified below that are specifically related to the formation of diastereomers of the adduct corresponding to **15**, the chemistry of cationic complexes of chiral derivatives of DIPHOS, notably of DIPAMP (**5**) and CHIRAPHOS (**6**), and the mechanisms of catalysis by these complexes of the hydrogenation of prochiral substrates such as MAC parallel closely those of the corresponding DIPHOS complexes (*6, 14*). When the mechanistic scheme of Fig. 1 is extended to catalysts containing such chiral ligands, it must be modified in accordance with Fig. 2 to accommodate the formation of diastereomeric forms of the adduct corresponding to **15** (**15′** and **15″**) and of the further reaction intermediates. Invoking the plausible assumptions (i) that the oxidative addition of H_2 and subsequent steps in the catalytic cycle are irreversible (supported by the absence of isotopic exchange with unreacted substrate when D_2 is used instead of H_2), and (ii) that the observed *cis*-addition of H_2 to the coordinated olefin is *endo*, that is, to the Rh-coordinated face, permits the stereochemistry of the product to be correlated with that of the adduct diastereomer from which it is derived in accord with Fig. 2, namely, the *N*-acetyl-(*R*)-phenylalanine ester product from **15′** and the *S*-product from **15″**.

Two possible limiting interpretations may be accorded to the origin of the enantioselection in such systems, namely:

1) The prevailing product chirality is determined by the preferred mode of initial binding of the substrate to the catalyst; that is, the predominant enantiomer of the product arises from the predominant diastereomer of the catalyst-substrate adduct.

2) The predominant enantiomer of the product arises from the minor diastereomer of the catalyst-substrate adduct by virtue of the much higher reactivity of the latter, compared with that of the predominant diastereomer, toward H_2.

Because the formation and dissociation of the catalyst-substrate adducts, and hence interconversion and equilibration of their diastereomeric forms, is rapid (compared with their reactions with H_2) at ambient temperatures and pressures, these two alternatives are kinetically indistinguishable under the usual catalytic conditions. In the absence of definitive proof, several lines of indirect evidence originally were interpreted as favoring the first of the above alternatives (*15–18*). This interpretation also seemed more attractive on conceptual grounds since it corresponds to the familiar lock-and-key concept that has been invoked so widely to explain the characteristically high selectivities of enzymic catalysts.

To resolve this ambiguity, it was necessary to correlate the absolute configurations of the products with those of the intermediate catalyst-substrate adducts. This was first achieved for the hydrogenation of ethyl-(*Z*)-α-acetamidocinnamate (EAC, **13b**) catalyzed by the rhodium complex of *S*,*S*-CHIRAPHOS (**18**) in

$$\textbf{13b} \ + \ H_2 \ \xrightarrow{\text{[Rh(S,S-CHIRAPHOS)]}^+}$$

$$(3)$$

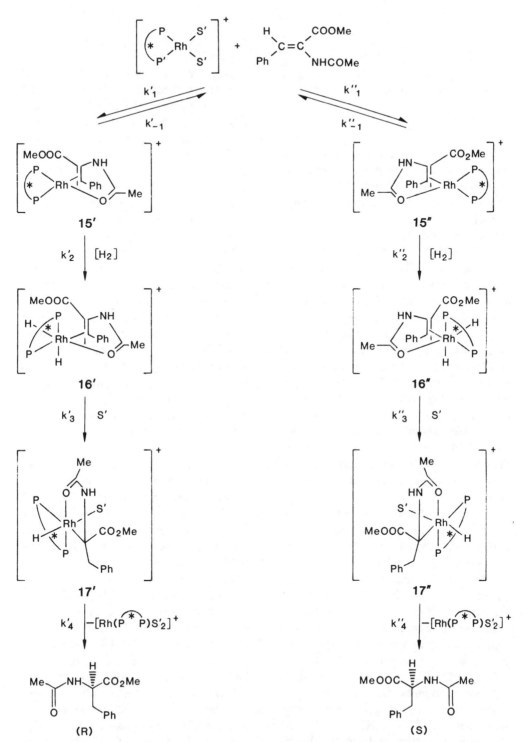

Fig. 2. Mechanistic scheme for the hydrogenation of a prochiral substrate (MAC) with a catalyst containing a chiral chelating diphosphine ligand (P̂*P, CHIRAPHOS or DIPAMP: S', methanol).

235

H3C   CH3
  H ⅲⅲC – C ◄ H
Ph₂P   PPh₂

S,S-CHIRAPHOS   (18)

accordance with reaction 3 (*14*). This reaction was shown to proceed with high enantioselectivity yielding N-acetyl-(*R*)-phenylalanine (> 95 percent enantiomeric excess).

The essential features of reaction 3 were found to parallel those of the corresponding [Rh(DIPHOS)]⁺–catalyzed reaction, as depicted by Fig. 1 (*14*). Formation of a [Rh(*S*,*S*-CHIRAPHOS)-(EAC)]⁺ adduct (**19**) analogous to **15** occurred with a similar equilibrium constant. The electronic spectrum and ³¹P NMR spectrum of **19** also were virtually identical to those of **15**. Only a single diastereomer of [Rh(*S*,*S*-CHIRAPHOS)-(EAC)]⁺ could be identified in solution by NMR and, hence, the other diastereomer (which is expected to exhibit a distinguishable NMR spectrum) must be present to the extent of less than 5 percent.

The rate law for reaction 3 was found to be similar to that for the [Rh(DIPHOS)]⁺–catalyzed reaction, namely, $-d[H_2]/dt = k_5[H_2][19]$, with $k_5 = 1.6 M^{-1}$ sec⁻¹; that is, only about 1/60 the corresponding value (k_2) for the [Rh(DIPHOS)]⁺–catalyzed reaction. This rate difference represents the only significant disparity revealed by the various solution measurements on the [Rh(DIPHOS)]⁺ and [Rh(*S*,*S*-CHIRAPHOS)]⁺ systems (*14*).

The structure of the predominant diastereomer of the [Rh(*S*,*S*-CHIRAPHOS)(EAC)]⁺ ion, determined by x-ray analysis of single crystals of the

perchlorate salt, is depicted in Fig. 3 (*14*, *19*) and is essentially identical to that previously determined for **15** (*11*). Of crucial significance in the present context is the finding that the C_α-re face of EAC is coordinated to the Rh atom. Addition of H₂ to this face, in accord with the mechanism deduced above (Fig. 1), would yield N-acetyl-(*S*)-phenylalanine ethyl ester. Instead it was found that the predominant product of reaction 3 (> 95 percent enantiomeric excess) was the *R* isomer (*3*, *14*).

We are, accordingly, led to the conclusion that it is not the preferred mode of initial binding of the prochiral olefinic substrate to the catalyst but, rather, differences in the rates of subsequent reactions of the diastereomeric catalyst–substrate adducts with H₂, that dictates the enantioselectivity of these catalyst systems. Apparently the minor diastereomer is sufficiently more reactive than the major one that it determines the predominant chirality of the product.

This conclusion also is consistent with, and serves to explain, the observation that, although the structural, equilibrium, and spectroscopic (electronic and NMR) properties of the predominant [Rh(*S*,*S*-CHIRAPHOS)(EAC)]⁺ diastereomer are virtually identical with those of the analogous [Rh(DIPHOS)(EAC)]⁺ complex, its apparent reactivity toward H₂ (k_5) is only about 1/60 of the latter (k_2). In contrast, the reactivity toward H₂ of the [Rh(*S*,*S*-CHIRAPHOS)]⁺ adduct of a simple olefin (1-hexene) actually was found to be about three times higher than that of the corresponding [Rh(DIPHOS)]⁺ adduct. This suggests that the low value of k_5 compared with k_2 is due not to the intrinsically lower reactivity of the *S*,*S*-CHIRAPHOS–derived catalyst but, rather, to the low concen-

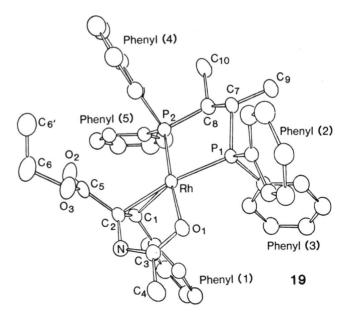

Fig. 3. Structure of the predominant diastereomer of [Rh(*S*,*S*-CHIRAPHOS)(EAC)]⁺.

19

tration of the "reactive" minor diastereomer of the EAC adduct (*14*).

Evidence for the same conclusion concerning the origin of the enantioselection in the [Rh(*R*,*R*-DIPAMP)]⁺–catalyzed hydrogenation of MAC and related substrates is provided by yet another line of evidence. With this catalyst, both diastereomers of the catalyst-substrate adduct can be detected in solution by NMR (although their absolute configurations cannot be assigned) (*6, 15*). At 25°C, the equilibrium ratio of the two diastereomers is about 11:1. Since this is also approximately the ratio of the two enantiomers derived from the hydrogenation of such solutions, the conclusion originally was drawn that the enantioselectivity of the reaction was determined by the ratio of the diastereomeric adduct precursors, the prevailing chirality of the product being that derived from the major diastereomer (*15*). However, by having H_2 react with a solution containing such a mixture of diastereomers of

[Rh(*R*,*R*-DIPAMP)(MAC)]⁺ at low temperatures (around –40°C), where the interconversion of diastereomers is frozen out, it was found that only the minor diastereomer reacts directly with H_2 (to form initially the hydridoalkyl complex corresponding to **17** and then, by reductive elimination, the hydrogenated product (*6, 16*). The subsequent slow reaction of the major diastereomer occurred at a rate independent of the H_2 concentration and approximating the rate of dissociation of the adduct. Thus, consistent with the high enantioselectivity of the reaction (> 95 percent *S* amino acid ester), product formation appears to proceed predominantly through the minor diastereomer in accordance with reaction 4 (*6*).

Kinetic measurements (*6*) on the [Rh(*R*,*R*-DIPAMP)]⁺–catalyzed hydrogenation of MAC in methanol at 25°C yielded the following values for the parameters defined in Fig. 2 (where *k'* and *k"* refer to the major and minor [Rh(*R*,*R*-

Table 2. Effect of H_2 pressure on the optical yield of the [Rh(diphosphine)]$^+$–catalyzed hydrogenation of (Z)-α-benzamidocinnamic acid (20).

H_2 pressure (atm)	Optical yield (percent enantiomeric excess)		
	S,S-BPPM	R,R-DIOP	R,R-DIPAMP
1	83.8 (R)	55.2 (R)	
5	62.3 (R)	8.4 (R)	63.6 (S)
20	21.2 (R)	0.5 (S)	29.9 (S)
50	4.7 (S)	4.9 (S)	
100	8.4 (S)		

[Rh(DIPAMP)(MAC)]$^+_{major}$ ⟶

[Rh(DIPAMP)]$^+$ + MAC ⟶

[Rh(DIPAMP)(MAC)]$^+_{minor}$ $\xrightarrow{H_2}$

[Rh(DIPAMP)(MAC)H$_2$]$^+_{minor}$ ⟶

[Rh(DIPAMP)(MACH)H]$^+_{minor}$ ⟶

[Rh(DIPAMP)]$^+$ + S−MACH$_2$ (4)

DIPAMP)(MAC)]$^+$ diastereomers, respectively); $k'_1 = 5.3 \times 10^3 M^{-1}$ sec^{-1}; $k'_{-1} = 0.15$ sec^{-1}; $k'_2 = 1.1 M^{-1}$ sec^{-1}; $k''_1 = 1.1 \times 10^4 M^{-1}$ sec^{-1}; $k''_{-1} = 3.2$ sec^{-1}; $k''_2 = 6.4 \times 10^2 M^{-1}$ sec^{-1}; K'_1(eq) $= k'_1/k'_{-1} = 3.5 \times 10^4 M^{-1}$; K''_1(eq) $= k''_1/k''_{-1} = 3.3 \times 10^3 M^{-1}$. The approximately 580-fold higher reactivity of the minor diastereomer (k''_2/k'_2) more than offsets its lower concentration ([15"]/[15']) = 0.09) and results in a product ratio of about 60:1 (> 96 percent enantiomeric excess) in favor of the S-enantiomer which is derived from the minor diastereomer.

Other Consequences and Indirect Criteria

The demonstration of the origin of the enantioselection in the two cases considered above depended on either (i) the isolation and structural characterization of the catalyst-substrate adduct (for CHIRAPHOS), or (ii) the detection and monitoring of both diastereomers of the adduct in solution (for DIPAMP). In general, for other catalyst systems, neither of these circumstances may be realized, so that it is necessary to resort to less direct criteria to ascertain whether the same conclusions apply.

One such criterion relates to the dependence of the optical yield on the H_2 concentration. According to the interpretation of the origin of enantioselection deduced above, the reversibility of the initial step of catalyst-substrate adduct formation, through which interconversion of the diastereomeric adducts apparently occurs, should be reduced by increasing the rate of the subsequent H_2 oxidative addition step. Thus, at sufficiently high H_2 concentrations (when $k_2[H_2][15] \gg k_{-1}[15]$), the rate and stereochemistry of the reaction should become determined by the initial binding rates of the prochiral substrate to the catalyst (that is, by k'_1/k''_1). For systems of the type we have described, this predicts that the enantioselectivity should decrease, with the possibility of eventual reversal of predominant product chirality, with increasing H_2 pressure. This has been quantitatively confirmed for the [Rh(DIPAMP)]$^+$–catalyzed hydrogenation of MAC (6). Furthermore, such an inverse dependence of optical yield on the H_2 pressure, exemplified by the results in Table 2, has been observed for virtually all of the asymmetric hydrogenation catalysts that have thus far been examined (2, 6, 20, 21). This feature of the behavior of these systems limits the scope of achieving higher rates, while still maintaining high optical yields, by increasing the H_2 pressure. Insofar as this inverse dependence of optical yield is fairly general for the catalytic hydrogenation of α-acylaminoacrylic acid derivatives with catalysts containing a variety of chiral phosphine ligands, we conclude that our interpretation of the enantioselection in such reactions probably extends to this whole class of catalysts and substrates.

A related, and somewhat surprising, consequence of our analysis concerns the temperature dependence of the optical yield. According to our interpretation of the origin of the enantioselection, high enantioselectivity depends on rapid interconversion of the diastereomeric catalyst-substrate adducts compared with the rates of their reactions with H_2. Since the adduct dissociation step through which such interconversion occurs typically has a much higher activation enthalpy than reaction with H_2 (18.3 compared to 6.3 kcal/mole for 15), the diastereomer interconversion process should become "frozen out" at sufficiently low temperatures. This leads to the prediction that, provided the major diastereomer exhibits some reactivity toward H_2 (as it must if the enantiomeric excess is less than 100 percent), the optical yield may actually decrease with decreasing temperature. Such an unusual dependence of enantioselectivity on temperature has been reported, for example, for the hydrogenation of MAC and related substrates catalyzed by rhodium complexes containing several chiral phosphine ligands (6, 20, 21). In one particularly dramatic case, the optical yield increases from 0 to 60 percent enantiomeric excess in going from 0° to 100°C (21).

Origin of Stability and Reactivity Differences of Diastereomeric Adducts

The conclusion that we have reached concerning the origin of enantioselection in these systems implies very large differences in reactivity toward H_2 between the two diastereomeric forms of the catalyst-substrate adduct. With [Rh(S,S-CHIRAPHOS)(EAC)]$^+$, the minor (less stable) diastereomer must be at least 10^3 times as reactive toward H_2 at 25°C as the major (more stable) diastereomer in order to account for the observed optical yield. In the case of [Rh(R,R-DIPAMP)-(MAC)]$^+$, where both diastereomers are observed, the corresponding reactivity difference has been directly measured to be about 580. The origin of these marked differences in reactivity clearly constitutes an important aspect of the behavior of these systems.

It has been pointed out (18) that a general feature of the conformations of the diarylphosphine ligands that are common to these catalysts is the arrangement of the four phenyl groups in a chiral alternating "edge-face" array depicted schematically (20) for [Rh(R,R-DIPAMP)]$^+$ and also re-

20

239

vealed by the structure of Fig. 3. The enantiorecognition involved in the binding of prochiral substrates such as (Z)-α-acylaminocinnamic acid derivatives to such a chiral template and the differences in stability between the diastereomers of the resulting adducts (for example, of [Rh(R,R-DIPAMP)(MAC)]$^+$ or [Rh(S,S-CHIRAPHOS)(EAC)]$^+$) has been attributed to the "matching" of the two faces of the substrate to the chiral "template" formed by this array of phenyl rings. The chirality and rigidity of this template, in turn, are determined by the nature of the chiral centers in the phosphine ligand and of the backbone connecting the phosphorus atoms. This model, as reflected in the structure depicted in Fig. 3, appears to provide a satisfactory basis for interpreting the stability difference between the diastereomers of the initial catalyst-substrate adducts; for example, $\Delta G° = 1.4$ kcal/mole, $\Delta H° = 2.2$ kcal/mole, $\Delta S° = 2.7$ cal/mole °K for [Rh(R,R-DIPAMP)-(MAC)]$^+$ (6), where $\Delta G°$, $\Delta H°$, and $\Delta S°$ are the standard free energy change, the standard enthalpy change, and the standard entropy change, respectively.

The origin of the striking differences between the reactivities of the diastereomeric adducts toward H$_2$ is more difficult to understand. It is, of course, not unexpected that the less stable of a pair of diastereomers will exhibit the higher reactivity by virtue of its higher initial free energy. However, to account for the enantioselectivity of the reaction, the difference in reactivity must be much greater than the difference in stability (that is, the difference in equilibrium concentration) of the diastereomers, indeed at least 50 times greater to accommodate an optical yield of 96 percent enantiomeric excess. At this stage, the origin of this marked reactivity differ-

ence (corresponding to $\Delta\Delta G^{\ddagger} = 4$ kcal in the case of [Rh(R,R-DIPAMP)(MAC)]$^+$) is still a matter of speculation. A reasonable suggestion is that the reactivity difference has its origin in the stability difference of the diastereomers of the initial product of the oxidative addition of H$_2$, the relative stabilities of the diastereomeric products being opposite to that of the parent catalyst-substrate adducts. Thus, the greater stability of the diastereomer of the dihydride, RhH$_2$(R,R-DIPHOS)(MAC)]$^+$, derived from the less stable diastereomer of [Rh(R,R-DIPHOS)(MAC)]$^+$, enhances the driving force and rate of the reaction of the former with H$_2$, compared with the rate of the more stable diastereomer. The reaction profiles corresponding to this situation are depicted schematically in Fig. 4. Examination of space-filling models provides some support for such an interpretation and suggests that a systematic reason for the inverted stabilities of the diastereomers of the initial catalyst-substrate adducts and the dihydrides derived from them, may be the *trans*-disposition of the substrate and diphosphine chelate rings in the former case and *cis*-disposition in the latter (Fig. 4). However, the space-filling models of these complexes (particularly of **16**) are extremely crowded and conclusions based on such models must be considered to be of limited reliability. Efforts are being made to explore this theme further and more directly, by actually intercepting and characterizing the dihydride intermediates in question.

Concluding Remarks

This article has dealt with a relatively restricted class of asymmetric catalytic

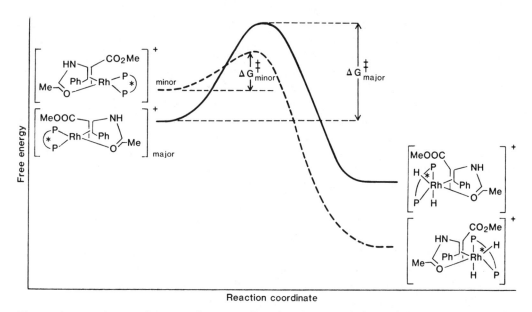

Fig. 4. Schematic reaction coordinate profiles for the enantiodetermining reactions of the diastereomeric [Rh(DIPAMP)(MAC)]$^+$ catalyst-substrate adducts with H$_2$ ($\Delta G^\ddagger$, the free energy of activation).

reactions, namely, the hydrogenation of α-acylaminoacrylic acid derivatives and related substrates, catalyzed by rhodium complexes containing chiral phosphine ligands. The scope of asymmetric catalysis by metal complexes is considerably more extensive, and other substrates as well as other reactions, such as hydroformylation (22), hydrosilylation (23), and olefin coupling (24), have been studied. While moderate optical yields, in some cases up to 70 percent, have been achieved in such reactions, only in one other case do these consistently approach the high enantioselectivities (> 90 percent enantiomeric excess) characteristic of the hydrogenation reactions discussed in this article. This case involves the epoxidation of allylic alcohols by *tert*-butylhydroperoxide, catalyzed by titanium alkoxides in the presence of a tartarate ester as the chiral component (25). The latter catalyst system, like those discussed in this article, exhibits high enantioselectivity only for a restricted class of substrates. Furthermore, in each case, the effective substrates are characterized by the presence, in addition to the reactive olefin site, of a neighboring binding site (a carbonyl and hydroxyl substituent, respectively) which, in combination with the olefin site can effect the anchoring of the substrate to the catalyst through chelation.

The studies described in this article constitute a detailed elucidation of the mechanism of such an asymmetric catalytic reaction. The most significant and surprising conclusion yielded by these studies is that, contrary to the view generally held previously, it is not the preferred mode of the initial binding of the prochiral substrate to the chiral catalyst, but rather overcompensating differences in the rates of the subsequent

241

reactions of the diastereomeric catalyst-substrate adducts that dominate the enantioselectivity of asymmetric catalytic hydrogenation. The predominant product enantiomer arises from the minor (less stable) diastereomer of the adduct, which frequently does not accumulate in sufficient concentration to be detected. In contrast, the major diastereomer, whose stability reflects the optimal fitting of the prochiral substrate to the chiral template of the catalyst and which is the principal species present under catalytic conditions, is unreactive and corresponds to a "dead-end" complex. Based on a variety of criteria, including the temperature- and H_2 pressure–dependence of the optical yield, this conclusion appears to be quite general for the variety of chiral catalysts identified for this class of reactions. It also finds parallels in the conclusions reached in studies on the mechanisms of other homogeneous catalytic reactions, for example, the [Rh(PPh₃)₃Cl]-catalyzed hydrogenation of olefins. In the latter case, it also was demonstrated that virtually none of the species that accumulate under the conditions of the catalytic reactions in sufficient concentrations to be detected actually are intermediates in the catalytic reaction (7). This serves to underscore the unreliable, and often misleading, mechanistic conclusions that may be derived by simply identifying the species present in a catalytic system unless such observations are coupled with kinetic and other appropriate (for example, stereochemical) evidence to define the role of such species in the reaction mechanism.

An important question that remains to be addressed is how widely our conclusion concerning the origin of enantioselection applies to other stereoselective catalyst systems. This question is particularly significant in the context of enzymic reactions where a contrary interpretation (embodied in the familiar lock-and-key concept) often is assumed; namely, that the stable adducts that are formed by optimal fitting of the substrate to the enzyme, and that so frequently are the focus of direct observation and characterization, are intermediates in the enzymic reactions. Since the formation of such adducts usually is rapid compared with the catalytic reaction, direct support for such a conclusion often is lacking, and the results of our studies on asymmetric hydrogenation raise serious questions about its widely accepted plausibility. The criteria for distinguishing between the alternative interpretations that need to be considered are not readily realized for many catalytic reactions, but the studies described in this article provide some leads as to how such criteria might be developed.

References and Notes

1. For reviews, see H. B. Kagan, *Pure Appl. Chem.* **43**, 401 (1975); J. D. Morrison and W. F. Masler, *Adv. Catal.* **25**, 81 (1976); P. Pino and G. Consiglio, in *Fundamental Research in Homogeneous Catalysis*, M. Tsutsui and R. Ugo, Eds. (Plenum, New York, 1977), p. 147; H. B. Kagan and J. C. Ciaud, *Top. Stereochem.* **10**, 175 (1978); D. Valentine, Jr., and J. W. Scott, *Synthesis* **1978**, 329 (1978); B. Bosnich and M. D. Fryzuk, *Top. Stereochem.* **12**, 119 (1981); H. Wynberg, *Rec, J. Roy. Netherlands Chem. Soc.* **100**, 399 (1981).
2. W. S. Knowles, M. J. Sabacky, B. D. Vineyard, *Adv. Chem. Ser.* **132**, 274 (1974); B. D. Vineyard, W. S. Knowles, M. J. Sabacky, G. L. Bachman, D. J. Weinkauff, *J. Am. Chem. Soc.* **99**, 5946 (1977).
3. M. D. Fryzuk and B. Bosnich, *J. Am. Chem. Soc.* **99**, 6262 (1977); P. A. MacNeil, N. K. Roberts, B. Bosnich, *ibid.* **103**, 2273 (1981).
4. I. Ojima, T. Kogure, N. Yoda, *J. Org. Chem.* **45**, 4728 (1980).
5. A. S. C. Chan and J. Halpern, *J. Am. Chem. Soc.* **102**, 838 (1980).
6. C. Landis, S. Okrasinski, J. Halpern, unpublished results.
7. J. Halpern, *Inorg. Chim. Acta* **50**, 11 (1981).
8. H. B. Kagan, *Ann. N.Y. Acad. Sci.* **333**, 1 (1980); L. Marko and J. Bakos, in *Aspects of Homogeneous Catalysis*, R. Ugo, Ed. (Reidel,

Dordrecht, Netherlands, 1981), vol. 3, p. 145; R. E. Merill, *Chemtech* **1981**, 118 (1981).

9. W. C. Christophel and B. D. Vineyard, *J. Am. Chem. Soc.* **101**, 4406 (1979); K. E. Koenig, G. L. Bachman, B. D. Vineyard, *J. Org. Chem.* **45**, 2362 (1980).

10. J. Halpern, D. P. Riley, A. S. C. Chan, J. J. Pluth, *J. Am. Chem. Soc.* **99**, 8055 (1977).

11. A. S. C. Chan, J. J. Pluth, J. Halpern, *Inorg. Chim. Acta* **37**, 2477 (1979).

12. B. R. James, *Homogeneous Hydrogenation* (Wiley, New York, 1969), and references therein.

13. J. Halpern, *Acc. Chem. Res.* **3**, 386 (1970).

14. A. S. C. Chan, J. J. Pluth, J. Halpern, *J. Am. Chem. Soc.* **102**, 5952 (1980).

15. J. M. Brown and P. A. Chaloner, *Tetrahedron Lett.* (1978), p. 1877; *J. Chem. Soc. Chem. Commun.* (1978), p. 321.

16. _____, *J. Chem. Soc. Chem. Commun.* (1980), p. 344.

17. I. Ojima and T. Kogure, *Chem. Lett.* **1978**, 1175 (1978); *ibid.* **1979**, 641 (1979).

18. W. S. Knowles, B. D. Vineyard, M. J. Sabacky, B. R. Stults, in *Fundamental Research in Ho-mogeneous Catalysis*, M. Tsutsui, Ed. (Plenum, New York, 1979), vol. 3, p. 537; W. E. Koenig et al., *Ann. N.Y. Acad. Sci.* **333**, 16 (1980).

19. P. S. Chua, N. K. Roberts, B. Bosnich, S. J. Okrasinski, J. Halpern, *J. Chem. Soc. Chem. Commun.* (1981), p. 1278.

20. I. Ojima, T. Kogure, N. Yoda, *Chem. Lett.* **1979**, 495 (1975); *J. Org. Chem.* **45**, 4728 (1980).

21. D. Sinou, *Tetrahedron Lett.* **22**, 2987 (1981).

22. P. Pino, G. Consiglio, C. Botteghi, C. Salomon, *Adv. Chem. Ser.* **132**, 295 (1974); C. F. Hobbs and W. S. Knowles, *J. Org. Chem.* **46**, 4422 (1981).

23. I. Ojima, K. Yamamoto, M. Kumada, in *Aspects of Homogeneous Catalysis*, R. Ugo, Ed. (Reidel, Dordrecht, Netherlands, 1977), vol. 3, p. 186.

24. B. Bodganovich, *Angew. Chem. Int. Ed. Eng.* **12**, 954 (1973).

25. T. Katsuki and K. B. Sharpless, *J. Am. Chem. Soc.* **102**, 5974 (1980); B. E. Rossiter, T. Katsuki, K. B. Sharpless, *ibid.* **103**, 464 (1981).

26. The research encompassed by this article was supported by a grant from the National Science Foundation.

Transition metal complexes have long been used to catalyze reactions of organic molecules (*1*). A relatively simple example is addition of molecular hydrogen to a carbon-carbon double bond in an alkene (olefin) to give an alkane with platinum metal used as the catalyst (*2*). This is a heterogeneous catalytic reaction since the catalyst (a solid) is in a different phase than the reactants (gas or solution). When the catalyst and reactants are in the same phase (usually in solution) the catalytic reaction is said to be homogeneous.

One of the most important and fundamental types of intermediates in a catalytic reaction involving organic molecules is one that contains a metal-carbon single bond, that is, a metal alkyl. A great many well-defined metal alkyl compounds have been isolated, and reactions that are steps in known homogeneous catalytic reactions have been studied in great detail. Yet there are still some reactions of metal alkyls that are not well understood. An important example is polymerization of ethylene (or propylene) by catalysts of the Ziegler-Natta type to give high molecular weight, crystalline polymers, a reaction that was discovered almost 30 years ago (*3*) and is believed to involve growth of an alkyl chain bound to a transition metal by "insertion" of the ethylene into the metal-alkyl (M–R) bond (Eq. 1).

$$M-R \xrightarrow{\quad CH_2=CH_2 \quad} M-CH_2-CH_2-R$$

$$\xrightarrow{\quad (n-1)CH_2=CH_2 \quad} M(CH_2CH_2)_nR \qquad (1)$$

Fischer's discovery of the carbene complex (*4, 5*) $(CO)_5W=C(OCH_3)(C_6H_5)$ in 1964 demonstrated that transition metals can form double bonds to carbon (*6*). At about the same time a remarkable

17

Catalysis by Transition Metals: Metal-Carbon Double and Triple Bonds

Richard R. Schrock

Science 219, 13–18 (7 January 1983)

unprecedented catalytic reaction called the olefin metathesis reaction was discovered. In this reaction an ill-defined molybdenum, tungsten, or rhenium catalyst (homogeneous or heterogeneous) rapidly establishes the equilibrium shown in Eq. 2 (R and R′ are alkyl groups) from either direction (7). About 5 years later it was proposed that the

$$2RCH{=}CHR' \rightleftarrows \quad RCH=CHR + \tag{2}$$
$$R'CH{=}CHR'$$

catalyst was a carbene complex and that the reaction consisted of random, reversible formation of all possible metallacyclobutane rings (for instance, Eq. 3). But the fact that complexes such as

$$M{=}CHR + RCH{=}CHR' \longrightarrow M{-}CHR$$
$$R'HC{-}CHR$$
$$\tag{3}$$
$$\longrightarrow M{\equiv}CHR' + RCH{=}CHR$$

$(CO)_5W{=}C(OCH_3)(C_6H_5)$ did not catalyze the metathesis of simple olefins suggested that the as yet uncharacterized metathesis catalysts were probably significantly different from those discovered by Fischer.

A tantalum alkylidene complex (8) discovered in 1974 (Eq. 4; Me = CH_3) was the first of many (9) that appeared to be fundamentally different from the carbene

$$Ta(CH_2CMe_3)_3Cl_2 + 2LiCH_2CMe_3 \xrightarrow{-2LiCl}$$
$$CMe_4 + (Me_3CCH_2)_3Ta{=}C\overset{H}{\underset{CMe_3}{\diagup}} \tag{4}$$

complexes discovered by Fischer. If we view the neopentylidene ligand as a dianion (8), it is isoelectronic with an oxo (O^{2-}) ligand and the metal is formally in its highest oxidation state, Ta(V). These types of molecules will react with alkenes as shown in Eq. 3. A few years

later related compounds containing a metal-carbon triple bond (alkylidyne complexes) were discovered. These alkylidyne complexes react with alkynes as shown in Eq. 5 and therefore catalyze the almost unknown alkyne metathesis

$$M{\equiv}CR' + RC{\equiv}CR \longrightarrow M{\equiv}CR + RC{\equiv}CR' \tag{5}$$

reaction. In this article I will discuss the preparation and structure of high-oxidation-state alkylidene and alkylidyne complexes, how they metathesize alkenes and alkynes, and what other reactions involving multiple metal-carbon double or triple bonds have been discovered recently. As we will see, even relatively old catalytic reactions such as alkene insertion into a metal-alkyl bond may actually involve alkylidene-like ligands.

All the transition metal complexes discussed here are well-defined compounds that catalyze reactions in solution. There is no good reason why the principles behind this chemistry should not extend to heterogeneous catalytic systems, since all the homogeneous catalytic reactions mentioned here are also catalyzed by heterogeneous catalysts.

Preparation and Structure of

Alkylidene and Alkylidyne Complexes

The first tantalum alkylidene complex was prepared by the reaction shown in Eq. 4. The unique aspect of this reaction is what is called α-hydrogen atom abstraction. In its simplest form α-hydrogen atom abstraction consists of migration of a hydrogen atom from the α-carbon atom (the one bound directly to the metal) of one alkyl group to another (Eq. 6). The α-hydrogen atom abstraction reaction has also been observed in complexes containing other alkyl ligands (such as benzyl, CH_2Ph; $Ph = C_6H_5$) that do not have any hydrogen atoms on their β-carbon atom. An alkylidene ligand can also be formed by removing a proton from the α-carbon atom in a cationic alkyl complex by adding some external base (Eq. 7).

$$\text{(6)}$$

$$\longrightarrow \quad \text{Ta=CHCMe}_3 + \text{CMe}_4$$

$$\text{(7)}$$

In $Ta(\eta^5\text{-}C_5H_5)_2(CH_3)(CH_2)$ the $Ta=CH_2$ bond length (2.03 angstroms) is approximately 10 percent less than the $Ta–CH_3$ bond length (2.25 Å), and the $Ta=C–H$ angle is $\sim 125°$.

An unusual structural feature of many tantalum alkylidene complexes is an inordinately large metal–C_α–C_β angle. Instead of $\sim 125°$ it can be as high as $175°$. An example is shown in Fig. 1. The reason is believed to be that the metal draws electron density from the C_α–H_α bond toward it and, in effect, partially forms a metal-carbon bond of multiplicity greater than two. As a result, the α-hydrogen atom is placed in a nonclassical bridging position between C_α and Ta, the tantalum-carbon "double bond" is too short (~ 1.90 Å), and the C_α–H_α bond is much longer (1.13 Å) than it should be (1.09 Å). The reason why the methylene ligand in the $Ta(\eta^5\text{-}C_5H_5)_2(CH_3)(CH_2)$ complex above is not distorted is that the metal has already achieved its maximum possible, 18-electron configuration (10).

The reaction shown in Eq. 8 was an attempt to prepare $W(CHCMe_3)(OCMe_3)_4$ by exchanging an oxo ligand on tungsten with a neopentylidene ligand

$$\text{(8)}$$

$$R = CMe_3; \quad L = PEt_3$$

on tantalum (in Eq. 8, $Et = CH_2CH_3$). The reaction took an entirely different course to give an oxo alkylidene complex (11). The most significant feature of tungsten oxo alkylidene complexes is that the neopentylidene ligand is not distorted significantly by the metal. For example, the $W=C_\alpha–C_\beta$ angle in $W(O)(CHCMe_3)(PEt_3)Cl_2$ is $\sim 141°$ (12), the smallest of any observed so far (Fig. 2). It is believed that the electron-deficient metal can acquire electron density more easily from the two nonbonded π-electron pairs on the oxo ligand than it can from the bonding electron pair between C_α and H_α in the neopentylidene ligand. In fact, the oxo ligand is one of the best π-electron donors known (13). Recently, it was found that two alkoxide ligands can take over the role of the oxo ligand (14) in trigonal bipyramidal molecules of the type $W(CHCMe_3)(OR)_2Cl_2$.

Alkylidyne complexes are formed by an α-hydrogen atom abstraction reaction (Eq. 9) which is related to the reaction that produces alkylidene complexes (Eq. 6). An important example is shown in Eq. 10 (15). Typical tantalum-carbon or tungsten-carbon triple bond lengths are ~ 1.80 Å, ~ 10 percent shorter again than double bond lengths. The $M\equiv C_\alpha$–

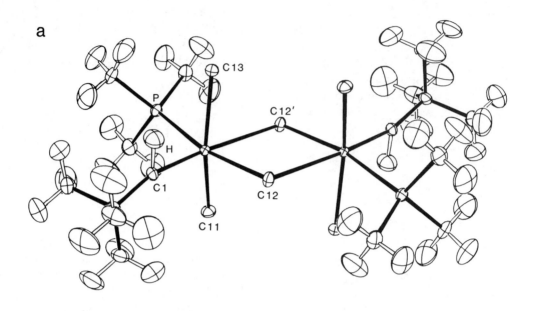

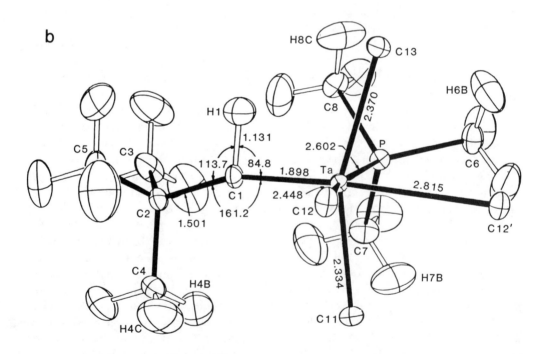

Fig. 1. (a) Computer-generated (ORTEP II) drawing of [Ta(CHCMe₃)(PMe₃)Cl₃]₂; (b) detail of half the structure. Except for H1, all hydrogen atoms have been omitted for clarity.

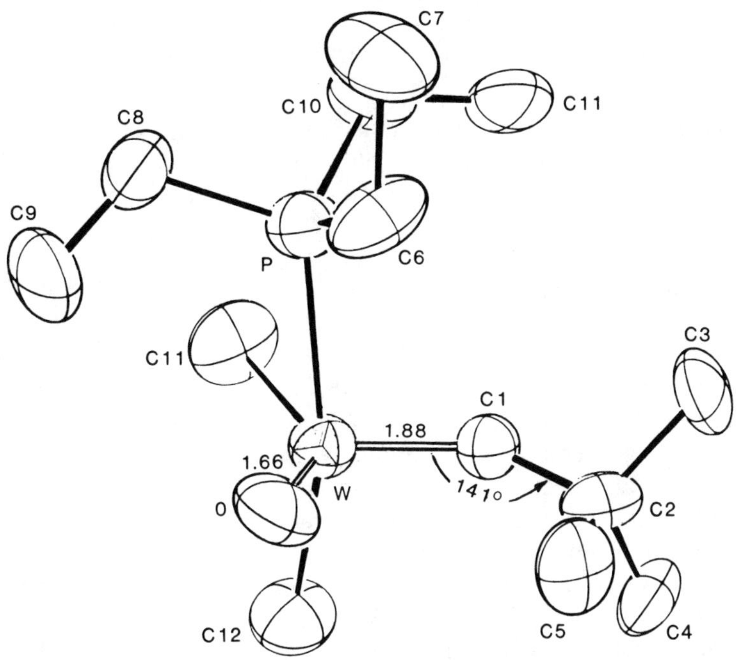

Fig. 2. ORTEP II drawing of W(O) (CHCMe₃) (PEt₃) Cl₂. All hydrogen atoms have been omitted for clarity.

C_β angle is usually within a few degrees of 180°, as it should be by analogy with an alkyne.

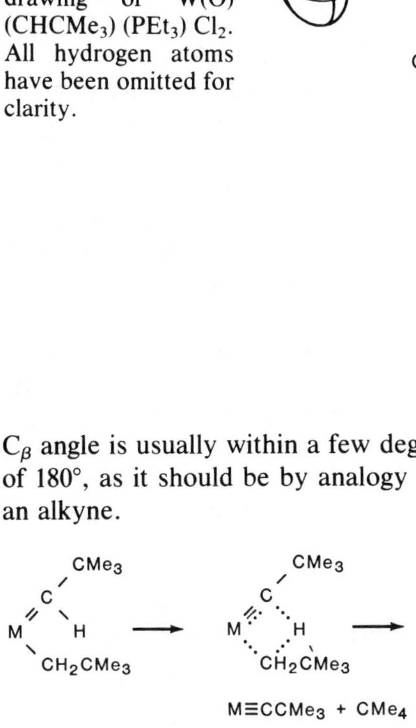

$$M \equiv CCMe_3 + CMe_4$$

$$WCl_6 + 6LiCH_2CMe_3 \longrightarrow$$ (10)
$$(Me_3CCH_2)_3W \equiv CCMe_3$$

An important reaction that is related to α-hydrogen atom abstraction is what has been called α-hydride elimination (*16*). Here the α-hydrogen atom migrates from an α-carbon atom to the metal, rather than to the α-carbon atom of another alkyl ligand. This can occur only when the complex that results is isoelectronic with other alkylidene or alkylidyne complexes. Therefore, the initial alkyl or

alkylidene complexes must first be reduced by two electrons. Examples are shown in Eqs. 11 and 12. Not surprisingly, α-hydride elimination is reversible; that is, Ta(CHCMe₃)(H) (PMe₃)₃Cl₂ is in equilibrium with Ta(CH₂CMe₃)(PMe₃)₃Cl₂.

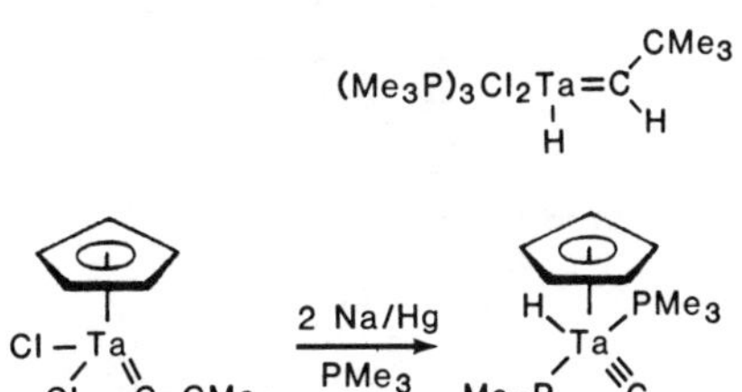

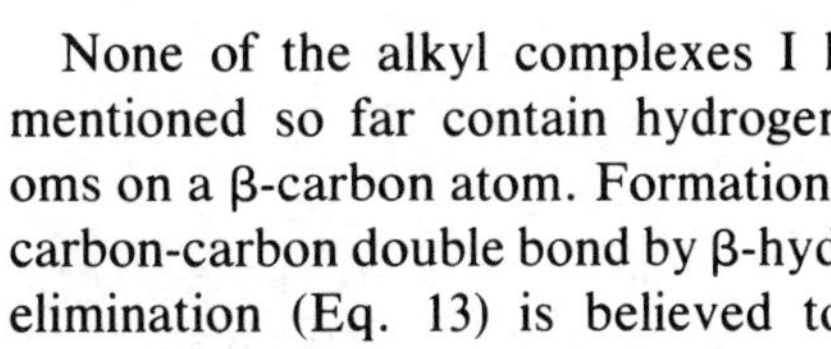

None of the alkyl complexes I have mentioned so far contain hydrogen atoms on a β-carbon atom. Formation of a carbon-carbon double bond by β-hydride elimination (Eq. 13) is believed to be

249

$$M-CH_2-CH_3 \longrightarrow M \overset{H}{\underset{CH_2}{\overset{|}{\underset{\diagdown}{\diagup}}}} \overset{CH_2}{\underset{}{}} \quad (13)$$

faster than formation of a metal-carbon double bond by α-hydrogen atom abstraction or α-hydride elimination.

Reaction of Alkylidene

Complexes with Alkenes

The niobium and tantalum alkylidene complexes whose reactions with alkenes have been studied most thoroughly are octahedral complexes, for example, $Ta(CHCMe_3)Cl_3(PMe_3)_2$ (17). Each member of the class of $M(CHR)X_3L_2$ complexes (M = Nb or Ta; R = CMe_3 or Ph; X = Cl or Br; L = an alkyl phosphine ligand) reacts with an alkene to give an intermediate metallacyclobutane complex (for instance, Eq. 14). Since β-hydrogen atoms are now present, a process related to that shown in Eq. 13 converts the metallacyclobutane ring into an olefin (Eq. 15). Another possible fate of the metallacyclobutane complex,

$$M=C\overset{CMe_3}{\underset{H}{\diagdown}} \;+\; CH_2=CH_2 \longrightarrow M\overset{\overset{CMe_3}{\overset{|}{C}}\diagdown H}{\underset{CH_2}{\overset{\diagup}{\diagdown}}}\overset{CH_2}{\underset{}{}} \quad (14)$$

$$M\overset{\overset{H}{\diagdown}\overset{CMe_3}{\diagup}}{\underset{CH_2}{\overset{C}{\diagup}}}\overset{CH_2}{\underset{}{}} \longrightarrow M\overset{\overset{H}{\diagdown}\overset{CMe_3}{\diagup}}{\underset{\overset{\|}{C}}{\overset{C}{}}}\overset{H}{\underset{H}{}} \quad (15)$$

$$\longrightarrow M\overset{H}{\underset{C}{\overset{|}{\underset{\diagdown}{\diagup}}}}\overset{CH_2CMe_3}{\underset{H}{}}$$

and a step that is the heart of the alkene metathesis reaction, is re-formation of

the other possible alkylidene complex, a methylene complex, by loss of t-butylethylene (Eq. 16). Evidently, rearrange-

$$\overset{H}{\underset{M}{\overset{\diagdown}{\overset{C}{\underset{CH_2}{\diagup\diagdown}}}}}\overset{CMe_3}{\underset{CH_2}{}} \longrightarrow Me_3CCH=CH_2 \;+\; M=CH_2 \quad (16)$$

ment of the metallacyclobutane ring is faster than cleavage of the ring. If we are interested in catalytically metathesizing alkenes we might begin by attempting to slow down the rate of rearrangement of the metallacyclobutane ring.

On the basis of some related work involving rearrangement of metallacyclopentane (MC_4) rings (18) we believed that replacing some chloride ligands with alkoxide ligands would yield that result. Indeed, $Ta(CHCMe_3)(OCMe_3)_2Cl(PMe_3)_2$ reacts with ethylene to give only cleavage products (Eq. 17). $Ta(CH_2)(OCMe_3)_2Cl(PMe_3)_2$ cannot be isolated because it decomposes readily (Eq. 18) by a process that is probably re-

$$Ta(CHCMe_3)(OCMe_3)_2ClL_2 \xrightarrow{\overset{excess}{C_2H_4}} \\ Me_3CCH=CH_2 + Ta(CH_2)(OCMe_3)_2ClL_2 \quad (17) \\ (L = PMe_3)$$

$$2Ta(CH_2)(OCMe_3)_2ClL_2 \longrightarrow \\ Ta(CH_2=CH_2)(OCMe_3)_2ClL_2 \quad (18) \\ + Ta(OCMe_3)_2ClL_2$$

lated to the bimolecular decomposition of $Ta(\eta^5\text{-}C_5H_5)_2(CH_2)(CH_3)$ to give, among other things, $Ta(\eta^5\text{-}C_5H_5)_2(CH_2=CH_2)(CH_3)$ (19). When a terminal alkene is added to $Ta(CHCMe_3)(OCMe_3)_2ClL_2$ the reaction proceeds largely as shown in Eq. 19. The Ta(CHR)

$$Ta(CHCMe_3)(OCMe_3)_2ClL_2 \xrightarrow{RCH=CH_2} \\ Me_3CCH=CH_2 + Ta(CHR)(OCMe_3)_2ClL_2 \quad (19)$$

250

(OCMe$_3$)$_2$ClL$_2$ complexes do not decompose bimolecularly nearly as fast as Ta(CH$_2$)(OCMe$_3$)$_2$ClL$_2$ does. Therefore they live long enough to react with more RCH=CH$_2$, and they can do so either in a "productive" (Eq. 20) or "degenerate" (Eq. 21) sense. But the methylene complex (Eq. 20) must still be formed in

$$Ta{=}C\overset{R}{\underset{H}{<}} + RCH{=}CH_2 \longrightarrow Ta\underset{\overset{|}{\underset{H}{C}}\overset{}{\underset{R}{\diagdown}}}{\overset{\overset{H\ R}{\diagdown/}}{\underset{|}{C}}\diagup\overset{}{H}}$$

$$\qquad\qquad (20)$$

$$\longrightarrow Ta{=}CH_2 + RCH{=}CHR$$

order to catalytically metathesize a terminal alkene to give ethylene and the internal alkene. Since the methylene complex decomposes, terminal alkenes cannot be metathesized catalytically.

$$Ta{=}C\overset{R}{\underset{H}{<}} + RCH{=}CH_2 \rightleftharpoons Ta\underset{\overset{|}{\underset{H}{C}}\overset{}{\underset{R}{\diagdown}}}{\overset{\overset{H\ R}{\diagdown/}}{\underset{|}{C}}}CH_2 \quad (21)$$

One might expect internal alkenes (RCH=CHR′) to be metathesized catalytically starting with Ta(CHCMe$_3$)(OCMe$_3$)$_2$ClL$_2$ since formation of Ta(CH$_2$)(OCMe$_3$)$_2$ClL$_2$ is not possible. This is the case, but, unfortunately, the catalytic activity is not long-lived. A new problem has arisen. The alkylidene ligands in the two intermediate complexes, Ta(CHR)(OCMe$_3$)$_2$ClL$_2$ and Ta(CHR′)(OCMe$_3$)$_2$ClL$_2$ (R may be CH$_3$; R′ = CH$_2$CH$_3$), contain β-hydrogen atoms. Although these alkylidene complexes do not decompose readily in a bimolecular reaction, the alkylidene ligands do rearrange readily to alkenes (Eq. 22). This β-hydride elimination re-

action is related to those shown in Eqs. 13 and 15.

$$Ta{=}C\overset{CH_2R}{\underset{H}{<}} \longrightarrow Ta{-}H\underset{\overset{|}{\underset{H\ H}{C}}\diagdown}{\overset{\overset{H\ R}{\diagdown/}}{\underset{|}{C}}} \quad (22)$$

The study of niobium(V) and tantalum(V) alkylidene complexes showed that, although certain complexes have some of the necessary properties to metathesize alkenes, bimolecular decomposition of alkylidene complexes, rearrangement of metallacyclobutane rings, and rearrangement of alkylidene ligands are three ways in which an alkylidene ligand is removed from the catalytic cycle and the potential catalytic alkene metathesis reaction thereby prevented. In retrospect, it is not surprising that tantalum and niobium alkylidene complexes do not metathesize alkenes well; niobium and tantalum do not form good metathesis catalysts in the classical "black-box" catalyst systems (7). Conversely, since tungsten is one of the three classical catalysts, it was gratifying to find that isolated tungsten(VI) alkylidene complexes do metathesize alkenes.

W(O)(CHCMe$_3$)L$_2$Cl$_2$ reacts with terminal alkenes in the presence of a trace of a Lewis acid such as AlCl$_3$ (20) to yield a mixture of the two possible alkylidene complexes that result from cleavage of the two possible initial metallacyclobutane rings (for instance, Eq. 23).

$$W(O)(CHCMe_3)L_2Cl_2 \xrightarrow{CH_2=CHMe}$$
$$W(O)(CHMe)L_2Cl_3 + W(O)(CH_2)L_2Cl_3 \quad (23)$$

Both the ethylidene complex and the methylene complex [the only isolable "high-oxidation-state" methylene complex besides Ta(η^5-C$_5$H$_5$)$_2$(CH$_2$)(CH$_3$)]

can be isolated. Therefore, the oxo alkylidene complex will slowly catalytically metathesize both internal and terminal alkenes. Complexes of the type $W(CHCMe_3)(OR)_2Br_2$ will also metathesize olefins, especially rapidly in the presence of aluminum halides (14). In each case it seems likely that the π-electron donor ligands (oxo or alkoxide, respectively) retard formal reduction of W(VI) to W(IV)—that is, they slow down the rate of rearrangement of intermediate tungstenacyclobutane rings (Eq. 15) and probably also slow down the rate of rearrangement of intermediate alkylidene complexes in which β-hydrogen atoms are present in the alkylidene ligand (as in Eq. 22). At the same time the rate of cleavage of intermediate tungstenacyclobutane rings to give an alkene and an alkylidene complex may increase markedly. Although aluminum halides are not required as cocatalysts (21), they increase the rate of alkene metathesis enormously, possibly by forming cationic alkylidene complexes. For example, addition of one or two equivalents of $AlCl_3$ to $W(O)(CHCMe_3)(PEt_3)_2Cl_2$ in polar, noncoordinating solvents such as dichloromethane or chlorobenzene yields $[W(O)(CHCMe_3)L_2Cl]^+AlCl_4^-$ and $[W(O)(CHCMe_3)L_2]^{2+}(AlCl_4^-)_2$ (L = PEt_3), some of the longest lived metathesis catalysts we have prepared.

It is now fairly certain that tungsten(VI) alkylidene complexes must be present in the classical black-box alkene metathesis systems. One of the remaining important questions concerning the classical alkene metathesis catalyst is how the initial alkylidene ligand forms. Although it now seems likely that there are many routes to alkylidene complexes, some form of α-hydrogen atom abstraction is almost certainly an important step in many, if not most, cases.

Reaction of Tungsten Alkylidyne Complexes with Alkynes

If W(VI) alkylidene complexes will metathesize alkenes, will W(VI) alkylidyne complexes react with alkynes to give new alkylidyne complexes (Eq. 24) and thereby catalyze the metathesis of alkynes (Eq. 25)? The only known homogeneous alkyne metathesis catalyst is inefficient and its composition is unknown (22). Alkylidyne complexes such as $(CO)_5BrW\equiv CPh$, one of numerous species prepared by Fischer over the last decade (23), do not catalyze the alkyne metathesis reaction. We should now not be surprised by that fact, since they contain W(IV), the "wrong" oxidation state.

$$M\equiv CR + RC\equiv CR' \longrightarrow \begin{matrix} & R \\ & | \\ M = C \\ | \; \; O \; \; | \\ C - C \\ | \quad\quad | \\ R' \quad\quad R \end{matrix} \longrightarrow$$

$$M\equiv CR' + RC\equiv CR \qquad (24)$$

$$2R'C\equiv CR \rightleftharpoons R'C\equiv CR' + RC\equiv CR \quad (25)$$

The first known tungsten(VI) alkylidyne complex (15), $(Me_3CCH_2)_3W\equiv CCMe_3$, serves as a starting material for preparing blue $[NEt_4][Cl_4W\equiv CCMe_3]$ (Eq. 26), and from it white $(Me_3CO)_3W\equiv CCMe_3$. The $(Me_3CO)_3W\equiv CCMe_3$ reacts rapidly with diphenyl acetylene as shown in Eq. 27 to give orange-red,

$$(Me_3CCH_2)_3W\equiv CCMe_3 \xrightarrow[NEt_4Cl]{3HCl}$$

$$[NEt_4][Cl_4W\equiv CCMe_3] \xrightarrow{3LiOCMe_3} \qquad (26)$$

$$(Me_3CO)_3W\equiv CCMe_3$$

$$(Me_3CO)_3W\equiv CCMe_3 + PhC\equiv CPh \longrightarrow \qquad (27)$$

$$(Me_3CO)_3W\equiv CPh + Me_3CC\equiv CPh$$

sublimable $(Me_3CO)_3W\equiv CPh$. Therefore, unsymmetrical alkynes such as $EtC\equiv CPh$ and $PrC\equiv CPh$ (where $Pr = CH_2CH_2CH_3$) are catalytically metathesized by $(Me_3CO)_3W\equiv CCMe_3$ (24). Some estimates of the initial rate of productive metathesis of several acetylenes are shown in Table 1. The most impressive is the metathesis of 3-heptyne. In neat 3-heptyne at 25°C it is estimated that on the order of one to ten metathesis steps per second are carried out by each molecule of catalyst. Some of the most active homogeneous catalysts known have turnover numbers of ~ 100 per second.

We found that metathesis of alkynes is the exception rather than the rule (25). When alkoxide ligands are not present a tungstenacyclobutadiene complex is formed which is relatively stable toward cleavage to reform an alkylidyne complex (Eq. 28, for example). Furthermore,

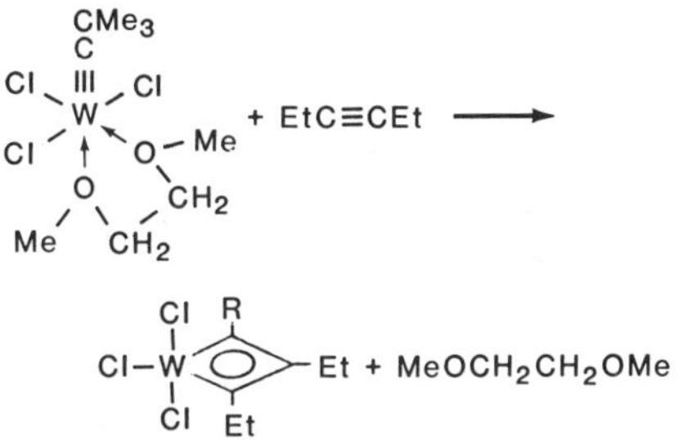

(28)

it readily reacts with a second equivalent of alkyne to give a cyclopentadienyl complex (Eq. 29, for example). The met-

al is thereby reduced from tungsten(VI) to tungsten(IV). Interestingly, alkyne metathesis activity in the system involving $(Me_3CO)_3W\equiv CR$ catalysts eventually does cease (in days), and when it does, catalytically inactive cyclopentadienyl complexes can be isolated. We can con-

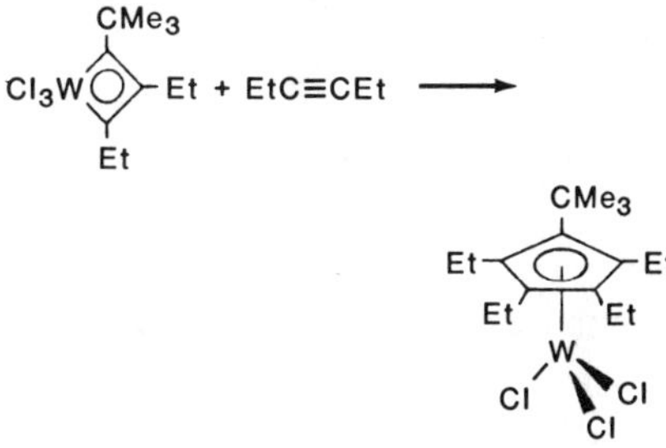

(29)

clude that alkoxide ligands slow down the rate of reduction of the metal [formation of W(IV) cyclopentadienyl complexes] and probably also enhance the rate of cleavage of the tungstenacyclobutadiene ring to give $(Me_3CO)_3W\equiv CR$ complexes. This role for t-butoxide ligands is analogous to their role (and that of the oxo ligand) in alkene metathesis, as discussed in the previous section.

Metathesis-Like Reactions of Other Triple Bonds

In thinking about the possibility of metathesis-like reactions of other $W\equiv X$ bonds, we noted that we had never observed $(Me_3CO)_3W\equiv CR$ to de-

Table 1. Productive metathesis of acetylenes in toluene at 25°C. Initial metathesis products formed rapidly (> 90 percent yield). The rate was measured after initial metathesis products had formed but while more than 90 percent of initial acetylene reactant remained. Unless otherwise noted, the error in k is on the order of ±10 percent (three to seven measurements).

Catalyst	Acetylene	k (M^{-1} sec^{-1})
$W(CCMe_3)(OCMe_3)_3$	$PhC\equiv CEt$	8.1×10^{-3}
$W(CCMe_3)(OCMe_3)_3$	$p\text{-}MeC_6H_4C\equiv CPh$	8.9×10^{-2}
$W(CCMe_3)(OCMe_3)_3$	$PrC\equiv CEt$	~ 3

compose to give $RC{\equiv}CR$ and $(Me_3CO)_3W{\equiv}W(OCMe_3)_3$, one of many dimeric compounds of molybdenum, tungsten, or rhenium that contain triple (or quadruple) bonds (26). We found this surprising for two reasons: metal-metal triple and quadruple bonds are thought to be extremely strong (and formation of them therefore thermodynamically quite favorable), and bimolecular decomposition of alkylidene complexes to give olefins is probably a major decomposition pathway in all homogeneous alkene metathesis reactions. Therefore, we attempted to add alkynes to $(Me_3CO)_3W{\equiv}W(OCMe_3)_3$ (27). This reaction was successful (Eq. 30). We were

$$(Me_3CO)_3W{\equiv}W(OCMe_3)_3 + EtC{\equiv}CEt \tag{30}$$

$$\xrightarrow[\text{fast}]{25^\circ C} 2(Me_3CO)_3W{\equiv}CEt$$

surprised, since compounds containing metal-metal triple bonds, including their reactions with alkynes, have been the subject of a good deal of research in the last decade (26). We soon found that of the several common compounds containing a metal-metal triple bond [including $(Me_3CO)_3Mo{\equiv}Mo(OCMe_3)_3$] none reacted with alkynes the way $(Me_3CO)_3W{\equiv}W(OCMe_3)_3$ does. Therefore it appears once again that the alkoxide ligand (or, more specifically, perhaps the t-butoxide ligand) is a rather special ligand. It is possible also that at least in this reaction tungsten is a rather special metal.

The observations above prompted two questions. First, do organonitriles, in which the carbon-nitrogen bond dissociation energy is approximately 215 kilocalories per mole (compared to a $C{\equiv}C$ bond energy of $\sim$ 200 kcal/mole in alkynes) also react with $W_2(OCMe_3)_6$? They do, as shown in Eq. 31. The nitrido

$$(Me_3CO)_3W{\equiv}W(OCMe_3)_3 + CH_3CN \xrightarrow[\text{fast}]{25^\circ C} \tag{31}$$

$$(Me_3CO)_3W{\equiv}CCH_3 + (Me_3CO)_3W{\equiv}N$$

ligand is isoelectronic with the alkylidyne ligand. Therefore it is not surprising to find that nitrido complexes analogous to alkylidyne complexes can be formed. Interestingly, nitrido complexes also will form (slowly) by the reaction shown in Eq. 32. Hence it appears that

$$(Me_3CO)_3W{\equiv}CR + RC{\equiv}N \longrightarrow \tag{32}$$

$$(Me_3CO)_3W{\equiv}N + RC{\equiv}CR$$

tungsten prefers to bond to the more electronegative atom (N). This finding is similar to the observation that a tantalum alkylidene complex can be converted into an isoelectronic imido complex, and the imido complex into an oxo complex (28), as shown in Eq. 33.

$$Ta{=}CHCMe_3 \xrightarrow[-PhCH=CHCMe_3]{+PhN=CHPh}$$

$$Ta{=}NPh \xrightarrow[-Me_2C=NPh]{+MeCMe\overset{O}{\parallel}} Ta{=}O \tag{33}$$

The second question is whether molecular nitrogen, whose $N{\equiv}N$ bond strength is $\sim$ 225 kcal/mole, reacts with $W_2(OCMe_3)_6$. At up to 100 atmospheres and 80°C it does not. Molecular nitrogen, unlike alkynes or nitriles, coordinates most strongly to relatively electron-rich metals in lower oxidation states (29), that is, it is a poor nucleophile. Therefore, the reaction of N_2 with $W_2(OCMe_3)_6$, like many reactions of N_2, probably failed for kinetic rather than thermodynamic reasons.

Ethylene Polymerization

I mentioned earlier in this article that tantalum neopentylidene hydride complexes can be prepared by reducing Ta(V) neopentyl complexes by two electrons. An example is shown in Eq. 11. Since a hydride ligand is rarely unreactive in organometallic chemistry, an interesting question is how alkylidene hydride complexes react with alkenes.

Green Ta(CHCMe$_3$)(H)L$_3$I$_2$ (L = PMe$_3$) in toluene reacts slowly with excess ethylene to give a pale green polymer (*30*). Protonolysis of this pale green polymer with HCl yields a white polymer, which was shown by field desorption mass spectroscopy to consist of primarily alkenes, C$_n$H$_{2n}$, where n = 50 to 100 and is both even and odd. The mechanism by which the polymers are believed to form is shown in Eq. 34. The odd-carbon polymers result from growth

$$
\begin{array}{c}
\overset{H}{\underset{H}{M{=}C}}\overset{R}{} + CH_2{=}CH_2 \longrightarrow \\[2ex]
\overset{H}{\underset{CH_2{-}CH_2}{M - C - R}}\overset{H}{} \longrightarrow MCH_2CH_2CH_2R \quad (34) \\[2ex]
\longrightarrow \overset{H}{\underset{H}{M{=}C}}\overset{CH_2CH_2R}{}
\end{array}
$$

of a polymer that contains the original C$_5$ fragment and termination of that growth by formation of an alkene. The even-carbon polymer is authentic polyethylene. The polymer mixture is pale green since it contains some "living" polymer, Ta[CH(CH$_2$CH$_2$)$_n$CH$_3$](H)L$_3$I$_2$, which is destroyed by HCl to give even-carbon alkanes, CH$_3$(CH$_2$CH$_2$)$_n$CH$_3$, which can be observed in small amounts in the mass spectrum.

One of the most interesting aspects of the mechanism shown in Eq. 34 is the last step, an α-elimination reaction to give the new alkylidene hydride complex. Such a proposal appears contrary to common sense, that is, "more favor-

able" β-hydride elimination to give an olefin hydride complex. But our results do not necessarily imply that β-elimination to give an alkene hydride complex is relatively slow. As shown in Eq. 35 it is possible that although K_2 is greater than K_1, k_1 is greater than k_2 ($K_1 = k_1/k_{-1}$, $K_2 = k_2/k_{-2}$)—that is, β-elimination

$$
\begin{array}{c}
\overset{H}{\underset{\underset{H}{\overset{|}{C}}\,\,R}{\overset{|}{Ta}\overset{CH_2}{\underset{-}{-}}}}\,\,\overset{k_1}{\underset{k_{-1}}{\rightleftharpoons}}\,\, TaCH_2CH_2R \\[3ex]
\overset{k_2}{\underset{k_{-2}}{\rightleftharpoons}}\,\, \overset{H}{\underset{\underset{H}{\diagdown}}{Ta{=}C}}\overset{CH_2R}{\diagup} \quad (35)
\end{array}
$$

could still be faster—if the olefin hydride complex also is relatively stable toward displacement of CH$_2$=CHR by ethylene under the reaction conditions employed here.

The alkylidene hydride catalyst closely resembles a Ziegler-Natta ethylene polymerization catalyst (*3*), except that the chain length with the tantalum catalyst is actually comparatively short. It has been proposed that alkylidene hydride complexes may be responsible for the stereospecific polymerization of propylene by Ziegler-Natta catalysts (*31*). However, an enormous amount of circumstantial evidence favors the accepted mechanism (*32*) of ethylene and propylene polymerization (Eq. 1), and growth of a polymer chain has been observed directly on a lutetium center (*33*). Although lutetium is, strictly speaking, not a transition metal, it resembles typical transition metals that are active in Ziegler-Natta systems.

Rather than attempt to choose between the two mechanisms, it should be pointed out that it is possible, on the basis of what I discussed earlier in this article, that an alkyl ligand in some alkyl complexes of titanium, vanadium, or chromium (typical Ziegler-Natta catalysts) may be severely distorted toward being an alkylidene hydride (Eq. 36). It would not be surprising if such an alkyl complex were to behave as an alkylidene hydride complex behaves. Such molecules will be difficult to observe since they may be some of the truly rapid alkene polymerization catalysts.

$$M\text{--}CH_2R \longrightarrow M\overset{H}{\cdots}CHR \qquad (36)$$

Conclusions

Metal-carbon double and triple bonds are natural and important features of the organometallic chemistry of metals in their highest possible oxidation state. Since molybdenum, tungsten, and rhenium form a large variety of oxo, imido, and nitrido complexes, high-oxidation-state alkylidene and alkylidyne complexes of these metals should be most numerous. High-oxidation-state complexes are active for methathesis-like catalytic reactions if the metal is surrounded with ligands that are compatible with the high oxidation state; rapid reduction of the metal and catalyst deactivation are thereby avoided. There seems to be some evidence that reactions of $M\equiv M$ bonds, $M\equiv C$ bonds, and $M\equiv N$ bonds are related; therefore, "heteroatomic" methathesis-like reactions of triple bonds are a good possibility.

References and Notes

1. C. N. Satterfield, *Heterogeneous Catalysis in Practice* (McGraw-Hill, New York, 1980); G. W. Parshall, *Homogeneous Catalysis* (Wiley-Interscience, New York, 1980); P. N. Rylander, *Organic Synthesis with Noble Metal Catalysts* (Academic Press, New York, 1973); R. Pearce and W. R. Patterson, *Catalysis and Chemical Processes* (Halstead, New York, 1981).

2. P. N. Rylander, *Catalytic Hydrogenation in Organic Syntheses* (Academic Press, New York, 1979).
3. J. Boor, Jr., *Ziegler-Natta Catalysts and Polymerizations* (Academic Press, New York, 1979).
4. The name derives from the fact that a carbene, $(C_6H_5)(CH_3O)C$:, which in the free state has only a limited lifetime (5), is "stabilized" by bonding to the metal.
5. M. Jones, Jr., and R. A. Moss, Eds., *Carbenes* (Wiley-Interscience, New York, 1973), vol. 1.
6. D. J. Cardin, B. Cetinkaya, M. F. Lappert, *Chem. Rev.* **72**, 575 (1972); F. A. Cotton and C. M. Lukehart, *Prog. Inorg. Chem.* **16**, 243 (1972).
7. N. Calderon, J. P. Lawrence, E. A. Ofstead, *Adv. Organometal. Chem.* **17**, 449 (1979).
8. When the carbene ligand contains only carbon and hydrogen, and especially if the metal is not in a relatively low oxidation state, the carbene ligand is more accurately called an alkylidene (CHR^{2-}) ligand.
9. R. R. Schrock, *Acc. Chem. Res.* **12**, 98 (1979); in *Reactions of Coordinated Ligands*, P. S. Braterman, Ed. (Plenum, New York, in press).
10. F. A. Cotton and G. Wilkinson, *Advanced Inorganic Chemistry* (Wiley-Interscience, New York, ed. 4, 1980).
11. J. H. Wengrovius and R. R. Schrock, *Organometallics* **1**, 148 (1982).
12. M. R. Churchill, J. R. Missert, W. J. Youngs, *Inorg. Chem.* **20**, 3388 (1981).
13. W. P. Griffith, *Coord. Chem. Rev.* **5**, 459 (1970).
14. J. Kress, M. Wesolek, J. Osborn, *J. Chem. Soc. Chem. Commun.* (1982), p. 514.
15. D. N. Clark and R. R. Schrock, *J. Am. Chem. Soc.* **100**, 6774 (1978).
16. N. J. Cooper and M. L. H. Green, *J. Chem. Soc. Dalton Trans.* (1979), p. 1121; J. D. Fellmann, H. W. Turner, R. R. Schrock, *J. Am. Chem. Soc.* **102**, 6608 (1980).
17. S. M. Rocklage, J. D. Fellmann, G. A. Rupprecht, L. W. Messerle, R. R. Schrock, *J. Am. Chem. Soc.* **103**, 1440 (1981).
18. S. J. McLain, J. Sancho, R. R. Schrock, *ibid.* **102**, 5610 (1980).
19. R. R. Schrock and P. R. Sharp, *ibid.* **100**, 2389 (1978).
20. R. Schrock, S. Rocklage, J. Wengrovius, G. Rupprecht, J. Fellmann, *J. Mol. Catal.* **8**, 73 (1980).
21. J. H. Wengrovius, R. R. Schrock, M. R. Churchill, J. R. Missert, W. J. Youngs, *J. Am. Chem. Soc.* **102**, 4515 (1980).
22. A. Mortreux, J. C. Delgrange, M. Blanchard, B. Labochinsky, *J. Mol. Catal.* **2**, 73 (1977); S. Devarajan, O. R. M. Walton, G. J. Leigh, *J. Organometal. Chem.* **181**, 99 (1977).
23. E. O. Fischer and U. Schubert, *J. Organometal. Chem.* **100**, 59 (1975).
24. J. H. Wengrovius, J. Sancho, R. R. Schrock, *J. Am. Chem. Soc.* **103**, 3932 (1981).
25. S. F. Pedersen, R. R. Schrock, M. R. Churchill, H. J. Wasserman, in preparation.
26. F. A. Cotton and R. H. Walton, *Multiple Bonds between Metal Atoms* (Wiley, New York, 1982); M. H. Chisholm, Ed., *Reactivity of Metal-Metal Bonds* (Symposium Series No. 155, American Chemical Society, Washington, D.C., 1981).

27. R. R. Schrock, M. L. Listemann, L. G. Sturge-off, *J. Am. Chem. Soc.* **104**, 4291 (1982).
28. S. M. Rocklage and R. R. Schrock, *ibid.*, p. 3077.
29. J. Chatt, J. R. Dilworth, R. L. Richards, *Chem. Rev.* **78**, 589 (1978).
30. H. W. Turner and R. R. Schrock, *J. Am. Chem. Soc.* **104**, 2331 (1982).
31. K. J. Ivin *et al.*, *J. Chem. Soc. Chem. Commun.* (1978), p. 604.
32. P. Cossee, *J. Catal.* **3**, 80 (1964).
33. P. L. Watson, *J. Am. Chem. Soc.* **104**, 337 (1982).
34. The work described in this article was supported by the National Science Foundation. I thank the students whose names appear in the references for their enthusiasm, dedication, and hard work.

The usual requirements for organic feedstocks in large-scale metal-catalyzed processes are in plentiful supply and reasonable reactivity toward transition metals. Alkenes (also called olefins or unsaturated hydrocarbons) fit these requirements well, since they are major constituents of petroleum and contain carbon-carbon double bonds, which are reactive toward a wide range of transition metal–based reagents (*1, 2*). From the point of view of quantity, alkanes (saturated hydrocarbons) would also be very attractive feedstocks for catalytic synthesis of organic molecules; they are major constituents of natural gas, petroleum, and coal liquefaction processes. However, they are much less often used in this way for a simple reason: alkanes are among the least reactive organic molecules. This unreactivity is a consequence of the high C–H bond energies in alkanes. Typical values are 98 and 95 kilocalories per mole for primary and secondary C–H bonds; in methane, the main constituent of natural gas, the C–H bond energy is 104 kcal/mole (*3*). This makes methane one of the most common but least reactive organic molecules in nature.

The large quantity and low reactivity of methane and higher alkanes has tantalized organic and organometallic chemists for several years. In 1968, when mechanistic theory in organometallic chemistry was just developing, one leader in this area stated that, "the development of successful approaches to the activation of carbon-hydrogen bonds particularly in saturated hydrocarbons, remains to be achieved and presently constitutes one of the most important and challenging problems in this whole field" (*4*). This feeling, shared by many workers in the area, prompted many

18

Activation of Alkanes with Organotransition Metal Complexes

Robert G. Bergman

Science 223, 902–908 (2 March 1984)

searches for clear-cut homogeneous metal-based alkane activation systems. However, only recently have soluble transition metal systems been found for which reaction with alkanes can be directly observed. These discoveries stimulated even greater interest in this problem. The purpose of this article is to summarize some of these discoveries, to discuss the advances they have led to in the fundamental understanding of reactions between alkanes and transition metal complexes, and to provide a general idea of the important and as yet unsolved questions that have been raised by this research.

Background

Alkanes are not completely unreactive. Free radicals capable of abstracting alkane hydrogen atoms [for example, chlorine atoms (Eq. 1)] have been known

$$Cl\cdot \ + \ R\text{-}H \longrightarrow HCl \ + \ R\cdot \quad (1)$$

for many years (5); these reactions lead to alkyl radicals, which may then be converted into other types of organic molecules (Eq. 2). In some cases, transi-

$$R\cdot \ + \ X\text{-}Y \longrightarrow R\text{-}X \ + \ Y\cdot \quad (2)$$

tion metals mediate the formation or reactions of free radicals (6). One problem with these systems is that they are either very unselective or, in more selective cases, tertiary hydrogens react much more readily than secondary or primary hydrogens. A typical example of a free-radical autoxidation, selective for tertiary hydrogens, is illustrated in Eq. 3.

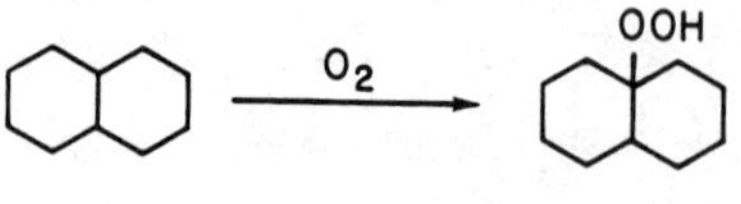

It is much more difficult to activate secondary and primary hydrogens in such processes, and only a very few systems capable of causing methane to undergo free-radical reactions are known (7). Some organic reagents react with alkanes by apparent nonradical mechanisms; examples are ozone, superacids, and fluorine (8).

In many laboratories, attention has been focused on a specific goal: the discovery of a soluble complex capable of inserting a metal center into the C–H bond of alkanes (Eq. 4). This is one

$$M \ + \ R\text{-}H \longrightarrow R\text{-}M\text{-}H \quad (4)$$

example of a general organometallic primary transformation called "oxidative addition" because the metal center is formally oxidized by two electrons in its conversion from starting complex to insertion product (9). During the past 10 years, many examples of intramolecular C–H oxidative addition [that is, the metal and reacting C–H bond are located in the same molecule (Eq. 5)] have been

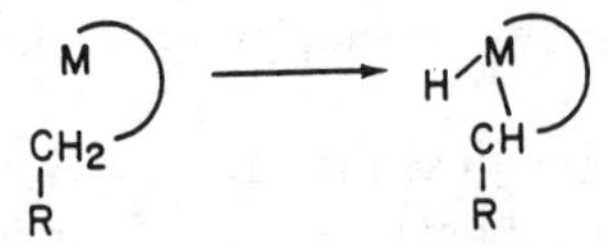

discovered (10). Some specific examples of this process are illustrated in Eqs. 6 to 8. The largest group involves so-called

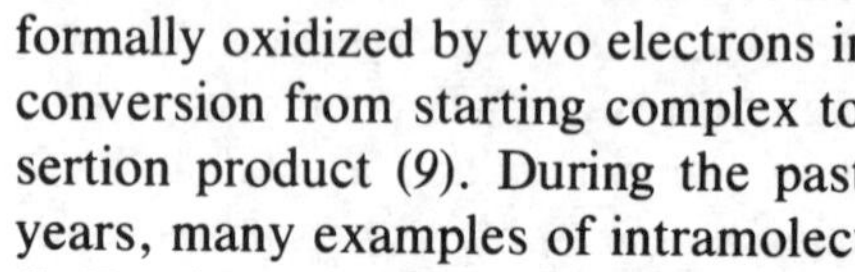

260

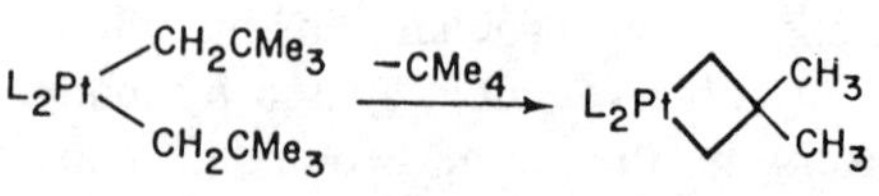

$$MCl_3 \xrightarrow{(t\text{-Bu})_2P-(CH_2)_5-P(t\text{-Bu})_2}$$

$$(M = Ir, Rh) \qquad (8)$$

orthometallation processes, in which insertion takes place into the C–H bond of an aromatic ring attached to an atom directly bound to the metal (Eq. 6). A few cases are also known of insertion into a C–H bond of an alkyl chain located in the same molecule as the metal (Eqs. 7 and 8). Clearly, proximity of the reacting C–H bond to the metal center is a critical factor favoring such cyclometallation processes. In attempts to extend these reactions to intermolecular cases, a few reactions have been found in which metal centers that are relatively electron-rich react with organic compounds having C–H bonds with low bond energy or high acidity (Eqs. 9 and 10). However, before

$$(M = Fe, Ru) \qquad (9)$$

the work reported here was initiated, all attempts to find a clear example of Eq. 4 involving simple alkanes had been unsuccessful.

A few examples of multistep metal-based transformations of alkanes are known. One example, the work of Shilov and his group in the Soviet Union (11), was reported in the late 1960's. These reactions involve both alkane hydrogen-

deuterium exchange and conversion to chlorides and acetates, catalyzed by soluble platinum salts. However, somewhat elevated temperatures (100° to 120°C) are required for these reactions. Little else is known about their mechanism, including whether they are truly homogeneous reactions. In much more recent work, Crabtree's group at Yale (12) and Felkin's at Gif-sur-Yvette in France (13)

$$L_2ReH_7 + \bigcirc + t\text{-BuCH=CH}_2 \longrightarrow$$

$$L_2ReH_2 \qquad + \; t\text{-BuCH}_2CH_3 \quad (11)$$

$$(L = PPh_3, PEt_2Ph)$$

$$[IrH_2S_2L_2]^+ + \bigcirc + t\text{-BuCH=CH}_2$$

$$\qquad + \; t\text{-BuCH}_2CH_3 \quad (12)$$

$$(L = PPh_3, S = H_2O \text{ or acetone})$$

have reported the novel iridium- and rhenium-induced dehydrogenation processes exemplified by Eqs. 11 and 12. Crabtree has provided convincing evidence that these reactions are homogeneous and has proposed that they are initiated by insertion of the metal center into an alkane C–H bond. However, this hypothesis has not yet been confirmed by direct observation of the oxidative addition in these systems.

Generation and Oxidative Addition Reactions of $(\eta^5\text{-}C_5Me_5)(PMe_3)Ir$

In connection with a project aimed at examining the products of hydrogenolysis of metal alkyls, we had occasion to prepare dihydridoiridium complex **2**

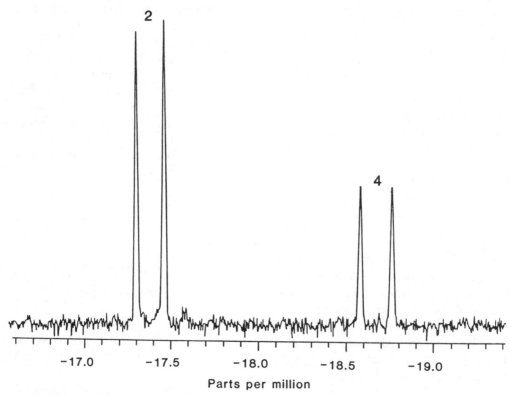

Fig. 1. Reproduction of the high-field portion of the proton nuclear magnetic resonance (NMR) spectrum of the mixture obtained during irradiation of Cp'(L)IrH$_2$ (complex **2**) with ultraviolet light in cyclohexane solvent. On the numbered axis at the bottom of the spectrum, the field increases toward the right. The position of each signal (that is, the chemical shift of the hydrogen atom producing that signal) is measured in parts per million distance from the single line observed for the internal standard tetramethylsilane, which appears at much lower field and is therefore not visible in this portion of the spectrum. Chemical shifts that appear at fields higher than tetramethylsilane are listed as negative numbers by convention. The lower-field doublet is due to the hydride signal for the starting dihydride **2**, and the higher-field doublet arises from the hydride signal for the C–H insertion product **4**.

by the reaction shown in Eq. 13. Many didhydridometal complexes undergo reductive elimination of H$_2$ upon irradiation (*14*). In complexes related to **2** (for example, Cp$_2$MoH$_2$; Cp $=$C$_5$H$_5$) the type of intermediates generated in such reactions have been shown to insert into at least some types of C–H bonds (see, for example, Eq. 10). We therefore decided to investigate the irradiation of **2**. We were encouraged to find that in benzene only **3**, the product of intermolecular C–H activation, was formed; no orthometallation product was observed. Benzene is reactive toward C–H activation, presumably because of involvement of its π-electrons in the insertion transition state. Our next step, therefore, was to attempt this reaction with somewhat less inherently reactive compounds. Exploratory experiments of this type are relatively simple because the presence of new metal hydride products can be detected by observing their characteristic resonances in the proton nuclear magnetic resonance (NMR) spectrum of the reaction mixture. The signals due to the metal-bound hydrogen atoms appear in the very-high-field region of the spectrum, and each signal is doubled as a result of spin-spin coupling between the hydride nucleus and the phosphorus nucleus also attached to the metal. Accordingly, we found that irradiation of 2 in acetonitrile gave H_2 and a new metal hydride product, and irradiation in tetrahydrofuran also gave new hydride products. We therefore proceeded immediately to alkanes. We found that irradiation of 2 in cyclohexane solvent gave a single new hydride in high yield. Figure 1 reproduces the high-field region of the proton NMR spectrum of the reaction mixture during irradiation. It shows the two-line pattern centered at -17.38 parts per million caused by the starting material, and the additional two-line pattern centered at -18.67 caused by the presence of the new hydride product. A similar result was obtained in neopentane solvent.

These products are the hydridoalkyl complexes 4 and 5 shown in Eq. 14. The new complexes are extremely hydrophobic and are therefore very difficult to obtain pure. However, they can be isolated and purified, with some loss of material, and characterized fully by proton NMR, carbon-13 NMR, and infrared

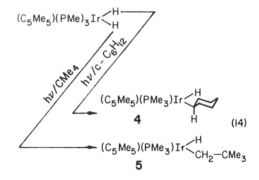

$$(14)$$

spectroscopic techniques. Furthermore, they can be converted to the corresponding bromoalkyliridium derivatives (see below), which can be characterized both by spectroscopic and elemental analysis techniques. The structure of 4 was confirmed recently by an x-ray diffraction study. Analogous C–H activation behavior in the ultraviolet irradiation of $(C_5Me_5)Ir(CO)_2$ was discovered independently and reported soon after our initial paper by Hoyano and Graham (15).

Mechanism of the C–H Oxidative Addition Reaction

The conventional mechanism for the C–H insertion is shown in Fig. 2. It is first assumed that, upon irradiation, an excited state of dihydride 2 is formed. This rapidly extrudes H_2, leaving behind the reactive, coordinatively unsaturated fragment 6. Complex 6 then inserts into a C–H bond via transition state 7, leading to the hydridoalkyliridium complexes 4 and 5.

Although this mechanism is reasonable, supporting evidence for it was needed, especially since more complicated mechanisms involving free radicals have been established in certain other oxidative addition reactions (16). Many

$(C_5Me_5)(PR'_3)IrH_2 \xrightarrow[-H_2]{h\nu}$

2 (R' = Me)

$[(C_5Me_5)(PR'_3)Ir]$

6

$\downarrow$ RH

$\left[(C_5Me_5)(PR'_3)Ir\overset{\cdots H}{\underset{\cdots R}{:}}\right]^{\ddagger}$

7

$\downarrow$

$(C_5Me_5)(PR'_3)Ir\overset{H}{\underset{R}{<}}$

4 (R = C$_6$H$_{11}$, R' = Me)

5 (R = CH$_2$_t_-Bu,
         R' = Me)

$(C_5Me_5)(PMe_3)Ir\overset{H}{\underset{H}{<}} \xrightarrow[\text{(1:1)}]{\underset{CMe_4/C_6D_{12}}{h\nu}}$

$Cp'(L)Ir\overset{H}{\underset{CH_2\text{-}t\text{-}Bu}{<}}$ $Cp'(L)Ir\overset{D}{\underset{C_6D_{11}}{<}}$

A   **B**

- -

$Cp'(L)Ir\overset{CH_2\text{-}t\text{-}Bu}{\underset{D}{<}}$ $Cp'(L)Ir\overset{C_6D_{11}}{\underset{H}{<}}$

C   **D**

$[Cp' \equiv C_5Me_5]$

Fig. 2 (left). Proposed mechanism for the photochemical conversion of dihydride **2** to C–H insertion products. Fig. 3 (right). Schematic illustration of the crossover experiment (see text) carried out by irradiating dihydride **2** in a mixture of neopentane (CMe$_4$) and completely deuterated cyclohexane (C$_6$D$_{12}$).

radical mechanisms proceed through a step in which the radical R· has independent existence. Thus the individual R and H groups located in each molecule of product **4** or **5** may not necessarily have been bound together in the starting hydrocarbon molecule. Information about this was obtained from the "crossover experiment" summarized in Fig. 3. We first irradiated **2** in 1:1 mixture of neopentane and cyclohexane. This established that the two hydrocarbons have similar reactivity toward **6**, although the C–H bond in neopentane is slightly more reactive. Next, irradiation of **2** in a 1:1 mixture of neopentane and the deuterium-labeled alkane cyclohexane-d$_{12}$ was carried out. As shown in Fig. 3, the products of this reaction were the hydri-

doneopentyl- and the deuterio-(perdeuterocyclohexyl)iridium complexes **A** and **B**, with very small amounts of contamination from the crossed products **C** and **D**. Therefore, the R and H groups remain associated with one another during the process that converts the hydrocarbon to a hydridoalkyliridium complex.

This experiment rules out reactions involving radicals having a free existence and finite lifetime; however, it is more difficult to rule out processes proceeding predominantly through solvent-caged radical pairs. Our best evidence against such intermediates comes from relative reactivity studies. Hydrogen abstraction reactions of radicals favor C–H bonds with low bond energies; very strong bonds are nearly always inert (5). Cyclopropane may therefore be used as a diagnostic substrate since its C–H bond energy is 106 kcal/mole, even stronger than that of methane (3). Nevertheless, as shown in Eq. 15, irradiation of 2 in

$$2 \xrightarrow[-35°C]{h\nu/\Delta}$$

$$(C_5Me_5)(PMe_3)Ir \underset{\overset{|}{8}}{\overset{H\cdots}{<}} \overset{CH_2}{\underset{CH_2}{<}} \quad (15)$$

liquid cyclopropane at $-35°C$ leads only to the C–H insertion product 8. Thus, addition to the C–H bond is favored even over insertion into the relatively weak C–C bond, which would have led to a four-membered iridium-containing ring compound. In view of this result, a radical-cage mechanism seems extremely unlikely for these reactions.

Functionalization Reactions

A primary goal of this work has been the conversion of the hydridoalkyliridium complexes to functionalized organic molecules. Unfortunately, many of the reagents that might be used to carry out such functionalizations promote only reductive elimination, simply regenerating the starting hydrocarbon. One solution to this problem is conversion of the hydrides to the corresponding bromoalkyliridium complexes 10 and 11. This can be done smoothly at room temperature by treating the hydrides with the common reagent $CHBr_3$, as shown in Eq. 16. The bromine-containing prod-

$$(C_5Me_5)(PMe_3)Ir \overset{H}{\underset{R}{<}} \xrightarrow[-CH_2Br_2]{Br_3CH}$$

$$\begin{array}{ll} 4 & R = C_6H_{11} \\ 5 & R = CH_2CMe_3 \end{array}$$

$$(C_5Me_5)(PMe_3)Ir \overset{Br}{\underset{R}{<}} \quad (16)$$

$$\begin{array}{ll} 10 & R = C_6H_{11} \\ 11 & R = CH_2CMe_3 \end{array}$$

ucts are somewhat easier to handle than their hydride precursors, and they do not easily undergo reductive elimination. As a result, functionalization of these materials has been more successful. The most convenient method we have found so far is summarized in Eq. 17. Treatment of

$$R-HgCl + (Me_5C_5)Ir \overset{Br}{\underset{Cl}{<}} \quad (17)$$
$$\underset{12}{\big|I_2} \qquad\qquad 13$$

$$R-I + HgClI$$

bromoalkyliridium complex 10 or 11 with mercuric chloride leads to the corresponding alkylmercurial 12 and bromochloroiridium complex 13. Conversion of 12 to organic halides can then be carried out with I_2 or Br_2 at room temperature. Thus overall conversion of an alkane to alkyl halide can be effected in good yield under mild conditions, in a process that completely avoids free radical intermediates.

Selectivity Studies

We next turned our attention to studies of the selectivity of the reactive inter-

Ring systems

2.6   1.6   1.0   0.09   4.7

Fig. 4. Relative rates of reaction with Cp′IrL of one C–H bond in each of the cyclic organic molecules illustrated.

mediate $Cp'(PMe_3)Ir$ (6) ($Cp' = C_5Me_5$; Fig. 2). Irradiation of $Cp'(PMe_3)IrH_2$ (2) in the presence of solvents having different types of C–H bonds allowed the reactive intermediate $Cp'(PMe_3)Ir$ to react competitively with those bonds. At low conversion, the ratio of the products in such an experiment is directly equal to the ratio of rate constants for insertion into each type of C–H bond. Two types of selectivity experiments were carried out: those involving competition of the intermediate for C–H bonds in different molecules (intermolecular selectivity) and competition for different types of C–H bonds located in the same molecule (intramolecular selectivity).

Intermolecular selectivities can be established readily because the individual oxidative addition products can be independently generated. Each selectivity experiment was carried out with solvents consisting of mixtures of two liquid hydrocarbons. First the dihydride 2 was irradiated in each pure solvent, and the NMR spectrum of the product was recorded. Then the irradiation was carried out in a mixture (usually 1:1) of the two solvents, and the ratio of the products was measured by repetitive integration of the hydride resonances in the highfield region of the spectrum. The rate ratio was calculated by correcting the ratio of product yields for the number of hydrogens available in each molecule. In this way, a neopentane-to-cyclohexane

rate ratio of 1.4 and a cyclopropane-to-cyclohexane ratio of 2.6 were obtained. Similarly, we were able to determine the relative rates of attack by $Cp'Ir(PMe_3)$ on one C–H bond in each of the molecules in Fig. 4, with attack on cyclohexane set as 1.0. The results are interesting and, as far as we know, unprecedented: reactivity seems to vary with ring size, smaller rings showing higher reactivity toward C–H insertion and larger rings showing lower reactivity. The physical reason for this is not yet clear but may be related partially to the greater steric accessibility of the C–H bonds in the smaller rings, where the two carbon atoms attached to the reacting carbon are somewhat more tightly "tied back" than they are in the larger systems. A competition experiment with a cyclohexane-benzene mixture showed that the aromatic C–H bonds were 4.0 times as reactive as the cyclohexyl.

Intramolecular selectivities presented a more difficult problem. Reaction of $Cp'(PMe_3)Ir$ (6) with unsymmetrical hydrocarbons (that is, those having C–H bonds in different chemical environments) in general leads to several structurally different hydridoalkyliridium complexes. Because of the chemical similarity of these materials, it was not feasible to separate and identify them individually. Fortunately, however, NMR experiments on a wide range of C–H oxidative additions using symmetrical hydrocarbons (such as those summarized in the preceding paragraph) showed that the NMR line position (chemical shift) due to a particular metal-bound hydride in the hydridoalkylmetal products correlated quite reliably with the type of substitution at the carbon attached to the metal. That is, secondary hydridoalkyl complexes (for example, those formed from ring compounds such as cyclohexane and cyclopentane, having general structure H–M–CHR$_2$) exhibit hydride chemical shifts at relatively high field, whereas primary hydridoalkyl complexes (H–M–CH$_2$R) invariably exhibit hydride signals at significantly lower field in the proton NMR. Oxidative addition reactions carried out on simple unsymmetrical systems confirm that this dependence holds up in more complex molecules; for example, generation of **6** in *n*-propane, as predicted, gives two new hydride signals, one in the lower-field "primary" and one in the higher-field "secondary" region of the spectrum. On the basis of this correlation, these signals can be assigned with some confidence to oxidative addition products **14** and **15**, respectively (Eq. 18). The results of ex-

$$Me_5C_5)Ir(PMe_3)H_2 \xrightarrow[\text{CH}_3\text{CH}_2\text{CH}_3]{h\nu}$$

2

$$\underset{Me_3P}{\overset{(Me_5C_5)}{\diagdown}}\underset{\diagup}{\overset{\diagup}{Ir}}\underset{H}{\overset{CH_2CH_2CH_3}{\diagdown}} \quad +$$

14

$$\underset{Me_3P}{\overset{(Me_5C_5)}{\diagdown}}\underset{\diagup}{\overset{\diagup}{Ir}}\underset{H}{\overset{CH{\overset{CH_3}{\diagdown}}_{CH_3}}{\diagdown}} \quad (18)$$

15

periments such as this one indicate that in simple acyclic systems such as *n*-propane, *n*-pentane, and 2-methyl-2-butane, **6** demonstrates measurable "antiradical" selectivity; it inserts into primary C–H bonds about twice as fast as into secondary, and insertion into tertiary C–H bonds is apparently so slow that we have never detected hydride signals due to this type of product.

Reductive Elimination and Thermochemical Studies

Regeneration (reductive elimination) of hydrocarbon R–H from hydridoalkyliridium complexes such as **4** and **5** can be induced by various oxidizing agents and Lewis acids. However, it can also be induced thermally. For example, heating hydridocyclohexyl complex **4** (see Eq. 14) to temperatures above 100°C in the presence of a solvent other than cyclohexane causes elimination of cyclohexane and (through the intermediacy of **6**) formation of the new hydridoalkyl complex resulting from insertion into a C–H bond of the solvent. This reaction proved to be especially useful in enhancing the effective primary-secondary selectivity of the C–H oxidative addition process. This is illustrated with the sequence of reactions carried out with *n*-pentane (Fig. 5). Irradiation of **2** in *n*-pentane gives four new hydride resonances in the NMR spectrum, one (that due to the major product **16**) in the primary and three in the secondary region of the spectrum. Three, rather than two, secondary products are observed because the insertion of iridium into the hydrogens at C-2 of this hydrocarbon can give rise to the two structurally distinguishable diastereomers shown in Fig. 5. Heating this mixture of hydridoalkyl complexes to 110°C in pentane causes disappearance of the resonances assigned to the secondary C–H insertion products and a corresponding increase in the resonances due to the primary insertion product **16**. This occurs because this temperature is sufficient to cause reversible reductive elimination from the secondary hydridoalkyl products; thus the product that is thermodynamically most stable (that is, the primary complex **16**) is

Fig. 5. The C–H insertion products formed on irradiation of dihydride **2** in *n*-pentane at 6°C and heating the resulting mixture to 110°C in the same solvent.

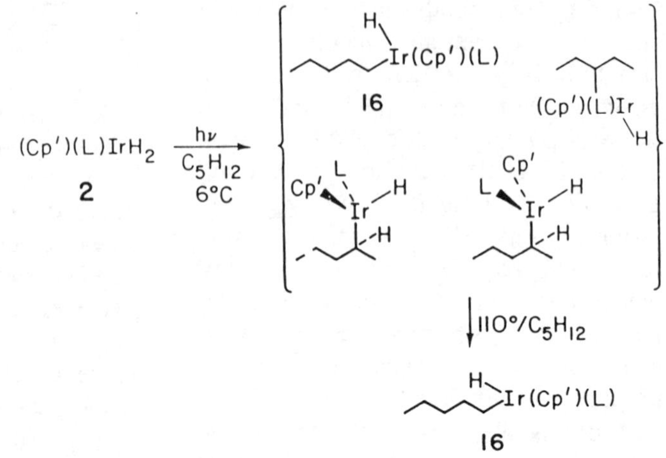

formed. This is an extremely important result from a practical point of view because it suggests that a combination of photolysis followed by thermal equilibration will allow exclusive primary functionalization of linear alkanes. Supporting evidence for the reductive elimination-readdition mechanism has been obtained by heating the mixture of hydridopentyliridium complexes with cyclohexane instead of *n*-pentane, as shown in Fig. 6. Under these conditions, the secondary products again disappear

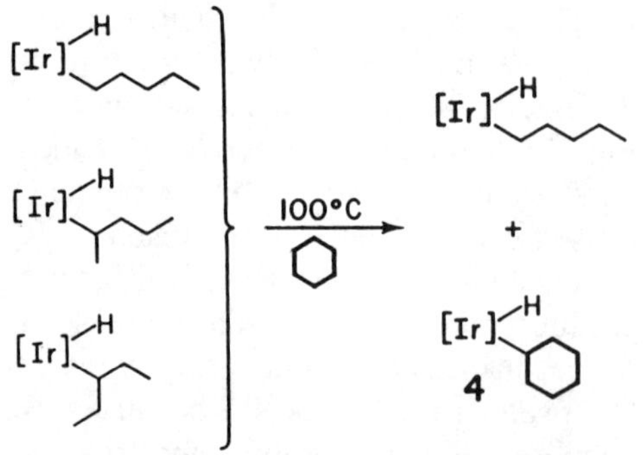

but are converted into the cyclohexylhydridoiridium complex **4** instead of additional 1-pentylhydridoiridium complex **16**.

Because different hydridoalkyliridium complexes are interconverting reversibly in these experiments, it is in principle possible to measure the equilibrium constants for the reactions and thus to obtain a quantitative measure of the relative iridium-carbon bond strengths involved. We have done this for the *n*-pentyl–cyclohexyl system. The reversible reaction under consideration and the equilibrium constant associated with it are illustrated in Eqs. 19 and 20. The

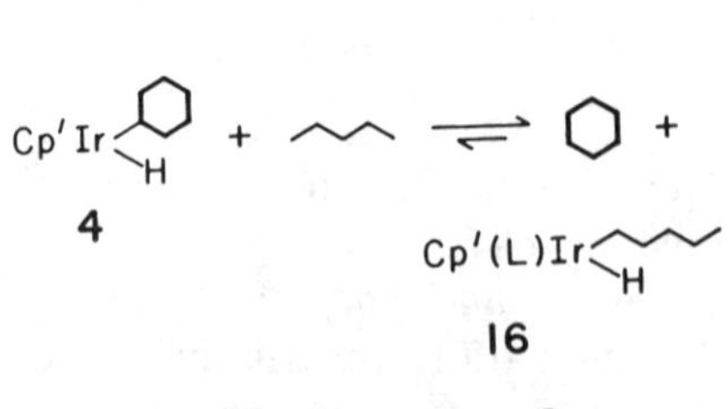

$[Ir] \equiv (\eta^5-C_5Me_5)Ir(PMe_3)$

Fig. 6. Illustration of the result of heating the products formed on irradiation of **2** in *n*-pentane to 100°C in cyclohexane solvent.

$$K_{eq} = \frac{[\bigcirc]\left[Cp'(L)Ir\diagdown_H\diagup\diagup\diagup\right]}{[\diagup\diagup][Cp'(L)Ir\diagdown_H\bigcirc]} =$$

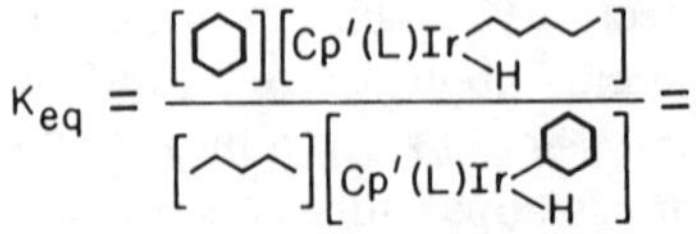

$$\frac{91.5}{8.5} \cdot \frac{50}{50} = 10.8 \quad (20)$$

equilibrium was studied by first dissolving the mixture of four hydridopentyliridium products illustrated in Fig. 5 in a 91.5:8.5 mixture of cyclohexane and *n*-pentane. This solution was then heated. At 120°C, reductive elimination and disappearance of the 2-pentyliridium complexes occurred. Heating to 140°C then brought the more stable complexes **4** and **16** into interconversion with one another via reversible reductive elimination–oxidative addition. In this solvent mixture and at this temperature, the system reached equilibrium after about 50 hours; 50 percent of the equilibrium mixture was **4** and 50 percent was **16**, and there were no detectable 2- or 3-pentyliridium complexes. With these equilibrium concentrations and the relative molar concentrations of the two solvents, the equilibrium constant was calculated to be 10.8, as shown in Eq. 20. This corresponds to a free energy difference of about 2 kcal/mole at 140°C. The primary C–H bond in *n*-pentane is about 4 kcal/mole stronger than the secondary C–H bond in cyclohexane (*3*). Using this number and the reaction free energy, and assuming a negligible entropy difference between reactants and products, we estimate the primary metal-carbon bond in the *n*-pentyliridium complex **16** to be about 6 kcal/mole stronger than the secondary metal-carbon bond in **4**. Clearly the secondary metal-carbon bonds in the 2-pentyliridium complexes are even weaker than that in **4**; we do not yet know the reason for this unusual result (*17*).

Activation of Methane

In considering various possible alkane substrates for C–H activation, the small-est hydrocarbon, methane (CH$_4$), is one of the most intriguing. The reasons for this, as mentioned earlier, are methane's very high bond dissociation energy (104 kcal/mole) (*3*), and its presence as a major constituent in natural gas. The fact that cyclopropane, despite its even higher bond energy, reacts readily with Cp'(PMe$_3$)Ir suggested that methane should be similarly reactive. However, there are serious technical problems in carrying out the irradiation of Cp'(PMe$_3$)IrH$_2$ in methane. At normal pressures, such low temperatures are required to liquefy methane that the dihydride is insoluble in it. Attempts to find an inert solvent in which to carry out the methane reaction were not successful. For example, irradiation of dihydride **2** in perfluoroalkane solvents (in which it is only slightly soluble) under 4 atmospheres of CH$_4$ gave only decomposition to intractable materials, in contrast to the observations of Hoyano and Graham (*18*) and Watson (*19*) [Eqs. 21 (R = CH$_3$) and 22]. Photolysis in cyclooctane, a "slow" substrate for C–H insertion of

$$(C_5Me_5)(CO)_2Ir \xrightarrow[R-H]{h\nu} (C_5Me_5)(CO)Ir{<}^{R}_{H}$$

$$\xrightarrow{CCl_4} (C_5Me_5)(CO)Ir{<}^{R}_{Cl} \quad (21)$$

$$(C_5Me_5)_2Lu{-}CH_3 + {}^{13}CH_4 \longrightarrow$$

$$(C_5Me_5)_2Lu{-}{}^{13}CH_3 + CH_4 \quad (22)$$

the presumed intermediate Cp'(PMe$_3$)Ir, gave only the previously observed hydridocyclooctyl complex.

It occurred to us that methane activation could be achieved thermally, under the reversible conditions described in the preceding section, by taking advantage of the presumption that the hydrido-methyl complex **17** would be thermodynamically more stable even than primary alkyl complexes such as **2**. It turned out that application of this idea gave successful results. Thus, as shown in Eq. 23, heating hydridocyclohexyl complex **4**

in cyclooctane solvent in a sealed Pyrex vessel under 20 atm of CH_4 at temperatures between 140° and 150°C for 14 hours led to a 58 percent yield of hydridomethyl complex **17** along with 8 percent of dihydride **2**. As with the other hydridoalkyl complexes, this material was quite sensitive, and therefore its identification by proton NMR was confirmed by conversion to the corresponding chloromethyl complex **18**, as well as by independent synthesis of **17** from dichloride **1** (via **18**) as shown in Eq. 23. In the methane thermal C–H oxidative addition experiment, we assume that the hydridocyclohexyl complex **4** and the corresponding hydridocyclooctyl complex are formed reversibly but do not build up because of their thermodynamic instability relative to the hydridomethyl complex **17**; that is, the hydridomethyl complex **17** is the "thermodynamic sink" for the system (*17*).

Preliminary Results on C–H Oxidative Addition with Rhodium Complexes

We prepared the rhodium complex **19**, analogous to dihydride **2**, and found that it, too, participates in C–H oxidative addition reactions (Fig. 7). Irradiation of **19** at room temperature leads only to an unidentified black substance. However, irradiation at −50°C produces new hydrides, again by observation of their absorptions in the high-field region of the low-temperature proton NMR spectrum. Preliminary studies have indicated that the C–H insertion reactions of the proposed intermediate Cp′RhL (Cp′ = C_5Me_5; L = PMe_3) shown in Fig. 7 are successful (*20*). With the exception of the hydridophenyl complex, warming the reaction solutions to −10°C causes reductive elimination of hydrocarbon and deposition of the black decomposition product. However, conversion of the hydridoalkyl complexes to the corresponding halides can be carried out at low temperature, and these products appear to be more stable than the hydrides.

Conclusions

We have found an example of the intermolecular reaction shown in Eq. 4, in which a soluble metal complex undergoes oxidative addition into the C–H bonds of alkanes, leading to hydridoalkylmetal complexes. Our most extensive work has been done with iridium, but preliminary results indicate that the analogous rhodium complexes undergo similar C–H insertions, although the products are considerably less stable. We have found it possible to convert the insertion products into organic halides, indicating that overall conversion of alkanes into functionalized organic mole-

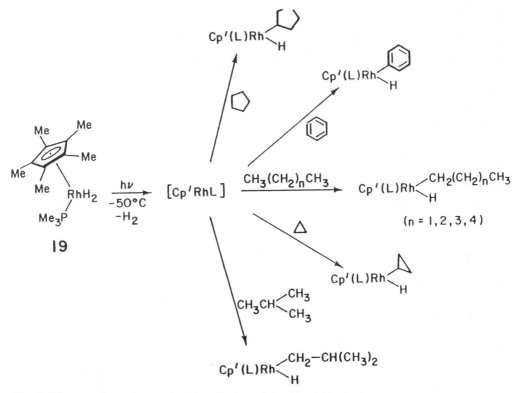

Fig. 7. The reactions observed on irradiation of the dihydridorhodium complex **17** at −50°C in various organic solvents.

cules is feasible. Selectivity studies have shown that the reactive iridium complex favors insertion into aromatic, primary, and small-ring C–H bonds.

Development of catalytic processes will be more challenging. The difference between a catalytic and noncatalytic sequence of reactions is that the former is cyclic, so that the catalyst is regenerated in the final step of the sequence and can be automatically reused. Development of such a cyclic process, like methane activation, does not require substantial new fundamental information but still faces difficult practical roadblocks. One of these is the fact that reagents which convert the C–H insertion product to functionalized organic molecules often

react irreversibly with the critical intermediate Cp′IrL, thus shutting down the cycle. However, research may uncover reagents that do not cause this problem.

On the fundamental side, we do not yet understand why the systems discussed here seem to favor intermolecular addition, whereas other metal complexes either favor intramolecular cyclometallation or do not react with unactivated C–H bonds at all. The search for additional systems that also engage in intermolecular addition should provide information that will be useful in answering this question. We also do not yet understand the physical basis for the selectivities we have observed. Finally, our results raise questions about the supposed requirement for a very-electron-rich metal center to induce C–H activation. We have found that replacing PMe$_3$ with the more electron-withdrawing phosphine P(OMe)$_3$ gives a system that also effects oxidative addition. Independent work in Graham's laboratory, illustrated in Eq. 21, demonstrates that even a CO ligand is not electron-withdrawing enough to prevent C–H insertion (15, 18). Finally, Watson's recent discovery of the reaction illustrated in Eq. 22 shows that methane (and other small hydrocarbons) will react with organometallic lutetium— a metal center normally considered to be very electrophilic (19). Intermolecular reactions between metal centers and alkanes are now being uncovered at an increasingly rapid rate and with metals having a wide range of electronic character. It seems likely that several different mechanisms for alkane oxidative addition will eventually be identified (21).

References and Notes

1. G. W. Parshall, *Homogeneous Catalysis* (Wiley, New York, 1980); P. Wiseman, *An Introduction to Industrial Organic Chemistry* (Applied Science Publishers, London, ed. 2, 1979).
2. H. A. Witcoff and B. G. Reuben, *Industrial Organic Chemicals in Perspective* (Wiley, New York, 1980), parts I and II.
3. S. W. Benson, *Thermochemical Kinetics* (Wiley, New York, ed. 2, 1976), p. 309.
4. J. Halpern, *Discuss. Faraday Soc.* **46**, 7 (1968). For reviews of the C–H activation area, see A. E. Shilov and A. A. Shteinman [*Coord. Chem. Rev.* **24**, 97 (1977)]; D. E. Webster [*Adv. Organomet. Chem.* **15**, 147 (1977)]; G. W. Parshall [*Acc. Chem. Res.* **8**, 113 (1975)]; A. E. Shilov [*Activation of Saturated Hydrocarbons by Transition Metal Complexes* (Reidel, Dordrecht, 1984)].
5. C. Walling, *Free Radicals in Solution* (Wiley, New York, 1957); E. S. Huyser, *Free Radical Chain Reactions* (Wiley, New York, 1970); W. A. Pryor, *Free Radicals* (McGraw-Hill, New York, 1966).
6. J. T. Groves, *Adv. Inorg. Biochem.* **1**, 119 (1979); _____, T. E. Nemo, R. S. Myers, *J. Am. Chem. Soc.* **101**, 1032 (1979); J. T. Groves and M. Van Der Puy, *ibid.* **98**, 5290 (1976); C. K. Chang and M.-S. Kuo, *ibid.* **101**, 3413 (1979); J. T. Groves and W. J. Kruper, Jr., *ibid.*, p. 7613; C. L. Hill and B. C. Schardt, *ibid.* **102**, 6374 (1980); J. A. Sofranko, R. Eisenberg, J. A. Kampmeier, *ibid.*, p. 1163.
7. See (2), part I, pp. 110ff.
8. A. L. J. Beckwith and T. Duong, *J. Chem. Soc. Chem. Commun.* (1978), p. 413; Z. Cohen, E. Keinan, Y. Mazur, T. Haim Varkony, *J. Org. Chem.* **40**, 2141 (1975), and references cited therein; N. C. Deno, E. J. Jedziniak, L. A. Messer, M. D. Meyer, S. G. Stroud, E. S. Tomezsko, *Tetrahedron* **33**, 2503 (1977).
9. J. P. Collman and L. S. Hegedus, *Principles and Applications of Organotransition Metal Chemistry* (University Science Books, Mill Valley, Calif., 1980), pp. 176–258.
10. See, for example, M. I. Bruce, *Angew. Chem. Int. Ed. Engl.* **16**, 73 (1977); P. Foley and G. M. Whitesides, *J. Am. Chem. Soc.* **101**, 2732 (1979); H. D. Empsall, E. M. Hyde, R. Markham, W. S. McDonald, M. C. Norton, B. L. Shaw, B. Weeks, *J. Chem. Soc. Chem. Commun.* (1977), p. 589; R. A. Andersen, R. A. Jones, G. Wilkinson, *J. Chem. Soc. Dalton Trans.* (1978), p. 446; H. Werner and R. Werner, *J. Organomet. Chem.* **209**, C60 (1981); J. W. Rathke and E. L. Muetterties, *J. Am. Chem. Soc.* **97**, 3272 (1975).
11. N. F. Gol'dshleger, M. B. Tyabin, A. E. Shilov, A. A. Shteinman, *Zh. Fiz. Khim.* **43**, 2174 (1969). For a review of this work, see A. E. Shilov and A. A. Shteinman [*Coord. Chem. Rev.* **24**, 97 (1977)].
12. R. H. Crabtree, J. M. Mihelcic, J. M. Quirk, *J. Am. Chem. Soc.* **101**, 7738 (1979); R. H. Crabtree, M. F. Mellea, J. M. Mihelcic, J. M. Quirk, *ibid.* **104**, 107 (1982).
13. D. Baudry, M. Ephritikhine, H. Felkin, *J. Chem. Soc. Chem. Commun.* (1980), p. 1243; *ibid.* (1982), p. 606.
14. G. L. Geoffrey and M. S. Wrighton, *Organometallic Photochemistry* (Academic Press, New York, 1979).
15. A. H. Janowicz and R. G. Bergman, *J. Am. Chem. Soc.* **104**, 352 (1982); *ibid.*, **105**, 3929 (1983); J. K. Hoyano and W. A. G. Graham, *ibid.* **104**, 3723 (1982).
16. See, for example, J. A. Labinger, R. J. Braus, D. Dolphin, J. A. Osborn [*J. Chem. Soc. Chem. Commun.* (1970), p. 612]; F. R. Jensen and B. Knickel [*J. Am. Chem. Soc.* **93**, 6339 (1971)]; J. A. Labinger and J. A. Osborn [*Inorg. Chem.* **19**, 3230 (1980)].
17. These results have been discussed in a recent manuscript (M. J. Wax, J. M. Stryker, J. M. Buchanan, C. A. Kovac, R. G. Bergman, *J. Am. Chem. Soc.*, in press).
18. J. K. Hoyano and W. A. G. Graham, *ibid.*, in press.
19. P. L. Watson, *ibid.* **105**, 6491 (1983).
20. R. A. Periana and R. G. Bergman, in preparation. Similar work has been reported independently by another group [W. D. Jones and F. J. Feher, *Organometallics* **2**, 562 (1983)].
21. J. K. Hoyano and W. A. G. Graham, *J. Am. Chem. Soc.* **104**, 3723 (1982).
22. I would like to acknowledge the important experimental and intellectual contributions made to the research described here by the graduate students and postdoctoral fellows whose names are listed in the references. This work was supported by the Director, Office of Energy Research, Office of Basic Energy Sciences, Chemical Sciences Division of the U.S. Department of Energy under contract No. DE-AC03-76SF00098.

The interconversion of different forms of energy has been of central importance in science and technology. Just as the practical application of heat engines, electric generators and motors, and storage batteries led to the development of the fields of thermodynamics and electrochemistry in the 19th and 20th centuries, so the problem of utilizing solar energy for the direct production of electricity and fuel has become a field of much current interest and has encouraged new fundamental investigations of the interactions of light, electron flow, and chemical reactions at electrode surfaces in electrochemical cells.

Sunlight in the near infrared, visible, and near ultraviolet regions has considerable energy (about 0.9 to 3.2 electron volts per photon, or about 87 to 308 kilojoules per mole) and intensity and could provide a significant contribution to our electrical and chemical resources if efficient and inexpensive systems utilizing readily available materials could be devised for the conversion process. Indeed the fossil fuels we largely depend on for our current energy needs presumably originated from biological photosynthesis, and even now the burning of wood represents an important part of the world energy supply. However, biological photosynthesis is relatively inefficient in terms of conversion of solar energy to dry fuels, with overall field efficiencies of about 1 percent (*1*). The fabrication of artificial photosynthetic systems for the conversion of abundant materials (for example, H_2O and CO_2) to fuels (for example, H_2 and CH_3OH), directly or by electrolysis or the production of electricity, is clearly an important goal (*2, 3*). Moreover, photochemical reactions could be employed to replace other energy-consuming chemical processes, for example, for pollution abatement or in

19

Photoelectro-chemistry

Allen J. Bard

Science 207, 139–144 (11 January 1980)

273

chemical synthesis. One of the more promising approaches to the design of such systems involves the application of photoelectrochemical cells or powder catalysts made of semiconductor materials. In this article I describe the basic principles of such systems, review some of their applications, and touch briefly on the relevancy of the concepts developed for these systems to such diverse topics as photography, electrochromic displays, and chemical evolution.

Basic Principles

The conversion of light to electrical or chemical (redox) energy results from light in the visible region acting as an electron pump. The absorption of a photon by an atom or molecule pumps an electron from a lower orbital to a higher one (Fig. 1A). The wavelength of light that causes such a transition is that with an energy equal to or greater than the difference in energies of the two orbitals, E_g. The result is an electron-hole (e^-h^+) pair formed by this intramolecular pumping in species, S. This produces an excited state, denoted S* (see Fig. 1A). If the e^-h^+ pair can be separated so that the e^- flows to a suitable acceptor species, A,

$$S^* + A \rightarrow S^+ + A^- \qquad (1)$$

or an electron from a suitable donor, D, fills h^+

$$S^* + D \rightarrow S^- + D^+ \qquad (2)$$

then the light energy has been stored, at least for a short time, as redox chemical energy. The reaction of S^+ and A^- or S^- and D^+ is spontaneous (or "downhill") and capable of liberating energy (that is, is exoergic). If e^- is pumped through a wire, it has been converted to an electrical current flow. However, excited states are very short-lived (typically lasting from nanoseconds to milliseconds in liquids) and the e^-h^+ pairs frequently recombine very quickly with the captured light energy degraded to heat or, sometimes with the emission of a photon, as in phosphorescence. To utilize the light in a form other than heat, one must achieve separation of the e^-h^+ pair before recombination. This separation can be promoted by an electric field (that is, a difference in electrical potential) or a "chemical field" (that is, a difference in chemical potential, as occurs in the presence of A or D). Unless the back reaction between S^+ and A^- or between S^- and D^+ is slow, the back electron transfer to produce S and A or S and D, respectively, will occur very quickly, and again this immediate intermolecular e^-h^+ recombination will usually result in heat production. If the chemical energy is to be stored in the form of oxidized and reduced species, either the energy of activation for this recombination process must be high, so that the back reaction is slow, or the oxidized and reduced products must have been formed at some distance from one another so that they can be separated before reacting.

Light absorption occurs in a similar way in a semiconductor (Fig. 1B). In semiconductors (for example, silicon, titanium dioxide, cadmium sulfide, gallium arsenide, and a wide range of other inorganic, organometallic, and organic substances) the orbitals are merged into a nearly filled valence band and a nearly vacant conduction band separated by the energy gap, E_g. When a semiconductor is immersed in a solution, charge transfer occurs at the interface because of the difference in the tendency of the two

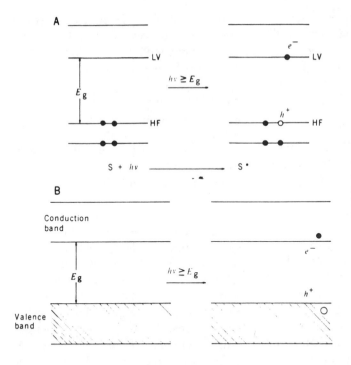

Fig. 1. (A) Electronic orbitals and light absorption in an atom or molecule, S. (B) Semiconductor bands and electron-hole pair formation on light absorption.

phases to gain or lose electrons (that is, difference in electron affinity or electrochemical potential of the two phases). The net result is the formation of an electrical field at the surface of the semiconductor to a depth of 5 to 200 nanometers. The direction of this electric field depends on the relative electron affinities of the semiconductors and solution. For an n-type semiconductor, which is one that is doped with a donor species so that some electrons are in the conduction band, the field frequently forms in the direction from the bulk of the semiconductor toward the interface (Fig. 2A). Thus, if an e^-h^+ pair forms in this region of the semiconductor (the space charge region) because light is absorbed at the interface, the electron moves toward the bulk of the semiconductor and the hole moves toward the surface. Thus the electric field that forms spontaneously at the interface accomplishes the e^-h^+ separa-

tion. The electric field within the semiconductor is represented, as in Fig. 2A, by a bending of the bands. In such diagrams, where the energy levels represent electronic energy, electrons move spontaneously "downhill" and holes, "uphill." After separation, what is the fate of e^- and h^+? If the solution contains a species, D, which has an energy level (that is, a solution redox potential) above that of the photogenerated hole at the surface, the electron transfer reaction

$$D + h^+ \rightarrow D^+ \qquad (3)$$

can occur. The excited electron, which can have an energy approaching that of the conduction band edge, can be transferred through a wire connected to the semiconductor to a second, nonphotoactive electrode (for example, made of carbon or a metal) where some oxidized form, O, can be reduced

$$O + e^- \rightarrow R \qquad (4)$$

The combination of the n-type semiconductor and the inert electrode immersed in the electrolyte solution comprises a photoelectrochemical cell (Fig. 2, B and C) in which light promotes the overall reaction

$$O + D \rightarrow R + D^+ \qquad (5)$$

For example, if O and D were both water, Eq. 5 would represent the photodecomposition of the water to H_2 and O_2

$$2H_2O \rightarrow O_2 + 2H_2 \qquad (6)$$

Although this interesting and potentially useful reaction has been demonstrated in such photoelectrochemical cells, its efficiency with the semiconductor materials employed so far (for example, TiO_2 and $SrTiO_3$) has been too low, for light that is characteristic of the solar spectrum, to be of practical interest. The reaction in Eq. 5 could be carried out by conventional electrochemical methods in cells such as those used to electrolyze water or to produce chlorine; in these an external electrical source drives the reaction in the uphill direction. In the photoelectrochemical cell it is light that pumps the electrons and provides the needed energy. Note that the photoelectrochemical cell provides both the required field for $e^- h^+$ separation as well as the considerable spacial separation of the products R and D^+ which are potentially capable of reacting with one another.

Cells constructed with p-type semiconductors can be described in a similar way (Fig. 3). A p-type semiconductor is one that is doped with an acceptor impurity to produce holes in the valence band. When a p-type material contacts a solution with a redox couple at an energy within the band gap of the semiconductor (Fig. 3A), electronic equilibration

again produces a space charge region with the field now pointing toward the bulk semiconductor. Photogenerated $e^- h^+$ pairs in the space charge region will again separate; in this case electrons move to the interface, where an acceptor species in solution, A, can be reduced, while holes move to the interior of the semiconductor (Fig. 3, B and C). Therefore, light promotes photoreductions at p-type materials and photooxidations at n-type ones.

This separation of the $e^- h^+$ pair in the electric field at the semiconductor-solution interface is very similar to that which occurs at the p-n junction of solid-state photovoltaic cells (for example, of Si or GaAs), which are the familiar solar cells used in space and some terrestrial applications. However, several important differences should be noted. Although the solid-state cells pump electrons through an external circuit, no actual chemistry occurs. In the liquid junction cells, since electrons and holes themselves are usually not very stable in the solution medium, their transport through the liquid phase depends on the occurrence of redox reactions at the electrode surfaces. This provides a means of direct conversion of radiant energy to storable chemicals. The occurrence of chemical reactions at the electrode surface also implies the possibility of the semiconductor itself reacting. Thus photogenerated holes produced at an n-type semiconductor may cause oxidation of the semiconductor surface, producing a blocking layer or dissolution of the electrode, as well as the solution species. Stability of the semiconductor electrodes under irradiation is thus of major concern in photoelectrochemical cells. However, stability can be attained by suitable choice of the solution redox

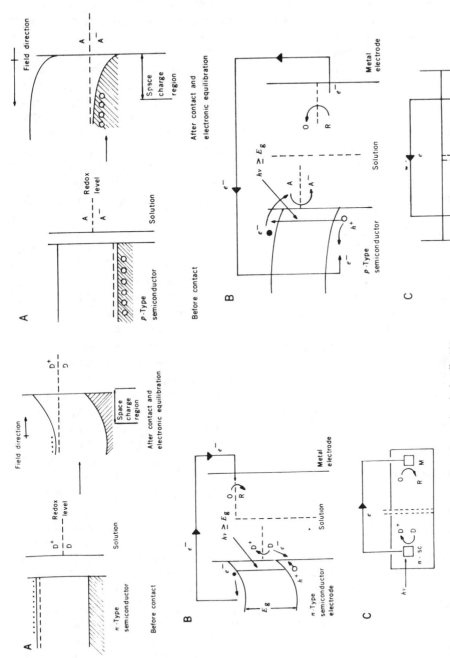

Fig. 2 (left). *n*-Type semiconductor photoelectrochemical cell. (A) Formation of space charge on immersion in solution with redox couple D/D⁺. (B) Electron flow under irradiation with solution containing species D and O. (C) Cell configuration.   Fig. 3 (right). *p*-Type semiconductor photoelectrochemical cell. (A) Formation of space charge on immersion in solution with redox couple A/A⁻. (B) Electron flow under irradiation with solution containing species A and R. (C) Cell configuration.

couple or solvent and by modification of the electrode surface. An important advantage of the liquid junction cells, compared to the solid-state devices, is that the junction is produced very easily by immersing the semiconductor in solution. Moreover, light is absorbed right at the interface, so that the photogenerated carriers must move only a short distance before they react. In the Si(p-n) cell the light must pass through the layer of p-type material and most of the photogenerated carriers must diffuse to the junction where separation occurs. Under such circumstances expensive single crystal material is needed, otherwise recombination of the diffusing species at grain boundaries occurs before their separation, which leads to loss of efficiency. This is much less of a problem at the semiconductor-liquid interface, and polycrystalline materials (produced by vapor deposition, chemical surface treatment, or simply by pressing powdered materials into pellets) have been used successfully. Several reviews that describe the phenomena at such interfaces in more detail and discuss a number of such systems are available (4-7).

Photoelectrochemical Cells

Several different types of photoelectrochemical cells can be devised (4). In the liquid junction photovoltaic (or regenerative) cell the aim is the production of an electrical current flow without net changes occurring in either the electrolyte solution or the electrode materials. These cells, which are the photoelectrochemical equivalent of solid-state solar cells, utilize a single redox couple in the solution. For example, if in the cell of Fig. 2B only the redox couple D/D$^+$ was present, the photodriven oxidation (D + $h^+ \rightarrow$ D$^+$) would occur at the n-type semiconductor electrode while the reverse reaction would occur at the metal electrode (D$^+$ + $e \rightarrow$ D). A number of cells of this type with different semiconductor materials, redox couples, and solvents have been described. The choice of the redox couple is important, because it is a factor in fixing the operating voltage of the cell and often serves to stabilize the semiconductor from photodecomposition. For example, a cell with an n-CdS electrode will not operate for very long in a solution lacking a suitable redox couple, because during irradiation the photogenerated holes at the surface cause decomposition of the CdS (CdS + 2$h^+ \rightarrow$ Cd^{2+} + S) rather than promoting water oxidation, and the semiconductor surface becomes covered with sulfur. In the presence of a high concentration of sulfide or selenide in the solution, however, preferential oxidation of these species occurs, and the CdS no longer decomposes (8). The photovoltaic cells that show the highest solar efficiencies are those using single crystal or epitaxial n-GaAs in an alkaline selenide electrolyte; solar power conversion efficiencies of 12 to 14 percent have been reported in these cells (9, 10).

Perhaps more interesting from the viewpoint of the storage of solar energy are cells in which there is a net chemical reaction. In photoelectrosynthetic cells the overall reaction (for example, see Eq. 5) is driven in an uphill direction (that is, $\Delta G^0 > 0$ for the reaction) by the light and a portion of the solar energy is stored as chemical energy in the products (R and D$^+$). For practical, long-term storage of fuels, the starting materials should be inexpensive, readily available materials (such as H$_2$O, CO, CO$_2$, or N$_2$).

In an alternative arrangement the products can be used nearby, for example, in a storage battery or fuel cell arrangement where the chemical energy of the products is converted to electricity at the same time as O and D are regenerated. Photoelectrosynthetic cells have been reported, for example with n-SrTiO$_3$ or n-TiO$_2$ for the decomposition of water to hydrogen and oxygen (11–13) or with p-GaP for reduction of CO$_2$ to formaldehyde and methanol (14). In most cases an external electrical bias had to be provided (so that the processes were "photoassisted electrolyses" rather than purely light-driven reactions) and the solar power efficiencies were below 1 percent. Such cells could also be used for carrying out other reactions that produce chemically useful products. For example, the photooxidation of chloride has been demonstrated on n-TiO$_2$ (although again with a rather low efficiency) (15, 16). Thus chlorine, which is currently produced throughout the world in large amounts with the expenditure of electrical energy, could be photosynthesized by the reaction

$$2H^+ + \tfrac{1}{2}O_2 + 2Cl^- \rightarrow H_2O + Cl_2 \quad (7)$$

In photocatalytic cells the light is used to drive a reaction in a downhill direction (that is, $\Delta G^0 < 0$). In this case the radiant energy is not stored as chemical energy but is instead used to overcome the energy of activation of the process. Such a cell could be of interest for the decomposition of waste materials. For example, the oxidation of cyanide ion by the reaction

$$CN^- + \tfrac{1}{2}O_2 \rightarrow OCN^- \quad (8)$$

is spontaneous thermodynamically but is kinetically very slow. However, in a photocatalytic photoelectrochemical cell

CN$^-$ can be photooxidized at an n-TiO$_2$ electrode with O$_2$ reduced at a suitable cathode, so that the net reaction in Eq. 8 is accomplished (15). The decomposition reactions for many organic substances to smaller molecules are also spontaneous. For example, although acetic acid is usually considered to be a very stable substance, the liquid (ℓ) to gas (g) reaction

$$CH_3CO_2H(\ell) \rightarrow \tfrac{1}{2}C_2H_6(g) + CO_2(g) + \tfrac{1}{2}H_2(g) \quad (9)$$

is downhill ($\Delta G^0 = -18.4$ kJ/mole). This decomposition reaction (referred to as the Photo-Kolbe reaction by analogy to the well-known nonphotoactivated electrochemical decomposition of carboxylic acids) has also been carried out in a photoelectrochemical cell with an n-TiO$_2$ photoanode (17, 18).

Heterogeneous Photosynthesis and Photocatalysis

The concepts that emerge from studies of these types of photoelectrochemical cells with semiconductor electrodes can be applied to the design of systems in which semiconductor particles or powders are used for similar purposes. For example, we have used powders of TiO$_2$ with platinum dispersed on the surface for a number of studies. Each particle can be pictured as a "short-circuited" photoelectrochemical cell, where the semiconductor electrode and metal counter electrode have been brought into contact (Fig. 4). Irradiation of such particulate systems still involves the e^-h^+ pair formation and surface oxidation and reduction reactions found in the cells but without external current flow. Although powders are obviously much simpler to use, the advantage of the large separation between the oxidation and reduction sites found in the electrochemical cells is missing. However, the distance between such sites on particles is still probably large compared to those found with photochemical redox reactions in solutions, where the products of the charge transfer reaction are in close proximity within the same solvent cage.

Although the particles have been depicted in Fig. 4 as semiconductor and metal systems, in fact the untreated powders alone can often serve as photocatalytic materials. For example, the reaction of Eq. 8 can be carried out on bare TiO_2 (anatase) powders (19). What is required is that the reaction (in this case the reduction of oxygen) occur readily at the unilluminated TiO_2 surface. The behavior of electrodes in the photoelectrochemical cells can serve as a useful guide to the design of the catalyst particles.

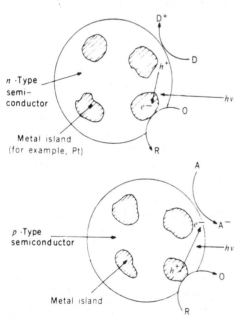

Fig. 4. Representation of semiconductor particulate systems for heterogeneous photocatalysis or photosynthesis (equivalent to the cells in Figs. 2 and 3).

cles. The rate of a reaction occurring at an electrode is measured by the current flowing through that electrode (since current flow is equivalent to charge transferred across the interface or amount of product formed per unit of time). The energy required to carry out this electrode reaction is represented by the electrode potential needed to produce this current flow. The behavior of an electrochemical cell is thus characterized by current-potential (i-V) curves such as those shown in Fig. 5A. These are typical of the curves found with cells of the type in Fig. 2. The oxidation of D occurs more easily on the illuminated semiconductor (Fig. 5, curve 1) than on the platinized electrode (curve 2) because the light energy promotes the electron transfer so that it takes place at less positive potentials than those corresponding to its reversible oxidation. The reduction of oxygen can occur at either the semiconductor (curve 3) or the platinum (curve 4). On the particles at steady state the net rate of the oxidation of D and reduction of oxygen must be the same. In terms of the i-V curves, this occurs when the anodic current has the same magnitude as the cathodic one. For the case illustrated in Fig. 5, one would predict that the platinized semiconductor would show a larger reaction rate (line a) as compared to the bare semiconductor (line b) because of the greater ease of reduction of oxygen on platinum. This application of i-V curves to the design of photocatalyst particles is analogous to treatments of corrosion in terms of mixed potentials and electrode kinetics (20, 21). In those cases the curves are usually presented in the form of log $|i|$ plotted against V and the intersection of the anodic and cathodic lines determines the rate of the process; this method of presentation for photoprocesses at particles is shown in Fig. 5B.

A number of heterogeneous photosynthetic and photocatalytic processes involving semiconductor particulates have been described (4, 22, 23); a representative sampling is shown in Table 1. Some of the same reactions that have been carried out in photoelectrochemical cells, such as the oxidation of CN^-, SO_3^{2-}, and $CH_3CO_2^-$, can also be accomplished at semiconductor powders suspended in solution. The products of the reactions may be different, however. Let us consider the oxidation of acetic acid. By application of spin trapping techniques with observation of the intermediates by electron spin resonance spectroscopy, we have demonstrated that methyl radicals ($CH_3 \cdot$) are intermediates in the photooxidation at TiO_2 (24). Thus the elementary step in the oxidation of acetate is probably

$$CH_3CO_2^- + h^+ \rightarrow CH_3 \cdot + CO_2 \quad (10)$$

where h^+ is a photogenerated hole at the semiconductor surface (or an adsorbed hydroxyl radical). In the divided photoelectrochemical cell the methyl radicals dimerize to form ethane (17, 18). At the platinized TiO_2 powders, however, reduction of protons to hydrogen atoms ($H \cdot$) adsorbed on the platinum sites can occur. Although some formation of ethane is observed, methane is the major product at the powders, probably because at reasonable light intensities the reaction of $CH_3 \cdot$ with $H \cdot$ is more probable than coupling (25). The decarboxylation of a number of other acids to form the alkane at irradiated platinized TiO_2 has also been reported. The photogenerated radical intermediates in these processes have also been shown to be useful in the initiation of polymerization reactions (26). The advantage in this method of initiation is that the initiator radical, and the subsequent endgroup of the polymer, can be selected without regard to the light-absorption properties of the initiator.

When the solution contains ions of metals that are deposited at potentials less negative than those required for proton reduction, these ions may play the role of species, O, and be preferentially reduced at the semiconductor. This method has been applied to the deposition of platinum, copper, and other metals on TiO_2 powders, and may be useful in the preparation of catalysts (27). Indeed, the platinized TiO_2 particles used in the decarboxylation reactions were prepared by just this process. This technique has advantages over more conventional chemical methods of catalyst preparation in that the low temperatures used and the control over the extent of platinum deposition by the light flux allow the production of very small clusters of platinum on the semiconductor surface. The light-induced deposition of metals is clearly related to photographic processes (28). Fundamental studies of the behavior of semiconductor electrodes and powders can provide insight into the physical and chemical processes occurring during image formation and may also be useful in the fabrication of new types of electrochromic devices for use in displays. Photoprocesses at TiO_2 and ZnO may also be implicated in the chalking of paints with these pigments (29); chalking occurs in exterior paints when the organic polymer binder is degraded and the pigment is exposed on the surface.

Perhaps the area of application of greatest potential utility, but also the one with the most serious difficulties, is that

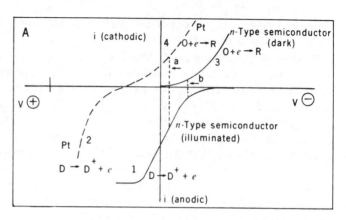

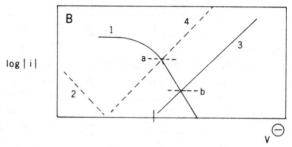

Fig. 5. The application of current-potential curves in the design of photoelectrochemical systems. (A) Representative curves for the system in Fig. 2. Solid lines indicate behavior at n-type semiconductor (curves 1 and 3); broken lines, at Pt metal (curves 2 and 4); a, operating point for semiconductor/Pt system; b, operating point for bare semiconductor system. (B) Curves showing log current plotted against potential for the system in (A).

of photoelectrosynthesis of useful chemicals from inexpensive and abundant materials (for example, H_2 from H_2O, CH_3OH from CO_2, and NH_3 from N_2). Several experiments (see Table 1) demonstrate the feasibility of such processes but so far both the quantum efficiency (molecules of product produced per photon absorbed) and solar efficiency (molecules of product produced per incident solar photon) have been rather low (usually less than 1 to 2 percent). These low efficiencies can be ascribed to problems caused by recombination of products and intermediates as well as the use of materials whose band gap is too large to utilize effectively the solar spectrum. Recent considerations of the photodecomposition of water by semiconductor systems or by other means with sunlight suggest that when the needed energy and driving forces are considered, a single

light-absorbing system necessitates the use of a semiconductor with a rather large band gap (2.5 to 3.0 eV) (2, 4). The efficient utilization of solar energy for water-splitting may thus require the strategy of green plants, that is, the use of two photosystems with the absorption of the two photons of lower energy per electron transferred.

We have discussed the possibility that such photoprocesses on semiconductors, presumably taking place on naturally occurring powders, played a role in early chemical evolution on earth (30). Irradiation of mixtures of NH_3, methane, and water in the presence of platinized TiO_2 with either a xenon lamp or sunlight was shown to produce amino acids. This experiment is thus similar to those demonstrating the production of amino acids in electric discharges or by ultraviolet light (31, 32), but demonstrates that light

Table 1. Representative heterogeneous photocatalytic and photosynthetic processes at semiconductor powders.

Reaction	Powder	Comments	Reference
Liquid phase			
$O_2 \rightarrow H_2O_2$	ZnO, others	Oxidation of organic additives	(22)
$RCO_2H \rightarrow RH + CO_2$	Pt/TiO_2	$R = CH_3, C_2H_5, \ldots$	(25)
$M^{+n} \rightarrow M$	TiO_2, WO_3	$M = Pt, Pd, Cu, \ldots$	(27, 33)
$CN^- \rightarrow NCO^-$	TiO_2, ZnO, CdS	Reduction of O_2	(19)
$CO_2 \rightarrow CH_3OH, HCHO$	TiO_2, CdS. . . SiC, GaP		(34)
$H_2O \rightarrow H_2 + O_2$	$Pt/TiO_2, Pt/SrTiO_3$		(11, 35)
$O_2 \rightarrow O_2^-\cdot$	CdS		(36)
Gas phase			
$CO \rightarrow CO_2$	ZnO		(22, 23)
$N_2 \rightarrow NH_3$	TiO_2(Fe-doped)		(37)
$RH \rightarrow$ ketones	TiO_2		(23, 38)
Quinoline $\rightarrow$ radical cation	TiO_2		(39)

within the solar spectrum is capable of promoting such a reaction. In the absence of the semiconductor such processes would not be possible, because none of the components of the mixture absorb radiation in the visible and near ultraviolet region. This type of process for the formation of amino acids from components of the primordial atmosphere is attractive, because it provides a means for their continuous production over long periods of time, thus providing nutrients for early heterotrophic organisms with energy inputs not very different from those on the earth now. Although platinized TiO_2 is an unlikely candidate for such a process under natural conditions, many other semiconductor powders, such as WO_3, Fe_2O_3, and ZnO that have been used for photocatalytic processes, might also be capable of such an inorganic photosynthesis. The role of similar processes as a nonbiological source of oxygen in the early atmosphere might also be considered.

Conclusions

The basic principles of photoelectrochemical cells and the photoprocesses occurring on semiconductor materials have been established and studies describing new materials and reaction schemes are appearing with increasing frequency. Although more efficient systems utilizing inexpensive and readily available semiconductors are still needed, the field of photoelectrochemistry has opened new areas in electrochemical research and has provided new insight into the interactions of light, electricity, and chemical reactions in a number of different processes.

References and Notes

1. D. O. Hall, in *Solar Power and Fuels*, J. R. Bolton, Ed. (Academic Press, New York, 1977), pp. 27–51.
2. J. R. Bolton, *Science* **202**, 705 (1978).
3. M. S. Wrighton, in *McGraw-Hill Yearbook of Science and Technology* (McGraw-Hill, New York, 1978), pp. 11–17.
4. A. J. Bard, *J. Photochem.* **10**, 59 (1979).
5. A. J. Nozik, *Annu. Rev. Phys. Chem.* **29**, 189 (1978).
6. R. Memming, in *Electroanalytical Chemistry*, A. J. Bard, Ed. (Dekker, New York, 1979), pp. 1–84.
7. H. Gerischer, in *Physical Chemistry—An Advanced Treatise*, H. Eyring, D. Henderson, W. Jost, Eds. (Academic Press, New York, 1970), pp. 463–542.
8. B. Miller and A. Heller, *Nature (London)* **262**, 680 (1976); G. Hodes, D. Cahen, J. Manassen, *ibid.* **260**, 312 (1976); R. N. Noufi, P. A. Kohl, A. J. Bard, *J. Electrochem. Soc.* **125**, 375 (1978); A. J. Bard and M. S. Wrighton, *ibid.* **124**, 1706 (1977); G. Hodes, J. Manassen, D. Cahen, *Nature (London)* **261**, 403 (1976); B. Miller, A. Heller, M. Robbins, S. Menezes, K. C. Chang, J. Thompson, Jr., *J. Electrochem. Soc.* **124**, 1019 (1977); M. S. Wrighton, A. B. Bocarsly, J. M. Bolts, A. B. Ellis, K. D. Legg, in *Semiconductor Liquid-Junction Solar Cells*, A. Heller, Ed. (Electrochemical Society, Princeton, N.J., 1977), vol. 77–3, p. 138; A. B. Ellis, S. W. Kaiser, M. S. Wrighton, *J. Am. Chem. Soc.* **98**, 1635 (1976); *ibid.*, p. 6418; *ibid.*, p. 6855; A. B. Ellis, S. W. Kaiser, J. M. Bolts, M. S. Wrighton, *ibid.* **99**, 2839 (1977); J. M. Bolts, A. B. Ellis, K. D. Legg, M. S. Wrighton, *ibid.*, p. 4826; A. Heller, K. C. Chang, B. Miller, in *Semiconductor Liquid-Junction Solar Cells*, A. Heller, Ed. (Electrochemical Society, Princeton, N.J., 1977), vol. 77–3, p. 54; *J. Electrochem. Soc.* **124**, 697 (1977).
9. B. A. Parkinson, A. Heller, B. Miller, *J. Electrochem. Soc.* **126**, 954 (1979).
10. R. Noufi and D. Tench, *ibid.*, in press.
11. M. S. Wrighton, A. B. Ellis, P. T. Wolczanski, D. L. Morse, H. B. Abrahamson, D. S. Ginley, *J. Am. Chem. Soc.* **98**, 2774 (1976).
12. J. G. Mavroides, J. A. Kafalas, D. F. Kolesar, *Appl. Phys. Lett.* **28**, 241 (1976).
13. T. Watanabe, A. Fujishima, K. Honda, *Bull. Chem. Soc. Jpn.* **49**, 355 (1976).
14. M. Halmann, *Nature (London)* **275**, 115 (1978).
15. S. N. Frank and A. J. Bard, *J. Am. Chem. Soc.* **99**, 4667 (1977).
16. T. Inoue, T. Watanabe, A. Fujishima, K. Honda, *Chem. Lett.* **1977**, 1073 (1977).
17. B. Kraeutler and A. J. Bard, *J. Am. Chem. Soc.* **99**, 7729 (1977).
18. _____, *Nouv. J. Chim.* **3**, 31 (1979).
19. S. N. Frank and A. J. Bard, *J. Phys. Chem.* **81**, 1484 (1977).
20. R. F. Steigerwald, *Corrosion* **24**, 1 (1968).
21. C. Wagner and W. Traud, *Z. Elektrochem.* **44**, 391 (1938).
22. T. Freund and W. P. Gomes, *Catal. Rev.* **3**, 1 (1969).
23. M. Formenti and S. J. Teichner, *Specialists Periodical Reports* (Chemical Society, London, in press), vol. 2, chap. 4.
24. B. Kraeutler, C. D. Jaeger, A. J. Bard, *J. Am. Chem. Soc.* **100**, 4903 (1978).
25. B. Kraeutler and A. J. Bard, *ibid.*, p. 2239; *ibid.*, p. 5985.
26. B. Kraeutler, H. Reiche, A. J. Bard, R. G. Hocker, *J. Polym. Sci. Polym. Lett. Ed.* **17**, 535 (1979).
27. B. Kraeutler and A. J. Bard, *J. Am. Chem. Soc.* **100**, 4317 (1978).
28. See, for example, E. J. DeLorenzo, L. K. Case, E. M. Stickles, W. A. Stamoulis, *Photogr. Sci. Eng.* **13**, 95 (1969) and references therein.
29. S. P. Pappas and R. M. Fischer, *J. Paint. Technol.* **46**, 65 (1974).
30. H. Reiche and A. J. Bard, *J. Am. Chem. Soc.* **101**, 3127 (1979).
31. See, for example, S. L. Miller, *Science* **117**, 528 (1953); *J. Am. Chem. Soc.* **77**, 2351 (1955).
32. S. L. Miller and H. C. Urey, *Science* **130**, 245 (1959).
33. M. S. Wrighton, P. T. Wolczanski, A. B. Ellis, *J. Solid State Chem.* **22**, 17 (1977); H. Reiche, W. W. Dunn, A. J. Bard, *J. Phys. Chem.* **83**, 2248 (1979).
34. T. Inoue, A. Fujishima, S. Konishi, K. Honda, *Nature (London)* **277**, 637 (1979); M. Halmann and B. Aurian-Blajeni, *Proceedings of the Photovoltaic Energy Conference, West Berlin, April 1979*, in press.
35. A. V. Bulatov and M. L. Khidekel, *Izv. Akad. Nauk SSSR, Ser. Khim.* **8**, 1902 (1976).
36. J. R. Harbour and M. L. Hair, *J. Phys. Chem.* **81**, 1791 (1977).
37. G. N. Schrauzer and T. D. Guth, *J. Am. Chem. Soc.* **99**, 7189 (1977).
38. M. Formenti, F. Juillet, P. Meriaudeau, S. J. Teichner, *Chem. Technol.* **1**, 680 (1971); M. Formenti, F. Juillet, S. J. Teichner, *Bull. Soc. Chim. France* **1976**, 1031 (1976).
39. K. M. Sancier and S. R. Morrison, *Surf. Sci.* **83**, 29 (1979).
40. We thank the National Science Foundation, Robert A. Welch Foundation, and the Office of Naval Research for support.

Part V

Organic Synthesis

Since the 3rd century and for more than a thousand years thereafter chemistry has been thought of as a complicated, hard-to-predict science. Efforts to improve even a part of its unpredictable character are said to have borne fruit first of all in the success of the "electronic theory." This was founded mainly by organic chemists, such as Fry, Stieglitz, Lucas, Lapworth, and Sidgwick, then brought to a completed form by Robinson and Ingold, and developed later by many other chemists (1). In the electronic theory, the mode of migration of electrons in molecules is noted and is considered from various points of view. For that purpose, a criterion is necessary with respect to the number of electrons that should originally exist in an atom or a bond in a molecule. Therefore, it can be said to be the concept by Lewis of the sharing of electrons that has given a firm basis to the electronic theory (2).

In the organic electronic theory, the chemical concepts such as acid and base, oxidation and reduction, and so on, have been conveniently utilized for a long time. Furthermore, there are terms centering more closely around the electron concept, such as electrophilicity and nucleophilicity, and electron donor and acceptor, both being pairs of relative concepts.

One may be aware that these concepts can be connected qualitatively to the scale of electron density or electric charge. In the electronic theory, the static and dynamic behaviors of molecules are explained by the electronic effects that are based on nothing but the distribution of electrons in a molecule.

The mode of charge distribution in a molecule can be sketched to some extent by the use of the electronegativity concept of atoms through organic chemical

20

Role of Frontier Orbitals in Chemical Reactions

Kenichi Fukui

Science 218, 747–754 (19 November 1982). Copyright © 1982 by the Nobel Foundation. This chapter is the basis of the lecture delivered by Kenichi Fukui in Stockholm on 8 December 1981 when he received the Nobel Prize in Chemistry, which he shared with Roald Hoffmann.

experience. At the same time, it is given foundation, made quantitative, and supported by physical measurements of electron distribution and theoretical calculations based on quantum theory.

The distribution of electrons or electric charge—with either use the result is unchanged—in a molecule is usually represented by the total numbers (generally not integer) of electrons in each atom and each bond, and it was a concept easily acceptable even to empirical chemists as having a tolerably realistic meaning. Therefore, chemists employed the electron density as a fundamental concept to explain or to comprehend various phenomena. In particular, for the purpose of promoting chemical investigations, researchers usually rely on the analogy through experience, and the electron density was very effectively and widely used as the basic concept in that analogy.

When the magnitude of electron density is adopted as the criterion, the electrostatic attraction and repulsion caused by the electron density are taken into account. Therefore, it is reasonable to infer that an electrophilic reagent will attack the position of large electron density in a molecule while a nucleophilic reaction will occur at the site of small electron density. In fact, Wheland and Pauling (3) explained the orientation of aromatic substitutions in substituted benzenes along these lines, and theoretical interpretations of the mode of many other chemical reactions followed in the same fashion.

However, the question why one of the simple reactions known from long before, the electrophilic substitution in naphthalene, for instance, such as nitration, yields α-substituted derivatives predominantly was not so easy to an-swer. That was because, in many of such unsubstituted aromatic hydrocarbons, both the electrophile and the nucleophile react at the same location. This point threw some doubt on the theory of organic reactivity, where the electron density was thought to do everything.

Concept of Frontier Orbital Interactions

The interpretation of this problem was attempted from variously different angles by many people. Above all, Coulson and Longuet-Higgins (4) took up the change of electron density distribution under the influence of approaching reagent. The explanation by Wheland (5) was based on the calculation of the energy required to localize electrons forcibly to the site of reaction. But I myself tried to attack this problem in a way which was at that time slightly unusual. Taking notice of the principal role played by the valence electrons in the case of the molecule formation from atoms, only the distribution of the electrons occupying the highest energy π orbital of aromatic hydrocarbons was calculated. The attempt resulted in a better success than expected, obtaining an almost perfect agreement between the actual position of electrophilic attack and the site of large density of these specified electrons as exemplified in Fig. 1 (6).

The "orbital" concept, which was established and developed by many scientists, such as Pauling, Slater, Mulliken, Roothaan, Löwdin, Hückel, and Parr, had until then been used to construct the wave function of a molecule, through which molecular properties were usually interpreted (7). It seemed that the electron distribution in an orbital was directly connected to chemical observations and this fact was certainly felt to be interesting by many chemists.

But the result of such a rather "extravagant" attempt was by no means smoothly accepted by chemists in general. That paper received a number of controversial comments. This was in a sense understandable, because, for lack of my experiential ability, the theoretical foundation for this conspicuous result was obscure or rather improperly given. However, it was fortunate for me that the article on the charge-transfer complex of Mulliken (8) was published in the same year as ours.

The model of Mulliken *et al.* for protonated benzene was very helpful (9). Our work in collaboration with Yonezawa, Nagata, and Kato provided a simple and pointed picture of theoretical interpretation of reactions (10), as well as the "overlap and orientation" principle proposed by Mulliken with regard to the orientation in molecular complexes (11). After the electrophilic substitution, the nucleophilic substitution was discussed, and it was found that in this case the lowest energy vacant orbital played this particular part (12). In reactions with radicals, both of the two orbitals mentioned above became the particular orbitals.

There was no essential reason to limit these particular orbitals to π orbitals, so that this method was properly applied not only to unsaturated compounds but also to saturated compounds. The applicability to saturated compounds was a substantial advantage in comparison with many theories of reactivity which were then available only for π electron compounds. The method displayed its particular usefulness in the hydrogen abstraction by radicals from paraffinic hydrocarbons, the S_N2 and E2 reactions in halogenated hydrocarbons, the nucleophilic abstraction of α-hydrogen of olefins, and so forth (13).

These two particular orbitals, which act as the essential part in a wide range of chemical reactions of various compounds, saturated or unsaturated, were referred to under the general term of "frontier orbitals," and abbreviated frequently by HOMO (highest occupied molecular orbital) and LUMO (lowest unoccupied molecular orbital).

In this way, the validity of the theory became gradually clearer. The vein of ore discovered by chance was found to be more extensive than expected. But it was attributed to the role of the symmetry of particular orbitals pointed out in 1964 with regard to Diels-Alder reactions (14) that the utility of our studies was further broadened. It was remarked that, as is seen in Fig. 2, the symmetries of HOMO and LUMO of dienes and those of LUMO and HOMO of dienophiles, respectively, were found to be in a situation extremely favorable for a concerted cyclic interaction between them.

This signified the following important aspects. First, it pointed out a possible correlation between the orbital symmetry and the rule determining the substantial occurrence or nonoccurrence of a chemical reaction, which may be called

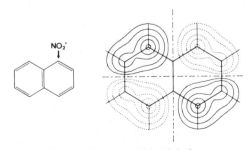

Fig. 1. Nitration of naphthalene.

the "selection rule," in common with the selection rule in molecular spectroscopy. Second, it provided a clue to discuss the question concerning what was the "concertedness" in a reaction which forms a cycle of electrons in conjugation along the way.

In 1965 Woodward and Hoffmann proposed the stereoselection rules that are established today as the "Woodward-Hoffmann" rules (15, 16). An experimental result that was developed in Havinga's [and colleagues] important paper (17) was extended. It is only after the remarkable appearance of the brilliant work by Woodward and Hoffmann that I became fully aware that not only the density distribution but also the nodal property of the particular orbitals have significance in such a wide variety of chemical reactions. In fact, we previously studied the $(4n + 2)$ rule proposed by Hückel (18) and noticed that the sign of the bond order in the highest energy electron orbital of an open-chain conjugation should be closely related to the stabilization of the corresponding conjugated rings (19). We did not imagine, however, on that occasion that the discussion might be extended to the so-called Möbius-type ring-closure (20).

If we take into consideration the HOMO-LUMO interactions between the fragments of a conjugated chain divided into parts (21), the frontier orbital theory can yield selection rules that are absolutely equivalent to those obtained from the principle called "the conservation of orbital symmetry" by Woodward and Hoffmann. One point that I may stress here is, as was pointed out by Fujimoto, Inagaki, and myself (22), that the electron delocalization between the particular orbitals interprets definitely in terms of orbital symmetries the formation and breaking of chemical bonds which, I believe, should be a key for perceiving chemical reaction processes.

In the cycloaddition of butadiene and ethylene shown in Fig. 2, both the interaction between the HOMO of diene and the LUMO of dienophile and that between the LUMO of diene and the HOMO of dienophile stabilize the interacting system. If one is interested in the local property of interaction, however, it is possible to recognize the clear distinction between the roles of the two types of orbital interactions. The HOMO of ethylene and the LUMO of butadiene are both symmetric with regard to the symmetry plane retained throughout the course of cycloaddition. This signifies that each of the carbon atoms of ethylene are bound to both of the terminal carbons of butadiene. The chemical bonding between the diene and dienophile thus generated may be something like the one in a loosely bound complex, for example, protonation to an olefinic double bond. On the contrary, the HOMO of butadiene and the LUMO of ethylene are antisymmetric. The interaction between these orbitals leads, therefore, to two separated chemical bonds, each of which combines a carbon atom of ethylene and a terminal carbon atom of butadiene. Needless to say, it is the interaction between the HOMO of diene and the LUMO of dienophile that is of importance for the occurrence of concerted cycloaddition (22).

In this way, it turned out in the course of time that the electron delocalization between HOMO and LUMO generally became the principal factor determining the easiness of a chemical reaction and the stereoselective path, irrespective of intra- and intermolecular processes, as illustrated in Fig. 3. In addition to our

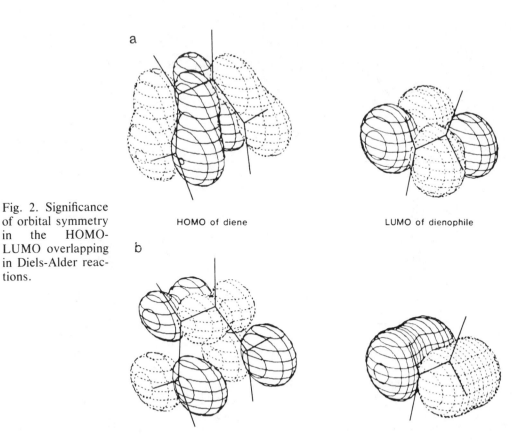

a

HOMO of diene

LUMO of dienophile

b

LUMO of diene

HOMO of dienophile

Fig. 2. Significance of orbital symmetry in the HOMO-LUMO overlapping in Diels-Alder reactions.

own school, a number of other chemists made contributions. I want to refer to several names that are worthy of special mention.

First of all, the general perturbation theory of the HOMO-LUMO interaction between two molecules was built up by Salem *et al.* (*23–25*). One of Salem's papers (*25*) was in line with the important theory of Bader (*26*), which specified the mode of decomposition of a molecule or a transition complex by means of the symmetry of the normal vibration. Furthermore, Pearson (*27*) investigated the relation between the symmetry of reaction coordinates in general and that of HOMO and LUMO.

The discussion so far may seem to be an overestimation of these selected orbitals, HOMO and LUMO. This point was ingeniously modified by Klopman (*28*). He carefully took into account the factors to be considered in the perturbation theory of reacting systems and classified reactions into two cases: the one was the "frontier-controlled" case in which the reaction was controlled by the particular orbital interaction, and the other was the "charge-controlled" case, where it was controlled by the electrostatic interaction of charges. This classification was conveniently used by many people. In this context the review articles of Herndon (*29*) and of Hudson (*30*)

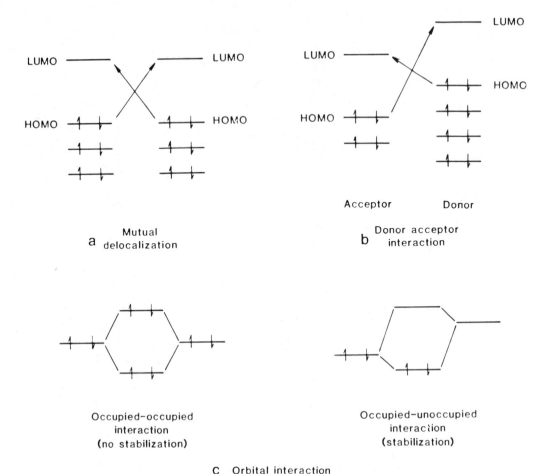

LUMO ——   —— LUMO

HOMO —⊣⊢—   —⊣⊢— HOMO

—⊣⊢—   —⊣⊢—

—⊣⊢—   —⊣⊢—

a Mutual delocalization

LUMO ——

LUMO ——   —⊣⊢— HOMO

HOMO —⊣⊢—   —⊣⊢—

—⊣⊢—   —⊣⊢—

—⊣⊢—

Acceptor   Donor

b Donor acceptor interaction

Occupied–occupied
interaction
(no stabilization)

Occupied–unoccupied
interaction
(stabilization)

C   Orbital interaction

Fig. 3. The mode of interaction between orbitals of two molecules.

appeared to be very useful. The names of Coulson (*4*) and Dewar (*31*) should also be noted here as those who contributed to the development of reactivity theories.

Returning to the subject again, let us assume that two molecules approach each other and orbital overlapping takes place. The perturbation theory (*32*) of this sort of interaction indicates that the larger the orbital overlapping is and the smaller the level separation of two overlapping orbitals is, the larger is the contribution of the orbital pair to the stabili-

zation of an interacting system. Accordingly, at least at the beginning, a reaction will proceed with a mutual nuclear configuration that is most favorable for the HOMO-LUMO overlapping.

Now let us suppose an electron flow from the HOMO of molecule I to the LUMO of molecule II. In each molecule the bonds between the reaction center—the place at which the orbitals overlap with those of the other molecule—and the remaining part of the molecule are weakened. On this occasion, in molecule I the bonds which are bonding in HOMO are weakened and those antibonding in HOMO are strengthened, while in molecule II the bonds which are antibonding in LUMO are weakened and those bonding in LUMO are strengthened. Consequently, the HOMO of molecule I particularly destabilizes as compared with the other occupied orbitals, and the LUMO of molecule II discriminatively stabilizes among unoccupied orbitals, so that the HOMO-LUMO level separation between the two molecules is decreased. Such a circumstance is clearly understandable in Fig. 2.

The following tendency is further stressed. When the bond weakenings specified above have arisen, the HOMO and the LUMO tend to become more localized at these weakened bonds in each molecule. Besides, the weakening of the bonds between the reaction center and the remaining part causes an increase of the amplitudes of HOMO and LUMO at the reaction centers, resulting in a larger overlapping of HOMO and LUMO (33). Such a trend of the characteristic change in the orbital pattern is made numerically certain by actual calculations. The role of interaction between HOMO and LUMO turns out in this way to become more and more important as the reaction proceeds.

A series of studies on chemical interactions was attempted in which the interaction of reactants was divided into the Coulomb, the exchange, the polarization, and the delocalization interactions, and their magnitude of contribution to the interaction energy was quantitatively discussed (32, 34). The interactions discussed were the dimerization (35) and the addition to ethylene (36) of methylene and the dimerization of BH_3 (37), and also several donor-acceptor interactions—BH_3-NH_3 (38), BH_3-CO (39),

NH_3-HF (40), and the like. The method was applied also to reactions of radicals, such as the abstraction of a methane hydrogen by methyl radical, the addition of methyl radical to ethylene (41), and recombinations, disproportionations, and self-reactions of two radicals (42). In these calculations, the configuration analysis proposed originally by Baba et al. (43) was also utilized conveniently. We could show numerically the mode of increase of the electron delocalization from HOMO to LUMO along with the proceeding of reaction, the increasing weight of contribution of such a delocalization to the stabilization of the reacting system, the driving force of the reaction in terms of orbital interactions, and so on.

The question "Why HOMO and LUMO solely determine the reaction path?" was one which I very frequently received from the audiences in my lectures given in the past in different places. The discussion so far made here is thought to correspond, at least partly, to that answer. But one may not adhere so strictly to the HOMO and LUMO. In one-center reactions like substitutions, which the orbital symmetry has nothing to do with, any occupied orbitals which are very close to HOMO should properly be taken into account (12). In large paraffin molecules a number of HOMO's (high-lying occupied MO's), and furthermore as will shortly be referred to later, in metal crystals, even "HOMO-band" must be taken along the line of reactivity argument. If HOMO or LUMO happens to be inadequate owing to its extension, the symmetry, or the nodal property, the next orbital should be sought. One of the simplest examples of such an instance is the protonation of pyridine. In this case, the nitrogen lone-pair orbital is not

HOMO, but the addition of proton to the nitrogen lone-pair so as not to disturb the π conjugation would evidently be more advantageous than the addition to higher occupied π orbitals that may intercept the π conjugation. Thus, the reason why proton dare not add to the positions of large amplitude of π HOMO in this case will easily be understood. It is not completely satisfactory to dispose of a disagreement between the HOMO-LUMO argument and the experimental fact formally as an *exception* to the theory. A so-called exception does possess its own reason. To investigate what the reason is will possibly yield a novel finding.

The HOMO-LUMO interaction argument was recently pointed out (44) to be in an auxiliary sense useful for the interpretation of the sign of a reaction constant and the scale of a substituent constant in the Hammett rule (45), which has made an immeasurable contribution to the study of the substituent effect in chemical reactivity. In the cyclic addition, like Diels-Alder reactions and 1,3-dipolar additions, the relative easiness of occurrence of reactions, various subsidiary effects, and interesting phenomena like regioselection and periselection were interpreted with considerable success simply by the knowledge of the height of the energy level of HOMO and LUMO, the mode of their extension, their nodal structure, and others (46). I defined these in a mass: the *"orbital pattern."*

Other topics that have been discussed in terms of HOMO-LUMO interactions are thermal formation of excited states (47), singlet-triplet selectivity (48), the chemical property of biradicals and excited molecules (49), the interaction of the central atom and ligands in transition metal complexes (50), the interaction of three or more orbitals (51), and others. Inagaki *et al.* included in the theory the polarization effect in HOMO and LUMO due to the mixing in of other orbitals and gave an elucidation for a number of organic chemical problems that were not always easy to explain. The unique stereoselection in the transannular cross-bond formation, the lone-pair effect, the *d* orbital effect, and the orbital polarization effect due to substituents were the cases (52).

As was partly discussed above, the method of orbital interaction was applied not only to the ground electronic state but to the excited states, giving an explanation of the path of even complicated photochemical isomerizations (13, 21). In a majority of cases, the HOMO and the LUMO of the ground-state molecule were also found to be the essential orbitals. Even the ground-state reaction of a strong electron acceptor (or donor) causes a mixing in of an ionized electron configuration or an excited electron configuration in another molecule. In consequence, a partial HOMO-HOMO or LUMO-LUMO interaction, which would be trivial if there were no influence of the acceptor (or donor), becomes important in stabilizing the interacting system (47).

The problems so far discussed have been limited to chemical reactions. However, the HOMO-LUMO interaction must also come into relation with other chemical phenomena in almost the same mechanism—with the exception of one different point that they usually do not bring about so remarkable a change in the nuclear configuration as in the case of chemical reactions. Now let us examine the possibility of applying the theory to so-called "aromaticity"—one of the simplest, but the hardest-to-interpret problems. There seem to be few problems so annoying to theoreticians as the explanation of this chemically classical concept. I greatly appreciate the contribution of Dewar's theory (53, 54) based on a quantitative energy values argument. Here, however, I want to give a qualitative comment through a totally different way of consideration.

It is easily ascertained (55, 55a) in Fig. 4 that in benzene, naphthalene, phenanthrene, and similar compounds any virtual division of the molecule into two always produces the parts in which their HOMO and LUMO overlap in-phase at the two junctions. But these circumstances are not seen in anthracene which is usually looked upon as one of the typical representatives of aromatic compounds. Hosoya (55a) pointed out from the comparison with phenanthrene indicated in Fig. 4, that the ring growth of type 2 was less stable than that of 1. It is

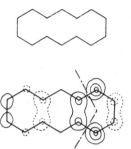

1   **2**

well known that anthracene occasionally exhibits a reactivity of olefin-like additions.

In view of the so-called Hückel's $(4n + 2)$ rule mentioned above, an anthracene molecule has 14 π electrons and fulfills the stability condition for "aromaticity." Actually, if one considers a molecule of anthracene with the two inside bonds deleted, it is really seen that the HOMO and the LUMO of the two parts overlap in an in-phase manner at both of the junctions:

In this way, it is understood that the two bonds that were deleted above exerted a certain unfavorable influence for aromaticity. Such an influence bears a close resemblance to that of impurity scattering in the wave of a free electron moving in a metal crystal.

This discussion seems to be a digression but, as a matter of fact, it relates to the essential question as to how an electron in a molecule can delocalize. As will be mentioned later, Anderson (56) solved the question of how an electron in a random system can localize. In a molecule, there are potential barriers between atoms which should be got over by the aid of a certain condition to be satisfied, in order for an electron to move around it freely. Although the question of how valence electrons can delocalize in a molecule may have not yet been solved satisfactorily under the condition of unfixed nuclear configuration, the in-phase relation of HOMO and LUMO at the junctions of the two parts of the molecule seems to be at least one of the conditions of intramolecular delocalization of electrons.

Generally speaking, the electron delocalization gives rise to a stabilization due to "conjugation," which is one of the old chemical concepts. If so, similar stabilization mechanisms must be chemically detected in other systems besides aromatic compounds. The discussion of this

Benzene

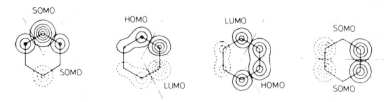

Naphthalene

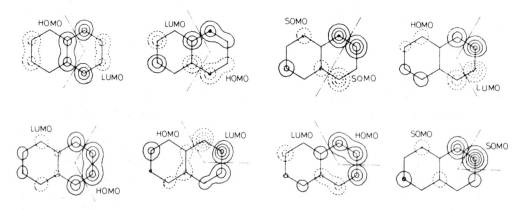

Phenanthrene

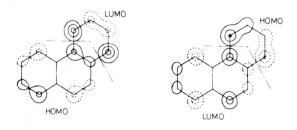

Anthracene

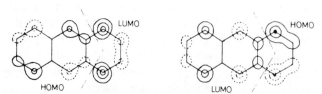

Fig. 4. The HOMO-LUMO phase relationship in virtual division of aromatic hydrocarbons. SOMO, a singly occupied MO of a radical.

delocalization stabilization at the transition state or on the reaction path was nothing but the reactivity theory hitherto mentioned.

The term "delocalizability" was attached to the reactivity indices we derived (10), and our reactivity theory itself was sometimes called "delocalization approach" (14). The "hyperconjugation" of various sorts is explained in the same manner. The stabilization due to homo-aromaticity or bicycloaromaticity of Goldstein (57), the stability in spirocycles, pericycles (58), "laticycles," and "longicycles" of Goldstein and Hoffmann (59), that of spirarenes of Hoffmann et al. (60), and so on, are all understood as examples of the stabilization due to the delocalization between HOMO and LUMO, although other explanations may also be possible.

You may be doubtful to what extent such a qualitative consideration is reliable. In many cases, however, a considerably accurate nonempirical determination of the stable conformation of hydrocarbon molecules (61, 62) results in a conclusion qualitatively not much different from the expectation based on the simple orbital interaction argument mentioned above.

Chemical Reaction Pathways

It has already been pointed out that the detailed mechanism of a chemical reaction was discussed along the reaction path on the basis of the orbital interaction argument. For that purpose, however, it is required that the problem as to how the chemical reaction path is determined should have been solved. The method in which the route of a chemical reaction was supposed on the potential energy surface and the rate of the reaction was evaluated by the aid of a statistical-mechanical formulation was established by Eyring (63). Many people wrote papers where the rate expression was derived from wave-mechanics with the use of the potential energy function. Besides, the problem of obtaining the trajectory of a given chemical reaction with a given initial condition was treated by Wang and Karplus (64).

The center line of the reaction path, so to speak, the idealized reaction coordinate—which I called "intrinsic reaction coordinate" (IRC) (65)—seemed to have been, rather strangely, not distinctly defined until then. For that reason, I began with the general equation that determines the line of force mathematically (34, 66, 67). Although my papers themselves were possibly not very original, they turned out later to develop in a very interesting direction (68–74). These papers opened the route to calculate the quasistatic change of nuclear configuration of the reacting system which starts from the transition state proceeding to a stable equilibrium point (66). I termed the method of automatic determination of the molecular deformation accompanying a chemical reaction as "reaction ergodography" (34, 67). This method was applied to a few definite examples by Kato and myself (67) and by Morokuma et al. (72, 73). Those examples were: abstraction and substitution of methane hydrogen by hydrogen atom (67), nucleophilic replacement in methane by hydride anion (72), and isomerization of methylcarbylamine to acetonitrile (73). All of these reactions thus far treated are limited to the simplest cases, but there seems to be no principal difficulty in extending the applicability to larger systems. Once IRC was determined in this way, the driving force of a chemical reaction was analyzed on the basis of the orbital interaction argument (66).

In the reacting system with no angular momentum it is possible to obtain the IRC by the use of the space-fixed Cartesian coordinate system. All of the calculated examples mentioned above belong to this case. However, in the reaction in which rotational motion exists, it is required to discuss the IRC after separating the nuclear configuration space from the rotational motion (74–77). For that purpose, it is essential to derive the general classical Hamiltonian of the reacting systems and then to separate the internal motion which is determined only by the internal coordinates. The nuclear configuration space thus separated out is in general a Riemannian space. The classical Lagrangian form to be obtained in that process of constructing the Hamiltonian is used to derive the IRC equation in the presence of rotational motion. It is thus understood that the rotational motion of the reacting system generally causes a deviation of IRC (74).

Once the method of determining unique reaction pathways is obtained, the next problem we are concerned with is to see if the calculated pathways are interpreted in terms of the frontier orbital interactions. A method referred to as the "interaction frontier orbitals" or "hybrid molecular orbitals" has been developed very recently by Fujimoto and myself in order to furnish a lucid scheme of frontier orbital interactions with the accuracy of nonempirical calculations now and in the future (78–80). By properly including contributions of other MO's than the HOMO and the LUMO, we realized in terms of orbital diagrams how ingenious the empirically estab-

lished chemical concepts—"reaction sites" and "functional groups"—and the empirical concept of reaction pathways could be. In Fig. 5 the HOMO of styrene is compared with its interaction frontier orbital for protonation to the olefinic double bond. The latter is seen to be localized very efficiently in the frontier of chemical interaction. The double bond is evidently the functional unit in this case. Innovation of the frontier orbital concept will be continued by young people to make it useful for one of our ultimate targets: theoretical design of molecules and chemical reactions.

Frontier Orbitals in Related Fields

Theoretical treatments of the property of solid crystals, or chemisorption on a solid surface, appear to have been hitherto almost monopolistically treated by the methodology of physics. But the orbital pattern technique has also advanced gradually in this field.

The "cluster approach" (81, 82), in which a portion of the metal crystal is drawn out as the form of a cluster of atoms and its catalytic actions or other properties are investigated, has contributed to the development of the orbital pattern approach, because the physical methods mentioned above can hardly be applied to such sizable systems. It is expected that, if clusters of various sizes and various shapes are studied for assessing the characteristic feature of their HOMO's (high-lying occupied MO's) and LUMO's (low-lying unoccupied MO's), then the nature of chemisorption and catalytic action, the mode of surface chemical reaction, and several related subjects of interest can be investigated theoretically.

As is the case of molecular interactions in usual chemical reactions, only the HOMO- and LUMO-bands lying in the range of several electron volts near the Fermi level can participate in the adsorption of molecules and surface reactions on solid crystals. You may recollect here that, in the BCS (Bardeen-Cooper-Schrieffer) theory of superconductivity, too, only the HOMO's and LUMO's in close proximity to the Fermi surface can be concerned in forming electron pairs as the result of interaction with lattice vibration. In the case of solid catalysts mentioned above, the discrimination of particular orbitals and electrons from the others have made the situation much easier.

Consider a system composed of a regular repetition of a molecular unit, for instance, a one-dimensional high polymer chain or a one-dimensional lattice, in which a certain perturbation is imposed at a definite location. Sometimes it is convenient to discuss the influence of this perturbation by transforming the or-

bitals belonging to the HOMO-band to construct the orbitals localized at that place. One such technique was proposed by Tanaka, Yamabe, and myself (83). This method is expected to be applied in principle to a local discussion of such problems as the adsorption of a molecule on the two-dimensional surface of catalysts, surface reactions, and related matters. This approach may be called a little more chemical than the method in which is used the function of local density of states (84) or similar ones, in that the former can be used for the argument of the reactivity of molecules on a catalyst surface in terms of the phase relationship of localized orbitals.

What are called low-dimensional semiconductors and some superconductors have also been the objects of application of the orbital argument. In these studies, the dimerization of S_2N_2 to S_4N_4 (85) and the high polymerization to $(SN)_x$ (86) were discussed, and the energy band structure of $(SN)_x$ polymer chain was analyzed to investigate the stable nuclear

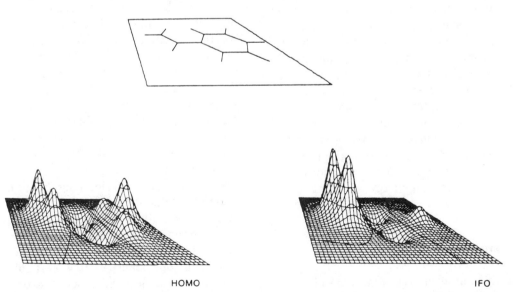

HOMO   IFO

Fig. 5. A comparison of HOMO and the interaction frontier orbital for protonation in styrene.

arrangement and the mode of interchain interactions (87).

The modern technique in solid-state physics to interpret the interesting characteristic behaviors of noncrystalline materials, in particular of amorphous materials, in which the nuclear arrangements were not regular, was certainly striking. Anderson showed generally that in a system of random lattice the electron localization should take place (56). Mott, stating in his 1977 Nobel lecture that he thought it was the first prize awarded for the study of amorphous materials, answered the question, "How can a localized electron be conducted?" with the use of the idea of hopping. Here, too, the HOMO-LUMO interaction—in this case the consideration of spin is essential—would play an important part.

Here in a few words, I want to refer to the meaning and the role of virtual orbitals. The LUMO, which has been one of the stars in orbital arguments hitherto discussed, is the virtual orbital which an external electron is considered to occupy to be captured by a molecule to form an anion. Virtual orbitals always play an essential part in producing metastable states of molecules by electron capture (88). To discuss such problems generally, Tachibana et al. (89) systematized the theory of resonant states from the standpoint of complex eigenvalue problem. The idea of resonant states will take a principal part in chemical reactions, particularly in high-energy reactions which will be developed more in the future.

Prospect

In introducing above a series of recent results of the studies carried out mainly by our group, I have ventured to make those things the object of my talk which are no more than my prospective insight and are not yet completely established. This is just to stimulate, by specifying what are the fields I believe promising in the future, the intentional efforts of many younger chemists in order to develop them further.

In my opinion, quantum mechanics has two different ways of making contributions in chemistry. One is the contribution to the nonempirical comprehension of empirical chemical results just mentioned. However, we should not overlook another important aspect of quantum mechanics in chemistry. That is the promotion of empirical chemistry from the theoretical side. But, also for this second purpose, as a matter of course, reliable theoretical foundations and computational methods are required. The conclusions of theories should be little affected by the degree of sophistication in approximations adopted.

On the other hand, for theoreticians to make the second contribution, the cases where predictions surpassing the experimental accuracy are possible by very accurate calculations are for the present limited to those of a very few, extremely simple molecules. In order to accomplish this object in regard to ordinary chemical problems, it becomes sometimes necessary to provide qualitative theories that can be used even by experimental chemists. If one can contribute nothing to chemistry without carrying out accurate calculations with respect to each problem, one cannot be said to be making the most of quantum mechanics for the development of chemistry. It is certainly best that the underlying concepts are as close to experience as possible, but the sphere of chemical experience is steadily expanding. Quantum chemistry has then to perform its duty by furnishing those concepts with the theoretical basis in order to make them chemically available and serviceable for the aim of promoting empirical chemistry.

Even the same atoms of the same element, when they exist in different molecules, exhibit different behaviors. The chemical symbol H even seems to signify atoms of a completely different nature. In chemistry, this terrible individuality should never be avoided by "averaging," and, moreover, innumerable combinations of such atoms form the subject of chemical research, where it is not the "whole assembly of compounds of different kinds but each individual kind of compounds" that is of chemical interest. On account of this formidable complexity, chemisty possesses inevitably one aspect of depending on the analogy through experience. This is in a sense said to be the fate allotted to chemistry, and the source of a great difference in character from physics. Quantum chemistry, too, so far as it is chemistry, is required to be useful in promoting empirical chemistry as mentioned before.

Finally, I want to mention, out of a sense of gratefulness, the names of many people in our group who have been walking on the same road as mine since my first paper (1952) on quantum chemistry, particularly Drs. T. Yonezawa, C. Nagata, H. Kato, A. Imamura, K. Morokuma, T. Yamabe, and H. Fujimoto; and also I cannot forget the names of younger doctors mentioned in the text who made a contribution in opening new circumstances in each field. Among them, Professor T. Yonezawa was helpful in performing calculations in our 1952 paper, and also, it is to be mentioned with appreciation that the attractive title "frontier orbitals" of my lecture originated from the terminology I adopted in that paper by the suggestion of Professor H. Shingu, who kindly participated in that paper as an organic chemist to classify the relevant experimental results. Furthermore, many other collaborators are now distinguishing themselves in other important fields of chemistry, which, however, have not been the object of the present lecture. It was the late Professor Yoshio Tanaka of the University of Tokyo and Professor Masao Horio of Kyoto University who recognized the existence and significance of my early work in advance of others. I owe such a theoretical work, which I was able to carry out in the Faculty of Engineering, Kyoto University, and moreover in the Department of Fuel Chemistry, to the encouragement and kind regard of Professor Shinjiro Kodama, who nurtured the Department. It was the late Professor Gen-itsu Kita, my life-teacher, and the founder of this department, who made me enter into chemistry, one of the most attractive and promising fields of science, and led me to devote my whole life to it. For all of these people no words of gratitude can by any means be sufficient.

References and Notes

1. For instance, see C. K. Ingold, *Structure and Mechanism in Organic Chemistry* (Cornell Univ. Press, Ithaca, N.Y., 1953).
2. For instance, see G. N. Lewis, *Valence and the Structure of Atoms and Molecules* (Chemical Catalog Co., New York, 1923).
3. G. W. Wheland and L. Pauling, *J. Am. Chem. Soc.* **57**, 2086 (1935).
4. C. A. Coulson and H. C. Longuet-Higgins, *Proc. R. Soc. (London) Ser. A* **191**, 39 (1947); *ibid.* **192**, 16 (1947).
5. G. W. Wheland, *J. Am. Chem. Soc.* **64**, 900 (1942).
6. K. Fukui, T. Yonezawa, H. Shingu, *J. Chem. Phys.* **20**, 722 (1952).

7. For instance, see R. G. Parr, *The Quantum Theory of Molecular Electronic Structure* (Benjamin, New York, 1963), and references cited therein.
8. R. S. Mulliken, *J. Am. Chem. Soc.* **74**, 811 (1952).
9. L. W. Picket, N. Muller, R. S. Mulliken, *J. Chem. Phys.* **21**, 1400 (1953).
10. K. Fukui, T. Yonezawa, C. Nagata, *Bull. Chem. Soc. Jpn.* **27**, 423 (1954); K. Fukui, H. Kato, T. Yonezawa, *ibid.* **34**, 1112 (1961).
11. R. S. Mulliken, *Recl. Trav. Chim. Pays-Bas* **75**, 845 (1956).
12. K. Fukui, T. Yonezawa, C. Nagata, H. Shingu, *J. Chem. Phys.* **22**, 1433 (1954).
13. For instance, see K. Fukui, *Theory of Orientation and Stereoselection* (Springer, Berlin, 1970).
14. K. Fukui, in *Molecular Orbitals in Chemistry, Physics and Biology*, P.-O. Löwdin and B. Pullman, Eds. (Academic Press, New York, 1964), p. 513.
15. R. B. Woodward and R. Hoffmann, *Angew. Chem.* **81**, 797 (1969); *The Conservation of Orbital Symmetry* (Academic Press, New York, 1969); see papers in *Orbital Symmetry Papers*, H. E. Simmons and J. F. Bunnett, Eds. (American Chemical Society, Washington, D.C., 1974).
16. R. B. Woodward and R. Hoffmann, *J. Am. Chem. Soc.* **87**, 395 (1965).
17. E. Havinga, R. J. de Kock, M. P. Rappoldt, *Tetrahedron* **11**, 276 (1960); E. Havinga and J. L. M. A. Schlatmann, *ibid.* **15**, 146 (1961).
18. E. Hückel, *Z. Phys.* **70**, 204 (1931); *ibid.* **76**, 628 (1932).
19. K. Fukui, A. Imamura, T. Yonezawa, C. Nagata, *Bull. Chem. Soc. Jpn.* **33**, 1501 (1960).
20. E. Heilbronner, *Tetrahedron Lett.* **1964**, 1923 (1964).
21. K. Fukui, *Acc. Chem. Res.* **4**, 57 (1971).
22. H. Fujimoto, S. Inagaki, K. Fukui, *J. Am. Chem. Soc.* **98**, 2670 (1976).
23. L. Salem, *ibid.* **90**, 543, 553 (1968).
24. A. Devaquet and L. Salem, *ibid.* **91**, 3743 (1969).
25. L. Salem, *Chem. Br.* **5**, 449 (1969).
26. R. F. W. Bader, *Can. J. Chem.* **40**, 1164 (1962).
27. R. G. Pearson, *Symmetry Rules for Chemical Reactions* (Wiley, New York, 1976), and references cited.
28. G. Klopman, *J. Am. Chem. Soc.* **90**, 223 (1968); *Chemical Reactivity and Reaction Paths* (Wiley, New York, 1974).
29. W. C. Herndon, *Chem. Rev.* **72**, 157 (1972).
30. R. F. Hudson, *Angew. Chem. Int. Ed. Engl.* **12**, 36 (1973).
31. For instance, see M. J. S. Dewar and R. C. Dougherty, *The PMO Theory of Organic Chemistry* (Plenum, New York, 1975), and many papers cited therein; M. J. S. Dewar, *Tetrahedron* **58** (part I), 85 (1966).
32. K. Fukui and H. Fujimoto, *Bull. Chem. Soc. Jpn.* **41**, 1989 (1968).
33. _____, *ibid.* **42**, 3392 (1969).
34. K. Fukui, in *The World of Quantum Chemistry*, R. Daudel and B. Pullman, Eds. (Reidel, Dordrecht, 1974), p. 113.
35. H. Fujimoto, S. Yamabe, K. Fukui, *Bull. Chem. Soc. Jpn.* **45**, 1566 (1972).
36. _____, *ibid.*, p. 2424.
37. S. Yamabe, T. Minato, H. Fujimoto, K. Fukui, *Theor. Chim. Acta* **32**, 187 (1974).

38. H. Fujimoto, S. Kato, S. Yamabe, K. Fukui, *J. Chem. Phys.* **60**, 572 (1974).
39. S. Kato, H. Fujimoto, S. Yamabe, K. Fukui, *J. Am. Chem. Soc.* **96**, 2024 (1974).
40. S. Yamabe, S. Kato, H. Fujimoto, K. Fukui, *Bull. Chem. Soc. Jpn.* **46**, 3619 (1973); *Theor. Chim. Acta* **30**, 327 (1973).
41. H. Fujimoto, S. Yamabe, T. Minato, K. Fukui, *J. Am. Chem. Soc.* **94**, 9205 (1972).
42. T. Minato, S. Yamabe, H. Fujimoto, K. Fukui, *Bull. Chem. Soc. Jpn.* **51**, 1 (1978); T. Minato, T. Yamabe, M. Miyake, M. Tanaka, H. Kato, K. Fukui, *Bull. Chem. Soc. Jpn.* **51**, 682 (1978).
43. H. Baba, S. Suzuki, T. Takemura, *J. Chem. Phys.* **50**, 2078 (1969).
44. O. Henri-Rousseau and F. Texier, *J. Chem. Educ.* **55**, 437 (1978).
45. L. P. Hammett, *Physical Organic Chemistry* (McGraw-Hill, New York, 1940).
46. K. N. Houk, *Acc. Chem. Res.* **8**, 361 (1975), and references cited.
47. S. Inagaki, H. Fujimoto, K. Fukui, *J. Am. Chem. Soc.* **97**, 6108 (1975).
48. See I. Fleming, *Frontier Orbitals and Organic Chemical Reactions* (Wiley, New York, 1976); T. L. Gilchrist and R. C. Storr, *Organic Reactions and Orbital Symmetry* (Cambridge Univ. Press, London, ed. 2, 1979).
49. K. Fukui and K. Tanaka, *Bull. Chem. Soc. Jpn.* **50**, 1391 (1977).
50. K. Fukui and S. Inagaki, *J. Am. Chem. Soc.* **97**, 4445 (1975), and many papers cited therein.
51. S. Inagaki, H. Fujimoto, K. Fukui, *J. Am. Chem. Soc.* **98**, 4693 (1976).
52. S. Inagaki and K. Fukui, *Chem. Lett.* **1974**, 509 (1974); S. Inagaki, H. Fujimoto, K. Fukui, *J. Am. Chem. Soc.* **98**, 4054 (1976).
53. M. J. S. Dewar, *The Molecular Orbital Theory of Organic Chemistry* (McGraw-Hill, New York, 1969).
54. _____, *Angew. Chem.* **83**, 859 (1971).
55. K. Fukui, *Kagaku To Kogyo* **29**, 556 (1976).
55a. H. Hosoya, *Symposium on Electron Correlation in Molecules* (Research Institute for Fundamental Physics, 18 December 1976).
56. P. W. Anderson, *Phys. Rev.* **109**, 1492 (1958); _____ and N. F. Mott, see *1977 Nobel Lectures in Physics* (Elsevier, New York, 1978).
57. M. J. Goldstein, *J. Am. Chem. Soc.* **89**, 6357 (1967).
58. H. E. Simmons and T. Fukunaga, *ibid.*, p. 5208.
59. M. J. Goldstein and R. Hoffmann, *ibid.* **93**, 6193 (1971).
60. R. Hoffmann, A. Imamura, G. Zeiss, *ibid.* **89**, 5215 (1967).
61. See W. J. Hehre and J. A. Pople, *ibid.* **97**, 6941 (1975).
62. W. J. Hehre, *Acc. Chem. Res.* **8**, 369 (1975).
63. For instance, see S. Glasstone, K. J. Laider, H. Eyring, *The Theory of Rate Processes* (McGraw-Hill, New York, 1941), and references cited therein.
64. I. S. Y. Wang and M. Karplus, *J. Am. Chem. Soc.* **95**, 8060 (1973), and references cited therein.
65. K. Fukui, *J. Phys. Chem.* **74**, 4161 (1970).
66. _____, S. Kato, H. Fujimoto, *J. Am. Chem. Soc.* **97**, 1 (1975).
67. S. Kato and K. Fukui, *ibid.* **98**, 6395 (1976).
68. A. Tachibana and K. Fukui, *Theor. Chim. Acta* **49**, 321 (1978).

69. _____, *ibid.* **51**, 189, 275 (1979).
70. _____, *ibid.*, p. 5.
71. K. Fukui, *Recl. Trav. Chim. Pays-Bas* **98**, 75 (1979).
72. B. D. Joshi and K. Morokuma, *J. Chem. Phys.* **67**, 4880 (1977).
73. K. Ishida, K. Morokuma, A. Komornicki, *ibid.* **66**, 2153 (1977).
74. K. Fukui, A. Tachibana, K. Yamashita, *Int. J. Quantum Chem.*, in press.
75. W. H. Miller, N. C. Handy, J. E. Adams, *J. Chem. Phys.* **72**, 99 (1980); S. K. Gray, W. H. Miller, Y. Yamaguchi, H. F. Schaefer III, *ibid.* **73**, 2732 (1980).
76. K. Fukui, *Int. J. Quantum Chem.*, in press.
77. _____, *Acc. Chem. Res.*, in press.
78. _____, N. Koga, H. Fujimoto, *J. Am. Chem. Soc.* **103**, 196 (1981).
79. H. Fujimoto, N. Koga, M. Endo, K. Fukui, *Tetrahedron Lett.* **22**, 1263, 3427 (1981).
80. H. Fujimoto, N. Koga, K. Fukui, *J. Am. Chem. Soc.*, in press.
81. For instance, see K. H. Johnson, in *The New World of Quantum Chemistry*, B. Pullman and R. G. Parr, Eds. (Reidel, Dordrecht, 1976), p.

317, and references cited therein.
82. H. Kobayashi, H. Kato, K. Tarama, K. Fukui, *J. Catal.* **49**, 294 (1977); H. Kobayashi, S. Yoshida, H. Kato, K. Fukui, K. Tarama, *Surf. Sci.* **79**, 189 (1979).
83. K. Tanaka, T. Yamabe, K. Fukui, *Chem. Phys. Lett.* **48**, 141 (1977).
84. For instance, J. R. Schrieffer, in *The World of Quantum Chemistry*, B. Pullman and R. G. Parr, Eds. (Reidel, Dordrecht, 1976), p. 305; J. B. Danes and J. R. Schrieffer, *Int. J. Quantum Chem.* **10S**, 287 (1976).
85. K. Tanaka, T. Yamabe, A. Noda, K. Fukui, H. Kato, *J. Phys. Chem.* **82**, 1453 (1978).
86. K. Tanaka, T. Yamabe, K. Fukui, H. Kato, *ibid.* **81**, 727 (1977).
87. K. Tanaka, T. Yamabe, K. Fukui, *Chem. Phys. Lett.* **53**, 453 (1978).
88. See S. Ishimaru, T. Yamabe, K. Fukui, H. Kato, *J. Phys. Chem.* **78**, 148 (1974); S. Ishimaru, K. Fukui, H. Kato, *Theor. Chim. Acta* **39**, 103 (1975).
89. A. Tachibana, T. Yamabe, K. Fukui, *J. Phys. B.* **10**, 3175 (1977); *Adv. Quantum Chem.* **11**, 195 (1978).

The quality of our lives today has benefited from successes in organic chemistry. Our food, clothing, shelter, and health care have been significantly improved by the creation of new materials based on carbon compounds. Safer and more effective agrichemicals such as pesticides and herbicides enhance the quality and quantity of our food supply. The miraculous polymers have been the basis of new materials that have increased the durability, versatility, and beauty of the clothes we wear and the homes we live in. New drug discoveries contribute significantly to life preservation and extension and an improved quality of living. Can we expect such contributions to continue at the same or perhaps an even greater rate in the future?

In order to address such a question, we must look at the present in order to extrapolate to the future. At a fundamental level, where do the challenges lie and what methods are being developed to meet these challenges? Four general areas may be recognized: structures, reactions, techniques, and concepts. In highlighting some of the specific advances in these areas in a field as broad as organic chemistry in a very brief overview, only a few of the many different exciting developments can be treated. There are undoubtedly developments not included that are as significant as the few illustrations presented here.

Structures

Historically, organic chemistry focused on isolating compounds from nature. Structure determination became the first step in understanding. However, the methods available, which relied on systematic degradation and correlations,

21

Sculpting Horizons in Organic Chemistry

Barry M. Trost

Science 227, 908–916 (22 February 1985)

required relatively large amounts of material; thus only abundantly occurring natural products were accessible. With the great advances that have been made in spectroscopy and separations science, a whole new world has begun to emerge. As it has become possible to detect, isolate, and characterize ever smaller components in nature, the importance of small organic molecules for a myriad of purposes has become evident.

Understanding the control mechanisms for the maturation of insects and for their behavior has led to an entire area called insect chemistry. Probing the physiology of insects revealed the presence of microgram amounts of hormones that regulate the maturation process. The availability of the new separations and spectroscopic techniques allowed the identification of two simple molecules—ecdysone, 1 (1), and juvenile hormone, 2 (2)—and led to the idea of using insect growth regulants as an approach

1

2

to control of insect populations. Present in even smaller quantities are constituents that control insect behavior. These

organic molecules, called pheromones (3), regulate behavior ranging from sex to defense and offer opportunities for defense against insect infestations ranging from early warning devices to terminating the populations.

New directions continue to emerge. Studies of mammalian physiology have revealed the role in metabolism of the arachidonic acid cascade (see Fig. 1A) (4). The role of the leukotrienes in control of human metabolism suggests pharmaceutical applications, such as in the control of inflammation. Another branch of the same metabolic cascade involves a family of human hormones, the prostaglandins (5). The pervasive influence of these hormones on the action of cells remained undiscovered until recently because of the limits of detection in analytical instrumentation. As instruments become still more refined, new factors will be discovered.

Our analytical tools allow precise definition of molecules of enormous complexity. In the life sciences, biomacromolecules such as the enzymes and polynucleic acids come to mind. Their primary structure is not a major problem today. However, other molecules from nature can constitute a major challenge. Palytoxin (see Fig. 1B), a highly toxic metabolite from algae, is a molecule of this type that is now in the realm of approachable problems (6). This molecule, with its 64 asymmetric centers and 7 double bonds, has so many possible geometric and stereoisomers that if a single molecule of each permutation were present in a flask, nearly 1 mole of compound would be present. In seeking to define the precise structure of this compound, we need to pick out a single molecule from the entire mole of molecules. The weaknesses of our current

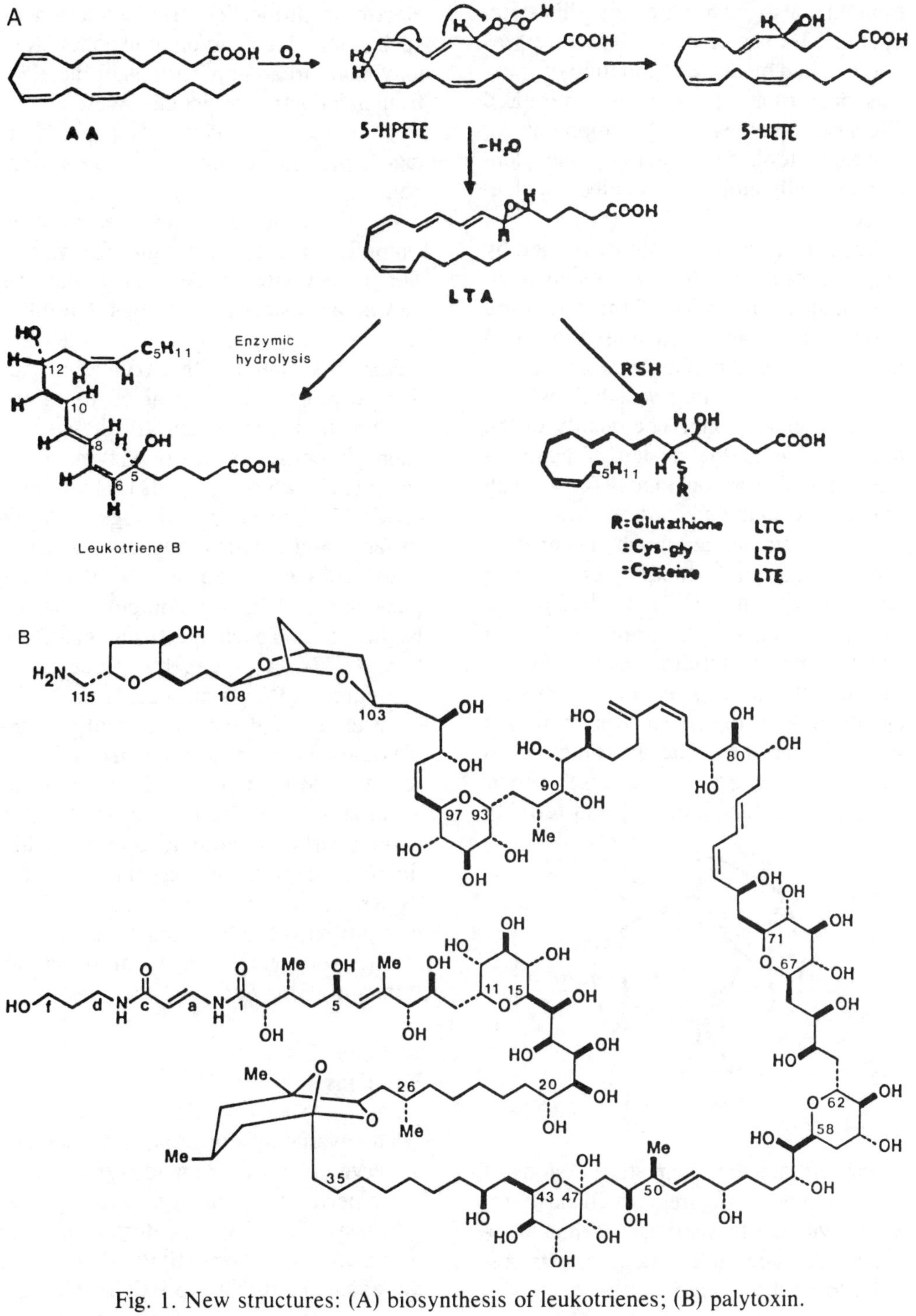

Fig. 1. New structures: (A) biosynthesis of leukotrienes; (B) palytoxin.

tools are also highlighted by this compound. The secret of its structure was not revealed by any combination of spectroscopic tools. It ultimately required chemical synthesis. Development of structural tools for noncrystalline, conformationally mobile molecules is a challenge.

Organic chemistry is not restrained by the limits of nature. Only man's imagination defines the limits of the structural varieties that can be created. What factors influence the choice from among these almost unlimited possibilities? Esthetics is clearly one. The beauty of the diamond led to the design of adamantane—a hydrocarbon model for a small unit of the diamond carbon lattice (7). Besides stimulating much fascinating fundamental understanding of bonding and reactivity, its design also had practical consequences. A simple analog of adamantane, amantidine, has proved to be one of the first effective antiviral agents and a beneficial treatment for parkinsonism (8). The platonic solids capture the imagination because of their beauty. Dodecahedrane (9) and tetrahe-

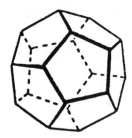

Dodecahedrane

drane (10) are the chemists' versions of these geometric figures which have recently yielded to synthesis. What properties such beautiful lattices possess remain to be discovered.

electronic properties. The evolution of such compounds into organic conductors may have broad applications in the electronics industry. Molecular switches, as well as having obvious weight advantages, promise to open up a new time domain.

Even the world of enzymes may be simplified into small organic fragments. The active sites of enzymes may be viewed as surfaces or cavities that influence a reacting substrate. While rate factors associated with enzymatic reactions capture a great deal of attention, perhaps more significant are the stereochemical consequences of enzymatic reactions. Crown compounds (11) and cavitands (12) are simple analogs of these surfaces and cavities in what is referred to as host-guest chemistry. Figure 2 depicts a chiral crown compound which begins to approach the behavior of enzymes (13). It is capable of selective recognition of (D)-amino acids to a sufficient extent that their preferential complexation allows a racemic mixture to be resolved. More significantly, it serves as a chiral surface for the formation of a carbon-carbon bond in a conjugate addition (also depicted in Fig. 3) and led to a 99 percent enantiomeric excess at −78°C in a partially catalytic reaction. Are we at the very early stages of designing "unnatural enzymes"?

Reactions

The availability of organic substances to serve our needs depends critically on the repertoire of reactions available for synthesis. The obvious interdependence of structure and reactivity determines the advances that are possible. The leg-

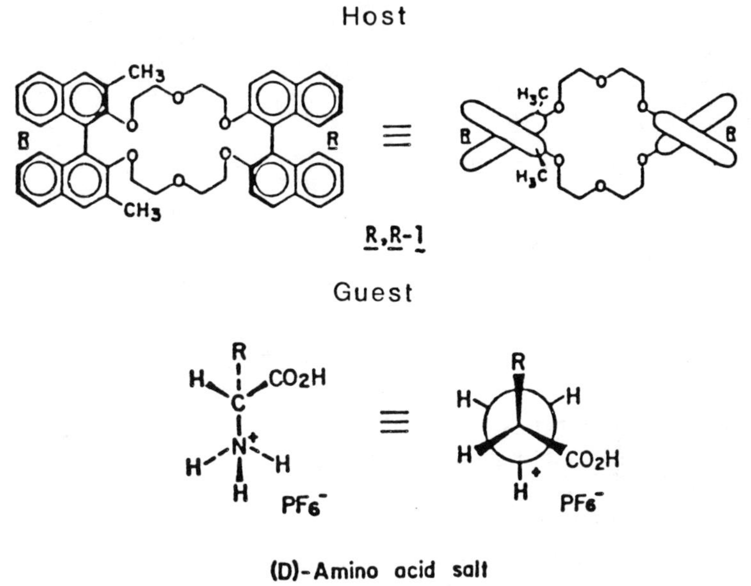
Host
R,R-1
Guest
PF6⁻
(D)-Amino acid salt

Fig. 2. New structures: chiral crown compounds.

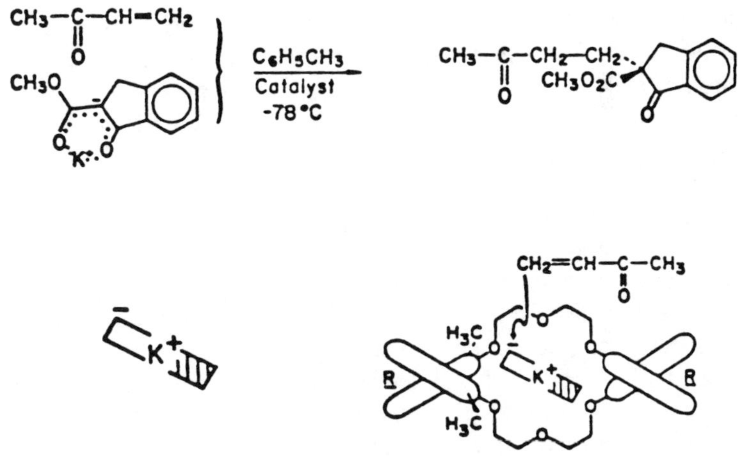
C6H5CH3
Catalyst
-78 °C

endary Diels-Alder reaction illustrates the evolutionary aspect of reaction development. Although the reaction was extensively described in the early part of this century and its impact was recognized by a Nobel Prize in 1950, the 1970's became the decade of recognition of its power in complex synthesis (see Fig. 3). Nevertheless, new faces of this classical reaction continue to be revealed. A most recent one is the reaction of heteroatom unsaturated molecules

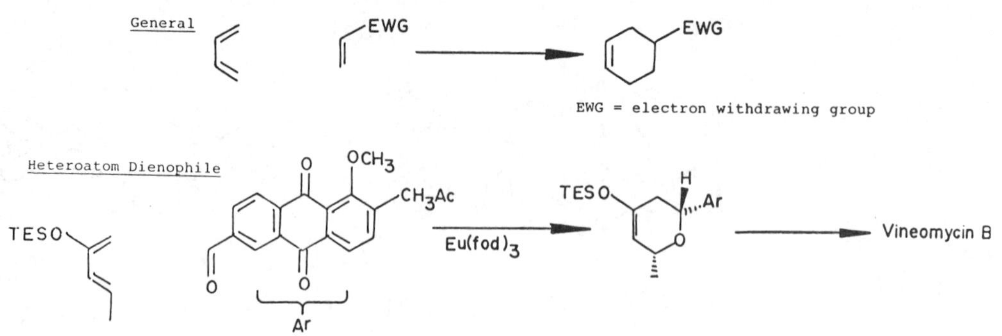

Fig. 3. Diels-Alder reaction: new faces.

with certain dienes, especially in the presence of lanthanides as catalysts. This makes the designed synthesis of some carbohydrates, especially those which are attached to other molecules by C–C bonds, a simpler procedure. Practical targets of this approach are antibiotics and antitumor agents, exemplified by vineomycin B_2 (see Fig. 3) (*14*) and, even more important, analogs of such molecules designed for improved therapeutic properties.

The power of the Diels-Alder reaction is due to its chemo-, regio-, and diastereoselectivity and its rapid creation of molecular complexity. The development of cycloadditions to rings of other than six members may extend these benefits to the synthesis of other ring sizes. Five-membered rings are of particular interest from both a biological and a theoretical point of view. An attractive possibility is the cycloaddition of trimethylenemethane (TMM), which conceptually is conveniently designated as a 1,3-dipole as in Fig. 4 although it exists as a 1,3-diradical. The resultant methylenecyclopentanes possess a useful array of functionality with which to mold the initial adducts into the ultimate targets. Chemistry involving such a reactive in-

termediate appears feasible with the use of transition metal templates (*15*). This leads to a strategy for synthesis of the antitumor, antiviral, and antifungal compound brefeldin A which resolves the stereochemical problems (*15b*).

Research on transition metal templates in catalytic reactions has suggested that these easily tailored templates may become the "chemists' enzymes." However, organotransition metal complexes are also useful stoichiometric templates. The "esoteric" carbene complexes of tungsten and chromium show remarkable behavior in joining an aromatic ring, an acetylene, and carbon monoxide to form naphthoquinones (see Fig. 5) (*16*). The use of chromium complexes in this way allows facile elaboration of the therapeutically important antitumor compounds of the anthracyclinone variety such as daunomycin (*17*).

The wealth of unusual chemistry revealed by merging carbon with the other hundred or so elements has only recently begun to be appreciated. The maturing of this field is not even in sight. Nevertheless, it will not be the only mechanism for discovering new reactions. The means by which nature creates complex molecules, discovered through studies of

310

biosynthesis and metabolism, suggest new pathways which we refer to as biomimetic. One of the most dramatic illustrations of this approach evolved from the discovery of squalene as the precursor of the tetracyclic steroids (18). The enzyme system responsible for squalene cyclization creates an electrophilic initiator, an epoxide, and then folds the linear chain to generate the beautiful steroid framework. Figure 6 shows that, with a proper initiator and a suitable terminator, a simple proton successfully promotes the polyolefin cyclization and translates a biosynthetic notion into a practical synthesis of the anti-inflammatory corticosteroids (19). As our understanding of how nature does things increases, new opportunities for reaction design will undoubtedly arise.

Techniques

Advances in a science closely parallel the development of new tools. Separations science, specifically chromatography, has had a great effect on the prac-

tice of organic chemistry and the types of problems that can be broached, such as the study of bioregulators present in very small amounts in complex mixtures. There will be further advances in designing selective absorbents and improved detection systems, and new concepts in separations science undoubtedly remain to be discovered which will make it possible to tackle problems that are now unapproachable.

Related to separation techniques for stable molecules is the development of methodology for isolating reactive intermediates. Matrix isolation permits the encapsulation of an unstable species in a frozen matrix at temperatures approaching absolute zero (20). Species which have been postulated as intermediates on the basis of indirect methods cannot only be generated but also directly observed. Direct study of these intermediates will lead to better control of their reactivity and thus to improved reactions and even new ones. The possibility of detecting previously unknown

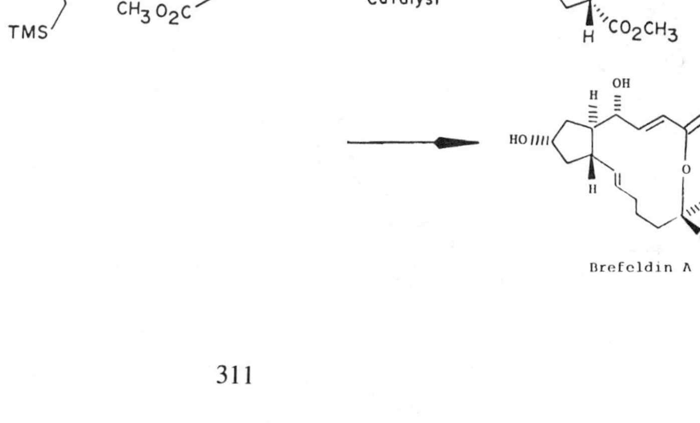

Fig. 4. Catalysis: cycloaddition to cyclopentanes.

311

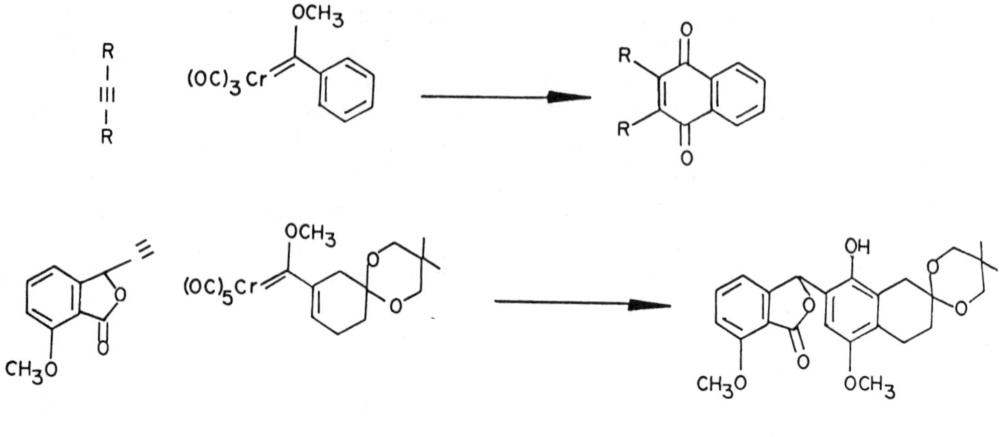

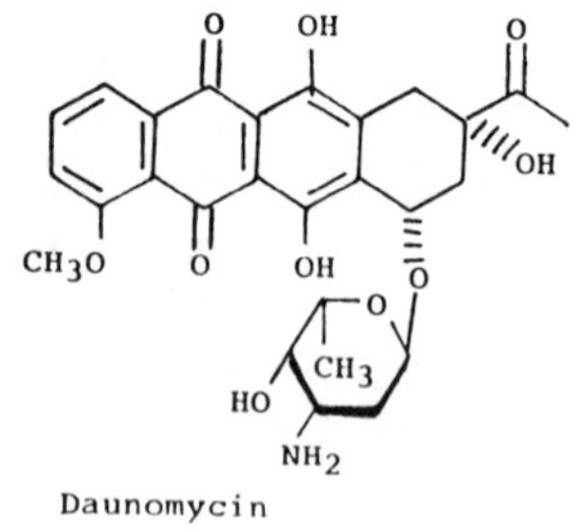

Fig. 5. Stoichiometric metal reactions: chromium synthesis of naphthoquinones.

Daunomycin

types of reactive intermediates is particularly exciting.

Organic chemistry has largely relied on thermal reactions for synthesis. The electronically excited state offers another dimension of reactivity and another world of reactions. Thermally forbidden reactions are sometimes photochemically allowed. Thus, while cycloaddition of two olefins to give a four-membered ring fails thermally, it succeeds photochemically, and this is one of the few syntheti-

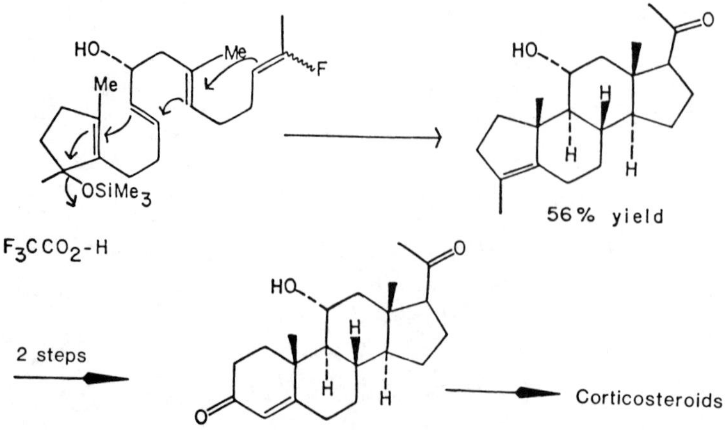

56% yield

F_3CCO_2-H

2 steps

Corticosteroids

Fig. 6. Biomimetic reactions: steroid synthesis.

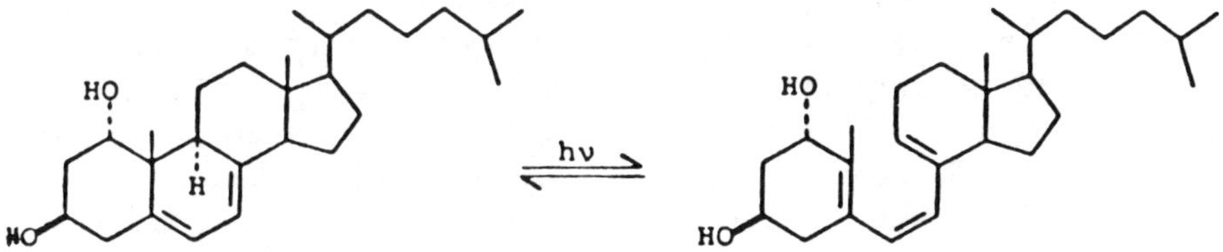

1α-Hydroxyprevitamin D₃   1α-Hydroxyprovitamin D₃

Fig. 7. Photolysis with lasers.

cally important photochemical processes (21). Controlling photoreactions appears to constitute one of the difficulties. The therapeutically important vitamin D metabolites rely on photoconversion of the previtamin to the provitamin form, as shown in Fig. 7 for 1-α-hydroxy series. The use of lasers with precisely tuned wavelengths of incident radiation generates the specific excited state to effect this transformation and minimizes undesired side reactions (22). The enhanced control doubled the efficiency of this commercially important process.

At times, the obvious appears to have been overlooked. Much of synthesis involves condensing two reacting species and forming one molecule from two. A decrease in the volume and normally a negative volume of activation accompanies such a process. The simple mechanical operation of high pressure should and does facilitate such condensations, as shown in Fig. 8 in a projected synthe-

sis of the antitumor quassinoids (23). The absence of any additional reactants and the lower temperatures frequently possible prevent decomposition pathways from superseding, as happens with acid or base catalysis. Greater accessibility of equipment for ultrahigh pressure and better reactor design will enhance the role of this technique in synthesis.

Clearly, other techniques remain to be explored. Effects of magnetic fields (24), of solid states, of molten salts, and of weightlessness, to name a few, can only be speculated upon. Such unexplored techniques may open up new dimensions in performing and studying reactions.

Concepts

The correlation of a vast number of observations into unified concepts or theories not only provides a better basis for understanding the existing facts but

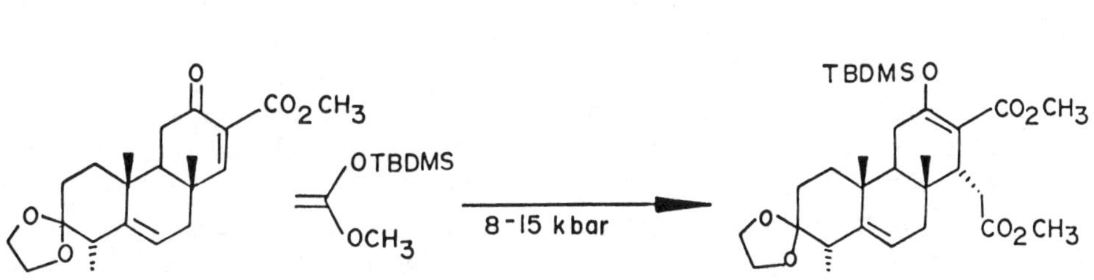

Fig. 8. Reaction under ultrahigh pressure.

313

also serves as a basis for making predictions. The concepts of aromaticity that evolved in the 1930's are still sources of animated debate and of inspiration for research directions. The concepts of orbital symmetry and frontier orbital control of reactions, enunciated less than 20 years ago, almost immediately profoundly changed the thinking of chemists. An avalanche of new ideas resulted and will continue to grow as further experimental data develop. Most important, our ability to rationally propose and experimentally implement the synthesis of unusual unsaturated molecules or the generation of new reactions has moved large steps forward.

Most areas in experimental organic chemistry have not enjoyed the benefits of such sweeping theories. In part, this may be due to the extremely small effects that we attempt to correlate. In this sense, organic chemistry is an exercise in perturbation theory. We must concern ourselves with many small effects and deduce the ultimate resultant. Defining and quantifying these small effects is a major endeavor.

A good illustration is provided by the idea of stereoelectronic effects as applied to saturated systems. The ability of saturated single bonds to transmit electronic information has usually been ignored. The "anomeric" effect, a propensity for an alkoxy group to prefer the normally less favorable axial position in a tetrahydropyran ring, appeared to be restricted to carbohydrates (Eq. A in Fig. 9 (25). The tendency to attribute its behavior to the special properties of the carbohydrate system delayed the development of a fundamental understanding. An appreciation of the general implications of this basic phenomenon has arisen from the observation that a t-butyl group, long considered to be an immovable object, an anchor, in terms of its propensity to be equatorial in a six-membered ring, can be forced into an axial position (Eq. B in Fig. 9) (26). The transmission of the electronic effects of the lone pairs on the ring heteroatoms to the axial C–O bond in Eq. A, a stabilizing effect, and the axial C–Li bond in Eq. B, a destabilizing effect, account for such observations. A growing understanding of this unusual effect and the uncovering of many related examples led to its predictable application in reaction design (25). It has already been applied in the preparation of important highly oxygenated natural (ionophores) and unnatural (crown and cryptand) products.

Almost any reaction reveals additional areas of little or no understanding. This primitiveness is revealed by the fact that as central and widely used (sometimes confused with understood) a reaction as the addition of nucleophiles to a carbonyl group remains a mystery. The observation that acetylide anions add preferentially in the normally "sterically more hindered" axial fashion (Fig. 10) (27) suggests the existence of unrevealed but significant concepts.

One strategy for trying to evaluate the resultant of many small effects is exemplified by the field of conformational analysis. The orientation of parts of molecules to each other which vary by only simple bond rotations that form different isomers (conformers) frequently determines the chemical or biological behavior of the molecules. In certain systems, such as six-membered rings, the molecules are relatively rigid and well understood. However, noncyclic systems and large ring molecules, of particular practical importance, are conformationally mobile.

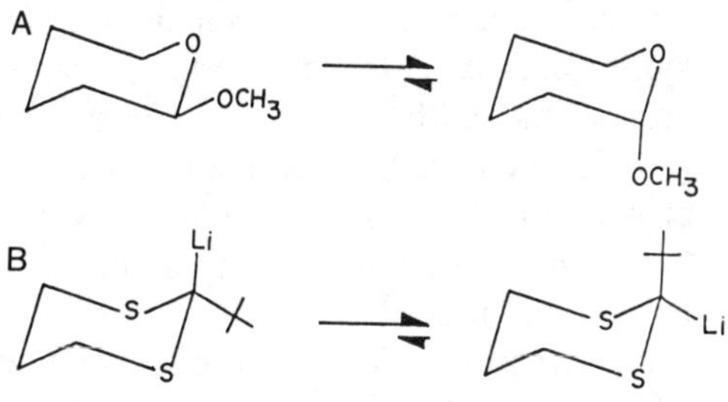

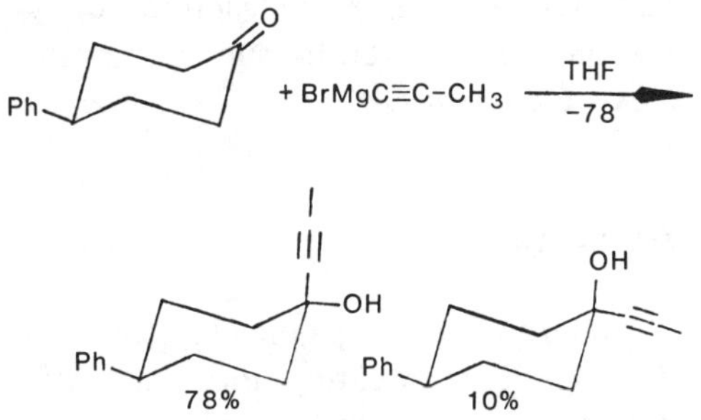

Fig. 9 (left). Stereoelectronic factors.   Fig. 10 (right). Stereoelectronics: axial selectivity.

Molecular mechanics, the empirical computation of conformational energy, combined with computer graphics constitutes molecular modeling (28). This technique has greatly enhanced our ability to perceive the exact shapes of molecules. Nevertheless, it is not the panacea. Using such models in a synthesis of the polyether antibiotic lysocellin, as de-

picted in Fig. 11, the prediction suggests formation of the *cis* alkylated product; experimentally, alkylation of **3**, Fig. 11, produces **4** with one alkylation proceeding as predicted but the second one not (29). Obviously, refinements are required before this approach becomes more generally useful. Nevertheless, the beginnings offer a great deal of anticipa-

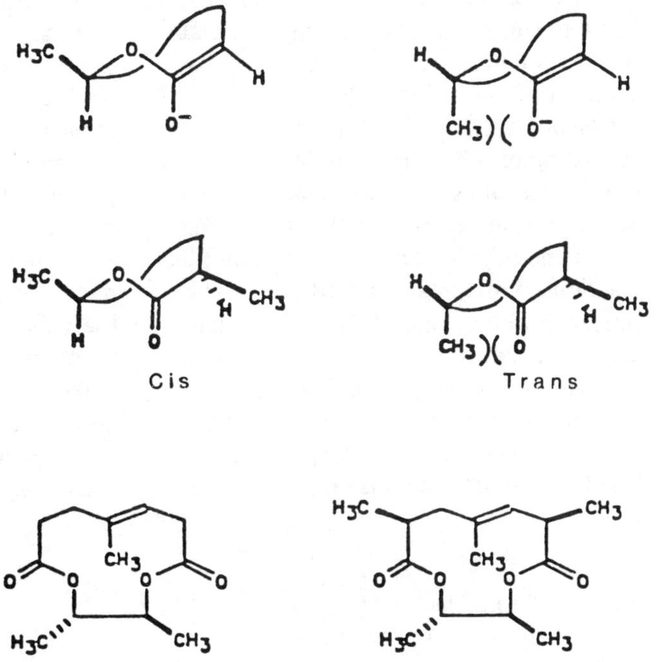

Fig. 11. Computer modeling—conformational analysis.

Cis   Trans

3   4

315

tion that an impasse in structural analyses may be removed by such computer-based techniques.

Selectivity

Underlying the major challenges in designing and creating more complex molecules for a multitude of uses is selectivity. In defining strategies and reactions to construct complex molecules we require synthetic methods that can (i) perform a wanted structural change and none other (that is, be chemoselective), (ii) orient the reacting partners in a correct fashion (be regioselective), (iii) create the correct orientations of the various parts of the molecule with respect to each other (be diastereoselective), and (iv) enable the formation of a molecule of one-handedness or a mirror image isomer (be enantioselective). Such extraordinary demands are exciting challenges. Some recent examples demonstrate that such challenges can be met but also demonstrate how far we have to go.

Chemoselectivity. If you have a target consisting of a hundred moving parts of similar size and shape, how can you hit a single preselected part without harming any of the others? Trying to carry out a reaction involving a hundred atoms is a similar problem and defines the challenge of chemoselectivity. An approach involves rediscovery of existing knowledge to offer new insight and applications. A renaissance in radical chemistry beautifully illustrates this concept. Radi-

cals, extremely important reactive intermediates because of their role in polymer synthesis, have the property of being compatible with most polar functional groups such as hydroxyl and carbonyl groups. In contrast, carbanion chemistry, a more commonly employed synthetic technique for forming C–C bonds, is incompatible with such groups—a fact which has led to the idea of "protecting" these functional groups. Unfortunately, protection results in synthetic inefficiency. Consider the cyclization of 5 in Fig. 12 (*30*). Any type of carbanion reaction would (i) lead to carbonyl rather than olefin addition or (ii) require protection of the carbonyl group. Quite the contrary is true for radical reactions. Indeed, direct cyclization with C–C bond formation and, most important, without recourse to protecting groups occurs upon exposure of 5 to a radical-generating agent. A particularly efficient synthesis of the fragrance ingredient seychellene results.

Transition metal complexes, which we referred to earlier as the chemists' enzymes, represent a tremendous stride in our quest for chemoselectivity. An important example is the catalyzed cross-coupling reaction, illustrated in Fig. 13 in a synthesis of the antifungal antibiotic pyrenophorin (*31*). Here, in contrast to the classical approaches, we adjust the reactivity of our carbanion chemistry to make it compatible with carbonyl groups. Clearly, we must avoid the use of "crutches" such as protecting and activating groups if we are to achieve full

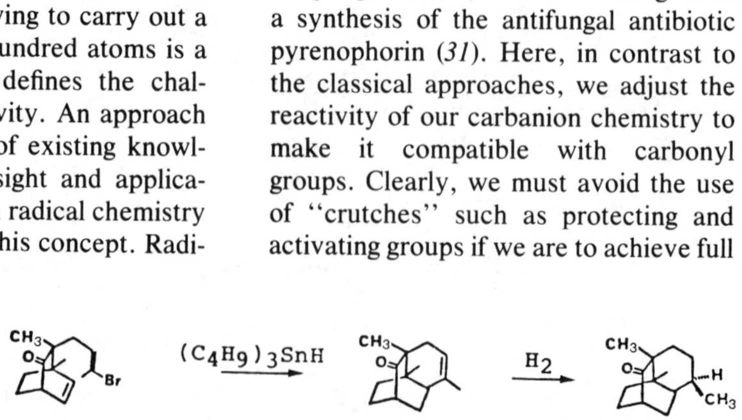

Fig. 12. Chemoselectivity: radical intermediates.

316

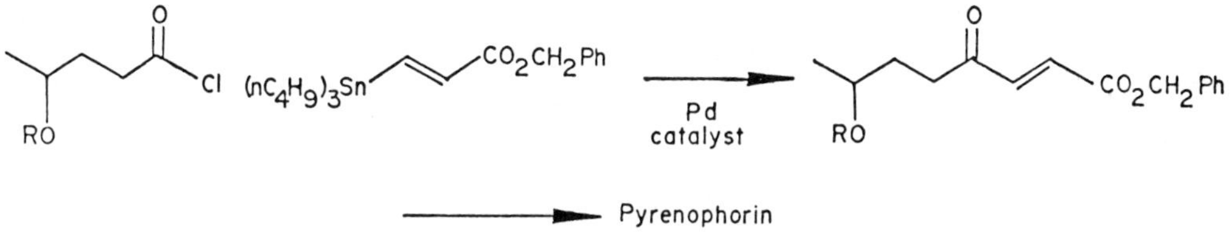

Fig. 13. Chemoselectivity: transition metal catalysts.

synthetic efficiency. Whether such a goal is practical remains to be determined.

Regioselectivity. An ability to adjust the orientation in which two reacting partners approach one another can increase our synthetic flexibility, although sometimes it is not so important. For example, in O versus C alkylation of enolates (Fig. 14), C alkylation usually dominates but it is also normally the desired product. In contrast, directing allylic alkylation to give either of the two regioisomeric products at will is desir-

able and important (Fig. 15). The transition metal complexes reveal their versatility in allowing us to titrate the reaction with full control by varying the electronic and steric nature of the catalyst template (*32*).

At times, introduction of an easily removed substituent permits regiocontrol. The discovery of the other elements of the periodic table by organic chemists has revealed a number of elements which are of value in such a task. Silicon, so far, appears to be a very powerful one.

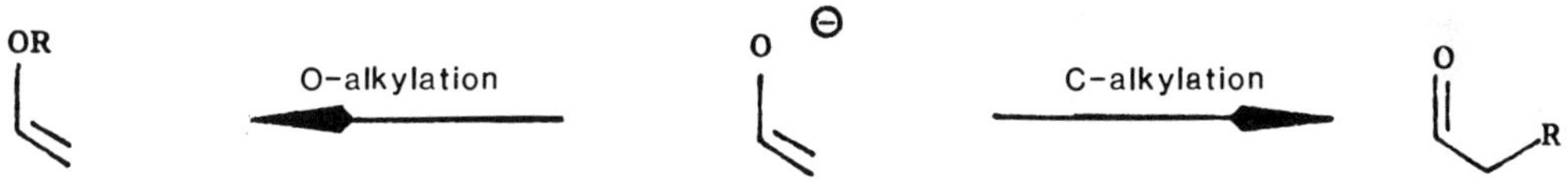

Fig. 14. Regioselectivity: O versus C alkylation.

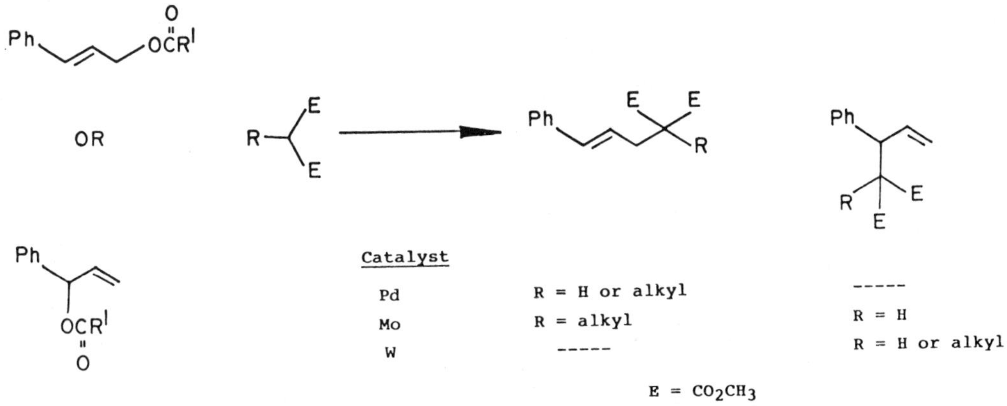

Catalyst		
Pd	R = H or alkyl	-----
Mo	R = alkyl	R = H
W	-----	R = H or alkyl

E = CO_2CH_3

Fig. 15. Regioselectivity: allylic alkylation.

317

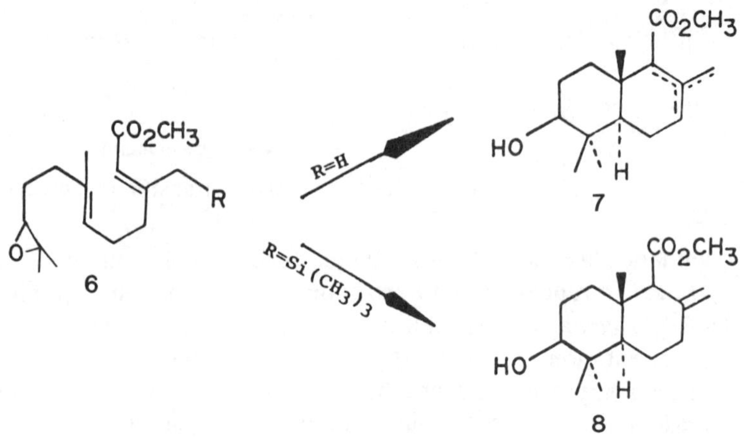

Fig. 16. Regioselectivity control element: silicon.

We discussed the biomimetic polyolefin cyclization as a useful approach to forming complex molecules from simple ones (Fig. 6). Figure 16 discloses a limitation, the lack of regioselectivity in the termination by proton loss, in which every possible olefin results (**6→7**). The simple expedient of employing an allyl silane as the terminator gives a unique product (**6→8**), even though it is the thermodynamically unfavorable one (*33*). Our ability to control regioselectivity remains

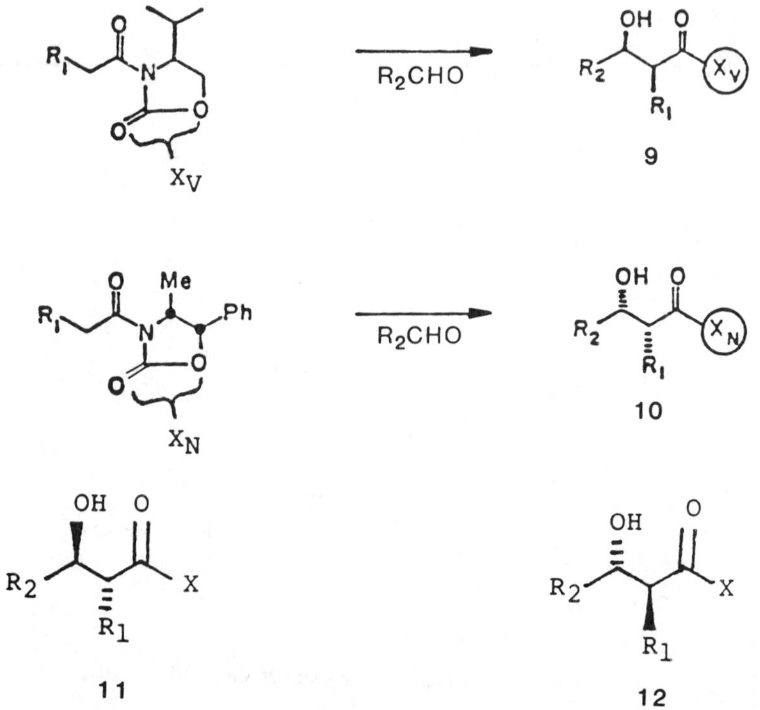

Fig. 17. Diastereoselectivity: aldol-type reaction.

highly limited. Nevertheless, the idea that we are powerless to do so must be put aside.

Diastereoselectivity. Nature has evolved elaborate mechanisms for molding the shape of its objects. Man does not share nature's success. Molding the shape of organic molecules, or stereocontrol, remains a formidable challenge for the chemist. Our limited success is understandable if we realize that the delicate interplay of many opposing forces, few of which can be quantified at present, determines the experimental outcome. Even so, impressive studies have recently been made.

The aldol reaction, one of the classical reactions in organic chemistry, reveals new faces in its ability to serve the chemist. As Fig. 17 discloses, with an appropriate choice of reaction conditions, such as metal reactant and substrate, *syn*-type products such as **9** and **10** dominate over *anti*-type products such as **11** and **12** (*34*). Such diastereoselectivity easily translates into enantioselectivity. In the examples of Fig. 17, X_V and X_N are optically active and transfer their stereochemistry to the products **9** and **10** in such a way that (ignoring the parts that correspond to X_V and X_N) these two compounds are mirror image

or optical isomers or enantiomers. By simple hydrolysis, we obtain enantiomerically pure products. We refer to X_V and X_N as chiral auxiliaries.

Another facet of this approach derives from the expanding world of organoboranes. While the use of such compounds for formation of C–H and C–O bonds is well recognized, their role in forming C–C bonds may ultimately be even more important. The conformational rigidity associated with the transition states for boron-mediated additions to carbonyl groups serves to enhance stereochemical control.

The aldol reaction, already discussed, is one reaction that is susceptible to the use of boron chemistry for such a purpose (*35*). Replacement of the magnesium of allyl Grignard reagents with boron (Fig. 18) is another example (*36*). The fact that the two stereochemical centers in the product **13** are not directly bonded to each other permits their separation by simple hydrolysis to the fragrance ingredient artemesia alcohol, **14**, in enantiomerically pure form. Thus, the diastereoselectivity of the carbonyl addition translates into enantioselectivity in the final product. A great deal of progress in diastereoselectivity still results from the empirical or "try and see"

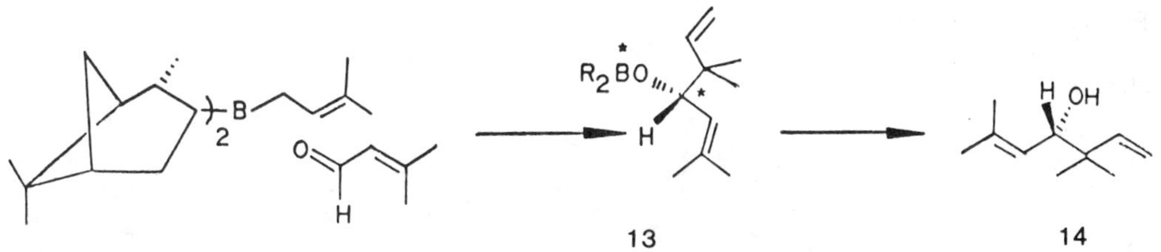

>96% de

Fig. 18. Diastereoselectivity: allylboranes (de, diastereomeric excess).

approach. The transition to rational design will become possible only with better understanding.

Enantioselectivity. If you take a glove from a box containing equal numbers of right-handed and left-handed gloves, you have a 50:50 chance of drawing a right-handed glove. Is there a way to increase that probability to 100 percent? This is the type of problem that the chemist confronts in trying to control enantioselectivity.

Once again, transition metal templates come into use. Previously, control of stereochemistry appeared to be possible only in the realm of enzymes. Now, however, abiological catalysts for asymmetric induction are beginning to emerge. Figure 19 illustrates a benefit of wine other than the obvious physical one. The metal titanium, in an environment of tartarates, constituents of wine, selectively delivers an oxygen atom to one of the enantiotopic faces of the olefin of an allyl alcohol (*37*). Unfortunately, very few cases of asymmetric metal cata-

lysts exist and only the formation of C–O or C–H bonds has so far succumbed to asymmetric induction.

Transition metal catalysts need not be the only substances used for this purpose. Figure 20 depicts a sculpted host used for asymmetric induction. In examining the source of chiral recognition at the active site of enzymes, we discuss many small effects such as hydrogen bonding and π stacking. By use of an optically active phase-transfer catalyst (**17**), a dramatic asymmetric alkylation is achievable (**15→16** in Fig. 20) in a process directed toward the uricosuric agent indacrinone (*38*). In retrospect, enzyme-like interactions such as that depicted in **18** may account for this extraordinary result. The simplicity of the catalyst but the enzyme-like experimental result begin to strip away some of the mysticism we associate with enzymes.

Nevertheless, enzymes, modified enzymes, or whole cells will undoubtedly play an increasing role in synthesis. The challenge of manipulating an enzyme by

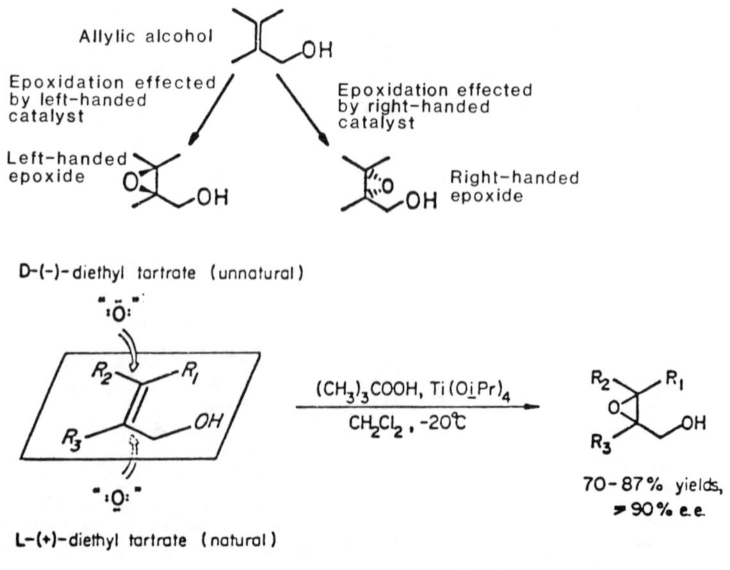

Fig. 19. Enantioselectivity: abiological catalysts.

320

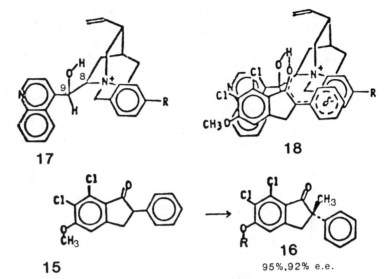

Fig. 20. Enantioselectivity: phase transfer catalysis.

17

18

15

$\longrightarrow$

16
95%, 92% e.e.

using biotechnology to modify enzyme selectivity or create new types of reactivity is exciting. However, potential pitfalls must be recognized. In a synthesis of L-carnitine, a compound with therapeutic use as an antihyperlipoproteinemic agent, the yeast reduction of the β-ketoester **19** was expected to give **20** but, surprisingly, gave a mixture of **20** and **21** (see Fig. 21). Careful investigative work showed that this lack of enantioselectivity was due to the presence of multiple oxidoreductases with opposing stereochemical effects (*39*). Substrate manipulation by changing the nature of an ester modifies the relative activity of the competing enzymes and restores high enantioselectivity.

Enantiocontrol remains most primitive. The fact that the world we live in is chiral and nature is particularly sensitive to enantioisomerism stresses the importance of the challenge.

Conclusion

At what stage of development does organic chemistry stand? Its successes of the past have led some to believe that it is mature. The idea persists that we can synthesize anything. The major flaw in this view is that it fails to recognize effectiveness and practicality. Clearly, we cannot synthesize with the directness with which nature manipulates polyfunc-

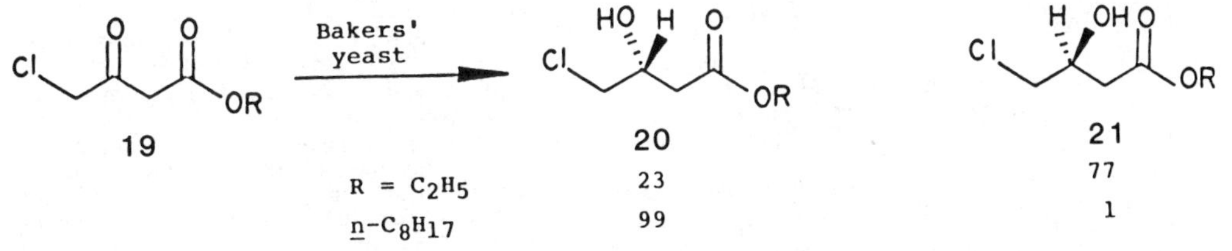

19

Bakers' yeast $\longrightarrow$

20

21

R = C_2H_5	23	77
$\underline{n}$-C_8H_{17}	99	1

Fig. 21. Enantioselectivity: microbial methods.

321

tional molecules. In fact, there is considerable evidence that we are still in our infancy and only beginning to learn to walk. The advances that have been made, although they are impressive, reveal how much further we need to go.

The relationship between chemistry and biology must enter any discussion of the future. Almost by definition, molecular problems of the life sciences are integral parts of organic chemistry. Some of the discussion in this article shows the prominent position such problems have in contemporary organic research and surely will have in the future. With the development of molecular biology as a subdiscipline of biology, the boundaries between these sciences have begun to vanish. While some speak of interdisciplinary research between these areas, perhaps we should speak of co- or even unidisciplinary research as the objectives of the researchers seem to merge. Regardless of tradition, the best definition of any science is what the individuals who call themselves practitioners of that science are doing.

References and Notes

1. K. Nakanishi, *Pure Appl. Chem.* **25**, 167 (1971).
2. B. M. Trost, *Acc. Chem. Res.* **3**, 120 (1970).
3. J. M. Brand, J. C. Young, R. M. Silverstein, *Fortschr. Chem. Org. Naturst.* **37**, 1 (1979); B. A. Leonhardt and M. Beroza, *ACS Symp. Ser.* **190** (1982).
4. B. Samuelsson, *Pure Appl. Chem.* **53**, 1203 (1981); E. J. Corey, *Experentia* **38**, 1259 (1982); T. Scheinman and J. Ackroyd, *Leukotriene Syntheses. New Class of Biologically Active Compounds Including SRSA* (Raven, New York, 1984).
5. B. Samuelsson, *Recent Prog. Horm. Res.* **34**, 239 (1978); K. C. Nicolaou, G. P. Gasic, W. E. Barnette, *Angew. Chem. Int. Ed. Engl.* **17**, 293 (1978).
6. Y. Kishi, in *Selectivity—a Goal for Synthetic Efficiency*, W. Bartmann and B. M. Trost, Eds. (Verlag Chemie, Weinheim, 1984), pp. 99–119.
7. R. C. Fort, Jr., and P. V. R. Schleyer, *Chem. Rev.* **64**, 277 (1964).
8. W. L. Davis *et al.*, *Science* **144**, 862 (1964).
9. L. A. Paquette, *Proc. Natl. Acad. Sci. U.S.A.* **79**, 4495 (1982).
10. G. Maier, S. Pfriem, U. Schafer, K.-D. Malsch, R. Matusch, *Chem. Ber.* **114**, 3965 (1981).
11. D. J. Cram, *Science* **219**, 1177 (1983).
12. J. M. Lehn, *Stud. Phys. Theor. Chem.* **24**, 181 (1983); *Pure Appl. Chem.* **52**, 2441 (1980).
13. D. J. Cram and J. M. Cram, in *Selectivity—a Goal for Synthetic Efficiency*, W. Bartmann and B. M. Trost, Eds. (Verlag Chemie, Weinheim, 1984), pp. 43–64.
14. S. J. Danishefsky, B. J. Uang, G. Quallich, *J. Am. Chem. Soc.* **106**, 2453 (1984).
15. (a) B. M. Trost, *Chem. Soc. Rev.* **11**, 1419 (1982); (b) ——, J. Lynch, P. Renaut, D. Steinman, unpublished results.
16. K. H. Dotz, *Angew. Chem. Int. Ed. Engl.* **23**, 587 (1984).
17. W. D. Wulff and P. C. Tang, *J. Am. Chem. Soc.* **106**, 434 (1984).
18. E. E. van Tamelen, J. D. Willett, R. B. Clayton, K. E. Lord, *ibid.* **88**, 4752 (1966); E. J. Corey and W. E. Russey, *ibid.*, p. 4750.
19. W. S. Johnson, T. A. Lyle, G. W. Daub, *J. Org. Chem.* **47**, 161 (1982).
20. G. C. Pimentel, *Angew. Chem. Int. Ed. Engl.* **14**, 199 (1975); I. R. Dunkin, *Chem. Soc. Rev.* **9**, 2 (1980).
21. P. Margaretha, *Top. Curr. Chem.* **103**, 1 (1982); S. W. Baldwin, *Org. Photochem.* **5**, 123 (1980).
22. W. G. Dauben and R. B. Phillips, *J. Am. Chem. Soc.* **104**, 5780 (1982).
23. C. H. Heathcock, C. Mahaim, M. F. Schlecht, T. Utawanit, *J. Org. Chem.* **49**, 3264 (1984).
24. N. J. Turro, *Proc. Natl. Acad. Sci. U.S.A.* **80**, 609, (1983).
25. P. Deslongchamps, *Stereoelectronic Effects in Organic Chemistry* (Pergamon, Oxford, 1983).
26. E. L. Eliel, A. A. Hartmann, A. G. Abatjoglou, *J. Am. Chem. Soc.* **96**, 1807 (1974).
27. I. Fleming and N. K. Tenett, *J. Organomet. Chem.* **264**, 99 (1984); G. Stork and J. M. Stryker, *Tetrahedron Lett.* **24**, 4887 (1983).
28. U. Burkert and N. L. Allinger, *ACS Monogr.* **177** (1982).
29. W. C. Still, in *Selectivity—a Goal for Synthetic Efficiency*, W. Bartmann and B. M. Trost, Eds. (Verlag Chemie, Weinheim, 1984), pp. 263–279.
30. G. Stork, *ibid.*, pp. 281–298.
31. J. W. Labadie, D. Tueting, J. K. Stille, *J. Org. Chem.* **48**, 4634 (1983).
32. B. M. Trost and M.-H. Hung, *J. Am. Chem. Soc.* **106**, 6837 (1984).
33. R. J. Armstrong and L. Weiler, *Can. J. Chem.* **61**, 214 (1983); *ibid.*, p. 2530.
34. D. A. Evans, M. D. Ennis, T. Le, N. Mandel, G. Mandel, *J. Am. Chem. Soc.* **106**, 1154 (1984).
35. S. Masamune *et al.*, *ibid.* **103**, 1566 (1981).
36. H. C. Brown and P. K. Jadhav, *Tetrahedron Lett.* **25**, 1215 (1984).
37. K. B. Sharpless, S. S. Woodard, M. G. Finn, in *Selectivity—a Goal for Synthetic Efficiency*, W. Bartmann and B. M. Trost, Eds. (Verlag Chemie, Weinheim, 1984), pp. 377–390.
38. U.-H. Dolling, P. Davis, E. J. Grabowski, *J. Am. Chem. Soc.* **106**, 446 (1984).
39. C. J. Sih and C.-S. Chen, *Angew. Chem. Int. Ed. Engl.* **23**, 570 (1984).
40. I thank the National Science Foundation and the National Institutes of Health for their generous support of our programs. I thank Professor Samuel J. Danishefsky for providing a copy of his address to the AAAS on 25 May 1984 on an overview of organic chemistry, which contributed some of the examples used here.

The organic constituents of plants, animals, and microorganisms—from the most primitive to the most highly developed—are composed of molecules that are chiral (1). Life processes on a molecular scale take place between chiral molecules in a chiral environment; presumably, this has been so since primordial biotic times (2). How did the selection (or production) of one enantiomer of a chiral pair of molecules take place in a prebiotic world where presumably ordinary chemical processes had no statistical preference for selecting (or producing) one isomer over the other? For example, if an achiral (1) amino acid precursor 1 is reduced in a symmetrical environment with a symmetrical reagent (an achiral reaction), a 50:50 mixture of the two enantiomeric (3) forms, 2-(S) and 2-(R) of the amino acid, results. The stereochemical situation is depicted in Fig. 1. The right and left faces of sym-

metrical structure 1 (Fig. 1) are indistinguishable to a symmetrical reagent such as hydrogen; therefore, attack upon either face is equally probable. Presumably this was the situation in prebiotic times before chiral molecules existed on earth (2). However, these faces are not the same to a chiral reagent such as an enzyme. When structure 1 is viewed from the vantage point of the H₂ on the left, the NH, COOH, and CH₃ groups

22

Current Status of Asymmetric Synthesis

Harry S. Mosher and James D. Morrison

Science 221, 1013–1019 (9 September 1983)

are arranged in a clockwise sequence; however, when viewed from the vantage point of the H_2 on the right, the same three groups occur in a counterclockwise sequence. These are designated prochiral faces. A chiral reagent attacking the prochiral faces of this symmetrical substrate can recognize the difference, thereby giving rise to unequal amounts of the products from the two modes of attack. The difference may vary from essentially zero to 100 percent; for instance, if 45 percent of 2-(R) and 55 percent of 2-(S) are formed, then there is a 10 percent excess of the 2-(S) enantiomer, namely, a 10 percent enantiomeric excess. In the reduction of such a substrate by an enzyme system, the perfect chiral reagent, one side will be attacked to the virtual exclusion of the other, leading to a 100 percent enantiomeric excess. Natural processes have resulted in the evolution of enzymes that for the most part produce only the amino acids corresponding to the 2-(S) configuration.

Because of the predominance of chiral organic molecules in nature, we must develop the best ways of obtaining these compounds so that their properties and reactions can be studied, and so that structural and stereochemical variations that do not occur in nature can be prepared. Chiral substances may be obtained in many ways (4): they may be isolated from a plant, animal, or microbial source; synthesized from a chiral substance isolated from nature; or obtained from a racemic mixture by use of a chiral resolving agent to accomplish the separation. In this article, we discuss the process of asymmetric synthesis, whereby an achiral starting material is converted into a chiral product by a reaction involving a chiral reagent.

Before 1965, asymmetric synthesis was generally considered a rather esoteric subject; recently it has undergone an exponential growth as reflected by publications and patents. In the 5-year period 1971 to 1976, *Chemical Abstracts* listed 47 entries under "Asymmetric synthesis." In the 5-year period, 1976 to 1981, there were more than 940 such entries (5). This is more than the total number of references found in the monograph on asymmetric synthesis (4) covering the literature from the turn of the century to 1970.

We believe that few recent basic breakthroughs have been made in the principles governing the stereochemistry of asymmetric synthesis. These principles were present in an embryonic form in the early 1950's with the contributions of Barton (6), Doering (7), Mosher (8), Cram (9), Prelog (10), and others. However, spectacular progress has been made in the application of stereochemical principles to organic synthesis. The most innovative advancements have been in the applications of metallo-organic chemistry in which the central metal atom, along with a coordinated chiral ligand or ligands, is used to guide and closely orient the stereochemical course of the reaction (11–15). Examples of such reactions are the homogeneous hydrogenations catalyzed by rhodium-chiral phosphine (11–15), the chirally modified lithium aluminum hydride reductions (16); the peroxidic allyl alcohol oxidations catalyzed by titanium alkoxide and diethyl tartrate (17); and alkylations with organometal derivatives of internally coordinated chiral reagents such as oxazolines (18), oxathianes (19, 20), or chiral proline derivatives (21). Asymmetric synthesis must now be considered on an equal basis with other available methods as a practical approach for obtaining any specified chiral compound.

The subject of asymmetric synthesis has been extensively reviewed (*4, 22–28*). Therefore, we present in this article a limited number of examples that have been chosen because they represent an optimistic assessment of the current status of asymmetric synthesis. Many of these examples came from papers presented at the United States–Japan joint conference on asymmetric reactions and processes, held at Stanford University in the summer of 1981 (*27*). Our objective is to point out how recent developments have influenced the perspective of organic chemists with respect to the use of asymmetric synthesis of the preparation of chiral compounds.

Asymmetric Homogeneous

Hydrogenations

Soluble cationic rhodium complexes of chiral tertiary phosphines (RR'R"P* symbolized by P*) catalyze asymmetric hydrogenations of unsaturated substrates. Such hydrogenations of α-acylaminoacrylic acids (3) in suitable cases have produced amino acid derivatives (5) approaching 100 percent enantiomeric purity.

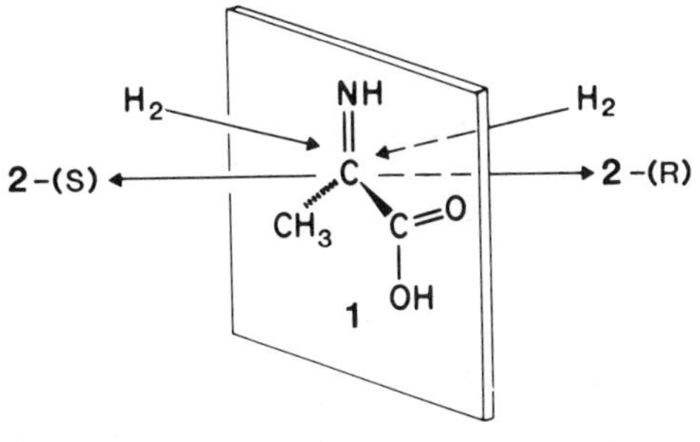

$$O$$
$$\parallel$$
RCNH   COOH
  \   /
   C
   ‖
   C
  /   \
Ar   H
3
Achiral

$$\xrightarrow[\substack{P^*-Rh-P^* \\ [\text{Chiral catalyst}]}]{H_2}$$

4

$$O \quad H \quad COOH$$
$$\parallel \quad | \quad \vdots$$
R—C—N►C◄H   and   H►C◄NHCR
        | |
      H—C—H   H—C—H
        | |
        Ar   Ar
     5-(S)   **5-(R)**
              Chiral

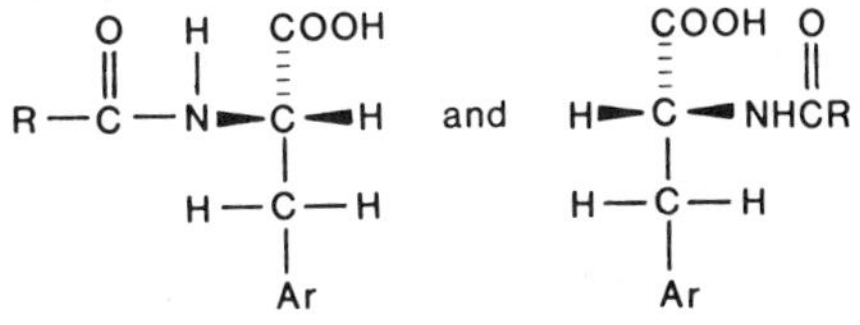

More than a decade ago, germinal papers by Knowles and Sabacky (*29*) and Horner *et al.* (*30*) on phosphorus ligands chiral at phosphorus, and later by Morrison *et al.* (*31*) and Kagan and Dang (*32*) on phosphorus ligands chiral at carbon, triggered an explosion of interest in the potential of such systems (*11–15*). Vineyard *et al.* (*33*) at Monsanto Chemical Company developed a commercial synthesis of L-dopa (L-3,4-dihydroxyphenylalanine, **7**, a compound used in the treatment of Parkinson's disease), using hydrogenation by a rhodium catalyst incorporating the chiral diphosphine ligand (*R,R*)-DIPAMP (**8**). This industrially

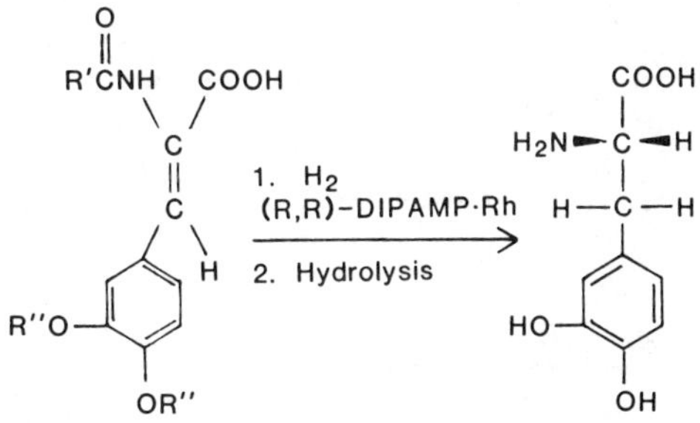

Fig. 1. Idealized model for attack of the reagent (H_2) on the prochiral faces of the carbon-nitrogen double bond in substrate **1**.

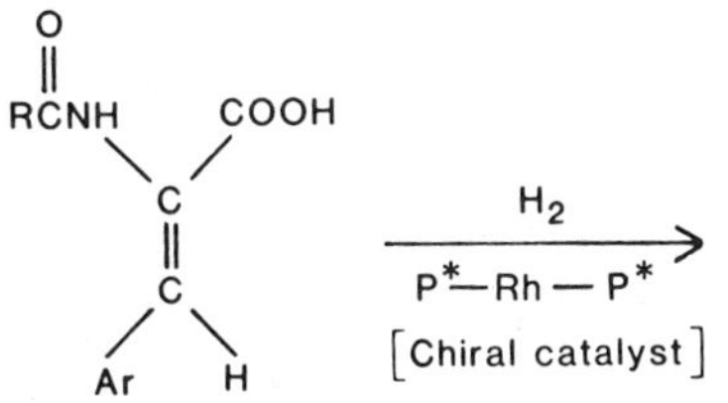

$$O$$
$$\parallel$$
R'CNH   COOH
   \   /
    C
    ‖
    C
   / \
R"O   H

|
OR"
6

$$\xrightarrow[\substack{1. \ H_2 \\ (R,R)\text{-DIPAMP·Rh} \\ 2. \ \text{Hydrolysis}}]{}$$

COOH
    |
H₂N►C◄H
H—C—H

HO

|
OH
7
L-3,4-Dihy-
droxyphenyl-
alanine
(L-dopa)

325

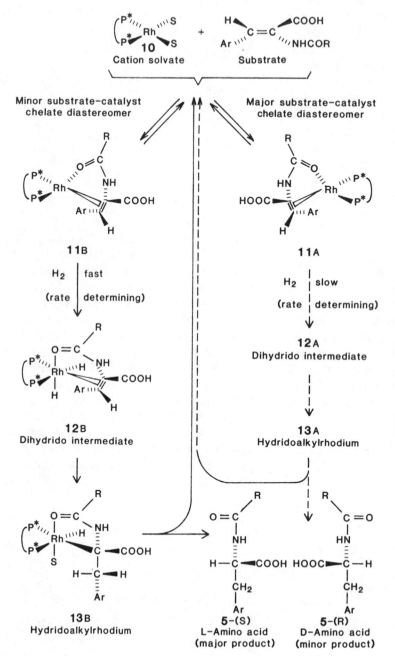

Fig. 2. Mechanism for asymmetric homogeneous hydrogenation with the rhodium chiral diphosphine ligand DIPAMP (**8**) represented by P*—P*. The solvent is designated *S*; *Ar* stands for an aryl group such as phenyl. Model according to Halpern (*12*).

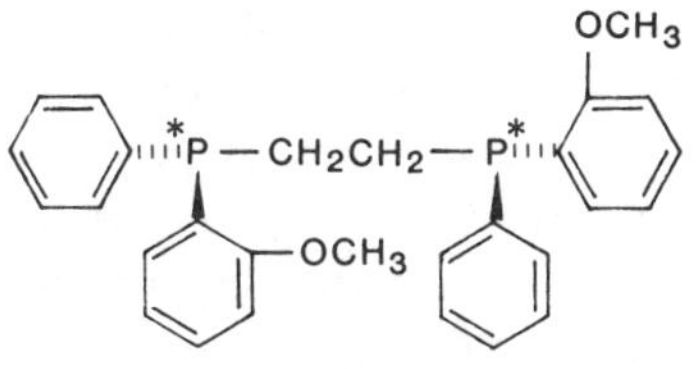

8, (R,R)-DIPAMP

successful application of asymmetric homogeneous hydrogenation—and others that can be imagined—stimulated a vast amount of activity in this area (14). More than 100 chiral phosphine ligands have been tested with various substrates; several such ligands are now commercially available. A large amount of a chiral product can be made from an achiral substrate by investing only a small amount of an active chiral catalyst. Even though the catalyst is relatively expensive, great economic leverage can be derived in suitable processes. This system is being used commercially to produce isotopically labeled amino acids. It is especially attractive for the preparation of rare D-amino acids and other chiral compounds for which alternative biochemical methods may not be suitable.

How does a diphosphine ligand such as 8 achieve this remarkable stereoselectivity? The currently accepted mechanism, represented in Fig. 2, has been deduced from kinetic, stereochemical, and x-ray structural evidence (12). Two diastereomeric catalyst-substrate chelates, 11A and 11B, are formed. The major chelate, observed in solution by nuclear magnetic resonance (NMR) is 11A, from which the D-amino acid 5-(R) should result. However, the major product is the L-amino acid 5-(S). The minor diastereomeric complex, 11B, which is in equilibrium with the major diastereomer, 11A, and which is present in too small an amount to be observed by NMR, ultimately leads to the L-amino acid derivative 5-(S). This happens because the subsequent rate-determining hydrogenation step shown in the reaction pathway on the left in Fig. 2 is much faster for the minor (11B) than for the major (11A) complex. After the initial equilibrium, the reaction path to the major product (Fig. 2) includes (i) the rate-determining addition of hydrogen to give the dihydrorhodium complex 12B, (ii) the transfer of one hydrogen to carbon with formation of the chiral carbon-rhodium bond to give the hydridoalkyl rhodium intermediate 13B, and (iii) the final irreversible transfer of hydrogen to carbon with retention of configuration to give the L-amino acid derivative, 5-(S). The enantiomeric 5-(R) product is formed to a small extent by a much slower reaction from the major complex 11A by the pathway shown on the right in Fig. 2. The high stereoselectivity occurs because of the differences in steric fits, and therefore energies of activations, between the competing, rate-determining, diastereomeric transition states. The central rhodium atom organizes the reactants, chiral ligands, and solvent in a closely packed, sterically demanding array in intermediate 12, which determines the chiral discrimination of the asymmetric reduction. Higher H_2 pressures may reduce the stereoselectivity by changing the relative concentrations of intermediate 12 and its diastereomer (15). The mechanistic details for chiral catalysts with monophosphine ligands versus diphosphine ligands may not be the same (13).

The chiral phosphine-rhodium asymmetric hydrogenation systems represent a triumph of both synthetic and mechanistic chemistry. In ideal cases, it produces chiral products with enzyme-like efficiency.

Asymmetric Epoxidation

The counterpart to asymmetric hydrogenation is asymmetric epoxidation. Henbest (34) in 1965 pioneered the use of chiral monoperoxycamphoric acid (14) to produce chiral epoxides but with an asymmetric bias of 5 percent or less; Pirkle and Rinaldi (35) in 1977 described improvements that gave asymmetric epoxidations up to 9 percent enantiomeric excess. The following equations illustrate this reaction for conversion of styrene (15) to styrene oxide (16) with an 8 percent excess of the (S)-(+) isomer. Presumably the low stereoselectivity results from too great a distance between

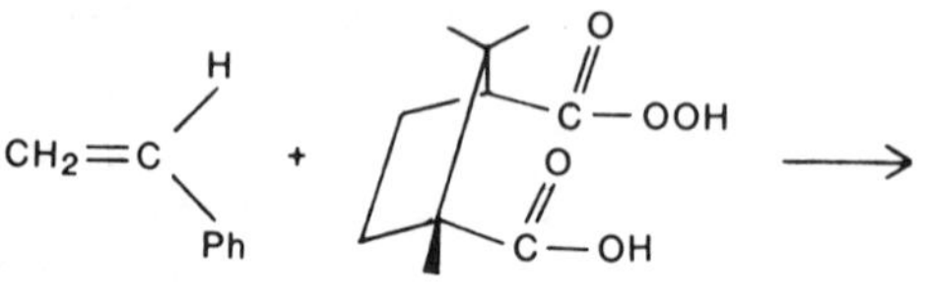

15   **14**

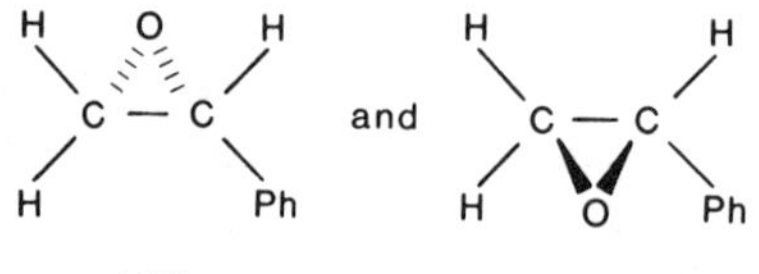

46%   54%
(R)(−)-**16**   (S)(+)-**16**

the chiral-inducing centers and the peroxidic bond so that there is little steric discrimination during attack on one prochiral face of the olefin versus the other.

Asymmetric epoxidations with hydrogen peroxide and *tert*-butyl hydroperoxide catalyzed by chiral phase-transfer agents such as benzyl quinidinium salts, were studied by Wynberg and others (36) and found to be moderately successful. Sharpless and his students, using vanadi-

um complexes (37), and Yamada *et al.* (38) using molybdenum complexes, independently reported the first metal-catalyzed asymmetric epoxidations. Additional examples were reported by Otsuka and his students (39) who, for example, described the treatment of squalene with a mixture of *tert*-butyl hydroperoxide, molybdenum oxide chelate, and the chiral inducing agent (+)-diisopropyl tartrate. This resulted in a high chemical yield of biogenetically important (S)-2,3-squalene epoxide in 14 percent enantiomeric excess. Studies on allylic alcohols were pursued by Sharpless and his students (40–43), who ultimately devised successful epoxidation, based on both organic and inorganic stereochemistry, which has given impressive results. In the Sharpless reaction, an allylic alcohol (17) is treated with a mixture of commercially available titanium tetra-isopropoxide, *tert*-butyl hydroperoxide, and (+)- or (−)-diethyl tartrate to give an epoxy carbinol (18) in high chemical yield. So far, most reactions have given more than

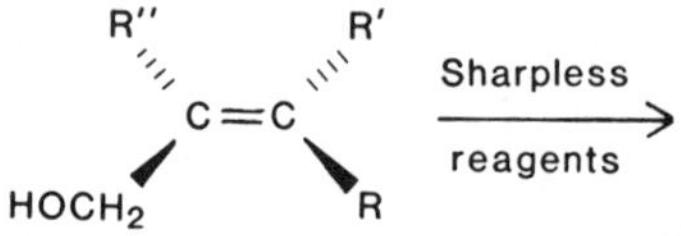

17

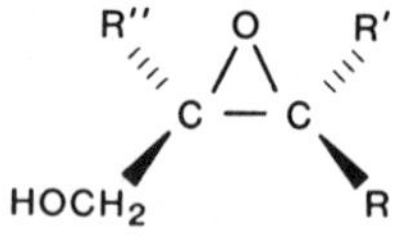

18

90 percent enantiomeric excess, with many giving better than 95 percent enantiomeric excess. An equivalent amount of the titanium-tartrate reagent is used when the allylic alcohol has one or two substituents on the double bond, but tri- and tetra-substituted derivatives are so reactive that only a catalytic amount of

the reagent is necessary. This catalytic asymmetric epoxidation is under intensive study as a practical method of introducing chirality into the intermediates for the synthesis of various commercially important biologically active compounds such as antibiotics, drugs, and insect pheromones (44). The following synthesis of the gypsy moth pheromone (+)-disparlure (21), starting with the achiral allylic alcohol 19, illustrates one of its commercial applications.

$$
\begin{array}{ccc}
\underset{\textbf{19}}{\ce{H\bond{...}C(CH2OH)=C(H)(CH2)9CH3}} & \xrightarrow[\text{reagent}]{\text{Sharpless}} & \underset{\textbf{20}\ (91\%\ \text{e.e.})}{\text{epoxide 20}}
\end{array}
$$

19 → Sharpless reagent → **20** (91% e.e.)

Several steps → → → → **21** (+)-Disparlure

The Sharpless allylic epoxidation catalyst begins to approach the stereoselectivity of an enzyme system and surpasses most enzymes in the variety of substrates it will accept. How does it work? It is clear that the titanium metal serves to bind and organize the epoxidizing agent (*tert*-butyl hydroperoxide), the chiral-inducing agent (diethyl tartrate), and the substrate (an allylic alcohol) in such a way that one face of the allylic alcohol is greatly favored over the other. Sharpless has proposed an empirical model to correlate the observed stereochemical results. Regardless of the nature of the substituents, when (S,S)-$(-)$-diethyl tartrate is used as the inducing

agent, the oxygen of the epoxidizing agent is inserted onto the top face of the double bond for the orientation shown in Fig. 3. The use of natural enantiomeric (R,R)-$(+)$-diethyl tartrate results in preferential attack from the opposite side, with formation of the epoxide with reversed stereochemistry. A theoretical model of the transition state for this epoxidation was provisionally proposed (42) (Fig. 4A). The reagent is now considered to be a dimer (43); a transition state model incorporating this dimer structure is shown in Fig. 4B. Both models depict bonding of the allylic oxygen to a central titanium atom that organizes the epoxidizing agent (*tert*-butyl hydroperoxide), the chiral-inducing agent $[(R,R)$-$(+)$-diethyl tartrate], and the substrate into a compact structure that favors attack on the bottom face of the allylic double bond. The use of enantiomeric (S,S)-$(-)$-diethyl tartrate changes the combined electronic and steric effects so that attack on the opposite side of the allylic bond is favored, with formation of the enantiomeric product.

This may be the most innovative reaction introduced into organic chemistry in modern times. Masamune and Sharpless have collaborated to synthesize the four

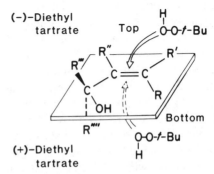

Fig. 3. Stereochemical model for Sharpless asymmetric epoxidation reaction (t-*BuOOH*, *tert*-butyl hydroperoxide).

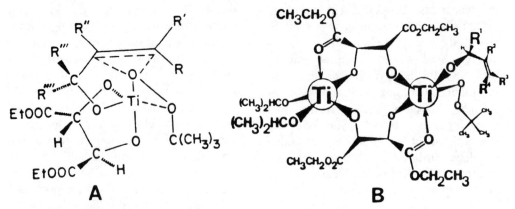

Fig. 4. Postulated organization of substrate (allylic alcohol), epoxidizing agent (*tert*-butyl hydroperoxide), chiral-inducing agent [(*R,R*)-(+)-diethyl tartrate], and titanium in the Sharpless reagent; (A) monomer model (*42*); (B) dimer model (*43*).

D-pentoses with this basic reaction followed by stereo-controlled opening of the epoxide ring (*41*). All eight isomeric L-hexoses (*43*) were then synthesized within a few months, an accomplishment that by older methods might take years.

Chelation-Controlled Addition Reactions

Recent efficient asymmetric addition reactions have produced various types of chiral carbinols, α-hydroxy carbonyl compounds, acids, hydroxy acids, and their derivatives, all of which have become important in the construction of complex chiral molecules. These asymmetric syntheses have been constructed upon the solid foundation of knowledge accumulated and correlated by Cram beginning in 1952 (*9*) and expanded upon by others during the intervening years. These earlier results have been reviewed (*4*) and are summarized by what are known as Cram's rules. The stereoselectivity of these addition reactions is greatly increased when there is a functional group, such as methoxy, to which the organometallic reagent can chelate. Chelation freezes out free rotation and in-

creases rigidity, thereby accentuating and localizing steric effects. Recent application and expansion of this principle has led to several important, highly stereoselective, asymmetric syntheses.

Oxathiane system. In the system developed by Eliel (*19, 20*) (Fig. 5) a chiral oxathiane, **22**, which is prepared in three steps from the readily available natural (+)-pulegone, is converted to the anion that undergoes electrophilic substitution at C-2 when treated with an aldehyde to give the product **23**. Essentially a single configuration at C-2 is produced because the anion assumes an equatorial orientation for stereoelectronic reasons. Mixed configurations are formed at the adjacent secondary carbinol center of the product; however, oxidation in the next step to **24** destroys the chirality at this center. Because the chelation of magnesium is much greater to oxygen than to sulfur, as shown in **25**, the R′ group of the Grignard reagent preferentially attacks the front face of the carbonyl group on the side of the ring oxygen. Therefore, the diastereomer produced in 90 to 98 percent enantiomeric excess has the configuration shown in **26**. Treatment with

330

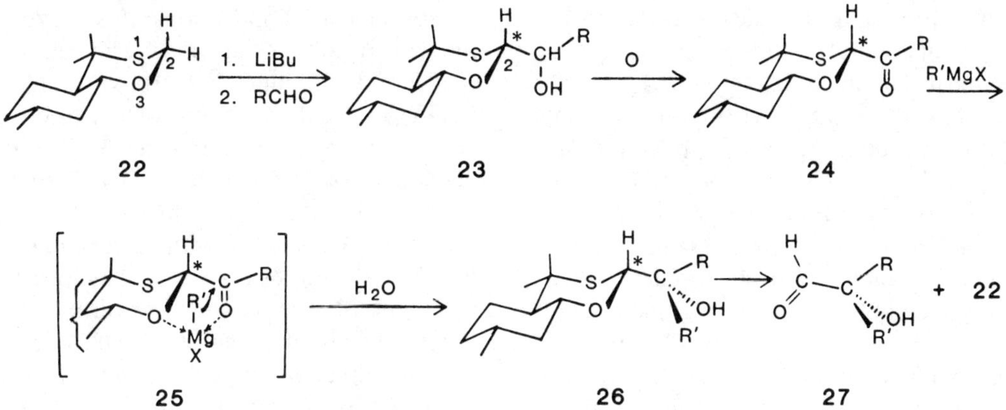

Fig. 5. A series of α-hydroxy aldehydes has been made by the Eliel oxathiane asymmetric synthesis (*19, 20, 27*) with enantiomeric purities of 90 to 98 percent.

thiocyanate and silver ions regenerates a precursor to the chiral-inducing pulegone derivative **22** and liberates the chiral hydroxy aldehyde **27** with an enantiomeric purity the same as that of the diastereomeric precursor **26**. Since the order of introducing R and R' can be reversed, either enantiomer may be synthesized at will. The hydroxy aldehyde can either be oxidized to the correspond-

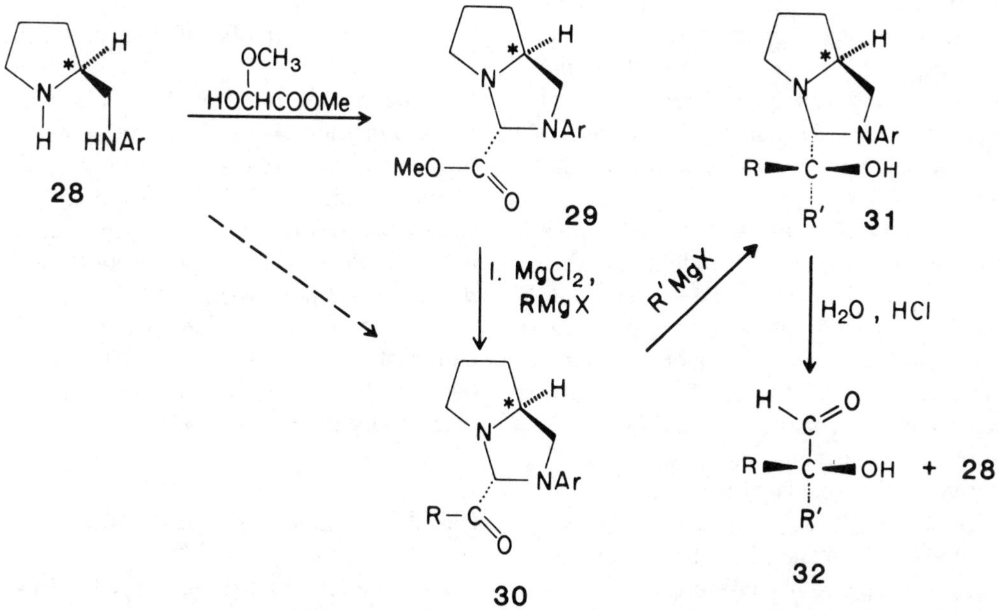

Fig. 6. The Mukaiyama chiral diamine asymmetric synthesis. By introducing the R and R' groups in the opposite sequence, the enantiomeric α-hydroxy aldehyde **32** will be formed.

ing chiral acid or reduced to the corresponding chiral glycol. Since there is an almost unlimited choice of R and R' groups, this asymmetric synthesis is one of wide-ranging potential application. An example of such an application is the synthesis of (R)-mevalactone (20).

Aminal system, Mukaiyama's chiral aminal system (21) (Fig. 6) bears a certain resemblance to the oxathiane synthesis, but with nitrogen replacing oxygen and sulfur in the ring. The chiral diamine **28**, which is available from L-proline in three steps, can be converted into aminals **29** and **30**. Steric hindrance (rather than stereoelectronic control, as in the oxathiane system) directs the substituent at C-2 so that one enantiomer is formed to the virtual exclusion of the other. Treatment of **29** with a suitable Grignard reagent and **30** with another Grignard reagent, followed by hydrolysis, regenerates the chiral-inducing agent **28** and forms the α-hydroxy aldehyde **32** in enantiomeric excess of 84 to 98 percent. As in the oxathiane case, diverse R groups are available. This chiral aminal synthesis has been engineered in several different ways (21) and has been used to synthesize several natural products (27, 45).

Oxazoline system. Extensive studies by Meyers (18) and his students on chelation control in oxazoline systems has had a profound influence on developments in the field of asymmetric synthesis during the last decade. The oxazoline template is a versatile chiral adjuvant; its applications are among the most imaginative in the annals of asymmetric synthesis.

The basic oxazoline chiral-inducing agent is prepared in two steps from commercially available (1S,2S)-1-phenyl-2-amino-1,3-propanediol (**33**) (Fig. 7). In one of the early applications, this oxazoline (**34**; R, alkyl or aryl) was allowed to react with lithium diisopropyl amide (LDA) to yield a mixture of chelation-stabilized E and Z aza-enolates **35-E** and **35-Z** (5:95 ratio when R is CH_3). Treatment of the aza-enolate mixture (reaction with **35-Z** shown) with electrophiles (shown as R'X) results in the substituted oxazoline **36**, which can be hydrolyzed to give the chiral substituted acid **37** and the O-methyl derivative **38**. Oxazoline **34** is resynthesized from **38**, and the cycle is ready to be repeated.

The two crucial structural elements that determine the stereochemical control of this oxazoline system are the methoxy group and the *trans*-phenyl substituent. As shown in **35-Z**, the methoxy group holds the enolate in a rigid conformation by chelation with the lithium while the phenyl group blocks the top face of the oxazoline ring. These factors force attack by the electrophilic R' group from the bottom face.

Numerous variants of the chirality transfer accomplished by this system have been reported (18, 27, 45). These reports include: 3-hydroxy- and 3-methoxy acids, 2-hydroxy-, 2-methoxy-, and 2-chloro acids, 3,3-dialkyl- and 2,3,3-trialkylpropanoic acids, 2-substituted butyrolactones and valerolactones, phthalides, thiolepoxides, dihydropyridines, and binaphthyls. A specific application of the oxazoline asymmetric synthesis is the preparation of the European pine sawfly pheromone (45).

Chiral Aldol Condensation Reactions

The aldol condensation appears frequently in both biosynthetic and laboratory synthetic sequences for the con-

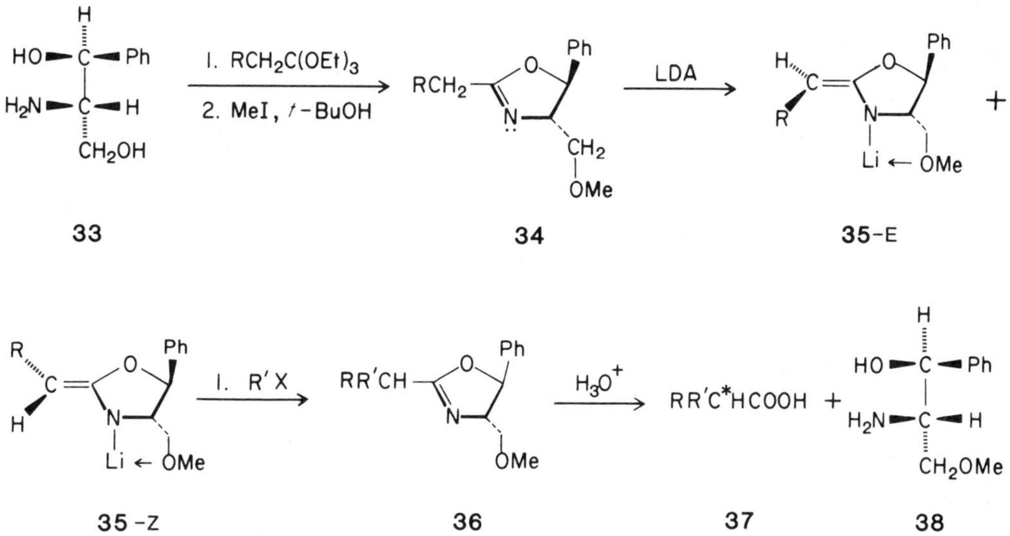

33 → **34** → **35-E** +

35-Z → **36** → **37** + **38**

Fig. 7. Asymmetric alkylation with the chiral oxazoline system. *LDA*, lithium diisopropyl amide [LiN(*i*-pr)$_2$].

struction of complex molecules. The detailed facets of its stereochemistry make it potentially a powerful reaction for the synthesis of natural products, but realization of this potential has been elusive. Complexities of the reaction rendered it one of the most challenging problems in synthetic organic stereochemistry. Cram (*9*) made important contributions to the stereochemistry of the aldol condensation in the early 1950's; however, the diastereomer selection of this reaction is just now being brought under satisfactory synthetic control, based primarily on the work of Evans (*46*), Masamune (*47*), and Heathcock (*48*).

A simple crossed aldol condensation between the carbonyl compounds **39** and **40** leads to structure **41**, which with two chiral centers, exists in four isomeric forms. If the G group in product **41** is hydrogen (or can be replaced by hydrogen), then another aldol condensation yields structure **44** with four contiguous chiral centers, which can exist in 16

isomeric forms. The extreme stereochemical complexities of the reaction are obvious. The isomers of **41** are *erythro*- or *threo*-type diastereomers, each of

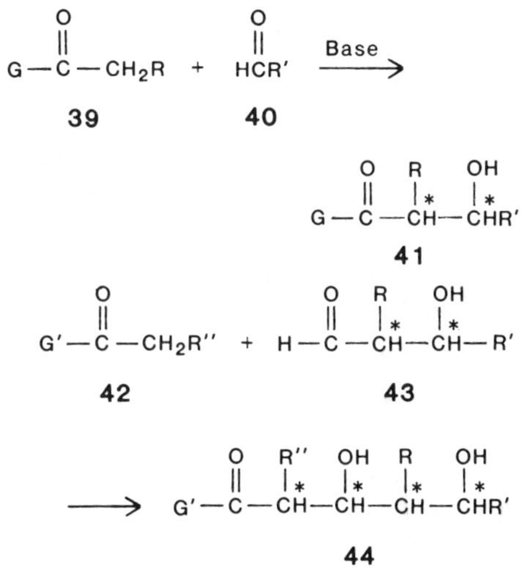

39 **40**

41

42 **43**

44

which exists in either *d* or *l* enantiomeric forms. Thus two types of stereochemical control are involved: diastereomer selection to give predominantly either the *erythro* or *threo* form (control of the

333

relative configuration of the two chiral centers) and enantiomer selection to give predominantly either the *d* or *l* isomer of the *erythro* or *threo* form (control of the absolute configuration of the two chiral centers).

The intricacies of the advances that have been made in the aldol condensation prevent an adequate treatment of its stereochemistry here. Accordingly we will make only a few general comments. If either of the carbonyl compounds **39** or **40** is already optically active by virtue of a chiral R group (that is, **39** or **40** has been obtained optically active by some independent method), then the enantioselectivity (control of absolute stereochemistry) is built into the reaction. However, there is need for a breakthrough in the asymmetric synthesis of the aldol condensation itself. Perhaps an effective and general chiral catalyst for the aldol condensation can be developed as has been for the allylic alcohol epoxidations. That this is not impossible is shown by the unique intramolecular aldol asymmetric synthesis catalyzed by L-proline discovered by Hajos and Parrish (*49*). Alternatively, it may be that a general effective reagent for preparing chiral enolates, involved as intermediates in the aldol condensation, can be developed. If so, then complete stereochemical control of this reaction would have been achieved.

Asymmetric Reduction

An available, widely applicable, chiral reducing agent for the conversion of unsymmetrical ketones **45** into the corresponding chiral secondary carbinols **46**, with enantiomeric purities of better than 90 percent of either desired configura-

tion, would be of great value (Fig. 8). Extensive research has resulted in several successful reagents (*4, 26, 27, 50, 51*) but none can be considered ideal. However, the results to date are sufficiently encouraging to allow the prediction that the problem will be successfully resolved in time.

The easiest chiral reducing agents to prepare and handle are those (**49**) generated from the reaction of lithium aluminum hydride (**47**) with a chiral ligand (R*QH; **48**). Examples of the most successful of such chiral ligands are the amino alcohol **50** (*50*), the diamine **51** (*21*), and the diol **52** (*51*). The reagent (*50*) from Darvon alcohol **50** is especially favorable for reduction of acetylenic ketones (*52*); chemical yields of 70 to 96 percent and stereochemical yields from 72 to 84 percent enantiomeric excess are observed. An example of the use of this reagent, as well as the borane reagent (*53*) from α-pinene and borobicyclo[3.3.1]nonane (*53*), is given in the polyene cyclization reaction illustrated in Fig. 9.

Polyene Cyclization

The potential of asymmetric synthesis as applied to steroids is being explored intensively. One of the most interesting systems is that of polyene cyclization which has been the subject of study by several groups, most successfully by Johnson and his co-workers (*54, 55*). In the example shown in Fig. 9 the polyenediol **56** has two chiral centers, one at C-3 and the other at C-11 (according to steroid numbering). The one at C-3 is destroyed during the cyclization and apparently has no stereochemical influence on the reaction, but the other at C-11 carries

334

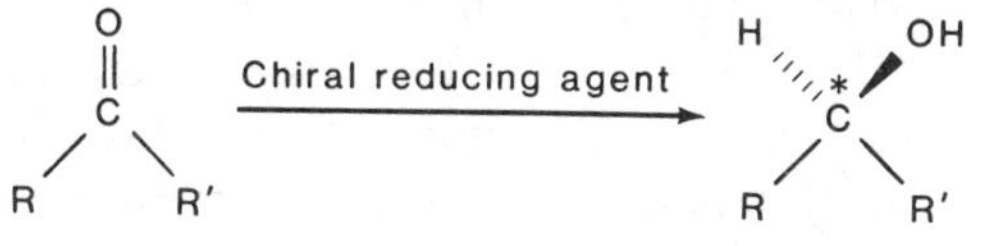

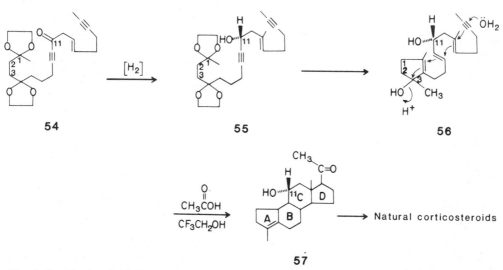

45 → (Chiral reducing agent) → **46**

LiAlH₄ + x R*QH ⟶ LiAl(QR)ₓH₄₋ₓ

47   **48**   **49**

R*QH = chiral alcohol and/or amine

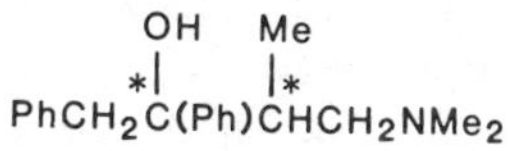

50 Darvon alcohol

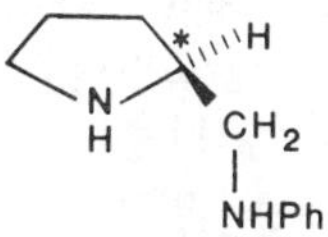

51 2-(Phenyl-aminomethyl)pyrrolidine

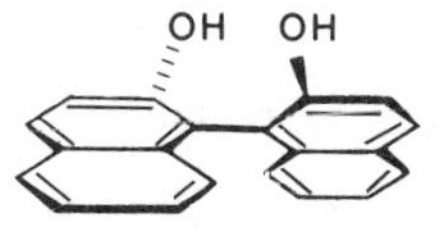

52 2,2'-Binaphthol

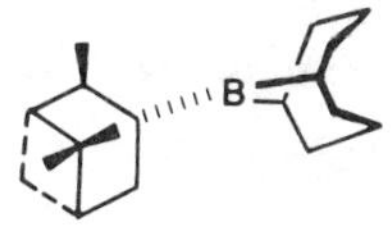

53 (+)-2-Pinanyl-borobicyclo-[3.3.1]nonane

Fig. 8. Chiral reagents for the asymmetric reduction of ketones to secondary alcohols (*26*).

through unchanged in the product. When the starting polyenediol **56** is a racemate, the product is a racemic mixture of the natural and unnatural enantiomers. When the pure 11-(*R*) enantiomer **56** is cyclized, the product **57** has the desired natural steroid configuration. Thus, once the chiral intermediate is at hand, the subsequent cyclization, forming six addi-

tional chiral centers, is directed to give only the one desired enantiomeric product. Accordingly, an efficient method of obtaining the chiral polyenediol **56** or its precursor **55** is needed. Classical resolution of **55** failed in this case, but asymmetric reduction of the precursor ketone **54** was successful (*54*). The LiAlH₂(OR*)₂ reagent made from **50** was

54 → [H₂] → **55** → **56**

56 (CH₃COH / CF₃CH₂OH) → **57** → Natural corticosteroids

Fig. 9. Synthesis of chiral secondary carbinol **55** by asymmetric reduction of achiral ketone **54**, followed by polyene cyclization to the steroid precursor, **57**.

335

used to reduce **54** in 78 percent chemical yield to give a 91:9 mixture of **55** and its C-11 enantiomer. A precursor acetylenic ketone that is related to **54** was reduced with the chiral (+)-α-pinene-9BBN reagent **53** (*53*) to give product with a 98.5-to-1.5 ratio of enantiomers (*55*).

Immobilized Enzymic Asymmetric Synthesis

In 1978, the Tanabe Seiyko Company in Japan went into commercial production of L-aspartic acid by a continuous process in which a buffered ammonia and fumaric acid solution was passed over a selected strain of *Eschericha coli* cells that were immobilized in a polysaccharide gel (*56*). The potential use of immobilized enzymes and immobilized

Fumaric acid

L-Aspartic acid

whole cells for production of chiral natural products seems limited only by economic factors and man's ingenuity in harnessing these natural systems. Furthermore, they serve as models of synthetic chiral catalyst systems not yet realized.

We have omitted many innovative and important developments in this general treatment of the current status of asymmetric synthesis. We are sorry that we

have slighted so much good work; for example, one obvious omission has been the many applications of hydroboration in this field which is a subject by itself (*53, 57*). We must also point out the article by Trost (*58*) in which stereochemical control of reactions, including enantioselectivity, is discussed.

We believe that the foregoing examples support the thesis that the science of asymmetric synthesis has come of age. Asymmetric synthesis must be systematically considered in any synthetic strategy aimed at the formation of chiral compounds.

References and Notes

1. Chiral molecules have the property of handedness [chiral from the Greek χειρ (*cheir*), meaning hand]. Chiral structures, like right versus left hands, have the same connections (connectivity); they are mirror images of each other, but they are not identical because they are not superposable on each other. An achiral structure lacks the property of handedness. A structure that is achiral but on further substitution becomes chiral is called prochiral.
2. W. A. Bonner, in *Exobiology*, C. Ponnamperuma, Ed. (North-Holland, Amsterdam, 1972), pp. 117–181.
3. The two enantiomeric forms of the amino acid, **2**-(*S*) and **2**-(*R*), have the same connectivity. However, they are not identical because of the way they are connected in three-dimensional space. To be identical, two structures must be superposable: **2**-(*S*) and **2**-(*R*) are not. As written **2**-(*S*) has the NH₂ group below the plane of the page, whereas **2**-(*R*) has it above. Reorientation of structure **2**-(*R*) by rotating 180° around the horizontal axis through the central carbon atom shows that **2**-(*S*) and **2**-(*R*) are mirror-image structures. Nonsuperposable mirror-image structures are known as enantiomers. A 50:50 mixture of enantiomers is called a racemate and is achiral in the aggregate.
4. J. D. Morrison and H. S. Mosher, *Asymmetric Organic Reactions* (Prentice-Hall, Englewood Cliffs, N.J., 1971; American Chemical Society reprint edition, 1976).
5. These numbers may not be a quantitatively accurate reflection of the research activity in this area because of both more efficient indexing by Chemical Abstracts Service and because of a change in their subject classifications, but the qualitative conclusion is valid.
6. D. H. R. Barton, *Experientia* **6**, 316 (1950).
7. W. Doering and R. W. Young, *J. Am. Chem. Soc.* **72**, 631 (1950).
8. H. S. Mosher and E. La Combe, *ibid.*, pp. 3994 and 4991.

9. D. J. Cram and F. A. Abd Elhafez, *ibid*. **74**, 5828 and 5851 (1952).
10. V. Prelog, *Helv. Chim. Acta* **36**, 308 (1953).
11. G. W. Parshall, *Science* **208**, 1221 (1980).
12. J. Halpern, *ibid*. **217**, 401 (1982).
13. J. D. Morrison, W. F. Masler, M. K. Neuberg, *Adv. Catal*. **25**, 81 (1976).
14. B. Bosnich and M. D. Fryzuk, *Top. Stereochem*. **12**, 119 (1981).
15. I. Ojima, T. Kogure, N. Yoda, *J. Org. Chem*. **45**, 4728 (1980).
16. J. D. Morrison, E. R. Grandbois, S. I. Howard, *ibid*., p. 4229.
17. T. Katsuki and K. B. Sharpless, *J. Am. Chem. Soc*. **102**, 5974 (1980).
18. A. I. Meyers, *Acc. Chem. Res*. **11**, 375 (1978); also see (*27*), pp. 83–98.
19. E. L. Eliel and J. E. Lynch, *Tetrahedron Lett*. **22**, 2855 (1981).
20. E. L. Eliel and S. K. Soai, *ibid*., p. 2859; also see (*27*), pp. 37–54.
21. T. Mukaiyama, *Tetrahedron* **37**, 4111 (1981); also see (*27*), pp. 21–36.
22. J. W. Scott and D. Valentine, Jr., *Science* **184**, 943 (1974).
23. Y. Izumi and A. Tai, *Stereo-Differentiating Reactions* (Kodansha, Tokyo, and Academic Press, New York, 1977).
24. H. B. Kagan and J. C. Fiaud, *Top. Stereochem*. **10**, 175 (1978).
25. D. Valentine and J. W. Scott, *Synthesis* (1978), p. 329.
26. J. W. ApSimon and R. R. Seguin, *Tetrahedron* **35**, 2797 (1979).
27. E. Eliel and S. Otsuka, Eds., *Asymmetric Reactions and Processes* (American Chemical Society Symposium Series, No. 185, American Chemical Society, Washington, D.C. 1982).
28. J. D. Morrison, Ed., *Asymmetric Synthesis* (Academic Press, New York, in press).
29. W. S. Knowles and M. J. Sabacky, *Chem. Commun*. (1968), p. 1445.
30. L. Horner, H. Siegel, H. Büthe, *Angew. Chem*. **80**, 1034 (1968).
31. J. D. Morrison, R. E. Burnet, A. M. Aguiar, C. J. Morrow, C. Phillips, *J. Am. Chem. Soc*. **93**, 1301 (1971).
32. H. B. Kagan and T. P. Dang, *ibid*. **94**, 6429 (1972).
33. B. D. Vineyard, W. S. Knowles, M. J. Sabacky, G. L. Bachman, D. J. Weinkauff, *ibid*. **99**, 5946 (1977).
34. H. B. Henbest, in *Organic Reaction Mechanisms* (Special Publication No. 19, Chemical Society, London, 1965), pp. 83–92.
35. W. H. Pirkle and P. L. Rinaldi, *J. Org. Chem*. **42**, 2080 (1977).
36. J. C. Hummelen and H. Wynberg, *Tetrahedron Lett*. (1978), p. 1089.
37. R. C. Michaelson, R. E. Palermo, K. B. Sharpless, *J. Am. Chem. Soc*. **99**, 1990 (1977).
38. S. Yamada, T. Mashiko, S. Terashima, *ibid*., p. 1988.
39. K. Tani, M. Hanafusa, S. Otsuka, *Tetrahedron Lett*. (1979), p. 3017.
40. B. E. Rossiter, T. Katsuke, K. B. Sharpless, *J. Am. Chem. Soc*. **103**, 464 (1981).
41. T. Katsuki *et al*., *J. Org. Chem*. **47**, 1373 (1982).
42. S. S. Woodard, thesis, Stanford University (1981).
43. S. Y. Ko, A. W. M. Lee, S. Masamune, L. A. Reed III, K. B. Sharpless, F. J. Walker, *Science* **220**, 949 (1983).
44. S. Bystrom, H. E. Högberg, T. Norin, *Tetrahedron* **37**, 2249 (1981).
45. Y. Sakito *et al*., *Chem. Lett*. (1979), p. 1027; *ibid*. (1980), p. 1223.
46. D. A. Evans, J. V. Nelson, T. R. Taber, *Top. Stereochem*. **13**, 1 (1982).
47. S. Masamune, S. Mori, D. Van Horn, D. W. Brooks, *Tetrahedron Lett*. (1979), p. 1665.
48. C. H. Heathcock, *Science* **214**, 395 (1981); compare references therein and (*27*), pp. 55–72.
49. Z. G. Hajos and D. R. Parrish, *J. Org. Chem*. **39**, 1615 (1974).
50. S. Yamaguchi and H. S. Mosher, *ibid*. **38**, 1870 (1973).
51. R. Noyori, I. Tomino, Y. Tanimoto, *J. Am. Chem. Soc*. **101**, 3129 (1979).
52. W. S. Johnson, R. S. Brinkmeyer, V. M. Kapoor, T. M. Yarnell, *ibid*. **99**, 8341 (1977); R. S. Brinkmeyer and V. Kapoor, *ibid*., p. 8339.
53. M. M. Midland, D. C. McDowell, R. L. Hatch, A. Tramontano, *ibid*. **102**, 867 (1980).
54. W. S. Johnson, *Bioinorg. Chem*. **5**, 51 (1976); *Acc. Chem. Res*. **1**, 1 (1968).
55. W. S. Johnson, B. Frei, A. S. Gopalan, *J. Org. Chem*. **46**, 1512 (1981).
56. I. Chibata, Ed., *Immobilized Enzymes, Research and Development* (Kodasha, Tokyo, and Halstead, New York, 1978).
57. H. C. Brown, *Chem. Eng. News* **59**, 24 (6 April 1981); _____ and P. K. Jadhav, *J. Org. Chem*. **46**, 4048 (1981).
58. B. M. Trost, *Science* **219**, 245 (1983).

337

One of the most vexing problems in the synthesis of complex organic compounds is control of relative stereochemistry. The problem arises from the fact that each tetracoordinate carbon may represent a center of chirality (1). Thus, any carbon that is bonded to four different substituents may exist in either "right-handed" or "left-handed" forms, and there are two stereoisomeric forms of a molecule that contains such a chiral element. In general, a molecule containing n chiral elements may exist in as many as 2^n stereoisomeric forms. For example, erythronolide A (1), the aglycone of the important antibiotic erythromycin A, has ten chiral centers and may exist in 1024 stereoisomeric forms (2). Compound 2, the aglycone of the polyether antibiotic septamycin, has 21 chiral centers and may exist in more than 2 million stereoisomeric forms.

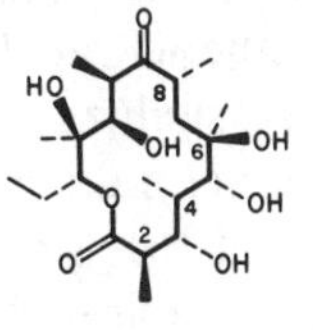

1

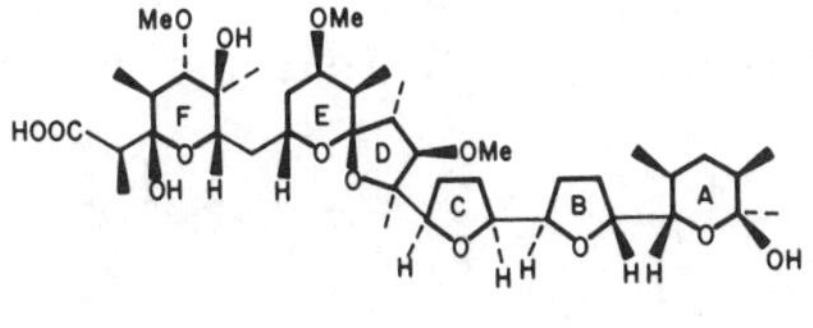

2

To be able to carry out efficient syntheses of complex molecules such as 1 and 2, chemists must be able to control the sense of chirality at each chiral center as it is introduced in the course of a synthesis. Traditionally, this has been accomplished by making use of the rather rigid structures of small-ring compounds. For example, in a synthesis of the tricyclic sesquiterpene alcohol cedrol

23

Acyclic Stereocontrol Through the Aldol Condensation

Clayton H. Heathcock

Science 214, 395–400 (23 October 1981)

339

(4) the correct relative stereochemistry at C-3 results from addition of methyllithium to ketone 3 because one face of the carbonyl group is shielded by one of the methyl groups at C-4 (3).

When the molecule to be synthesized does not have such a rigid structure, control of stereochemistry is much more difficult. A classic way in which chemists have approached this problem is to establish the sense of chirality of new asymmetric carbons by using small-ring intermediates, and then break open the ring to obtain an acyclic building block containing the desired asymmetric carbons. For example, in one synthesis of erythronolide A (1) the correct relative chirality at the five asymmetric centers in the C-2 to C-8 portion of the molecule was established by using intermediate 5,

which contains two six-membered rings (4). This material is manipulated by a series of steps including cleavage of the indicated carbon-carbon bond to obtain lactonic acid (6), a further intermediate in the preparation of 1.

Such indirect methods of achieving stereocontrol, although conceptually elegant, limit the options available to the synthetic chemist and often cause syntheses to be long and cumbersome. Consequently, many research groups have been investigating methods in which effective stereocontrol might be realized without resorting to such indirect meth-

ods (5). For the past 5 years, my research group at Berkeley and several other groups in the United States, Japan, and Germany have been reexamining one of the oldest organic reactions—the aldol condensation (Eq. 1). This venera-

ble process (6) has several desirable features which highly recommend it as a synthetic method. First, a new carbon-carbon bond is formed. Reactions in which carbon-carbon bonds are formed are especially important, relative to those involving simply the conversion of one functional group into another, since they allow us to build up larger molecules from small building blocks. Second, the aldol condensation provides a product having two useful functional groups, which can serve as reaction sites for further synthetic transformations. Finally, in a case such as that depicted in Eq. 1, two new chiral centers are created. If one of the reactants is chiral, the aldol product contains three chiral elements and there are $2^3 = 8$ possible stereoisomers (Eq. 2). Thus, the aldol con-

densation is potentially a powerful method for synthesizing polyfunctional compounds with many chiral centers, such as compounds 1 and 2, provided we can learn to control its stereochemical outcome.

The first significant breakthrough was the development of reagents that enable the chemist to control the relative chirality of the two new asymmetric carbons formed in an aldol condensation such as Eq. 1. Dubois and co-workers showed that in the reaction of a preformed lithium enolate with an aldehyde, there is a natural predisposition for *cis* enolates to give *erythro* aldols (Eq. 3) while *trans* enolates tend to give *threo* aldols (Eq. 4) (*7, 8*).

We capitalized on this important observation by finding that, when the substituent R in Eqs. 3 and 4 is large, the

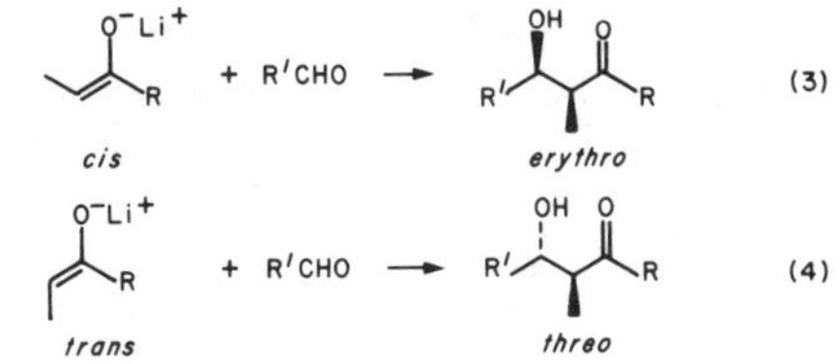

(3)

(4)

stereoselectivity observed is quite high (*9*). Thus, the preformed lithium enolate of the ethyl ketone **7** (a *cis* enolate) reacts with benzaldehyde in tetrahydrofuran (THF) to give virtually exclusively the *erythro* aldol **8**. Similar high stereoselectivity is seen in the reaction of ketone **7** with a variety of other aldehydes. On the other hand, the preformed *trans* eno-

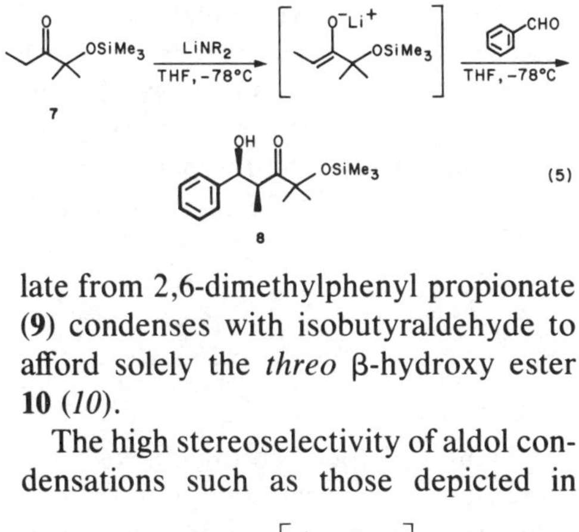

(5)

late from 2,6-dimethylphenyl propionate (**9**) condenses with isobutyraldehyde to afford solely the *threo* β-hydroxy ester **10** (*10*).

The high stereoselectivity of aldol condensations such as those depicted in

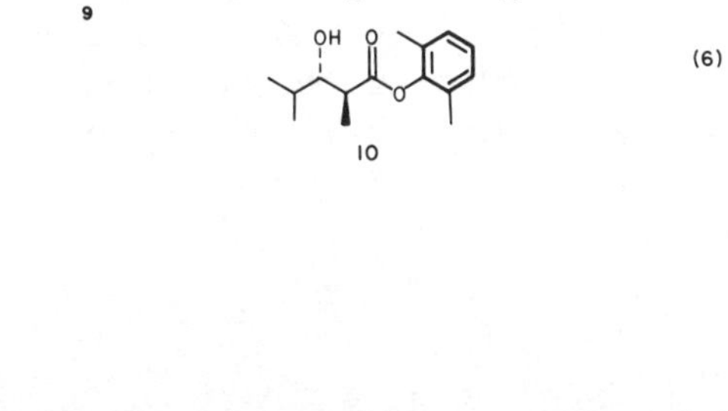

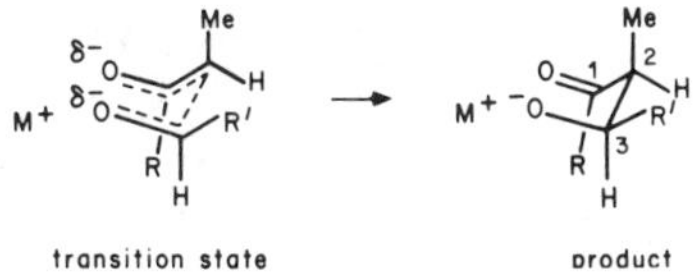

(6)

Fig. 1. Zimmerman transition state for the aldol condensation: *cis* enolate giving *erythro* aldol.

Eqs. 5 and 6 has usually been interpreted in terms of a transition state proposed by Zimmerman and Traxler (*11*) for a related reaction in 1957. The basic assumption of their proposal is that the new carbon-carbon bond is partially formed in the transition state for the reaction. Thus, there is partial negative charge on each of the two oxygens and it is reasonable that the assembly would be ordered so that these partial negative charges are both oriented in the general direction of the cation. This electrostatic ordering of the transition state results in a *cis* enolate giving an *erythro* aldol if the distance between R and R′ is maximized (Fig. 1). The alternative arrangement is shown in Fig. 2 (*12*). In this arrangement R and R′ are closer together. This arrangement is disfavored for simple steric reasons, particularly for enolates having large R groups (*13*).

Reagents such as **7** and **9** are useful because they allow the formation of either the *erythro* or the *threo* β-hydroxy carbonyl system and because the products may be converted into a variety of other useful intermediates. For example, the aldols formed from reagent **7** may easily be transformed into *erythro* β-hydroxy acids (Eq. 7) (*9*), *erythro* β-hydroxy aldehydes (Eq. 8) (*14*) or other *erythro* β-hydroxy ketones (Eq. 9) (*15*). In a similar manner, the *threo* aldols prepared from reagents such as **9** can be transformed into a variety of useful *threo* β-hydroxy carbonyl compounds.

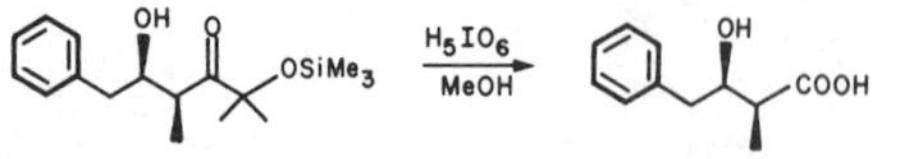

(7)

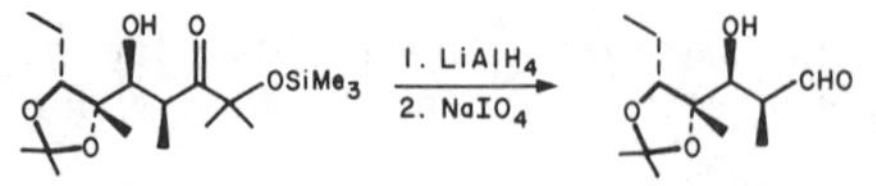

(8)

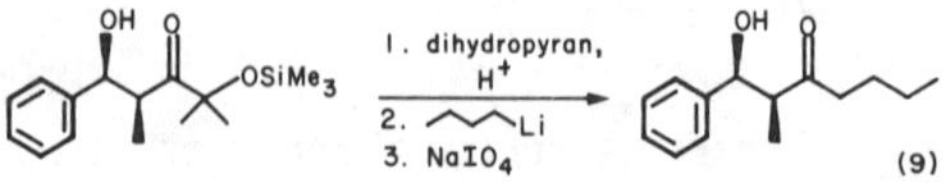

(9)

At Berkeley, we have concentrated on using preformed lithium enolates to accomplish kinetic stereoselection. Other research groups have investigated stereocontrol with enolates of other metals and also with thermodynamic selection (*16*). Masamune at Massachusetts Institute of Technology and Evans at Caltech and their respective co-workers find that highly stereoselective aldol condensations may be realized with boron enolates. For example, *S*-phenyl propanethioate may be converted in the *cis* enolate **10**, which gives *erythro* aldols of high stereochemical purity (Eq. 10) (*17*). On

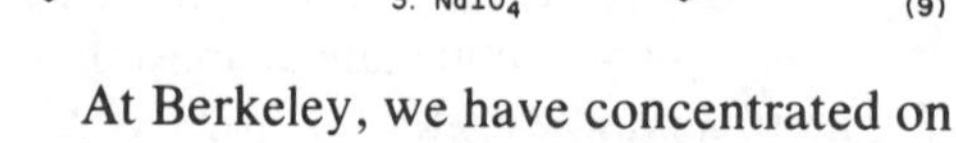

(10)

the other hand, *S-tert*-butyl propanethioate affords a *trans* enolate (**11**) which gives rise to *threo* aldols (Eq. 11) (*18*).

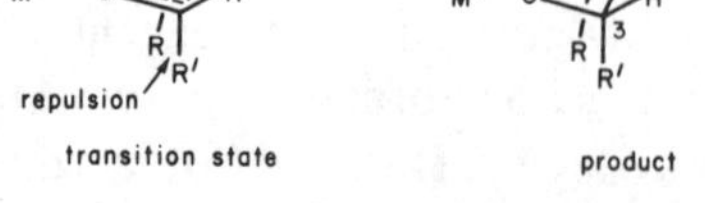

Fig. 2. Zimmerman transition state for the aldol condensation: *cis* enolate giving *threo* aldol.

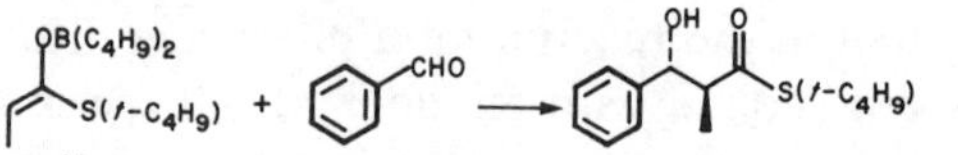

---

(11)

Progress has also been made in achieving thermodynamic stereoselection (*16*). The aldol condensation is easily reversible, and *erythro-threo* equilibration is often observed when an ethereal solution of the aldolate (the salt of the aldol) is allowed to stand for a period of time (Eq. 12) (*9*). The rate of this equilibration

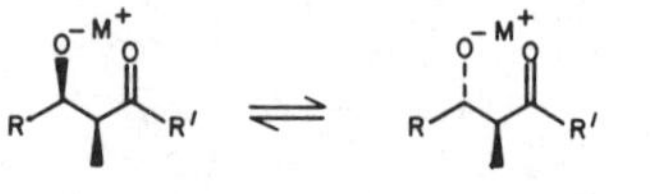

(12)

depends on a number of factors, including the nature of the cation, M. In general, cations that are strongly chelated by the two oxygens of the aldolate (B, Al, and to some extent Li) stabilize it and retard retroaldolization. Aldolates are more prone to undergo retroaldolization when the cation is one that is largely dissociated (K, Na, and R_4N). Thus, *erythro-threo* equilibration is a process that can often be promoted by a judicious choice of reaction conditions. It turns out that the *threo* isomers of many aldolates are substantially more stable than their *erythro* counterparts (for instance, Eq. 13) (*19*). Thus, equilibration

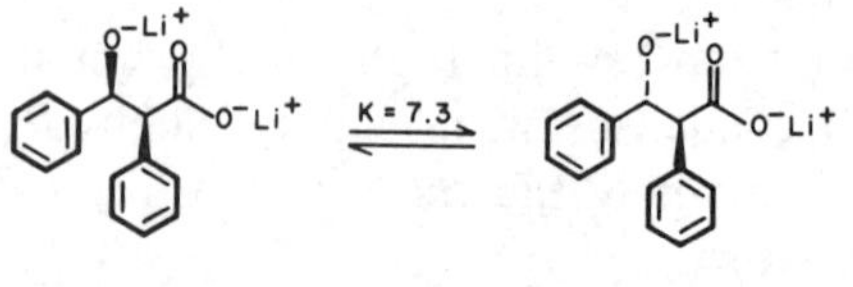

(13)

can sometimes be used to achieve *threo* stereoselection. In addition, House *et al.* (*20*) at Georgia Tech discovered that certain cations, notably Zn^{2+} and Mg^{2+}, actually enhance the thermodynamic *threo* selectivity in many cases. The House procedure has become one of the standard methods for achieving modest *threo* stereoselectivity. For example, it was used by Kishi and co-workers in a crucial aldol coupling of two intermediates in a synthesis of the polyether antibiotic lasalocid (Eq. 14) (*21*). Lasalocid, which has the *threo* configuration at the pertinent chiral centers, is obtained as the major isomer of a 40:10:7:3 mixture of stereoisomers.

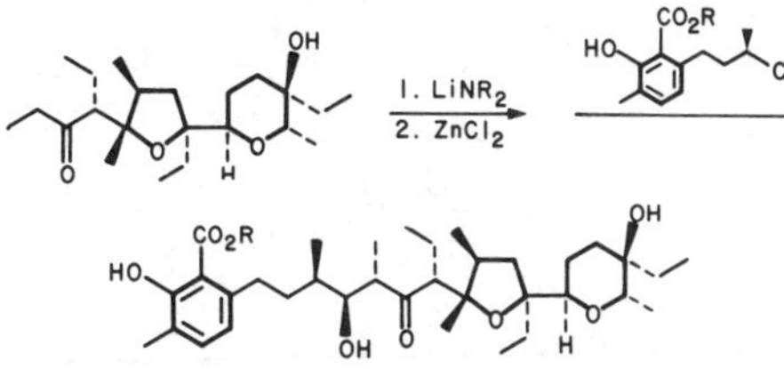

(14)

The ability to rather confidently control the relative configuration of the two chiral centers produced in an aldol condensation is only half a victory. A much more difficult problem presents itself when the aldehyde is chiral. As pointed out earlier, the product from such an aldol condensation (Eq. 2) has three chiral centers, and hence there are eight possible stereoisomers. Four of these stereoisomers are depicted in Eq. 15; the other four are the enantiomers (mirror-image stereoisomers) of the ones shown. Two of the isomeric products are *erythro* and two are *threo*, because in a chiral aldehyde the two faces of the carbonyl group are diastereotopic rather than enantiotopic (*22*). Of course, by using an *erythro*-selective or *threo*-selective reagent, we can select two of the four stereoisomers. For example, addition of

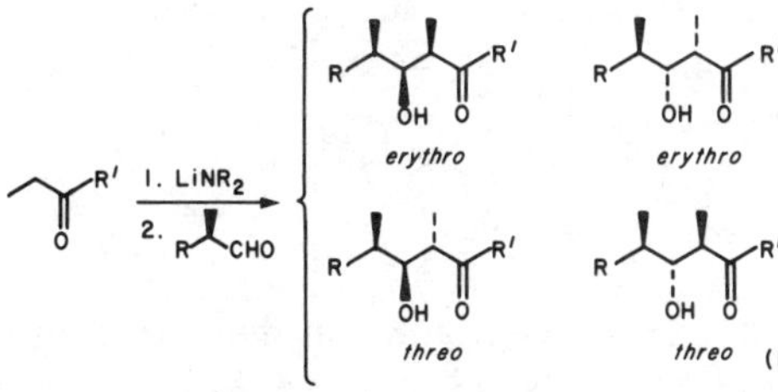

(15)

the enolate of ketone **7** with chiral aldehyde **12** provides a 2:1 mixture of two

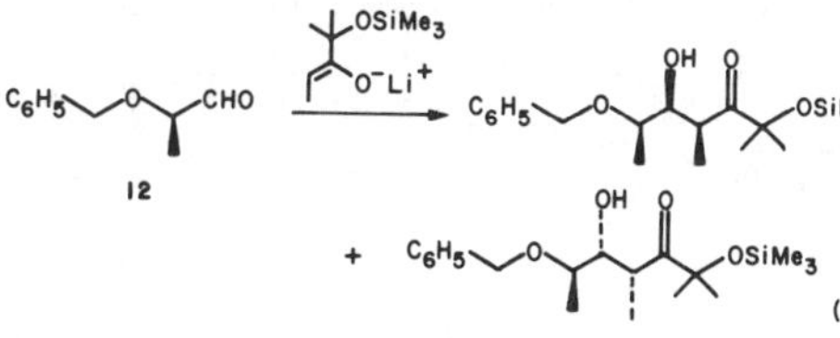

(16)

erythro aldols (Eq. 16) (*14*). The 2:1 ratio reflects the inherent difference in reaction rate when the enolate of ketone **7** attacks the two diastereotopic faces of the aldehyde; it is called the diastereoface preference of the aldehyde.

At present, the only effective way to influence the diastereoface preference in additions to such chiral aldehydes is a strategy called double stereodifferentiation (*23*). To understand how this method works, one must first realize that either reactant in an aldol condensation can be chiral and can show diastereoface selectivity. For example, as shown in Eqs. 17 and 18, assume that one has a chiral aldehyde that shows diastereoface preference of 5:1 in the indicated sense and a chiral enolate that also shows diastereoface preference of 5:1 as indicated. Now consider what happens when one allows the indicated enantiomer of the chiral aldehyde to react with the

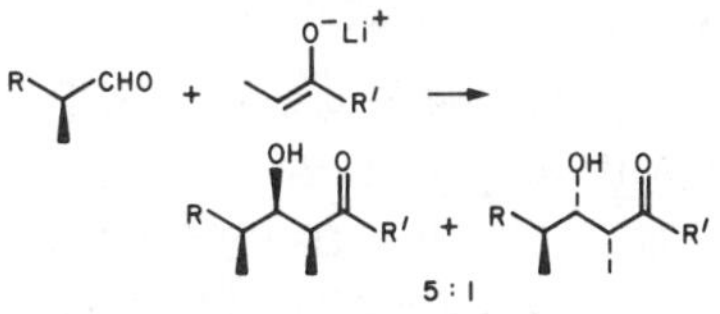

(17)

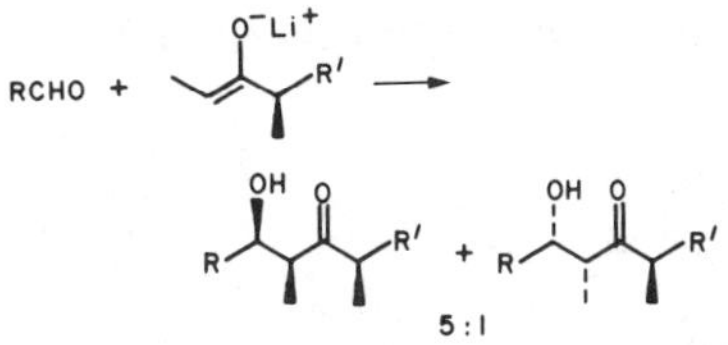

(18)

indicated enantiomer of the chiral enolate. As shown in Eq. 19, one of the

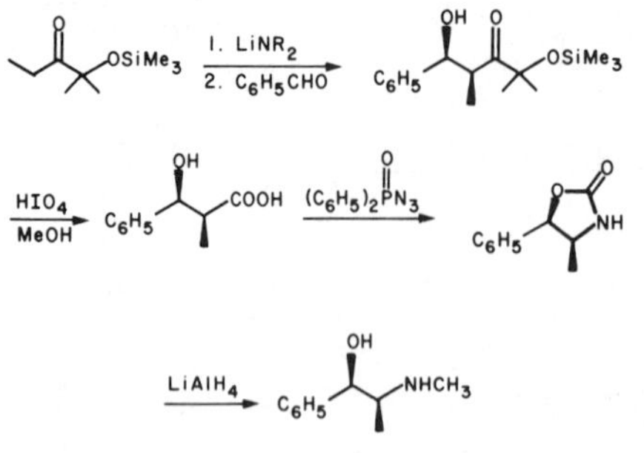

Fig. 3 (left). Synthesis of (±)-ephedrine.
Fig. 4 (right). Synthesis of (±)-blastmycinone.

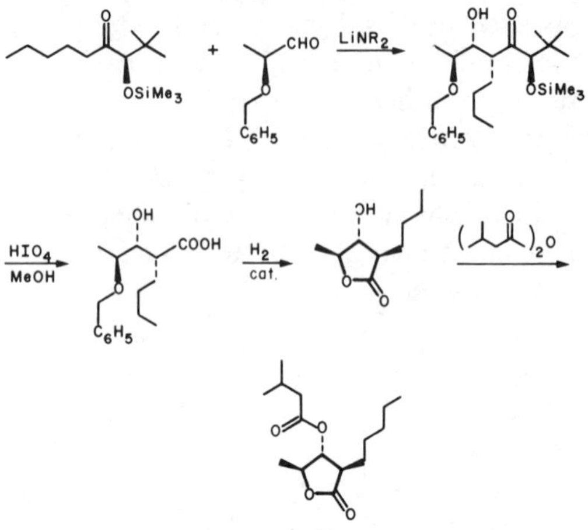

erythro aldols will be formed in much greater quantity than the other, since both chiral reactants are promoting the same sense of chirality at the newly created centers; to a first approximation, the two *erythro* aldols will be formed in a ratio of 25:1. Of course, if one allows the same enantiomer of the aldehyde to react with the other enantiomer of the enolate, the net diastereoface preference of the

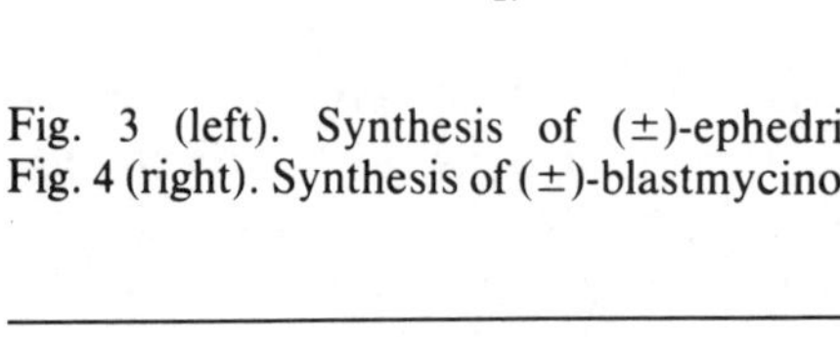

aldehyde should be even worse than in its reaction with a typical achiral enolate, since the two chiral reactants are now working at cross-purposes in inducing the sense of chirality at the two new centers (Eq. 20).

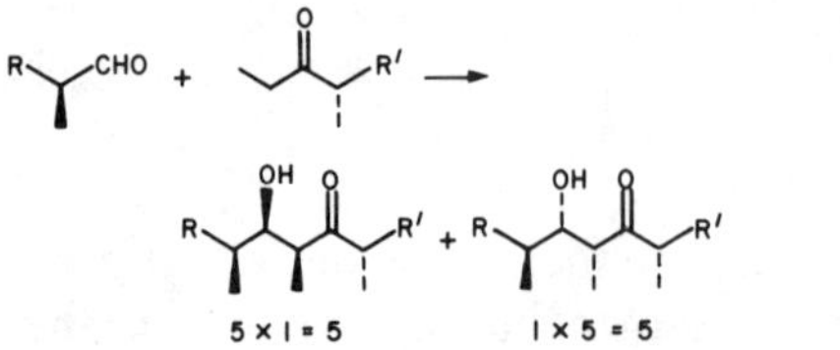

An example of double stereodifferentiation in an aldol condensation is shown in Eqs. 21 and 22. The condensation depicted in Eq. 21 represents an unproductive combination of ketone and aldehyde; the two *erythro* aldols are produced in a ratio of only 2:1. However, when the other enantiomer of the aldehyde is used, both reactants promote the same sense of chirality at the new centers. Productive double stereodifferentiation results and the two *erythro* aldols are produced in a ratio greater than 30:1.

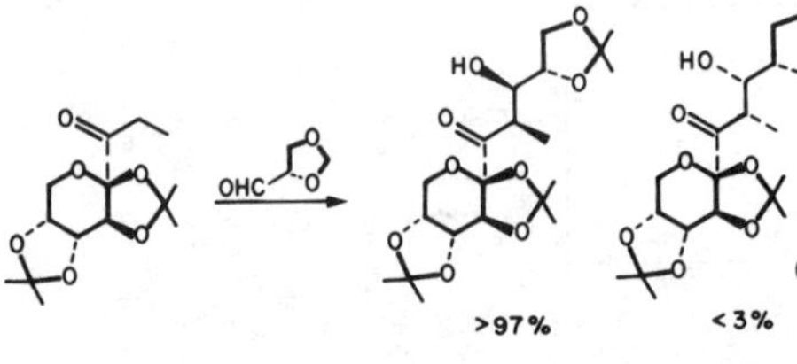

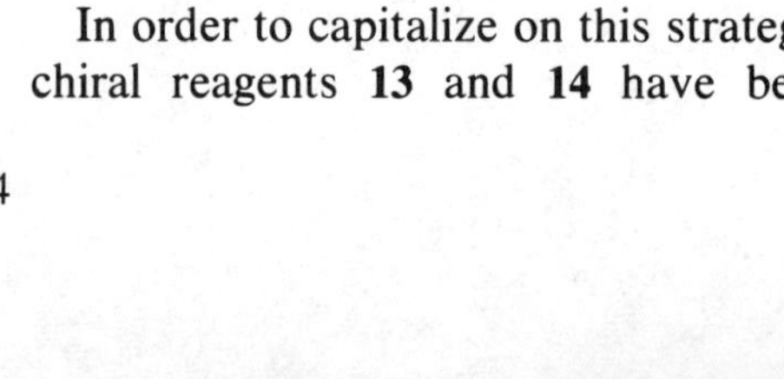

In order to capitalize on this strategy, chiral reagents **13** and **14** have been developed by my research group at Berkeley and by Masamune's group (23). A related class of highly selective reagents has recently been introduced by

13   14

Evans and co-workers (24). The Evans reagents are propionyl imides 15 and 16, which are derived from the readily available amino alcohols valinol and norephedrine. The value of the Evans reagents is

15   16

that the derived enolates show exceptional inherent diastereoface selectivity in aldol condensations. For example, the dibutylboron enolate derived from 15 reacts with benzaldehyde to give aldols 17 and 18 in a ratio of 332:1 (Eq. 23).

1 . C$_6$H$_5$CHO
2. OH$^-$, H$_2$O$_2$

(23)

332:1

17   18

These reagents are so demanding in promoting a given sense of chirality at the newly created chiral centers that they should totally overwhelm the modest inherent diastereoface preferences of most chiral aldehydes. Thus, compound 15 should react with the indicated enantiomer of 2-phenylpropanal to give almost completely aldol 19, even though the innate diastereoface preference of this aldehyde is for formation of 20 (Eq. 24).

C$_6$H$_5$—CHO   15

(24)

19   20

Up to the present, most of the research on aldol stereoselection has been aimed at understanding the factors responsible for stereoselectivity and developing reagents for controlling the *erythro-threo* and diastereoface preference problems. Application of the technique in the synthesis of complex structures has barely begun. However, some progress in this direction has been made. A simple example in which a stereoselective aldol condensation provides stereochemical control is the synthesis of the alkaloid ephedrine (21), outlined in Fig. 3 (9). This synthesis, which has a 71 percent overall yield, illustrates the power of the aldol condensation as a synthetic method for a variety of structural types, since ephedrine does not even contain the 1,3-keto alcohol functionality typical of an aldol. However, by using well-established functional group chemistry, it is easy to transform the COOH group into NHCH$_3$.

A second example is the synthesis of blastmycinone (22), a degradation product of the antibiotic antimycin A$_3$ (Fig. 4) (25).

For about 18 months, several of my co-workers have been developing a synthesis of erythronolide A (1) (26). Our goal is to use a series of stereoselective aldol condensations to form all of the necessary carbon-carbon bonds. Although the synthesis is not yet complete, we have successfully negotiated three of the five necessary aldol condensations and have prepared an intermediate that has six of the ten chiral centers of erythronolide A (Fig. 5). The three aldol condensations in Fig. 5 proceed with stereoselectivities of 15:1, 5.7:1, and 5.3:1, respectively. Thus, the overall stereochemical yield for the preparation of aldehyde 23 is 69 percent. Since 23 has six chiral centers, there are 64 possible stereoisomers. A stereorandom synthesis would give 23 (and its enantiomer) in a stereochemical yield of only 3 percent. It is interesting to note that intermediate 23 has the same six chiral centers as compound 6, an intermediate in one of the previous successful syntheses of compound 1.

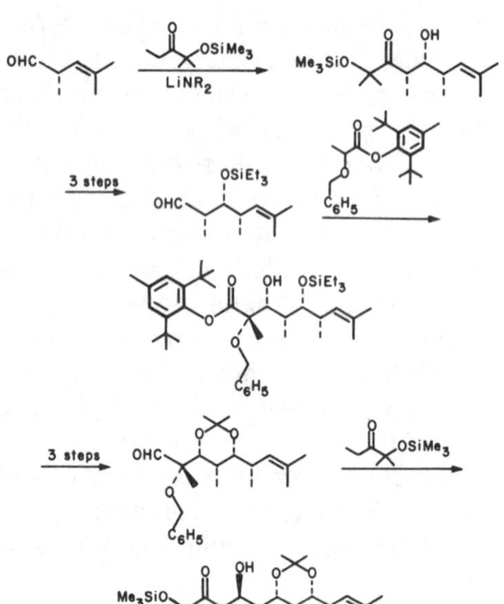

In this brief article, I think I have given a taste of the exciting progress that has been made in bending the aldol condensation to our will over the past 5 years. Much has been accomplished and much remains to be done. At the same time, other research groups are actively investigating many other organic reactions from the standpoint of achieving good stereocontrol in the synthesis of conformationally flexible systems. It is clear that this field will continue to be an active area of research for some time to come.

References and Notes

1. The terms chiral and chirality were first introduced by Lord Kelvin when he wrote: "I call any geometrical figure, or any group of points chiral, and say that it has chirality, if its image in a plane mirror, ideally realized, cannot be brought to coincide with itself" [*Lord Kelvin: Baltimore Lectures* (Clay, London, 1904)]. The terms have gained widespread use in chemistry since 1966 due to the landmark article "Specification of molecular chirality" [R. S. Cahn, C. K. Ingold, V. Prelog, *Angew. Chem. Int. Ed. Engl.* 5, 385 (1966)]. Cahn *et al.* define chiral and chirality in the following manner: "A model which has no element of symmetry except at most axes of rotation may be called chiral. Thus, chirality expresses the necessary and sufficient condition for the existence of enantiomers. Chirality means handedness, and, in our context, topological handedness." Because of the nature of three-dimensional space, there are three kinds of chiral element: centers, axes, and planes. The most common one is the center of chirality—the familiar asymmetric carbon.
2. In this article, the following conventions are used in depicting structures. The main skeleton of the molecule is written as a geometric figure so that each vertex and the end of each line represents an atom, which is carbon unless otherwise indicated. Hydrogens bonded to tetracoordinate carbons are usually omitted unless the omission would lead to confusion. Lines of normal density represent bonds that lie more or less in the plane of the page. Bold lines indicate bonds that project toward the viewer, and dashed lines represent bonds that project away from the viewer. Thus, each vertex bearing bold or dashed bonds corresponds to a chiral center.
3. G. Stork and F. H. Clarke, *J. Am. Chem. Soc.* 77, 1072 (1955); *ibid.* 83, 3114 (1961).
4. E. J. Corey, P. B. Hopkins, S. Kim, S. Yoo, K. P. Nambiar, J. R. Falck, *ibid.* 101, 7131 (1979).
5. For an excellent review of the field see P. A. Bartlett, *Tetrahedron* 36, 2 (1980).
6. R. Kane, *Ann. Phys. Chem.* 44, 475 (1838); *J. Prakt. Chem.* 15, 129 (1838). The aldol condensation takes its name from the trival name for 3-

Fig. 5. Stereoselective aldol approach to (±)-erythronolide A.

hydroxybutanal, the simplest product of the type (*a*ldehyde, alcoh*ol*).

7. J. E. Dubois and M. Dubois, *Chem. Commun.* (1968), p. 1567; J. E. Dubois and P. Fellman, *C. R. Acad. Sci. Ser. C* **274**, 1307 (1972); *Tetrahedron Lett.* (1975), p. 1225.
8. In describing enolates, *cis* and *trans* refer to the relationship of the enolate oxygen and the alkyl group attached to the adjacent carbon of the double bond. The convention used to describe the relative configuration of aldols is as follows: when the aldol is written in an extended (zigzag) manner, if the β-hydroxy group and the alkyl group at the α position both extend either toward the viewer or away from the viewer, this is an *erythro* stereoisomer. If one of these substituents projects away from the viewer and one toward the viewer, this is a *threo* stereoisomer. In reactions involving only achiral reactants, such as Eqs. 3 and 4, the indicated products are accompanied by an equal amount of the mirror-image stereoisomer (enantiomer) in each case. Normally, only one member of each pair of enantiomers is depicted, since we are usually concerned with the relative chirality of the two newly created chiral centers.
9. C. H. Heathcock, C. T. Buse, W. A. Kleschick, M. C. Pirrung, J. E. Sohn, J. Lampe, *J. Org. Chem.* **45**, 1066 (1980).
10. M. C. Pirrung and C. H. Heathcock, *ibid.*, p. 1727.
11. H. Zimmerman and M. Traxler, *J. Am. Chem. Soc.* **79**, 1920 (1957).
12. Deconvolution of perspective drawings such as those labeled "product" in Figs. 1 and 2 is often difficult, even for experienced chemists. These drawings are related to the planar, zigzag depictions used elsewhere in this article in the following way. For the product in Fig. 1, note that the chain of atoms C_1-C_2-C_3-R' lies in a plane. If we view this plane from the top we see

For the product in Fig. 2, we must first rotate about the C_2-C_3 bond so that the chain of atoms C_1-C_2-C_3-R' again lies in a plane:

Viewing this plane from the top we see

13. Other mechanistic interpretations have also been advanced. For instance, see J. Mulzer, M. Zippel, C. Brüntrup, J. Segner, J. Finke, *Justus Liebigs Ann. Chem.* (1980), p. 1108.
14. C. H. Heathcock, S. D. Young, J. P. Hagen, M. C. Pirrung, C. T. White, D. Van Derveer, *J. Org. Chem.* **45**, 3846 (1980).
15. C. T. White and C. H. Heathcock, *ibid.* **46**, 191 (1981).
16. Stereoselectivity in a reaction can result from either the kinetics or the thermodynamics of the process. Kinetic stereoselection means that the desired stereoisomer is formed faster than other stereoisomers that might be formed. Thermodynamic stereoselection operates when the desired stereoisomer is more stable than other stereoisomers with which it is in equilibrium.
17. D. E. Van Horn and S. Masamune, *Tetrahedron Lett.* (1979), p. 2229.
18. D. A. Evans, J. V. Nelson, E. Vogel, T. R. Taber, *J. Am. Chem. Soc.* **103**, 3099 (1981).
19. J. Mulzer, J. Segner, G. Brüntrup, *Tetrahedron Lett.* (1977), p. 4651.
20. H. O. House, D. S. Crumrine, A. Y. Teranishi, H. D. Olmsted, *J. Am. Chem. Soc.* **95**, 3310 (1973); see also (9).
21. T. Nakata, G. Schmid, B. Vranesic, M. Okigawa, T. Smith-Palmer, Y. Kishi, *J. Am. Chem. Soc.* **100**, 2933 (1978); see also R. E. Ireland, G. J. McGarvey, R. C. Anderson, R. Badouch, B. Fitzsimmons, S. Thaisrivongs, *ibid.* **102**, 6180 (1980).
22. Addition of a nucleophile to an achiral aldehyde gives a pair of enantiomers; hence, the two faces of the aldehyde carbonyl are said to have an enantiotopic relationship. However, addition of a nucleophile to the two faces of a chiral aldehyde produces two diastereomers (nonmirror-image stereoisomers). Thus, the faces of a chiral aldehyde are said to have a diastereotopic relationship.
23. C. H. Heathcock and C. T. White, *J. Am. Chem. Soc.* **101**, 7076 (1979); _____, J. J. Morrison, D. Van Derveer, *J. Org. Chem.* **46**, 1296 (1981); S. Masamune, S. A. Ali, D. L. Snitman, D. S. Garvey, *Angew. Chem. Int. Ed. Engl.* **19**, 557 (1980).
24. D. A. Evans, personal communication.
25. C. H. Heathcock, M. C. Pirrung, J. Lampe, C. T. Buse, S. D. Young, *J. Org. Chem.* **46**, 2290 (1981).
26. The co-workers are S. Young, J. Hagen, and U. Badertscher.
27. I owe a great deal to a talented group of students and postdoctoral associates who have done much of the experimental work on which this account is based. I also acknowledge my colleagues at other universities for valuable exchanges of information, particularly D. Evans of Caltech. Finally, I thank the U.S. Public Health Service for financial support for our aldol project under grant AI-15027.

At the turn of the century several reports appeared on the use of light as a reagent in organic synthesis. Ciamician's research efforts resulted in the discovery of the photochemically mediated [2 + 2] cycloaddition of an enone and an alkene to afford a cyclobutane (1). His colleague Paternò followed this report with a description of the first [2 + 2] photocycloaddition of an aldehyde to an alkene to provide an oxetane (2). During the ensuing 80 years organic chemists have come to recognize and routinely employ the photochemical method for effecting this cycloaddition reaction in complex syntheses of chiral molecules. The four-membered ring products are found to occur in such compounds as thromboxane A_2 (1) (3), the arachidonic acid metabolite that plays an important role in platelet aggregation and muscle constriction in humans, and the sesquiterpene essential oil caryophyllene (2) (4).

24

A feature of these photoproducts with far-reaching implications is that they serve as vehicles for subsequent chemical manipulations. It is this property that has imparted great significance to these chemical reactions in the field of organic synthesis. Evidence for this comes from the recent literature where one can find several reviews on the application of these reactions in organic synthesis (5–7).

[2 + 2] Photocycloadditions in the Synthesis of Chiral Molecules

Stuart L. Schreiber

Science 227, 857–863 (22 February 1985)

The purpose of this article is to report some developments from our laboratories that have employed the photochemical [2 + 2] cycloaddition reaction as a springboard for the synthesis of naturally occurring substances with significant physiological properties. Through chemical synthesis, samples of these important and often scarce molecules can be prepared, which may facilitate subsequent studies of their biological properties. Modification of their structure through synthesis can lead to new materials with altered properties. In this process of creating new substances we learn about the nature of chemical reactivity.

Synthesis of a Sex Pheromone of the American Cockroach

Isolation of periplanones A and B. In 1952 an important result in pheromonal research was reported by Roth and Willis. These workers were studying the reproductive behavior of the three most common species of cockroaches (*germanics, orientalis*, and *americana*) and found that the female of *Periplaneta americana* was able to sexually stimulate the male by olfactory communication (*8*). Many years of intensive investigations of the active component produced by the female American cockroach led to refinement of the techniques used for the isolation of the pheromone and of the bioassay to determine activity. Employing optimal conditions for pheromone biosynthesis and executing a direct extraction on mass scale, Persoons and coworkers were able to isolate and determine the structures of two active components termed periplanone A (pA) and periplanone B (pB) (*9–11*). In this work,

the alimentary tracts of 35,000 insects were dissected and the midguts were extracted. Another 20,000 insects were kept in a cage suitable for routine recovery of feces, which were extracted and added to the material isolated from the midguts. Silica gel chromatography separated the active compounds pB (200 μg) and pA (20 μg). Through a combination of spectroscopic analyses, primarily nuclear magnetic resonance spectroscopy (NMR), the structure of pB was established as that depicted in **3**. A tentative assignment of structure for pA based largely on mass spectral analysis was suggested and is shown as **4**.

3   4

In 1979 W. C. Still reported his landmark studies that culminated in the synthesis of pB (*12, 13*). Through conformational analysis of germacranoid intermediates, Still established the relative stereochemistry of pB and laid the groundwork for a powerful method of stereocontrol in organic synthesis based on conformational analysis of medium and large ring systems (macrocyclic stereocontrol).

With samples of pB made available through synthesis, breakthroughs in neurobiological studies of the pB olfactory response have been reported. For example, Sass (*14*) and Nishino and Manabe (*15*) have employed electrophysiological techniques to show that two separate receptors for pA and pB exist in the antennae of the male cockroach. Thus, a two-component attractant system has been developed by the American cockroach. These studies complement earlier work by M. Burrows and his co-workers, which resulted in the discovery of two sets of interneurons in the deutocerebrum of the male American cockroach brain that respond separately to either pA or pB (16). From the standpoint of the cockroach, pA and pB are not the same and both are important. The cockroach has developed two distinct systems to detect these pheromones and process the information they impart by separate paths. The reasons for this and the mechanism by which this is achieved remain a mystery.

Recent entomological studies have shown that pB exhibits sex attractant properties in addition to its previously recognized sex excitant properties (17, 18). The synthesis of the sex pheromones pA and pB and the design and synthesis of analogs could lead to further neurobiological studies and an understanding of the chemical basis for olfaction. A flexible synthesis might be modified to provide radioactively labeled pheromones or affinity-column analogs that could be employed in the isolation of the bioreceptor molecules. A greater understanding of the two-component attractant system might also lead to an optimal procedure for use of the sex pheromones as lures in baited traps.

[2 + 2] *Allene photocycloadditions.* Our studies of the synthesis of sex pheromones pA and pB have employed an allene photocycloaddition to a cyclohexenone as an entry point in each synthesis. These [2 + 2] photocycloaddition reactions illustrate the ability of this method to rapidly construct the carbon skeletons of both molecules. In this section an efficient synthesis of pB, which contains a ten-membered ring, will be detailed (19, 20).

The virtues of early and rapid skeletal construction in a synthesis have been appreciated by organic chemists for some time. In the periplanone B problem we recognized that regioselective photocycloaddition of an allene to a substituted cyclohexenone would set the stage for a rearrangement to form a ketone (called an oxy-Cope rearrangement) followed by a ring-opening rearrangement of the cyclobutene ring (scheme 1). The resultant product contains many structural similarities to the target system.

The ten-membered-ring ketone containing the *trans*-butadiene present in pB was constructed in five steps from the enone **5**. Four of the five steps employ light or heat as the reagent to mediate the transformations. The synthesis commenced with a [2 + 2] photocycloaddition of enone **5** and allene on a multigram scale. The regiochemical outcome of this reaction was anticipated on the basis of the pioneering studies of [2 + 2] photocycloadditions in the laboratories of Corey (21, 22). The reaction affords two stereoisomers (**6a** and **6b**) that need not be separated since the stereocenters at the ring-fusion carbons are returned to achiral trigonal centers in subsequent transformations and thus converge on the same end product. A Grignard addition of vinyl bromide to the ketone produced a divinylcyclohexanol **7**. Marvell and Whalley had first shown that compounds of this type serve as substrates for thermally induced oxy-Cope ([3.3] sigmatropic) rearrangements to provide ring expanded cyclodecenones (23). Presumably due to poor overlap of the orbitals in the 1,5-diene, **7** failed to undergo the desired transformation under thermolysis or metal-catalyzed (palladium acetate, mercuric trifluoroacetate) conditions. However, when the conditions

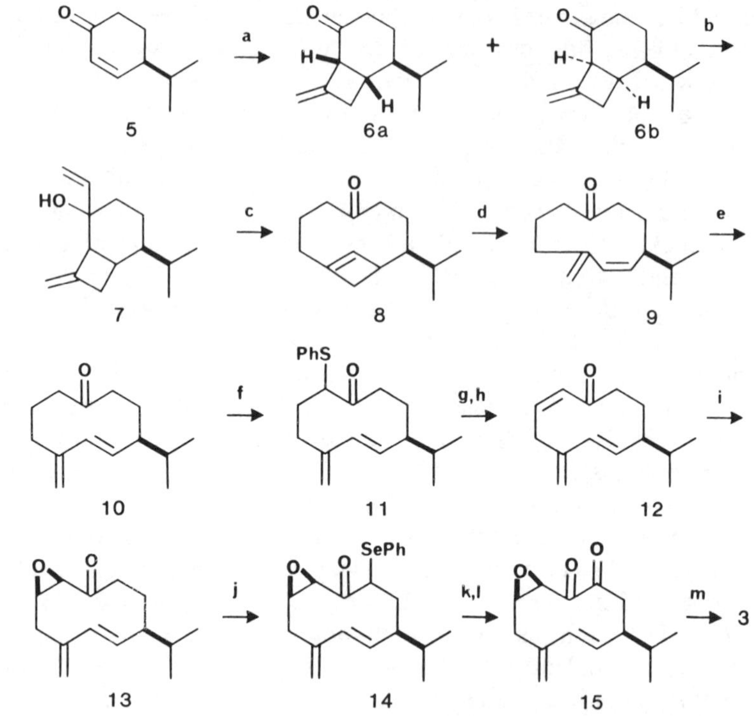

Scheme 1. The conditions and yields represented by the letters a to m are: (a) CH$_2$=C=CH$_2$, diethyl ether, hv (450-watt Hanovia lamp equipped with a Pyrex filter), −30°C, 72 percent yield (6a:6b = 2:1); (b) CH$_2$=CHMgBr, diethyl ether, −78°C, 63 percent yield; (c) KH (5 equivalents), 18-Crown-6, tetrahydrofuran, 60°C, 25 minutes, 100 percent yield; (d) toluene, 175°C, 20 hours, 77 percent yield; (e) benzene, hv (450-watt Hanovia lamp equipped with a Vycor filter), 82 percent yield; (f) tetrahydrofuran, LiN(Si(CH$_3$)$_3$)$_2$, −78°C, C$_6$H$_5$SSO$_2$C$_6$H$_5$, regioselectivity = 16:1; (g) aqueous methanol, NaIO$_4$, 71 percent yield from 7; (h) toluene, 110°C, 45 percent yield; (i) tetrahydrofuran, (CH$_3$)$_3$COOH, KH, 0°C, 83 percent yield (four to one mixture of β and α epoxides); (j) tetrahydrofuran, Li[N(i-C$_3$H$_7$)$_2$], −78°C, C$_6$H$_5$SeBr, 83 percent yield; (k) 30 percent H$_2$O$_2$, tetrahydrofuran, 97 percent yield; (l) tetrahydrofuran, acetic anhydride, CH$_3$CO$_2$Na, and then methanol, H$_2$O, K$_2$CO$_3$, 60 percent yield; (m) dimethyl sulfoxide, tetrahydrofuran, [(CH$_3$)$_3$S]I, (CH$_3$SOCH$_2$)Na, 62 percent yield.

of Evans and Golob were used for the anion-accelerated oxy-Cope rearrangement (24) a smooth transformation of 7 to the cyclobutene bridgehead olefin 8 took place. The product of this reaction is now a candidate for another thermally allowed rearrangement, the electrocyclic ring opening of a cyclobutene. Heating 8 at 175°C for 24 hours provided a 2:1 mixture of the *cis* and *trans* 1,3-diene isomers 9 and 10, which could be photoisomerized to a 15:1 mixture favoring the desired *trans* diene 10.

The photocycloaddition reaction that initiated the synthesis provided in one step a β,γ-unsaturated ketone that served as an ideal precursor to the divinylcyclohexanol required for ring expansion. The anion-accelerated oxy-Cope rearrangement produced the ten-membered ring with concomitant shift of the olefin into the four-membered ring. Here, the olefin was properly positioned for the electrocyclic ring opening that produced the required diene moiety. The cyclodecenone 10 was prepared in multigram quantities in five steps and converted to pB by a procedure that required eight additional steps. Interestingly, an electroantennagram bioassay indicated

10 was capable of initiating an electrical signal from a severed antenna of a male (but not female) American cockroach (25).

A solution to the problem of converting cyclodecenone 10 to pB was uncovered after we noted that enolization of the ketone and sulfenylative trapping of the resultant enolate could be made to proceed with regio- and stereocontrol. We believe the regiocontrol that is operating in the enolization reaction is a manifestation of what Still has termed macrocyclic stereocontrol (26). The formation of the enolate removes a serious transannular interaction between the methylene group adjacent to the ketone and an olefinic carbon of the diene.

The introduction of the phenylsulfenyl group provided a route to the cis enone 12 by pyrolytic sulfoxide elimination. Bis epoxidation of the cis enone occurred in a stereoselective manner analogous to that in the Still synthesis of pB (12). To complete the synthesis, the introduction of the requisite ketone functionality was achieved by means of a selena-Pummerer rearrangement.

Our studies of the chemistry and neurobiology of the American cockroach sex pheromones continue (25). We have prepared analogs of pB and are evaluating their activity. With these compounds in hand, the search for the pB receptor is under way. Studies in support of the synthesis of pA and isolation of the pA receptor are also in progress.

A Photochemical Equivalent to
Threo-Selective Aldol Condensation

The Paternò-Büchi photocycloaddition (7) of aldehydes to furan heterocycles was examined in detail by Sakurai and his co-workers and was shown to give rise to *exo*-substituted photoadducts with a high degree of regio- and stereocontrol (27, 28). We recognized that in this reaction the furan can be considered an equivalent to a Z-enolate and that the photoadduct serves as a type of protected aldol (29). Accordingly, the furan-carbonyl photocycloaddition is related to the stereoselective aldol condensation, a reaction of considerable importance in organic synthesis (30–34). Hydrolytic unmasking of the photoadduct yields aldol products with the *threo* configuration. A special feature of the photoadduct is that its cup-shaped geometry lends itself to a variety of reactions wherein electrophilic reagents add to the enol ether from the convex face. The resulting product can then be hydrolytically unmasked to afford an acyclic chain with a variety of substitution patterns. The identity of the substituents on the chain is determined by the selection of R groups on the furan and aldehyde and by the reaction conditions (scheme 2).

We have learned how to oxidize or reduce the dihydrofuran with stereocontrol, and we have been able to form carbon-carbon bonds at either the alpha or beta positions of the enol ethers. Having gained confidence in our ability to produce a variety of structural types, we have embarked on new studies with the objective of illustrating how this strategy can be applied to the synthesis of naturally occurring systems. Our approach to the synthesis of various target molecules using the photoaldol reaction was simplified by the information we received from other workers' studies of the application of the aldol reaction (or equivalent processes) to syntheses of chiral molecules (30–34). The following sections provide several illustrations.

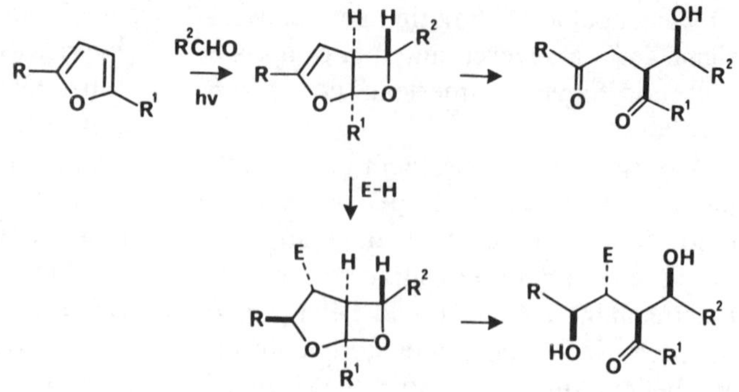

Scheme 2.

Antibiotics, antifungals, and myco-toxins. The growth of a mold colony from spores present in the air is not an uncommon occurrence. These molds, usually *Aspergillus* or *Penicillium* species, synthesize a variety of secondary metabolites that provide an effective means of fending off potential competitors from their food source. These metabolites have been classified by observing the type of organisms that respond to them in a deleterious manner. To drive away bacteria or other fungi the molds produce antibacterials or antifungals, and these compounds (when available from fermentation or synthesis) have been used by man to combat bacterial or fungal infections. These fungi are also capable of producing metabolic products, termed mycotoxins, which are toxic to mammals. As mycotoxin-producing molds tend to grow rapidly and aggressively, they can cause epidemic mycotoxicoses, several of which have had a massive impact on human communities (*35, 36*).

These compounds have for a long time attracted the attention of organic chemists. In some cases they may have important therapeutic value. A study of the relation between their structure and activity may contribute significantly to a mechanistic understanding of their mode of action. In the case of mycotoxins, this understanding might lead to a means of protection against their toxic effects.

In the following sections we discuss our studies of the furan-carbonyl photo-cycloaddition reaction as a method for stereocontrolled synthesis of two metabolites produced by *Aspergillus* species. Five-membered-ring heterocycles are a structural feature these two compounds share. A stereochemical feature they share is a *threo*-aldol embedded in their structures (or in the structures of logical synthetic precursors).

Avenaciolide: An antifungal metabolite from Aspergillus avenaceus. When grown in aqueous media *Aspergillus avenaceus* G. Smith produces a metabolite (avenaciolide, **16**) which possesses antifungal properties (*37–39*). At *p*H 3.5,

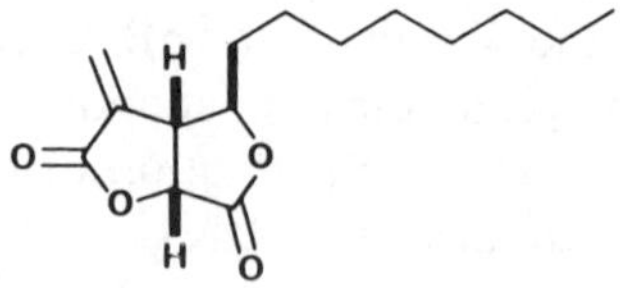

16

avenaciolide inhibits germination of spores of a wide range of fungi (for example, *Botrytis allii* and *Penicillium gladioli*) at concentrations of 1 to 10 μg/ml. This unusual bis-lactone is also weakly antibacterial.

354

Avenaciolide was first synthesized by Johnson in 1969 and has since been prepared by alternative routes by several research teams (40–48). We felt an efficient synthesis of avenaciolide would serve to illustrate an important feature of the furan-carbonyl photocycloaddition strategy (49): the photoaddition reaction can be easily carried out on a multigram scale with a high degree of stereocontrol so that its use as the first step in a synthesis can be recommended (scheme 3).

The photocycloaddition of nonanal to furan afforded a single photoadduct 17 in near quantitative yield on a 50-g scale. In this manner, two readily available starting materials were joined and two of the stereocenters in the target system were created. At this stage we required a method for the insertion of carbon monoxide into the acetal bond of the oxetane. Methods for such a transformation are known, especially those mediated by organometallic compounds of transition metals; however, we chose a slightly more circuitous but stereospecific method, as follows. Hydrogenation of 17 yielded 18, and then hydrolysis provided the *threo*-aldol 19, which exists as the corresponding butyrolactol. Comparison of 19 and 16 reveals the *threo*-aldol substructure of avenaciolide. The addition of vinyl magnesium bromide proceeded with 5:1 stereoselection to yield 20 and the corresponding allylic epimer, respectively. Although the stereochemistry at the allylic carbinol site of the major diastereomer is opposite to that required for the avenaciolide synthesis, we recog-

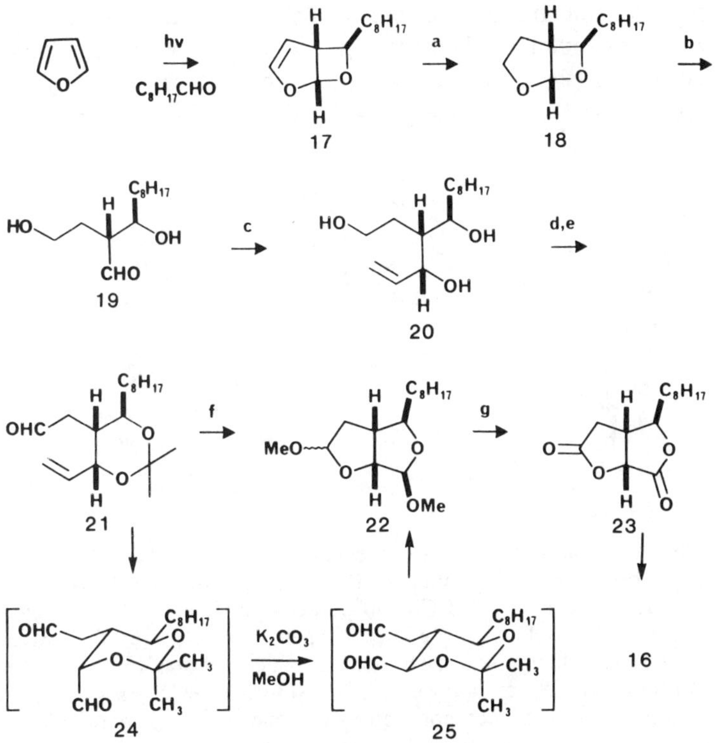

Scheme 3. The conditions and yields represented by the letters a to g are: (a) H_2, Rh-Al_2O_3, ethyl acetate, 97 percent yield; (b) $0.1N$ HCl and tetrahydrofuran (1:4), 96 percent yield: (c) CH_2=CHMgBr, tetrahydrofuran, room temperature, 80 percent yield; (d) acetone, $CuSO_4$, *p*-toluenesulfonic acid; (e) pyridinium chlorochromate, CH_3CO_2Na, CH_2Cl_2, 91 percent yield; (f) ozone, methanol, $-78°C$, and then $(CH_3)_2S$, room temperature, and then K_2CO_3, and then HCl, 31 percent yield; (g) *m*-chloroperbenzoic acid, $BF_3 \cdot O(C_2H_5)_2$, CH_2Cl_2, 80 percent yield.

355

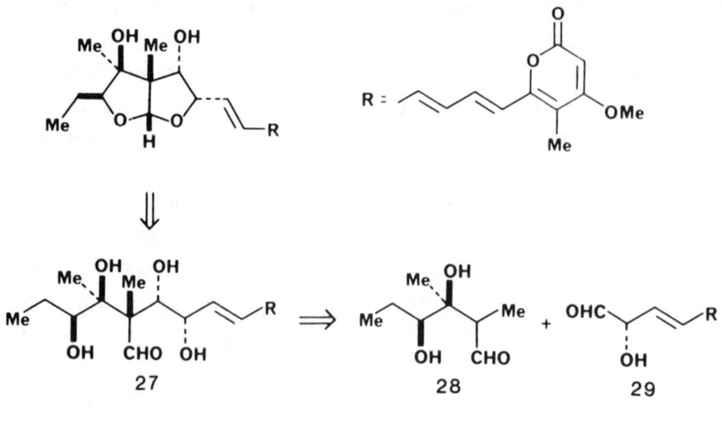

Scheme 4.

27   28   29

nized this could be corrected at a later stage.

We were able to engage the two secondary hydroxyls in the formation of an acetonide, leaving free the remaining primary (less hindered) hydroxyl. Oxidation to the aldehyde 21 set the stage for final skeletal construction and correction of the offending stereocenter. In a one-pot operation 21 was converted to the bismethoxy lactols labeled 22 (a 2:1 mixture of methoxy anomers) with complete inversion of configuration at the carbon bearing the allylic ether in 21. Ozonolysis and reduction of 21 gave rise to the axial aldehyde 24, which in a chair conformation suffers a severe 1,3-diaxial interaction with the axial methyl at the ketal carbon center. Due to the presence of the carbonyl group, this compound can transform into the all equatorially substituted isomer 25 upon base-cata-

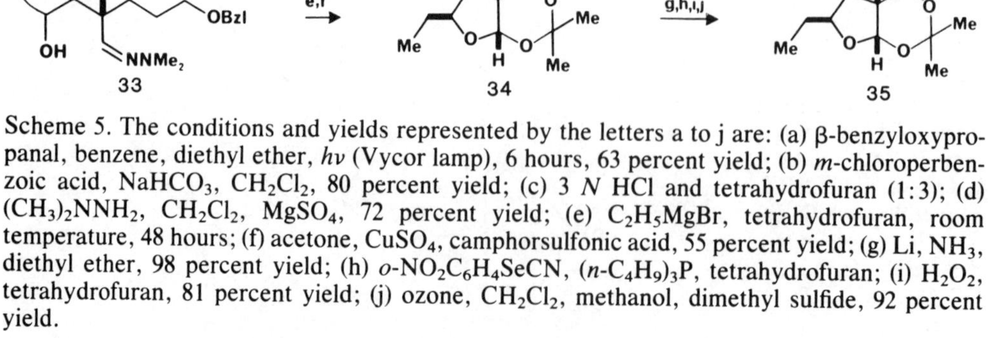

30   31   32

33   34   35

Scheme 5. The conditions and yields represented by the letters a to j are: (a) β-benzyloxypropanal, benzene, diethyl ether, hv (Vycor lamp), 6 hours, 63 percent yield; (b) m-chloroperbenzoic acid, NaHCO$_3$, CH$_2$Cl$_2$, 80 percent yield; (c) 3 N HCl and tetrahydrofuran (1:3); (d) (CH$_3$)$_2$NNH$_2$, CH$_2$Cl$_2$, MgSO$_4$, 72 percent yield; (e) C$_2$H$_5$MgBr, tetrahydrofuran, room temperature, 48 hours; (f) acetone, CuSO$_4$, camphorsulfonic acid, 55 percent yield; (g) Li, NH$_3$, diethyl ether, 98 percent yield; (h) o-NO$_2$C$_6$H$_4$SeCN, (n-C$_4$H$_9$)$_3$P, tetrahydrofuran; (i) H$_2$O$_2$, tetrahydrofuran, 81 percent yield; (j) ozone, CH$_2$Cl$_2$, methanol, dimethyl sulfide, 92 percent yield.

lyzed epimerization. Addition of potassium carbonate effected the desired transformation, and after equilibration was complete the mixture was acidified. These conditions resulted in the cleavage of the acetonide and formation of the skeleton of avenaciolide. This method of 1,3-stereochemical control was introduced by Stork and co-workers as part of their studies directed toward the synthesis of erythronolide A and was termed "ancillary stereocontrol" (50). A feature of this method is that the stereochemical outcome of the Grignard addition is inconsequential to the avenaciolide synthesis. Accordingly, the 5:1 mixture of stereoisomers produced in this step could be used, allowing greater material throughput.

Direct oxidation of **22** afforded the lactone **23** (15-g scale) which served as the penultimate compound in the Johnson synthesis of avenaciolide (41). Using Johnson's conditions for introduction of the α-methylene group in the A-ring butyrolactone, we were able to complete the synthesis (49).

Asteltoxin: A mycotoxin isolated from Aspergillus stellatus. Investigations of toxic maize cultures of *Aspergillus stellatus* resulted in the isolation and structural determination of a mycotoxin that was named asteltoxin, **26** (51). This mycotoxin is a member of a class of compounds distinguished by the presence of an α-pyronylpolyene side chain. Related

compounds such as the aurovertins and citreoviridin have been used extensively as inhibitors of oxidative phosphorylation. Recent studies have indicated that asteltoxin has a similar inhibitory effect on the adenosinetriphosphatase activity of *Escherichia coli* BF$_1$ (52). Extensive studies have been reported on the biological activity and biosynthesis of members of this class; however, before we initiated our studies we found no reports on the chemical synthesis of any of the members of this important class of compounds.

An analysis of the B-ring tetrahydrofuran of asteltoxin revealed the presence of a β-hydroxy acetal moiety. The corresponding aldehyde **27**, from an imagined hydrolysis of the bis(tetrahydrofuran), can be recognized as the product of a *threo*-aldol condensation of **28** and **29** or their equivalents in the indicated manner (scheme 4). Accordingly, our strategy was to carry out a first step *threo*-selective Paternò-Büchi photocycloaddition reaction and to use the cup-shaped geometry of the resultant molecule to facilitate introduction of the remaining peripheral stereocenters (53, 54).

Our synthesis began with a Paternò-Büchi photocycloaddition of 3,4-dimethylfuran and β-benzyloxypropanal on a 10-g scale (scheme 5). The *exo*-substituted photoadduct **31** was oxidized with *m*-chloroperbenzoic acid (MCPBA) to yield the β-hydroxytetrahydrofuran **32**. Hydrolysis afforded the desired *threo*-aldol, which exists as the monocyclic lactol. Protection of the more reactive free aldehyde produced the hydrazone **33**. Then, we were able to use the α-hydroxyl substituent to direct introduction of the ethyl side chain to the hemiacetal carbon. Chelation-controlled addition of ethyl magnesium bromide provided, after hy-

26

357

drolysis and internal protection, the acetonide **34**. Degradation of the β-benzyloxyethyl side chain to an aldehyde set the stage for final introduction of the pyronyltriene side chain.

Completion of the synthesis required the stereoselective addition of a 4-formyl-1,3-butadiene anion equivalent and a final crossed-aldol condensation-dehydration reaction with a suitable α-pyrone. These objectives were achieved by a process that is outlined in scheme 6. Corey's anion **36** (*55*) reacted smoothly with the aldehyde **35** to provide a 3:1

mixture of primary addition products. When the products stood at room temperature for several hours, a double [2.3] sigmatropic rearrangement took place, resulting in the formation of the pentadienyl sulfoxide **37** as the major stereoisomer. A Pummerer rearrangement and hydrolysis produced the dienal **39**, which could be coupled with the indicated pyrone by the crossed-aldol reaction to give **40**. Selective dehydration of the secondary hydroxyl on the side chain in the presence of the more hindered, unprotected secondary and tertiary hy-

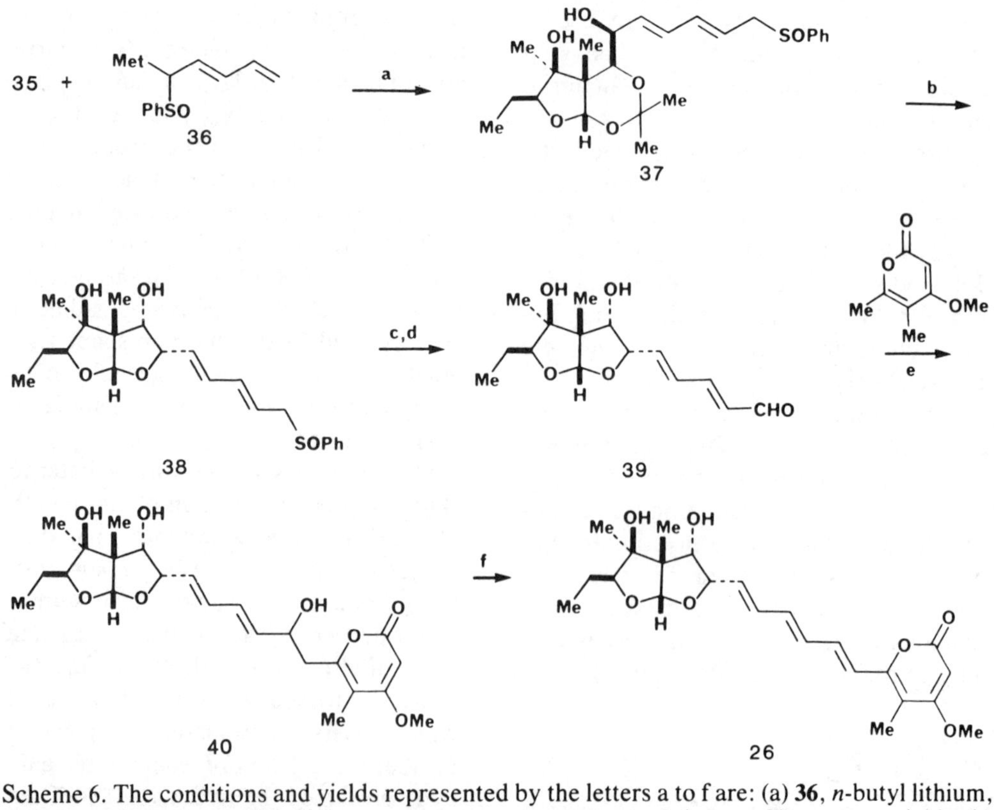

Scheme 6. The conditions and yields represented by the letters a to f are: (a) **36**, *n*-butyl lithium, −78°C, and then **35**, NH$_4$Cl in water, room temperature, 3 hours, 89 percent yield; (b) camphorsulfonic acid, CH$_2$Cl$_2$, 77 percent yield; (c) CF$_3$CO$_2$COCH$_3$, acetic anhydride, 2,6-lutidine; (d) HgCl$_2$, CaCO$_3$, CH$_3$CN and water (4:1), 60 percent yield; (e) pyrone, Li[N(*i*-C$_3$H$_7$)$_2$], hexamethylphosphoramide, tetrahydrofuran, −78°C, and then **39**, −78°C, 80 percent yield; (f) *p*-toluenesulfonyl chloride, dimethylaminopyridine, (C$_2$H$_5$)$_3$N, CH$_2$Cl$_2$, 82 percent yield.

droxyls on the ring system completed the synthesis of asteltoxin, **26**.

When compared to an authentic sample of natural asteltoxin, our synthetic material exhibits identical spectroscopic and chromatographic properties. Our material differs from natural asteltoxin in that it exists as a racemic mixture. We are presently studying a new synthetic route to asteltoxin that would result in the nonracemic natural product. However, the absolute configuration of asteltoxin, which had not been determined, was a problem that had to be dealt with in the early part of these studies. Our experiments that relate to this point are outlined in scheme 7.

Whereas the furan-carbonyl photocycloaddition results in a highly diastereoselective *threo*-aldol addition, the reaction proceeds with lack of facial selectivity for the chiral aldehydes that we have examined. For example, irradiation of R-glyceraldehyde acetal (readily available from D-mannitol) in the presence of 3,4-dimethylfuran afforded a nearly 1:1 mixture of **41** and the *threo* adduct resulting from opposite facial selectivity. These two compounds can be separated by silica gel chromatography. When **41** was treated with MCPBA in aqueous tetrahydrofuran, oxidation and ring opening ensued and the lactol **42** was isolated. Protection of the free aldehyde and chelation-controlled addition of ethyl magnesium bromide to the lactol proceeded with the same stereocontrol observed in the racemic synthesis. Treatment with acid effected the removal of the acetonide and caused cyclization to the bis(tetrahydrofuran) **43**. Ozonolytic degradation of natural asteltoxin and subsequent reduction with sodium borohydride afforded the same triol **43**. Comparison of the optical rotation values of our synthetic substance with the degradation product showed that these substances have the same absolute configuration (*56*). These results indicate the absolute configuration of asteltoxin is as shown in scheme 7.

We are presently studying a new method for the attachment of the α-pyronyltriene side chain to the synthetic intermediate **43**. In this manner we hope to complete a new synthesis of nonracemic asteltoxin in half the number of steps required in the racemic synthesis.

Mechanism of the furan-carbonyl photocycloaddition reaction. The lack of fa-

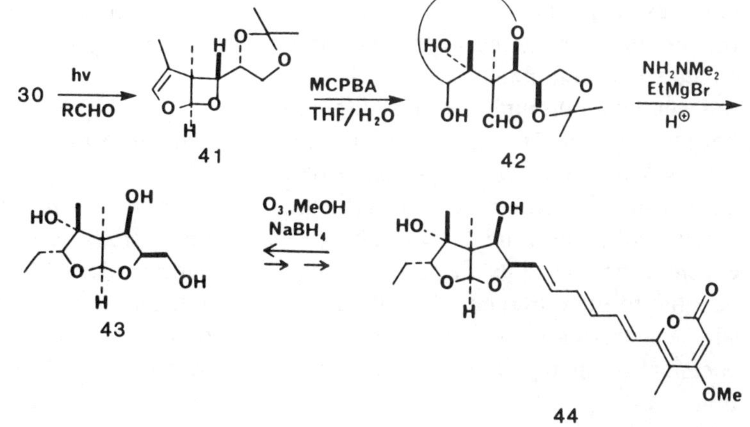

Scheme 7.

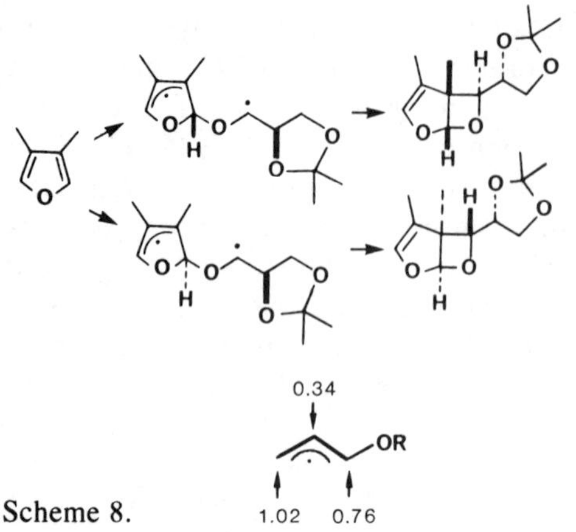

Scheme 8.

cial selectivity in the addition of the furan to an excited state of a chiral aldehyde is in contrast to the normal aldol reaction (30–34). This feature of the photoreaction suggests the operation of a mechanism that is insensitive to the substitution pattern of the chiral aldehyde. One such mechanism is depicted in scheme 8. Reaction between an excited aldehyde (singlet or triplet state) and the furan proceeds with initial carbon-oxygen bond formation to produce either of the two diradical species shown. A stereocenter adjacent to the carbonyl is now in a 1,4-relationship to the newly formed stereocenter at the acetal carbon and is expected to exert little influence as a stereocontrol device. Two stereocenters are present in these intermediates, but the stereocenter at the acetal carbon is expected to completely dictate the stereochemical outcome of the diradical bond formation. In each case ring closure will produce a *cis* ring fusion with an *exo*-substituted side chain, as in the reaction of achiral aldehydes. The stereocenter on the side chain is inconsequential to the outcome of diradical closure and has minimal influence, stereochemically, on the initial acetal formation.

The regiochemical alternative of diradical ring closure would provide the product of a [4 + 2] cycloaddition. From our proposed diradical intermediate there might appear to be little impediment to such a process since the product, as well as the corresponding transition state, is expected to contain less total strain energy. We believe ring closure in the observed sense (to provide the [2 + 2] adduct) is favored by virtue of the greater free valency at C_3 of the 1-oxyallyl radical. For example, a simple Hückel molecular orbital calculation of 1-methoxyallyl radical 44 provides the values of free valence index at the three carbon centers indicated in scheme 8 (57). This calculation indicates that bond formation with a carbon radical would be favored at the C_3 position, since bonding at C_1 would involve a greater loss of the π-bonding character present in the oxyallyl radical intermediate.

Our hypothesis for the mechanism of this reaction should be considered tentative. Nevertheless, it serves as a working model. The mechanism can be tested by examining the effect of furan substituents on the mode of selectivity (whether [2 + 2] or [4 + 2] addition is favored) and by examining the relative efficiency of various intramolecular cycloadditions. These experiments are ongoing in our laboratories.

Current Studies

The previous two examples illustrate the application of the Paternò-Büchi photocycloaddition of a furan and an aldehyde to the synthesis of compounds that contain five-membered-ring oxygen heterocycles. We are currently involved in several studies that employ this reac-

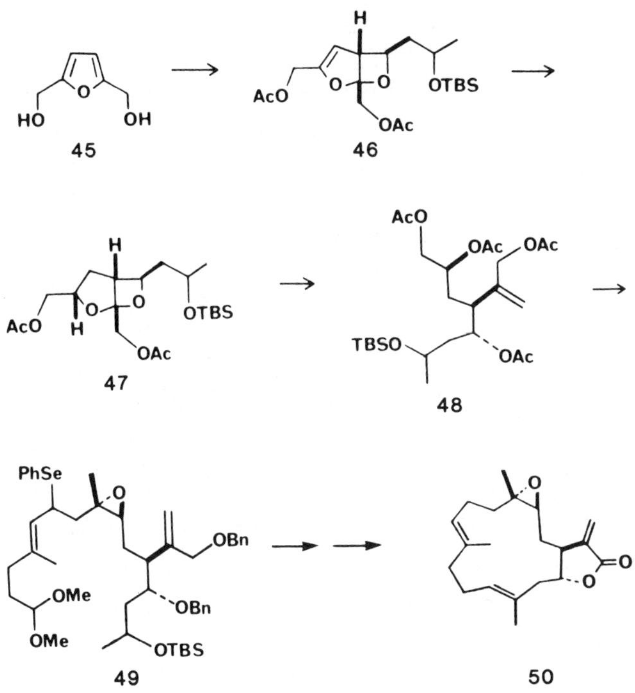

Scheme 9.

tion as an entry point for the synthesis of other types of compounds. For example, significant progress has been made in our studies directed towards the synthesis of members of the cembranolide class of diterpenes, which contain 14-membered rings. These compounds are represented by the structures of the antileukemic substances crassin acetate and isolobophytolide (58). A brief description of our studies in this area is given in scheme 9.

The acyclic chain **48** can be prepared in six steps from 2,5-bis(hydroxymethyl)furan, **45**. A photocycloaddition of this furan and β-*t*-butyldimethylsiloxybutanal on a 25-g scale provided, after acetylation, the photoadduct **46**. Hydrogenation from the convex face proceeded with complete stereocontrol to produce **47**. Hydrolysis, acetylation, and Wittig methylenation afforded **48**, which contains three of the four stereocenters

present in the target molecule, isolobophytolide. The *seco* derivative **49**, which contains all the carbons and stereocenters present in isolobophytolide, **50**, has been prepared in 14 steps from **48** (59). We are currently investigating methods for closure of the 14-membered ring and completion of the cembranolide synthesis.

In this brief account of photochemically mediated [2 + 2] cycloadditions in chiral molecule synthesis one strategy has been revealed that receives growing popularity. These reactions provide four-membered ring adducts that can be transformed in a controlled manner to advanced synthetic intermediates with skeletal and stereochemical complexities. As entry points in challenging syntheses, these reactions present the possibility for concise and efficient solutions to a variety of synthesis problems.

References and Notes

1. G. Ciamician and P. Silber, *Chem. Ber.* **41**, 1928 (1908).
2. E. Paternò and G. Chieffi, *Gazz. Chim. Ital.* **39**, 341 (1909). This reaction, now referred to as the Paternò-Büchi reaction, went largely unnoticed until G. Büchi and co-workers were able to confirm the oxetane structure as part of their investigations [G. Büchi, C. G. Inman, E. S. Lipinsky, *J. Am. Chem. Soc.* **76**, 4327 (1954)].
3. H. A. J. Carless and G. K. Fekarurhobo, *J. Chem. Soc. Chem. Commun.* (1984), p. 667.
4. E. J. Corey, R. B. Mitra, H. Uda, *J. Am. Chem. Soc.* **86**, 485 (1964).
5. S. W. Baldwin, in *Organic Photochemistry*, A. Padwa, Ed. (Dekker, New York, 1981), vol. 5, p. 123.
6. W. Oppolzer, *Acc. Chem. Res.* **15**, 135 (1982).
7. G. Jones, II, in *Organic Photochemistry*, A. Padwa, Ed. (Dekker, New York, 1981), vol. 5, p. 1.
8. L. M. Roth and E. R. Willis, *Am. Midl. Nat.* **47**, 66 (1952).
9. C. J. Persoons *et al.*, *Tetrahedron Lett.* (1976), p. 2055.
10. E. Talman, J. Verwiel, F. J. Ritter, C. J. Persoons, *Isr. J. Chem.* **17**, 227 (1978).
11. C. J. Persoons, P. E. J. Verweil, E. Talman, F. J. Ritter, *J. Chem. Ecol.* **5**, 219 (1979).
12. W. C. Still, *J. Am. Chem. Soc.* **101**, 2493 (1979).
13. M. A. Adams *et al.*, *ibid.*, p. 2495.
14. H. Sass, *J. Comp. Physiol.* **152**, 309 (1983).
15. C. Nishino and S. Manabe, *Experientia* **39**, 1340 (1983).
16. M. Burrows, J. Boeckh, J. Esslen, *J. Comp. Physiol.* **145**, 447 (1982).
17. Y. S. Chow and S. F. Wang, *J. Chem. Ecol.* **7**, 265 (1981).
18. T. R. Tobin, G. Seelinger, W. J. Bell, *ibid.*, p. 969.
19. S. L. Schreiber and C. Santini, *Tetrahedron Lett.* (1981), p. 4651.
20. S. L. Schreiber and C. Santini, *J. Am. Chem. Soc.* **106**, 4038 (1984).
21. E. J. Corey, J. D. Bass, R. Mahieu, R. B. Mitra, *ibid.* **86**, 5570 (1964).
22. P. E. Eaton, *Tetrahedron Lett.* (1964), p. 3695.
23. E. N. Marvell and W. Whalley, *ibid.* (1970), p. 509.
24. D. A. Evans and A. M. Golob, *J. Am. Chem. Soc.* **97**, 4765 (1975).
25. M. Lerner, personal communication.
26. W. C. Still and V. J. Novack, *J. Am. Chem. Soc.* **106**, 1148 (1984).
27. S. Toki, K. Shima, H. Sakurai, *Bull. Chem. Soc. Jpn.* **38**, 760 (1965).
28. K. Shima and H. Sakurai, *ibid.* **39**, 1806 (1966).
29. S. L. Schreiber, A. H. Hoveyda, H-J. Wu, *J. Am. Chem. Soc.* **105**, 660 (1983).
30. D. A. Evans, J. V. Nelson, T. R. Taber, *Top. Stereochem.* **13**, 1 (1982).
31. T. Mukaiyama, *Org. React.* **28**, 203 (1982).
32. C. H. Heathcock, *Science* **214**, 395 (1982).
33. _____, in *Comprehensive Carbanion Chemis-
try*, T. Durst and E. Buncel, Eds. (Elsevier, Amsterdam, 1983), vol. 2.
34. _____, in *Asymmetric Synthesis*, J. D. Morrison, Ed. (Academic Press, Orlando, Fla., 1984), vol. 3, p. 111.
35. J. G. Heathcote, M. O. Moss, D. H. Watson, W. H. Butler, J. E. Smith, *Chem. Ind. (London)* **15**, 530 (1984).
36. B. Franck, *Angew. Chem., Int. Ed. Engl.* **23**, 493 (1984).
37. D. Brookes, B. K. Tidd, W. B. Turner, *J. Chem. Soc.* (1963), p. 5385.
38. J. J. Ellis, F. H. Stodola, R. F. Vesonder, C. A. Glass, *Nature (London)* **203**, 1382 (1964).
39. D. Brookes, S. Sternhell, B. K. Tidd, W. B. Turner, *Aust. J. Chem.* **18**, 373 (1967).
40. W. L. Parker and F. Johnson, *J. Am. Chem. Soc.* **91**, 7208 (1969).
41. W. L. Parker and F. Johnson, *J. Org. Chem.* **38**, 2489 (1973).
42. J. L. Herrmann, M. H. Berger, R. H. Schlessinger, *J. Am. Chem. Soc.* **95**, 7923 (1973).
43. J. L. Herrmann, M. H. Berger, R. H. Schlessinger, *ibid.* **101**, 1544 (1979).
44. R. C. Anderson and B. Fraser-Reid, *ibid.* **97**, 3870 (1975).
45. H. Ohrui and S. Emoto, *Tetrahedron Lett.* (1975), p. 3675.
46. T. Sakai, H. Horikawa, A. Takeda, *J. Org. Chem.* **45**, 2039 (1980).
47. A. Murai, K. Takahashi, H. Takatsura, T. Masamune, *J. Chem. Soc., Chem. Commun.* (1981), p. 221.
48. F. Kido, Y. Tooyama, Y. Noda, A. Yoshikoshi, *Chem. Lett.* (1983), p. 881.
49. S. L. Schreiber and A. H. Hoveyda, *J. Am. Chem. Soc.* **106**, 7200 (1984).
50. G. Stork, I. Paterson, K. C. F. Lee, *ibid.* **104**, 4868 (1982).
51. G. J. Kruger, P. S. Steyn, R. Vleggaar, *J. Chem. Soc. Chem. Commun.* (1979), p. 441.
52. M. Satre, *Biochem. Biophys. Res. Commun.* **100**, 267 (1981).
53. S. L. Schreiber and K. Satake, *J. Am. Chem. Soc.* **105**, 6723 (1983).
54. _____, *ibid.* **106**, 4186 (1984).
55. E. J. Corey and D. J. Hoover, *Tetrahedron Lett.* (1982), p. 3463.
56. S. L. Schreiber and K. Satake, in preparation.
57. The free valence index provides an indication of the amount of π-bonding that is available at each of the atoms of the π-system.
58. For a stereochemical study of these two compounds, see J. A. Marshall, L. J. Karas, M. J. Coghlan, *J. Org. Chem.* **47**, 699 (1982).
59. S. L. Schreiber and A. H. Hoveyda, in preparation.
60. I thank C. Santini, K. Satake, and A. H. Hoveyda whose work made this account possible. I also thank my colleagues, especially J. Berson, M. R. Lerner, and W. E. Crowe, for their insight and exchange of information. This research was supported by the Institute for General Medical Sciences of the National Institutes of Health (GM-32527).

During the past 15 years, organosilicon chemistry has developed to the point where it is now a highly valued tool for the preparative organic chemist (*1–3*). From its early concerns primarily with silicone polymer fabrication (*4*), stereochemical phenomena (*5*), and functionalization of polar substances to enable gas chromatographic analysis (*6*), the field of organosilicon research has attained an increasingly important position in organic synthesis. The vigorous growth of this field has been due to the ever-increasing need for methods of forming carbon-carbon bonds under relatively controlled cationic and anionic conditions. Because it is the mildest metal, silicon fulfills this role and allows also for the implementation of additional reaction types. Furthermore, by the proper choice of reagents it is often possible to attain regioselectivity or stereoselectivity, that is, to obtain predominantly one positional or spatial isomer.

Some appreciation of the synthetic potential of silicon can be gained by comparing its properties with those of carbon, which is positioned above it in the periodic table. Chiefly because silicon is larger (atomic radius, 1.17 angstroms) and more electropositive (electronegativity, 1.74) than carbon (0.77 Å, 2.50), and possibly because silicon has empty $3d$ orbitals, Si–C bonds are weaker than C–C bonds (75 versus 83 kilocalories per mole, respectively). In contrast, single bonds from silicon to electronegative atoms such as oxygen (108 kcal/mole) and fluorine (129 kcal/mole) are notably strong. Consequently, while fluorine ion shows poor nucleophilicity for carbon, it readily attacks silicon.

In keeping with the position of silicon as a second-row element, bimolecular nucleophilic substitution in compounds

25

Silicon-Mediated Organic Synthesis
Leo A. Paquette

Science 217, 793–800 (27 August 1982)

of the type R$_3$SiX (where R is an aryl or alkyl group and X is the leaving group) occurs much more readily than in their carbon counterparts. This heightened reactivity carries over to carbon-functional silanes, where attack by oxygen and halogen nucleophiles is generally faster at silicon than at hydrogen. Also to be appreciated is the effect of silicon's greater electropositive character, which causes the sigma orbital associated with an Si–C bond to have a higher coefficient on carbon. Exceptional stabilization is therefore provided by overlap of such orbitals with the empty orbital of an adjacent carbocation center when an in-plane arrangement is adopted as in **1** (*7*). A carbon-metal (C–M) bond in the α position (**2**) is also favorably disposed for stabilization as the result of energy-lowering overlap of the filled C–M orbital with the *d* orbitals on silicon or the antibonding orbitals of the methyl-silicon (Me–Si) bond, both of which are empty (*8*). For related reasons, α-silyl cations of type **3** are appreciably destabilized despite an offsetting inductive effect (*9*).

These reactivity patterns, coupled with the low steric requirements of the trimethylsilyl group and the ability of Si–H bonds to add across olefinic and acetylenic linkages when appropriately catalyzed (hydrosilation), are the basis for the contributions of organosilicon chemistry to organic synthesis. The challenge has been to deploy the special chemical properties of silicon to achieve transformations heretofore more cumbersome or impossible to realize. In the following sections suitable examples of the power of this methodology are discussed.

Electrophilic Chemistry

The activation of carbonyl groups through conversion to silyl enol ethers has become a much utilized technique. Conditions have been developed to enable trapping of both kinetic and thermodynamic enolate anions such as **4** and **5** (DMF = dimethyl formamide) (*10*).

O-Silylation almost always predominates heavily (*11*). Regio-specificity in the generation of silyl enol ethers can also be attained by 1,4 addition to conjugated enones (*10, 12*), enone reduction by lithium in liquid ammonia (*13*), and reductive silylation of α-halo ketones (*14*); examples of these approaches are shown in the next group of reactions.

Although isomerically pure trimethylsilyl enol ethers can be transformed cleanly into structurally well-defined lithium enolates by reaction with methyllithium (for instance, in the replacement of Me$_3$Si in **6** as **6** → **7**) (*10, 15*), they are more often used directly with reagents that have reasonable electrophilicity. A valid assumption appears to be that silyl enol ethers have reactivities comparable to those of their derived enols. The key advantage to be gained is that one no longer deals with the complications which arise from keto-enol equilibria. A wider range of electrophilic processes is consequently made available; the formation of **8**, **9**, and **10** is exemplary (*16*). The facile generation of a quaternary center, as in **8**, is particularly noteworthy.

Upon thermal activation, silyl enol ethers can enter into vinylcyclopropane rearrangement, providing a useful means for cyclopentanone annulation, as in the formation of **14** (*20*). Through stereospecific Claisen rearrangement, olefinic acids have been produced which serve as precursors of insect pheromones, the antibiotic botryodiplodin (**15**), and other types of natural products (*21*).

The double bond of silyl enol ethers can also be used advantageously. Following cyclopropanation, as in **11**, ring opening can be achieved in acid or base, or by using various electrophilic agents to achieve selective monoalkylation (*17*). Smooth ring expansions can be achieved, as in **12**, by halocarbene additions (*18*). The preparation of muscone (**13**) by this method is an example of a protocol by which an inexpensive starting material can be conveniently transformed into a valued commodity (*19*).

While silyl enol ethers are usually inert toward Grignard reagents, coupling can be readily effected by the addition of a nickel catalyst (*22*). The technique represents a powerful cross-coupling method for olefin synthesis (*23*).

Aryl- and vinyl-silicon bonds are cleaved with unusual facility by electrophilic reagents because Si–C bonds stabilize a β-carbocation (see **1**) better than do C–H and C–C bonds. Attack by the electrophile (E$^+$) is consequently most often directed to the silicon-bearing carbon, as in **16**. With loss of the silyl

functionality, a product is formed with net substitution by the electrophile. This type of intermolecular reaction has been widely employed; some examples from the area of natural products chemistry include the formation of isoegomaketone (**17**) (*24*), nuciferal (**18**) (*25*), and estra(10)-trien-17-one (**19**) (*26*).

A highly diastereoselective condensation reaction occurs when the vinylsilane also forms part of an allylic boronate moiety, such as **20** (*27*). Controlled deoxysilylation of the resulting *threo* hydroxy silanes (note preferential loss of the boron functionality) provides a convenient synthesis of either *Z* or *E* terminal dienes.

Still more impressive are the recent advances in organic synthesis which have as their basis intramolecular variants of this concept. The vinylsilane-mediated spiroannulation sequence which effects facile overall conversion of **21** to spirodienone **22** (*28*) and the iminium ion-vinylsilane cyclization **23** to **24** which stereospecifically assembles the (*Z*)-6-alkylidene-indolizidine ring system known as dendrobatid toxin 251D are particularly noteworthy (*29*). Where silylacetylenes are concerned, free radical conditions can be applied to achieve addition across the triple bond. In the cyclization of **25**, the key step in a new synthesis of β-agarofuran (**26**), kinetically favored six-membered ring formation controls the regio-chemistry of C–C bond formation (*30*).

Like vinylsilanes, alkynylsilanes generally react with electrophiles to give substitution products (*31*). When alkynylsilanes are used as terminator groups in polyolefin cyclization, stabilization provided by the C–Si bond in **27** gives rise to a regioselectivity opposite to that provided by alkyl substituents (*32*). In intramolecular cyclization reactions, addition rather than substitution will be forced on the alkynylsilane if the ring being formed is small (compare **25** → **26**). When factors of this type are not at issue, normal reactivity returns (*33*).

The bifunctional acetylene **28** reacts with electrophiles in the sense shown in intermediate **29** (*34*). This reactivity pat-

tern indicates that cationic stability as in **1** is more favorable than that found, for example, in **27**. An identical pattern of behavior is followed by allylsilanes, loss of the silicon substituent occurring rapidly. As suggested by examples **30** to **33**, the range of electrophiles toward which allylsilanes are reactive is large (*35*). As a rule, allylsilanes are somewhat more nucleophilic than the corresponding hydrocarbons, while vinylsilanes usually show comparable reactivity.

As a consequence of their chemical versatility, allylsilanes have been applied in a broad spectrum of imaginative synthetic protocols. In the methylenecyclobutane **34**, reaction with electrophiles proceeds exclusively in the expected way with allyl shift to give a 1-substituted cyclobutene. Subsequent thermal ring opening provides a functionalized isoprene; the terpenoid tagetol (**35**) is an example (*36*). Comparable mechanistic considerations lead one to view silane **36** as an immediate precursor of 2,3-disubstituted butadienes. Because the reactivity of **36** is comparable to that of simpler allylsilanes, an efficient and simple route has been opened to such compounds as **37** (*37*). The in situ Claisen rearrangement which transforms enol ethers into 2-substituted 3-trimethylsilyl enones— for instance, **38** and **40**—provides a new and richly variable technique for the synthesis of annulated (**39**) and spirocyclic (**41**) systems (*38*). In one of the more elegant applications of this chemistry, disilane **42** was prepared by [2 + 2] photocyclization and treated with stannic chloride. Following Lewis acid-promoted elimination of one silyl substituent, which yields an allylsilane, cyclization to tricyclic hydroxy ketone **43** takes place (*39*).

367

As foreshadowed by **28**, propargylsilanes are readily transformed to allenes by electrophilic reagents. The chemical reactivity of **44** and **45** is prototypical (*40*). Allenylsilanes experience electrophilic attack regiospecifically at C-3, initially providing a vinyl cation. This species may lose the silyl substituent to yield an acetylene (**46**), undergo intramolecular capture with retention of the silane residue (**47**), or be sufficiently stabilized by interaction with the adjacent C-Si bond to be subject to 1,2-silyl migration (**48**) (*41*).

Although silylcyclopropanes show appreciable stability in the presence of a variety of chemical reagents, they are known to be subject to electrophilic substitution (*42*) and ring opening (*43*). Recent studies have shown that α-trimethylsilylcyclopropylcarbinyl cations, such as **49**, are unwilling to undergo elimination to the methylenecyclopropane (*44*).

The pyrolysis of structurally related vinylcyclopropanes does, however, lead smoothly to annulated vinylsilanes (**50**), which can be transformed under electrophilic conditions to functionalized cyclopentenes (**51**) (*45*).

Open-chain epoxysilanes are hydrolyzed in aqueous acid to aldehydes and ketones. When used in this manner, a vinylsilane can conveniently serve as a masked carbonyl group (*46*). In these reactions the carbonyl group appears on the carbon atom that originally carried the silicon substituent. This process can be cleanly reversed in cyclic epoxysilanes. A useful method for achieving 1,2-

carbonyl migration—for instance, **52** → **53**—has stemmed from this chemistry (*47*). Singlet oxygen can act on vinylsilanes with similar results (*48*).

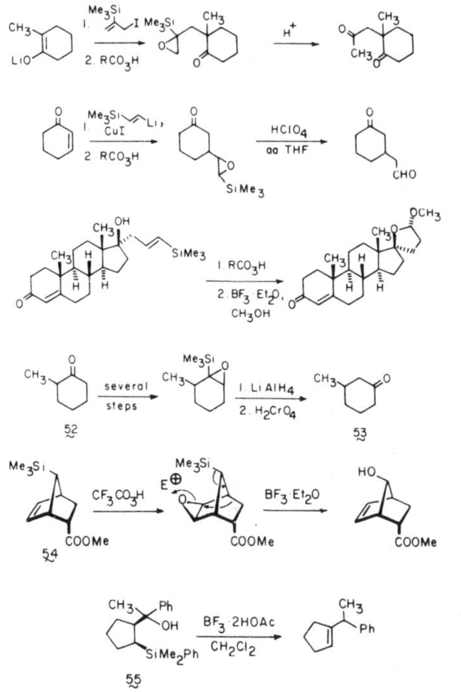

Finally, it is important to recognize that silicon substituents can direct the course of carbonium ion rearrangements from distal positions (*49*). The manner in which a silyl group can control the outcome of these reactions is illustrated for two examples (**54** and **55**).

Nucleophilic Chemistry

As noted earlier, silicon is able to stabilize adjacent carbon-metal bonds even though it is more electropositive than carbon. For this reason, formation of the Grignard reagent **56** (*50*) and α-deprotonation of a host of functionalized

silanes to produce α-lithiosilanes are now commonplace (*51*). Among reagents of this class, **57** and **58** have so far commanded the greatest attention because of their usefulness as nucleophilic carboxaldehyde and methyl ketone equivalents (*52*). Since allylsilanes are easily metallated, their anions have also found broad use in synthesis (*53*).

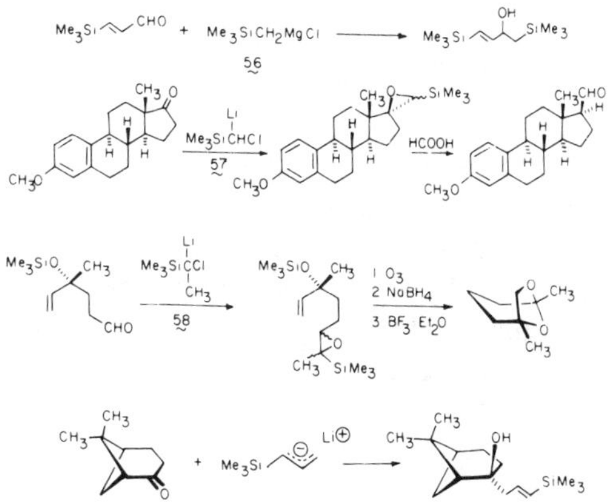

The ease with which (trimethylsilyl) diazomethane is deprotonated is noteworthy since the resulting carbonyl addition products are transformed readily into epoxysilanes (*54*). Alkyllithium reagents also remove the proton α to silicon in epoxysilanes and provide for convenient nucleophilic transfer of this heterocyclic moiety (see **59**) (*55*). Triphenylvinylsilane and related compounds enter into 1,2 addition with alkyllithiums to generate α-lithio-α-silyl nucleophiles (*56*). The utility of silylated vinyl ketones (**60**) in Robinson annulation reactions has been demonstrated. The trialkylsilyl group serves to retard enone polymerization and allow use of these building blocks under aprotic conditions with regiospecifically generated enolates (*10, 13, 57*).

369

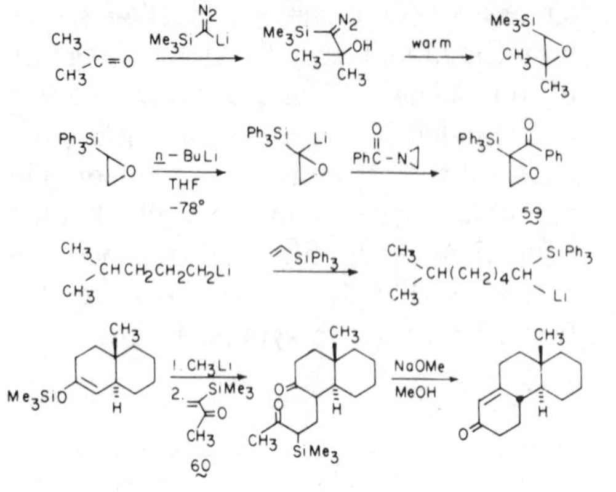

rangement, and hydrolysis (*59*). This methodology and anionic sigmatropy were elegantly coupled in the formation of aldehyde **63** from **62** (*60*). The silicon-substituted ylide **64** undergoes conjugate addition to enones to give silylcyclopropanes (*61*). The organometallic produced by reaction of (trimethylsilyl)propargyl bromide with aluminum amalgam condenses with carbonyl compounds to give allenic products **65** (*62*).

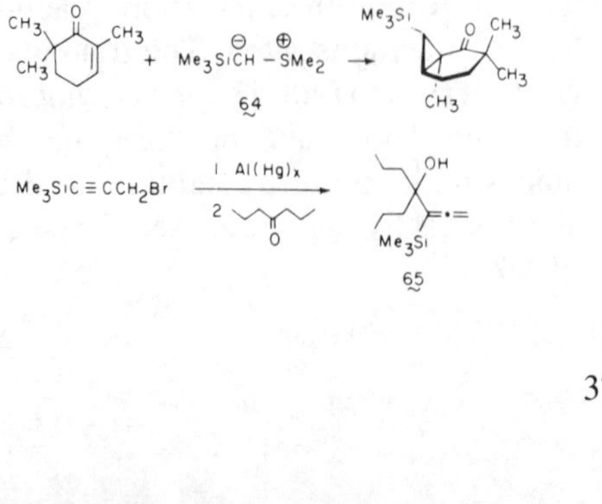

Carbenoid Chemistry

Cuprous chloride–promoted decomposition of (trimethylsilyl)diazomethane in olefins produces silylcyclopropanes with retention of alkene stereochemistry (*63*). More recently, the combination of lithium 2,2,6,6-tetramethylpiperidide and (chloromethyl)trimethylsilane has evolved as a functionally similar (*64*) but not identical (*43*) carbene source. Other known routes to silylcarbenes involve the decomposition of organomercury reagents such as $(Me_3SiCCl_2)Hg$ (*65*) and insertion of atomic carbon into Si–H bonds (*66*). Detailed studies of (trimethylsilyl)(carboethoxy)carbene (*67*), (trimethylsilyl)phenylcarbene (*68*), (trimethylsilyl)carbene (*69*), and bis(trimethylsilyl)carbene (*70*) have also been reported. The pyrolysis of 1-(trimethylsilyl)-1-alkanols (such as **66**) as a method of carbene formation holds considerable promise (*71*).

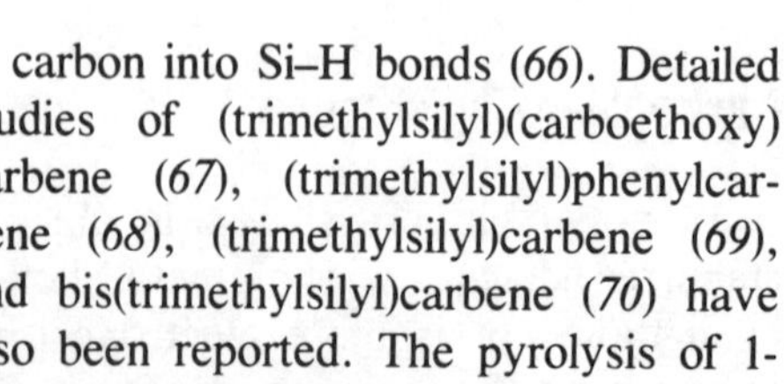

High levels of 1,2-asymmetric induction are attained when methyllithium is added to a vinylsilane which is further activated by a phenylsulfonyl group and has an allylic alkoxide as an existing chiral center (*58*).

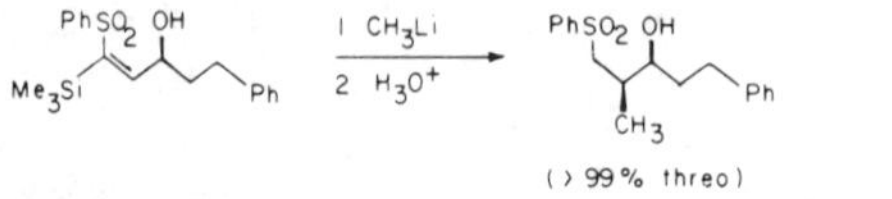

The ease with which **61** can be deprotonated and alkylated is valued because the adducts are convertible to aldehydes by sequential oxidation, thermal rear-

Elimination Reactions

Although alkene-forming 1,2-elimination reactions of β-functional organosilanes have been known for more than three decades, applications to organic synthesis have appeared only recently. The rapid development of this area appears to be due principally to (i) the demonstration that these eliminations can proceed under very mild conditions with fluoride ion (72, 73); (ii) the discovery that β-hydroxyalkylsilanes undergo elimination under both acidic and basic conditions (74), most notably with high and opposite stereoselectivities (see 67) (75); and (iii) the realization that 1,3- and 1,4-elimination reactions are feasible.

The most useful alkene-forming process is the Peterson olefination reaction, which involves the condensation of carbonyl compounds with α-silyl organometallics (76). As exemplified by the preparation of β-gorgonene (68), the scheme is closely allied to the Wittig reaction but is more powerful because of the higher nucleophilicity and smaller bulk of the silicon reagent relative to a phosphorus ylide. If the silicon-bearing carbon does not carry a carbanion-stabilizing group, β-hydroxysilanes are generally isolated and must be separately treated with acid (77) or base (78). Otherwise, a fast elimination reaction occurs.

The usefulness of fluoride ion–promoted β-dehalosilylation has been amply demonstrated. Allenes, allene oxides, and cyclopropenes have been prepared (73). The smooth elimination of β-silyl-sulfones in the presence of F⁻ (79) has been applied to cycloadducts of 70 with the result that synthetic dienophile equivalents of HC≡CH, HC≡CD, DC≡CD, RC≡CH, and RC≡CD are now available (80).

Special attention has been given to β-(trimethylsilyl)ethoxy protecting groups for both amines (81) and carboxylic acids (82). Derivatives such as 71 and 72 are stable under a wide variety of conditions, but may be readily cleaved by fluoride ion without racemization occurring. The exceptional strength of the Si–F bond has been used to advantage in converting the silyl-substituted tropylium ion 73 to cycloheptatrienylidene (83).

Imaginative applications of this synthetic methodology have recently appeared. Use of the trimethylsilyl group

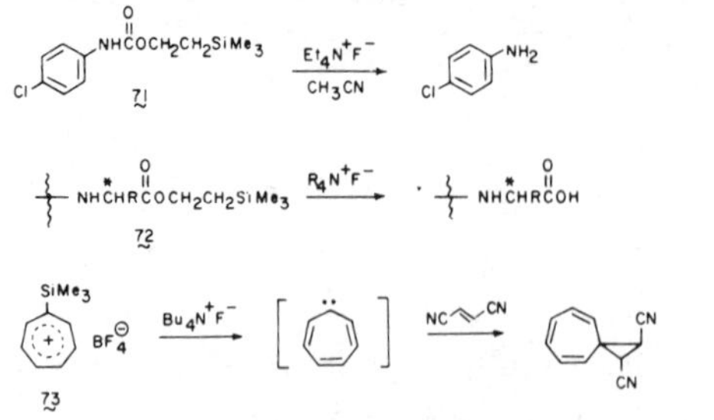

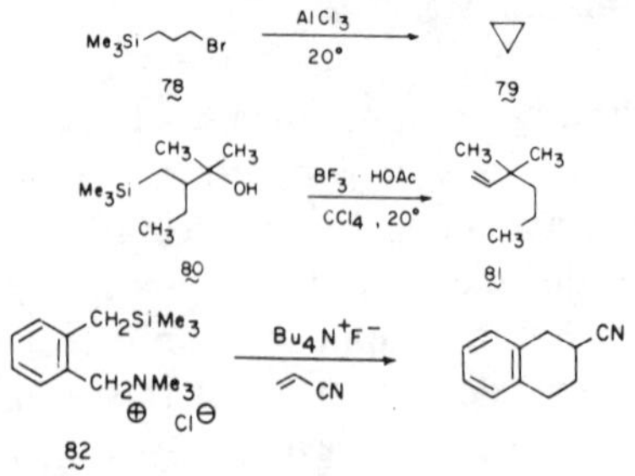

Fluoride ion–induced 1,4 eliminations of activated benzylic systems such as **82, 83,** and **84** proceed under very mild conditions to liberate highly reactive *o*-quinonemethides, which, under the proper conditions, can be trapped inter- or intramolecularly (*89*). This stratagem now serves as the basis for the stereoselective synthesis of carbocyclic and heterocyclic natural products.

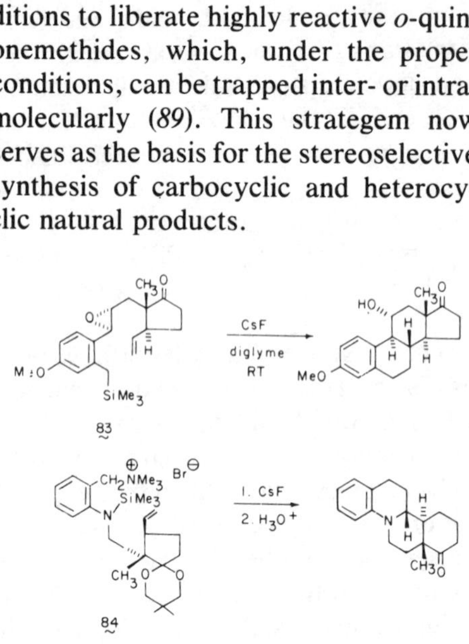

as a vinyl substituent in **74** allows for stereoselective epoxidation. Subsequent removal of the silicon residue with fluoride ion proceeds with retention of stereochemistry (*84*). Whereas photocycloaddition reactions of 2-cyclopentenone to terminal olefins show virtually no regioselectivity, this complication is nicely circumvented through incorporation of a trimethylsilyl group at C-2. Desilylation of the head-to-tail cycloadducts (such as **75**) delivers isomerically clean bicyclic ketones (*85*). The first stage of a general three-carbon ring expansion sequence illustrated by the conversion of **76** to **77** involves silane-mediated allyl anion generation (*86*).

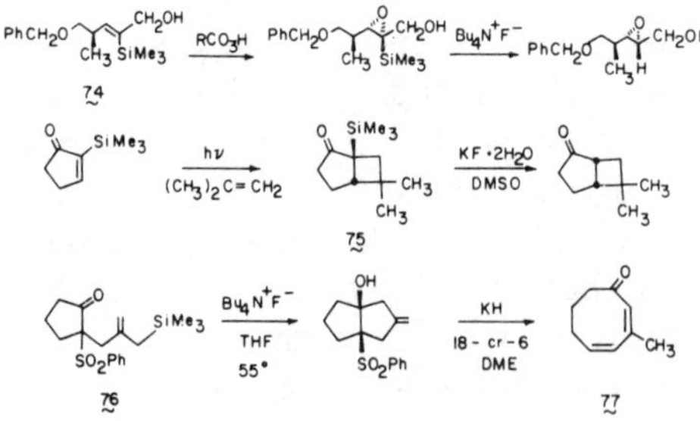

Although several 1,3 eliminations of a silyl group are known—for example, **78** → **79** (*87*)—the more common observation is β-alkyl or hydrogen migration with ultimate loss of the silyl substituent (compare **55**) (*88*).

Synthetic Applications of Me₃SiX Reagents

Monofunctionalized trimethylsilyl compounds have emerged as utilitarian reagents for effecting a diverse array of chemical transformations. Cyanotrimethylsilane, for example, condenses readily with aldehydes and ketones to produce silylated cyanohydrins (**85**) (*90*). When the nitrile group in such adducts is reduced, β-aminoethanols result and Tiffeneau-Demjanov ring expansions become possible (*91*). Since conversion to the corresponding anion (in aldehyde derivatives) can also be accomplished, these nucleophiles serve well as acyl anion equivalents (*92*).

Iodotrimethylsilane, an equally versatile reagent, readily cleaves esters (*93*), lactones (*94*), carbamates (*95*), and aryl ethers (*96*) under conditions presumed to be neutral. Furthermore, alcohols are converted to alkyl iodides (*97*), sulfoxides to sulfides (*98*), and epoxides to silylated iodohydrins (*99*) in its presence. If the latter reaction is performed in the presence of a tertiary amine base, allylic alcohols result (*100*). There are also conditions that lead directly from epoxide to alkene with an exceptionally high yield (*101*).

Azidotrimethylsilane behaves analogously (*102*), primary amines being produced if the adducts are treated with lithium aluminum hydride (*103*). The condensation of methylthiotrimethylsilane with aldehydes and ketones likewise proceeds readily to deliver dithioacetals, regioselectively if desired (*104*).

The foregoing examples should give some idea of the exciting and highly promising nature of organosilicon-mediated organic synthesis. This field of research has been rapidly gaining momentum and its continued growth can be confidently forecast.

References and Notes

1. For general references, see C. Eaborn, *Organosilicon Compounds* (Academic Press, New York, 1960); C. Eaton and R. W. Bott, in *Organometallic Compounds of the Group IV Elements*, vol. 1, *The Bond to Carbon*, A. G. MacDiarmid, Ed. (Dekker, New York, 1968), part 1, pp. 105–536; A. W. P. Jarvie, *Organomet. Chem. Rev. Sect. A* **6**, 153 (1979).
2. Some early reviews of organosilicon chemistry are summarized by S. S. Washburne, *J. Organomet. Chem.* **83**, 155 (1974).
3. Excellent reviews of most recent developments are given by S. S. Washburne, *ibid.* **123**, 1 (1976); D. Seyferth, Ed., *New Applications of Organometallic Reagents in Organic Synthesis* (Elsevier, Amsterdam, 1976); P. F. Hudrlik, in *ibid.*, vol. 1, p. 127; R. Calas and J. Dunoguès, in *ibid.*, vol. 2, p. 277; E. W. Colvin, *Chem. Soc. Rev.* **7**, 15 (1978); T.-H. Chan, *Acc. Chem. Res.* **10**, 442 (1977); ____ and I. Fleming, *Synthesis* (1979), p. 761; I. Fleming, in *Comprehensive Organic Chemistry*, D. H. R. Barton and W. D. Ollis, Eds. (Pergamon, New York, 1979), vol. 3D; W. C. Groutas and D. Felker, *Synthesis* (1980), p. 861; P. Magnus, *Aldrichimica Acta* **13**, 43 (1980); R. Calas, *J. Organomet. Chem.* **200**, 11 (1980); L. Birkofer and O. Stuhl, *Top. Curr. Chem.* **88**, 33 (1980); L. A. Paquette, *Isr. J. Chem.* **28**, 128 (1981).
4. W. Noll, *Chemistry and Technology of Silicones* (Academic Press, New York, 1968).
5. L. H. Sommer, *Stereochemistry, Mechanism, and Silicon* (McGraw-Hill, New York, 1965).
6. A. E. Pierce, *Silylation of Organic Compounds* (Pierce Chemical Co., Rockford, Ill., 1968).
7. C. Eaborn, T. A. Emokpae, V. I. Sidorov, R. Taylor, *J. Chem. Soc. Perkin Trans. 2* (1974), p. 1454; T. G. Traylor, H. J. Berwin, J. Jerkunica, M. L. Hall, *Pure Appl. Chem.* **30**, 599 (1972).
8. D. J. Peterson, *J. Org. Chem.* **33**, 780 (1968); G. A. Gornowicz and R. West, *J. Am. Chem. Soc.* **90**, 4478 (1968).
9. F. K. Cartledge and J. P. Jones, *Tetrahedron Lett.* (1971), p. 2193; M. A. Cook, C. Eaborn, D. R. M. Walton, *J. Organomet. Chem.* **29**, 389 (1971); Y. Apeloig, P. von R. Schleyer, J. A. Pople, *J. Am. Chem. Soc.* **99**, 1291 (1977).
10. G. Stork and P. F. Hudrlik, *J. Am. Chem. Soc.* **90**, 4462 (1968); *ibid.*, p. 4464; H. O. House, M. Gall, H. D. Olmstead, *ibid.* **36**, 2361 (1971).
11. G. L. Larson and L. M. Fuentes, *ibid.* **103**, 2418 (1981).
12. J. W. Patterson, Jr., and J. H. Fried, *J. Org. Chem.* **39**, 2406 (1974); E. S. Binkley and C. H. Heathcock, *ibid.* **40**, 2156 (1975); T. Mukaiyama, K. Saigo, O. Takazawa, *Chem. Lett.* (1976), p. 1033.
13. L. A. Paquette, G. D. Annis, H. Schostarez, J. F. Blount, *J. Org. Chem.* **46**, 3768 (1981).

14. G. C. Joshi and L. M. Pande, *Synthesis* (1975), p. 450.
15. G. Stork and B. Ganem, *J. Am. Chem. Soc.* **95**, 6152 (1973).
16. T. H. Chan, I. Paterson, J. Pinsonnault, *Tetrahedron Lett.* (1977), p. 4183; M. T. Reetz and W. F. Maier, *Agnew. Chem. Int. Ed. Engl.* **17**, 48 (1978); G. M. Rubottom, J. M. Gruber, G. M. Mong, *J. Org. Chem.* **40**, 3427 (1975); E. Nakamura and I. Kuwajima, *J. Am. Chem. Soc.* **99**, 961 (1977).
17. J. M. Conia, *Pure Appl. Chem.* **43**, 317 (1975).
18. L. Blanco, P. Amice, J. M. Conia, *Synthesis* (1981), p. 289.
19. G. Stork and T. L. Macdonald, *J. Am. Chem. Soc.* **97**, 1264 (1975).
20. S. A. Monti, F. G. Cowherd, T. W. McAninch, *J. Org. Chem.* **40**, 858 (1975); B. M. Trost and M. J. Bogdanowicz, *J. Am. Chem. Soc.* **95**, 290 (1973).
21. R. E. Ireland and J. P. Vevert, *J. Org. Chem.* **45**, 4259 (1980).
22. T. Hayashi, Y. Katsuro, M. Kumada, *Tetrahedron Lett.* (1980); p. 3915.
23. G. D. Crouse and L. A. Paquette, *J. Org. Chem.* **46**, 4272 (1981).
24. J.-P. Pillot, B. Bennetau, D. Dunoguès, R. Calas, *Tetrahedron Lett.* (1980), p. 4717.
25. K. Yamamoto, J. Yoshitake, N. T. Qui, J. Tsuji, *Chem. Lett.* (1978), p. 859.
26. R. L. Funk and K. P. C. Vollhardt, *J. Am. Chem. Soc.* **99**, 4058 (1977).
27. D. J. S. Tsai and D. S. Matteson, *Tetrahedron Lett.* (1981), p. 2751.
28. S. D. Burke, C. W. Murtiashaw, M. S. Dike, S. M. M. Strickland, J. O. Saunders, *J. Org. Chem.* **46**, 2400 (1981).
29. L. E. Overman and K. L. Bell, *J. Am. Chem. Soc.* **103**, 1851 (1981).
30. G. Büchi and H. Wüest, *J. Org. Chem.* **44** 546 (1979).
31. P. Casara and B. W. Metcalf, *Tetrahedron Lett.* (1978), p. 1581.
32. W. S. Johnson, T. M. Yarnell, R. F. Myers, D. R. Morton, *ibid.*, p. 2549.
33. K. Utimoto, M. Tanaka, M. Kitai, H. Nozaki, *ibid.*, p. 2301.
34. P. Bourgeois and G. Mérault, *C. R. Acad. Sci.* **273**, 714 (1971); *J. Organomet. Chem.* **39**, C44 (1972).
35. I. Fleming and D. Marchi, Jr., *Synthesis* (1981), p. 560.
36. S. R. Wilson, L. R. Philips, K. J. Natalie, Jr., *J. Am. Chem. Soc.* **101**, 3340 (1979).
37. M. Montury, B. Psaume, J. Goré, *Tetrahedron Lett.* (1980), p. 163.
38. I. Kuwajima, T. Tanaka, K. Atsumi, *Chem. Lett.* (1979), p. 779.
39. M. Ochiai, M. Arimoto, E. Fujita, *J. Chem. Soc. Chem. Commun.* (1981), p. 460.
40. T. Flood and P. E. Peterson, *J. Org. Chem.* **45**, 5006 (1980); J. Pornet, *Tetrahedron Lett.* (1981), p. 455; R. Schmid, P. L. Huesmann, W. S. Johnson, *J. Am. Chem. Soc.* **102**, 5122 (1980).
41. R. L. Danheiser, D. J. Carini, A. Basak, *J. Am. Chem. Soc.* **103**, 1604 (1981).
42. M. Grignon-Dubois, J. Dunoguès, R. Calas, *Synthesis* (1976), p. 737; *Tetrahedron Lett.* (1976), p. 1197.
43. R. G. Daniels and L. A. Paquette, *J. Org. Chem.* **46**, 2901 (1981).
44. L. A. Paquette, K. A. Horn, G. J. Wells, *Tetrahedron Lett.* (1982), p. 259.
45. L. A. Paquette, G. J. Wells, K. A. Horn, T.-H. Yan, *ibid.*, p. 263.
46. G. Stork and E. Colvin, *J. Am. Chem. Soc.* **93**, 2080 (1971).
47. W. E. Fristad, T. R. Bailey, L. A. Paquette, *J. Org. Chem.* **45**, 3028 (1980).
48. _____, R. Gleiter, M. C. Böhm, *J. Am. Chem. Soc.* **101**, 4420 (1979).
49. I. Fleming and J. P. Michael, *J. Chem. Soc. Chem. Commun.* (1978), p. 245; I. Fleming and S. K. Patel, *Tetrahedron Lett.* (1981), p. 2321.
50. T. H. Chan, E. Chang, E. Vinokur, *Tetrahedron Lett.* (1970), p. 1137.
51. D. Seebach, D. Enders, B. Renger, *Chem. Ber.* **110**, 1852 (1977).
52. C. Burford, F. Cooke, E. Ehlinger, P. D. Magnus, *J. Am. Chem. Soc.* **99**, 4536 (1977).
53. E. Ehlinger and P. D. Magnus, *ibid.* **102**, 5004 (1980).
54. U. Schollkopf and H.-I. Scholz, *Synthesis* (1976), p. 271.
55. J. J. Eisch and J. E. Galle, *J. Am. Chem. Soc.* **98**, 4646 (1976).
56. L. Cason and H. Brooks, *J. Org. Chem.* **19**, 1278 (1954); T. H. Chan and E. Chang, *ibid.* **39**, 3264 (1974).
57. R. K. Boeckman, Jr., *J. Am. Chem. Soc.* **95**, 6867 (1973).
58. M. Isobe, M. Kitamura, T. Goto, *Tetrahedron Lett.* (1980), p. 4727.
59. P. J. Kocienski, *ibid.*, p. 1559; D. J. Ager and R. C. Cookson, *ibid.*, p. 1677; D. J. Ager, *ibid.* (1981), pp. 587 and 2803.
60. P. J. Kocienski, *J. Chem. Soc. Chem. Commun.* (1980), p. 1096.
61. F. Cooke, P. Magnus, G. L. Bundy, *ibid.* (1978), p. 714.
62. R. G. Daniels and L. A. Paquette, *Tetrahedron Lett.* (1981), p. 1579.
63. D. Seyferth, A. W. Dow, H. Menzel, T. H. Flood, *J. Am. Chem. Soc.* **90**, 1080 (1969); D. Seyferth, H. Menzel, A. W. Dow, T. H. Flood, *J. Organomet. Chem.* **44**, 279 (1972).
64. R. A. Olofson, D. H. Hoskin, K. D. Lotts, *Tetrahedron Lett.* (1978), p. 1677.
65. D. Seyferth and E. M. Hanson, *J. Am. Chem. Soc.* **90**, 2439 (1968); *J. Organomet. Chem.* **27**, 19 (1971).
66. P. S. Skell and P. W. Owen, *J. Am. Chem. Soc.* **94**, 1578 (1972).
67. W. Ando *et al.*, *ibid.* **101**, 6393 (1979).
68. W. Ando, A. Sekiguchi, T. Hagiwara, T. Migita, *J. Chem. Soc. Chem. Commun.* (1974), p. 372; *ibid.* (1975), p. 145.
69. W. Ando, A. Sekiguchi, T. Migita, *Chem. Lett.* (1976), p. 779.
70. T. J. Barton and S. K. Hoekman, *J. Am. Chem. Soc.* **102**, 1584 (1980).
71. A. Sekiguchi and W. Ando, *J. Org. Chem.* **45**, 5286 (1980).
72. R. F. Cunico and E. M. Dexheimer, *J. Am. Chem. Soc.* **94**, 2868 (1972); *J. Organomet. Chem.* **59**, 153 (1973); R. F. Cunico and Y.-K. Han, *ibid.* **105**, C29 (1976).
73. T. H. Chan and W. Mychajlowskij, *Tetrahedron Lett.* (1974), p. 171; T. H. Chan, M. P. Li,

W. Mychajlowskij, D.N. Harpp, *ibid.*, p. 3511; T. H. Chan and D. Massuda, *ibid.* (1975), p. 3383.

74. F. C. Whitmore, L. H. Sommer, J. Gold, R. E. Van Strien, *J. Am. Chem. Soc.* **69**, 1551 (1947); H. Gilman and R. A. Tomasi, *J. Org. Chem.* **27**, 3647 (1962).

75. P. F. Hudrlik, D. Peterson, R. J. Rona, *J. Org. Chem.* **40**, 2263 (1975).

76. H. Taguchi, K. Shimoji, H. Yamamoto, H. Nozaki, *Bull. Chem. Soc. Jpn.* **47**, 2529 (1974).

77. P. F. Hudrlik, A. M. Hudrlik, A. Nagendrappa, T. Yimenn, E. T. Zellers, E. Chin, *J. Am. Chem. Soc.* **102**, 6894 (1980).

78. P. F. Hudrlik and D. Peterson, *Tetrahedron Lett.* (1974), p. 1133; *J. Am. Chem. Soc.* **97**, 1464 (1975).

79. P. J. Kocienski, *Tetrahedron Lett.* (1979), p. 2649.

80. L. A. Paquette and R. V. Williams, *ibid.* (1981), p. 4643.

81. L. A. Carpino, J.-H. Tsao, H. Ringsdorf, E. Fell, G. Hettrich, *J. Chem. Soc. Chem. Commun.* (1978), p. 358.

82. P. Sieber, *Helv. Chim. Acta* **60**, 2711 (1977); H. Gerlach, *ibid.*, p. 860.

83. M. Reiffen and R. W. Hoffman, *Tetrahedron Lett.* (1978), p. 1107.

84. I. Hasan and Y. Kishi, *ibid.* (1980), p. 4229; T. H. Chan, P. W. K. Lau, M. P. Li, *ibid.* (1976), p. 2667.

85. J. S. Swenton and E. L. Fritzen, Jr., *ibid.* (1979), p. 1951.

86. B. M. Trost and J. E. Vincent, *J. Am. Chem. Soc.* **102**, 5680 (1980).

87. L. H. Sommer, R. E. Van Strien, F. C. Whitmore, *ibid.* **71**, 3056 (1949).

88. I. Fleming, *Bull. Soc. Chim. Fr. No. II-7* (1981).

89. Y. Ito, S. Miyata, M. Nakatsuka, T. Saegusa, *J. Am. Chem. Soc.* **103**, 5250 (1981).

90. D. A. Evans, G. L. Carroll, L. K. Truesdale, *J. Org. Chem.* **39**, 914 (1974); D. A. Evans and L. Y. Yong, *ibid.* **42**, 350 (1977).

91. R. W. Thies and E. P. Seitz, *ibid.* **43**, 1050 (1978).

92. D. A. Evans, K. M. Hurst, L. K. Truesdale, J. M. Takacs, *Tetrahedron Lett.* (1977), p. 2495.

93. M. E. Jung and M. A. Lyster, *J. Am. Chem. Soc.* **99**, 968 (1977), T. L. Ho and G. A. Olah, *Synthesis* (1977), p. 417.

94. G. A. Olah, S. C. Narang, B. G. B. Gupta, R. Malhotra, *J. Org. Chem.* **44**, 1247 (1979).

95. M. E. Jung and M. A. Lyster, *J. Chem. Soc. Chem. Commun.* (1978), p. 315; R. S. Lott, V. S. Chuhan, C. H. Stammer, *ibid.* (1979), p. 495.

96. M. E. Jung and M. A. Lyster, *J. Org. Chem.* **42**, 3761 (1977).

97. M. E. Jung and P. L. Ornstein, *Tetrahedron Lett.* (1977), p. 2659; T. Morita, S. Yoshida, Y. Okamoto, H. Sakurai, *Synthesis* (1979), p. 379.

98. G. A. Olah, S. C. Narang, B. G. B. Gupta, R. Malhotra, *ibid.*, p. 61.

99. H. Sakurai, K. Sasaki, A. Hosomi, *Tetrahedron Lett.* (1980), p. 2329; O. Ceder and B. Hansson, *Acta Chem. Scand. Ser. B* **30**, 574 (1976).

100. G. A. Kraus and K. Frazier, *J. Org. Chem.* **45**, 2579 (1980); M. R. Detty, *ibid.*, p. 924.

101. R. Caputo, L. Mangoni, O. Neri, G. Palumbo, *Tetrahedron Lett.* (1981), p. 3551.

102. H. Vorbrüggen and K. Krolikiewicz, *Synthesis* (1979), p. 35.

103. E. P. Kyba and A. M. John, *Tetrahedron Lett.* (1977), p. 2737.

104. D. A. Evans, L. K. Truesdale, K. G. Grimm, S. L. Nesbitt, *J. Am. Chem. Soc.* **99**, 5009 (1977).

Part VI

Chemistry of Life Processes

Several million organic compounds with a wide variety of shapes, sizes, and properties have been synthesized. Some of these compounds are rigid at ordinary temperatures in their nonsolid states, but the majority are capable of assuming a variety of shapes through rotations of their various parts around covalent single bonds. Because chemical bonds radiate outward from nuclei, most organic compounds have convex surfaces composed of hydrogen atoms bound to carbon, oxygen, or nitrogen. A few simple compounds such as benzene and cyclobutane have flat or nearly flat surfaces. Like stones, very few organic compounds have concave surfaces of any size. If organic compounds that contain enforced (rigid) cavities are to be designed and prepared, they must be composed of units that are concave on parts of their surfaces. The virtual nonexistence of such units until recently is related to the nonexistence of cavity-containing compounds.

I use the class name cavitands for synthetic organic compounds that contain enforced cavities of dimensions at least equal to those of the smaller ions, atoms, or molecules. Among the latter, Li^+ has a diameter of about 1.48 angstroms, He a diameter of 2.6 angstroms, and H_2 is about 2 by 2.5 angstroms. The interior surfaces of cavitands can have spherical, hemispherical, ellipsoidal, oblong, collar, or any other designable and synthesizable shapes.

The cyclodextrins are the most studied of the naturally occurring compounds that contain rigid cavities. They are cyclic oligomers of the 1,4-glucopyranoside unit. They are torus-shaped and have cavities that are enforced by the rigidity of the six, seven, or eight monosaccharide units. Their internal dimensions are

26

Cavitands: Organic Hosts with Enforced Cavities

Donald J. Cram

Science 219, 1177–1183 (11 March 1983)

large enough to embrace aryl units. Much interesting enzyme modeling has been done with cyclodextrins and their derivatives (1). Other important cavities in the biological world include the binding sites of enzymes and the troughs in the helices of RNA and DNA, which thus far have no counterparts in synthetic organic chemistry.

Why should cavitands interest scientists? One of the supreme challenges to the organic chemist is to design and synthesize compounds that simulate the properties of the working parts of evolutionary chemistry. Complexation between substrate and enzyme, or between inhibitor and receptor site, is a central feature in catalysis and regulation of biological processes. Cooperativity between catalyzing functional groups in enzyme systems is possible only if those groups are held in positions that converge on a substrate-binding site, usually located in a cavity. The design, synthesis, and study of cavity-containing organic compounds is a principal theme in what we term host-guest complexation chemistry.

A host is a compound with binding sites arranged to converge on the same point, line, or surface. A guest is a compound or ion with binding sites that diverge from a point, line, or surface. When a host and a guest have complementary binding sites and steric requirements, complexation occurs. A complex is composed of two or more distinct molecules held together by noncovalent forces in a definable structural relationship. The binding can result from any combination of hydrogen bonding, ion pairing, metal ion to ligand attractions, π-acid to π-base attractions, van der Waals attractions, and solvent-liberation driving forces. Hosts can contain cavi-

ties that are rigid or that are developed by reorganization of the hosts during the process of complexation. Thus cavitands are particular kinds of hosts—those whose cavities are enforced and exist prior to complexation.

The first section of this article describes a cavitand whose rigid organization vastly enhances its ability to bind guests with high selectivity. The second describes larger cavitands that contain moving parts. The third section traces the evolution of an approach to another family of cavitands whose syntheses are in progress, and the fourth describes a saddle-shaped cavitand containing two clefts of substantial size. The final section indicates how cavitands might be chemically manipulated to produce desired physical and chemical properties.

Importance of Preorganization of Hosts to Their Complexing Powers

Among hosts that bind metal cations, the most thoroughly investigated have been Pedersen's chorands (or crowns) (2), particularly 18-crown-6 (1) and its derivatives. The crystal structures of 1 and other chorands (3) show that they do not contain cavities because one or two methylene groups turn toward the general center of the molecule to fill potential intramolecular voids. The crystal structure of 18-crown-6-K^+ (2) shows that the complex contains a cavity formed and filled by the potassium ion during complexation. Other chorands behave similarly (3).

Although less conformationally mobile than the chorands, Lehn's cryptands (4) also fill their own cavities when not complexed, and are reorganized by the guest during complexation. The complexation of [2.2.2]cryptand (3) with K^+ provides an example. Crystal structures reveal that two inward-turned CH_2 groups fill the cavity in free host 3 that is occupied by K^+ in 4 (3).

Unlike the chorands or cryptands, whose cavities are developed by the complexing guest, the spherands contain cavities in their uncomplexed state which become filled upon complexation (3, 5). Thus the spherands are rigidly organized for complexation during their synthesis rather than during their complexation. In spherand 5, the six oxygens are octahedrally arranged with their 24 unshared electrons lining a cavity enforced by the rigidity of the benzene rings and by the spatial requirements of the methoxy groups. The ring system inhibits all but small rotations about the aryl-aryl bonds. The cumulative effect of three methyls protruding from each face of the best plane of the macroring, coupled with the flanking benzene rings, prevents all but small rotations about the Ar–O bonds (3, 5). In projection, formulas 5 and 6 are shaped like snowflakes.

Host 5 readily complexes Li^+ and Na^+ and is the strongest known binder of these ions (6). Other ions such as K^+, Rb^+, Cs^+, Mg^{2+}, and Ca^{2+} are completely rejected by 5. Since no electron pairs on the oxygens of 5 can turn outward, ions are either fully complexed by encapsulation or not complexed at all. Thus 5 shows very high structural recognition in complexation. The crystal structures of 5, 6, and sodiospherium complex 7 (not shown) are drawn in 8, 9, and 10, respectively. Clearly visible are the enforced cavity in 8 and the filled cavities in 9 and 10 (5). The cavity diameter in 8 is 1.62 Å, the Li^+-filled cavity diameter of 9 is 1.48 Å, and the Na^+-filled cavity diameter of 10 is 1.75 Å.

Thus when 5 becomes complexed to form 6 or 7, very little reorganization occurs. In Corey-Pauling-Koltun (CPK) molecular models (7), the metal ions of 6 and 7 are barely visible because they are buried in a skin of C–H bonds.

A measure of the importance of preorganization to complexation is found in a comparison of the binding free energies of spherand 5 with those of its unorganized open-chain relative 11. Compound 11 differs constitutionally from 5 only in the sense that two hydrogen atoms terminate the chain of 11 at the point where an aryl-aryl bond completes the ring system of 5. Whereas spherand 5 has a single conformation organized for binding, compound 11 has 1024 conformations, only two of which provide the octahedral arrangement of oxygens needed for fully cooperative binding. In chloroform at 25°C, spherand 5 binds sodium picrate by > 13 kilocalories per mole more, and lithium picrate by >> 16 kilocalories per mole more than does 11 (6). Normally, the anisyl unit of 5 and 11 is regarded as a poor ligand for binding metal ions.

Larger Cavitands That Contain Moving Parts

Collar-shaped cavitands have been prepared whose enforced cavities have variable shapes and sizes because they contain inward-directed, mobile substituents. The most studied example is spherand 12, whose eight oxygens have an enforced, square antiprismatic arrangement. The methyl groups of 12 are much less congested than the corresponding methyls of its smaller analog, 5. As a result, each methyl in 12 can turn inward toward the center of the cavity,

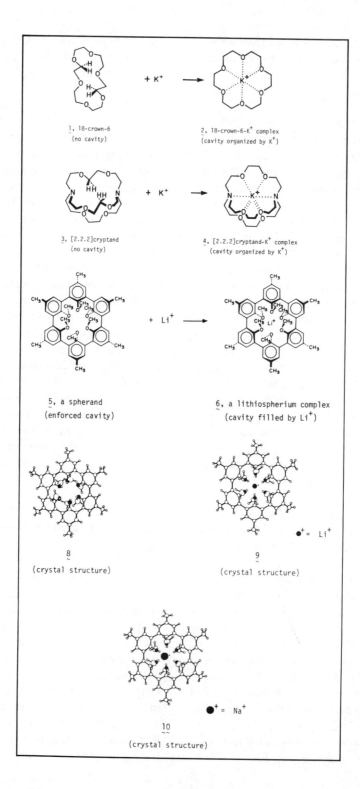

1, 18-crown-6
(no cavity)

2, 18-crown-6-K⁺ complex
(cavity organized by K⁺)

3, [2.2.2]cryptand
(no cavity)

4, [2.2.2]cryptand-K⁺ complex
(cavity organized by K⁺)

+ Li⁺ →

5, a spherand
(enforced cavity)

6, a lithiospherium complex
(cavity filled by Li⁺)

8
(crystal structure)

9
(crystal structure)

●⁺ = Li⁺

10
(crystal structure)

●⁺ = Na⁺

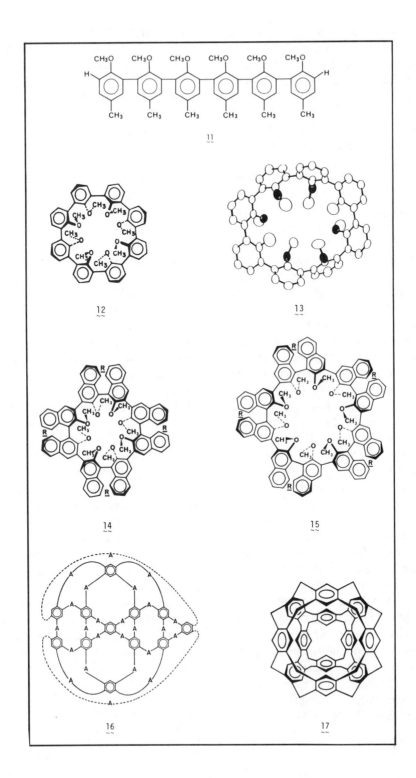

or away from the center between the oxygens (as in the drawing of **12**), thus extending the cavity. In molecular models of **12** with all eight methyl groups turned inward, the cavity is cylindrical and is lined by the hydrogens of the eight methyl groups. The cylinder has a diameter of about 3 Å and a length equal to that of diacetylene ($\sim$ 8 Å), whose model just fits into the cylinder. With all the methyl groups facing outward, the molecular model of **12** has a cavity just large enough to accommodate a model of chair cyclohexane. A sphere with the diameter of Cs^+ (3.4 Å), when inserted in this model of **12**, barely reaches the eight oxygens when the aryl-aryl dihedral bond angles are minimized (8).

The crystal structure of **12** is shown in **13** (9). As expected, the eight oxygens have an alternating up-down arrangement. Of the eight methyl groups, two that are adjacent and the two that are opposite these extend inward, two others opposite one another turn upward, and the remaining two, which are also opposite, turn downward. The resulting cavity is egg-shaped, $\sim$ 4 Å in its longer and $\sim$ 2.5 Å in its shorter dimension (9).

At room temperature, the proton nuclear magnetic resonance spectrum of **12** shows a singlet for its methyl protons, which indicates that the methyls rapidly equilibrate between their possible positions. Such flexibility allows this host, while rigidly maintaining a cavity, to vary the size, shape, and lining of that cavity, depending on the size, shape, and surface polarity of the guest.

Spherand **12** complexes cesium picrate in chloroform with a binding free energy of 11.8 kcal mole^{-1}. This is 4 kcal mole^{-1} greater than the binding free energy observed for **12** complexing potassium picrate and 6 kcal mole^{-1} greater than for **12** complexing lithium picrate. Probably all eight oxygens ligate Cs^+ simultaneously but cannot reach the smaller ions (9).

Spherands **14** and **15** have also been prepared (10). They are composed of chiral binaphthyl units and possess D_4 and D_5 symmetry, respectively. In molecular models, **14** resembles **12** except that the dihedral angles within the binaphthyl units of **14** must be substantially greater than those within the biphenyl units of **12** (3). Consequently, the enforced cavity of **14** must be larger than that of **12**, which explains why **12** binds cesium picrate less well (by 3.5 kcal mole^{-1}) than does **14**. Host **14** crystallizes with 1 mole of cyclohexane (10). A molecular model of cyclohexane neatly occupies the cavity of a model of **14**. However, a crystal structure of the complex is not yet available.

In models, larger spherand **15** possesses an enforced arrangement of oxygens complementary to the ten hydrogens of ferrocene. With all the methyl groups of the host turned outward, a model of ferrocene beautifully fills the minimum cavity available, with each of the ten guest hydrogens touching an oxygen of the host. The model dimensions of the cavity of this conformation range from about 6 by 8 Å to about 7.2 by 8.4 Å, depending on the aryl-aryl dihedral angles. With all of the methyl groups in the model turned inward, the cylindrical cavity is about 4.8 Å in diameter and 7.2 Å in length. A model of chair cyclohexane nicely fills this cavity. The complexing properties of **15** have not yet been examined.

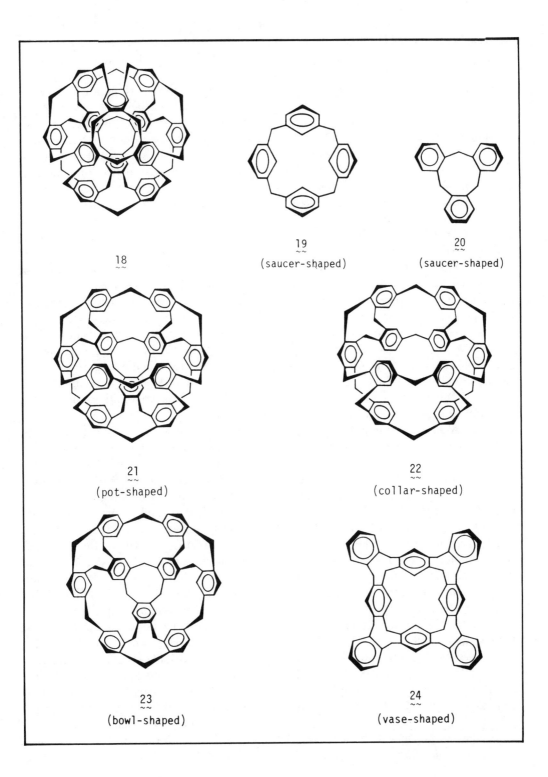

18

19
(saucer-shaped)

20
(saucer-shaped)

21
(pot-shaped)

22
(collar-shaped)

23
(bowl-shaped)

24
(vase-shaped)

Approaches to a New Family of Cavitands

Hosts are yet to be prepared whose cavities can imprison simple organic compounds. These hosts suggest a variety of interesting questions. (i) Can the closing of the shells of such cavitands be templated by solvent or by solvated ion pairs? (ii) When organic (or inorganic) guests are encapsulated by such cavitands, how do the physical properties of the complex differ from those of the host and guest taken separately? (iii) What kinds of organic reactions can be carried out on encapsulated guests? (iv) Can cavitands be prepared with "pores" in their "skins" that allow the entrance and departure of certain guests from their interior, but forbid passage to others? In other words, can cavitands be made that resemble cells? (v) What types of synthetic strategies are required to bring such cavitands to hand and to make their solubility properties manipulable?

Examination of many possible structures by use of molecular models led us to prototype structure **16**, in which A stands for simple bridging groups such as CH_2, O, S, NH, C=O, CH_2–CH_2, O–CH_2–O, o-C_6H_4, m-C_6H_4, and p-C_6H_4. Such compounds are close to being spherical, and their parts are rigidly distributed in three dimensions. In order to show their bonding sequences, we resort to Mercator projection formulas such as **16**. Gerhardus Mercator was a Flemish geographer who in the 16th century invented the Mercator projection maps of the world for navigation. The spirit of our use of a drawing such as **16** is that it is useful for molecular navigation.

A model of **16**, in which each A is a CH_2 group, is almost strain-free. The model is spherical in shape, about the size of a small melon, and has a cavity about the size of a large orange. It contains 12 benzene rings and 24 CH_2 groups, and each benzene ring has two hydrogens *para* to one another. Formulas **17** and **18** are different representations of this hypothetical compound, which is composed of two kinds of concave units (**19** and **20**) that repeat on the surface of the sphere. Unit **19** is composed of four *meta*-attached benzenes, and **20** of three *ortho*-attached benzenes, each linked through CH_2 groups. In **17**, the viewer looks into the sphere through the center of an expanded **19** unit; in **18**, through the tiny opening of a greatly expanded **20** unit. In models of the structure, these two different kinds of holes in the skin of the sphere are very small. Only those described by the **19** units are large enough to allow passage of the smallest of chemical entities, such as e⁻, H⁺, H_2, or Li⁺.

Removal of one or more benzene units with their attached CH_2 groups from models of **18** (or **17**) produces simpler cavitands, with ample room for entrance to and escape from the cavities by guests. For example, omission of the benzene located at 12 o'clock in the drawing of **18** along with its four attached CH_2 groups gives a pot-shaped entity. Formula **21** represents the resulting structure, with the viewer looking into the "pot" off center from its top. Omission of the benzene located at 6 o'clock and its four CH_2 groups from **21** gives **22**, which is shaped like a collar. Elimination of the top three benzene rings and their attached CH_2 groups from **18** results in **23**, which is bowl-shaped. If the top four benzene rings and their attached CH_2 groups are subtracted from **17**, vase-shaped **24** is produced. Extension of

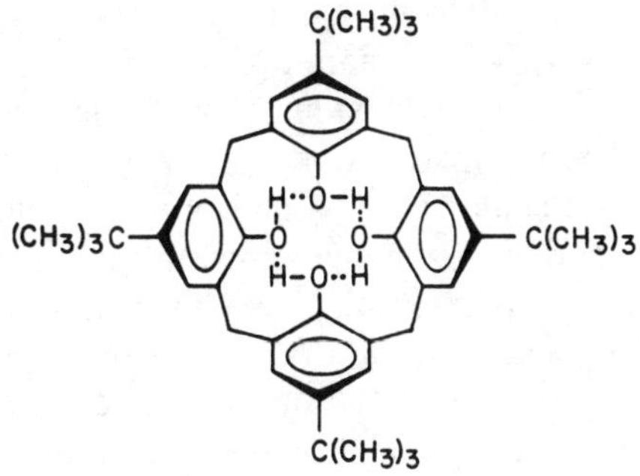

25

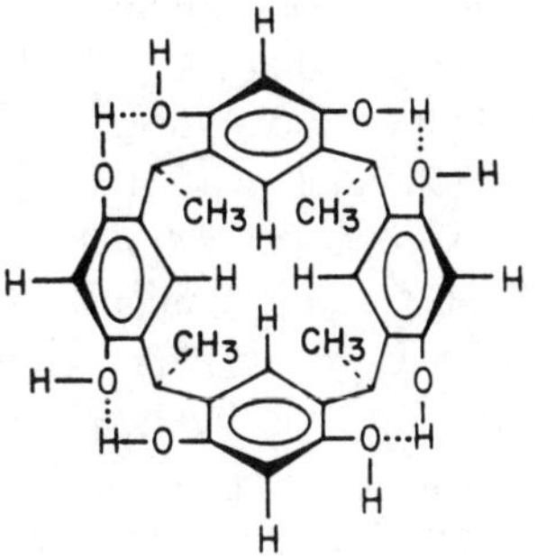

26

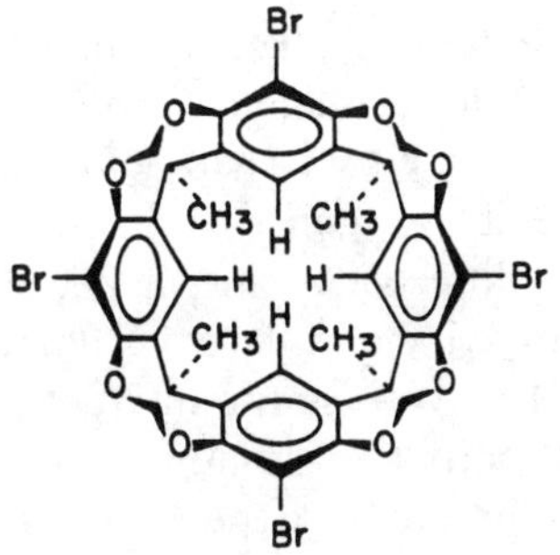

27

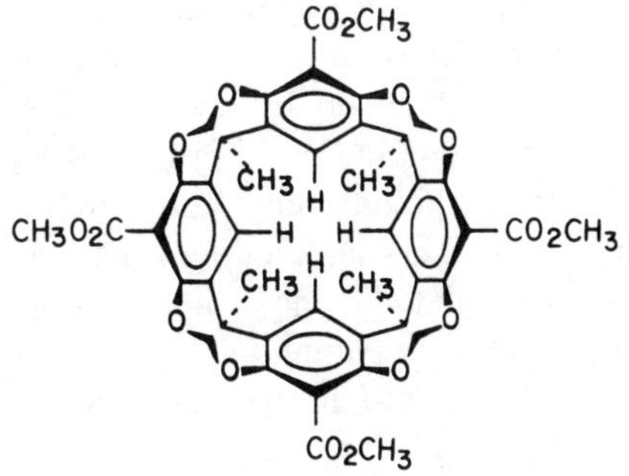

28

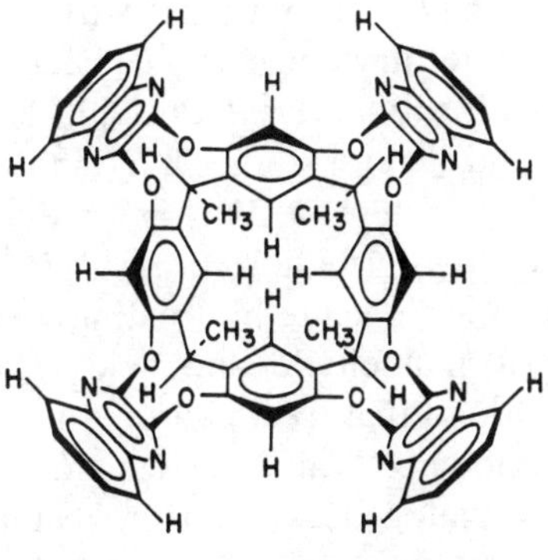

29

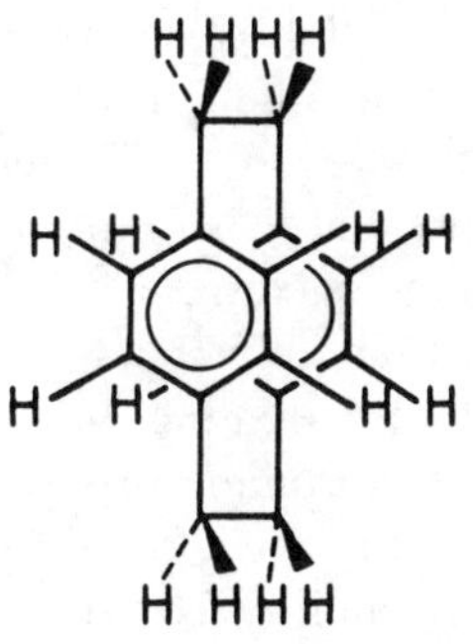

30

such processes ultimately produces **19** or **20**, which in their concave conformations are saucer-shaped.

Formally, we have shown how hypothetical molecules with the shapes of spheres, pots, collars, bowls, or vases can be constructed from assemblies of saucers, which in turn are constructed from three or four benzene rings and methylene groups. How close have organic chemists come to preparing these compounds or ones like them?

Derivatives of simple units **19** and **20** have been reported, and are easily prepared. For example, Gutsche *et al.* (*11*) reported that treatment of 4-*tert*-butylphenol with formaldehyde and base gave **25**. The compound equilibrates between a number of different conformations, only the saucer-shaped one of which has been formulated. The other conformations have been eliminated by converting all of the aryl-OH to aryl-OSi(CH$_3$)$_3$ groups, which are too large to pass through the center of the ring (*12*). The authors call such compounds calixarenes (*11, 12*).

Following the lead of Erdtman *et al.* (*13*), Högberg (*14*) reported that resorcinol and acetaldehyde in an appropriately acidic medium produced stereoisomerically pure **26** in 57 percent yield. This compound was found to exist as an equilibrating mixture of two conformations, only one of which has been formulated (*14*). By treating **26** with bromine and the tetrabromide product with base and ClCH$_2$Br, we closed four rings to give **27**, which was converted to **28** through metallation and carbonation (*15*). Both **27** and **28** can exist only in the shallow bowl conformation shown because of the constraints of the multiple ring systems.

Compound **26** also served as our starting material for the construction of the tall, vase-shaped compound **29**. When mixed with four equivalents of 2,3-dichloro-1,4-diazanaphthalene and excess potassium hydroxide, **26** was converted to **29** by closing four new rings (*15*).

Molecular models and proton nuclear magnetic resonance spectral measurements indicate that **29** exists in either the compact structure drawn or an extended structure. At 25°C and above, **29** is exclusively in the compact structure in which the four diazanaphthalene groups occupy planes perpendicular to the plane of the eight oxygens. At −100°C and below, the compound exists exclusively in the extended structure in which the four diazanaphthalene groups occupy planes more closely parallel to the plane of the eight oxygens. As a result of their drastic differences in shape, the two structures have different numbers and kinds of solvent-solute interactions, and consequently their relative stabilities are very temperature-dependent. The molecular model of the compact structure has a cavity of sufficient size to contain one model of [2.2]paracyclophane or 12 models of water (*15*).

The hexamethoxy derivative of **20**, cyclotriveratrylene (**31**), has been identified by Erdtman *et al.* (*13*) and Lindsey (*16*), and several studies have shown that the ring system of **31** exists only in the saucer conformation drawn (*16*). The compound is made by treating veratrole with formaldehyde and acid (*16*). We have found that treatment of 3-methoxy-4-bromobenzyl alcohol with acid produces analog **32**, which on metallation and carbonation gives **33** (*17*).

These three saucer-shaped units and their analogs provide good starting points for the syntheses of a variety of cavitands. An example is cavitand **34**, whose elegant synthesis has just been reported by Gabard and Collet (*18*).

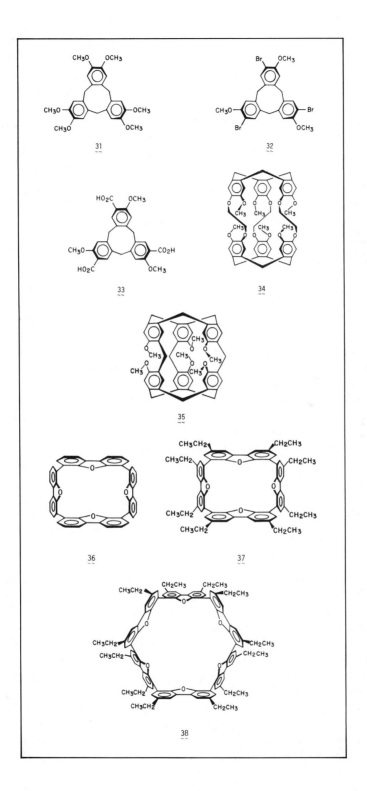

31

32

33

34

35

36

37

38

389

Compound **31** is chiral and has D_3 symmetry. One of our target cavitands is **35**, which in molecular models must be chiral and has D_3 symmetry.

The sizes of the cavities and perforations in molecular models of **34** and **35** provide interesting comparisons. Experimentally, cavitand **34** crystallizes with one molecule of chloroform, which escapes while a mass spectrum is being taken (*18*). In models, the gaps in the skin of **34** are just barely large enough in the most advantageous conformations to allow entrance or egress of a single molecular model of chloroform. Up to six molecules of water (in models) can be inserted with ease into the cavity of a model of **34**. The cavity of a model of **35** can contain only one molecular model of water. The largest organic compound whose model can be encapsulated by **35** is CH_4. The holes in the skin of a model of **35** are too small to allow either H_2O or CH_4 to enter or depart from cavity without severe strain. However, models of H_2, H–F, or LiF can enter and depart without much difficulty.

Cavitands with Rectangular or Hexagonal Cross Sections

Examination of molecular models of possible cavitands containing rigid clefts led to the design of compound **36**. It is composed of four dibenzofuran units bound directly to one another in a cycle. Since these units are rigid and planar, the model has only a low order of conformational freedom, which can be exercised by rotating the aryl groups a few degrees around all of the aryl-aryl bonds in concert with one another. The largest cavity in models of **36** has a rectangular or square cross section. The compound belongs to the point group D_{2d}. The overall shape of the model is that of a high western saddle, and it contains two long, cleft-shaped cavities approximately 10.8 Å long, 3.4 Å wide, and 4.3 Å deep. Attempts to prepare **36** led to material that, although sublimable at very high temperatures, was essentially insoluble even under severe conditions in all solvents tried. To increase the solubility we prepared compound **37**, whose eight ethyl groups made it subject to dissolution and purification (*19*). The compound has been characterized.

Cavitand **38** was obtained as a by-product of **37**. In spite of its high molecular weight (1332), the compound was easily purified and characterized by ordinary techniques. In molecular models the substance has a sizable enforced cavity whose shape depends somewhat on the dihedral angles between the planes of the attached aryl rings. These six angles vary in concert with one another in a way similar to those in a collapsible wine rack. The largest cavity in models of **38** has a hexagonal cross section. The compound belongs to point group D_{3d}, since it contains a C_3 axis, three C_2 axes, and three mirror planes. The dimensions of the cavity are approximately 11 by 7 by 7 Å. It neatly embraces molecular models of seven benzene rings, one lying in the plane of the six oxygens, three stacked in planes perpendicular to one side of the first benzene, and three similarly stacked on the other side. Alternatively, molecular models of two molecules of p-$CH_3C_6H_4C_6H_4CH_3$-p completely occupy the cavity.

The cavity of **38** can partially collapse. Total collapse is inhibited by two sets of oxygens of the dibenzofuran units running into one another. The resulting minimum cavity has an irregular hexagonal cross section. Its dimensions are 11 by 9 by 3.5 Å. This cavity snugly embraces models of four benzenes stacked in specialized ways. The orientations of the ethyl groups affect the cavity sizes in only minor ways. They tend to line the cavity surfaces in the areas not lined by aromatic rings.

Unlike the other characterized cavitands, **37** and **38** are more soluble in toluene and xylene than in chloroform. Molecular model fitting indicates that the space in cavitands **37** and **38** can be more thoroughly occupied by aromatic solvent molecules whose bulk is distributed largely in two dimensions than by those whose bulk is distributed largely in three dimensions, such as chloroform. It will be interesting to discover whether a general correlation exists between solubility and complementarity in shapes of cavities and solvents. Preliminary results show that both **37** and **38** complex a variety of solvents in the crystalline state.

Envisioned Modifications of Cavitands

Some of the kinds of cavitands whose structures have been suggested here are potential hosts for binding and orienting guests. Others are hydrocarbons that offer only π-electron or relatively weak van der Waals forces to attract possible guests. Complexations involving such hosts are expected to be greatest with water as the solvent and with guests that are also lipophilic. Many of the suggested cavitands could be rendered water-soluble by introduction of several $ArSO_3^-$ or $ArCH_2\overset{+}{N}(CH_3)_3$ groups appropriately placed in the compounds so that micelle formation is avoided. For example, introduction of eight $CH_2CH_2\overset{+}{N}(CH_3)_3$ groups into the framework of **37** in place of the eight CH_2CH_3 groups should provide water solubility as well as opportunities for ion pairing with appropriate guests. Water solubility opens the way for use of hydrophobic binding (solvent liberation driving forces for complexation). Thus both solubility and binding site manipulation are subject to molecular design.

One of the incentives for the exploration of host-guest complexation chemistry is the expectation that organic catalysts of the future will combine binding and orientation with cooperativity between catalytic functional groups. Many of the cavitands suggested here have aryl-hydrogen bonds that are potentially substitutable with catalytic groups such as carboxyl, hydroxyl, amino, or imidazole. The direct attachment of such functional groups to aryl rings tends to make their location relatively free of conformational mobility. I believe that structural recognition in complexation and catalysis will depend on the availability of hosts whose cavity sizes and shapes and catalytic sites are subject to a minimum of ambiguity.

References and Notes

1. For example, see M. L. Bender and M. Komiyama, *Cyclodextrin Chemistry* (Springer, Berlin, 1978); F. Cramer, M. L. Bander, R. Breslow, E. T. Kaiser, and I. Tabushi have all made important contributions to this field.
2. C. J. Pedersen, *Synthetic Multidentate Macrocyclic Compounds*, R. M. Izatt and J. J. Christensen, Eds. (Academic Press, New York, 1978), chapter 1.
3. D. J. Cram and K. N. Trueblood, *Top. Curr. Chem.* **98**, 43 (1981).
4. J.-M. Lehn, *Struct. Bonding (Berlin)* **16**, 1 (1973).
5. K. N. Trueblood, C. B. Knobler, E. Maverick, R. C. Helgeson, S. B. Brown, D. J. Cram, *J. Am. Chem. Soc.* **103**, 5594 (1981).
6. G. M. Lein and D. J. Cram, *J. Chem. Soc. Chem. Commun.* (1982), p. 301.
7. CPK molecular models are space-occupying and are based on crystal structures of the organic

compounds important in biological processes. Except when compounds are severely compressed or have abnormal bond angles or bond lengths, CPK models are very useful in predicting molecular shape. They are models referred to throughout this article.

8. M. P. deGrandpre and D. J. Cram, unpublished results.
9. We thank K. N. Trueblood for this information in advance of publication.
10. R. C. Helgeson, J.-P. Mazaleyrat, D. J. Cram, *J. Am. Chem. Soc.* **103**, 3929 (1981).
11. C. D. Gutsche, B. Dhawan, K. H. No, R. Muthukrishnan, *ibid.*, p. 3782.
12. C. D. Gutsche and J. A. Levine, *ibid.* **104**, 2652 (1982).
13. H. Erdtman, F. Haglid, R. Ryhage, *Acta Chem. Scand.* **18**, 1249 (1964).
14. A. G. S. Högberg, *J. Org. Chem.* **45**, 4498 (1980).
15. D. J. Cram, J. R. Moran, S. Karbach, *J. Am. Chem. Soc.* **104**, 5826 (1982).
16. A. S. Lindsey, *J. Chem. Soc.* (1965), p. 1685; A. Lüttringhaus and K. C. Peters, *Angew. Chem. Int. Ed. Engl.* **5**, 593 (1966); A. Collet *et al.*, *J. Chem. Soc. Perkin Trans.* **1**, 1630 (1981).
17. S. J. Keipert and D. J. Cram, unpublished work.
18. J. Gabard and A. Collet, *J. Chem. Soc. Chem. Commun.* (1981), p. 1137.
19. R. C. Helgeson, M. Lauer, D. J. Cram, *ibid.*, in press.
20. I thank the U.S. Public Health Service for grant GM 12640 and the National Science Foundation for grant NSF CHE 81-09532, which supported this research. This article is dedicated to H. Erdtman on the occasion of his 80th birthday.

Beyond molecular chemistry based on the covalent bond lies supramolecular chemistry based on molecular interactions—the associations of two or more chemical entities and the intermolecular bond (*1–3*).

Molecular interactions are the basis of the highly specific processes that occur in biology, such as a substrate binding to an enzyme or a receptor, the assembling of protein complexes, intermolecular reading of the genetic code, signal induction by neurotransmitters, and cellular recognition. The correct manipulation of the energetic and stereochemical features of the noncovalent, intermolecular forces (electrostatic forces, hydrogen bonding, van der Waals forces, and so forth) within a defined molecular architecture should allow the design of artificial receptor molecules capable of binding substrate species strongly and selectively, forming supramolecular entities, so-called supermolecules, of well-defined structure and function. One might say that supermolecules are to molecules and the intermolecular bond what molecules are to atoms and the covalent bond (*4*).

In order for the receptor to "recognize" a potential substrate and bind to it, the two species must complement each other both in size and shape (geometry) and binding sites (energy). This extends Emil Fischer's "lock and key" concept from steric fit to other, intermolecular, properties.

Receptor chemistry, therefore, may be considered generalized coordination chemistry. It extends the purpose of designed organic complexing agents from the coordination of transition metal ions, for which they were first used, to the coordination of all kinds of substrates: cationic, anionic, and neutral

27

Supramolecular Chemistry: Receptors, Catalysts, and Carriers

Jean-Marie Lehn

Science 227, 849–856 (22 February 1985)

species of an inorganic, organic, or biological nature.

In addition to binding sites, the receptor may bear reactive sites that transform the bound substrate, which would make the receptor a molecular reagent or catalyst. If it is fitted with lipophilic groups that allow it to dissolve in a membrane, it may act as a molecular carrier. Thus, the functional properties of a supermolecule cover molecular recognition, catalysis (transformation), and transport (translocation) (Fig. 1) (4).

Macropolycyclic Receptors

Molecular recognition requires that a receptor and its substrates be in contact over a large area. Thus, artificial receptors must contain intramolecular cavities sufficiently large to allow substrate inclusion as well as structural elements that endow the three-dimensional framework with planned geometric and dynamic features (a balance between flexibility and rigidity). This leads to concave, hollow molecules of defined architecture that can bind substrate species by multiple noncovalent interactions.

Macropolycyclic architectures, in principle, meet the requirements. Being large (macro) and highly connected (polycyclic), they are suitable for the construction of molecules containing the cavities, clefts, and pockets that provide the framework for arrangement of binding sites, reactive groups, and bound species.

The binding of a substrate by a macropolycyclic receptor forms an inclusion complex, a cryptate, in which the substrate is contained inside the molecular cavity (crypt) of the ligand (cryptand). Although we introduced cryptates as a class of cation-inclusion complexes (5), they may be considered a general type of compound, independent of the nature of the receptor and substrate. As our work progressed to include several classes of macropolycyclic structures—macrocycles, macrobicycles, and cylindrical and spherical macrotricycles (Fig. 2)—the initial studies of macrobicyclic cationic cryptates expanded into the study of the structures and functions of supermolecules, covering the design of artificial molecular receptors, catalysts, and carriers. These species may be either biomimetic or abiotic since they may either

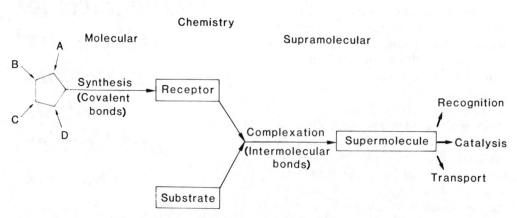

Fig. 1. From molecular to supramolecular chemistry.

394

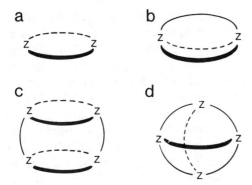

Fig. 2. Some macropolycyclic structures: (a) macrocyclic, (b) macrobicyclic, (c) cylindrical macrotricyclic, and (d) spherical macrotricyclic.

binding the spherical alkali cations (AC) and alkaline-earth cations (AEC).

Indeed, macrobicyclic ligands such as **1** through **3** form cryptates [M^{n+} ⊂ cryptand], **4**, by inclusion of metal cations inside the molecule. The optimal cryptates of AC and AEC have stabilities

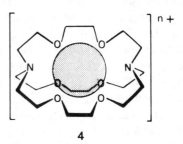

1 m=0, n=1
2 m=1, n=0
3 m=n=1

several orders of magnitude higher than those of either the natural or synthetic macrocyclic ligands. They show pronounced selectivity as a function of the size complementarity between the cation and the intramolecular cavity, a feature termed spherical recognition. As the bridges of the macrobicycle are length-

serve as models of biological systems and reactions or provide access to non-biological systems or processes displaying the efficiency and selectivity of biological ones.

This review will mainly cover molecular recognition and the design of molecular receptors, going from monotopic receptors, which possess a single site for substrate complexation, to polytopic receptors, which contain several binding subunits. Some of the results cited will demonstrate the molecular catalysis and transport properties of supermolecules (6).

Spherical Recognition: Cationic

Cryptates of Macrobicyclic Ligands

The coordination chemistry of alkali cations developed only about 15 years ago after the discovery that natural (7) or synthetic (8, 9) macrocycles and macrobicycles (1, 5, 10, 11) are powerful ligands. Whereas macrocycles define a two-dimensional, circular hole, macrobicycles define a three-dimensional, spheroidal cavity, particularly well suited for

$$4$$

ened from cryptand **1** to **3**, the most strongly bound ion becomes, respectively, Li^+, Na^+, and then K^+. Cryptand **3** also displays a higher selectivity for Sr^{2+} and Ba^{2+} than for Ca^{2+}. Suitable structural modifications can result in a high AC to AEC (M^+ to M^{2+}) selectivity (1, 5, 10, 11).

Cryptands **1** through **3** thus function as receptors for spherical cations. Their special complexation properties result from their macropolycyclic nature and define a cryptate effect characterized by high stability and selectivity, slow exchange rates, and efficient shielding of the bound ion from the environment.

Numerous other macrobicyclic cryptands and cryptates have been obtained. Replacing the oxygen sites in **1** through **3** with sulfur or nitrogen sites yields cryptands that show marked preference for transition metal ions and that allow highly selective complexation of toxic heavy metals such as cadmium, lead, and mercury. Cryptation markedly affects the redox properties of the enclosed metal ion (*12*) and may stabilize uncommon oxidation states such as Eu^{II} (*12a*).

Tetrahedral Recognition by Macrotricyclic Cryptands

Selective binding of a tetrahedral substrate requires the construction of a receptor molecule with a tetrahedral recognition site. This may be achieved by positioning four suitable binding sites at the corners of a tetrahedron and linking them with six bridges. Such a structure, formally a cylindrical macrotricycle, has been realized in the spherical cryptand **5**, which contains four nitrogens located at the corners of a tetrahedron and six oxygens located at the corners of an octahedron, as shown by **6** (*11*).

Indeed, **5** binds the tetrahedral NH_4^+ cation exceptionally strongly and selectively (as opposed to K^+) (*13*), forming an ammonium cryptate, designated by $[NH_4^+ \subset 5]$ and shown as **7**. This complex presents a high degree of structural and binding complementarity between the substrate, NH_4^+, and receptor, **5**. The ammonium ion fits into the cavity of **5**, and it is held by a tetrahedral array of $^+N–H \ldots N$ hydrogen bonds and by electrostatic interactions with the six oxygens. The strong binding in **7** results in an effective pK_a for the NH_4^+ that is about six units higher than that of free NH_4^+. This illustrates how large may be the changes in substrate properties brought about by binding. Similar changes may take place when substrates bind to enzyme-active sites and to biological receptors.

The remarkable protonation features of **5** led to the formulation of the diprotonated species as the water cryptate, $[H_2O \subset 5–2H^+]$, **8**, in which the water molecule accepts two $^+N–H \ldots O$

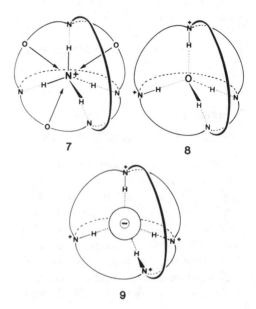

7   8

5

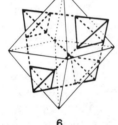

6

9

bonds from the protonated nitrogens and donates O–H . . . N bonds to the unprotonated ones (*2, 11*). The second protonation of **5** is facilitated by the substrate; it is considered a "positive cooperativity" effect, which is mediated by H_2O as the effector. When **5** is tetraprotonated, it forms the chloride cryptate [$Cl^- \subset$ **5**–$4H^+$], **9**, in which the included anion is bound by four ^+N–H . . . X^- hydrogen bonds (*2, 11, 14*).

The spherical macrotricycle **5** is thus a receptor molecule possessing a tetrahedral recognition site that binds the substrates in a tetrahedral array of hydrogen bonds, as in **7**, **8**, and **9**. It illustrates the molecular engineering required in abiotic receptor chemistry.

Chemical Applications of Cryptates

The strong binding of AC's and AEC's by neutral cryptands of types **1** through **3** has led to numerous applications both in pure and applied chemistry. Cryptate formation transforms a small metal cation into a large, spheroidal, organic cation about 10 Å in diameter, a sort of super-heavy AC or AEC. This allows the study of ionic solvation and makes the bound ions more difficult to reduce. The stability of the cryptates and the large distance imposed by the thick organic-ligand shell between the enclosed cation and the environment (both the anion and the solvent) have many physical and chemical consequences (*11, 15–19*).

Cryptate counterions are able to stabilize unusual species such as alkalides, as in ([$Na^+ \subset$ **3**]Na^-) (*17*), electrides, as in ([$M^+ \subset$ cryptand]e^-) (*17*), and anionic clusters of the heavy posttransition metals, as in ([$K^+ \subset$ cryptand]$_2$ Pb_5^{2-}) (*18*).

Cryptation promotes the solubilization of salts and dissociation of ion pairs, resulting in strong anion activation. It thus markedly increases the rate of numerous reactions, such as those involving the generation of strong bases, nucleophilic substitutions, carbanion reactions, alkylations, rearrangements, anionic polymerizations, and phase transfer catalyses. It may even change their course. Conversely, cryptate formation inhibits reactions in which cation participation (electrostatic catalysis) plays an important role. Thus, cryptands are powerful tools for studying the mechanism of ionic reactions that involve complexable metal cations. Their effect on a reaction is a criterion for ascertaining the balance between anion activation and cation participation under a given set of conditions (*16*).

Cryptands, either alone or fixed on a polymer support, have been used in many processes, including selective extraction of metal ions, solubilization, isotope separation, decorporation of radioactive or toxic metals, and cation-selective analytical methods (*19*). A number of patents have been granted for such applications.

Anion Receptor Molecules and

Anion Coordination Chemistry

In view of the fundamental role played by anions in chemical as well as biological processes, the binding of anions by organic ligands would be expected to provide a multitude of novel structures with properties of wide significance. However, it has received little attention in comparison with cation coordination, which has been the subject of numerous studies. Only in recent years has anion coordination chemistry been developed as a new area of coordination chemistry (*2, 14, 20*). This arose from the design of anion receptor molecules of various types, especially macrocyclic and macropolycyclic polycations capable of forming strong and selective complexes with inorganic anions and with negatively charged functional groups (such as carboxylate and phosphate) of organic or biological substrates. The binding strength and selectivity of the receptors are provided by electron-deficient interaction sites (like the positively charged sites of the polyammonium and polyguanidinium cations, which may establish multiple $^+N–H \ldots X^-$ bonds), suitably arranged around an intramolecular cavity of a shape and size adapted to the anionic substrate to be bound (*14, 20–23*).

Polyammonium macrocycles of various ring sizes (for instance, **10** through **12**) act as anion receptors towards organic polycarboxylates, displaying stabilities and selectivities that result from both electrostatic and structural effects (*21*). The binding of complex anions of transition metals, such as the hexacyanides $M(CN)_6^{n-}$, markedly affects their redox and photochemical properties (*24, 25*). The strong complexation of adenosine mono-, di-, and triphosphates (AMP, ADP, and ATP) is particularly significant in view of their role in bioenergetics. It presents the possibility of devising molecular catalysts and carriers for these substrates. Substances like **10** and **11** are cyclic analogs of biological polyamines and could thus interact with biomolecules; indeed several macrocyclic polyamines induce efficient polymerization of actin.

Protonated macrobicyclic diamines form katapinates by inclusion of halide

ions (*22*). Tetraprotonated macrotricycles, such as **5–4H$^+$**, are geometrically suitable receptors for spherical anions and form anion cryptates with halides. Thus, **5–4H$^+$** yields a chloride cryptate [Cl$^-$ ⊂ **5–4H$^+$**], **9**, of high stability, and it shows a high selectivity for chloride over bromide, but it does not complex other types of anions (*14*).

The hexaprotonated form of the ellipsoidal cryptand Bis-Tren, **13**, binds various monoatomic and polyatomic anions and extends the recognition of anionic substrates beyond the spherical halides (*23*). The strong and selective binding of the linear, triatomic anion N_3^- results from its complementarity to the receptor **13–6H$^+$**. In [N_3^- ⊂ **13–6H$^+$**], **14**, the substrate is held inside the cavity by two pyramidal arrays of $^+N–H \ldots N^-$ hydrogen bonds, each of which binds one of the two terminal nitrogens of N_3^-.

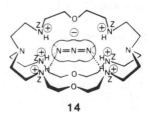

14

398

The noncomplementarity between the ellipsoidal 13–6H$^+$ and the spherical halides results in much weaker binding and appreciable distortions of the ligand, as seen in the crystal structures of the cryptates 15 where the bound ion is F$^-$, Cl$^-$,

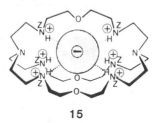

15

or Br$^-$. In these compounds, the F$^-$ is bound by a tetrahedral array of hydrogen bonds; and the Cl$^-$ and the Br$^-$, by an octahedral array. Thus, 13–6H$^+$ is a molecular receptor for the recognition of linear triatomic species of a size compatible with the size of the molecular cavity.

A cryptate effect is observed for anion complexes as well as cation complexes. In general, an increase in cyclic order from acyclic to macrocyclic to macrobicyclic significantly increases the stability and selectivity of the anion complexes formed by polyammonium ligands.

In addition, in 14, the receptor is built from two protonated tripodal subunits of the tren type, N(CH$_2$CH$_2$NH$_2$)$_3$, located at each pole of the molecule, which cooperate in substrate binding. This is a feature of coreceptor molecules, which will be discussed below.

Macrocyclic Receptors for

Ammonium Ions

Macrocyclic polyethers and aza-polyethers selectively bind primary ammonium ions by anchoring the –NH$_3$$^+$ into the circular cavity with three $^+$N–H . . . X (X = O, N) hydrogen bonds (2, 6, 8, 26, 27). In view of the role of such substrates in both chemistry and biology, we sought a derivative that would yield stronger complexes than those of the parent macrocycles and bear functional groups for further modification. Thus, the chiral, tetrafunctional macrocycle in 16 was devised, which, by attachment of lateral substituents to the central core, led to molecular receptors with a variety of binding properties (27).

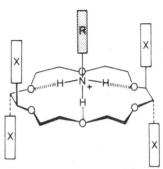

16a X = CO$_2$$^-$
  b X = CONHR
  c X = 2 x CO$_2$$^-$, 2 x CONHR
     (syn or anti isomer)
  d X = CONMe$_2$

The tetracarboxylate, 16a, forms the strongest metal-ion and ammonium complexes of any crown ether. It displays marked selectivity in favor of primary ammonium ions against more highly substituted ones (central discrimination). Of special interest is its selective binding of biologically active ions such as nor-adrenaline and norephedrine with respect to their N-methylated derivatives, adrenaline and ephedrine.

Varying the side groups, X, in 16b affects the interactions (electrostatic, lipophilic, H-bonding, and charge transfer) between the side groups and the R group of the centrally bound substrate. This affects both the stability and selectivity of the complexes, which is called lateral discrimination, and allows the receptor-substrate interactions in biological systems to be modeled, for instance, the interaction between nicotinamide and tryptophane (28).

The structural features of **16** and its remarkable binding properties make it an attractive unit for the construction of macropolycyclic multisite receptors, molecular catalysts, and carriers for membrane transport. Such extensions require separate handling of the side groups, as in the face- and side-discriminated derivatives **16c** (29).

Coreceptor Molecules

Coreceptors are defined as polytopic receptor molecules combining two or more discrete binding subunits within the same macropolycyclic architecture (30). In terms of the general functions of supramolecular systems, recognition, catalysis, and transport, they may act as coreceptors, cocatalysts, or cocarriers whose subunits cooperate for, respectively, the complexation, transformation, or translocation of either several singly bound substrates or a multiply bound, polyfunctional substrate. Depending on the subunits, such coreceptors may bind metal ions, organic molecules, or both. Their ability to perform multiple recognition provides entry into higher forms of molecular behavior, such as cooperativity, allostery, regulation, and communication (signal transfer).

**Dinuclear and Polynuclear
Metal-Ion Cryptates**

Macropolycyclic ligands incorporating two or more binding subunits for metal ions form dinuclear or polynuclear cryptates in which the distance and arrangement of the cations held inside the molecular cavity may be controlled through ligand design. They allow the study of cation-cation interactions (magnetic coupling, electron transfer, redox, and photochemical properties) as well as the inclusion of bridging substrates to yield cascade complexes, which are of interest for bioinorganic modeling and multicenter-multielectron catalysis.

Depending on the nature and number of binding subunits and of connecting bridges used as building blocks, a variety of macropolycyclic structures may be envisioned. Ditopic ligands that contain two units, which may be chelating, tripodal, or macrocyclic, bind two metal ions to form dinuclear cryptates of various types (Fig. 3). Combining three or four such subunits leads to tritopic and tetratopic metal-ion receptors. Dissymmetric ligands that contain subunits with "hard" and "soft" binding sites yield complexes in which the bound ions act as either redox or Lewis-acid centers. Representatives of these types of ligands and complexes have been obtained and studied. Only a few will be illustrated here (31).

Dinuclear cooper (II) cryptates of macrocyclic ligands (for example, **12**) or macrobicyclic ligands (for example, **13**) containing bridging groups (imidazolato, hydroxo, or azido) display antiferromagnetic or ferromagnetic coupling between the ions and bear a relation to dinuclear sites of copper proteins. Lateral macrobicycles are dissymmetric by design; thus, monoelectronic reduction of the Cu(II) bound to the [12]-N_2S_2 macrocyclic subunit in the bis-Cu(II) cryptate **17**, gives a mixed-valence Cu(I)-Cu(II) complex. Macrotricycle **18** also forms a dinuclear

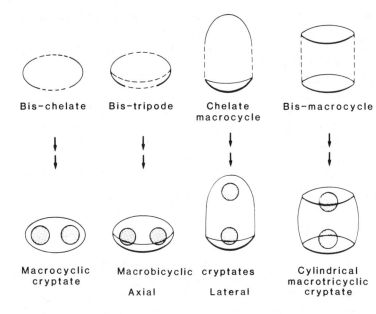

Bis-chelate	Bis-tripode	Chelate macrocycle	Bis-macrocycle

Fig. 3. Some dinuclear cryptates of macropolycyclic cryptands resulting from connection of chelating, tripodal, and macrocyclic subunits.

Macrocyclic cryptate

Macrobicyclic cryptates

Axial   Lateral

Cylindrical macrotricyclic cryptate

Cu(II) cryptate, which acts as a dielectronic receptor and exchanges two electrons in a single electrochemical wave.

trinuclear complex **20** containing a [tris-Cu(II), bis-μ_3-hydroxo] group in the cavity is formed by a tritopic, tris-ethylenediamine ligand (*33*). These few examples may at least have shown how rich the field of polynuclear metal cryptates is, both in structures and properties. Their chemical reactivity and use in catalysis have barely been studied up to now.

17

18

19

Polytopic receptors have the ability to assemble metal ions and bridging species within their molecular cavity to form "cluster cryptates." The bis-chelating macrocycle **12** gives complex **19** in which a triply bridged [Rh(CO)$_3$Rh]$^{2+}$ unit is built inside the ligand cavity (*32*). A

20

401

Diammonium Cryptates of Macrotricyclic Coreceptors

When two binding subunits are located at the poles of a coreceptor molecule, the complexation of a difunctional substrate will depend on the complementarity of the distance between the two binding sites in the receptor and the distance between the two corresponding functional groups in the substrate (*30*). Such linear recognition by ditopic coreceptors has been achieved for both dicationic and dianionic substrates, diammonium and dicarboxylate ions, respectively; it corresponds to the binding modes illustrated in **21** and **22**.

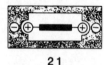

21

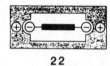

22

The cylindrical macrotricycles **23a** through **23d**, which contain two macrocyclic subunits capable of binding –NH$_3^+$ groups, yield molecular crypt-

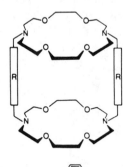

23a R=

b R=

c R=

d R=

ates with diammonium ions. In the supermolecules thus formed, **24**, the substrate is located in the central molecular

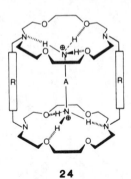

24

cavity of the coreceptor and anchored by each terminal –NH$_3^+$ group to an [18]-N$_2$O$_4$ macrocycle with three hydrogen bonds, as confirmed by the crystal structure of [$^+$H$_3$N–(CH$_2$)$_5$–NH$_3^+$ ⊂ **23d**], **25** (*34*).

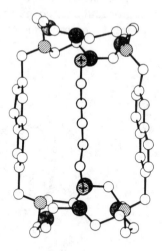

25

The $^+$H$_3$N–(CH$_2$)$_n$–NH$_3^+$ substrate that is preferentially bound has a length complementary to the length of the molecular cavity, with $n = 4$, 5, and 7 for **23a**, **23b**, and **23c**, respectively, Thus, ligand **23b** discriminates between cadaverine ($n = 5$) and putrescine ($n = 4$) cations. Similar effects may be expected for polyamines of various chain lengths such as ornithine, lysine, diammonium dipeptides, or spermine. Different selectivity sequences have been obtained with other macrotricycles (35), and triply bridged cylindrical receptors show a high degree of selectivity in forming their diammonium cryptates, **26** (36).

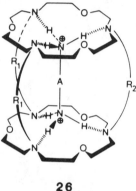

26

These results are evidence for structural complementarity of receptors and substrates being the basis for linear molecular recognition.

Carbon-13 nuclear magnetic resonance studies of complexes of the type **24** indicate that complementary receptor-substrate pairs display similar molecular motions, a dynamic fit, in addition to their steric fit. Thus, complementarity between components of a supramolecular species expresses itself in both structural and dynamic properties (37).

Ditopic Coreceptors for

Dicarboxylate Anions

The two hexaammonium macrocycles **27a** and **27b** possess the features of ditopic coreceptor molecules for dianionic substrates, since they contain two triammonium chelating subunits that may

serve as binding sites for a negatively charged group, as shown in **28**. They complex dicarboxylates with a selectivity that depends on the chain length of the substrates, ^-O_2C–$(CH_2)_m$–CO_2^-. The preferential binding of substrates with $m = 2$ and 3 by **27a** and of substrates with $m = 5$ and 6 by **27b** corresponds to an equivalent increase in length of the polymethylene chains separating binding subunits in both the substrates and the receptors. Receptors **27a** and **27b** also bind biological dicarboxylates of compatible chain lengths, respectively, amino acid and dipeptide dicarboxylates (38).

As in the binding of diammonium substrates to macrotricyclic coreceptors, this chain length selection describes a linear recognition process, based on structural complementarity in a ditopic binding mode, **20**. In both cases, the receptor acts as a discriminating sensor of molecular length.

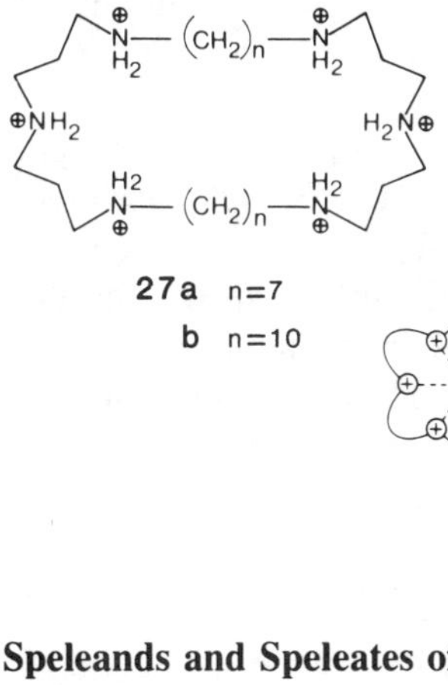

27a n=7
b n=10

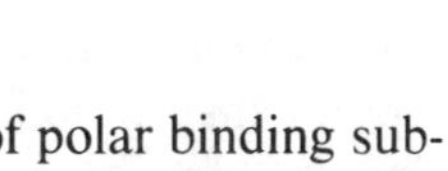

28

Speleands and Speleates of

Molecular Cations

The combination of polar binding subunits with more or less rigid, apolar shaping components provides amphiphilic, macropolycyclic coreceptors of cryptand type, termed speleands, that form speleates by substrate inclusion (30, 39, 40).

The macrocyclic speleand **29** combines two tartaric acid units with two diphenylmethane groups (40). It strongly binds a range of molecular cations by electrostatic and hydrophobic effects. It not only yields, with primary ammonium ions, more stable complexes than the common polyether macrocycles (except **16a**), but also it strongly binds secondary, tertiary, and quaternary ammonium substrates. Among the latter, the complexation of acetylcholine is of special interest: it provides one specific answer to the general question of how acetylcholine can be bound and sheds light on the type of interactions that may play a role in biological acetylcholine receptors.

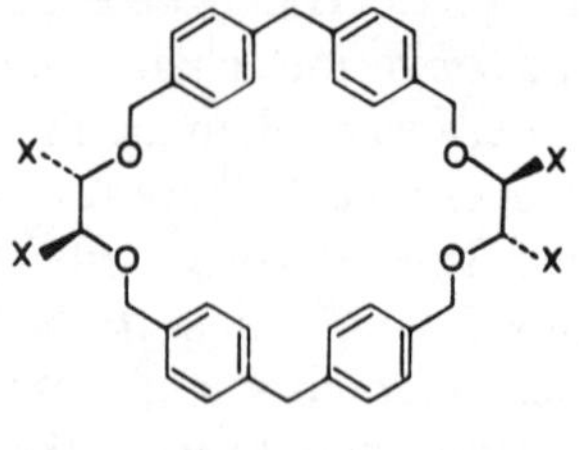

The macropolycyclic speleand **30a** combines an [18]-N_3O_3 macrocyclic binding subunit with a cyclotriveratrylene shaping component (39). Its tight intramolecular cavity allows inclusion of the CH_3–NH_3^+ ion, which results in the speleate [CH_3–$NH_3^+ \subset$ **30a**], shown as **30b**.

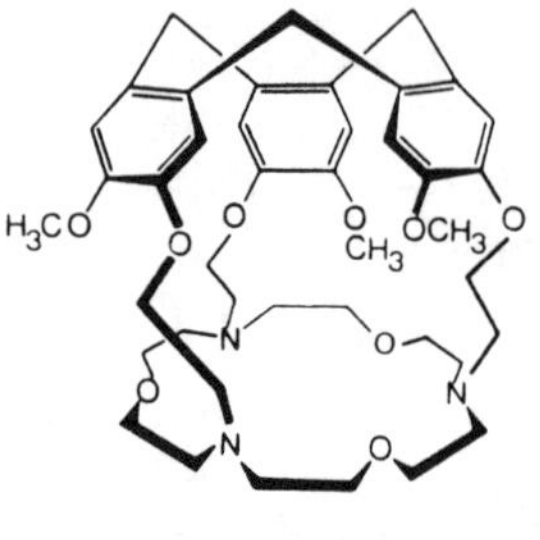

Numerous other combinations of (polar) binding units and (apolar) architectural components may be imagined, making speleands attractive for the design of novel, efficient molecular receptors for anions and neutral substrates as well as for cations.

Metalloreceptors and
Mixed-Substrate Supermolecules

Metalloreceptors are heterotopic coreceptors that contain substrate-selective binding subunits for the complexation of both metal ions and organic species within the same superstructure.

29 X=CO_2^-

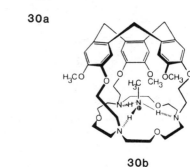

30a

30b

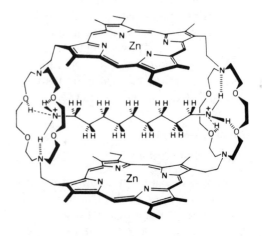

31

404

Such substances have been obtained by introducing one or two porphyrin or α,α'-bipyridine metal ion binding units as bridges in the macrotricycles of type **23a** through **23d**. These compounds may complex diammonium ions, as in **24**, as well as metal ions (*30, 31, 41*). Simultaneous binding of $^+H_3N-(CH_2)_9-NH_3^+$ and of two Zn^{2+} cations by such a bis-porphyrin metalloreceptor yields the mixed-substrate supermolecule **31**. Complexation of several metal ions gives polynuclear cryptates. Many variations are conceivable involving the subunits as well as the overall macropolycyclic architecture. The simultaneous complexation of organic and inorganic substrates offers the opportunity to induce or adjust physical and chemical interactions and reactions (metallocatalysis) between the metal-centered reactive sites and the co-bound molecular substrates. It could also mimic essential features of metalloenzymes. For instance, in species like **31**, activation of the internally bound organic substrate by the metal-porphyrin sites may be envisioned. Also by binding an effector species, the macrocyclic units could act as regulation sites for external interaction with the metal-porphyrin centers, to the point of exerting allosteric control and cooperativity.

Supramolecular Catalysis

Molecular receptors bearing appropriate functional groups may bind a substrate, react with it, and release the products. This would be called supramolecular catalysis, catalysis within a supermolecule (Fig. 4). The design of efficient and selective molecular catalysts may give mechanistic insight into the elementary steps of catalysis, provide new types of chemical reagents, and provide models, "artificial enzymes," that reveal factors which contribute to enzyme catalysis (*42, 43*).

Enhanced rates were observed for hydrogen transfer from 1,4-dihydropyridyl side chains attached to the macrocyclic ammonium receptor **16** to bound pyridinium substrates, as shown in **32**. This intracomplex reaction was inhibited by addition of a complexable cation that displaced the substrate (*44*).

Ester cleavage has been induced with the tetra-(L)-cysteinyl derivative of **16**, which binds *p*-nitrophenyl (PNP) esters of amino acids and reacts with the bound species, releasing PNP at various rates. This intracomplex reaction displays substrate selectivity with marked rate enhancement in favor of bound dipeptide esters such as glycylglycine-

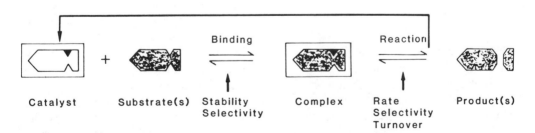

Fig. 4. Schematic representation of the supramolecular catalysis process.

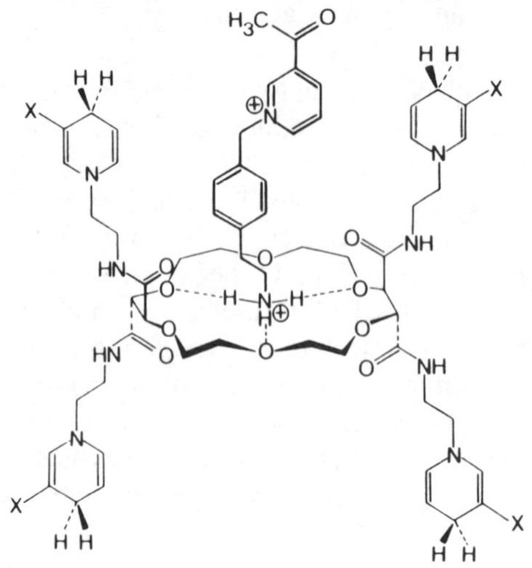

32

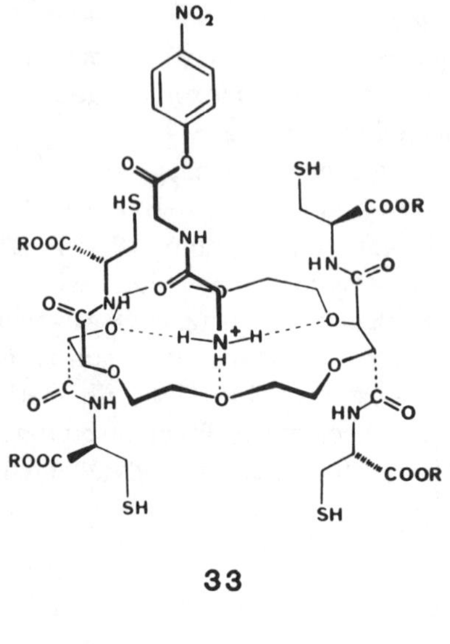

33

The development of anion receptor molecules allows molecular catalysis to be performed on anionic substrates of chemical and biochemical interest. Thus, macrocyclic polyamines were found to catalyze ATP hydrolysis, protonated [24]-N_6O_2, **12**, being particularly efficient (46). It strongly binds ATP and markedly accelerates its hydrolysis over a wide pH range, probably through a combination of acid, electrostatic, and nucleophilic catalysis. The latter reaction proceeds with an N-phosphoryl intermediate, which is subsequently hydrolyzed, a process reminiscent of the enzyme-catalyzed reaction, which gives **12** proto-adenosinetriphosphatase-type activity.

Recent work showed that when **12** is used to catalyze acetylphosphate hydrolysis, appreciable amounts of pyrophosphate are formed. Again an N-phosphorylated macrocycle intermediate is generated, which is apparently capable of transferring its phosphoryl group to a phosphate substrate (probably held in proximity), thus achieving the synthesis of inorganic pyrophosphate. This was confirmed by nuclear magnetic resonance spectroscopy and isotopic (^{18}O) labeling experiments. Macrocycle **12** may be considered to display protokinase activity in this process (47).

These systems possess a number of properties that supramolecular catalysis should display, that is, properties of an abiotic enzyme (protoase): selective substrate binding, reaction within the supramolecular complex, rate acceleration, inhibition by species competing for the binding site, structural selectivity, and chiral recognition.

Numerous other processes may be imagined. Of particular interest is the development of catalysts capable of realizing synthetic reactions, bond-making

OPNP in complex **33**. It is inhibited by a complexable metal cation, such as K^+, and it shows high chiral recognition between the enantiomeric dipeptide esters (Gly-(L)-Phe-OPNP and Gly-(D)-Phe-OPNP, the former reacting more than 50 times faster than the latter (45).

rather than bond-breaking processes. For this, the presence of several binding and reactive sites within the molecular catalyst is essential. Thus, coreceptors open the way to the design of artificial molecular cocatalysts of the ligase, metallocatalyst, and enzyme-coenzyme types, which act on two or more cobound and spatially oriented substrates. In the pyrophosphate generation mentioned above, **12** displays such a cocatalysis function by mediating bond formation between two species.

Transport Processes and Carrier Design

One of the initial motivations of our work with cryptates was to investigate the ability of such complexes to transport cations through membranes (*1, 5*). This led us to explore the chemistry of transport and the design of transport effectors (*42, 48*).

Among the different transport mechanisms, facilitated diffusion, or carrier-mediated transport, consists of the transfer of a substrate across a membrane with the assistance of a carrier molecule (Fig. 5). This cyclic process may be considered as a physical catalysis that effects a translocation on the substrate just as chemical catalysis effects a transformation. The carrier is the transport catalyst, which strongly increases the rate of passage of the substrate with respect to its free diffusion, controls its selectivity, and allows the transport to be coupled to other processes such as proton cotransport (a proton pump) or electron cotransport. The active species is the carrier-substrate supermolecule; and transport is thus one of the basic functional features of supramolecular systems, together with recognition and catalysis.

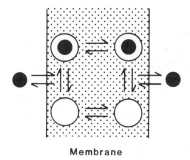

Membrane

Fig. 5. Schematic representation of carrier-mediated transport through a membrane. Closed circles represent substrates; open circles represent carriers; and closed circles within open circles represent complexes.

Depending on internal factors, such as ligand structure or ligand-cation pairing, and external factors, such as the nature of the counterion, the nature of the membrane, or the concentrations of the substrates and carriers, macrobicyclic cryptands of type **1** through **3** were capable of selectively transporting alkali cations, even under conditions where natural or synthetic macrocycles show little activity. The transport rates and the selectivity of the transport for the various alkali cations can be varied by modifying the structure of the cryptand (*49*). Both experimental results and kinetic analysis indicate that there is an optimal binding ability for the highest transport efficiency. Molecular cations, like physiologically active, primary ammonium ions, are transported selectively by macrocyclic polyethers.

Anion carriers may, in principle, be derived from anion receptors. However, this area has been comparatively little explored, although it promises significant developments, for example, the selective transport of organic and biological anions.

The ability to design and to set up coupled transport processes may be clearly illustrated by two examples.

Electron-cation symport has been achieved in a two carrier process in which the electrons and K^+ ions are transported in the same direction, pumped by a redox gradient. This transport is mediated simultaneously by an electron carrier, a nickel complex, and a selective cation carrier, a polyether macrocycle (*50*).

A striking regulation of the selectivity of Ca^{2+} and K^+ transport by *p*H has been achieved with both isomers of the lipophilic dicarboxylate-dicarboxamide macrocyclic carriers of type **16c**. The process involves competitive Ca^{2+} and K^+ symport coupled to back-transport of protons in a *p*H gradient (a proton pump). It displays a change from preferential K^+ transport to preferential Ca^{2+} transport as a function *p*H (*51*). These results demonstrate how appropriate design provides carriers that perform a given function.

Cation passage through membranes may occur through a transmembrane channel rather than a mobile carrier. A solid state model of a K^+ channel is provided by the crystal structure of the KBr complex of **16d** (*52*), and attempts to synthesize a channel structure on the basis of such macrocyclic subunits are under way.

Transport studies provide the means for effector design, analysis of the elementary steps and mechanisms of transport, coupling transport to chemical potentials, energy and signal transduction, models of biological transport processes, and so on. There are a variety of possible applications, for instance, in separation and purification, in batteries, or in systems for artificial photosynthesis. Again,

the multiple sites in coreceptors should allow the design of cocarriers capable of transporting several substrates with the transport coupled to electrochemical gradients.

Conclusion

The chemistry of artificial molecular receptors, catalysts, and carriers has produced supramolecular species that effect molecular recognition, catalysis and transport. It has provided insight into the elementary interactions and processes on which these functions are based, and it has an increasing impact on organic, inorganic, and biological chemistry as well as on other fields of science. This chemistry has led to the elaboration of numerous novel structures and made available new properties of abiotic as well as biomimetic interest. Molecules belonging to a number of structural categories other than those described above, have been studied [for example, cyclodextrins (*43, 53*), cyclophanes (*43*), calixarenes (*54*), cavitands (*9*), cryptophanes (*55*), and catenands (*56*)] (*57*) and many more may be imagined, assisted by the refinement of methods for theoretical molecular design (*58*). The driving force and the selection process of the chemical evolution of these artificial systems rest in the chemist's creative imagination and synthetic (molecular and supramolecular) power.

In combination with molecular layers, membranes and vesicles (*59*), receptors, catalysts, and carriers are necessary elements in the elaboration of chemical microreactors and artificial cells. In another perspective, they may play a key role in the development of an intriguing although still rather futuristic area, which may be termed "chemionics," the design of components, circuitry, and systems for signal and information treatment at the molecular level (*3*). These "molecular devices" may be defined as structurally organized and funtionally integrated chemical species built on supramolecular architectures and operating via electrons ("molecular electronics"), inorganic or organic ions ("molecular ionics") or photons ("molecular photonics") (*60*). Such prospects offer challenging opportunities for continuously expanding on the words of Marcelin Berthelot in 1860: "La Chimie crée son objet."

References and Notes

1. J. M. Lehn, *Struct. Bonding (Berlin)* **16**, 1 (1973).
2. _____, *Pure Appl. Chem.* **50**, 871 (1978).
3. _____, *Leçon Inaugurale* (Collège de France, Paris, 1980).
4. A number of the basic concepts and definitions that led to the formulation of supramolecular chemistry have been introduced in the course of the design of receptors for the spherical alkali cations (*1*). General presentations may be found in (*1–3*) as well as in (*30, 31, 42, 48*). The term "Übermolekeln" (supermolecules) was introduced in the mid-1930's to describe entities of higher organization resulting from the association of two or more coordinatively saturated species; see K. L. Wolf, F. Frahm, H. Harms, *Z. Phys. Chem. Abt. B* **36**, 17 (1937); K. L. Wolf, H. Dunken, K. Merkel, *ibid.* **46**, 287 (1940); K. L. Wolf and R. Wolff, *Angew. Chem.* **61**, 191 (1949). We consider that a supermolecule results from binding of substrate to a receptor. This terminology conveys the sense of biological receptor-substrate interactions, with their highly defined structural and functional properties. Furthermore, it is easily converted from one language to another. The "inclusion compound" and "host-guest" designations also cover species that exist only in the solid state (also termed clathrates) and are not supermolecules; see, for instance, F. Cramer, *Einschlussverbindungen* (Springer-Verlag, Berlin, 1954); J. E. D. Davies, W. Kemula, H. W. Powell, N. O. Smith, *J. Inclusion Phenom.* **1**, 3 (1983).
5. B. Dietrich, J. M. Lehn, J. P. Sauvage, *Tetrahedron Lett.* (1969), p. 2885; *ibid.*, p. 2889.
6. The present account surveys various aspects of our own work. Numerous reports and reviews describe other aspects of this field; for instance, see R. C. Hayward, *Chem. Soc. Rev.* **12**, 285 (1983).
7. Yu. A. Ovchinnikov, V. T. Ivanov, A. M. Skrob, *Membrane Active Complexones* (Elsevier, New York, 1974); B. C. Pressman, *Annu. Rev. Biochem.* **45**, 501 (1976).
8. C. J. Pedersen and H. K. Frensdorff, *Angew. Chem. Int. Ed. Engl.* **11**, 16 (1972); C. J. Pedersen, *J. Am. Chem. Soc.* **89**, 7017 (1967).
9. D. J. Cram, *Science* **219**, 1177 (1983).
10. J. M. Lehn and J. P. Sauvage, *J. Am. Chem. Soc.* **97**, 6700 (1975).
11. J. M. Lehn, *Acc. Chem. Res.* **11**, 49 (1978).
12. J. P. Gisselbrecht and M. Gross, *Adv. Chem. Ser.* **201**, 111 (1982).
12a. E. L. Yee, O. A. Gansow, M. J. Weaver, *J. Am. Chem. Soc.* **102**, 2278 (1980).
13. E. Graf, J. P. Kintzinger, J. M. Lehn, J. LeMoigne, *ibid.* **104**, 1672 (1982).
14. E. Graf and J. M. Lehn, *ibid.* **98**, 6403 (1976).
15. A. I. Popov and J. M. Lehn, *Coordination Chemistry of Macrocyclic Compounds*, G. A. Melson, Ed. (Plenum, New York, 1979).
16. J. M. Lehn, *Pure Appl. Chem.* **52**, 2303 (1980).
17. J. L. Dye, *Angew. Chem. Int. Ed. Engl.* **18**, 587 (1979).
18. J. D. Corbett, S. C. Critchlow, R. C. Burns, *ACS Symp. Ser.* **232**, 95 (1983).
19. I. M. Kolthoff, *Anal. Chem.* **51**, 1R (1979).
20. J. L. Pierre and P. Baret, *Bull. Soc. Chim. Fr.* (1983), p. 367.
21. B. Dietrich, M. W. Hosseini, J. M. Lehn, R. B. Sessions, *J. Am. Chem. Soc.* **103**, 1282 (1981).
22. C. H. Park and H. E. Simmons, *ibid.* **90**, 2431 (1968).
23. B. Dietrich, J. Guilhem, J. M. Lehn, C. Pascard, E. Sonveaux, *Helv. Chim. Acta* **67**, 91 (1984).
24. F. Peter, M. Gross, M. W. Hosseini, J. M. Lehn, *J. Electroanal. Chem. Interfacial Chem.* **144**, 279 (1983).
25. M. F. Manfrin *et al.*, *J. Chem. Soc. Chem. Commun.* (1984), p. 555.
26. D. J. Cram and J. M. Cram, *Acc. Chem. Res.* **11**, 8 (1978).
27. J. P. Behr, J. M. Lehn, P. Vierling, *Helv. Chim. Acta* **65**, 1853 (1982).
28. J. P. Behr and J. M. Lehn, *ibid.* **63**, 2112 (1980).
29. _____, D. Moras, J. C. Thierry, *J. Am. Chem. Soc.* **103**, 701 (1981).
30. J. M. Lehn, in *Biomimetic Chemistry*, Z. I. Yoshida and N. Ise, Eds. (Elsevier, New York, 1983), pp. 163–187. Earlier work had shown that receptors of this type form complexes with fluorescent substrate molecules: J. M. Lehn, J. Simon, J. Wagner, *Angew. Chem. Int. Ed. Engl.* **12**, 578 (1973); *ibid.*, p. 579.
31. J. M. Lehn, *Pure Appl. Chem.* **52**, 2441 (1980); in *IUPAC Frontiers in Chemistry*, K. J. Laidler,
32. J. P. Lecomte, J. M. Lehn, D. Parker, J. Guilhem, C. Pascard, *J. Chem. Soc. Chem. Commun.* (1983), p. 296.
33. B. Dietrich, J. Comarmond, J. M. Lehn, R. Louis, *ibid.* (1985), p. 74.
34. F. Kotzyba-Hibert, J. M. Lehn, P. Vierling, *Tetrahedron Lett.* (1980), p. 941; C. Pascard, C. Riche, M. Cesario, F. Kotzyba-Hibert, J. M. Lehn, *J. Chem. Soc. Chem. Commun.* (1982), p. 557.
35. N. F. Jones, A. Kumar, I. O. Sutherland, *J. Chem. Soc. Chem. Commun.* (1981), p. 990.
36. F. Kotzyba-Hibert, J. M. Lehn, K. Saigo, *J. Am. Chem. Soc.* **103**, 4266 (1981).

37. J. P. Kintzinger, F. Kotzyba-Hibert, J. M. Lehn, A. Pagelot, K. Saigo, *J. Chem. Soc. Chem. Commun.* (1981), p. 833.
38. M. W. Hosseini and J. M. Lehn, *J. Am. Chem. Soc.* **104**, 3525 (1982).
39. J. Canceill *et al.*, *Helv. Chim. Acta* **65**, 1894 (1982).
40. M. Dhaenens *et al.*, *J. Chem. Soc. Chem. Commun.* (1984), p. 1097.
41. A. D. Hamilton, J. M. Lehn, J. L. Sessler, *ibid.* (1984), p. 311.
42. J. M. Lehn, *Pure Appl. Chem.* **51**, 979 (1979).
43. R. Breslow, *Science* **218**, 532 (1982); R. M. Kellogg, *Topics Curr. Chem.* **101**, 111 (1982); C. Sirlin, *Bull. Soc. Chim. Fr. II* (1984), p. 5; I. Tabushi and K. Yamamura, *Topics Curr. Chem.* **113**, 145 (1983); Y. Murakami, *ibid.* **115**, 107 (1983).
44. J. P. Behr and J. M. Lehn, *J. Chem. Soc. Chem. Commun.* (1978), p. 143.
45. J. M. Lehn and C. Sirlin, *ibid.* (1978), p. 949.
46. M. W. Hosseini, J. M. Lehn, M. P. Mertes, *Helv. Chim. Acta* **66**, 2454 (1983).
47. M. W. Hosseini and J. M. Lehn, *J. Chem. Soc. Chem. Commun.* (1985), p. 1155.
48. J. M. Lehn, in *Physical Chemistry of Transmembrane Ion Motions*, G. Spach, Ed. (Elsevier, New York, 1983), pp. 181–206.
49. M. Kirch and J. M. Lehn, *Angew. Chem. Int. Ed. Engl.* **14**, 555 (1975).
50. J. J. Grimaldi and J. M. Lehn, *J. Am. Chem. Soc.* **101**, 1333 (1979).
51. A. Hriciga and J. M. Lehn, *Proc. Natl. Acad. Sci. U.S.A.* **80**, 6426 (1983).
52. J. P. Behr, J. M. Lehn, A. C. Dock, D. Moras, *Nature (London)* **295**, 526 (1982).
53. M. L. Bender and M. Komiyama, *Cyclodextrin Chemistry* (Springer-Verlag, Berlin, 1978).
54. C. D. Gutsche, *Topics Curr. Chem.* **123**, 1 (1984).
55. J. Gabard and A. Collet, *J. Chem. Soc. Chem. Commun.* (1981) p. 1137; J. Canceill, L. Lacombe, A. Collet, *C.R. Acad. Sci. Ser. B* **289**, 39 (1984); *J. Am. Chem. Soc.* **107**, 6993 (1985).
56. C. O. Dietrich-Bucheker, J. P. Sauvage, J. M. Kern, *J. Am. Chem. Soc.* **106**, 3043 (1984).
57. F. Vögtle, Ed., *Topics Curr. Chem.* **98** (1981); *ibid.* **101** (1982).
58. G. Wipff, P. A. Kollman, J. M. Lehn, *J. Mol. Struct.* **93**, 153 (1983).
59. J. H. Fendler, *Membrane Mimetic Chemistry* (Wiley, New York, 1982).
60. For the incorporation of macrocyclic molecules into organized, liquid crystalline phases, see the "tubular mesophases": J. M. Lehn, J. Malthete, A.-M. Levelut, *J. Chem. Soc. Chem. Comm.* (1985), p. 1794.
61. I thank my co-workers, whose skill and dedication allowed us to realize the work surveyed in this article. Their names appear either in the references listed or in the original publications cited in the review articles.

Organic chemistry is only in part a "natural science." Much of the field is concerned with creating and understanding chemical substances and reactions other than those existing in nature. Thus the role of synthesis greatly exceeds that in other sciences. Most of the compounds and reactions examined are created by the chemists themselves. Chemistry is in this sense akin to electrical engineering, with its similar concern for understanding the behavior of humanly fabricated systems.

The inspiration for the design of new substances can come from considering natural compounds. For example, synthetic medicinals with structures related to those of important metabolites have been created. The motivation for such work is in part the desire to understand the properties of the natural substances by putting them into a general context, but the principal interest is in the creation of substances with novel properties. Chemistry is also concerned with reactions, that is, chemical transformations. The chemical reactions occurring in the natural world are almost exclusively the reactions of life. Thus it is reasonable that chemists have been interested in imitating and generalizing the biochemical reactions of nature. Since most such reactions are catalyzed by enzymes, efforts have been made to produce artificial enzymes.

Natural enzymes are proteins of considerable size, made up of hundreds of amino acids. They catalyze biochemical reactions by binding one or two small molecules into an active site, with catalytic groups of the enzyme held nearby so as to interact effectively with the bound substrate molecules. The result is reactions of great speed—velocities 10 billion times the speed of the uncata-

28

Artificial Enzymes
Ronald Breslow

Science 218, 532–537 (5 November 1982)

411

lyzed reaction are not uncommon. At least as remarkable is the typical selectivity of enzyme-catalyzed processes. Enzymes can select and bind one particular molecule among the many that are present in solution in a cell. Then the enzyme can selectively catalyze a reaction in one section of the molecule, which might not be the one most prone to simple chemical attack. This catalyzed reaction is normally under strong geometric control by the enzyme, so that products will be formed with particular spatial arrangements, including chirality ("handedness").

For many practical purposes, learning how to imitate the selectivity of enzyme-catalyzed reactions is more important than learning how to achieve their velocity. Simple chemical processes do not generally produce high yields of a single desired product that is free of impurities from side reactions. The achievement of good selectivity has been a major aim of work on artificial enzymes.

Selective Chlorinations

The cyclodextrins have excited much interest as possible components of artificial enzymes (1). They are doughnut-shaped molecules made up of glucose units, with an interior cavity whose size and shape is determined by the number of glucose units that make up the ring. In α-cyclodextrin (cyclohexaamylose, made up of six glucose units) the almost cylindrical cavity is approximately 7 angstroms deep and 5 angstroms in diameter. In β-cyclodextrin (cycloheptaamylose, seven glucose units) the depth is the same but the diameter is 7 angstroms, and in γ-cyclodextrin (eight glucose units) it is again 7 angstroms deep, but 9 angstroms in diameter.

Cyclodextrins are soluble in water, because of the many hydroxyl groups of glucose that rim the cavity. However, since the cavities themselves are hydrophobic (poorly solvated by water), the cyclodextrins have the ability to extract small organic molecules out of water solution and bind them into the cavities. This is similar to the ability of enzymes to bind substrates into interior cavities. Cyclodextrin binding is also selective for molecules that have the correct shape and hydrophobic character.

In free solution, the chlorination of anisole (methoxybenzene) normally yields a mixture of two products, *ortho*-chloroanisole (1) and *para*-chloroanisole (2) (Fig. 1). When α-cyclodextrin is present, the anisole binds into the cavity. Molecular models suggested that in the complex it should be difficult to chlorinate the anisole so as to form *ortho*-chloroanisole, since the ortho position is buried inside the cavity. This was the experimental finding: no *ortho*-chloroanisole was formed (2). At the same time, *para*-chloroanisole (2) was produced by a new chemical pathway, catalyzed by a hydroxyl group of α-cyclodextrin. By this path (Fig. 1), a chlorine atom was delivered to the bound anisole in the complex. The chlorination system catalyzed by α-cyclodextrin thus shows many enzyme-like properties. The cyclodextrin selectively binds particular molecules. It then selectively produces a single product by a catalyzed reaction within the complex. Finally, the product *para*-chloroanisole is released and a new anisole substrate molecule binds to start the cycle again.

Anisole can also be chlorinated by the enzyme chlorinase. However, with this enzyme, a mixture of products 1 and 2 is formed. In this instance, the α-cyclodextrin system shows a more typical en-

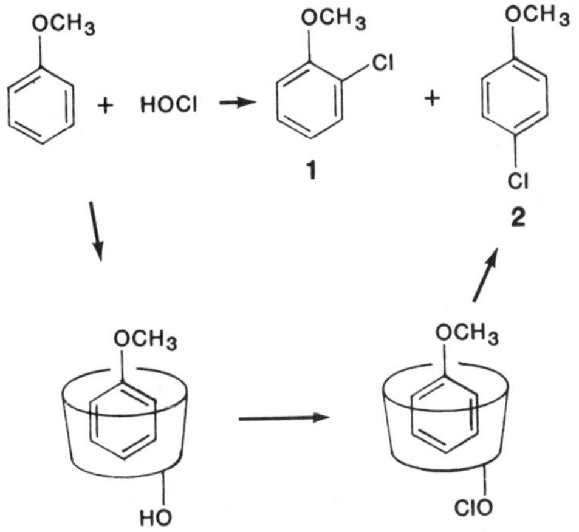

Fig. 1. The random chlorination of anisole in solution is converted to a selective process by a cyclodextrin catalyst, which binds the anisole in a cavity and delivers chlorine under geometric control.

zyme-like selectivity than does the enzyme itself. As this shows, there are many chemical reactions of interest for which no enzyme exists, or an enzyme exists that catalyzes the reaction only poorly. In such cases artificial enzymes can play an important role.

A more challenging problem is the development of highly selective reactions of large organic species such as steroid molecules. Reaction of a large hydrocarbon structure (for example those in Fig. 2) in simple chemical processes would normally be selective only for reactions involving one of the reactive groups of the molecule, and attack on the unreactive C–H groups is usually random, leading to hopeless mixtures. By contrast, enzymes are known that can selectively attack unreactive C–H bonds because of the geometry of the enzyme-substrate complex. Since steroids have great practical importance as medicinal compounds, we have tried to learn how to imitate this selective enzyme-catalyzed chemistry.

The first problem was to show that simple geometrical control could indeed be achieved, leading to selective functionalization. A template molecule carrying an iodine atom was attached to a steroid (3) (Fig. 2). When chlorination was performed, a chlorine became attached to the iodine atom and thus held so that it could reach and attack only one C–H group, the one shown. The 46 other hydrogens of the steroid were not attacked. This led, in a few simple steps, to the formation of a chlorinated steroid (4) in which the chlorine was selectively placed on a carbon (C-17) at the opposite end of the molecule from the carbon to which the template had been attached (C-3). We have described such reactions as "remote functionalizations" (3).

This process is catalyzed by the template species. It is faster than chlorination of a steroid without the template group, and it shows geometrically directed selectivity. Furthermore, the ester linkage that attaches the template to the steroid can be hydrolyzed so that the catalyst group can be reused. The direct chemical linkage between the substrate and the catalytic group in this instance contrasts with the simple binding in the anisole system described above or with that in enzymes.

Different selectivities can be achieved by changing the geometry of the templates. Thus with the shorter template in compound 5, the chlorination has been directed to the hydrogen on a carbon (C-9) in the middle of the molecule. This template was also highly selective for the hydrogen shown among the 47 in the steroid; in particular, it afforded none of compound 4. The C-9 halogenation is useful in the production of corticosteroids, and we have used it in the synthesis of cortisone from a more available steroid (4). Still another template

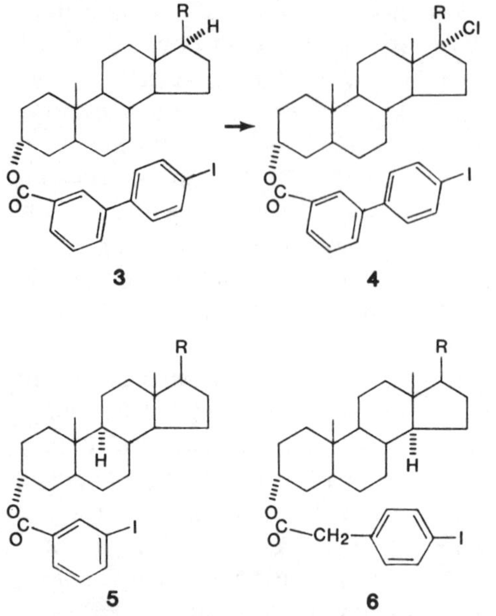

Fig. 2. The steroid, 3-α-cholestanol, can be selectively catalytically chlorinated at any one of the three hydrogens shown by different templates whose geometry determines the point of selective attack.

(in 6) has let us perform a selective chlorination at C-14, useful for the preparation of cardiac-active steroids. However, close geometric control is needed, as predicted from molecular models or computer modeling. Templates with incorrect geometries yield mixtures of products, directing the attack to several positions.

The imitation of the geometric control used to achieve selective enzymatic reactions of steroids has led to useful selective chemical processes. Even more utility, as well as a better imitation of enzymes, will come when such template control can be achieved with simple binding interactions, not chemical linkages, between substrates and catalysts. [This has now (1985) been achieved.]

Mimics of Transaminase

Pyridoxal phosphate and the related pyridoxamine phosphate are the coenzymes for many enzymatic transformations involving amino acids. Of these, the reactions of most general importance are the transaminations by which amino acids are synthesized from keto acids. The transaminase enzymes that catalyze this process perform two reactions. In the first, an α-keto acid reacts with enzyme-bound pyridoxamine phosphate to form an imine structure (7) (Fig. 3). This imine then undergoes isomerization to another imine, which hydrolyzes to form the amino acid and a new coenzyme, pyridoxal phosphate (the actual enzymatic reaction is a bit more complicated than simple hydrolysis). The process then goes over the same path in the reverse direction (all the steps of Fig. 3), but with a different amino acid. The result is to regenerate the pyridoxamine form of the coenzyme, with the formation of a new keto acid. Since the total reversible path converts keto acid 1 and amino acid 2 to amino acid 1 and keto acid 2, with overall preservation of the structure of the coenzyme, the coenzyme is a catalyst (along with the enzyme).

Many elegant studies (5) have demonstrated that transamination (and other biochemical reactions involving these coenzymes) can be catalyzed in simple chemical systems by pyridoxal, pyridoxamine, or related compounds in the absence of enzymes. These nonenzymatic reactions of the coenzymes with amino or keto acids are slower than the enzyme-catalyzed processes, and they are also much less selective. We have designed more complicated catalysts to try to achieve better rates and selectivity.

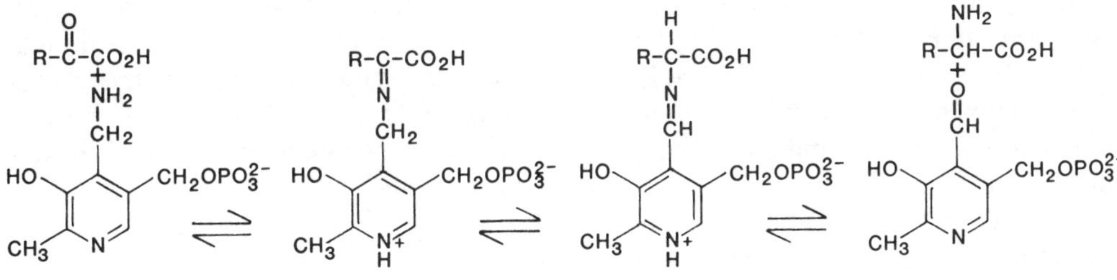

7

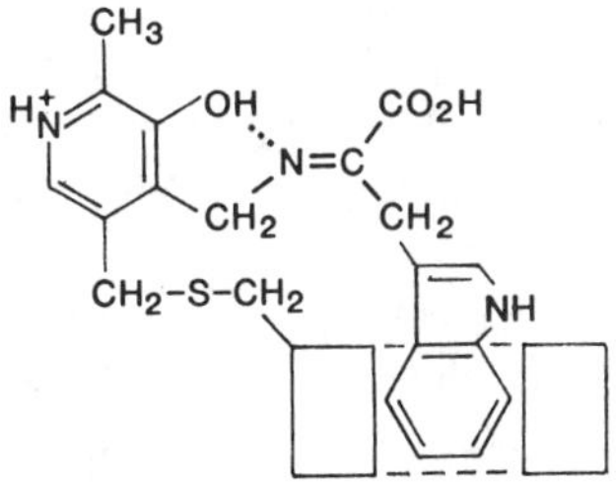

Fig. 3. Amino acids are produced by the transamination mechanism shown, involving the coenzyme pyridoxamine phosphate. A nonenzymatic transamination is performed by an artificial enzyme in which pyridoxamine has been attached to cyclodextrin. This artificial enzyme shows selectivity among substrates and produces optically active amino acids.

The selectivity of the enzyme-catalyzed processes results in part from the ability of an enzyme to bind particular substrates, and not others. Furthermore, within the selectively formed complexes there are additional catalytic groups of the enzyme able to reach particular atoms, and these impose additional selectivity. Because these groups are in general not symmetrically arranged, they can produce optically active amino acids by adding a proton to one face of an intermediate rather than to the other one. If all of these types of selectivity could be achieved within artificial enzyme-like systems, it might be possible to use them in chemical syntheses of important amino acid derivatives.

In one approach to this objective, we have constructed an artificial transaminase enzyme that combines the pyridoxamine-pyridoxal coenzyme system with a cyclodextrin-binding group (6). The compound was constructed to show selectivity for amino acid derivatives carrying aromatic side chains that would

bind into the cyclodextrin cavity. The desired selectivity was achieved. When the pyridoxamine catalyst carrying a cyclodextrin ring was allowed to react with a mixture of simple pyruvic acid and indolepyruvic acid (Fig. 3), the indole derivative was preferentially bound to the catalyst and selectively converted to the amino acid tryptophan. There was a 50-fold preference for the formation of tryptophan in competition with the formation of alanine from pyruvic acid, although with simple pyridoxamine carrying no cyclodextrin-binding group the competition led to essentially equal reactivities of the two compounds. Thus, a random reaction with respect to two substrates of the simple coenzyme has been converted to a selective reaction by the addition of a binding group.

An additional feature is that the cyclodextrin system is optically active, being composed of glucose units. When an amino acid is produced by our pyridoxamine-cyclodextrin system it is being formed in an asymmetric enviroment and can, in principle, be created in an optically selective fashion. We found that the reaction of phenylpyruvic acid to form phenylalanine did indeed show an optical preference, forming five times as much of the natural L-enantiomer as of the D-enantiomer (6). With simple pyridoxamine alone, no such optical selectivity is seen, since it is a completely achiral molecule. These systems are catalytic; in the presence of a second amino acid they can perform the entire transamination cycle repeatedly, turning one keto acid into an amino acid while catalyzing the reverse process for a second amino acid.

The enzymatic reaction also differs from the simple chemistry of the coenzymes themselves in the presence of additional catalytic groups in the enzyme. We have been studying mimics of this situation by adding catalytic groups to pyridoxamine and pyridoxal that can catalyze the removals and additions of protons required in the overall mechanism for transamination and other related processes (7). We find that we can attach chains carrying basic groups at the same position in pyridoxamine to which we attached the cyclodextrin. Some of these compounds are much more effective than simple pyridoxamine as catalysts for transamination. The most effective catalysts are those with a flexible chain that permits the basic group to reach both the carbon from which a proton must be removed and the other carbon to which a proton must be delivered in the reactions of Fig. 3. This is consistent with the finding for the enzyme that the same basic group is removing the proton from one carbon and carrying it to the other. Work is still under way to produce a catalyst that combines the pyridoxamine-pyridoxal

coenzyme with a binding group to create substrate selectivity and an additional catalytic group to perform the proton transfers. Such a species could be a very selective and effective artificial transaminase. [This was achieved in 1985.]

Mimics of Ribonuclease

Ribonuclease, which catalyzes the hydrolytic cleavage of RNA, is one of the better understood enzymes (8). The cleavage occurs in two steps (Fig. 4) that are closely related in mechanism. In the first step, a hydroxyl group of RNA itself attacks the phosphorus atom, converting the original phosphate diester of RNA into a cyclic phosphate diester group. This results in a break in the RNA chain. In the second step the enzyme catalyzes an attack on this cyclic phosphate diester by water, producing a phosphate monoester on C-3 of the ribose sugar unit or RNA, while regenerating the hydroxyl group at C-2 of that ribose unit. Thus in the overall hydrolytic cleavage of the RNA chain, the first step involves the cleavage and the second step involves the hydrolysis of an intermediate product.

The enzyme catalyzes both steps by closely related mechanisms. In the first reaction, a basic catalytic group of the enzyme removes the proton from the attacking hydroxyl group, and an acidic catalytic group of the enzyme puts a proton onto the leaving oxygen atom. In the second step, the proton is put back on the C-2 oxygen of ribose, which is now the leaving group, while a basic catalytic group of the enzyme removes the proton from an attacking water molecule. Thus in both steps, a basic and an acidic catalyst are needed, the base as-

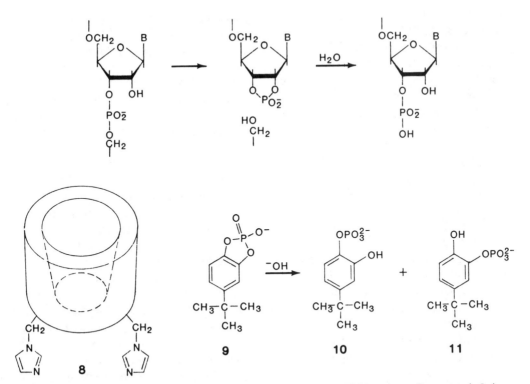

Fig. 4. The sequence by which ribonuclease hydrolyzes RNA (top). Compound **8** is a cyclodextrin bisimidazole artificial enzyme that catalyzes the hydrolysis of substrate **9** by an enzyme-like mechanism. Under simple hydrolysis, **9** yields a mixture of **10** and **11**.

sisting the removal of a proton from an attacking oxygen atom while the acid group is putting a proton onto the departing oxygen atom.

In ribonuclease, these two catalytic groups are both imidazole rings of histidine-12 and histidine-119. Catalysis occurs when one of them is present as the free imidazole ring, capable of acting as a base catalyst, and the other one is present as the protonated imidazolium ring, able to act as an acid catalyst. Thus the enzyme shows a rate maximum near pH 6. At a lower pH both rings are protonated and the needed basic catalytic group is absent, whereas at high pH, both rings are unprotonated and the acidic catalyst is no longer available. There is an addi-

tional important functional group in the enzyme, the ammonium group of lysine-41. This binds the phosphate anion of RNA at which the catalyzed reaction is to occur, and the hydrogen bond established makes the phosphate more reactive and thus serves a direct catalytic function.

We have produced two types of mimics of ribonuclease. In the first series, the principal catalytic groups of ribonuclease were mounted on a cyclodextrin framework so as to cause binding of the substrate and to bring about catalyzed reactions in the complex. By the use of rigid reagents it is possible to activate pairs of hydroxyl groups in cyclodextrin selectively (9). Using such selectively activated cyclodextrins, we have prepared cyclodextrin derivatives carrying two imidazole rings to see if they could mimic the catalytic reaction by which one of them acts as a base and the other one, in its protonated form, acts as an acid group to catalyze phosphate cleavages (10, 11).

A cyclodextrin bisimidazole (compound 8) has two of the correct catalytic groups to imitate ribonuclease, but the binding site is not optimized for the binding of RNA itself. We have therefore used these cyclodextrin-based systems to examine other substrates whose geometry is optimal for binding and reactions with our catalyst. Artificial enzymes would be expected to have different selectivities from natural enzymes because of their different geometries. This novel selectivity is one of the important reasons for preparing such catalysts.

Molecular models suggested that cyclodextrin bisimidazoles should be able to catalyze the hydrolysis of compound 9, a cyclic phosphate based on *tert*-butylcatechol. Such a cyclic phosphate resembles the intermediates in the cleavage of RNA, and the process we are examining is then very similar to the second step in the reaction of ribonuclease with RNA itself. We found that this was indeed the case. The cyclodextrin derivative 8 was an effective catalyst for the hydrolysis of the cyclic phosphate 9, and by the same mechanism as that used by the enzyme. This catalyst also showed a rate optimum near *p*H 6, indicating that one of the imidazoles was acting as a base catalyst while the other one, in its protonated form, was acting as an acidic catalyst. Perhaps even more striking is the fact that the catalyst showed a selectivity closely related to

that expected for an enzyme-like mechanism.

When the simple cyclic phosphate 9 is submitted to vigorous chemical conditions, it hydrolyzes to form two products, 10 and 11, because attack by water can lead to departure of either of the oxygen atoms of the cyclic phosphate. By contrast, when this same substrate undergoes the much faster hydrolysis catalyzed by 8, it forms chiefly product 11. This selectivity is caused by the geometry of the catalyst-substrate complex. In the attack by water on a cyclic phosphate ester, it is expected that the attacking water oxygen atom will come in on exactly the opposite side of phosphorus from the leaving oxygen atom. That is, the geometry of the activated complex at the transition stage for the reaction has five groups around the phosphorus (the original four plus the attacking water oxygen) arranged in the form of a trigonal bipyramid. The three oxygens that are not attacking or leaving are arranged in an equilateral triangle around the phosphorus, while the attacking and leaving oxygens are arranged above and below the plane of this triangle so as to form a 180° alignment of O–P–O. Molecular models of this mechanism for the cleavage of substrate 9 bound into the cavity of catalyst 8 indicate that this geometry is possible only for one oxygen as leaving group, delivering the water and ejecting that oxygen so as to form the observed product 11. If instead, a molecular model is constructed in which water is aligned with the other P–O bond, in a mechanism aimed at producing product 10, the water is too far away from the cyclodextrin catalyst and cannot be delivered by one of the imidazole groups. Thus as Fig. 5 shows, the observed selectivity of this process is precisely what is expected from the geometry of the system.

This geometry can be changed by constructing a new catalyst in which the catalytic groups are further from the cavity. This was done with catalyst **12** carrying two imidazole rings attached to cyclodextrin through a short chain. Molecular models show that this species can reach out to a water molecule attacking substrate **9** so as to form product **10** and should be able to catalyze an alternative selective cleavage of the same substrate. This was indeed the case. This new catalyst **12** uses a mechanism similar to that of **8**, with pH optimum indicating that the catalyst must be present with one group in the basic form and the other one in the protonated acidic form. How-

ever, this catalyst is able to perform the alternative mode of cleavage of substrate **9**, producing product **10**. Figure 5 shows that this is to be expected from the geometry of the system, as judged from molecular models.

The catalytic function of the protonated imidazole in this new system **12** is apparently not to protonate the leaving group, but to protonate the phosphate anion, thus resembling the catalytic function of lysine-41 in ribonuclease itself. This can be determined from a detailed consideration of the pH-rate profile and its comparison with the titration curve for the catalyst. However, the selectivity is a reflection of the location of the base catalytic groups in catalysts **8** and **12**, respectively. Cyclodextrins with

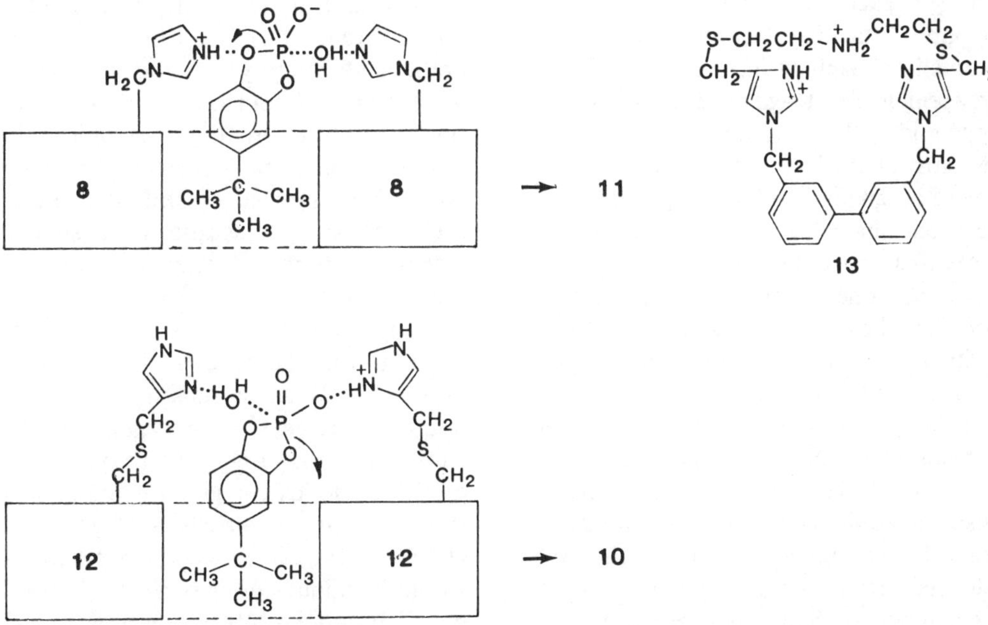

Fig. 5. The geometry of the complexes is such that artificial enzyme **8** hydrolyzes substrate **9** to afford mainly **11**, while artificial enzyme **12** converts the same substrate mainly to **10**. Both use ribonuclease-like mechanisms. Compound **13** is another artificial enzyme that catalyzes the hydrolysis of RNA itself.

only one catalytic group, acting as a base catalyst (*11*), have been prepared. These are less effective because of the loss of the additional acidic catalytic function, but they show the expected selectivity. A cyclodextrin monoimidazole related to compound **8** gives mainly product **11**, whereas a similar monoimidazole derivative related to catalyst **12** gives mainly product **10**.

Although artificial enzymes like **8** and **12** combine catalytic and binding groups and show enzyme-like mechanisms, with enzyme-like selectivities for a substrate (**9**) that fits the cavity, it would be of interest to develop artificial enzymes that are effective at cleaving RNA itself. We have approached this problem by considering geometrical questions related to those outlined above. In the transition state for the first step of the ribonuclease reaction, in which a hydroxyl group is attacking the phosphate to form a cyclic phosphate intermediate, the arrangement must again be a trigonal bipyramid with the attacking and leaving groups aligned at 180°. Furthermore, these attacking and leaving groups must be coordinated to the catalytic imidazoles that are adding or subtracting protons, and one of the equatorial oxygens on the phosphorus must be ion-paired with the lysine-41 ammonium ion. We thus set out to construct a molecule that would carry three such catalytic groups—two imidazoles and an ammonium ion—in the correct position to establish the required coordination. Molecular model building made it clear that the desired arrangement could be achieved in compound **13**, a structure that looked synthetically accessible. We have synthesized this molecule (*12*) and find that it is indeed a catalyst for the cleavage of RNA. Selective catalysts based on this structure may become useful biological tools.

Enzyme-Like Reactions with Very High Velocities

In all of the cases we have been discussing the reactions have been accelerated by the catalysts and in most of the cases have resulted in interesting new selectivities. However, the rate accelerations are modest by enzymatic standards. For instance, in the transamination mimic, the binding to a cyclodextrin group led to a 50-fold acceleration. While this acceleration is enough to make the reaction highly selective, because it occurs only for some substrates and not others, it does not produce a very large velocity. Similarly, in the cyclodextrin ribonuclease mimics, fairly high selectivities were observed, but the rate accelerations for hydrolysis of the substrate **9** were at most 50-fold as a result of bifunctional catalysis by the two imidazole groups. If artificial enzymes are to be considered truly successful, they must produce rate accelerations of similar magnitudes to those achievable with natural enzymes.

Many research groups have studied reactions of simple cyclodextrin complexes (*1*). When a substrate is bound into the cyclodextrin cavity, parts of the substrate are held near hydroxyl groups of the cyclodextrin glucose units, and this proximity might lead to high intracomplex reaction rates. Such reactions would be mimics of enzymatic reactions in which a bound substrate is attacked by an enzymatic hydroxyl group. For instance, in proteases such as chymotrypsin, attack by the hydroxyl group of the amino acid serine on an amide bond of a bound substrate leads to the formation of an ester. Part of the substrate becomes attached to the serine hydroxyl while the other part has been broken off and can depart. In a second step, the intermediate ester is then hydrolyzed, breaking off the other piece of the original peptide and regenerating the enzyme. This two-step sequence is reminiscent of a similar sequence for ribonuclease, described above, except that with the serine proteases, the hydroxyl group is derived from the enzyme, not from the substrate. Other enzymes also use a serine group to attack carbonyl and phosphoryl groups. Thus, early studies on cyclodextrin chemistry focused on an imitation of this first step, namely, attack by a hydroxyl group on a bound substrate.

Attack by a hydroxyl group within the same molecule can be very rapid (13). An intramolecular reaction, with a well-placed hydroxyl attacking a neighboring group, can occur at rates 1 million or more times as fast as the rate of attack on a substrate by solvent molecules that completely surround the substrate, but are not directly attached to it. Such very fast reactions by groups within the same molecule reflect the entropy advantages of an intramolecular attack (14). These examples suggested that the very high reaction rates seen in many enzymatic processes, compared with simple chemical reactions between unassociated reagents, similarly reflect the entropy advantages of chemical reactions between species bound together in a tight complex. For this reason, it was disturbing that in all of the examples (1) of cyclodextrin-promoted reactions, the accelerations did not exceed 300-fold.

When an ester substrate is bound into cyclodextrin and is then attacked by a cyclodextrin hydroxyl group to make a new ester, the rate of this attack by the cyclodextrin hydroxyl can be compared with the rate of attack by a solvent water hydroxyl under the same conditions in the absence of the cyclodextrin. Because, over the normal region, both the cyclodextrin process and the simple hydrolysis reaction have the same dependence on base, the relative reaction rate ratio is independent of pH. When the comparison is done, it turns out that for many bound ester substrates, attack by cyclodextrin is only 100-fold or less faster than simple hydrolysis of the corresponding substrate. The best rate acceleration with β-cyclodextrin when we started our studies was 250-fold, achieved with the substrate m-tert-butylphenyl acetate which had been tailored to fit the cavity well (15).

Molecular models show that when the tert-butylphenyl group of this substrate binds into the cyclodextrin cavity, the ester carbonyl group of the acetate can be in contact with one of the hydroxyl groups of the cyclodextrin. Thus, good proximity seems to be present. However, molecular model building also indicates that when the hydroxyl group attacks the ester carbonyl, the normal geometry of the resulting chemical bonds requires that the tert-butylphenyl group be pulled significantly up out of its most stable binding position. Thus, the prediction from these models was that the binding of the substrate is better than the binding of the transition state for the intracomplex reaction in which the hydroxyl is attacking the carbonyl group. Ideally an enzyme should bind the transition state more strongly than it binds the starting material (16); loss of binding in the course of the chemical reaction should seriously slow the reaction rate. We therefore set out to design systems in which not only the substrate but also the transition state would be well bound in order to optimize the velocity effect of complexing.

One approach to this is to modify the cyclodextrin cavity. We have prepared several derivatives of cyclodextrin carrying groups that can penetrate part of the way into the cavity to produce an intrusive floor (17). Studies of the binding of various substrates to the modified cavity show that this intrusion does occur; the cyclodextrin open cavity is turned into a closed pocket that is shallower than the cavity of the unmodified system. Under these circumstances, the original substrate must bind less deeply, although the binding is still strong because the new floor on the molecule contributes to the overall binding interaction with the substrate. As expected from this, the velocity of reaction of a substrate with the modified cyclodextrins was improved because the new binding geometry is closer to that required for the transition state. Increases in rate by an additional factor of 10 or so were achieved with this kind of modification. In another approach (17), we constructed substrate molecules carrying additional projecting groups that prevented them from binding deeply into the cyclodextrin cavity. Again this change brought the geometry of binding of the substrate much closer to the geometry required for the transition state, and the reaction rate with a cyclodextrin increased to 4000 times the reaction rate with solvent. However, the most dramatic improvements (17) occurred with novel substrate classes whose geometries were optimized so that there would be little loss of binding in proceeding to the transition state from the bound starting material.

One set of substrates was based on the adamantane nucleus (17). A reaction rate 15,000 times the rate with solvent at the same pH was observed for the reaction of one such substrate with the cyclodextrin. The substrate carried some projecting groups to guarantee that it would adopt the productive geometry rather than an alternative inactive complex. The most striking rate accelerations were seen with substrates based on the ferrocene nucleus. Ferrocene (the sandwich compound between two cyclopentadienyl rings and an iron atom) fits β-cyclodextrin well and is strongly bound into the cavity. Since substituents come off at essentially right angles to the axis of the ferrocene system, we expected (from models) that the transition state for attack by hydroxyls would also be strongly bound. In one substrate based on a simple ferrocene nucleus, the attack by a hydroxyl within the complex was 750,000 times as fast as simple attack by solvent under the same conditions (17). Improved fused-ring derivatives of the ferrocene system, with better geometric definition, have led to accelerations of as much as 6 million–fold (18), and in one instance, the optical activity of the cyclodextrin nucleus led to a selection of one of the two mirror-image isomers of the substrate in the ratio 65 to 1.

A 6 million–fold acceleration, achieved by complexing a substrate to put it near cyclodextrin hydroxyl groups, is quite substantial. In fact, the reaction is performed in a mixed organic-water solvent, which imitates the interior of an enzyme. The rate is 150 million times the rate of an uncatalyzed reaction in water (the comparison used for enzymes themselves). The enzyme chymotrypsin achieves much less than this rate acceleration in attack on esters, although most enzymes operating on their natural substrates achieve somewhat higher rate accelerations. However, the simple reaction between cyclodextrin and a bound substrate does not involve any of the other catalytic groups normally present in an enzyme. If such catalytic groups are placed correctly in enzyme mimics, so as to assist the intracomplex reactions, even larger accelerations can be expected. Furthermore, in the best substrate examined so far, the transition state for the chemical reaction seems to be almost as well bound as the substrate. A more interesting class of substrates would be those in which the transition state is bound more strongly than the substrate, mimicking the situation believed to exist for many enzymes. In such cases, accelerations even larger than those already achieved can be expected.

Conclusions

The field of the synthesis and study of artificial enzymes is in a sense in its infancy. Most of the work with molecular complexing has involved the use of cyclodextrins, and only recently has attention begun to focus on new artificial cavities that could bind substrates in other ways (19) and with different kinds of selectivity. However, it is already clear that with appropriate molecular design it will be possible to achieve very large rate accelerations, comparable to those typical of enzymatic processes. More exciting, the application of the principles learned from the study of enzymatic reactions permits us to design novel chemical processes that can achieve desirable selectivity of the sort not otherwise available outside of biochemistry. The selective, accelerated reactions achieved with artificial enzymes

that mimic the natural catalysts have the potential to play an important role in chemical synthesis, and such substances may even prove to have therapeutic utility. Thus one can look forward to continued vigorous growth of this field.

References and Notes

1. For a review, see M. L. Bender and M. Komiyama, *Cyclodextrin Chemistry* (Springer-Verlag, Berlin, 1978). See also I. Tabushi, *Acc. Chem. Res.* **15**, 66 (1982). These publications describe the pioneering work in this field by F. Cramer, M. L. Bender, and others, which made our studies possible.
2. R. Breslow and P. Campbell, *J. Am. Chem. Soc.* **91**, 3085 (1969); *Bioorg. Chem.* **1**, 140 (1971); R. Breslow, H. Kohn, B. Siegel, *Tetrahedron Lett.* (1976), p. 1645.
3. R. Breslow, *Acc. Chem. Res.* **13**, 170 (1980).
4. _____, B. B. Snider, R. J. Corcoran, *J. Am. Chem. Soc.* **96**, 6792 (1974); R. Breslow, R. J. Corcoran, B. B. Snider, R. J. Doll, P. L. Khanna, R. Kaleya, *ibid.* **99**, 905 (1977).
5. Reviewed by A. E. Braunstein, in *The Enzymes*, P. D. Boyer, Ed., (Academic Press, New York, ed. 3, 1973), vol. 9, chap. 10.
6. R. Breslow, M. Hammond, M. Lauer, *J. Am. Chem. Soc.* **102**, 421 (1980).
7. S. Zimmerman and A. Czarnik, unpublished data.
8. Reviewed by F. M. Richards and H. W. Wyckoff, in *The Enzymes*, P. D. Boyer, Ed. (Academic Press, New York, ed. 3, 1973), vol. 9, chap. 24.
9. I. Tabushi, K. Shimokawa, N. Shimizu, H. Shirakata, K. Fujita, *J. Am. Chem. Soc.* **98**, 7855 (1976).
10. R. Breslow, J. Doherty, G. Guillot, C. Lipsey, *ibid.* **100**, 3227 (1978).
11. R. Breslow, P. Bovy, C. Lipsey Hersh, *ibid.* **102**, 2115 (1980).
12. R. Corcoran and M. Dolson, unpublished data.
13. For a review, see A. J. Kirby and A. R. Fersht, *Prog. Bioorg. Chem.* **1**, 1 (1971).
14. M. I. Page and W. P. Jencks, *Proc. Natl. Acad. Sci. U.S.A.* **68**, 1678 (1971).
15. R. L. Van Etten, J. F. Sebastian, G. A. Clowers, M. L. Bender, *J. Am. Chem. Soc.* **89**, 3242 (1967).
16. L. Pauling, *Am. Sci.* **36**, 58 (1948).
17. R. Breslow, M. F. Czarniecki, J. Emert, H. Hamaguchi, *J. Am. Chem. Soc.* **102**, 762 (1980).
18. R. Breslow and G. Trainor, *ibid.* **103**, 154 (1981); G. Trainor, unpublished data.
19. Y. Chao, G. R. Weisman, G. D. Y. Sogah, D. J. Cram, *J. Am. Chem. Soc.* **101**, 4948 (1979); J.-M. Lehn, inaugural address, Collège de France, 1980; Y. Murakami, J. Sunamoto, K. Kano, *Chem. Lett.* (1973), p. 223.
20. Supported by the National Institutes of Health and the National Science Foundation. I thank my co-workers, some of whom are named in the references, for experimental and intellectual contributions.

A major objective of biochemists continues to be the elucidation of the structural basis of enzyme function and activity. The understanding of the relation between enzyme structure and activity has been facilitated by x-ray crystallographic studies, chemical modification experiments, kinetic investigations, and other related techniques. These kinds of experimentation have yielded considerable information concerning two principal phenomena that underlie the activity of enzymes: substrate binding and the subsequent intracomplex catalysis.

The information on structures now available on enzymes, and knowledge of the pathways by which enzymatic catalysis occurs, have made it possible for the chemist to contemplate designing enzymes from their constituent amino acids and cofactors. The clear prediction of the pattern of the folding of long peptide chains into specific tertiary structures still poses many difficulties. Therefore, the construction of new enzymes through total synthesis still remains a goal. Nevertheless, a breakthrough has been made in a major aspect of the problem of enzyme design, the creation of new enzyme active sites, by a process that we term "chemical mutation" (1–7).

In the chemical mutation approach, the starting material is a natural protein that has folded to a stable conformation. Groups at or on the periphery of the active site are chemically modified to produce a "semisynthetic" enzyme having catalytic activity different from that of the original enzyme. An attractive feature of the chemical mutation process is that the wealth of x-ray structural information available for relatively simple enzymes makes possible a considerable degree of flexibility in the choice of

29

Chemical Mutation of Enzyme Active Sites

E.T. Kaiser and D.S. Lawrence

Science 226, 505–511 (2 November 1984)

the natural system in which the new catalytic group is introduced. Until now, much of the effort has been on the conversion of readily available enzymes of moderate molecular weight that are hydrolytic catalysts into modified enzymes capable of catalyzing other important reactions such as oxidation-reduction, decarboxylation, and transamination. We have shown that we can achieve the "chemical mutation" of enzyme active sites by the reaction of appropriate coenzyme analogs containing reactive functional groups with amino acid residues in or near, active sites of hydrolytic enzymes. With suitably chosen enzyme templates, appropriate coenzyme analogs can be covalently attached in a manner permitting the enzyme binding sites to remain accessible to organic substrates as shown in Eq. 1.

Active site residue
+
Reactive coenzyme analog
↓
Semisynthetic enzyme
containing covalently
bound coenzyme   (1)

The choice of an appropriate enzyme as the starting material for the preparation of a semisynthetic enzyme has been made with consideration to five criteria.

1) The enzyme to be used should be readily available in highly purified form.

2) The x-ray structure of the enzyme should be known.

3) The enzyme should have a suitably reactive amino acid functional group at or near the active site.

4) The covalent modification of the enzyme should result in a significant change in the activity characteristic of the native enzyme.

5) The attachment of the coenzyme analog should not cause the entry of substrates to the binding site to be blocked.

In our initial research on semisynthetic enzymes, we examined briefly the modification of the serine proteinase α-chymotrypsin at a residue on the periphery of the active site Met[192] (7). In most of our work, however, we have modified a residue directly at the active site. The cysteine proteinase papain appeared to be an excellent candidate for this type of modification. Both the x-ray structure of the enzyme (8–11) and solution studies with peptide substrates have shown that papain contains an extended groove in the vicinity of the active site residue Cys[25]. Therefore, it appeared possible that the sulfhydryl group of this residue could be alkylated with a coenzyme analog, while the binding region for potential substrates would remain accessible. Not only would the hydrolytic activity of papain be lost when the cysteine residue would be modified and thus allow the facile monitoring of the modification reaction, but also the introduction of the coenzyme analog might permit a potential substrate to bind.

The choice of a suitable coenzyme analog to act as a modifying agent was based primarily on three criteria.

1) The coenzyme should have the potential to act as a catalyst when bound to an enzyme active site without a requirement for specific functional groups of amino acid residues in the enzyme to participate in the catalytic act.

2) Model building should indicate that the placement of the coenzyme analog is compatible with the spatial requirements of the enzyme template. In other words, there should be a reasonable likelihood that the coenzyme analog, when cova-

lently bound to the active site, would remain in close proximity to the substrate binding site without blocking it.

3) Model building should indicate that the covalently bound coenzyme should be capable of interacting with a potential substrate in a productive fashion.

In principle, future research on semisynthetic enzymes may not require adherence to the first criterion mentioned for the choice of the coenzyme analog modifying agent. However, in the early phases of this research, it was questionable whether appropriate modification with coenzyme analogs at enzyme active sites could lead to effective enzymes with new catalytic activities. Therefore, it seemed prudent not to try to build complex cases where the precise geometry of the interaction of functional groups on the enzyme with the coenzyme might be crucial to the development of a successful catalyst.

Flavins were chosen as the modifying agents used in the preparation of the first semisynthetic enzymes because of their known general catalytic versatility. Even model flavins can be quite effective catalysts, and therefore it seemed likely that flavoenzymes could be generated by chemical modification of an enzyme like papain without a requirement for the involvement of specific amino acid functional groups in the enzyme for the catalytic act to occur. Furthermore, flavins are known to catalyze many diverse transformations, suggesting that the preparation of semisynthetic flavoenzymes with different types of catalytic activity would be a distinct possibility.

The design of specific flavin analogs for use in the preparation of a semisynthetic enzyme was based on the x-ray diffraction studies of covalent papain-inhibitor complexes by Drenth and col-

leagues (12). These papain derivatives were obtained from the reaction of chloromethyl ketone peptide substrate analogs with the sulfhydryl of Cys^{25}. In each case the carbonyl oxygen of what had originally been the chloromethyl ketone group was positioned near two potential hydrogen-bond donating groups, the backbone NH of Cys^{25} and a side chain NH of Gln^{19}. We thought that building in the possibility of a similar interaction in the case of the flavin modifying agents might serve to constrain the covalently bound flavin moiety to the interior of the enzyme near the substrate binding site. The flavopapains 1C to 3C, prepared by alkylation of the sulfhydryl of Cys^{25} and having the potential for this interaction, were constructed according to the second criterion for the choice of the coenzyme modifying agent, namely, that the structure of the coenzyme analog be closely compatible with the geometry of the enzyme template. In contrast, flavopapains 4C and 5C do not have a carbonyl group attached to the flavin ring system near the alkylation site and, therefore, could not be held via hydro-

427

4

Tetraacetylribose

5

	X
A	H
B	Br
C	S-papain

Kinetic and Stereochemical Studies on the Flavopapains

The kinetic data for the oxidation of dihydronicotinamides by flavopapains **1C** to **5C** were measured primarily under aerobic conditions with excess substrate. Under these conditions the accumulation of the dihydroflavin product was not observed for **2C**, **4C**, or **5C**. However, recent findings with the species **1C** indicate that even under aerobic conditions the buildup of the dihydroflavin species produced by reduction of the flavin ring is readily observed (6). The general scheme which is postulated to apply to the oxidation of the dihydronicotinamides by flavopapains **1C** to **5C** is illustrated in Eqs. 3A and 3B. For those cases where no buildup of dihydroflavins is seen under aerobic conditions, the rate of formation of nicotinamide is independent of oxygen (that is, $k_o[O_2] >> k_{cat}$). Since dihydroflavin buildup is observed aerobically in the case of flavopapain **1C** under readily accessible conditions, the effect of oxygen on the rate of nicotinamide formation cannot be neglected. If we analyze the kinetics of reaction under aerobic conditions according to the rate expression shown in Eq. 3C, the meaning of the apparent k_{cat} and K_m (Michaelis constant) values will depend on the relative rates of the formation of dihydroflavin in the enzymatic reduction step and of its reaction with oxygen.

gen bonding to the active site in the manner proposed for flavopapains **1C** to **3C**. In view of this difference, there may not be a driving force for positioning the flavin moiety in flavopapains **4C** and **5C** near the substrate binding region. On this basis the efficiency of flavopapains **4C** and **5C** as enzymatic catalysts would not be expected to be particularly high. As shall be seen, our experimental work demonstrated the correctness of these predictions.

The oxidation of N^1-alkyl-1,4-dihydronicotinamides by the flavopapains **1C** to **5C** (Eq. 2) is the process that we have examined most carefully. Our model building suggested that various N^1-alkyl-1,4-dihydronicotinamides could be comfortably ensconced within the binding pocket of the semisynthetic enzymes.

(2)

Furthermore, in at least some instances, such as flavopapains **1C** and **2C**, the dihydronicotinamide substrates could be placed in close proximity to the flavin group, indicating that hydrogen transfer might be facilitated.

$$E_{ox} + NRNH \underset{K_m}{\overset{}{\rightleftharpoons}} ES \xrightarrow{k_{cat}} EH_{2red} + NRN \quad (3A)$$

$$EH_{2red} + O_2 \xrightarrow{k_0} E_{ox} + H_2O_2 \quad (3B)$$

$$v = \frac{k_{cat}[E]_0[NRNH]}{K_m + [NRNH]} \quad (3C)$$

428

E_{ox} is the oxidized form of flavopapain, EH_{2red} is the reduced form of flavopapain, ES is the Michaelis complex, NRNH is dihydronicotinamide, NRN is nicotinamide, and k_o and k_{cat} are the rate constants.

In order to assess the effectiveness of the catalytic action of the semisynthetic enzymes produced by covalently attaching flavins **1B** to **5B** to papain's active site, we also studied the kinetics of the model reactions illustrated in Eq 4. Under aerobic conditions and when the N^1-alkyl-1,4-dihydronicotinamide was present in substantial excess (usually, however, at concentrations appreciably less

than 0.01M), pseudo first-order kinetics were seen for the model reactions. Under these circumstances, since the pseudo first-order rate constant k_{obs} depends directly on the flavin concentration, the

second-order rate constant for the reaction is easily calculated (Tables 1 and 2).

Flavopapains 4C and 5C. In the oxidation of N^1-alkyl-1,4-dihydronicotinamides, flavopapains **4C** and **5C** exhibited a practically insignificant (threefold) rate enhancement over the corresponding reactions of the model compounds **4A** and **5A** and did not display saturation kinetics (*1*). The flavins in the modified enzymes **4C** and **5C** are attached to papain via a thioether bridge. As already mentioned, unlike the species **1C** and **2C**, they do not contain a carbonyl function attached to the flavin at position 8 of the ring system. Since the carbonyl group is predicted to be involved in properly aligning the cofactor relative to the substrate in the enzyme active site, it is not surprising that **4C** and **5C** are poor catalysts.

Flavopapain 2C. Flavopapain **2C** produced from the alkylation of the Cys^{25} residue of papain by the 7-bromoacetyl-substituted flavin **2B** was the first effective semisynthetic enzyme to be prepared (*1, 2*). In the oxidation of N^1-alkyl-1,4-dihydronicotinamides by **2C**, the k_{cat}/K_m values seen are one to two orders of magnitude larger than the second-order rate constants observed for the corresponding model reactions. Furthermore, under conditions of substrate in excess, the semisynthetic enzyme **2C** exhibits saturation kinetics even at a relatively low substrate concentration, a phenomenon not observed in the corresponding model reactions of compounds like **2A**.

Table 1 lists the rate parameters for the oxidation of dihydronicotinamides by flavopapain **2C** and by the model flavin 7-acetyl-10-methylisoalloxazine, **2A**. For comparison, the kinetic parameters in the cases of five naturally occur-

Table 1. Kinetic parameters for the oxidation of dihydronicotinamides by 7-acetylflavopapain **2C** and 7-acetylflavin **2A**.

Substrate*	Parameters for reactions			
	Enzymatic†			Model
	K_m (M)	k_{cat} (sec^{-1})	k_{cat}/K_m (M^{-1} sec^{-1})	k_2 (M^{-1} sec^{-1})
NBzNH	1.9×10^{-4}	0.64	3,370	185
NEtNH	1.3×10^{-4}	0.72	5,500	853
NPrNH	1.0×10^{-4}	0.81	8,100	845
NHxNH	0.42×10^{-4}	0.44	10,500	843

*NBzNH, N^1-benzyl-1,4-dihydronicotinamide; NEtNH, N^1-ethyl-1,4-dihydronicotinamide; NPrNH, N^1-propyl-1,4-dihydronicotinamide; NHxNH, N^1-hexyl-1,4-dihydronicotinamide.   †Rate constants for the enzymatic reaction were measured at 25°C in 0.1M tris-HCl containing 0.001M EDTA (or in buffer solutions passed through Chelex-100), pH 7.5, 0 to 5 percent ethanol (by volume).

ring flavoenzymes are given in Table 3. It can be seen that the rate constant k_{cat}/K_m for the oxidation of N^1-hexyl-1,4-dihydronicotinamide by flavopapain **2C** is larger than the corresponding rate constant for the oxidation of reduced nicotinamide adenine dinucleotide (NADH) by old yellow enzyme and is comparable to the value of k_{cat}/K_m displayed by glucose oxidase. However, the semisynthetic enzyme's efficiency as a catalyst is somewhat lower than two of the other naturally occurring flavoenzymes shown and much lower than that seen for NADH dehydrogenase. Our results then show that flavopapain **2C**, while not an exceptional catalyst, is a moderately effective flavoenzyme, comparable in activity to a number of the naturally occurring enzymes.

Although an extensive search for the best substrate has not been carried out, some trends in the selectivity toward substrates exhibited by **2C** are nevertheless apparent. Flavopapain **2C** contains an extended hydrophobic binding region. Therefore, it is not surprising that there is an increase in k_{cat}/K_m as the N^1-alkyl group of the dihydronicotinamide increases in chain length. It is probably for

Table 2. Kinetic parameters for the oxidation of dihydronicotinamides by flavopapain **1C** and 8-acetylflavin **1A**. The measurements were made at pH 7.5 in 0.1M tris-HCl, 0.1 mM EDTA containing 0.1 percent ethanol at 25°C.

Substrate	Parameters for reactions			
	Enzymatic			Model
	K_m (M)	k_{cat} (sec^{-1})	k_{cat}/K_m (M^{-1} sec^{-1})	k_2 (M^{-1} sec^{-1})
NBzNH	2.7×10^{-6}	0.093	33,800	170
NPrNH	0.81×10^{-6}	0.048	58,700	878
NHxNH	0.12×10^{-6}	0.067	570,000	917
NADH	340×10^{-6}	0.0073	21	5

Table 3. Kinetic parameters for several naturally occurring flavoenzymes.

Enzyme	Source	K_m (M)	k_{cat} (sec^{-1})	k_{cat}/K_m (M^{-1} sec^{-1})
NADH-specific FMN oxidoreductase (13)	B. Harveyi	47.5×10^{-6}	15.5	3.26×10^5
NADPH-specific FMN oxidoreductase (13)	B. Harveyi	40.0×10^{-6}	34.0	8.50×10^5
Old yellow enzyme (14)	Yeast	1100×10^{-6}	0.67	6.1×10^2
NADH dehydrogenase (15)	Bovine heart			$\sim 10^8$
Glucose oxidase (16)				1.05×10^4

the same reason that NADH, having a relatively hydrophilic N^1 substituent, is a very poor substrate for **2C**.

Flavopapain 1C. Flavopapain **1C** is the most efficient semisynthetic enzyme constructed to date (5). It can show rate enhancements of nearly three orders of magnitude relative to the corresponding model reactions and displays saturation kinetics in the oxidation of dihydronicotinamides under substrate in excess conditions. The rate parameters for the oxidation of the dihydronicotinamides by flavopapain **1C** are illustrated in Table 2. The k_{cat}/K_m for the oxidation of N^1-hexyl-1,4-dihydronicotinamide by **1C** is either larger than or equal to the corresponding rate parameter for four of the five naturally occurring flavoenzymes listed in Table 3. Indeed, this semisynthetic enzyme approaches the activity displayed by all but the most efficient flavin-containing oxidoreductases known.

Flavopapain 3C. In marked contrast to flavopapains **1C** and **2C**, flavopapain **3C** is an extremely poor catalyst for the oxidation of the N^1-alkyl-1,4-dihydronicotinamides (6). The k_{cat}/K_m value for the oxidation of N^1-benzyl-1,4-dihydronicotinamide by **3C**, for example, is 41 M^{-1} sec^{-1} at pH 7.5 and 25°C, a value less

than the second-order rate constant, 64 M^{-1} sec^{-1}, seen for the corresponding model reaction with flavin **3A**. The enormous difference between the catalytic behavior of flavopapain **1C**, an excellent oxidoreductase, flavopapain **2C**, a moderately effective catalyst, and flavopapain **3C**, a poor catalyst, illustrates the dependence of the enzymatic catalytic efficiency on the proper positioning of the isoalloxazine moiety in the active site. Clearly, the catalytic efficiency of the semisynthetic enzyme produced by chemical modification of the active site of an enzyme is exquisitely sensitive to the proper design of the new catalytic group introduced.

Anaerobic Studies and Their Mechanistic Consequences

While we have not, as yet, focused on the mechanisms by which the flavopapains react with their substrates in oxidation-reduction reactions, some information about the pathway followed by dihydronicotinamides has come from studies under anaerobic conditions with flavopapain **2C**. Stopped-flow spectrophotometry was used to study the reaction of flavopapain **2C** with N-benzyl-1,4-dihydronicotinamide under anaerobic conditions. When the substrate was present in excess, the reaction displayed biphasic kinetic behavior. We calculated the apparent rate constants for each phase from the experimental data, assuming consecutive first-order kinetics and interpreting the results according to the pathway shown in Eq. 5.

$$E_{ox} + NBzNH \underset{K_s}{\rightleftharpoons} ES \xrightarrow{k_2}$$

$$ES' \xrightarrow{k_3} EH_{2red} + NBzN \qquad (5)$$

where NBzNH is N^1-benzyl-1,4-dihydronicotinamide.

A logical explanation for the biphasic kinetic phenomena is that a labile intermediate is formed during the course of the reaction. This intermediate is shown as ES' in Eq. 5. The collapse of ES' to the product corresponds to the slower phase of the reaction, and the apparent first-order rate constant calculated for this phase does not show a dependency on the substrate concentration. The formation of the intermediate corresponds to the initial, faster phase of the reaction.

The flavin moiety in 2C exhibits an ultraviolet-visible spectrum which deviates from that of the corresponding free flavin. There is a significant long wavelength absorption present in the spectrum of 2C (with tailing beyond 600 nm) which suggests that a charge-transfer interaction may exist between the flavin and an aromatic amino acid in the active site. This kind of long wavelength tailing is also seen with the reduced flavopapain-N-benzylnicotinamide mixture. However, when the ES' intermediate is observed by stopped-flow spectral measurements in the oxidation of N^1-benzyl-1,4-dihydronicotinamide under anaerobic conditions, the ultraviolet-visible spectrum seen contains far less long

wavelength tailing than either the starting flavopapain or the reduced flavopapain-product mixture. This suggests that the charge-transfer complex between the flavin moiety and the aromatic amino acid is disrupted prior to the redox reaction with NBzNH. Although alternative explanations have been examined (2), these considerations led us to propose the pathway described below for the oxidation of dihydronicotinamides by the related enzyme 1C.

Figure 1a represents the resting state of the flavoenzyme 1C. The carbonyl group of the acetyl substituent on the flavin ring system is within hydrogen bonding distance of the backbone NH of Cys^{25} and the side chain NH of Gln^{19}. The flavin entity itself is participating in a charge-transfer complex with the indole side chain of Trp^{26}.

According to our analysis, the initial fast kinetic phase seen in the anaerobic oxidation of N^1-benzyl-1,4-dihydronicotinamide corresponds to a rapid formation of the Michaelis complex illustrated in Fig. 1b; it is followed by generation of the ES' intermediate illustrated in Fig. 1c. The dihydronicotinamide substrate is bound within the long hydrophobic cavity of the enzyme (Fig. 1b). In the ES' intermediate (Fig. 1c), the charge-transfer complex existing between the flavin moiety of flavopapain 1C and an aromatic amino acid in the enzyme's active site (presumably Trp^{26}) is disrupted. The disruption of this complex should lead, as a consequence, to the observed decrease in the long wavelength absorption seen for the oxidized enzyme. In other words, in the reaction of flavopapain 2C with N^1-benzyl-1,4-dihydronicotinamide the k_2 step seen kinetically corresponds to the formation of the ES' species illustrated in Fig. 1c in which the flavin moiety

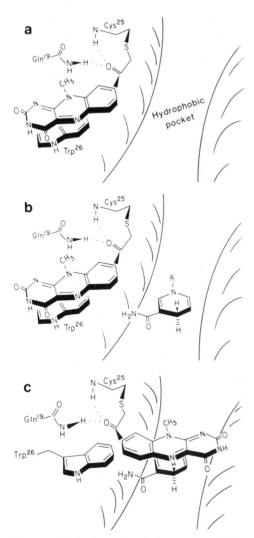

Fig. 1. (a) Active site of the semisynthetic enzyme **1C**. The acetyl side chain of the flavin moiety is hydrogen bonded to the Gln[19] and Cys[25] backbone. The flavin is participating in a charge transfer complex with Trp[26]. (b) Michaelis complex. The dihydronicotinamide is embedded within the hydrophobic groove of the flavoenzyme. (c) ES′ intermediate. The flavin-Trp[26] charge-transfer complex has been disrupted and the flavin now lies directly over the nicotinamide substrate. The pro-R hydrogen is shown as the species being transferred to the N-5 position of the flavin. This corresponds to the same transfer preference found for the oxidation of labeled NADH by flavopapain **2C** (3).

has moved to a distinctly different environment from the one it had occupied in the Michaelis complex ES shown in Fig. 1b. In the step leading from the structure of Fig. 1b to that of Fig. 1c, the flavin remains in the oxidized state while the substrate dihydronicotinamide remains reduced. After the realignment of the flavin moiety has taken place giving ES′, the redox reaction in which hydrogen transfer occurs from the dihydronicotinamide to the flavin takes place in the k_3 step, which does not show a rate dependence on the concentration of the substrate.

The postulated formation of the ES′ intermediate (Fig. 1c) also could explain the relative kinetic behavior of flavopapains **1C**, **2C**, and **3C**. Model building indicates that, when the flavin attached to the enzyme via position 8 of the ring system (that is, **1C**) undergoes the rotation step leading to ES′, the N-5 atom of the flavin ring system is in a highly favorable alignment to receive a hydride or hydride equivalent from the substrate. However, when the flavin is attached to the enzyme through position 7 of the ring system (**2C**), this orients the N-5 atom in a manner not quite as advantageous as that in the 8-substituted species **1C**. Finally, the N-5 atom in the 6-acetyl substituted flavin species **3C** is far removed from where we expect the reactive position of the bound substrate to be. For this reason, the low catalytic efficiency of flavoenzyme **3C** is understandable. However, since we have not as yet carried out extensive anaerobic kinetic studies on the reactions of N^1-alkyl-1,4-dihydronicotinamides with the flavopapains **1C** and **3C**, the hypothesis just discussed must be regarded as reasonable but not firmly established. Indeed, our results indicate, as already described, that under aerobic conditions reduction of the flavin moiety of flavopapain **1C** by various N^1-alkyl-1,4-dihydronicotinamides in excess proceeds rapidly relative to the subsequent reaction of the reduced flavin with oxygen to regenerate the oxidized flavin species. This contrasts with the case of flavopapain **2C** and underscores the importance of carrying out detailed anaerobic kinetic comparisons between the behavior of flavopapains **1C** and **2C**.

Stereochemical Studies

In view of the asymmetry of the environment of the binding groove of the flavopapains, it is reasonable to expect selectivity in the abstraction of hydrogen from the C-4 prochiral center of the N^1-alkyl-1,4-dihydronicotinamides. Furthermore, it was also anticipated that the semisynthetic enzymes would be able to discriminate between the enantiomers of various dihydronicotinamides and related derivatives containing chiral centers. Considerable experimental support has been obtained for both of these predictions.

The stereochemical consequences of hydrogen transfer from the dihydronicotinamide substrate to the flavin moiety can be elucidated with the use of 1,4-dihydronicotinamides stereospecifically labeled at the C-4 position with deuterium (17). However, when our stereochemical work was undertaken, there were no methods available to label stereospecifically model compounds like N^1-alkyl-1,4-dihydronicotinamides in a simple fashion. Consequently, it was necessary to study the reaction of flavopapain with the relatively poor substrate NADH which can be labeled stereospe-

cifically with deuterium at either the 4A or 4B positions (scheme 1). The results of experiments in which flavopapain **2C** was the catalyst for the oxidation of NADH and the corresponding deuterated derivatives are shown in Table 4.

Flavopapain **2C** exhibits a substantial preference for abstraction of the 4A (pro-R) hydrogen of NADH. The stereoselectivity exhibited by flavopapain **2C** is presumed to be the consequence of a difference in the rates of hydrogen transfer from the A and B sides of the 1,4-dihydronicotinamide ring as illustrated in scheme 1. It should be mentioned here that a complicating factor in our study of the oxidation of deuterated NADH derivatives is the possibility that the product NAD^+ and the reactant NADH might undergo nonstereospecific exchange of the C-4 hydrogen (18). While we used NADH concentrations less than 0.5 mM in order to minimize exchange under our reaction conditions, there is a possibility that such exchange might still complicate the product ratio results shown in Table 4. Ideally, by a calculation combining the results of the rate

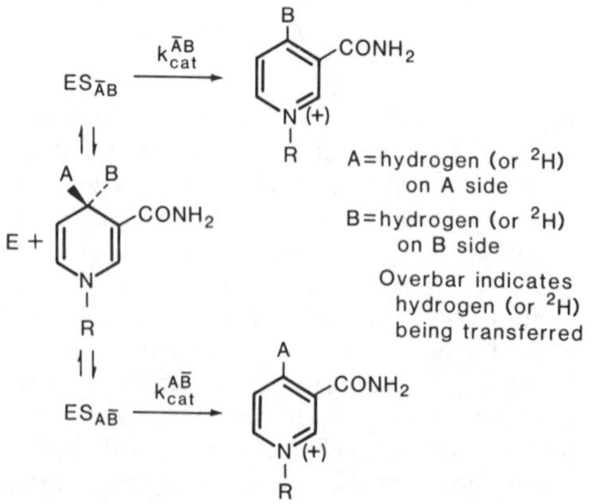

Scheme 1. Kinetic scheme for hydrogen transfer from the A and B faces of the dihydronicotinamide.

Table 4. Oxidation of NADH and deuterated NADH derivatives by flavopapain 2C. Product ratios and rate parameters.

NADH derivative	[4-^{2}H]-NAD$^+$/NAD$^+$	k_{cat}/K_m (M^{-1} sec^{-1})
NADH		68.1
[4A-^{2}H]NADH	0.47	17.3
[4B-^{2}H]NADH	7.33	43.6
[4AB-^{2}H$_2$]NADH		3.2

measurements with the various dihydronicotinamides (undeuterated, mono-deuterated, and dideuterated) and the results of product determination on the amount of hydrogen or deuterium transfer, the ratio for the rate of hydrogen transfer from the A side to that from the B side could be calculated. However, a meaningful solution was not obtained in this way, possibly because of complications from the nonstereospecific exchange of the product NAD$^+$ and the reactant NADH species. As an alternative, the ratio of the rate constants for hydrogen transfer from the A face to the B face was estimated from the relative values of $(k_{cat}/K_m)_{H\overline{D}}$ and $(k_{cat}/K_m)_{D\overline{H}}$. (The hydrogen or deuterium being transferred is denoted by the overbar.) Our results indicate that hydrogen transfer from the A face occurs at a rate approximately sevenfold higher than the corresponding transfer of hydrogen from the B face.

In the case of flavoenzyme 1C its ability to discriminate between enantiomers of dihydronicotinamide containing chiral centers was also studied briefly. The results of these experiments are provided in Table 5 (6).

The enantiomers of the 1,4-dihydronicotinamide 6 which possess a chiral center at the secondary carbon of the alkyl substituent attached to the N^1-position are oxidized at comparable rates. Model building suggests that the substrate can orient its N^1-substituent into the hydrophobic groove of the enzyme and, thus, away from the active site where hydride transfer occurs. If this is, indeed, the case, then it is not surprising that flavopapain 1C fails to distinguish between enantiomeric dihydronicotinamides which contain the chiral center on the N^1-substituent.

In contrast, when the kinetic behavior for the oxidation of substrates 7 and 8 was determined, appreciable chiral discrimination was observed. In both cases the L isomer reacts faster than its D counterpart (as judged by k_{cat}/K_m) by a factor of approximately 2. Clearly, studies on the chiral selectivity of the flavopapains have not yet been extensive. However, the findings with flavopapain 1C which shows some selectivity in discriminating between the D and L isomers of compounds 7 and 8, even though these compounds do not have the chiral center at the reactive function, and the observation with flavopapain 2C that there is approximately a sevenfold preference for removal of the 4A hydrogen over the 4B hydrogen at the prochiral center are very encouraging.

Oxidation of Dithiols by Flavopapain 2C

The oxidation of dithiols by flavins has been studied extensively. The generally accepted mechanism (scheme 2) involves a rate determining nucleophilic attack of a thiolate ion on the C-4A position of the flavin ring (19, 20). In our laboratory the oxidation of dithiothreitol (DTT), dl-dihydrolipoic acid, and dl-dihydrolipoamide by flavopapain 2C and by the corresponding flavin 2A has been

Table 5. Kinetic parameters for the oxidation by flavopapain 1C of dihydronicotinamides containing chiral centers. Kinetic measurements were carried out at pH 7.5 and 25°C.

Substrate	Isomer	k_{cat} (sec^{-1})	K_m (M)	k_{cat}/K_m (M^{-1} sec^{-1})	Ratio of rate constants L/D
6	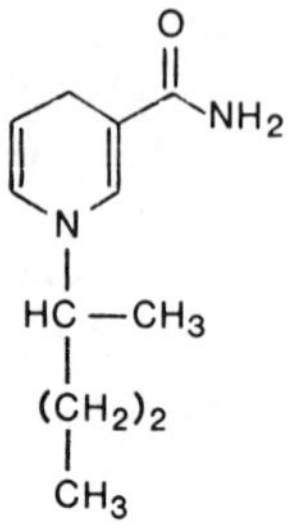 D/L	0.052	3.8×10^{-7}	137,000	1
7	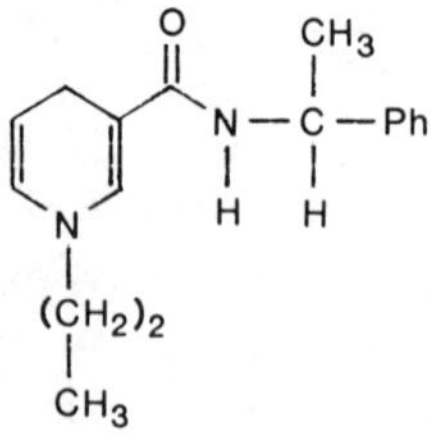 D	0.09	19×10^{-6}	5,100	
	L	0.07	6×10^{-6}	12,000	2.4
8	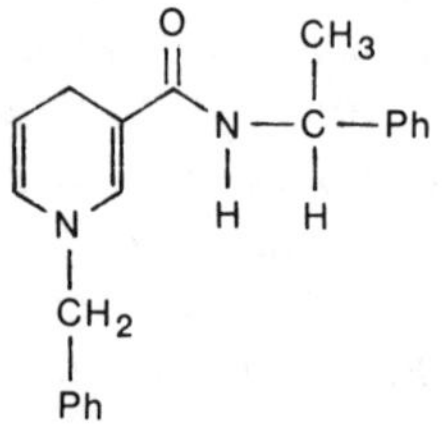 D	0.08	31×10^{-6}	2,200	
	L	0.08	22×10^{-6}	3,800	1.7

investigated (*4*), and the kinetic results obtained are shown in Table 6.

In the cases of all three dithiols, the semisynthetic enzyme reacts at a rate faster than that of the corresponding model reaction (comparing k_{cat}/K_m for the enzymatic reaction to the second-order rate constant, k_2, for the model reaction). Furthermore, in spite of the lack of stereoselectivity for the enzymatic reactions, the rate enhancements seen in these cases increases as the hydrophobicity of the substrate increases. This trend, similar to that observed in the oxidation of dihydronicotinamides, is undoubtedly a consequence of the presence of the hydrophobic binding groove in the semisynthetic flavoenzyme which

Table 6. Kinetic parameters for the oxidation of dithiols by flavopapain **2C** and 7-acetylflavin **2A**.

Substrate	Reaction		Rate enhance-ment
	Enzymatic k_{cat}/K_m (M^{-1} sec^{-1})	Model k_2 (M^{-1} sec^{-1})	
Dithiothreitol*	3.86	0.99	3.9
Dihydrolipoic acid†	6.70	0.84	8.0
Dihydrolipoamide‡	21.0	1.21	17.4

*Rate constants were measured at 25°C at pH 7.5. †Same conditions as for * except pH 7.3. ‡Same conditions as for * except with 3 to 6 percent dimethyl sulfoxide (by volume).

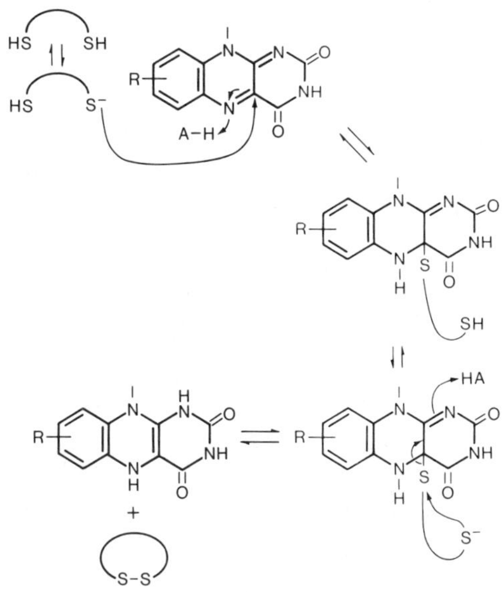

Scheme 2. Postulated mechanism of flavin-catalyzed oxidation of dithiols.

helps bring the reacting dithiols to the proximity of the flavin group.

Other Ways to Construct

Semisynthetic Enzymes

In 1966 the conversion by chemical methods of the serine residue in the active site of subtilisin to a cysteine residue was reported by Neet and Koshland (*21, 22*) and by Polgar and Bender (*23, 24*). The resultant "thiolsubtilisin" did not possess proteinase activity although a number of naturally occurring hydrolytic enzymes like papain contain histidine and cysteine as active site residues. Indeed, the semisynthetic species thiolsubtilisin showed significant activity as a hydrolytic catalyst against only the most highly activated substrates, such as nitrophenyl esters and acyl imidazole derivatives. Since the thiolsubtilisin was first reported, "thioltrypsin" (*25*) and "hydroxypapain" (*26*) have been prepared and these semisynthetic enzymes show a similar lack of catalytic activity. The exact reasons for the diminished catalytic activity of these semisynthetic enzymes relative to their natural counterparts are not known. A possible explanation is that in a given enzyme active site environment there is a requirement that the acidic and basic catalytic groups have just the right difference in their ionization constants in order to maximize the rates of proton transfer required for the catalytic act to occur efficiently. In other words, while the ionization constants of the sulfhydryl of cysteine and the imidazole of histidine may have just the right separation in the active site environment of papain, this may not be the case in the environment of the active site of thiolsubtilisin.

In another approach to semisynthetic enzymes, Wilson and Whitesides have reported the use of the biotin binding protein, avidin, as a chiral template (27). The complex 9 is a moderately active hydrogenation catalyst that does not exhibit any enantioselectivity. However, in the presence of avidin, 9 catalyzed the hydrogenation of α-acetamidoacrylic acid to (S)-N-acetylalanine, giving a 34 percent enantiomeric excess of this isomer.

Royer has attempted to generate semisynthetic enzymes from immunoglobulins (28). The specificity of an antibody for its antigen or hapten is well known. Royer utilized this specificity to label the binding site of antibodies to a dinitrophenyl hapten with a dinitrophenyl derivative of histidine p-nitrophenyl ester. Unfortunately, the modified proteins obtained failed to display any rate enhancement for the catalysis of the hydrolysis of p-nitrophenyl acetate. This lack of catalytic activity was ascribed to a misorientation of the bound substrate.

Finally, recombinant DNA technology has provided a means to obtain site-specific mutations at the active site of an enzyme. Restriction enzymes are used to excise specific gene segments, and then chemically synthesized oligonucleotides containing the information for the desired amino acid substitution are introduced. This methodology has resulted in the expression of a β-lactamase in which the active site serine has been replaced by a cysteine. The resultant "thiol-β-lactamase" exhibits reduced but detectable activity (29). The use of site-specific mutagenesis in combination with the "chemical mutation" methodology described in this article should increase enormously the range of possibilities in the design of semisynthetic enzymes.

References and Notes

1. H. L. Levine, Y. Nakagawa, E. T. Kaiser, Biochem. Biophys. Res. Commun. 76, 64 (1977).
2. H. L. Levine and E. T. Kaiser, J. Am. Chem. Soc. 100, 7670 (1978).
3. _____, ibid. 102, 343 (1980).
4. H. E. Fried and E. T. Kaiser, ibid. 103, 182 (1981).
5. J. T. Slama, S. R. Oruganti, E. T. Kaiser, ibid., p. 6211.
6. J. T. Slama, C. H. Radziejewski, E. T. Kaiser, ibid., in press.
7. E. T. Kaiser, H. L. Levine, T. Otsuki, H. E. Fried, R. M. Dupeyre, Adv. Chem. Ser. 191, 35 (1980).
8. A. N. Glazer and E. L. Smith, in The Enzymes, P. D. Boyer, Ed. (Academic Press, New York, 1971), vol. 3, p. 501.
9. J. Drenth, J. N. Jansonius, R. Koekoek, H. M. Sween, B. G. Wolthers, Nature (London) 218, 929 (1968).
10. J. Drenth, J. N. Jansonius, R. Koekoek, B. G. Wolthers, Adv. Protein Chem. 25, 79 (1971).
11. R. E. Dickerson and I. Geiss, The Structure and Action of Proteins (Benjamin, Menlo Park, Calif., 1969), p. 86.
12. J. Drenth, K. H. Kalk, H. M. Swen, Biochemistry 15, 3731 (1976).
13. F. Jablonski and M. DeLuca, ibid. 16, 2932 (1977).
14. D. J. T. Porter and H. J. Bright, J. Biol. Chem. 255, 7362 (1980).
15. T. P. Singer, in Biological Oxidations, T. P. Singer, Ed. (Interscience, New York, 1968), p. 339.
16. Q. H. Gibson, B. E. P. Swoboda, V. J. Massey, J. Biol. Chem. 239, 3927 (1964).
17. L. J. Arnold, K. You, W. S. Allison, N. O. Kaplan, Biochemistry 15, 4844 (1976).
18. M. F. Powell and T. C. Bruice, J. Am. Chem. Soc. 104, 5834 (1982).
19. Y. Yokoe and T. C. Bruice, ibid. 97, 450 (1975).

20. L. E. Loechler and T. C. Hollocher, *ibid.*, p. 3235.
21. K. E. Neet and D. E. Koshland, *Proc. Natl. Acad. Sci. U.S.A.* **56**, 1606 (1966).
22. K. E. Neet, A. Nanci, D. E. Koshland, *J. Biol. Chem.* **243**, 6392 (1968).
23. L. Polgar and M. L. Bender, *J. Am. Chem. Soc.* **88**, 3153 (1966).
24. _____, *Adv. Enzymol.* **33**, 381 (1970).
25. H. Yokosawa, S. Ojima, S. Ishii, *J. Biochem. (Tokyo)* **82**, 869 (1977).
26. P. I. Clark and G. Lowe, *Eur. J. Biochem.* **84**, 293 (1978).
27. M. E. Wilson and G. M. Whitesides, *J. Am. Chem. Soc.* **100**, 306 (1978).
28. G. P. Royer, *Adv. Catal.* **29**, 197 (1980).
29. I. S. Sigal, B. G. Harwood, R. Arentzen, *Proc. Natl. Acad. Sci. U.S.A.* **79**, 7157 (1982).
30. This research was supported in part by NSF grants APR 72-03577, AFR 77-14529, DAR 79-10245, and CHE 82-18637.

The rational design and construction of biologically active peptides and polypeptides is now an attainable goal through the applications of the tools of modern chemistry (*1*). Foremost among these tools is the technique of solid phase peptide synthesis (*2*). No longer is it necessary for a chemist interested in preparing a peptide 20 or 30 amino acids in length to spend a substantial portion of his career in its construction. Rather, through the judicious application of modern purification and analytical techniques such as high-performance liquid chromatography (HPLC), it is now possible for a graduate student in the course of his thesis studies to prepare as many as five to ten peptides of this size and to characterize their physical and biological properties. Because of these technological advances, it is feasible not only to propose structural hypotheses for the construction of biologically active peptides and polypeptides but also to test thoroughly the experimental aspects of these proposals.

Although we have increased our understanding of tertiary structure in recent years, we have not yet reached the point at which the folding of a peptide with a given amino acid sequence into a tertiary structure can be predicted with confidence. In contrast, a solid foundation has been laid for the prediction of amino acid sequences that form certain types of secondary structures. The recognition of the importance of these secondary structures has led us to a new approach to the design of biologically active peptides such as hormones.

A host of biologically important peptides composed of 10 to 50 amino acids are devoid of well-defined tertiary structure. As a rule, they lack disulfide bonds, they are linear (*3*), and their conforma-

30

Amphiphilic Secondary Structure: Design of Peptide Hormones

E.T. Kaiser
and F.J. Kézdy

Science 223, 249–255 (20 January 1984)

tion depends entirely on their environment; they can assume completely different secondary structures in water, in a detergent micelle, or in trifluoroethanol or other organic solvents. Many of these peptides play very specific roles as hormones, cofactors, signals for membrane translocation, and the like. The high activity and the specificity of their action imply a well-defined structure traditionally associated with ligand-enzyme and ligand-receptor interactions, although their behavior in aqueous solution gave no evidence for any predominant structure that could be associated with their biological activity. Probing for the "active site" of many of these molecules by selective chemical modification gave different results from those seen in similar experiments with enzymes; it appeared that all parts of the molecule were essential for high activity, although in some cases a number of amino acids could be modified without a major change in activity. The requirement that most of the molecule—not only the part which comes into direct contact with the receptor—be present argues against the hypothesis that the biologically active conformation would be imposed on the peptide by the complementary tertiary structure of the macromolecule which is the target of the peptide; that is, the receptor or the antibody upon which the peptide acts. Indeed, most ligand-macromolecule interactions that have been studied are limited to the specific interaction of only a few amino acid residues.

Many peptide hormones act at membranes that represent a characteristic, anisotropic environment due to the amphiphilic nature of the dividing line between the aqueous solution and the lipid bilayer of the membrane. It is then possible that the amphiphilic environment could impose a secondary structure on the peptide. If such a membrane-directed secondary structure exists, it should be characterized by the amphiphilic distribution of the individual amino acid side chains. One face of the peptide should be occupied preferentially by lipophilic side chains (4), such as Ile, Leu, Val, Phe, Trp, Tyr, and Met, whereas all hydrophilic residues, such as Asp, Glu, Arg, and Lys should be located at the opposite face. The free energy contribution of locating the amino acid residues in their optimal microenvironment is considerable, yielding additional stability (as much as 1 to 3 kcal/mole) per amino acid residue. Thus, secondary structures which would only have marginal or no stability in an aqueous solution could very well become the particular conformation of the peptide once located in the membrane. Most important, secondary structures that are energetically too unfavorable to exist in aqueous solution could be easily generated in an amphiphilic environment, provided that full advantage is taken of the hydrophilic and lipophilic interactions of the amino acid side chains. Thus, one could hypothesize that the biologically active conformation of peptides acting in membranes could be any of the sterically allowed secondary structures, namely, right- and left-handed α-helix, β-pleated sheet, 3,10-helix, or π-helix.

The search for the possibility of the occurrence of amphiphilic secondary structures in peptides needs only very simple tools. Once the amino acid sequence is known, regular alternation of the lipophilic and hydrophilic residues indicates the possibility of an amphiphilic β-sheet. The possibility of amphiphilic α- or π-helices is detected by the construction of an "Edmundson wheel" (5) of the appropriate pitch and noting the segregation of the hydrophilic and lipophilic residues on opposite faces of the cylinder circumscribing the helix. Alternatively, the helix can be projected in a two-dimensional fashion, which then shows more distinctly the interactions among side chains along the axis of the helix as well as the occurrence of the hydrophilic and lipophilic domains. Figure 1, A to C, shows the application of such constructions to a peptide with a potential pleated sheet, an α-helix, and a π-helix. The prediction of the occurrence of possible amphiphilic structures can then be further affirmed by calculating the probability of secondary structures by one or another of the semiempirical methods, such as the Chou-Fasman method (6) and the calculation of the free energy changes associated with the transfer of lipophilic residues into a nonaqueous environment (7).

Once the possible occurrence of an amphiphilic secondary structure is established, numerous experimental ap-

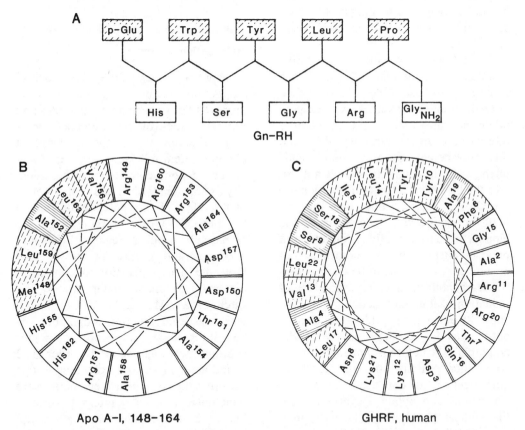

Fig. 1. Graphical method for searching for amphiphilic secondary structures. (A) β-Pleated sheet conformation of gonadotropin releasing hormone. (B) Axial projection of a potential α-helical conformation of residues 148 to 164 of apolipoprotein A-I. Residues in the white areas are hydrophilic. Those in the areas containing dashed lines are lipophilic. (C) Axial projection of a potential π-helical conformation of residues 1 to 22 of human growth hormone releasing factor. The segregation of the hydrophilic and lipophilic residues is shown.

proaches are needed to establish the existence of such a conformation. Indirect evidence can be derived from the fact that the peptide would bind readily to single bilayer phospholipid vesicles and would form stable monomolecular layers at the air-water interface (8). Also, the peptides should form readily "micellar structures," that is, small oligomers formed with a well-defined number of monomers, where the lipophilic domains are strongly interacting. The secondary structures of such assemblies can then be probed by conventional methods, that is, nuclear magnetic resonance spectroscopy, optical rotatory dispersion, or circular dichroism.

The ultimate confirmation of active secondary amphiphilic structures can come from the use of peptide synthesis. Indeed, in many instances one can postulate that most amino acids in the peptide assume only a structural role and their replacement by other amino acids of the same lipophilicity should not impair the biological activity of the peptide (9). On the contrary, one could imagine that the amino acid sequence of the biologically active peptide is dictated not only by structural considerations but also by considerations of biocompatibility, degradability, specificity, and lack of induction of immunoreaction. Thus, the naturally occurring peptide might be suboptimal as far as the stability of the amphiphilic structure is considered. One could then optimize the structural requirements at the expense of other considerations and obtain a synthetic peptide with a higher biological activity than the naturally occurring peptide.

A brief review of the early work in our laboratories which forms the basis of this approach seems in order. We began our work by searching for peptide systems where secondary structural features might dominate the biological activity. From the outset it was clear that such systems were likely to be found among the peptides that bind to phospholipids, membranes, or other amphiphilic surfaces. There appeared to be a good possibility that the peptides that bind to such surfaces, which have the characteristic of being hydrophobic on one face and hydrophilic on the other, might have complementary amphiphilic secondary structures.

Synthetic Studies on Apolipoproteins and Peptide Toxins

Significant clues to the importance of amphiphilic secondary structures to the properties of surface-active peptides and proteins came from work on apolipoproteins. Examination of the sequence of apolipoprotein A-I (apo A-I), the principal polypeptide constituent of high density lipoprotein (HDL) and of the molecular models of this polypeptide, led to the proposal (10) that there might be amphiphilic α-helical regions throughout a considerable portion of the apo A-I molecule. This suggestion was amplified by analyses made independently by Fitch (11) and McLachlan (12) in which it was suggested that the amphiphilic α-helical regions had a repeating character and that they might be on the order of 22 amino acids in length. According to this picture, the α-helical segments were punctuated at regular intervals by proline or glycine residues acting as helix breakers.

Model building of the amphiphilic α-helical structures revealed to us that they would provide a very reasonable mode for the interaction of surface active peptides like apo A-I with phospholipid vesicles and other amphiphilic surfaces. When a model of apo A-I was constructed we saw that the helices could be placed on the surface of an HDL particle in such a way that the axes of the helices were roughly tangential to the particle surface. The model showed that the hydrophobic faces of the helices could penetrate into the particle in the spaces between the phospholipid head groups, allowing these faces to come into contact with the long hydrophobic chains of the phospholipids. The helices then had their hydrophilic faces oriented toward the aqueous environment. Whereas this picture appeared to be attractive, an experimental test had to be performed to determine whether amphiphilic α-helices were induced in the apo A-I molecule in its binding to phospholipid in the HDL particle. At the point when we began, the physical chemical methodology for examining the secondary structures of peptides bound to membrane surfaces, while undergoing development, did not appear to us to provide the most incisive approach toward attacking the problem. Rather, it appeared to us that a more definitive picture might come from a synthetic organic study of the amphiphilic α-helix and its role in apolipoprotein binding to phospholipid surfaces.

If it is argued that a secondary structural feature like an amphiphilic α-helical region is crucial to the biological activity of a peptide such as apo A-I, then a fundamental prediction which emerges is that this activity ought to be simulated by a peptide with a radically different amino acid sequence but designed to have essentially the same secondary structure (13). To test this basic hypothesis, we designed a 22 amino acid peptide which represented an "idealized" ver-

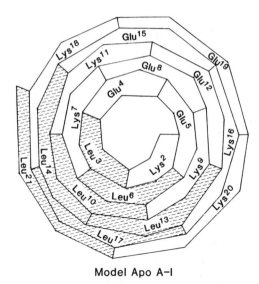

Model Apo A-I

Fig. 2. Two-dimensional axial projection of the α-helical conformation of a model peptide designed to simulate the properties of apolipoprotein A-I.

sion of the amphiphilic α-helical regions of apo A-I (Fig. 2). The peptide was constructed primarily of three types of amino acid residues with a high potential to participate in α-helical structures: Leu, a hydrophobic amino acid; Lys, a hydrophilic residue positively charged in the pH region near neutrality; and Glu, a negatively charged hydrophilic residue (9). The 22 amino acid model peptide was designed to be capable of forming an amphiphilic α-helix with a hydrophobic face covering about one-third of the surface and the hydrophilic face covering the remaining two-thirds. Most importantly, it was designed to have minimal homology to any of the amphiphilic α-helical regions of the naturally occurring protein. The synthetic model peptide was shown to mimic successfully the properties of apo A-I, a 243 amino acid peptide (9, 14). The model bound comparably to phospholipid surfaces, showed related α-helical character in water and water-trifluoroethanol solutions and showed similar aggregation behavior. Also, the model peptide showed appropriate biological activity, acting as a good activator of the enzyme lecithin:cholesterol acyltransferase (E.C. 2.3.1.43), an important function of apo A-I (14).

A peptide in which an amphiphilic secondary structural region was attached to a simple "active site" was the next most complex system examined (15). The bee venom toxin, melittin, an activator of the enzyme phospholipase A_2 and an effective hemolytic agent, contains in its major form a peptide segment about 20 amino acids in length, which has the potential to form an amphiphilic α-helix when bound to an appropriate surface. It contains a proline residue part way through the sequence, which should cause a kink in the helix. The helical region is attached to a hexapeptide segment which contains a cluster of positive charges (16). When the hexapeptide segment is removed from melittin, the remaining 20 amino acid segment, while capable of binding to erythrocytes, does not lyse them (17). Accordingly, a model was constructed in which the 20 amino acid segment starting at the amino terminus was designed to have relatively little homology to the melittin sequence itself but to have a comparable hydrophobic-hydrophilic balance and a somewhat accentuated potential for forming an α-helix (15). In this case, the helical portion was constructed with Leu residues as the hydrophobic groups, and because it appeared that the charge was not an important factor in the helix, Gln and Ser residues were chosen as the hydrophilic groups (Fig. 3). The hexapeptide portion attached at the COOH terminus to the 20

amino acid peptide had the same sequence as that in natural melittin. The 26 amino acid model peptide constructed in this fashion proved to be both a highly effective lytic agent and a satisfactory activator of phospholipase A_2 (18).

Amphiphilic Secondary Structures in Hormones

Our studies on apolipoproteins and toxins showed that the analysis of the structure of surface active peptides in terms of amphiphilic secondary structure yielded important new information. In view of the limited number of possible secondary structures and, as suggested by others, because of the likelihood that various peptide hormones might have α-helical regions, the question arose whether or not the amphiphilic secondary features might play a major role in hormone structure (15, 19). When we considered the structures of peptide hormones and the vast amount of data on this topic, it seemed to us that these compounds might be classified into three general categories. First, there are systems such as the opioid peptide [Met[5]]-enkephalin where the whole hormone consists of a relatively short sequence of amino acids, an "active site" containing specific amino acids required for eliciting biological activity. In a second category of hormones consisting of peptides where the structures are rather complex and where, as in the case of insulin, multiple disulfide bonds may be present, tertiary structure is very important. In a third category, however, the peptides either have no disulfide bridge or have only a single bridge, and quite frequently have a length in the vicinity of 10 to 50 amino acids. In this last category, it

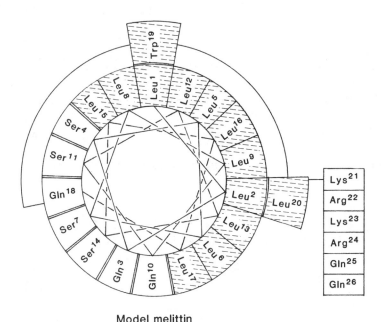

Fig. 3. Model melittin. Axial projection of the α-helical conformation of a model peptide designed to have melittin-like activity.

Model melittin

seemed quite likely that binding to the amphiphilic environment of a receptor might require the presence of an amphiphilic secondary structure in the peptide for the active site to be in the proper geometry to interact with the receptor. When we set out to search for such systems, it soon became rapidly obvious that there were many peptide hormones where sizable regions of the molecule could be forming amphiphilic secondary structures in appropriate environments (19).

The amphiphilic structural regions of most of the hormones that we have examined up to now are α-helical in nature. However, as will be discussed below, we have found systems where, we believe, the amphiphilic structures are π-helices or β-pleated sheets. The peptide hormones containing amphiphilic α-helical regions are treated first.

Calcitonin

We have proposed that calcitonin, a peptide with potent hypocalcemic activity, consists of an "active site" amino terminal heptapeptide containing a disulfide loop between Cys^1 and Cys^7, an amphiphilic α-helical region from residues 8 through 22, and a hydrophilic COOH-terminal region from residues 23 through 32 (Fig. 4A for the sequence of salmon calcitonin, SCT-I) (20). To test this hypothesis, a 32 amino acid peptide, MCT-I (Fig. 4A), designed to retain the structural characteristics of the amphiphilic α-helical region, but having an amino acid sequence from residues 8 through 22 with minimal homology to any naturally occurring calcitonin, was synthesized. The axial projections of the amphiphilic α-helical regions of SCT-I and MCT-I (Fig. 4B) show that the resi-

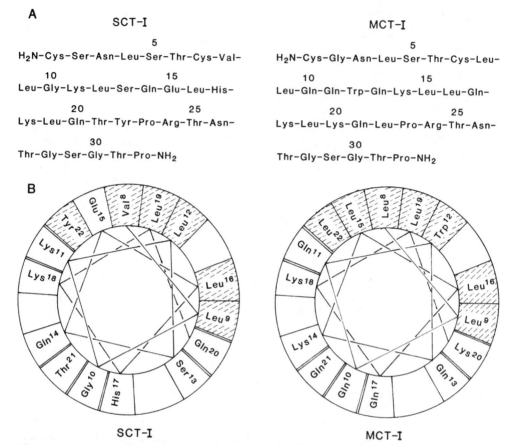

Fig. 4 (A). Amino acid sequences of salmon calcitonin (SCT-I) and a model peptide designed to have calcitonin-like activity (MCT-I) (20). (B) Axial projections of the potential α-helical conformation of residues 8 to 22 in SCT-I and MCT-I.

dues are segregated into opposing hydrophilic and hydrophobic faces. The physical and biological properties of MCT-I were studied and compared to those of SCT-I. In aqueous solution, the model peptide MCT-I was slightly more α-helical than SCT-I, and in an amphiphilic environment—the air-water interface—it formed a somewhat more stable monolayer. Most important, MCT-I, while somewhat less effective than SCT-I, specifically binds to calcitonin receptors in rat brain particulate fractions and has potent hypocalcemic activity, as mea-

sured using a rat bioassay. Taken together, the results obtained provide strong evidence that the region from residues 8 to 22 of calcitonin has a primarily structural role, interacting in the amphiphilic α-helical form with the amphiphilic environment of the calcitonin receptor (20).

We have recently undertaken work on a new calcitonin model, MCT-II (Fig. 5) (21). This model is basically similar to MCT-I in the region 8 to 22 except for the change of Trp[12] to Leu and of Leu[22] to Tyr. The amphiphilic α-helical segment in MCT-II is expected to be in

```
      ┌──────S──────S──────┐
H₂N-Cys-Ser-Asn-Leu-Ser-Thr-Cys-Leu-

         10
Leu-Gln-Gln-Leu-Gln-Lys-Leu-Leu-Gln-

         20
Lys-Leu-Lys-Gln-Tyr-Pro-Arg-Thr-Asn-

            30
Thr-Gly-Ser-Gly-Thr-Pro-NH₂
```

Fig. 5. Amino acid sequence of second model peptide designed to have high calcitonin-like activity (MCT-II) (21).

```
                        5
H₂N-Ser-Gln-Glu-Pro-Pro-Ile-Ser-Leu-Asp-

    10                15
Leu-Thr-Phe-His-Leu-Leu-Arg-Glu-Val-

      20              25
Leu-Glu-Met-Thr-Lys-Ala-Asp-Gln-Leu-

         30                35
Ala-Gln-Gln-Ala-His-Ser-Asn-Arg-Lys-

            40
Leu-Leu-Asp-Ile-Ala-NH₂
```

Fig. 6. Amino acid sequence of ovine CRF.

closer structural analogy than that of MCT-I to the corresponding region in the very potent naturally occurring calcitonin, SCT-I. Although our experiments on MCT-II are not complete, it appears that in its ability to bind to calcitonin receptors and in its hypocalcemic activity it is approximately equivalent to the salmon calcitonin, despite the substantial nonhomology between the two peptides in the α-helical region. Our experiments thus demonstrate that the use of the secondary structural analysis of peptide hormones enables us to produce new peptide sequences that have activities comparable to the most effective naturally occurring peptide hormones both in vitro and in vivo.

Corticotropin Releasing Factor

Corticotropin releasing factor (CRF), a hormone produced by the hypothalamus, increases the rate of secretion of corticotropin by the pituitary gland. Recently, Vale et al. (22) determined the primary structure of ovine CRF (Fig. 6). The helical potential of CRF has been analyzed by the method of Chou and Fasman (6), indicating that two large sections of the molecule have very pro-

nounced α-helical potential. Axial projection of the amino acid sequence of the two regions of high helical potential, using an Edmundson wheel, shows that the hydrophilic and hydrophobic residues are segregated on opposite sides of the cylindrical helix. The hydrophobic domain is nearly twice as large as the hydrophilic one.

We have examined (23) the tendency of CRF to assume an amphiphilic secondary structure when it is exposed to amphiphilic environments, such as the air-water interface and the surfaces of unilamellar phospholipid vesicles. In aqueous solution CRF exists predominantly as a random coil. When the peptide is at concentrations greater than $10^{-5}M$, it shows a tendency to self-aggregate with a concurrent slight increase in the apparent α-helical content as measured by circular dichroism. As expected from its potential amphiphilic structure, CRF binds readily to the surface of single bilayer egg phosphatidylcholine vesicles. This binding fits a simple Langmuir isotherm with the following parameters: dissociation constant $K_d = (1.3 \pm 0.6) \times 10^{-7}M$ and capacity at saturation $N = (1.1 \pm 0.1) \times 10^{-2}$ mole of peptide per mole of phospholipid. At the air-water interface CRF also forms an insoluble monolayer composed of monomers of the hormone with a molecular area which suggests the presence of a compact secondary structure. In addition, this compactness is indicated by the low compressibility of the monolayer. Indeed, the collapse pressure, 19.0 ± 0.1 dynes per centimeter, of the monolayer suggests that the amphiphilicity of CRF approximates that of the plasma apolipoproteins which, as we have seen, comprise a class of proteins of the most pronounced amphiphilic character. Thus, the binding of CRF to the cell membrane appears to be accompanied by the induction of an α-helical secondary structure and it is the predominantly helical form which is the biologically active conformation of the peptide (23). The design of models for this peptide hormone along the same lines as our earlier studies on the apolipoproteins, melittin, and calcitonin should help elucidate the structure of CRF further.

β-Endorphin

In an approach similar to those for calcitonin and CRF, we have analyzed (19) the structure of β-endorphin, a 31 amino acid peptide hormone with potent opiate activities. Examination of the linear amino acid sequence of β-endorphin has led us to propose that three separate structural regions can be distinguished: a very specific opiate recognition site in residues 1 to 5 that is identical to [Met⁵]enkephalin; a hydrophilic spacer region in residues 6 to 12 and a 16-residue sequence between Pro[13] and Gly[30], capable of forming an amphiphilic α- or π-helix with half of its surface hydrophobic and the hydrophilic residues being neutral or basic. As in the

cases of calcitonin (20) and CRF (23), the hydrophobic domain in the proposed α-helical region of β-endorphin is continuous. However, in contrast to calcitonin and CRF, where this domain lies straight along the length of the helix, in the α-helical form of β-endorphin the hydrophobic domain would twist along the length of the helix. As an alternative, the COOH-terminal region might be in a π-helical form and, in this case, the hydrophobic domain would lie straight along the length of the helix (Fig. 7) (19). At present, we are not sure which of these models for β-endorphin is the correct one, and since most of our earlier studies were with α-helices, we have until now concentrated our efforts primarily on testing α-helical models for the COOH-terminal region of the hormone.

Several model peptides have been synthesized and tested to evaluate the validity of the amphiphilic helical structural

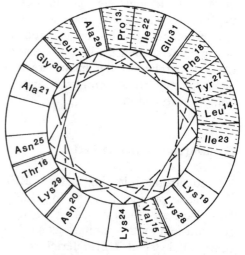

Beta-endorphin, 13-31

Fig. 7. Axial projection of a potential π-helical region of β-endorphin showing the segregation of the hydrophobic and hydrophilic residues.

450

hypothesis for β-endorphin (19, 24–26). These models include a peptide in which, in addition to the construction of an amphiphilic helical region from residues 13 to 31 with a sequence having minimal homology to that of β-endorphin, a hydrophilic spacer region from residues 6 to 12 with minimal homology to the equivalent region of the natural hormone was built (24). In a very stringent test for the structure of β-endorphin, we prepared and examined a peptide designed to be a "negative" model (26). The latter peptide retains almost all of the proposed features of β-endorphin, the only difference being that in an α- or π-helical conformation of the region 13 to 31, the amphiphilic character present in the previous models is minimized. Our results with this model, as well as with the earlier models which contained amphiphilic helices, provide evidence for the importance of an amphiphilic helical structure in β-endorphin residues 13 to 31, which determines the resistance to proteolysis of the natural hormone and which makes some contribution to the interactions of the hormone with the opiate receptors δ and μ. Indeed, the amphiphilicity of the COOH-terminal helical structural region was found to be essential for high opiate activity on the rat vas deferens (ε receptors), whereas no such structural requirement appears to be necessary for interaction with the opiate receptors on the guinea pig ileum (26).

Although our results do not allow us to choose between the α- or π-helical models for the COOH-terminal region of β-endorphin, our work illustrates the power of the structural analysis we have developed in conjunction with our synthetic approach in establishing the requirements for receptor binding and biological activity in pharmacologically complex situations like that encountered with β-endorphin.

Growth Hormone Releasing Factor

The π-helix is not a structure that is commonly encountered in globular proteins, probably because it is not compact and would be a metastable structure under the conditions where most x-ray crystal structure determinations are performed. Nevertheless, examination of molecular models of various peptide hormones leads us to propose the possibility that amphiphilic π-helices may play a significant role in the binding of these molecules to membrane surfaces. Although the π-helical structures may not be very stable under conditions where close packing is crucial, the proposal that such structures may be formed when peptide hormones bind to membrane surfaces is reasonable. Examination of the region 1 to 29 of human growth hormone releasing factor (hGRF) shows that this segment has an excellent potential to form an amphiphilic π-helix, as illustrated for the sequence 1 to 22 in Fig. 1C. We have shown that hGRF forms a very stable monolayer at the air-water interface (27). Furthermore, it binds tightly to unilamellar phosphatidylcholine vesicles. These observations are consistent with the hypothesis that hGRF forms an amphiphilic secondary structure, most likely a π-helix, in amphiphilic environments such as the air-water interface and phospholipid surfaces. Measurements of the physical properties of a synthetic peptide designed to have the structural features proposed for hGRF would allow us to test further the concept of the π-helix.

451

Glucagon

The binding of glucagon to hepatic plasma membrane receptors is the first event in the action of this hormone (28). Linkage of the occupied receptor to the catalytic entity of adenylyl cyclase results in elevation of the cytoplasmic cyclic adenosine monophosphate (AMP) and in the resulting modulation of cell metabolism (29). The binding of glucagon to hepatocytes and plasma membrane vesicles has been analyzed quantitatively in terms of two noninteracting receptor populations (30). We have been engaged in preparing and characterizing several models for glucagon, including one in which the COOH-terminal region is an amphiphilic α-helix in which the aromatic residues of the natural sequence have been preserved but the remainder of the helical region has been "idealized" (Fig. 8). When we employed rat hepatocytes, this model peptide was found to mimic (i) the activity of glucagon in binding to receptors, (ii) the elevation of cyclic AMP, and (iii) the inhibition of the incorporation of carbohydrate into glycogen (31). Binding of glucagon itself to the rat hepatocytes is consistent with the presence of two separate populations of hormone binding sites with the association constants of 57 pM and 41

```
                        5
H2N-His-Ser-Gln-Gly-Thr-Phe-Thr-Ser-

      10                    15
Asp-Tyr-Ser-Lys-Tyr-Leu-Asp-Ser-Arg-

      20                    25
Arg-Leu-Gln-Glu-Phe-Leu-Gln-Trp-Ala-

Leu-Gln-Thr
```

Fig. 8. Amino acid sequence of glucagon analog.

nM. In contrast, we found that a single sigmoidal binding curve was obtained for the model peptide, extending over a concentration range of only two orders of magnitude. Even at a 50 μM concentration of the analog only 50 percent inhibition of the binding of ^{125}I-labeled glucagon to hepatocytes was observed. Our data fit a scheme where the model peptide competes with the binding of ^{125}I-labeled glucagon to only one of the two receptor populations present on rat hepatocytes. In other words, the binding of the model peptide obeys a single thermodynamic equilibrium with the dissociation constant of 1.2 μM. To establish which of the two glucagon receptor types was responsible for the binding of the model peptide, we studied the inhibition of ^{125}I-labeled glucagon binding by unlabeled glucagon in the presence of the model peptide at $10^{-5}M$. Under such conditions, the binding curve for glucagon became a single sigmoidal one which extended over less than two orders of magnitude of competitor concentration. The dissociation constant thus obtained for the binding of glucagon was $K_d = 46$ nM, indicating that the hormone was interacting with its low affinity receptor. Thus, the use of the model peptide provides conclusive support for the hypothesis that there are two distinct, homogeneous populations of glucagon receptors in the rat hepatocytes.

The model peptide has been very useful in helping us to understand the interaction of glucagon with its receptors. These results, together with those previously obtained with β-endorphin, show the power of the structural approach we have adopted not only in testing the conformational requirements of a peptide to act as a hormone but also in permitting us to determine what parts of the molecule are crucial for binding to particular receptors in those cases where multiple receptors exist.

Conclusions

The results presented in this article provide a glimpse of new approaches that peptide synthesis offers for the elucidation of basic problems in biochemistry. Direct extension of the present methods of design should yield answers to a host of important questions. To cite a few: What is the relation between the extent of the hydrophobic domains and the biological activity? Is the size of the secondary structural domain critical or can one extend it without any interference with the receptor macromolecule? Can one replace the structural amphiphilic domain with a non–peptide moiety of comparable amphiphilic properties? Models designed with these questions in mind should also possibly yield new classes of molecules with interesting biochemical and pharmacological properties.

Further extension of our studies should involve the design of hitherto unobserved secondary structures of peptides. Judicious use of the stabilizing forces provided by the medium should allow one to stabilize energetically unfavorable conformations of the peptide bond, such as dihedral bond angles and even cis-peptide bonds which, in turn, should generate novel secondary structures.

On the basis of the available data it appears that peptides interact with their receptor macromolecules by a mechanism which is similar to that of information-bearing molecules of singular tertiary structure. In particular, the number of functional groups interacting with the receptor must be small and most of the amino acid components serve a purely structural role, namely, the proper positioning of the ligand functions.

References and Notes

1. E. T. Kaiser and F. J. Kézdy, *Proc. Natl. Acad. Sci. U.S.A.* **80**, 1137 (1983).
2. G. Barany and R. B. Merrifield, in *The Peptides*, E. Gross and J. Meienhofer, Eds. (Academic Press, New York, 1980), vol. 2, p. 3.
3. See, for example, T. Blundell and S. Wood, *Annu. Rev. Biochem.* **51**, 123 (1982).
4. Abbreviations for the amino acid residues are: Ala, alanine; Arg, arginine; Asn, asparagine; Asp, aspartic acid; Cys, cysteine; Gln, glutamine; Glu, glutamic acid; Gly, glycine; His, histidine; Ile, isoleucine; Leu, leucine; Lys, lysine; Met, methionine; Phe, phenylalanine; Pro, proline; Ser, serine; Thr, threonine; Trp, tryptophan; Tyr, tyrosine; Val, valine.
5. M. Schiffer and A. B. Edmundson, *Biophys. J.* **7**, 121 (1967).
6. P. Y. Chou and G. D. Fasman, *Annu. Rev. Biochem.* **47**, 251 (1978).
7. C. Edelstein, F. J. Kézdy, A. M. Scanu, B. W. Shen, *J. Lipid, Res.* **20**, 143 (1979).
8. D. J. Kroon, J. P. Kupferberg, E. T. Kaiser, F. J. Kézdy, *J. Am. Chem. Soc.* **100**, 5975 (1978).
9. D. Fukushima et al., *ibid.* **101**, 3703 (1979).
10. J. P. Segrest et al., *FEBS Lett.* **38**, 247 (1974).
11. W. M. Fitch, *Genetics* **86**, 623 (1977).
12. A. D. McLachlan, *Nature (London)* **267**, 465 (1977).
13. D. Fukushima et al., *Ann. N.Y. Acad. Sci.* **348**, 365 (1980).
14. S. Yokoyama, D. Fukushima, F. J. Kézdy, E. T. Kaiser, *J. Biol. Chem.* **255**, 7333 (1980).
15. W. F. DeGrado, F. J. Kézdy, E. T. Kaiser, *J. Am. Chem. Soc.* **103**, 679 (1981).
16. E. Habermann, *Science* **177**, 314 (1972).
17. E. Schröeder, K. Lübke, M. Lehmann, I. Beitz, *Experientia* **27**, 764 (1971).
18. W. F. DeGrado, G. F. Musso, M. Lieber, E. T. Kaiser, F. J. Kézdy, *Biophys. J.* **37**, 329 (1982).
19. J. W. Taylor, R. J. Miller, E. T. Kaiser, *J. Am. Chem. Soc.* **103**, 6965 (1981).
20. G. R. Moe, R. J. Miller, E. T. Kaiser, *ibid.* **105**, 4100 (1983).
21. G. R. Moe, in preparation.
22. W. Vale, J. Spiess, C. Rivier, J. Rivier, *Science* **213**, 1394 (1981).
23. S. H. Lau, J. Rivier, W. Vale, E. T. Kaiser, F. J. Kézdy, *Proc. Natl. Acad. Sci. U.S.A.* **80**, 7070 (1983).
24. J. W. Taylor, R. J. Miller, E. T. Kaiser, *Mol. Pharmacol.* **22**, 657 (1982).
25. J. W. Taylor, R. J. Miller, E. T. Kaiser, *J. Biol. Chem.* **258**, 4464 (1983).
26. J. Blanc, R. J. Miller, E. T. Kaiser, *ibid.* **258**, 8277 (1983).
27. S. H. Lau, private communication.
28. V. Bonnevie-Nielsen, K. S. Polonsky, J. J. Jaspan, A. H. Rubenstein, T. W. Schwartz, H.

S. Tager, *Proc. Natl. Acad. Sci. U.S.A.* **76**, 3189 (1979), and references therein.

29. E. M. Ross and A. C. Gilman, *Annu. Rev. Biochem.* **49**, 533 (1980).

30. V. Bonnevie-Nielsen and H. S. Tager, *J. Biol. Chem.* **258**, 11313 (1983).

31. G. F. Musso, E. T. Kaiser, F. J. Kézdy, H. S. Tager, *Abstracts, Eighth American Peptide Symposium, Tuscon, Ariz., May, 1983.*

32. We thank our collaborators and faculty colleagues whose contributions are cited in this article and without whose enthusiastic work and helpful discussions this paper could not have been written. The work presented here was supported by Public Health Service Program Project HL-18577 and by a grant from The Dow Chemical Company Foundation.

In recent years nuclear magnetic resonance (NMR) spectroscopy has been used extensively to study the mode of action of various enzymes with both inhibitors and substrates. So far, reporter nuclei, including ^{15}N, ^{13}C, ^{31}P, and ^{1}H, have been observed either at or near the active site of several enzymes. These studies have been reviewed elsewhere (1–3) and will be referred to in our discussion, which is concerned with the reactions of ^{13}C-enriched inhibitors and substrates with enzymes. Until recently, ^{13}C has not received much attention in relation to enzyme mechanism, in part because of certain inherent disadvantages. These include a low natural abundance (1.1 percent) compared to ^{13}P (100 percent) and ^{1}H (100 percent) and a relative insensitivity to the NMR experiment resulting from the small gyromagnetic ratio (γ) of ^{13}C, which directly governs the size of the nuclear response to electromagnetic excitation (4). To overcome these problems, the site or sites of interest in any system must be artificially enriched (either synthetically or biosynthetically) in ^{13}C to levels approaching 100 percent. The ultimate benefit from such tedious and time-consuming operations is that ^{13}C becomes the nucleus of choice in monitoring enzyme-substrate (or enzyme-inhibitor) interactions since its wide range of chemical shifts ($\approx$220 ppm) makes it an exquisitively sensitive probe of molecular structure.

With few exceptions, the use of ^{13}C NMR spectroscopy in studying enzyme reactions has been directed toward monitoring the fate of a specifically enriched carbonyl function in a substrate or inhibitor. This is a consequence of the large chemical shift change (Δ) brought about by substitution α to this group (Δ ~2 to

31

Studying Enzyme Mechanism by ^{13}C Nuclear Magnetic Resonance

Neil E. Mackenzie, J. Paul G. Malthouse, and A.I. Scott

Science 225, 883–889 (31 August 1984)

30 ppm) or by direct covalent attachment of the carbonyl (Δ ~100 ppm) to the enzyme. There are other advantages of the carbonyl function (and quaternary carbons in general). (i) Full nuclear Overhauser effect (5) proton decoupling can be maintained by low decoupler settings with concomitant minimization of dielectric heating. (ii) Since chemical shift anisotropy (CSA) is the predominant relaxation mechanism at high field strengths, the values of the longitudinal relaxation times are reduced (6), allowing rapid data acquisition without saturation. (iii) The line widths of the observed signals convey important information (see below) and are smaller for quaternary carbons relaxing by the CSA mechanism than for carbons having directly bonded protons, which have a predominantly dipolar relaxation mechanism (6, 7). The gain in sensitivity resulting from the use of very high magnetic fields (up to 500 MHz) is offset to some extent by increased line width at high field strength.

Almost all enzymatic processes utilize multistep reactions during catalysis, and the characterization of each of these stages is essential if one is to understand enzyme mechanism at the molecular level. Earlier investigators have used spectrophotometric methods for characterizing enzyme-catalyzed reactions, whereas inhibitors that form stable "transition state analogues" have been examined by other techniques that require long periods of data accumulation, for example, x-ray analysis. However, the advent of NMR and its subsequent use in enzymology are beginning to provide a novel and penetrating probe for elucidating enzyme mechanism by directly characterizing intermediates formed in catalysis. The studies of "transition state analogues"

allow access to the unique properties of enzymes that enable them to stabilize labile intermediates and therefore achieve their remarkable catalytic efficiency. We discuss first the structures of enzyme-inhibitor adducts of three examples of proteases—trypsin, papain, and carboxypeptidase—and compare the results of direct observation by ^{13}C NMR with the structural information inferred by more classical techniques. We then review the powerful combination of NMR and cryoenzymology. The development of this technique, whereby enzyme-catalyzed reactions are studied at sub-zero temperatures in aqueous organic solvents (cryosolvents), allows such reactions to be slowed down and can prolong the lifetime of enzyme-substrate intermediates (8, 9).

Proteases

The proteases are classified according to the nature of their functional groups. For example, the thiol and serine proteases have reactive cysteine thiol and serine hydroxyl groups, respectively, whereas the acid proteases have a catalytically essential carboxylate group. The latter function (and others) are combined with metal (for example zinc) in the carboxypeptidases, which are metalloenzymes. It is generally assumed that, for both the thiol and serine proteases, catalysis proceeds via tetrahedral and acyl intermediates as shown in Scheme 1 (10). As we shall see later, ^{13}C NMR allows the direct observation, identification, and characterization of such species, providing direct confirmation of the structure and existence of such intermediates in enzyme catalysis.

Scheme 1 shows the generally accepted mechanism for the hydrolysis of an

1. Enzyme–substrate complex

His-57
Asp-102
Ser-195

2. Tetrahedral intermediate

Ser-195

3. Acyl enzyme

Leaving Group

Ser-195

4. Acyl enzyme

Ser-195

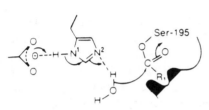

5. Tetrahedral intermediate

Ser-195

6. Enzyme–product complex

Ser-195

Scheme 1.

amide function by a serine protease and is representative of the mechanism for all the hydrolyses except that the active-site nucleophile OH of serine (see **1** in Scheme 1) can be replaced by the groups described above. To confirm this mechanistic pathway, the visualization and rigorous characterization of productive tetrahedral intermediates and acyl enzymes by ^{13}C NMR is the ultimate goal, and we now discuss the progress made so far, particularly with the serine and thiol proteases.

Serine Proteases: Inhibitor Complexes of Tetrahedral Geometry

Henderson (*11*) was the first to suggest that the tetrahedral intermediates (**2** and **5** in Scheme 1) are specifically stabilized at the active site of a serine protease via hydrogen bonding of this anionic species. Although transition-state stabiliza-

457

tion (12) of a tetrahedral intermediate may account for much of the catalytic efficiency of proteases, the stoichiometric accumulation of such a species would not result in efficient catalysis. Hence, the detection of such an intermediate during catalysis is unlikely under normal conditions of physiological pH and temperature.

Inhibitors that have the potential to form tetrahedral adducts with the thiol and serine proteases, and to behave as transition state analogues (13, 14), include both synthetic and natural inhibitors usually containing a carbonyl group at the scissile position. For example, the potency of both soybean trypsin inhibitor (STI) and pancreatic trypsin inhibitor (PTI) was first attributed to their ability to bind as tetrahedral adducts of trypsin at the scissile C=O center (15). However, later x-ray crystallographic evidence (16) suggested that a tetracovalent inhibitor complex was not formed, although the refined data indicated distortion of the bound C=O group from planarity. The latter view was confirmed by ^{13}C NMR experiments in which both STI and PTI were prepared with the appropriate carbonyl carbon enriched in ^{13}C and neither showed any significant perturbation (<1.0 ppm) of the carbonyl resonance in the inhibitor complex with trypsin (17–19). A fully tetrahedral adduct would have displayed an upfield shift (Δ) of ~100 ppm.

In the domain of synthetic inhibitors, chloromethylketone derivatives of specific substrates are potent irreversible covalent inhibitors of the serine proteases, alkylating the active-center histidine at N-2 (see Schemes 1 and 2). X-ray crystallographic studies (20) led to the suggestion that, in addition to the above alkylation, there was also nucleophilic attack by the active-center serine hydroxyl to form a hemiketal, which is stereochemically analogous to the tetrahedral intermediate purported to occur during catalysis.

In ^{1}H NMR studies (21) of chymotrypsin inhibited by chloromethylketones, attempts have been made to characterize the hydrogen bond between the histidine and aspartate residues (Scheme 1) involved in the catalytic triad of these enzymes (Ser-195, His-57, and Asp-102). These studies showed that the exchange rate of the N-1 proton and the chemical shift upon pH titration are both reduced relative to the native enzyme. It was also calculated that the pK_a (negative logarithm of the acidity constant) of this proton was ≈8.4, possibly reflecting ionization of a tetrahedral adduct, with the histidine-aspartate hydrogen bond remaining intact (Scheme 2). Alkylation of N-2 would be expected to raise the pK_a of the active-center histidine (pK_a >10?) as the N-1 proton is no longer freely available to solvent.

To test these suggestions trypsin was inhibited with the highly specific reagent N^α - carbobenzyloxylysylchloromethyl - ketone (22) (RCOCH$_2$Cl), labeled in the ketone carbon with 90 percent ^{13}C, for it was predicted that a tetrahedral adduct, if formed, would be directly observable by this technique. In aqueous solution (Fig. 1A), RCOCH$_2$Cl exists as a mixture of the ketone (δ = 204.7 ppm) and its hydrate (δ = 95.4 ppm) (23). At pH 3.2 no alkylation or inhibition of the enzyme is observed and the spectrum of [^{13}C=O]RCOCH$_2$Cl is unperturbed (Fig. 1, B and C), showing that at low pH there is no detectable binding or tetrahedral adduct formation prior to alkylation. At pH 6.9 there is rapid, irreversible inhibition, and resonances due to both

458

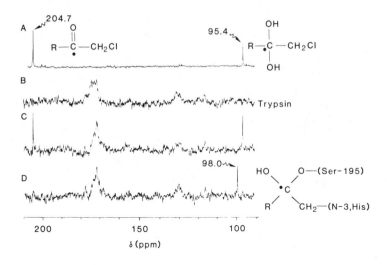

Fig. 1 (left). The ^{13}C NMR spectra of (A) N^α-carbobenzyloxylysyl-chloromethylketone (RCOCH$_2$Cl); • = ^{13}C, 90 percent enrichment; 47.6 mM (by weight); 1 mM HCl; D$_2$O, 16.7 percent (by volume); volume, 0.6 ml; 1660 accumulations; pH 3.1; (B–D) 20 mM sodium phosphate; 12.5 percent (by volume) D$_2$O; 10,000 accumulations; trypsin, 0.28 mM (concentration of fully active enzyme); volume, 8 ml. The RCOCH$_2$Cl concentrations, pH, and enzyme activities were as follows: (B) 0.00 mM, 3.2, 100 percent; (C) 0.38 mM, 3.2, 100 percent; (D) 0.37 mM, 6.9, 0.06 percent. All spectra were accumulated on a Bruker WM300 WB spectrometer at 75.4 MHz for ^{13}C nuclei. Fig. 2 (right). Structure of pepstatin (1) and a synthetic analogue (2) enriched in ^{13}C in the ketonic function.

[^{13}C=O]RCOCH$_2$Cl and its hydrate are replaced by a single resonance at 98.0 ppm (Fig. 1D), an indication that the ^{13}C-enriched carbonyl of the inhibitor is tetrahedrally coordinated in the inhibitor-enzyme adduct. Denaturation (23) of the trypsin led to reappearance of a carbonyl resonance (205.5 ppm) and a decrease in the intensity of the resonance at 98.0 ppm. This demonstrates that the tetrahedral adduct formed by the attack of the serine hydroxyl on the inhibitor carbonyl and characterized by the resonance at 98.0 ppm requires an intact trypsin structure.

The chemical shift (98.0 ppm) of the tetrahedral adduct increased with increasing pH (pK_a 7.9) to 102.1 ppm.

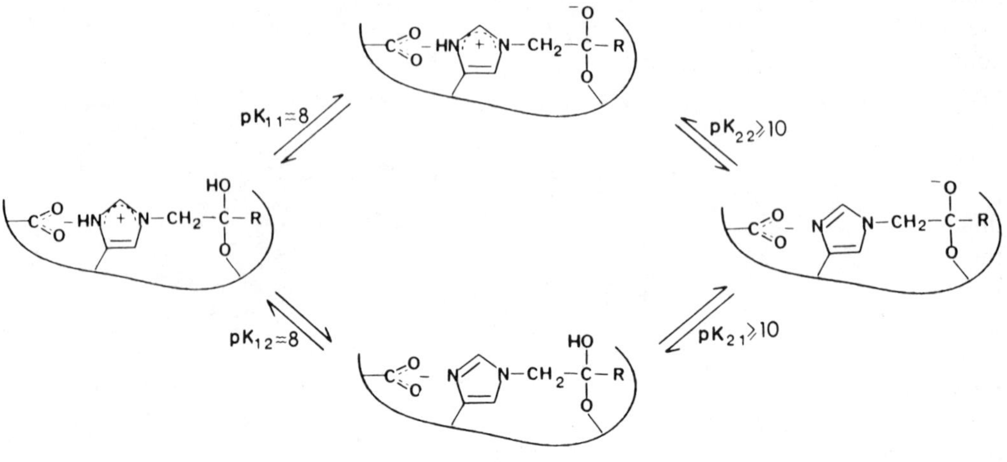

$$K_{11} \gg K_{21} \quad K_{12} \gg K_{22} \quad K_I = K_{11} + K_{12} \quad K_{II} = \frac{K_{22}K_{21}}{K_{22} + K_{21}}$$

Scheme 2. K_I and K_{II} are the experimentally observed molecular dissociation constants, which are related to the group dissociation constants (K_{11}, K_{12}, K_{21}, and K_{22}).

Studies with model compounds (24) suggest that the β-shift observed on deprotonation of alkylated imidazoles is too small to account for the Δ of the ^{13}C-enriched carbon in the enzyme adduct. A pK_a of 7.9 is in reasonable agreement with that (8.4) described in the ^{1}H NMR studies of Robillard and Shulman (21) and suggests that this pK_a (pK_{11} in Scheme 2) does indeed reflect the ionization of a tetrahedral adduct (Scheme 2). It has also been found that with negatively charged inhibitors (25) the histidine-aspartate hydrogen-bonded proton does not titrate; this has led to the suggestion (26) that possibly the tetrahedral adduct can be stabilized by interaction with the protonated histidine. Such an interaction would explain the low pK_a value of the tetrahedral adduct and would also raise the pK_a of the active-center histidine, normally at a value of ≈7. A small fraction of the histidine would therefore ionize with a pK_a (K_{12} in Scheme 2) of ≈8 and the remainder (K_{22} in Scheme 2)

with a pK_a >10, while conversely a small fraction of the tetrahedral adduct would ionize (K_{21} in Scheme 2) with a pK_a >10 and the remainder with a pK_a (K_{11} in Scheme 2) of ~8. Incorporation of ^{13}C-enriched histidine (at the C-2 of the imidazole ring) into α-lytic protease has shown that in the native enzyme the histidine is protonated with a pK_a of ~7 (2, 27, 28). However, in the presence of an aldehydic inhibitor the chemical shift of the carbon-enriched imidazole (of histidine) titrates cooperatively with pK_a values of 7 and 5.5 and the histidine is neutral at pH 8.0 (29). Fastrez (30) has argued that this suggests that an anionic tetrahedral adduct is not stabilized by interaction with the protonated histidine. It has, however, been shown by x-ray crystallography (31) that aldehydes binding to α-lytic protease cause large movements of the active-center histidine, which is consistent with the cooperative ionization observed in the ^{13}C NMR study by Hunkapillar et al. (29). In chloromethylketone inhibitor complexes and presumably in true enzyme-substrate complexes in which the leaving group prevents such movement, histidine-57 is held firmly in place.

In the above experiments with [^{13}C=O]RCOCH$_2$Cl and trypsin, it is difficult to discount the possibility that the resonance at 98.0 ppm could result from hydration of the carbonyl of the covalently bound inhibitor. Clearly, however, both the neutral and the ionic tetrahedral species are stabilized by trypsin, as no resonance near 205 ppm is detectable (23, 32). This confirms that trypsin can stabilize both the neutral and anionic tetrahedrally coordinated adducts.

The line width of the resonance at 98.0 ppm in the pH range 3.75 to 7.0 (7 to 10 Hz) is close to the value expected for a quaternary carbon at 75 MHz on a protein of molecular weight 24,000 (6).

Papain and Pepsin

N-acetylphenylalanylglycinal is an extremely potent inhibitor of papain (14) and is presumed to form stable hemithioacetals which resemble tetrahedral intermediates. Inhibition of papain with N-acetyl-L-phenylalanyl[1-^{13}C]glycinal (δ = 200.9 ppm) has made possible the direct observation of hemithioacetal formation by ^{13}C NMR spectroscopy (33). The lower electronegativity of sulfur compared to oxygen results in decreased deshielding of the enriched carbon in the thiohemiacetal relative to the hydrated aldehyde. As a result, the papain thiohemiacetal resonance at δ = 75 ppm is clearly resolved from the hydrate resonance (δ = 88.2 ppm). Earlier ^{1}H NMR studies (34, 35) involving aldehydic inhi-

bition of papain showed by cross-saturation that indeed the free aldehyde and not the hydrate is the inhibiting species of this reaction. Scale expansion of the signal at 75 ppm revealed that it was composed of two resonances, one at 75.02 ppm and another at 74.68 ppm. The former resonance can be selectively removed by titration of the free enzyme with 2,2'-dipyridyldisulfide (36) at pH 7.0. This allowed the ^{13}C-^{1}H coupling constant (J) values to be assigned as 155 Hz and 160 Hz for the low-field and high-field signals, respectively. Both the chemical shift values and the separation of the papain hemithioacetal resonances are very similar to those of the diastereoisomeric hemithioacetals formed chemically from N-acetyl-L-phenylalanyl[1-^{13}C]glycinal and L-cysteine, which display resonances at 75.8 and 76.8 ppm.

The line widths of the papain thiohemiacetal resonances are approximately 50 Hz, from which, if we assume predominantly dipolar relaxation, the average rotational correlation time (τ_r) is estimated at 36 nsec. This value lies in the range expected for a protein of molecular weight 23,406; for example, myoglobin, molecular weight 17,600, τ_r = 19 nsec; hemoglobin, molecular weight 64,500, τ_r = 47 nsec (7). The aldehyde inhibitor is therefore rigidly held to the protein with no significant librational freedom. These results illustrate the value of ^{13}C NMR spectroscopy as a direct stereochemical probe of enzyme-inhibitor adducts. The fact that only one diastereoisomer is observed in the tetrahedral adduct formed with the ketonic inhibitor adducts of trypsin (23) indicates a much greater enzymatic stereospecificity in this case.

A similar situation has been found in inhibition studies of pepsin, an acid protease. Pepstatin (isovaleryl-Val-Val-Sta-Ala-Sta) (**1** in Fig. 2) is a specific and potent inhibitor ($K_I = 5 \times 10^{-11}M$) (*37, 38*) of this enzyme, and its effectiveness led to the suggestion that the 3*S*-hydroxyl of the internal statin residue acts as an analogue of a tetrahedral intermediate (*39, 40*) formed during catalysis.

A ketone analogue (**2** in Fig. 2) of pepstatin was also found to be an effective inhibitor of pepsin (*41*) ($K_I = 5.6 \times 10^{-9}M$) and was prepared with the ketonic function enriched in carbon-13 [$\delta = 204.2$ ppm; CHCl$_3$-d (85 percent); CH$_3$OH-d (15 percent)] (*42*). On inhibition of pepsin with structure **2** in Fig. 2 a new single resonance at 99 ppm was observed and assigned to a tetrahedral adduct. Evidence that this resonance arises from an enzyme-bound species was derived by addition of the more effective inhibitor, pepstatin, which caused the resonance at 99 ppm to disappear and to be replaced by one at 204.2 ppm. As in the case of trypsin, the remote possibility remains that the enzyme could specifically stabilize the ketone hydrate, although the spectrum of ketone displaced by pepstatin gave no evidence of a hydrate resonance.

Carboxypeptidase

A recent ^{13}C NMR study (*43*) has been performed on this last subset of the proteases to establish the distribution of charge in the enzyme-inhibitor complex formed between carboxypeptidase A and the powerful reversible inhibitor 2-benzylsuccinate. To this end, the inhibitor was enriched with ^{13}C in the two carboxyl groups, alternatively. Thus, upon inhibition with both 2-benzyl[1-^{13}C]succinate and 2-benzyl[4-^{13}C]succinate, in two separate experiments, these carbox-

ylate resonances remained intact in the inhibition complex although moved to lower fields. Titration of these resonances showed that they were both ionized in the bound inhibitor complex. However, from the magnitude of the chemical shift in the case of inhibition with 2-benzyl[4-^{13}C]succinate it was inferred that one of the oxygens of the 4-carboxylate group is in coordination with the active-site zinc and that the other projects directly toward the protonated γ-carboxylate group of Glu-270 and forms a hydrogen bond.

Aldolase: Detection of a Carbinolamine

Schiff base formation between the ε-amino group of a lysine residue of an enzyme and the carbonyl group of a substrate is common, for example, in fructose diphosphate aldolase (*10*), which mediates the condensation of dihydroxyacetone phosphate with glyceraldehyde phosphate. However, ^{13}C NMR studies (*44*) with the pseudosubstrates [^{13}C=O]hydroxyacetone phosphate and [^{13}C=O, ^{2}H]hydroxyacetaldehyde phosphate (**3** in Fig. 3) showed that the former binds noncovalently (chemical shift differences of ≤0.1 ppm) and, most interestingly, that the latter forms a carbinolamine adduct (**4** in Fig. 3) ($\delta = 79.7$ ppm) at the active site.

The Quest for True Intermediate Structure: Cryoenzymology

The detection and characterization of a productive acyl intermediate by ^{13}C NMR during catalysis with natural substrates at ambient temperatures is not possible at present, because there is an inherent lack of sensitivity in ^{13}C NMR spectroscopy. Therefore, the lifetimes of these intermediates must be extended into the domain of the NMR experiment. One approach is to use synthetic substrate analogues which deacylate slowly. Another is to utilize low-temperature cryoenzymological techniques to extend the lifetimes of intermediates.

For example, nonspecific dithioester substrates give rise to dithioacylpapains that have been directly observed by ultraviolet (45) and resonance Raman spectroscopy (46). These dithioacylpapains are thought to be analogous to the thioacylpapains that are believed to occur during catalysis of natural ester substrates. The acylation of papain by N-cinnamoylimidazole causes the ultraviolet spectrum to be red-shifted by 20 nm relative to the model (S)-trans-cinnamolycysteine, and only on denaturation of the "acylated papain" did the spectrum resemble that of the model compound (47).

Since the ^{13}C resonances of the carbonyl carbon of thioesters are shifted (Δ 20 to 30 ppm) upfield relative to their oxygen analogues ($\delta \approx$ 165 to 185 ppm), ^{13}C NMR spectroscopy should allow the direct monitoring of the formation and decay of a thioacyl intermediate. Using [$^{13}C=O$]N-benzoylimidazole (δ = 168.7 ppm), we were able to observe directly a thioacyl intermediate at δ = 195.9 ppm in the presence of papain under the

R = papain

231 to 276 minutes

186 to 231 minutes

141 to 186 minutes

96 to 141 minutes

51 to 96 minutes

6 to 51 minutes

195.9 ppm

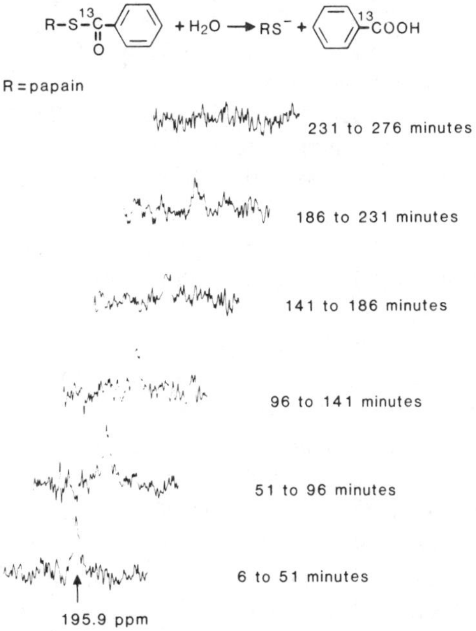

Fig. 4. Reaction of 1.7 mM papain (72 percent active enzyme), 5.4 mM potassium chloride, 25 percent (by volume) dimethyl sulfoxide, 0.1M sodium formate buffer (pH 4.1), and 23.6 mM benzoylimidazole. [$^{13}C=O$]Benzoylimidazole was added at 0°C; after 15 minutes the reaction was cooled to −6°C, and the NMR data acquisition commenced 6 minutes after the reaction was initiated. Spectra represent 10,000 accumulations recorded sequentially starting 6, 51, 96, 141, 186, and 231 minutes after adding benzoylimidazole (48). The spectral range shown is 192 to 200 ppm.

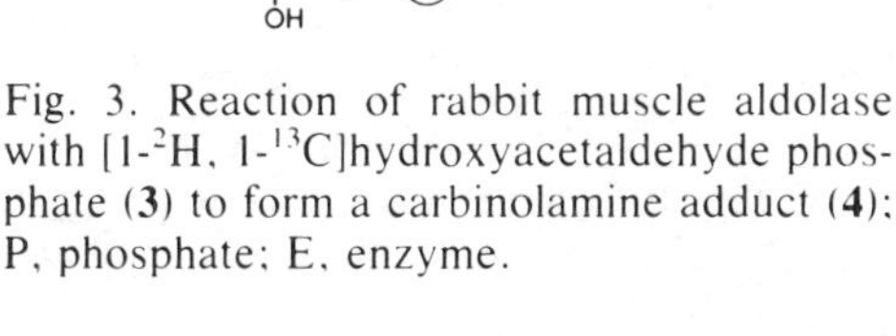

cryoenzymological conditions of −6°C in 25 percent aqueous dimethyl sulfoxide (48) (Fig. 4). Moreover, the thioacyl species is clearly a productive intermediate since the decrease in its signal intensity was accompanied by an increase in the product resonance and by release of free enzyme (half-life, ≈96 minutes) determined by titration of its thiol group. The line width of the resonance at 195.9 ppm was 25 ± 5 Hz.

Similar experiments with chymotrypsin and the nonspecific substrate [$^{13}C=O$]p-nitrophenylacetate (δ = 170.4

Fig. 3. Reaction of rabbit muscle aldolase with [1-^{2}H, 1-^{13}C]hydroxyacetaldehyde phosphate (3) to form a carbinolamine adduct (4): P, phosphate; E, enzyme.

463

ppm) have allowed observation of an acyl adduct (δ = 174.0 ppm) by ^{13}C NMR spectroscopy (49).

The trypsin-catalyzed hydrolysis (Fig. 5) of the highly specific substrate N^{α}-carbobenzyloxy-L-lysine-p-nitrophenyl-ester (Z-lys-pNP) has been studied in detail under cryoenzymological conditions by both spectrophotometry (50) and ^{13}C NMR spectroscopy (51). The kinetic data from both techniques (50, 51) confirm that the kinetics and mechanism under cryoenzymological conditions are essentially the same as those determined at ambient temperatures by rapid reaction techniques (52). The hydrolysis of [1-^{13}C]Z-lys-pNP (S in Fig. 5A, δ = 173.6 ppm) by trypsin was monitored by the decrease in intensity of this signal and the increase in the signal arising from the product (P$_2$ in Fig. 5A, δ = 177.7 ppm) at $-21°$C in 41 percent aqueous dimethyl sulfoxide (Fig. 5). The continued formation of product (P$_2$) after all the substrate has been consumed (Fig. 5B) provides indirect evidence for an enzyme-bound intermediate whose line width is much greater than that of the free substrate or product and is therefore not directly observable in the individual ^{13}C NMR spectra of Fig. 5. Improvement of the signal-to-noise ratio by summation of sets of the individual spectra in Fig. 5 made it possible to use difference techniques and allowed direct observation (Fig. 6B) at $-21°$C of a resonance (δ = 176.5 ppm) that was assignable to the acyl enzyme on the basis of its chemical shift and the kinetics of its breakdown-formation. Experiments at $-1.5°$C (Fig. 6A) showed that the acyl intermediate was much more readily detected at this higher temperature, the resonance being clearly resolved in individual spectra as a result of the smaller line width of the resonance at $-1.5°$C compared to that at $-21°$C (22 ± 3 Hz and 100 ± 10 Hz, respectively). This illustrates one of the main difficulties of a combined NMR–cryoenzymological approach in which the line width of enzyme-bound species increases dramatically as the cryosolvent viscosity increases on lowering the temperature.

Coenzyme Enrichment: ^{13}C-Enriched Coenzymes and Cofactors

An early and cogent example is provided by the experiments of Ghisla et al. (53), using flavin mononucleotide enriched in ^{13}C at the 4a position. On addition to bacterial luciferase in 20 percent aqueous ethylene glycol and reduction with dithionite, two signals were observed, one at 103 ppm arising from bound reduced flavin mononucleotide and one at 104 ppm for the free reduced flavin. On cooling to $-15°$C, molecular oxygen was added, whereupon the spectrum showed a sharp signal at 137 ppm due to reoxidized flavin and a broad signal, at least 50 Hz wide, arising from the 4a oxygen adduct of the coenzyme-enzyme complex.

Recently an elegant ^{13}C NMR study (54) on the hydroxylation of phenylalanine (Phe) to tyrosine (Tyr) by phenylalanine hydroxylase (PAH), an enzyme that requires a pteridine cofactor, has been described. The reduced cofactor, in the tetrahydro form, combines with molecular oxygen to form an active hydroxylating species. Unlike the tightly bound flavin coenzymes (see above), the pterin cofactor dissociates from the enzyme at the end of each catalytic cycle and is subsequently regenerated, an event that alleviates, to some extent, the problems

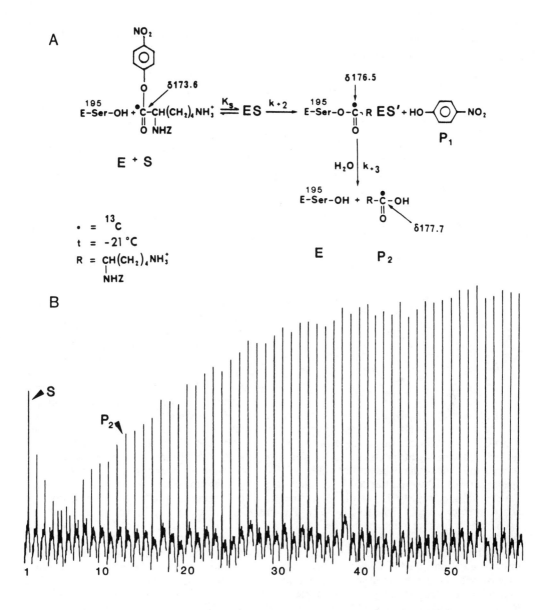

Fig. 5. (A) Reaction of [1-^{13}C]Z-lys-pNP, 0.83 mM; active trypsin, 0.7 mM; 1 mM HCl (apparent pH, 3.2); and 40 percent (by volume) dimethyl sulfoxide (sample volume, 10 ml; sample temperature, −21°C); NHZ, carbobenzloxy-. The reaction was initiated by the addition of [1-^{13}C]Z-lys-pNP; after mixing (≈1 minute) NMR data acquisition (B) commenced within 3 minutes. Each spectrum (1 to 58) represents 10,000 accumulations (time per spectrum, 41 minutes) recorded sequentially (51). The spectral width was 0 to 220 ppm, but only the region from 172 to 180 is shown. [Reprinted by permission from the *Biochemical Journal*, vol. 219, p. 437. © 1984 The Biochemical Society, London.]

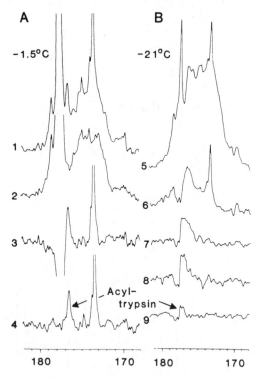

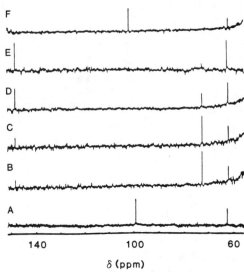

Fig. 6 (left). The ^{13}C NMR spectra of the reaction of (A) [1-^{13}C]Z-lys-pNP, 1.8 mM; active trypsin, 0.64 mM; 1 mM HCl (apparent pH, 3.02); and 40 percent (by volume) dimethyl sulfoxide (sample volume, 10 ml; sample temperature, -1.5°C). The reaction was initiated as described in Fig. 5, and spectra 1 to 15 (not shown) were recorded sequentially in the manner of Fig. 5B. Each spectrum resulted from 5000 accumulations (time per spectrum, 20.5 minutes). A 10-Hz exponential weighting factor was used. Spectra 1 and 2 represent the sums (25,000 accumulations of spectra 1 to 5 and spectra 11 to 15, respectively). Spectrum 3 is the difference between spectra 1 and 2. Spectrum 4 is spectrum 1 from which natural-abundance trypsin and product have been subtracted. (B) The free induction decay spectra from which the spectra in Fig. 5B were obtained were added together in groups of ten. Spectrum 5 represents the sum of spectra 2 to 11 (of Fig. 5B). Spectra 6 to 9 represent progressive subtractions of added spectra 12 to 21, 22 to 31, 32 to 41, and 49 to 58 from which product and natural-abundance trypsin resonances had been subtracted.   Fig. 7 (right). The ^{13}C NMR spectra of the 4a-^{13}C intermediates described in the text and Scheme 3 (54) at -30°C (the signal at 61.5 ppm is due to 20 mM tris buffer): (A) [4a-^{13}C]6-methyltetrahydropterin (structure 5 in Scheme 3) prior to addition to PAH and phenylalanine; (B) spectrum acquired after mixing with PAH, time = 19 to 54 minutes (100 scans); (C) time = 64 to 170 minutes (260 scans); (D) time = 228 to 258 minutes (200 scans); (E) time = 360 to 400 minutes (100 scans); (F) solution after 580 minutes, allowed to warm to 30°C for 3 hours (300 scans). [R. A. Lazarus, C. W. DeBrosse, S. J. Benkovic, *Journal of the American Chemical Society* **104**, 6869 (1982), reprinted with permission. © 1982 American Chemical Society.]

that arise from the large line widths of tightly bound enzyme intermediates.

In the event, 6-methyltetrahydropterin (**5** in Scheme 3) was synthesized with C-

4a enriched to 90 atom percent and added to an activated PAH solution (at 0°C) saturated with oxygen. It was necessary to perform the initial intermediate of the

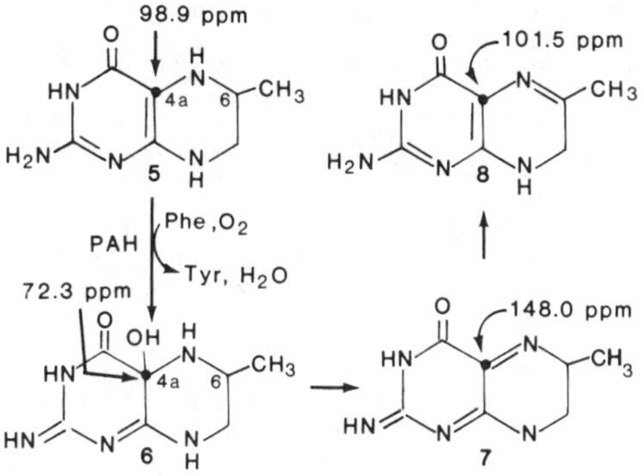

98.9 ppm

101.5 ppm

72.3 ppm

148.0 ppm

PAH

Phe, O₂

Tyr, H₂O

Scheme 3 (54). [R. A. Lazarus, C. W. De-Brosse, S. J. Benkovic, *Journal of the American Chemical Society* **104**, 6869 (1982), reprinted with permission. © 1982 American Chemical Society.]

reaction at 0°C because of its instability at higher temperatures (half-life = 2.0 minutes at 23.2°C). After 3.5 minutes at 0°C, methanol was added to give a 40 percent aqueous methanol solution while cooling to −30°C. Figure 7 shows the ^{13}C NMR spectra of the tetrahydropterin and of subsequent reaction products after the addition of oxygen and dehydration (Scheme 3). It was concluded that C-4a was hydroxylated (**5 → 6** in Scheme 3) to give a resonance at 72.3 ppm. Dehydration of **6** gave **7** with a C-4a resonance at 148.0 ppm. Other structural possibilities such as ring-opened oxenoid species were ruled out since the spectra depicted in Fig. 7 were also accumulated with ^{1}H coupling. The continuous observation of coupling constants between C-4a and H-6 proved that the pterin ring system remained intact.

Another coenzyme, pyridoxal phosphate, is a cofactor in a wide variety of enzyme-mediated reactions (55, 56). Researchers have used ^{13}C NMR to examine the binding of ^{13}C-enriched pyridoxal phosphate to D-serine dehydratase (molecular weight 46,000) and L-glutamate decarboxylase (subunit molecular weight 50,000) (57). In aqueous solution [4′,5′-$^{13}C_2$]pyridoxal phosphate shows two sharp resonances at δ = 196.8 and 88.4 ppm (for the 4′-aldehyde carbonyl and its hydrate, respectively) and one at δ = 62.8 ppm (for the 5′-methylene carbon). On binding to D-serine dehydratase, two resonances not found in the native enzyme are observed at 167.7 and 62.7 ppm with line widths of 24 ± 3 and 48 ± 3 Hz, respectively. The former resonance has been assigned to a Schiff's base formed between the 4′-^{13}C aldehyde group of pyridoxal phosphate and an ε-amino function of a lysine residue of the enzyme. The 5′-methylene carbon remains virtually unaltered save for line-broadening. The twofold increase in line width for the carbon-5′ signal compared with the carbon-4′ signal is to be expected for methylene and methine carbons. The τ_r of the enzyme calculated from both these line width values was 16 ± 2 nsec. This is significantly less (see above) than expected for a ^{13}C nucleus rigidly bound to a protein of molecular weight 46,000 and suggests that the cofactor has some degree of motional freedom.

The failure to detect any ^{13}C pyridoxal phosphate resonances on binding with L-glutamate decarboxylase presumably results from the fact that this enzyme is a hexamer (effective molecular weight 300,000). In this case, the bound pyridoxal phosphate does not have any detectable librational freedom.

Conclusions and Prognosis

From the discussion presented here it is clear that the stage is now set for intensive research into the mechanism of enzyme action with cryoenzymology. Although most of the enzymes discussed above lie in the molecular weight range of 20,000 to 35,000, an upper figure of ~50,000 probably represents the maximum convenient size for ^{13}C NMR studies on currently available superconducting instruments with substrates and inhibitors. The use of ^{13}C-enriched coenzymes has added another powerful tool which not only broadens the range of accessible enzymes but also appears to be independent of enzyme molecular weight, especially in those cases where free coenzyme can be observed. For example, the use of ^{13}C-enriched pyridine nucleotides, which has been restricted in the past to the in vivo determination of intracellular coenzyme pools and their redox status (58), can be extended to a wide range of enzymes that use nicotinamide adenine dinucleotide and its reduced form as cofactors.

Finally, we can expect exciting developments in the application of ^{13}C solid-state NMR spectroscopy to structural problems in enzyme complexes, a field which, although still in its infancy, already shows considerable promise (59).

References

1. O. Jardetzky and G. C. K. Roberts, *NMR in Molecular Biology* (Academic Press, New York, 1981), p. 417.
2. T. A. Steitz and R. G. Shulman, *Annu. Rev. Biophys. Bioeng.* **11**, 419 (1982).
3. K. Kanamari and J. D. Roberts, *Acc. Chem. Res.* **152**, 35 (1983).
4. T. C. Farrar and E. D. Becker, *Pulse and Fourier Transform NMR* (Academic Press, New York, 1971).
5. J. H. Noggle and R. E. Schirmer, *The Nuclear Overhauser Effect* (Academic Press, New York, 1971).
6. R. S. Norton, A. O. Clouse, R. Addleman, A. Allerhand, *J. Am. Chem. Soc.* **99**, 79 (1977).
7. E. Oldfield, R. S. Norton, A. Allerhand, *J. Biol. Chem.* **250**, 6368 (1975).
8. P. Douzou, *Cryobiochemistry* (Academic Press, New York, 1977).
9. A. L. Fink and S. J. Cartwright, *CRC Rev. Biochem.* 145 (1981).
10. C. Walsh, *Enzyme Reaction Mechanisms* (Freeman, San Francisco, 1979).
11. R. Henderson, *J. Mol. Biol.* **54**, 341 (1970); _____, C. S. Wright, G. P. Heis, D. M. Blow, *Cold Spring Harbor Symp. Quant. Biol.* **36**, 63 (1971).
12. R. Wolfenden, *Acc. Chem. Res.* **5**, 10 (1972).
13. J. Kraut, *Annu. Rev. Biochem.* **46**, 331 (1977).
14. J. O'C. Westerick and R. Wolfenden, *J. Biol. Chem.* **247**, 8195 (1972).
15. R. M. Sweet, H. T. Wright, J. Ganin, C. H. Chathia, D. M. Blow, *Biochemistry* **13**, 4212 (1974); A. Rhulmann, D. Kukla, P. Schwager, K. Bartels, R. Huber, *J. Mol. Biol.* **77**, 417 (1973).
16. R. Huber, D. Kukla, W. Bode, P. Schwager, K. Bartels, J. Deisenhofer, N. Steigemann, *J. Mol. Biol.* **89**, 73 (1976).
17. M. W. Hunkapillar, M. D. Forgac, E. H. Yu, J. H. Richards, *Biochem. Biophys. Res. Commun.* **87**, 25 (1979).
18. M. W. Baillargeon, M. Laskowski, D. E. Neves, M. A. Porubcan, R. E. Santinini, J. L. Markley, *Biochemistry* **19**, 5703 (1980).
19. R. Richardz, H. Tschesche, K. Wüthrich, *ibid.*, p. 5711.
20. T. L. Poulos, R. A. Alden, S. T. Frier, J. J. Burktoft, J. Kraut, *J. Biol. Chem.* **251**, 1097 (1976).
21. G. Robillard and R. G. Shulman, *J. Mol. Biol.* **86**, 519 (1974).
22. J. R. Coggins, W. Kray, E. Shaw, *Biochem. J.* **138**, 579 (1974).
23. J. P. G. Malthouse, N. E. Mackenzie, A. S. F. Boyd, A. I. Scott, *J. Am. Chem. Soc.* **105**, 1685 (1983).
24. W. U. Primrose, thesis, Edinburgh University (1984).
25. G. Robillard and R. G. Shulman, *J. Mol. Biol.* **86**, 541 (1974).
26. A. A. Kossiakoff and S. A. Spencer, *Biochemistry* **20**, 6462 (1981).
27. M. W. Hunkapillar, S. H. Smallcombe, D. R. Whitaker, J. H. Richards, *ibid.* **12**, 4732 (1973).
28. W. W. Bachovchin, R. Kaiser, J. H. Richards, J. D. Roberts, *Proc. Natl. Acad. Sci. U.S.A.* **78**, 7323 (1981).
29. M. W. Hunkapillar, S. H. Smallcombe, J. H. Richards, *Org. Magn. Reson.* **7**, 262 (1975).
30. J. Fastrez, *Eur. J. Biochem.* **135**, 339 (1983).
31. M. W. G. James, A. R. Sielecki, G. D. Brayer, L. T. J. Delbaerie, *J. Mol. Biol.* **144**, 43 (1980).
32. A. I. Scott, unpublished results.
33. M. P. Gamcsik *et al.*, *J. Am. Chem. Soc.* **105**, 6324 (1983).
34. R. Bendall, I. L. Cartwright, P. I. Clark, G. Lowe, D. Nurse, *Eur. J. Biochem.* **79**, 201 (1977).
35. P. I. Clark, G. Lowe, D. Nurse, *Chem. Commun.* **1977**, 451 (1977).
36. K. Brocklehurst and G. Little, *Biochem. J.* **128**, 471 (1972).
37. R. J. Workman and D. W. Burkett, *Arch. Biochem. Biophys.* **194**, 157 (1979).
38. D. H. Rich, E. T. O. Sun, E. J. Ulm, *J. Med. Chem.* **23**, 27 (1980).
39. J. Marciniszyn, J. A. Hartsuck, J. Tang, *J. Biol. Chem.* **251**, 7088 (1976).

40. G. R. Marshall, *Fed. Proc. Fed. Am. Soc. Exp. Biol.* **35**, 2494 (1976).
41. D. H. Rich, A. S. Bapani, M. S. Bernatowicz, *Biochem. Biophys. Res. Commun.* **104**, 1127 (1982).
42. D. H. Rich, M. S. Bernatowicz, P. G. Schmidt, *J. Am. Chem. Soc.* **104**, 3535 (1982).
43. A. R. Palmer, P. D. Ellis, R. Wolfenden, *Biochemistry* **21**, 5056 (1982).
44. B. D. Ray and E. T. Harper, *J. Am. Chem. Soc.* **105**, 3731 (1983).
45. G. Lowe and A. Williams, *Biochem. J.* **96**, 189 (1965).
46. P. R. Carey and A. C. Storer, *Acc. Chem. Res.* **16**, 455 (1983).
47. M. L. Bender and L. J. Brubacher, *J. Am. Chem. Soc.* **86**, 5333 (1964); *ibid.* **88**, 5871 (1966).
48. J. P. G. Malthouse, M. P. Gamcsik, A. S. F. Boyd, N. E. Mackenzie, A. I. Scott, *ibid.* **104**, 6811 (1982).
49. C. H. Niu, H. Shindo, J. S. Capen, *ibid.* **99**, 3161 (1977).
50. J. P. G. Malthouse and A. I. Scott, *Biochem. J.* **215**, 555 (1983).
51. N. E. Mackenzie, J. P. G. Malthouse, A. I. Scott, *ibid.* **219**, 437 (1984).
52. E. Antonini and P. Ascenzi, *J. Biol. Chem.* **256**, 12449 (1981).
53. S. Ghisla *et al.*, *Proc. Natl. Acad. Sci. U.S.A.* **75**, 5860 (1978).
54. R. A. Lazarus, C. W. DeBrosse, S. J. Benkovic, *J. Am. Chem. Soc.* **104**, 6869 (1982).
55. A. E. Braunstein, in *The Enzymes*, P. D. Boyer *et al.*, Eds. (Academic Press, New York, ed. 2, 1960), part A, vol. 2, p. 113.
56. L. Davis and D. E. Metzler, in *The Enzymes*, P. D. Boyer *et al.*, Eds. (Academic Press, New York, ed. 3, 1972), vol. 7, p. 33.
57. M. H. O'Leary and J. R. Payne, *J. Biol. Chem.* **251**, 2248 (1976).
58. C. J. Unkefer, R. M. Blazer, R. E. London, *Science* **222**, 62 (1983).
59. N. E. Mackenzie and A. I. Scott, *Chem. Commun.* **635** (1985).

Coenzyme B_{12} (5'-deoxyadenosylcobalamin, abbreviated $AdCH_2$-B_{12}), whose structure is depicted in Fig. 1, serves as a cofactor for various enzymatic reactions (Table 1). [For comprehensive accounts of the chemistry and biochemistry of vitamin B_{12}, see (1, 2).] A common feature of these reactions, depicted by Eq. 1, is the 1,2-interchange of a hydrogen atom and another substituent [X = OH, NH_2, C(=O)SCoA, C(=CH_2)COOH or CH(NH_2)COOH] on adjacent carbon atoms of the substrate (3). Among these reactions, the methylmalonyl–coenzyme A (CoA) mutase rearrangement (Eq. 2) (4), whose mechanism is discussed in this article, is distinctive in that it plays a role in mammalian systems, whereas the other reactions occur only in microorganisms.

$$- \overset{X}{\underset{|}{\underset{|}{C}}_1} - \overset{H}{\underset{|}{\underset{|}{C}}_2} - \longrightarrow - \overset{H}{\underset{|}{\underset{|}{C}}_1} - \overset{X}{\underset{|}{\underset{|}{C}}_2} - \qquad (1)$$

$$\underset{CH_3CHCOOH}{\overset{\overset{O}{\overset{\|}{CSCoA}}}{\underset{|}{}}} \longrightarrow CoASC-CH_2CH_2COOH \qquad (2)$$

Enzymatic studies have provided convincing evidence for the essential features of the mechanistic scheme depicted by Fig. 2 (5–7). This mechanism encompasses the following sequence of steps: (i) enzyme-induced homolytic dissociation of the cobalt-carbon bond of coenzyme B_{12} to generate cob(II)alamin (also designated vitamin B_{12_r}) and a 5'-deoxyadenosyl radical (abbreviated $AdCH_2\cdot$), (ii) abstraction of a hydrogen atom (8) from the substrate to generate a substrate radical and 5'-deoxyadenosine ($AdCH_3$), (iii) rearrangement of the resulting substrate radical (either directly or through additional intermediate steps) to the corresponding product radical,

32

Mechanisms of Coenzyme B_{12}-Dependent Rearrangements

Jack Halpern

Science 227, 869–875 (22 February 1985)

and (iv) abstraction of a hydrogen atom from $AdCH_3$ by the product radical (8) to complete the rearrangement reaction.

The evidence (5–7) for this mechanistic scheme includes (i) the demonstration, with the use of deuterium tracers, that the migrating H atom is scrambled with the two methylene H atoms of $AdCH_2$-B_{12}, suggesting that the three H atoms become equivalent during the course of the reaction (consistent with the intermediate formation of $AdCH_3$), (ii) the direct spectroscopic (electronic and electron paramagnetic resonance) observation of the free radical intermediates (B_{12_r} and a carbon-centered free radical) in certain coenzyme B_{12}–dependent rearrangements (notably ethanolamine ammonia lyase and diol dehydrase); and (iii) the demonstration, in at least one case, of the reversible formation of $AdCH_3$ during the course of the enzymatic reaction.

Figure 2 represents the minimal mechanistic scheme that accommodates the above observations. One important feature not specifically depicted by this scheme is that both the coenzyme and the substrate are bound to the enzyme during the course of the reaction. This binding presumably dictates the chemical selectivity and regioselectivity of the H abstraction steps and the stereospecificity of the H-X interchange [for example, retention of the configuration at C-2 for methylmalonyl-CoA mutase (4) and inversion for diol dehydrase (9)]. The mechanism also may be incomplete in that there may be additional steps and intermediates, such as intermediate hydrogen carriers in the H transfer steps, as well as additional steps and intermediates (for example, carbonium ions, carbanions, or substrate-derived organocobalt complexes) in the rearrangement

process itself (that is, the 1,2-migration of X).

In this article, the following themes that relate to the mechanistic scheme of Fig. 2 for coenzyme B_{12}–dependent rearrangements are discussed: (i) the cobalt-carbon bond dissociation energy of coenzyme B_{12}, (ii) the factors that influence cobalt-carbon bond dissociation energies and that may contribute to the enzyme-induced bond weakening and dissociation, (iii) the mechanism of the 1,2-migration of X, (iv) the role of the coenzyme and, in particular, of the cobalt atom, and (v) the role of the enzyme.

In addition to the enzymatic studies cited above, various studies on model systems (involving modeling of both the coenzyme and substrates) have been important in elucidating aspects of the chemistry of coenzyme B_{12} and the mechanisms of its rearrangement reactions (10). Examples of organocobalt compounds that have been widely in-

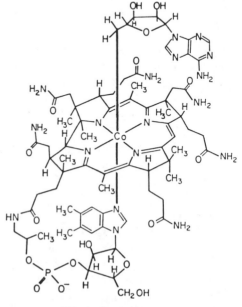

Fig. 1. Coenzyme B_{12}.

Table 1. Coenzyme B_{12}-Dependent Rearrangements*

Reaction	Enzyme

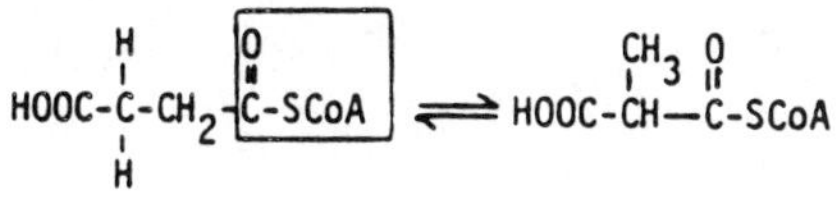

Reaction: HOOC-C(H)(H)-CH₂-[C(=O)-SCoA] ⇌ HOOC-CH(CH₃)-C(=O)-SCoA — Methylmalonyl-CoA mutase

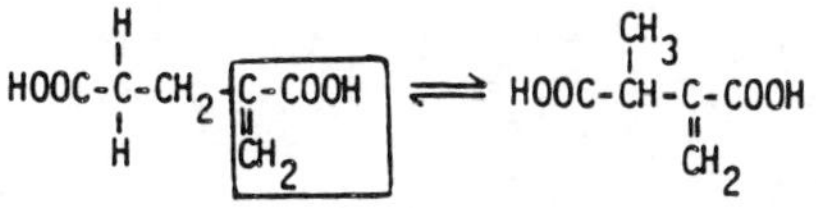

Reaction: HOOC-C(H)(H)-CH₂-[C(-COOH)(=CH₂)] ⇌ HOOC-CH(CH₃)-C(-COOH)(=CH₂) — α-Methyleneglutarate mutase

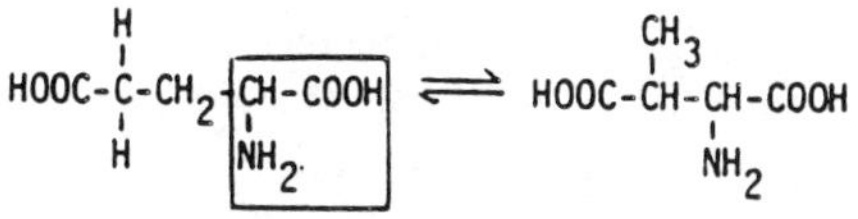

Reaction: HOOC-C(H)(H)-CH₂-[CH(COOH)(NH₂)] ⇌ HOOC-CH(CH₃)-CH(COOH)(NH₂) — Glutamate mutase

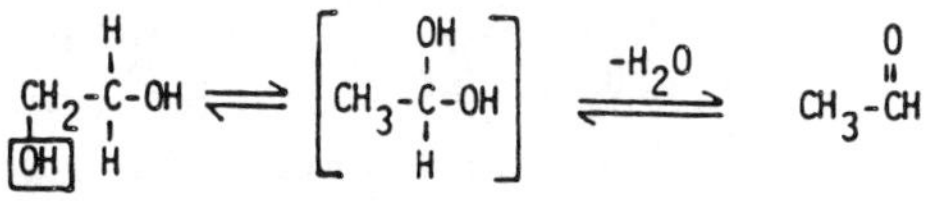

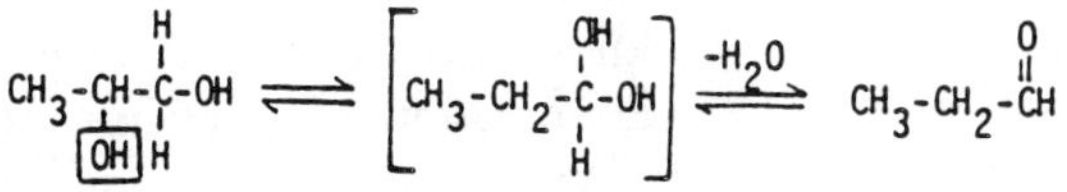

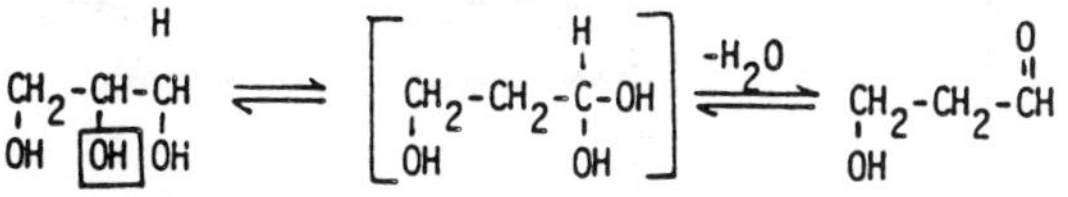

Reaction: Diol dehydrase

Reaction: Diol dehydrase

Reaction: Glycerol dehydrase

Reaction: Ethanolamine ammonia lyase

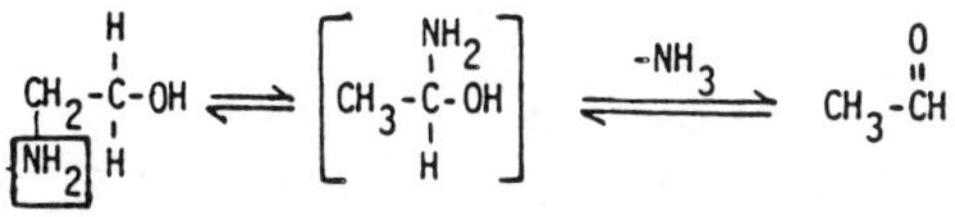

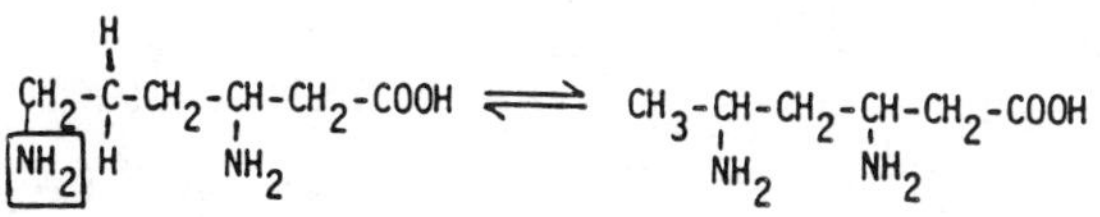

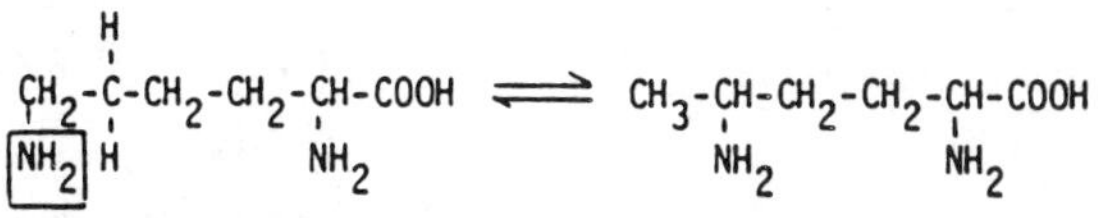

Reaction: L-β-Lysine mutase

Reaction: D-α-Lysine mutase

Reaction: Ornithine mutase

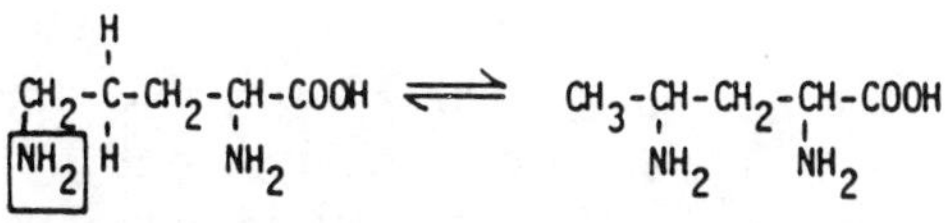

*The migrating group that interchanges with an adjacent hydrogen atom is shown enclosed in a rectangle.

473

voked as coenzyme B_{12} models in such studies include alkyl derivatives of bis-(dimethylglyoximato)cobalt (**1**) (abbreviated [RCo(DH)$_2$L], where DH$_2$ is dimethylglyoxime and L is an axial ligand such as water, pyridine, or a tertiary phosphine), and cobalt complexes of Schiff bases such as bis(salicyclaldehyde)phenylenediimine (**2**) (abbreviated [RCo(saloph)L]). With respect to properties that are recognizably relevant to the biochemical roles of coenzyme B_{12} (for example, redox properties and cobalt-carbon bond dissociation energies), the cobalt–Schiff base complexes appear to come closest to providing a model for coenzyme B_{12} (*10*).

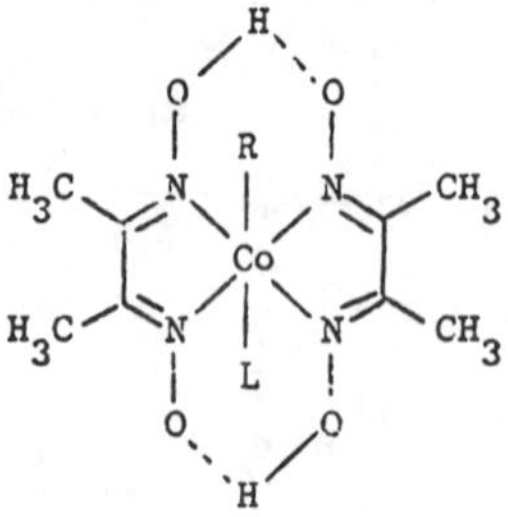

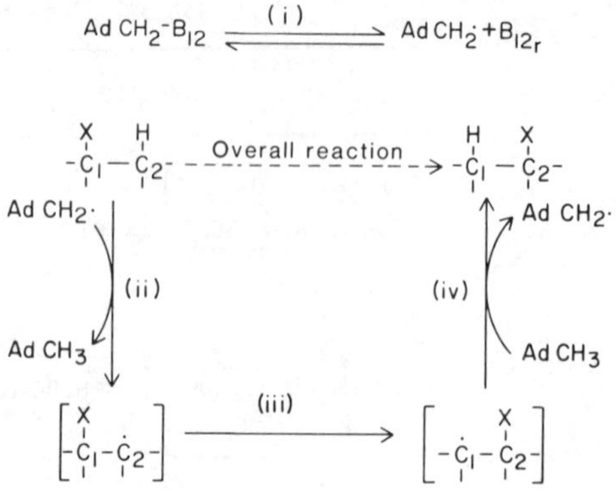

Fig. 2. Mechanistic scheme for coenzyme B_{12}–dependent rearrangements.

plies a very weak cobalt-carbon bond. A troublesome feature of this mechanistic scheme has been the absence until recently of precedents for such weak transition metal-alkyl bonds or, indeed, knowledge of transition metal-alkyl bond dissociation energies in general (*11*).

We have recently accomplished the measurement of the cobalt-carbon bond dissociation energy ($D_{\text{Co-CH}_2\text{Ad}}$) of coen-

$$AdCH_2\text{-}B_{12} \underset{k_{-3}}{\overset{k_3}{\rightleftharpoons}} AdCH_2\cdot + B_{12_r} \qquad (3)$$

$$AdCH_2\cdot + [Co^{II}(DH)_2(H_2O)] \xrightarrow{k_4}$$

$$[AdCH_2\text{-}Co(DH)_2(H_2O)] \qquad (4)$$

$$AdCH_2\text{-}B_{12} + [Co^{II}(DH)_2(H_2O)] \longrightarrow$$

$$[AdCH_2\text{-}Co(DH)_2(H_2O)] + B_{12_r} \qquad (5)$$

[RCo(DH)$_2$L]

1

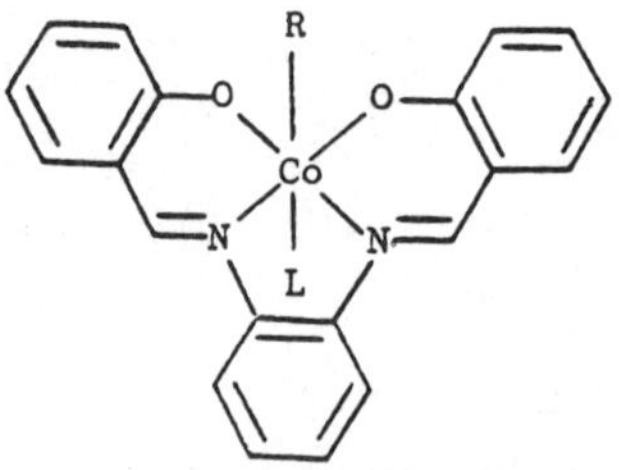

[RCo(saloph)L]

2

Cobalt-Carbon Bond Dissociation Energy of Coenzyme B_{12}

The role of coenzyme B_{12}, encompassed by the mechanism of Fig. 2, implies

zyme B_{12} by determining the kinetics of the bond dissociation process in aqueous solution. For this purpose, we used [Co^{II}(DH)$_2$(H$_2$O)] (which forms a stronger Co-C bond than B_{12_r}) to trap the AdCH$_2\cdot$ radical, in accord with Eqs. 3 to 5 (*12*).

Application of the steady-state approximation to this reaction sequence yields the rate law corresponding to Eq. 6. Fitting our kinetic data (12) to Eq. 6 yielded $k_3 = 1.0 \times 10^{-4}$ sec^{-1} and $k_{-3}/k_4 = 1.1$ at 100°C. Measurement of the temperature dependence of k_3 yielded (after correction for the equilibrium involving dissociation of the pendant axial 5,6-dimethylbenzimidazole ligand to form the unreactive "base-off" form) $\Delta H_3^{\ddagger} = 28.6$ kcal/mol and $\Delta S_3^{\ddagger} = 2$ cal mol^{-1} K^{-1}. Earlier demonstrations that the recombinations of B$_{12_r}$ with various free radicals including AdCH$_2 \cdot$ are diffusion-controlled (13) permit the value of $\Delta H_{-3}^{\ddagger}$ to be estimated as about 2 kcal/mol. These data yield a value of about 26 kcal/mol for the Co-C bond dissociation energy of coenzyme B$_{12}$ ($D_{\text{Co-CH}_2\text{Ad}} = \Delta H_3^{\ddagger} - \Delta H_{-3}^{\ddagger}$) in aqueous solution (14).

$$\frac{-d[\text{AdCH}_2\text{-B}_{12}]}{dt} =$$

$$\frac{k_3 k_4 [\text{AdCH}_2\text{-B}_{12}][\text{Co}^{II}(\text{DH}_2)(\text{H}_2\text{O})]}{k_{-3}[\text{B}_{12_r}] + k_4[\text{Co}^{II}(\text{DH})_2(\text{H}_2\text{O})]} \quad (6)$$

Factors Influencing Cobalt-Carbon

Bond Dissociation Energies

Although the Co-C bond of coenzyme B$_{12}$ is weak compared with typical covalent bonds in organic molecules, the value of k_3 at 30°C (that is, the rate constant for dissociation of the Co-C bond of the free coenzyme), calculated from the above values of $\Delta H_3^{\ddagger}$ and $\Delta S_3^{\ddagger}$, is only about 10^{-7} sec^{-1}. This is approximately 10^{-9} times ($\Delta \Delta G^{\ddagger} \sim 13$ kcal/mol) the values of the catalytic rate constants ($k_{\text{cat}} \sim 10^2$ sec^{-1}) that have been estimated for several coenzyme B$_{12}$–dependent enzymatic reactions, including methylmalonyl-CoA mutase (4, 9, 15). Even though this discrepancy will be

offset by the chain length of the catalytic cycle that is initiated by cleavage of the coenzyme Co-C bond, it is unlikely that the chain length is as high at 10^9. Thus, considerable further weakening of the Co-C bond, by interaction with the enzyme, appears to be required to achieve dissociation rates that are compatible with the enzymatic rates.

To identify the factors that might be responsible for this bond weakening and dissociation, we examined the influence of various electronic and steric parameters on the Co-C bond dissociation energies of some coenzyme B$_{12}$ model compounds including [R-Co(DH)$_2$L] (1) and [R-Co(salop)L] (2) (16). Values of $D_{\text{Co-R}}$ for [R-Co(salop)L] were deduced from kinetic measurements analogous to those described above for coenzyme B$_{12}$, but with n-C$_8$H$_{17}$SH used as the

$$[\text{C}_6\text{H}_5(\text{CH}_3)\text{CH-Co(DH)}_2\text{L}] \rightleftharpoons$$
$$[\text{Co}^{II}(\text{DH})_2\text{L}] + \text{C}_6\text{H}_5\text{CH=CH}_2 + \tfrac{1}{2}\text{H}_2 \quad (7)$$

$$\text{C}_6\text{H}_5\text{CH=CH}_2 + \tfrac{1}{2}\text{H}_2 \rightleftharpoons \text{C}_6\text{H}_5\overset{\cdot}{\text{C}}\text{HCH}_3 \quad (8)$$

$$[\text{C}_6\text{H}_5(\text{CH}_3)\text{CH-Co(DH)}_2\text{L}] \rightleftharpoons$$
$$[\text{Co}^{II}(\text{DH})_2\text{L}] + \text{C}_6\text{H}_5\overset{\cdot}{\text{C}}\text{HCH}_3 \quad (9)$$

radical trap (17). Values of the Co-C bond dissociation energies of a series of [C$_6$H$_5$(CH$_3$)CH-Co(DH)$_2$L] compounds were deduced either from kinetic measurements (18) or, in some cases, from measurements of equilibrium constants and of the enthalpy (ΔH_7^o) of the reversible reaction depicted by Eq. 7, in combination with published data for ΔH_8^o (−2.2 kcal/mol) (19). Thus, $D_{\text{Co-R}} \approx \Delta H_9^o \approx \Delta H_7^o + \Delta H_8^o \approx \Delta H_7^o - 2.2$ kcal/mol. The values of $D_{\text{Co-R}}$ resulting from these measurements are given in Table 2.

The Co-R bond dissociation energies listed in Table 2 range from 17 to 25 kcal/mol, approaching the value deduced above for coenzyme B_{12} itself. The influence of electronic factors is most clearly revealed by the values of D_{Co-R} for $[C_6H_5(CH_3)CH-Co(DH)_2L]$ where L is pyridine (py) or a para-substituted pyridine (so that steric influences are constant) (13). For this series, D_{Co-R} increases systematically with the basicity of the axial pyridine ligand, as depicted also by the plot of D_{Co-R} versus pK_a in Fig. 3. The probable explanation for this trend is that dissociation of the Co-C bond, in accordance with the generalized representation of Eq. 10, involves reduction of the cobalt—that is, a decrease in the formal oxidation state of cobalt from +3 to +2. Thus, more basic ligands are expected to stabilize the parent cobalt-(III)-alkyl relative to the cobalt(II) dissociation product and, hence, to increase the bond dissociation energy.

$$[L_5Co^{III}-R^-] \rightleftharpoons [L_5Co^{II}] + R\cdot \qquad (10)$$

Comparisons of some of the data in Table 2 reveal the importance of steric influences on Co-R bond dissociation energies. Thus the large (~7 kcal/mol) difference between the Co-C bond dissociation energies of $[CH_3CH_2CH_2-Co(saloph)py]$ and $[(CH_3)_3CCH_2-Co(saloph)py]$ presumably is due largely to steric factors. The important influence of steric factors also is revealed by the trend of values of D_{Co-R} for the series of $[C_6H_5(CH_3)CH-Co(DH)_2(PR_3)]$ compounds that exhibit a marked inverse dependence (Fig. 4) on the size [as measured by the "cone angle" (20)] of the phosphine ligand. Thus, D_{Co-R} ranges from about 24 kcal/mol for L = PMe_2Ph (cone angle 122°) to 17 kcal/mol for L = PPh_3 (cone angle 145°).

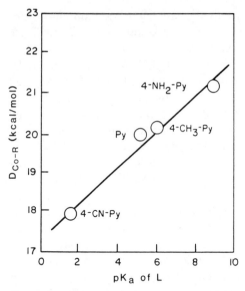

Fig. 3. Dependence of the Co-C bond dissociation energy of $[C_6H_5(CH_3)CH-Co(DH)_2L]$ on the pK_a of L (19).

Evidence for the importance of steric influences on Co-C bond stability also is provided by the results of x-ray structural determinations on a series of [R-Co(DH)₂L] compounds (Table 3) (21). These measurements reveal significant lengthening of the Co-C bond with increasing steric bulk of R and L, as well as significant sterically induced conformational distortions from planarity of the bis(dimethylglyoximate) structure, reflected in variations of the dihedral angle α between the planes of the two dimethylglyoxime ligands (3).

3

The structure of coenzyme B_{12}, as revealed by x-ray diffraction studies, shows evidence of similar steric crowding. The Co-C bond is quite long (2.05 angstroms) and the Co-C-C bond angle

Table 2. Cobalt-carbon bond dissociation energies.

Cobalt alkyl	$D_{\text{Co-R}}$ (kcal/mol)	Reference
[CH$_3$CH$_2$CH$_2$-Co(saloph)(py)]	25	(17)
[(CH$_3$)$_2$CH-Co(saloph)(py)]	20	(19)
[(CH$_3$)$_3$CCH$_2$-Co(saloph)(py)]	18	(19)
[C$_6$H$_5$CH$_2$-Co(saloph)(py)]	22	(19)
[C$_6$H$_5$(CH$_3$)CH-Co(DH)$_2$(4-NH$_2$-py)]	21	(19)
[C$_6$H$_5$(CH$_3$)CH-Co(DH)$_2$(4-CH$_3$-py)]	20	(19)
[C$_6$H$_5$(CH$_3$)CH-Co(DH)$_2$(py)]	20	(19)
[C$_6$H$_5$(CH$_3$)CH-Co(DH)$_2$(4-CN-py)]	18	(19)
[C$_6$H$_5$(CH$_3$)CH-Co(DH)$_2$(imidazole)]	21	(19)
[C$_6$H$_5$(CH$_3$)CH-Co(DH)$_2$(PMe$_2$Ph)]	24	(18)
[C$_6$H$_5$(CH$_3$)CH-Co(DH)$_2$(P(CH$_2$CH$_2$CN)$_3$)]	20	(18)
[C$_6$H$_5$(CH$_3$)CH-Co(DH)$_2$(PBu$_3^n$)]	21	(18)
[C$_6$H$_5$(CH$_3$)CH-Co(DH)$_2$(PEtPh$_2$)]	19	(18)
[C$_6$H$_5$(CH$_3$)CH-Co(DH)$_2$(PPh$_3$)]	17	(18)
Ado-CH$_2$-B$_{12}$	26	(12)

of 125° is much larger than the tetrahedral value of 109.5°, apparently reflecting repulsions between the 5′-deoxyadenosyl group and substituents on the corrin ring (16). Consistent with this is the identification of several close contacts (~3 Å) between atoms of the 5′-deoxyadenosyl group and atoms of the corrin ring and its substituents (22). In light of these considerations it seems highly likely that the enzyme-induced coenzyme Co-C bond weakening is due to steric influences—namely, an upward conformational distortion of the corrin ring that increases the steric repulsion of the 5′-deoxyadenosyl substituent and induces dissociation of the Co-C bond (22, 23). The results of the cited structural and bond dissociation studies on coenzyme B$_{12}$ model compounds support the plausibility of this view and suggest that only a modest distortion of the already crowded coenzyme molecule is sufficient to effect the necessary Co-C bond weakening.

Mechanism of Substrate Rearrangement

The least well understood and most controversial aspect of the mechanism of coenzyme B$_{12}$–dependent rearrangements continues to be the mechanism of the rearrangement step itself—that is, of the 1,2-migration of X (Eq. 11).

$$-\overset{\bullet}{\underset{|}{C}}_1 —\overset{\overset{X}{|}}{\underset{|}{C}}_2- \longrightarrow -\overset{\overset{X}{|}}{\underset{|}{C}}_1—\overset{\bullet}{\underset{|}{C}}_2- \qquad (11)$$

$$(S\cdot) \qquad\qquad (P\cdot)$$

The mechanism depicted by Fig. 2 implies that the rearrangement is triggered by H atom abstraction from the substrate to generate the substrate radical, S· but does not require that the rearrangement involving the 1,2-migration of X (to yield, ultimately, the product radical P·) actually occur at the free radical stage. Possible alternatives to such a direct rearrangement (that is, to Eq. 12a) include rearrangement via intermediate carbonium ions or carbanions (generated by oxidation or reduction of S· by B_{12_r}) or via an organocobalt intermediate formed by combination of S· with B_{12_r} (Eqs. 12b, 12c, and 12d), respectively).

Free radical rearrangement pathways. One problem with the proposal of rearrangement at the initially formed free radical stage (Eq. 12a) is that such a 1,2-migration in a free radical is precedented for only one of the coenzyme B_{12} substrates, α-methyleneglutarate, which involves migration of a substituted vinyl group [$-C(=CH_2)COOH$] (24). For the other B_{12} substrates, migration of X [$-OH$, $-NH_2$, $-C(=O)SCoA$ or $-CH(NH_2)COOH$] has not previously been observed in model free radicals. However, it should be noted that those coenzyme B_{12}–dependent reactions whose rates have thus far been reported are fairly slow ($k_{cat} \approx 10^2$ sec^{-1} for methylmalonyl-CoA mutase and diol dehydrase) (4, 9, 15). Radical rearrangement processes compatible with this time scale may well have escaped detection in earlier studies of free radical rearrragements, most of which were restricted to much shorter time scales.

Testing whether a given substrate radical, S· (or appropriate model thereof) would rearrange spontaneously on the time scale of coenzyme B_{12}–dependent reactions required generating the free radical unambiguously, under conditions in which its lifetime was fairly long ($\geq 10^{-2}$ second) and, preferably, susceptible to measurement and systematic variation. Furthermore, it seemed preferable to accomplish this in the absence of any cobalt complexes to eliminate the issue of possible cobalt participation in such rearrangements.

We have recently accomplished this for a radical $EtSC(=O)C(CH_3)-(CH_2\cdot)COOEt$ (5) that models the substrate radical of the methylmalonyl-CoA mutase reaction [that is, $CoASC-(=O)CH(CH_2\cdot)COOH$] (25). The procedure used was an adaptation of the one used earlier by Walling and Cioffari (26) to study the rearrangement of the 5-hexenyl radical. The model radical 5 was generated from the corresponding bromide 4 by reaction with n-Bu$_3$Sn· [generated by reaction of n-Bu$_3$SnH with 2,2'-azobisisobutyronitrile (AIBN)] and the competition between direct trapping with n-Bu$_3$SnH (k_t) to yield 6 and rearrangement (k_r), followed by trapping of the rearranged radical **7** to yield **8** as shown in Eq. 13, was monitored as a function of the initial n-Bu$_3$SnH concentration. Only the direct trapping product (**6**) and that resulting from 1,2-migration of the thioester group (**8**) were formed, and they were obtained together in essentially quantitative yield. No other products, notably that resulting from migration of the ester group, were detected.

According to the scheme of Eq. 2

$$\frac{d[6]}{d[8]} = \frac{k_t[n\text{-Bu}_3\text{SnH}]}{k_r} \qquad (14)$$

Fitting the results of our measurements to Eq. 14 in combination with published data for k_t (27) yielded the values k_r (60.5°C) = 24 sec^{-1}, $\Delta H_r^{\ddagger}$ = 13.8 kcal/mol, and $\Delta S_r^{\ddagger}$ = −11 cal mol^{-1} K^{-1}. The value of k_r at 30°C, calculated from these activation parameters, is 2.5 sec^{-1}. This is about 1/40 of the estimated value of k_{cat} for the methylmalonyl-CoA mutase reaction. However, this relatively modest rate difference could well be accommodated by the chemical and structural differences between the model radical **5** and the methylmalonyl-CoA mutase radical, as well as by effects of interaction of the (enzyme-bound) substrate with the enzyme—for example, hydrogen bonding to the sulfur atom or conformational influences.

Related experiments in which the carbanion corresponding to **5**—namely, EtSC(=O)C(CH$_3$)(CH$_2^-$)COOEt, was generated by reduction of **4** with sodium naphthenide, revealed that rearrangement, while rapid, was less selective than for the free radical, yielding products resulting from migration of both the thioester and ester groups, together with other unidentified products (25). The for-

mation of such a substrate carbanion [presumably by electron transfer between the initially formed B$_{12_r}$ and substrate radical (Eq. 12c)] is expected to be highly unfavorable and thus to constitute a much less likely pathway than the alternative free radical rearrangement process that now has been shown to be a chemically viable pathway.

As already noted, rearrangement at the free radical stage [involving 1,2-migration of the –C(=CH$_2$)COOH group] also seems likely for the α-methyleneglutarate mutase reaction, since analogous 1,2-migrations of vinyl groups in related radicals, via intermediate cyclopropyl-methyl radicals, are well documented (for example, Eq. 15) (28).

It is likely that 1,2-migration of the thioester group proceeds through the analogous cyclopropyloxy radical **9**.

EtS — O•

COOEt

9

Carbonium ion rearrangements. The facile 1,2-migration of saturated groups, such as OH, NH$_2$, and CH(NH$_2$)COOH, in free radicals is neither precedented nor is it expected on theoretical grounds.

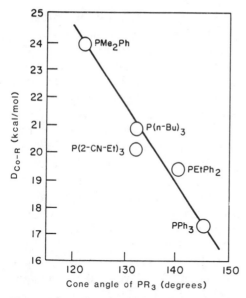

Fig. 4. Dependence of the Co-C bond dissociation energy of $[C_6H_5(CH_3)CH\text{-}Co(DH)_2(PR_3)]$ on the cone angle of PR_3 (18).

Acid-promoted dehydration of 1,2-diol radicals—for example, of 1,2-dihydroxyethyl to form acetaldehyde (Eq. 16)—has been observed (29) and has been invoked as a model of the dioldehydrase reaction (4, 30, 31). However, it is unlikely that such dehydration proceeds by 1,2-OH migration as does the enzymatic reaction [although it has been suggested that such migration may be accomplished at the radical stage through a sequence involving dehydration of the bound $CH_2(OH)\dot{C}HOH$ radical and rehydration of the resulting bound $\dot{C}H_2CHO$ radical (30)].

$$CH_2OH\dot{C}HOH \xrightarrow[-H_2O]{H^+} [\overset{+}{C}H_2\dot{C}HOH]$$

$$\xrightarrow{-H^+} [\dot{C}H_2\text{-}CHO] \xrightarrow{[H]} CH_3CHO \quad (16)$$

Since 1,2-migration of OH or NH_2 in carbonium ions is expected to be a facile process, the alternative possibility of mechanisms corresponding to Eq. 12b

warrants consideration. We have previously proposed such a mechanism for enthanolamine ammonia lyase (32), which is adapted in Fig. 5 for the corresponding diol dehydrase reaction. Steps i, ii, and vii of this mechanism correspond to steps i, ii, and iv of the mechanistic scheme of Fig. 2, and the other steps depict the carbonium ion rearrangement. Although the initial heterolytic cleavage of the coenzyme B_{12} Co-C bond to form the unstabilized primary $AdCH_2^+$ carbonium ion is highly unfavorable, the considerably enhanced stability of the oxocarbonium ion, $CH_2(OH)\text{-}\overset{+}{C}HOH$, could provide the necessary driving force for the proposed electron transfer (step iii) after the H-atom transfer (step ii). Conversely, the ensuing 1,2-OH shift and dehydration to form the unstable $\overset{+}{C}H_2CHO$ ion, would generate a driving force for reversal of the electron transfer step (that is, for step vi). Direct evidence for such a mechanistic scheme and, in particular, for the intermediacy of vitamin B_{12_s} (Co^I) in any coenzyme B_{12}–dependent reaction is at present lacking.

Rearrangement via organocobalt intermediates. Other mechanistic proposals have invoked the intermediacy of organocobalt adducts, arising from Co-C bond formation between B_{12_r} and the substrate radical (33–36). One such proposal for the diol dehydrase reaction, which derives some support from model experiments, is depicted by Eqs. 17 to 19 (34).

Assessment of the validity of this mechanistic proposal, as well as that of Fig. 5, must await further evidence. In this connection, recent model studies suggest that $Co\text{-}CH_2CHO$ is not an intermediate in the formation of CH_3CHO (30).

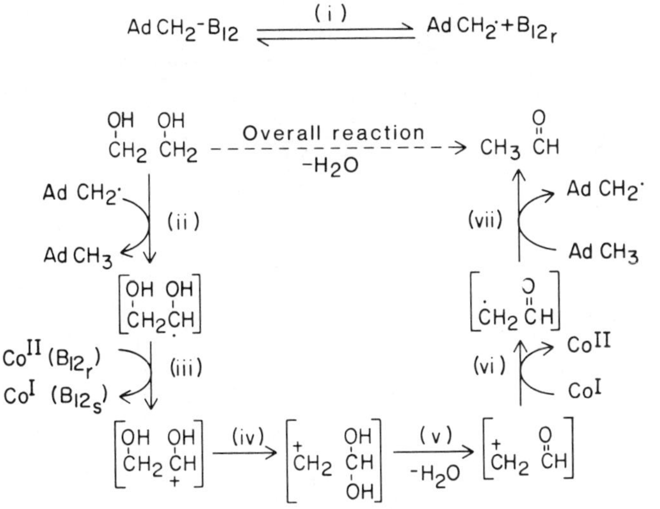

$$Ad\,CH_2\text{-}B_{12} \underset{\longleftarrow}{\overset{(i)}{\rightleftharpoons}} Ad\,CH_2^{\cdot} + B_{12r}$$

Fig. 5. Possible mechanism of the diol dehydrase rearrangement.

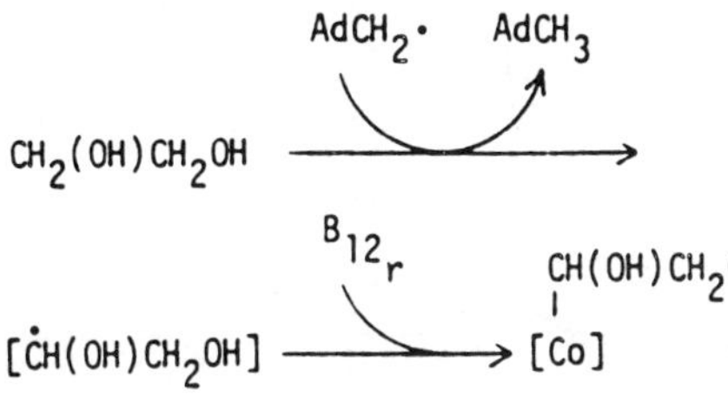

$$CH_2(OH)CH_2OH \quad (17)$$

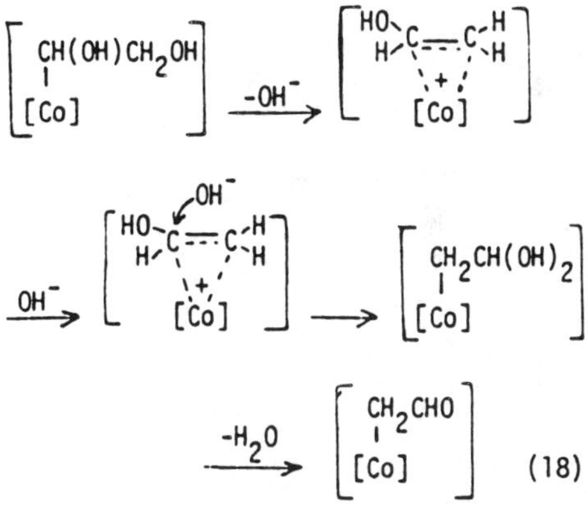

$$(18)$$

Table 3. Structural data for [R-Co(DH)₂L] compounds (21).

R	L	Co-C (Å)	Co-L (Å)	α (degrees)
CH₃	H₂O	1.990	2.058	−4
CH₃	Pyridine	1.998	2.068	+3.2
CH₃	PMe₃	2.015	2.294	+4.0
CH₃	PPh₃	2.026	2.418	+14.0
CH₂C(CH₃)₃	H₂O	2.044	2.056	−7
CH₂C(CH₃)₃	Pyridine	2.060	2.081	−5.2
CH₂C(CH₃)₃	PMe₃	2.084	2.316	−5
CH₂C(CH₃)₃	PPh₃	2.118	2.460	+2
CH(CH₃)₂	Pyridine	2.085	2.099	+4

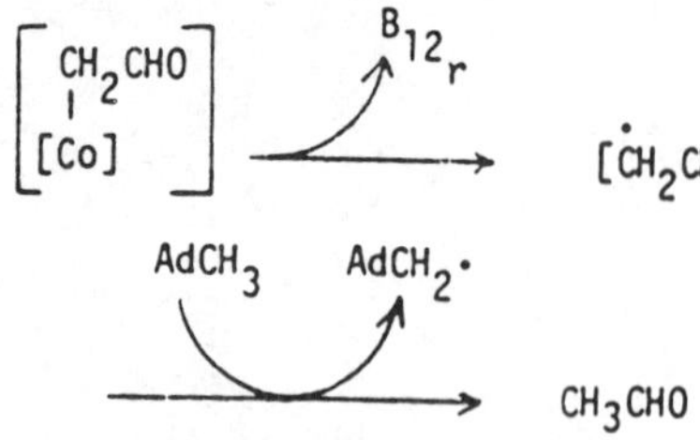
CH2CHO
[Co]
B12r
[ĊH2CHO]
AdCH3
AdCH2·
CH3CHO
(19)

Several model experiments, such as those depicted by Eqs. 20 to 22, have been reported, that attempt to probe the possible role of organocobalt intermediates in B_{12}-dependent rearrangements by generating organocobalt adducts containing cobalt-bonded groups intended to model those derived from coenzyme B_{12} substrates (*37–39*).

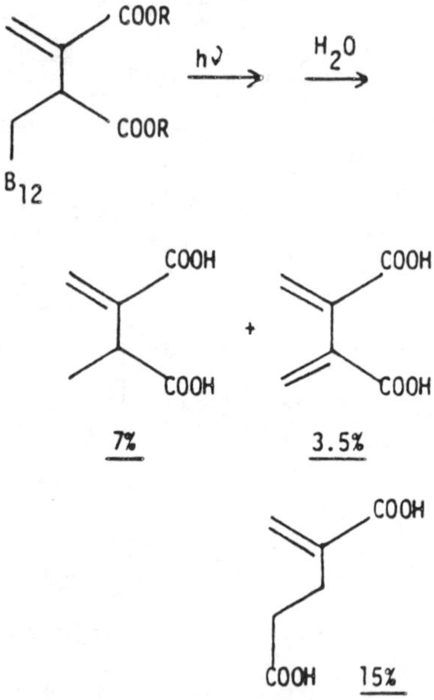
COOR
hν
H2O
COOR
B12
COOH
COOH
7%
COOH
COOH
3.5%
COOH
COOH
15%
(20)

Decomposition of such adducts under various conditions (photochemical, thermal, reductive) did, in some cases, lead to products that result from 1,2-migration, for example, of $-C(=CH_2)COOR$, $-COOEt$, and $-COSEt$ in the examples

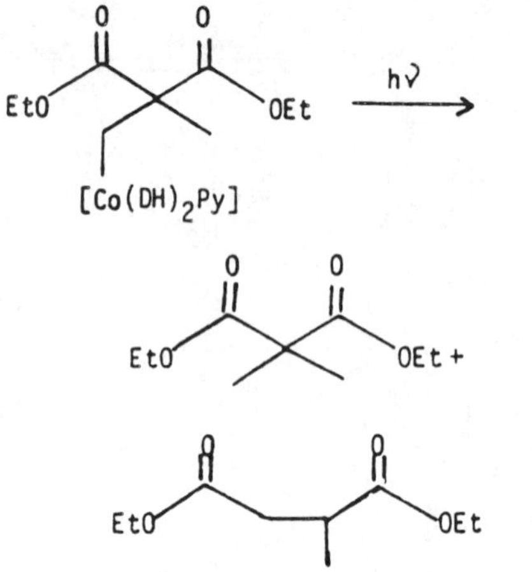
EtO
OEt
hν
[Co(DH)2Py]
EtO
OEt +
EtO
OEt
(21)

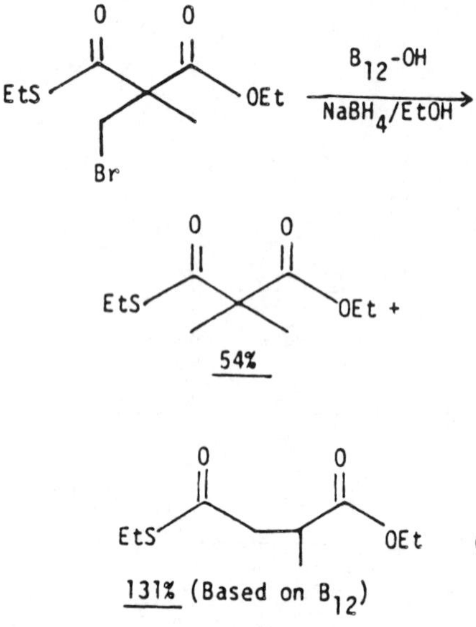
EtS
OEt
B12-OH
NaBH4/EtOH
Br
EtS
OEt +
54%
EtS
OEt
131% (Based on B12)
(22)

of Eqs. 20, 21, and 22, respectively (*31–39*). Although such observations have been interpreted as supporting the role of a substrate-derived organocobalt intermediate in related coenzyme B_{12}–dependent rearrangements (*37, 38*) this conclusion does not seem warranted. Thus, it is possible that the deliberately synthesized organocobalt compounds in these experiments serve simply as precursors of organic free radicals that are generated by homolysis of the cobalt-carbon bonds (for example, photochemically in Eqs. 20 and 21) or of carbanions generated under reducing conditions (Eq. 22).

Several other mechanistic proposals that have been advanced for coenzyme B_{12}–dependent rearrangements invoke substrate-derived organocobalt intermediates but do not include the cobalt-carbon homolysis step and the free radical intermediates that characterize the mechanistic scheme of Fig. 2 and its variants (40, 41). These proposals are not readily reconciled with the available evidence. Indeed, at this stage there is no convincing evidence, from either enzymatic or model system studies, for the formation of substrate-derived organocobalt adducts in coenzyme B_{12}–dependent rearrangements or for any role of such adducts in the rearrangement mechanisms.

Concluding Remarks

A combination of enzymatic and chemical studies, including studies on model systems, has resulted in considerable progress toward establishing the validity of the mechanistic scheme depicted by Fig. 2. At least for methylmalonyl-CoA mutase and α-methyleneglutarate mutase it seems likely that the substrate rearrangement step itself (the migration of X) occurs at the free radical stage. For other substrates the actual rearrangement mechanism remains to be elucidated.

The principal, if not the only, role of coenzyme B_{12} in these enzymatic processes appears to be that of a free radical precursor, a role that utilizes the weakness of the cobalt-carbon bond. The use of an organometallic molecule for this purpose seems appropriate since it is difficult to conceive of a stable organic

molecule that would undergo thermal dissociation under such mild conditions to generate a highly reactive primary radical. At this stage there is no convincing evidence that the coenzyme or the cobalt atom plays any other role—for example, that of mediating the rearrangement step itself.

The reversible cobalt-carbon bond dissociation of coenzyme B_{12} and related organocobalt compounds, depicted by Eq. 10, corresponds formally to an inner sphere redox process. There is at least a formal analogy between this process and the reversible binding of dioxygen by cobalt(II) and iron(II) complexes (such as myoglobin) as shown in Eqs. 23 and 24. This parallel is quite far-reaching and is reflected in trends in the dependence of the $Co-O_2$ (and presumably $Fe-O_2$) bond dissociation free energies and enthalpies that parallel those of Co-R bond dissociation energy trends, as revealed by a comparison of Figs. 3 and 6 (42, 43). Indeed, typical $Co-O_2$ bond dissociation energies in such reversible dioxygen carriers lie in the range 10 to 20 kcal/mol, which is not far from the range of typical cobalt-alkyl bond dissociation energies.

$$[L_5Co^{III}-O_2^-] \rightleftharpoons [L_5Co^{II}] + O_2 \quad (23)$$

$$[L_5Fe^{III}-O_2^-] \rightleftharpoons [L_5Fe^{II}] + O_2 \quad (24)$$

In the light of these considerations, the role of coenzyme B_{12} in biological systems may be described as that of a "reversible free radical carrier," analogous to the role of myoglobin or hemoglobin as a "reversible dioxygen carrier." Thus, coenzyme B_{12} fulfills its biochemical role by serving as a "free radical reservoir" from which 5'-deoxyadenosyl radicals are reversibly released

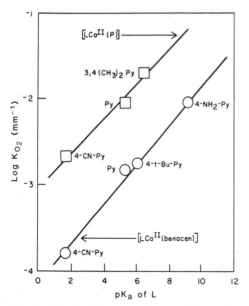

Fig. 6. Effect of the axial ligand (L) on the equilibrium constant for the reversible binding of O_2 to protoporphyrin IX ester cobalt(II) [LCo^{II}(P)] and N,N'-ethylenebis(benzyoylacetiminato)-cobalt(II) [LCo^{II}(benacen)] in toluene (42, 43).

under mild conditions, just as oxyhemoglobin serves as a reservoir for the storage and reversible release of dioxygen. Significant questions that warrant further attention relate to the alternative choices of cobalt and iron, as well as of the corrin and porphyrin ligand systems, for these parallel functions.

References and Notes

1. D. Dolphin, Ed., B_{12} (Wiley, New York, 1982), vols. 1 and 2.
2. B. Zagalak and W. Friedrich, Eds., *Vitamin B12* (de Gruyter, Berlin, 1979).
3. Another coenzyme B_{12}–dependent enzymatic reaction, that of ribonucleotide reductase, differs from these rearrangements in that it involves a net reduction of the substrate by an external reducing agent. Although free radical intermediates also have been implicated in this reaction, and the mechanism of the reaction is probably related to the mechanisms of the rearrangement reactions discussed in this article, the relation remains to be elucidated [R. L. Blakley, in (*1*), vol. 2, p. 382].
4. J. Retey, in (*1*), vol. 2, p. 357.
5. B. M. Babior, *Acc. Chem. Res.* **8**, 376 (1975).
6. B. T. Golding, in (*1*), vol. 1, p. 543, and references therein.
7. R. H. Abeles, in (*2*), p. 373, and references therein.
8. Such H-atom transfer from C-H to carbon radicals is well established to be a facile process for which the activation energy typically is in the range 7 to 10 kcal/mol [J. A. Kerr and A. F. Trotman-Dickenson, *Prog. React. Kinet.* **1**, 105 (1961)].
9. T. Toraya and S. Fukui, in (*1*), p. 233.
10. J. Halpern, in (*1*), p. 501, and references therein.
11. _____, *Acc. Chem Res.* **15**, 238 (1982).
12. _____, S.-H. Kim, T. W. Leung, *J. Am. Chem. Soc.* **106**, 8317 (1984).
13. J. F. Endicott and J. G. Ferandi, *ibid.* **99**, 243 (1977); J. F. Endicott and T. L. Netzel, *ibid.* **101**, 4000 (1979).
14. A related recent measurement in ethylene glycol solution, with tetramethylpiperidinyloxy (TEMPO) used as the radical trap, yielded a significantly higher value of $D_{AdCH_2-B_{12}}$, namely, 32 kcal/mol. The reason for this discrepancy, which may be solvent-related, is unclear [R. G. Finke and B. P. Hay, *Inorg. Chem.* **23**, 3041 (1984)].
15. B. M. Babior, in (*1*), vol. 2, p. 263.
16. P. G. Lenhert, *Proc. R. Soc. London, Ser. A* **303**, 45 (1968).
17. T. T. Tsou, M. Loots, J. Halpern, *J. Am. Chem. Soc.* **104**, 623 (1982).
18. F. T. T. Ng, G. L. Rempel, J. Halpern, *Inorg. Chim. Acta* **77**, L65 (1983).
19. J. Halpern, F. T. T. Ng, G. L. Rempel, *J. Am. Chem. Soc.* **101**, 7124 (1979); F. T. T. Ng, G. L. Rempel, J. Halpern, *ibid.* **104**, 621 (1982).
20. C. A. Tolman, *Chem. Rev.* **77**, 3131 (1977).
21. L. Randaccio, N. Bresciani-Pahor, P. J. Toscano, L. G. Marzilli, *J. Am. Chem. Soc.* **103**, 6347 (1981); N. Bresciani-Pahor, M. Calligaris, G. Nardin, L. Randaccio, *J. Chem. Soc. Dalton Trans.* (1982), p. 2549; M. F. Summers, P. J. Toscano, N. Bresciani-Pahor, G. Nardin, L. Randaccio, L. G. Marzilli, *J. Am. Chem. Soc.* **105**, 6259 (1983), and references therein.
22. J. P. Glusker, in (*1*), vol. 1, p. 23.
23. S. M. Chemaly and J. M. Pratt, *J. Chem. Soc. Dalton Trans.* (1980), pp. 2259, 2267, and 2274.
24. A. L. J. Beckwith and K. U. Ingold, in *Rearrangements in the Ground State and Excited States*, P. de Mayo, Ed. (Academic Press, New York, 1980), vol. 1, p. 161, and references therein.
25. S. Wollowitz and J. Halpern, *J. Am. Chem. Soc.* **106**, 8319 (1984).
26. C. Walling and A. Cioffari, *ibid.* **94**, 6059 (1972).
27. C. Chatgilialoglu, K. U. Ingold, J. C. Scaiano, *ibid.* **103**, 7739 (1981).
28. B. Maillard, D. Forrest, K. U. Ingold, *ibid.* **98**, 7024 (1976); A. Effio, D. Griller, K. U. Ingold, A. L. J. Beckwith, A. K. Serelis, *ibid.* **102**, 1734 (1980).
29. C. A. Walling and R. A. Johnson, *ibid.* **97**, 2405 (1979); B. C. Gilbert, J. P. Larkin, R. O. C. Norman, *J. Chem. Soc. Perkin Trans.* 2 (1972), p. 794.
30. R. G. Finke and D. A. Schiraldi, *J. Am. Chem. Soc.* **105**, 7605 (1983); _____, B. J. Mayer, *Coord. Chem. Rev.* **54**, 1 (1984).
31. B. T. Golding, T. J. Kemp, C. S. Sell, P. J. Sellars, W. R. Watson, *J. Chem. Soc. Perkin Trans.* 2 (1978), p. 839; B. T. Golding and L. Radom, *J. Am. Chem. Soc.* **98**, 6331 (1979); *J. Chem. Soc. Chem. Commun.* (1973), p. 939.

32. J. Halpern, *Ann. N.Y. Acad. Sci.* **239**, 2 (1974).
33. R. G. Eager, Jr., B. G. Baltimore, M. M. Herbst, H. A. Barker, J. H. Richards, *Biochemistry* **11**, 253 (1972).
34. R. B. Silverman, D. Dolphin, B. M. Babior, *J. Am. Chem. Soc.* **94**, 4028 (1972); R. B. Silverman and D. Dolphin, *ibid.* **95**, 1686 (1973); *ibid.* **98**, 4626 and 4633 (1976); D. Dolphin, A. R. Banks, W. R. Cullen, A. R. Cutler, R. B. Silverman, in (2), p. 575.
35. K. L. Brown and L. L. Ingraham, *J. Am. Chem. Soc.* **96**, 7681 (1974).
36. R. Hamilton and J. J. Rooney, *J. Mol. Catal.* **17**, 29 (1982); J. J. Rooney, *ibid.* **26**, 13 (1984); *J. Chem. Res. (S)*, 48 (1984).
37. P. Dowd, M. Shapiro, K. Kang, *J. Am. Chem. Soc.* **97**, 4754 (1975); P. Dowd and M. Shapiro, *ibid.* **98**, 3724 (1976); P. Dowd, in (2), p. 557, and references therein.
38. G. Bidlungmaier, H. Flohr, U. M. Kempe, T. Krebs, J. Retey, *Angew. Chem. Int. Ed. Engl.* **14**, 822 (1975); H. Flohr, W. Pannhorst, J. Retey, *ibid.* **15**, 561 (1976); *Helv. Chim. Acta* **61**, 1565 (1978); J. Retey, in (2), p. 439, and references therein.
39. A. I. Scott and K. Kang, *J. Am. Chem. Soc.* **99**, 1997 (1977); A. I. Scott, J. Kang, D. Dalton, S. K. Chung, *ibid.* **100**, 3603 (1978); A. I. Scott *et al.*, *Bioorg. Chem.* **9**, 227 (1980).
40. G. N. Schrauzer and R. W. Windgassen, *J. Am. Chem. Soc.* **89**, 143 (1967); G. N. Schrauzer and W. Sibert, *ibid.* **92**, 1022 (1970).
41. E. J. Corey, N. J. Cooper, M. L. H. Green, *Proc. Natl. Acad. Sci. U.S.A.* **74**, 811 (1977).
42. D. V. Stynes, H. C. Stynes, B. R. James, J. A. Ibers, *J. Am. Chem. Soc.* **95**, 1796 (1973).
43. M. Carter, D. P. Rillema, F. Basolo, *ibid.* **96**, 392 (1974).
44. The research encompassed by this article was supported by grants from the National Institutes of Health and the National Science Foundation.

Part VII

Chemistry of Materials

In 1980 an article on biomaterials documented the use of more than 40 different materials in more than 50 different medical and dental devices (*1*). A common characteristic of most of these biomaterials and devices, also discussed in a recent NIH conference (*2*) and a recent book (*3*), is their so-called "bio-inertness." However, it is now well established that no material implanted in living tissues is inert. All materials elicit a response from living tissues. Four types of response are possible: (i) if the material is toxic, the surrounding tissue dies; (ii) if the material is nontoxic and dissolves, the surrounding tissue replaces it; (iii) if the material is nontoxic and biologically inactive, a fibrous tissue capsule of variable thickness forms; and (iv) if the material is nontoxic and biologically active, an interfacial bond forms.

The purpose of this article is to discuss the current state of the science and development for clinical application of this last class of surface-active biomaterials. It is timely to do so for several reasons. First, there is increasing clinical evidence that the useful life of most implants made from inactive biomaterials is much shorter than the patient requires (*2*). Second, failure usually follows movement at the implant-tissue interface. Third, surface-active biomaterials are becoming more widely used in clinical applications, particularly musculoskeletal and dental applications. Fourth, a scientific understanding of the mechanisms of bonding between living and nonliving interfaces is emerging and has major implications for the biological sciences in general as well as for medical and dental surgery. It is essential that this class of biomaterials be understood and used properly if surface-active implants are to realize their potential for long-term stability.

33

Surface-Active Biomaterials

Larry L. Hench and June Wilson

Science 226, 630–636 (9 November 1984)

Four major categories of surface-active biomaterials have been developed during the past 15 years; dense hydroxylapatite (HA) ceramics, bioactive glasses, bioactive glass-ceramics, and bioactive composites. After many years of animal tests, clinical trials of all four types have begun, and some have been in progress for as long as 5 years.

Hydroxylapatite Ceramics

Hydroxylapatite materials have been used for implants in many forms, especially in dental applications. Denissen (4) reported that root-shaped HA implants, buried in contoured, fresh extraction sites in dogs, bonded in place without bone resorption around them. These, however, were not load-bearing implants, but were analogous to the alveolar ridge maintenance devices now in use. In 1983 de Putter et al. (5) showed that load-bearing, transmucosal implants of dense HA acted as ankylotic elements, similar to bone that is not load-bearing; that is, they did not acquire a periodontal ligament attachment. The gingival tissue response did resemble the natural interface, but chewing forces caused fatigue failure in the implants. The studies of Ogiso et al. (6), in which hemidesmosomes were identified, support the possibility of normal epithelial attachment to HA surfaces. However, de Putter et al. (5) concluded that apatite ceramics had no clinical potential in situations where forces other than compression play a role. Recent work by de Putter, de Groot, and others in the Netherlands (7) suggests that prestressing the HA implant will prevent fatigue fracture. When HA was implanted in the femur in experimental animals by Denissen (4),

the tissue reaction filled an oversized hole and the bone grew as a collar over protruding parts of the implant. This has been noted by other workers who used solid HA in skeletal models (8). The ability to fill an oversized implant site might provide a clinical advantage if the overgrowth of bone could be effectively understood and controlled.

In restoration of the bony conduction system and canal wall of the middle ear, HA in a combination of porous and dense forms has been used successfully by Grote (9). Where there is no demand for mechanical strength of the device, as in the middle ear, bony ingrowth into porous nonresorbable HA can provide a good functioning structure with integration of implant and host bone.

Hydroxylapatite has been used clinically in particulate form to augment the alveolar ridge (10) and in a variety of maxillofacial applications (11); it is especially effective when the particulates are mixed with autogenous bone. Data covering periods of 4 and 5 years show continued success with ridge augmentation (11). However, particulate HA, when used as a treatment for periodontal disease, has not yet fulfilled its early promise in long-term applications.

Hydroxylapatite has been used as a coating on other, mechanically stronger, materials to provide load-bearing implants. Ducheyne et al. (12) used an HA coating on porous stainless steel as a tooth replacement. The HA in this application was resorbable and allowed bone to be incorporated within the porous metal. The coating allowed good initial stabilization of the artificial tooth in the critical early weeks after implantation.

In reviewing the uses of hydroxylapatites it becomes clear that the material, which forms the mineral component of bone, is entirely biocompatible in its many forms but can have variable properties related to its method of preparation. It may be solid or particulate, as mentioned here; but it has also been reported in microporous (<5 μm) and macroporous (>100 μm) forms and with major or minimal resorbability. The factors governing resorbability have been reviewed by Jarcho (13), and it is clear that close attention to manufacturing procedures is essential for the production of HA implants that will behave predictably and reliably in vivo.

Surface-Active Glasses

All surface-active glasses under investigation today derive from materials developed at the University of Florida. These Bioglass (14) materials were the first man-made materials that formed a chemical bond with bone. Only certain compositions achieve the bone bond, and these are shown in Fig. 1 and discussed in a later section.

Potential clinical applications of implants made from or incorporating surface-active glass components exist in orthopedic, otolaryngological, dental, and maxillofacial surgery. The inherent mechanical weakness of the glass has divided these applications into two groups, those where mechanical strength is unimportant and those where it is critical. Implants in the first group include devices for maintenance of the alveolar ridges in mandible and maxilla and restoration of the ossicular chain in the middle ear. Clinical trials in both of these areas are in progress and to date have been successful. The ridge maintenance devices are cones made from a Bioglass composition, designated 45S5, which are

buried in a reamed extraction site. Middle ear prostheses made from the same 45S5 components are used to replace all or part of the ossicular chain in patients who have a history of chronic otitis media and contributing developmental defects. Implants made from Bioglass and those made from other materials now in clinical use have been compared by Merwin *et al.* (*14a*), who showed that a significant determinant of long-term stability of Bioglass devices is the presence of a soft-tissue bond between the implant and the tympanic membrane. This has not been seen with any other material. This bond prevents movement at that critical interface at an early stage and prevents the inflammation and scarring which are associated with loss of transmission and eventual extrusion. If the periosteum of the remaining ossicles can be retained at the interface with the Bioglass implant, a soft-tissue bond is induced there and provides an interface more closely resembling the natural one between undamaged ossicles. If the periosteum is removed, either accidentally

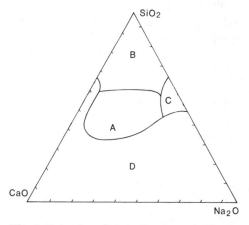

Fig. 1. Behavior of bioactive glass of different compositions. Region A, bone bonding; B, fibrous tissue encapsulation; C, dissolution; D, non–glass-forming. All compositions have a constant 6 percent P_2O_5 by weight.

or deliberately, the bonded interface will be an ankylotic one resembling that between allogenic ossicular implants, at present considered to be the best available for clinical use. In experiments with mice the soft tissue between Bioglass implants and the remaining ossicular chain and thin collagenous capsule around the implants has persisted apparently unaltered throughout the animals' lifetime. These results and data from other animal experiments reported recently by Merwin *et al.* (*14a*) suggest that modes of failure such as extrusion and scarring associated with other available materials, as reported by various authors (*9*), will not compromise long-term survival of these implants. In addition, the machinability and transparency of these bioactive glass implants allow ease and flexibility of use by the surgeon in the operating room.

The second set of applications requires the combination of bone-bonding activity of the bioactive glass with mechanical properties of a substrate. Bioglass coatings have been successfully applied to substrates of stainless steel, Vitallium (*15*), titanium metals, and high-density alumina ceramic. A composite of 45S5 Bioglass and stainless steel fibers has been produced (*16*) which can reproduce the mechanical properties of bone without compromising the integrity of the glass-metal interface. Extensive animal experiments (*3*) have shown that Bioglass coatings on orthopedic devices for noncement fixation should provide a generation of hip prostheses with a longer life than the 20 years available with conventional cement fixation. Load bearing can be achieved with dental implants made from coated alumina or coated metals. Smith (*17*) showed that coated-alumina devices have excellent

mechanical behavior. Japanese scientists have used stainless steel implants with Bioglass coatings to hold crowns. These functioned well in dogs (*18*) and are currently being used in clinical trials in Japan.

Considerable attention has been given to the use of Bioglass-coated implants of either alumina or metal in orthodontics. When a tooth is stressed it may be moved in the direction of the load. This movement is a consequence of osteoclastic activity in which bone is removed and osteoblastic activity in which bone is laid down under the influence of increased and decreased pressure, respectively. This is the basis of orthodontic movement of teeth, and there can be a complication if undesired movement is produced in the teeth which act as anchors. It is known that ankylosed teeth are not moved in this way (*19*).

Smith (*17*) inserted blade-shaped implants of alumina, coated with surface-active glass, into the alveolar ridge of rhesus monkeys. These animals had a rod protruding through the gingiva to which, after 9 weeks of healing, posts to hold a lingual arch appliance were attached. The anchors were then subjected to forces up to 950 g for several weeks with no movement of the implant and no change in adjacent bone. Smith suggested that the presence of connective tissue, as in the periosteum and periodontal membrane, is essential for the cellular involvement that causes teeth to move. Absence of this connective tissue in the essentially ankylosed pegs, as in ankylosed teeth, prevented this movement. Paige *et al.* (*20*), recognizing that useful orthodontic anchors would have to be small and simple in shape, showed that Smith's findings could apply to coated Vitallium wires as small as 2 mm in diameter. Turley *et al.* (*21*) put small coated-alumina implants into monkeys to produce maxillary expansion with equally good results, and Grey *et al.* (*22*) used Bioglass-coated Vitallium implants, only 1.6 mm in diameter, with equal success in rabbits. When these orthodontic pegs are removed, as they may be by rotational force, the gel layer shears and only scraps of biocompatible material are left in the healing wound area.

In all of these applications that require coating of bioactive glass on metal it is essential to consider the consequences of introduction of other chemicals into the glass. Recent experiments (*23*) showed that accidental introduction of aluminum into the glass during coating will prevent bonding. It has been suggested that the introduction of other metal oxides will either alter the reaction rate of the glass or facilitate coating procedures. However, Gross and Strunz (*24*), in the course of numerous comparative evaluations of a wide range of bioactive materials, found that the introduction of other oxides, particularly those of zirconium, titanium, and tantalum, can impair bone development (by disturbance of osteoblast metabolism), matrix vesicles function, and collagen deposition. These oxides should not be used in materials for bone and tooth replacement.

Glass-Ceramics

The surface-active, bone-bonding glass formulations can also be produced as glass-ceramics by nucleation and growth of crystals in the glass. Transparent glasses, which are monophase systems, become opaque glass-ceramics, which contain crystals within a glassy matrix. This transformation produces materials that are often mechanically stronger. However, the grain boundaries in a glass-ceramic may provide sites at which dissolution can occur (*25*).

One such material, Ceravital (*26*), which is based on the Bioglass formulations but has a lower alkali content, has been extensively tested in orthopedic, dental, maxillofacial, and otolaryngological applications. The surface-active glass-ceramic, used as a coating on a metal femoral head endoprosthesis, provided good noncement fixation in dogs for periods up to 20 weeks. Tests to failure of these implants postmortem showed that rupture lines occurred in the bone and not the interface. This provides confirmation that the interface is indeed stronger than the bone.

Good results were obtained when Ceravital as a bulk material was used for jaw augmentation of osteotomies in pigs. More than 60 percent of the surface was bonded to the surrounding bone after 1 year (*27*). Bunte and Strunz (*28*) performed a trial study with 12 implants in humans. Overall the results were very good, and all the implants were incorporated without irritation. However, subperiosteal implants gave poor results, as the thin layer of bone over the implant caused pain during loading, and they were removed.

Ceravital has been used successfully for middle-ear prosthetic devices. Reck and Helms (*29*) used it to restore the ossicular chain and reconstruct the bony posterior wall of the chamber in rabbits. As reported previously (*25*), small areas of lysis at the surface occurred where capillaries in the bone were adjacent to the glass-ceramic. These areas became filled with new bone; which continued to grow over the surface of the implant to a depth of 40 μm. Mucosa covered this bone, and the two layers prevented further lysis of the material for up to 2 years in animals. Where mucosa directly covered the implant the lysis sometimes progressed, and to prevent this bone pâté was required between the implant and soft tissue to ensure the development of bone at the interface. After 5 years of clinical use of Ceravital in patients, Reck and Helms concluded that for total ossicular chain reconstruction, prostheses made of this material give better results than preserved allogenic ossicles, previously considered to be the best available. Almost all middle-ear devices must be contoured in the operating room, and it has been shown clinically and experimentally that Ceravital glass-ceramic can be successfully contoured with standard operating room drilling equipment, albeit at a slower rate than bioactive glass. Ceravital devices were contoured by Babighian (30), who also confirmed the need for bone pâté to prevent extrusion through the soft tissue of the tympanic membrane. When the pâté was not used extrusion almost invariably occurred within a short time.

A bioactive glass-ceramic has been developed by Vogel and co-workers at the University of Jena in East Germany. Their objective is to produce a material that can be easily machined, and they use variations in composition to produce different properties in the material, including magnetic properties and bioactivity. The bioactive glass-ceramic, which is now undergoing preclinical testing, consists of a mica crystalline phase for machinability and an apatite crystalline phase for bioactivity in a residual glassy phase of unknown composition. Implantations of unloaded samples in

guinea pig bone showed satisfactory bonding after 16 weeks (31).

At the Kyoto Institute in Japan a new glass-ceramic, known as A/W ceramic, has been developed in a search for bioactive materials with sufficient strength to allow their use in load-bearing conditions. The material contains HA and wollastonite (a form of a calcium silicate) in a glassy phase of undetermined composition. Implantation of unloaded implants in rabbit tibiae showed good bonding at 8 weeks with a bonding strength greater than that of Bioglass, comparable to that of dense HA, but 70 percent of the value for bone. At 25 weeks, when only A/W and dense HA were compared, the relative strengths remained the same (32). The material was used clinically to provide spinal fusion in a patient for whom no autologous bone was available (33) and has been in place and functioning well for 2 years.

Surface-Active Composites

Many composite materials have been produced since the bone- and tissue-bonding abilities of bioactive materials were recognized. All natural tissues are themselves composites, and the combination of bioactivity and specific mechanical properties should allow the production of materials with properties selected for particular biomedical problems. This is an attractive theory, but in practice most of these composites have not been successful. The principal mode of failure is at the bond between matrix and filler, and many potential materials have failed at this interface under the action of tissue fluid and cellular enzymes. However, success has been achieved in some areas. When particulate HA is mixed with finely ground autologous bone the resulting mixture becomes a composite after implantation as physicochemical bonding takes place in vivo between the components, and this provides a material that is more satisfactory than either component alone in maxillofacial and dental applications (*10*).

A successful composite of 45S5 Bioglass and 316L stainless steel has been made (*16*) by bonding the glass onto a sintered stainless steel fiber matrix. This process results in a composite which has two continuous phases, rather than matrix and filler, but which has only the bioactive phase at the interface with bone. This material can be produced with variable mechanical properties for orthopedic applications.

A novel approach to providing a material that can be used to fix orthopedic prostheses, notably artificial hips, with the convenience of cement fixation allied with the greater long-term reliability of noncement fixation by bioactive materials, has been taken by the Leitz Company in the production of Palavital, a mixture of the conventional bone cement polymethylmethacrylate (PMMA) with Ceravital glass-ceramic particles and a small amount of glass fiber. This cement mixture polymerizes in situ, as does PMMA. After implantation, this bioceramic bone cement, while not strictly a composite, gives a combination of mechanical and physicochemical adhesion, with junctions of bone to the cement instead of the fibrous capsule usual with bone cements. Preliminary tests (*34*) showed the mechanical strength of Palavital to be comparable to that of other bone cements. Because of its bioactive component, this material may be superior in long-term behavior to other bone cements and may alleviate problems caused by loosening of prosthetic devices. More experimental data are needed to evaluate the contribution of the heat generated on polymerization of PMMA in situ to long-term loosening or long-term stability of the heterogeneous interface.

Interfacial Bonding

It is a common characteristic of bioactive implants that an interfacial bond forms between the smooth, nonporous surface of the material and adjacent tissue. The idea that chemical bonding could be achieved between a nonliving material and living tissue was first proposed to the U.S. Army Medical R&D Command by Hench in 1967, and evidence for a chemically bonded implant-tissue interface soon followed (*35*). Specially designed bioactive glasses containing Na_2O, CaO, P_2O_5. and SiO_2, termed Bioglasses, were shown to bond to rat femoral bone as early as 10 days postoperatively. The time dependence of hard-tissue bonding, strength of the bond, and proposed bonding mechanisms were soon described. A few years later these findings began to be confirmed with the same bioactive glasses and related compositions of surface-active glass-ceramics (*27, 28, 32, 36*).

Driskell *et al.* (*37*) made the first observation of apparent chemical bonding of bone to tricalcium phosphate ceramics. However, the most convincing evidence of this phenomenon for surface-active ceramics was provided by the studies of Jarcho and co-workers (*8, 13*) on bonding of dense HA implants. Subsequently, numerous investigators, using a variety of materials and animal models, showed that bone bonds to surface-active apatite ceramics (*10, 38*).

A unique feature of surface-active biomaterials is that the interfacial bond with bone is generally stronger than either the bone or the implant. Fracture almost never occurs at the bone-implant interface during mechanical testing (*3*). The strength of the bond therefore has not been measured. However, a series of studies with eight different models has established a lower limit of interfacial bond strength with surface-active glass implants (45S5 Bioglass). The maximum stress level sustained by a load-bearing segmental bone implant loaded in shear, in a torsional test, was calculated as 117 MPa. The femur fractured at this stress, which is 1 standard deviation below the average shear strength of normally healed fractured bone in the same animal (monkey) at 42 weeks. Eventual bone remodeling will increase the stress that the repaired bone can withstand and should lead to a measurable interfacial stress value equivalent to that of natural bone. The average interfacial stress measured for the load-bearing segmental bone model was 83 MPa. This is substantially more than the 3.2-MPa average shear stress calculated for the same 45S5 Bioglass material tested in the form of non–load-bearing cylinders in the cortices of dog femurs, assuming nominal contact area. This wide range of values is associated, in part, with the fact that the interfacial area of mineralized bone bonded to a surface-active implant increases with time and with the loads applied. An unloaded implant develops a bond much more slowly, just as it takes longer for an unloaded fracture to heal. Also, dehydration of the bonding zone can occur during testing and lead to shear failure within the surface gel layer.

Other studies of interfacial shear strengths of bioactive glasses and glass-ceramics also show wide ranges of results because of these variables (*32*).

Mechanisms of Bonding

The bonding mechanisms of surface-active glasses and glass-ceramics involve a complex combination of physicochemical and ultrastructural phenomena (*39–41*). At the microscopic level the bond between a surface-active glass and bone appears as a sharp interface, although a compositional gradient is present within the interfacial layer formed on the implant. Figure 2 shows the interface between bone and surface-active glass in a 4-week rat tibial implant. The areas labeled show bulk 45S5 Bioglass, an SiO_2-rich layer, a Ca,P-rich layer, and bone. Electron microprobe analysis and scanning electron microscopy–energy-dispersive x-ray analysis (SEM-EDXA) were used to measure this interfacial compositional sequence, which extended to a thickness of 70 μm at 4 weeks, 75 μm at 12 weeks, 97 μm at 52 weeks, and 268 μm at 128 weeks (the lifetime of a male Sprague-Dawley rat). The thickness of the Ca,P-rich zone remains relatively constant at 30 to 40 μm during this time period, whereas the silica-rich layer increases at a rate approximately proportional to $t^{1/2}$ due to continued exchange of Na^+ in the glass with H^+ ions from solution.

The Ca,P-rich layer forms instantaneously on the surface-active glasses, as shown by Auger electron spectroscopy of the 45S5 Bioglass composition after 1 hour in vitro (distilled water at 37°C) or in vivo (*40*). Initially the Ca,P-rich layer is amorphous, but it crystallizes into

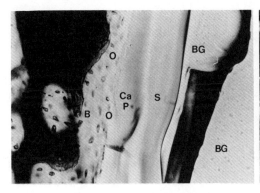 

Fig. 2 (left). Bonded interface between rat tibial bone and surface-active glass 30 days after implantation. BG, bulk 45S5 Bioglass; S, SiO₂-rich layer; CaP, Ca,P-rich layer; B, bone; O, osteocyte (×200).   Fig. 3 (right). Scanning electron micrograph of collagen fibers attached to a 45S5 Bioglass surface after exposure in vitro at 37°C for 10 days (×5000). [Photo courtesy C. G. Pantano]

mixed hydroxyl-carbonate apatite agglomerates within 7 to 10 days. When the apatite phase crystallizes in the presence of collagen fibers, in vitro or in vivo, the collagen becomes structurally integrated within the apatite agglomerates and vice versa (Fig. 3). The presence of mucopolysaccharides, such as chondroitin sulfate D, significantly enhances the physicochemical interaction between the crystallizing apatite layer and collagen. Between the collagen fibers and surface-active glass in vivo there is an amorphous zone 80 to 100 nm thick, which can be mineralized. This zone or seam apparently consists of extracellular ground substance which may contain mucopolysaccharides, glycoproteins, and various nectins, and provides a basis for attachment of collagen. Such a seam is present only at the interface of surface-active biomaterials bonded to bone.

After formation of the amorphous cementing zone on the Ca,P-rich layer, further steps in bone development and bonding are governed by osteoblasts in the implant area. Gross and Strunz (24) have shown that, at the bonding interface, osteoblasts provide (i) collagen and ground substance and (ii) matrix vesicles for primary mineralization. The sequence of events is consistent with present concepts of primary bone formation. Gross and Strunz summarized their findings: "Within the extracellular matrix and between small bundles of fibrils, matrix vesicles appear and display small, electron-dense, needle-like crystallites assumed to be apatite. After rupture of the vesicle membrane, calcifying fronts are formed. Often this process begins and is therefore more pronounced in the surroundings of the implant and the adjacent osteoblast, but may also start in the area around the osteoblast and then involve the surroundings of the already mineralized seam of amorphous cementing substance at the interface. Later on the whole area is mineralized, the osteocytes being rather evenly distributed and often arranged with their long axis parallel to the surface of the implant. This feature is found in different species—rats, dogs, pigs, chickens and humans—and provides the morphologic basis for the biomechanical quality of the bone bonding.''

Systematic studies by Hench *et al.* (*3, 39, 40*) and Gross and Strunz (*24, 25, 41*) have identified many of the compositional factors that can affect bone bonding to surface-active glasses. Substitution of 5 to 15 percent B_2O_3 for SiO_2 in the original 45S5 Bioglass formula (45 SiO_2, 6 P_2O_5, 24.5 CaO, and 24.5 Na_2O, in percent by weight) results in a more reactive composition, whereas replacement of varying proportions of CaO with CaF_2

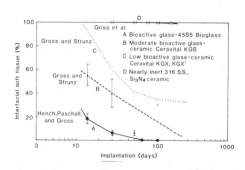

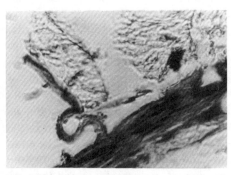

Fig. 4 (top). Time dependence of replacement of soft tissue with bone for various biomaterials (*3*).   Fig. 5 (bottom). Soft tissue attachment to partially decalcified 45S5 Bioglass after 8 weeks subcutis in rat (×250). Note greater adhesion than cohesion.

produces glasses with a wide range of surface activity and resistance to demineralization (*42*). Variations in Na_2O/CaO ratios and SiO_2/(Na_2O + CaO) ratios in the 45S5 Bioglass formula, with a constant 6.0 percent by weight P_2O_5, result in a compositional field (region A in Fig. 1) where bone bonding occurs in the rat within 30 days. Compositions in region B do not bond, those in region C resorb, and those in region D cannot be formed into glasses.

Gross and Strunz (*24*) have shown that a range of low-alkali (0 to 5 percent by weight) surface-active glass-ceramics (Ceravital) bond to bone; however, addition of Al_2O_3, Ta_2O_5, TiO_2, Sb_2O_3, or ZrO_2 tends to inhibit bone-bonding mechanisms at the interface. Addition of as little as 3 percent Al_2O_3 in the 45S5 formula seriously degrades bondability of the material (*23*). A series of commercial vitreous enamels showed similarly negative results (*41*). Often a seam of unmineralized osteoid tissue was present, indicating release of substances that impede steps of the mineralization process. Morphometric measurements also showed persistence of chondroid on the implant interface for these compositions, suggesting inhibition of cellular differentiation into osteoblasts. The inhibited cells did not switch from the production of metachromic ground substance and type II collagen to the production of type I collagen and organelles for mineralization. In contrast, bioactive glasses and glass-ceramics, without such inhibiting elements, do show bonding with osteoblasts, matrix vesicles, and normal mineralization at the bonding interface. Gross and Strunz (*41*) showed that compositions which bond release monophosphates at their interface, whereas nonbonding compositions release tri-, tetra-, or polyphosphates. The altered monophosphate and polyphosphate concentrations may influence local alkaline phosphatase concentrations and the formation or function of matrix vesicles. Glasses with a high P_2O_5 content also show adverse effects on tissue reactions and bonding; which may also be related to the interfacial monophosphate and polyphosphates developed in vivo (41).

These and other studies have identified a number of the surface chemical features essential for a stable interfacial bond with bone. In hard tissues, the central issue seems to be the relative competition between fibrogenesis and osteogenesis at the interface. Many factors, such as movement or infection, favor proliferation of the less highly differentiated fibroblasts with eventual capsule formation, whereas very specific conditions must be satisfied for osteogenesis to occur (3). Of course, this is exactly the situation in the repair of natural tissues. Recent studies (43) have indicated that attachment of osteogenic stem cells to a precursor acellular structure on an implant is necessary for differentiation to proceed and mineralizable bone matrix to be generated. The tissue culture findings, comparing CHO, NIL, and HeLa fibroblast cell lines with primary bone cell cultures, showed a factor of 3 to 10 decrease in rates of cell spreading and mitosis on the bioactive surfaces for cells with small fibronectin concentrations. In contrast, the time of spreading and division of osteoblast-like primary culture cells was equivalent on active and inert surfaces. These data indicate that a population of fibroblasts, such as that present at an implant interface in the first week of healing, will respond much more slowly on the bioactive surface, allowing attachment and

proliferation of osteoblasts to be favored.

A quantitative comparison of the relative percentages of soft tissue, osteoid, chondroid, and bone contact, or bone connection [based on figure 14.35 in (3)] shows an extensive amount of soft tissue in contact with "bioinert" implants after 2 weeks (Fig. 4). This soft-tissue capsule remains during the lifetime of the implant and is responsible for eventual movement and failure of the interface. In contrast, by the end of 2 weeks surface-active implants show substantially less soft tissue, with the quantity varying with relative surface activity of the implant. The relative proportions of bonded bone, chondroid, and osteoid tissue at the implant interface are dependent on composition.

Evidence for the bonding process for HA implants was included in a review by Jarcho (13). He pointed to the resemblance between the mechanisms of bonding of this material and the Bioglass range of materials. As acellular bone matrix from differentiating osteoblasts at the surface appears, there is a narrow amorphous electron-dense band 3 to 5 μm wide. Between this area and the cells, collagen bundles are seen. Bone mineral crystals have been identified in this "amorphous area." This is the earliest observation; as the site matures the bonding zone shrinks to a depth of only 0.05 to 0.2 μm [which agrees with Denissen's observations (4)]. The eventual picture is of normal bone attached through a very thin bonding layer to bulk implant. A consequence of this thin bonding zone is a very high gradient in elastic modulus at the bonding interface (3), which is a major difference between surface-active apatites and surface-active glasses.

Soft Tissue Bonding to Surface-Active Materials

The role of collagen in the bonding of surface-active materials to bone has been clearly demonstrated, and a similar effect may be demonstrated with Bioglass in connective tissue if the processing problems associated with relative movement at the interface can be solved. After decalcification of the glass in situ, 8 weeks after subcutaneous implantation in rats, histological sections showed collagen fibers adherent to the remnants of the glass where the fibers in the capsule were pulled apart before the bond between glass and fiber was broken (Fig. 5). A similar effect was observed when fibers of Bioglass were implanted in muscle and then pulled out of the tissue bed (44). Although adhesion between a bioactive solid material and soft tissue is not likely to be useful in load-bearing situations or where there is other mechanical stress, there are applications where such a bond is desirable—for example, when surface-active glasses are used for replacement of the ossicular chain and in transmucosal applications. For such implants the bond that attaches the prosthesis to the two remaining ossicles or to the gingiva will be a soft-tissue bond. This interface resembles the natural interface in both form and function.

Conclusions

There is now a wide range of surface-active implants made from glasses, glass-ceramics, ceramics, and composites. All of them develop a bond with tissues that prevents motion at the interface. The implants are used in dental, maxillofacial, otolaryngological, and orthopedic surgery, although their use as load-bearing devices will require improvements in strength and fatigue resistance. The rate of bonding and the strength and stability of the bond vary with the composition and microstructure of the bioactive material. The mechanism of bonding generally involves a bioactive acellular layer rich in calcium phosphate, mucopolysaccharides, and glycoproteins, which provides an acceptable environment for collagen and bone mineral deposition. The biologically active surfaces of these materials uniquely influence the behavior of different cell types, and an understanding of the mechanisms involved has broad implications for the life sciences as well as for the surgical repair of the musculoskeletal system.

References and Notes

1. L. L. Hench, *Science* **208**, 826 (1980).
2. J. Boretos and M. Eden, Eds., *Contemporary Biomaterials, Material and Host Response, Clinical Applications, New Technology, and Legal Aspects* (Noyes Press, Park Ridge, N.J., 1984).
3. L. L. Hench and E. C. Ethridge, *Biomaterials—An Interfacial Approach* (Academic Press, New York, 1982).
4. H. Denissen, thesis, Free University of Amsterdam (1979).
5. C. de Putter, K. de Groot, P. A. E. Sillevis Smitt, in *Clinical Applications of Biomaterials*, A. J. C. Lee, T. Albrektsson, P.-I. Brånemark, Eds. (Wiley, New York, 1982), p. 237.
6. M. Ogiso *et al.*, *J. Dent. Res.* **60A**, 419 (1981).
7. C. de Putter, K. de Groot, P. A. E. Sillevis Smitt, J. M. Van der Zel, *Trans. Soc. Biomater.* **6**, 27 (1983).
8. M. Jarcho *et al.*, *J. Mater. Sci.* **11**, 2027 (1976).
9. J. J. Grote, Ed., *Biomaterials in Otology* (Nijhoff, The Hague, 1984).
10. J. M. Kent, J. H. Quinn, M. F. Zide, M. Jarcho, *Abstracts of the Second Southern Biomedical Engineering Conference* (Southwest Research Institute, San Antonio, Texas, 1983), p. 27.
11. A. N. Cranin, in *Biomaterials '84, Transactions, Second World Congress on Biomaterials*, J. M. Anderson, Ed. (Society for Biomaterials, Washington, D.C., 1984), vol. 7, p. 324.
12. P. Ducheyne *et al.*, *J. Biomed. Mater. Res.* **14**, 225 (1980).
13. M. Jarcho, *Clin. Orthop. Relat. Res.* **157**, 259 (1981).
14. Bioglass is a registered trademark of the University of Florida.
14a. G. E. Merwin, J. Wilson, L. L. Hench, in (9), pp. 220–229.

15. Vitallium is a registered trademark of Howmedica, Inc.
16. P. Ducheyne and L. L. Hench, in *Biomaterials '82* (Advances in Biomaterials Series), G. D. Winter, D. Gibbons, H. Plenk, Eds. (Wiley, New York, 1982), p. 21.
17. J. R. Smith, *Am. J. Orthod.* **76**, 618 (1979).
18. N. Inoue *et al.*, *J. Jpn. Orthod. Soc.* **40** (No. 3), 291 (1981).
19. D. L. Mitchel and J. D. West, *Am. J. Orthod.* **68**, 404 (1975).
20. S. F. Paige, A. E. Clark, R. Costa, G. J. King, J. M. Waldron, *J. Dent. Res.* **59A**, 445 (1980).
21. P. Turley, P. A. Shapiro, B. C. Moffett, *Arch. Oral Biol.* **25**, 259 (1980).
22. J. B. Grey, M. E. Steen, G. J. King, A. E. Clark, *Am. J. Orthod.* **83** (No. 4), 311 (1983).
23. R. L. Folger, C. S. Kucheria, R. E. Wells, G. E. Gardiner, in *Biomaterials '84, Transactions, Second World Congress on Biomaterials*, J. M. Anderson, Ed. (Society for Biomaterials, Washington, D.C., 1984), vol. 7, p. 352.
24. U. M. Gross and V. Strunz, in *Clinical Applications of Biomaterials*, A. J. C. Lee, T. Albrektsson, P.-I. Brånemark, Eds. (Wiley, New York, 1982), p. 237.
25. U. M. Gross, J. Brandes, V. Strunz, I. Bab, J. Sela, *J. Biomed. Mater. Res.* **15**, 291 (1981).
26. Ceravital is a trademark of E. Leitz Wetzlar, West Germany.
27. U. M. Gross and V. Strunz, *J. Biomed. Mater. Res.* **2**, 46 (1978).
28. M. Bunte and V. Strunz, *J. Maxillofacial Surg.* **5**, 303 (1977).
29. R. Reck and J. Helms, in (9), pp. 230–241.
30. G. Babighian, in (9), pp. 242–247.
31. W. Holland, K. Naumann, W. Vogel, J. Gummel, *Wiss. Z. Freidrich Schiller Univ. Jena Math. Naturwiss. Reihe* **32**, 571 (1983).
32. T. Kokubo *et al.*, in *Biomaterials '84, Transactions, Second World Congress on Biomaterials*, J. M. Anderson, Ed. (Society for Biomaterials, Washington, D.C., 1984), vol. 7, p. 351.
33. T. Yamamuro, private communication.
34. F. Hahn, V. Strunz, J. Boese-Landgraf, *Adv. Biomater.* **4**, 95 (1982).
35. L. L. Hench, R. J. Splinter, W. C. Allen, T. K. Greenlee, *J. Biomed. Mater. Res. Symp.* **2**, 117 (1971).
36. B. A. Blencke, H. Bromer, K. K. Deutscher, *J. Biomed. Mater. Res.* **12**, 307 (1978).
37. T. D. Driskell, C. R. Hassler, V. J. Tennery, L. G. McCoy, W. J. Clarke, paper presented at the IADR (International Association for Dental Research) meeting, Chicago, 1973.
38. H. W. Denissen, B. V. Rjeda, K. de Groot, *J. Biomed. Mater. Res. Symp.* **2**, 188 (1978).
39. L. L. Hench, in *Fundamental Aspects of Biocompatibility*, D. F. Williams, Ed. (CRC Press, Boca Raton, Fla., 1981), chapter 4.
40. L. L. Hench and A. E. Clark, in *Biocompatibility of Orthopaedic Implants*, D. F. Williams, Ed. (CRC Press, Boca Raton, Fla., 1982), vol. 2, p. 129.
41. U. Gross and V. Strunz, *J. Biomed. Mater. Res.*, in press.
42. D. Spilman, J. Wilson, L. L. Hench, in *Biomaterials '84, Transactions, Second World Congress on Biomaterials*, J. M. Anderson, Ed. (Society for Biomaterials, Washington, D.C., 1984), vol. 7, p. 287.
43. T. L. Seitz, K. D. Noonan, L. L. Hench, N. E. Noonan, *J. Biomed. Mater. Res.* **16**, 195 (1982).
44. J. Wilson, G. H. Pigott, F. J. Schoen, L. L. Hench, *ibid.* **15**, 805 (1981).

501

Conducting organic materials have attracted the interest of scientists for about 20 years. The past 10 years have seen the level of activity grow rapidly as new and unanticipated phenomena were observed which were not found in nonconducting organic materials and were found only rarely in inorganic materials. The prediction that superconductivity (the complete absence of electrical resistance) might occur in organic solids at a technologically useful temperature (for instance, room temperature) also provided a stimulus for this field. So far, this prediction has not been fulfilled, although some organic materials have been found to be superconductors at very low temperatures ($T < 5$ K). The field is highly interdisciplinary involving chemists, materials scientists, and theoretical and experimental physicists, whose interaction has led to much new chemistry and physics.

The bulk of the research has been on three broad classes of materials: the crystalline charge transfer complexes or salts, polymers, and graphite and its intercalation compounds. The charge transfer (CT) salts and the conducting polymers have many features in common and therefore we shall limit our discussion to these two classes of materials. For example, they both exhibit anisotropic, quasi–one-dimensional properties; that is, the conductivity is much greater in one direction than in others. In conducting polymers the conductivity is greatest along the chain direction, whereas in the CT salts the anisotropy is determined by the stacking of the molecules in the crystal structure. Another common feature of these materials is that in the conducting state they are ionic. In the case of the polymers charge is transferred between the polymer chain and a

34

Conducting Organic Materials

R.L. Greene
and G.B. Street

Science 226, 651–656 (9 November 1984)

dopant, and in the case of the CT salts charge is transferred between the donor and acceptor components of the complex. In spite of the similarities between these two classes of materials, most of their novel properties are different and therefore we shall discuss them separately in this article.

It would be impossible to cover all the significant work that has been done on conducting organic materials in the space available for this review. Many other reviews and several books have been devoted to this topic. Some of the most recent reviews and conference proceedings are cited in (*1–6*), and an exhaustive bibliography of all the known reviews up to 1982 is given in (*3*). It is our intent here to give a brief introduction to the properties of the two main classes of conducting organic solids and to give an impression of some of the current active areas of research.

Charge Transfer Salts

The first stable organic materials which could conduct electricity like a metal were discovered in the early 1960's. They are materials in which charge is transferred between donor (cation) molecules (such as TTF) and acceptor (anion) molecules (such as TCNQ) to form a charge transfer compound (TTF-TCNQ). A few examples of donor and acceptor molecules which produce organic "metals" are shown in Fig. 1. A description of all the molecules which form the important conducting CT salts is given in (*2*). The electrical conduction arises because in the crystalline compound the large planar molecules can stack on top of each other like pancakes, which allows the transferred charge to move easily along the separate donor and acceptor stacks. In the language of solid-state physics, a conduction band is formed from the overlap of π orbitals on neighboring molecules along the stack. The partial filling of this band with electrons by charge transfer from donor to acceptor leads to the metallic levels of conductivity observed. The CT salts are quasi–one-dimensional metals because the orbital overlap is greater along the stacks of molecules and less between them, which makes the conductivity greater along the stack than in other directions. The best organic conductors have a conductivity of order 2000 ohm^{-1} cm^{-1} at room temperature, or about three orders of magnitude smaller than that of copper. Below room temperature, the metallic conduction is usually interrupted by a transition to an insulating state; in other words, a metal-insulator phase transition occurs. One of the major results of research on CT salts has been the elucidation of the nature of this metal-insulator phase transition as well as the development of chemical and physical techniques to suppress the transition in order to retain the metallic state to near absolute zero in temperature. An elementary review of this work which includes a historical perspective is given in (*1*).

Metal-insulator phase transitions. To understand the nature of the metal-insulator transition, we need to discuss some properties of one-dimensional metals. Physics in one dimension is different from that in higher dimensions and has been studied theoretically for many years. However, the synthesis of real materials (such as the CT salts) with quasi–one-dimensional properties has led to further understanding of one-dimensional metals. Consider a chain of

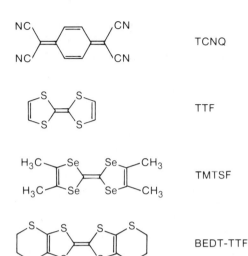

TCNQ

TTF

TMTSF

BEDT-TTF

Fig. 1. A few important donor and acceptor molecules are tetracyanoquinodimethane (TCNQ), tetrathiafulvalene (TTF), tetramethyltetraselenafulvalene (TMTSF), and bis(ethylenedithiolo)tetrathiafulvalene (BEDT-TTF).

molecules, separated by a distance "a," with one electron in the highest unoccupied orbital (Fig. 2a). If separation "a" is small enough to allow overlap of the molecular wave functions, the electrons can move freely along the stack and a metallic state is obtained. However, this one-dimensional metallic state is unstable against a variety of distortions which lead to metal-insulator phase transitions. These are shown in Fig. 2, b to d. When the electrons interact with the lattice vibrations it can become energetically favorable for the chain of molecules to distort so that the molecules dimerize, as shown in Fig. 2b. This gives rise to two electrons in the highest occupied orbital of each dimer and consequently to completely filled bands and an insulating state. This was first shown theoretically by R. Peierls and has since become known as a Peierls distortion. Sometimes the Coulombic repulsion between

electrons on the chain of molecules is greater than their interaction with the lattice. This repulsion causes the electrons to be confined to their individual molecules with their spins ordered in the alternating fashion shown in Fig. 2c, and one finds an insulating magnetic state. This state is sometimes called a spin density wave (SDW) state. Finally, the presence of an ordered array of anions with a periodicity "2a," shown in Fig. 2d, will cause a doubling of the original lattice (analogous to the situation represented by Fig. 2b), and this too can lead to an insulating state. In a strictly one-dimensional system, these phase transitions can occur only at absolute zero in temperature. However, the CT salts are three-dimensional crystals rather than single-chain filaments and therefore there is some electronic coupling between neighboring chains (hence the label quasi–one-dimensional materials). In this case, any phase transition will occur at a temperature (T_c) given roughly by the average of two energies, $T_\parallel$ and $T_\perp$, where $T_\perp$ characterizes the strength of the interchain interactions and $T_\parallel$ characterizes the intrachain interactions driving the phase transition. In the temperature region between $T_\parallel$ and T_c one expects to observe precursor (or fluctuation) effects associated with the transition.

Recent results. The synthesis of the CT salt TTF-TCNQ (Fig. 1) in the early 1970's led to a dramatic increase in the research on organic metals. This CT salt was much more conducting than the previously known CT salts and was one of the first clear examples of a Peierls transition. Attempts to eliminate this Peierls transition, which occurred at ~54 K, and thereby maintain the metallic state to lower temperatures, were unsuccessful.

During the later 1970's many analogous conducting compounds were found. A study of their properties and those of TTF-TCNQ led to a deeper understanding of the physical processes in quasi–one-dimensional materials (2–5).

In 1980, Bechgaard et al. (7) synthesized compounds of the form $(TMTSF)_2X$, where X is an inorganic anion such as PF_6^-, AsF_6^-, FSO_3^-, ReO_4^-, or ClO_4^-. These materials have given several significant new results and have accounted for 90 percent of the research activity on CT compounds since 1980. The major discoveries were the observation of superconductivity at low temperatures, the observation of SDW and anion-ordered insulating states, and the observation of a unique phase transition upon application of a magnetic field. Thus the $(TMTSF)_2X$ compounds and their derivatives exhibit a variety of physical properties that are not found in other organic metals. These materials have metal-insulator transitions between 10 and 200 K but under hydrostatic pressure become superconductors. A notable exception is the ClO_4^- salt, which is a superconductor ($T_c \sim 1.2$ K) at ambient pressure. The isostructural sulfur compounds $(TMTTF)_2X$ have lower conductivity and do not exhibit superconductivity. In contrast, the sulfur-containing molecule BEDT-TTF (Fig. 1) in the stoichiometry (BEDT-TTF)$_2X$ does show superconductivity (8) for X = ReO_4^- and X = I_3^-. The latter compound has the highest transition temperature ($T_c \sim 2.5$ K) observed at ambient pressure in an organic solid. Salts with different stoichiometries or crystal structures have metal-insulator transitions but no superconductivity. Magnetism (Fig. 2c) is found in salts with symmetrical anions (PF_6^-, AsF_6^-, and so on), while anion ordering (Fig. 2d) is found with nonsymmetrical anions (such as ReO_4^-). Why the magnetic (SDW) state is favored over the Peierls state in these materials is not well understood. The application of modest levels of hydrostatic pressure eliminates the metal-insulator transition and leads to a metallic state with superconductivity appearing at a low temperature. A typical pressure phase diagram is shown in Fig. 3. An understanding of exactly how pressure suppresses the metal-insulator transition is also lacking at present. The simplest explanation is that pressure squeezes the molecules closer together, increasing the interchain interaction to the point at which the metallic state is

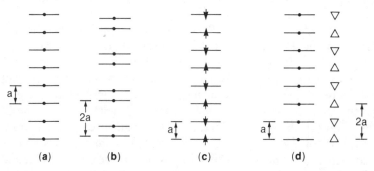

Fig. 2. Several possible configurations of a linear chain of molecules with one electron (●) in the highest orbital: (a) metal with uniform spacing of lattice constant "a"; (b) insulator with dimerization caused by Peierls transition; (c) SDW insulator with spin (↑) periodicity caused by Coulomb interactions; (d) insulator with periodicity caused by ordering of nonsymmetric anions (△).

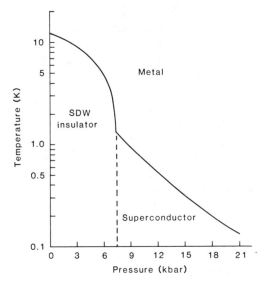

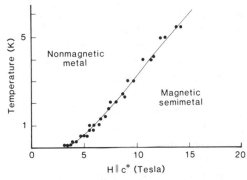

Fig. 3 (left). Pressure-temperature phase diagram of typical $(TMTSF)_2X$ salt showing the boundaries between the metallic, superconducting, and insulating states.   Fig. 4 (right). Phase diagram of $(TMTSF)_2ClO_4$ at ambient pressure for magnetic field (H) applied perpendicular to conducting plane.

more energetically favorable and the system becomes an anisotropic three-dimensional metal (9). Since superconductivity is a common property of three-dimensional metals, its occurrence in the $(TMTSF)_2X$ salts would then not be surprising. A different idea has been proposed by researchers in France (3). In their view, the $(TMTSF)_2X$ salts are highly one-dimensional, have a high $T_{||}$, and exhibit superconducting fluctuations at temperatures up to 40 K. Since the measured T_c for superconductivity is no more than 2.5 K, this proposal would suggest that by increasing the interchain coupling ($T_\perp$) one could obtain organic superconductors with transition temperatures comparable to or higher than those of the best superconductors known to date (Nb_3Ge has the highest known T_c, about 23 K). Unfortunately, almost all the experimental and theoretical evidence to date suggests that the organic superconductors are two- or three-dimensional rather than one-dimensional in their properties (9).

The third unusual property of the $(TMTSF)_2X$ salts is the occurrence of a magnetic field–induced transition at low temperature. The phase diagram for this transition (as determined by many different experiments) is shown in Fig. 4. The system, in this case $(TMTSF)_2ClO_4$, goes from a metallic, nonmagnetic state to a semimetallic, magnetic state as the magnetic field is increased. To our knowledge, there is no other solid (organic or inorganic) which has such a phase transition. This transition was first discovered in the PF_6^- salt in 1981 (10) and its possible origin is just now being clarified. The theory assumes that in the low-temperature metallic state the motion of the electrons is two-dimensional, that is, the electrons can move easily along the molecular stacks and can hop readily from stack to stack in one plane. When a magnetic field is applied along the third direction, the motion along the stack is not affected but the stack-to-stack motion is reduced until finally the electron is confined to a single stack. At this point, the electron motion is truly one-dimensional and a magnetic (SDW) instability can occur. This explains the essential physics of the transition but leaves out many subtle points, which cannot be discussed here but can be discovered by the interested reader (11). At magnetic fields and temperatures above the transition, unusual electronic properties are found, very similar to the quantum Hall effect observed in two-dimensional metals (12). The study of this magnetic field–induced state is now one of the most active areas of research in CT materials.

Future issues. Many conducting CT salts have been synthesized since the early 1960's [see (2) for a summary] but only a few of these have had properties sufficiently different to stimulate a detailed study. Through research on some of these, such as TTF-TCNQ, $(TMTSF)_2X$, and now the $(BEDT-TTF)_2X$ compounds, much has been learned about the properties of organic metals. However, there are still many gaps in our knowledge. It is not possible, in general, to predict or design the physical (or chemical) properties of a new CT crystal. The relation between crystal structure, stoichiometry, and molecular properties and the actual physical properties of the crystal are at best understood in a phenomenological way. The discovery of new materials with interesting properties is empirical in nature and depends more on the intuition of the materials scientist than on hard facts. Many other important issues are unresolved and will be the focus of future research. These include determining the interactions responsible for the superconductivity; attempting to raise the transition temperature to the level of good inorganic superconductors; understanding why the SDW is favored over the Peierls transition in the $(TMTSF)_2X$ salts; understanding the influence of the Coulomb interaction on the properties of the CT salts; determining which concepts of one-dimensional physics are relevant to the observed physical properties; and understanding in more detail the state of these materials at high magnetic fields and low temperatures, that is, above the phase transition shown in Fig. 4. Considerable effort will be required to make progress on resolving these still quite fundamental questions.

Conducting Polymers

In contrast to the CT salts, the widespread fascination with conducting polymers is clearly motivated by their perceived technological potential (13–15). However, as in the case of the CT salts, these materials have also presented a strong scientific challenge. This challenge has attracted a diverse community of chemists, physicists, and materials scientists who have not previously been actively interested in polymer science. The resulting involvements and contributions have gone a long way toward contradicting the view of some scientists that "there is no science in polymers." This prejudice has been nurtured by the difficulty in defining the structure of polymers with the precision associated with more crystalline materials. In reviewing the current status of the field of conducting polymers, we will emphasize the progress which has been made despite the difficulties in understanding the chemistry and physics of these materials. This progress has been achieved both in the understanding of the physics of electrical conductivity and in the control over the physical properties of the polymers. Because of their excellent physical and mechanical properties, plastics have pervaded most areas of our lives, replacing many other materials in a multitude of applications. However, because typical plastics are electrically insulating, they have not yet made an impact on the electronics industry except as dielectrics. In fact, electrically conducting or semiconducting applications remain the last great challenge for polymeric materials, although organic polymers, it should be pointed out, have been important as photoconductors for many years (13).

Poly(sulfur nitride): An intrinsically conducting polymer. A key development in the evolution of conducting polymers as we know them today was the discovery (16) that the inorganic polymer poly(sulfur nitride), $(SN)_x$, is a metal (that is, its electrical conductivity decreases with increasing temperature) and at low temperature becomes superconducting. This discovery was important because it provided an existence theorem for highly conducting polymers and stimulated the enormous amount of work necessary to synthesize other polymeric conductors. The metallic conductivity observed for $(SN)_x$ crystals is an intrinsic property of the crystals, arising from the presence of one unpaired electron associated with each sulfur-nitrogen unit; this unpaired electron can move under the effect of an applied field, carrying charge and giving rise to electrical conductivity.

Though the electronic properties of $(SN)_x$ were very exciting, its other physical properties did not lend themselves to commercialization (for instance, it was explosive and synthesized from explosive intermediates). However, $(SN)_x$ did provide the field with further insight which proved significant in the development of an entirely new class of conducting polymers. It was observed that the conductivity of $(SN)_x$ could be increased by reaction with bromide (15) (and similar oxidizing agents) to give an even more conducting polymer where the conducting entity was no longer a neutral polymer but a polymeric cation, charge neutrality being maintained by incorporating Br_3^- anions (formed by reduction of the Br_2) into the structure, as shown below.

$$(SN)_x + 3/2\, yBr_2 \rightarrow (SN)_x^{y+} + yBr_3^-$$

This has provided a model for all of the nonintrinsically conducting polymers which have been discovered subsequently.

Polyacetylene and the nonintrinsically conducting polymers. With the exception of $(SN)_x$, most polymeric systems are closed-shell systems with all of their electrons paired. Such an electronic configuration leads, as expected, to dielectric polymers. However, as was first demonstrated in the case of polyacetylene [often referred to as $(CH)_x$], an insulating polymer can be made to conduct by partial oxidation, which removes electrons from the system and facilitates conductivity (17). This process is referred to as doping by analogy with the doping of inorganic semiconductors. As we shall discuss later, this is a rather misleading analogy. Essentially, we are chemically converting the insulating neutral polymer into an ionic compound consisting of a polymeric cation and a counterion which is the reduced form of the oxidizing agent (Br_3^- in the case of bromine). In this sense, the doped polymers are similar to the CT salts discussed earlier.

In selecting potentially conducting polymer systems, it is important to choose a polymer which can be readily oxidized and which forms a cation rather than undergoing other chemistry. This accounts in part for the choice of π-bonded unsaturated polymers like $(CH)_x$. The π electrons can be relatively easily removed to give a polymeric cation without breaking the σ bonds which are primarily responsible for holding the polymer together. In contrast, with saturated polyethylene oxidation would remove σ electrons and lead to bond breaking and chemical decomposition. Even with π-bonded polymers, it is important to choose mild oxidizing agents which do not cause other disruptive chemistry such as addition across the carbon-carbon double bond. The latter reaction tends to be a problem when oxidizing $(CH)_x$ with Br_2.

Although $(CH)_x$ is clearly the most extensively studied and understood of the conducting organic polymers (18), this basic principle has been applied with success to an increasingly large number of other polymer systems. Conductivity is usually achieved by oxidation, but several polymer systems can also be reduced to give conducting polymeric anions. (Again, charge neutrality is preserved by incorporation of the cation derived from the reducing agent, for example, Li^+ in the case of reduction by Li.) Thus, both n-type (conduction by negative charge carriers) and p-type (conduction by positive charge carriers) materials can be prepared.

$$(CH)_x + 3/2\ yBr_2 \rightarrow (CH)_x^{y+} + yBr_3^-$$

$$(CH)_x + yLi \rightarrow (CH)_x^{y-} + yLi^+$$

In some cases, oxidation or reduction of the polymer can be achieved electro-chemically by subjecting the neutral polymer to the appropriate oxidizing or reducing voltage in an electrochemical cell (19). The charge on the polymer is then neutralized by a counterion from the electrolyte solution. Figure 5 shows the principal polymers which are or can be made to be electrically conducting. An important subgroup of nonintrinsically conducting polymers consists of those prepared by the electrochemical oxidation of some monomer which polymerizes at the anode of an electrochemical cell. These include polypyrrole, polythiophene, polyazulene, and polymers derived from other aromatic heterocyclic precursors (20).

Advantages of nonintrinsically conducting polymers. There are several at-

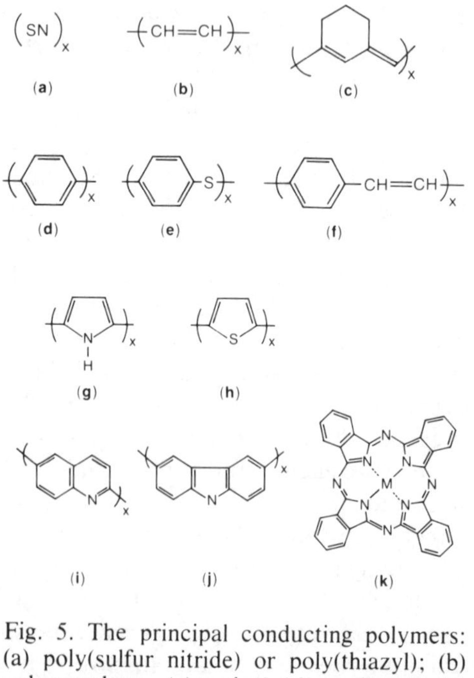

Fig. 5. The principal conducting polymers: (a) poly(sulfur nitride) or poly(thiazyl); (b) polyacetylene; (c) poly(1,6-heptadiene); (d) poly(p-phenylene); (e) poly(p-phenylene sulfide); (f) poly(p-phenylene vinylene); (g) polypyrrole; (h) polythiophene; (i) polyquinoline; (j) polycarbazole; (k) metallo-phthalocyanine.

510

tractive features of nonintrinsic conducting polymers. Their conductivity can be controlled in many cases by varying the extent of doping so that one system can span the whole conductivity range from insulator to semiconductor to metal. In some systems, doping can be achieved with more precision with electrochemical doping techniques, which also enable the conductivity of these systems to be switched between the conducting and insulating states by application of the appropriate voltage. This switching is usually accompanied by a color change, leading to efforts to develop electrochromic displays with conducting polymers (21). This electroactivity of conducting polymers is also the basis of the much more significant efforts which have gone into the development of batteries having conducting polymers as one or both of the electrodes. In the case of a $(CH)_x$ battery, where the polymer is used for both electrodes (22), the discharge reaction involves, for example, the reduction of polyacetylene perchlorate to neutral polyacetylene at one electrode and at the other electrode the oxidation of the lithium salt of the reduced form of polyacetylene to give neutral polyacetylene. In battery applications, the fibrous nature of $(CH)_x$ is advantageous in giving high current densities.

Advantage can also be taken of the counterion present in the nonintrinsically conducting systems to modify the physical properties of the parent polymer. For instance, the mechanical properties of polypyrrole are significantly improved by the incorporation of toluenesulfonate as the anion (23). Specific anions have also been incorporated into polypyrrole to prepare catalysts (24). The technological potential has led to considerable efforts to improve the me-

chanical properties and stability of conducting polymer systems. Currently, this has been most successful for polymers derived from aromatic heterocycles, especially the polyphthalocyanines (25, 26) and polypyrroles. Polyphthalocyanines, in particular, have been used as fillers in the high-strength polymer Kevlar to give conducting fiber composites of encouraging mechanical strength and stability (25). The polypyrroles can be made into tough, flexible plastics either by the appropriate choice of anion (as mentioned above), by forming graft copolymers, or by electrochemically depositing the polymer in the matrix of a swellable polymer such as polyvinyl alcohol or polyvinyl chloride (27).

The intense efforts to understand the chemistry and physics of conducting polymers have been hampered by the insolubility and limited crystallinity of these materials, characteristics which have made traditional methods of polymer characterization difficult. Progress has begun to be made toward solution-processible conducting polymers. Conducting films of poly(p-phenylene sulfide) (PPS) hexafluoroarsenate have been cast from an AsF_3-AsF_5 solvent (28). Several polymers which are soluble in their insulating state, such as PPS or polycarbazole (29), become insoluble when oxidized to a conducting state. Perhaps the most significant progress along these lines has been the Feast synthesis (30) of polyacetylene outlined in Fig. 6. This synthesis involves the formation of a soluble prepolymer from which polyacetylene can be prepared by heating. By applying a stress to the prepolymer film during the thermal conversion, it has been possible to obtain films of $(CH)_x$ which are extremely highly crystalline and close to theoretical densi-

Fig. 6 (left). Feast synthesis of polyacetylene from a soluble prepolymer: I, soluble prepolymer; II, polyacetylene; III, leaving group.   Fig. 7 (right). Comparison of acceptor doping of silicon and polypyrrole showing polaron and bipolaron formation.

ty, with the $(CH)_x$ chain oriented parallel to the direction of stress (*31*). Previous films of $(CH)_x$ were much less crystalline; thus these highly ordered films will be of great importance in resolving some of the problems associated with the char-

acterization of polyacetylene. Highly oriented films of poly(*p*-phenylene vinylene) have also been prepared from stressed prepolymers (*32*).

Carrier generation in conducting polymers: Polarons, bipolarons, and soli-

tons. It was initially assumed that the generation of charge carriers in nonintrinsically conducting polymers was analogous to that in classical inorganic semiconductors such as silicon or germanium. However, this assumption was challenged by the discovery that doped conducting polymers could display conductivity which was not associated with unpaired electron spins (*33*).

The generation of charge carriers in inorganic semiconductors is achieved by doping, for instance, silicon with part-per-million quantities of elements such as boron or lithium which are appropriate in size to fit into the silicon lattice either substitutionally or interstitially (Fig. 7). This process involves removing a neutral atom of silicon and substituting for it a neutral atom of, for example, boron which fits on the lattice site vacated by the silicon atom (*34*). Whereas the silicon atom had four valence electrons, the boron has only three, leading to the appearance of empty energy levels associated with the boron atoms slightly above the top of the silicon valence band. Electrons can be thermally transferred from the silicon to the boron levels, producing holes in the silicon valence band and giving rise to unpaired electrons, which can be detected by electron spin resonance (ESR) techniques.

The situation is very different when it comes to generation of charge carriers by doping of conjugated polymers. Doping of polymers is more appropriately seen as chemical modification because it involves incorporating into the polymer several percent of an oxidizing agent (acceptor) or a reducing agent (donor). The dopant perturbs the polymer extensively not only because of its large physical size, which does not allow it to fit substitutionally into the polymer lattice,

but also because of the extensive charge transfer which takes place between the polymer chain and the dopant, causing both to become ionic and leading to changes in the geometry of the chain. In the case of acceptor doping, such as iodine doping, the polymer becomes a cation and the iodine becomes an anion, I_3^-. The levels corresponding to the acceptor in the doped polymer, that is, the I_3^-, are not in the gap but quite deep below the top of the polymer valence band.

The conducting polymer chains, like organic molecules in general, adopt a different geometry in the ionized state than in the neutral state. This is a consequence of the fact that polymer lattices are much more readily distorted than, say, the stiffer three-dimensional silicon lattice. Thus, on doping the polymer, electronic energy levels associated with the geometry modifications are removed from the top of the valence band and the bottom of the conduction band and brought into the gap. This process widens the original gap between the conduction and valence bands but does not lead to holes in the valence band. Thus, it is clear that such holes are not the source of charge carriers. In fact, the source of the charge carriers is believed to be related to the charge transfer–induced geometric modifications of the polymer chains, which lead to the formation of radical cations (polarons) and dications (bipolarons) (*35–37*).

If, for example, the polypyrrole chain is oxidized by removal of an electron, the lattice distorts locally around the resulting positive charge which appears on the polymer chain, forming a polaron. This lattice distortion (shown in Fig. 7) involves switching of the positions of the carbon-carbon double and single bonds and requires elastic energy since the resulting structure is of higher energy. However, calculations (38) have shown that this is more than compensated for by the lower ionization energy of the distorted form of the chain. If a second electron is now removed from the same chain, calculations show that it is energetically favorable to remove it from the polaron to create a bipolaron (see Fig. 7) rather than a second polaron. Although doubly charged, the bipolaron is spinless and can be considered analogous to a Cooper pair in the Bardeen-Cooper-Schrieffer theory of superconductivity, which consists of a pair of electrons coupled together through a lattice vibration, that is, a phonon (39).

In the case of polypyrrole, these polarons and bipolarons extend over about four pyrrole rings and can become mobile (38). The presence of both these species has been detected in polypyrrole from ESR and optical experiments (40). Although bipolarons are stable relative to polarons, the latter can be seen at low and intermediate doping concentrations because the kinetics of their interconversion is slow (41). At low dopant concentrations, ESR measurements show that the number of spins increases with the conductivity, as would be expected if the charge carriers are polarons (40). However, at higher dopant concentrations, the ESR signal saturates and then decreases, consistent with the fact that polarons combine to form spinless bipolarons. No ESR signal is detected in the highly conducting state, indicating that mobile bipolarons are the charge carriers. This conductivity mechanism is very different from that typical of doped inorganic semiconductors or normal metals in that all the electronic bands are either completely filled or empty, a situation classically associated with the insulating state (38).

In the case of polymers such as trans-polyacetylene which have degenerate ground states (that is, two ground-state geometric structures of identical total energy), the bipolaron can dissociate into two isolated spinless charged species (Fig. 8), referred to as charged solitons. This charge separation does not cost energy and effectively lowers the Coulombic repulsion between the two charges. These charged solitons were, historically, the first postulated spinless carriers and were invoked to explain the spinless conductivity when it was first observed in trans-polyacetylene (33, 42). However, most conducting polymers have nondegenerate ground states; thus, the geometric distortion of the ground state produced on separating the bipolaron into two solitons requires energy. Because of this energy, the solitons are confined in pairs as bipolarons, and so the soliton should be viewed as a rather special case of the spinless carrier of importance for polymers with degenerate ground states. On the other hand, the

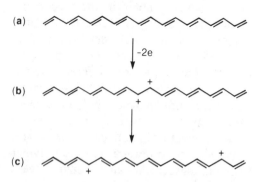

Fig. 8. Instability of bipolarons relative to charged solitons in trans-polyacetylene. (a) Removal of two electrons from polyacetylene to form (b) a bipolaron and dissociation of the bipolaron to form (c) two charged solitons.

polaron-bipolaron model is a very general model for conduction in all conjugated polymers with degenerate and nondegenerate ground states.

Conclusions

We have presented a brief overview of some of the activity in conducting organic materials. Obviously, some important results have been omitted, and we recommend to the reader the more detailed reviews given in the references. We have emphasized some common features of the two main classes of organic conductors: the charge transfer which leads to the ionic nature of both the conducting polymers and the charge transfer salts, and the quasi–one-dimensional nature of these materials which results in physical properties not found in inorganic isotropic materials. In the case of the polymers, we have discussed the role of polarons, bipolarons, and solitons. It is important to appreciate that most of the exciting phenomena found in these materials have not resulted from theoretical predictions but from the difficult, empirical efforts of synthetic chemistry. For the future, this is still the only viable approach, considering the minimal level of understanding of the correlations between chemical or crystalline structure and specific solid-state properties. Certainly, new materials will determine the scientific or technological future of this field. Perhaps too much has already been said of the potential commercial application of these conductors (13, 15, 43, 44). We shall not indulge in further speculation along these lines, but point out that so far the technological impact of these materials has been minimal. Though the chances of finding organic superconductors with transition temperatures greater than their inorganic counterparts seem remote, the future for scientific and technological progress must be viewed optimistically in the light of the unanticipated discoveries in this field during the past 10 years.

References and Notes

1. K. Bechgaard and D. Jerome, *Sci. Am.* **247**, 50 (July 1982).
2. M. Bryce and L. Murphy, *Nature (London)* **309**, 119 (1984).
3. D. Jerome and H. J. Schulz, *Adv. Phys.* **31**, 299 (1982), and references therein.
4. H. J. Keller, Ed., *Chemistry and Physics of One-Dimensional Metals* (Plenum, New York, 1977); J. T. Devresse, Ed., *Highly Conducting One-Dimensional Solids* (Plenum, New York, 1979).
5. J. S. Miller, Ed., *Extended Linear Chain Compounds* (Plenum, New York, 1981–1983), vols. 1–3.
6. Proceedings of the International Conference on Synthetic Metals, *J. Phys. (Paris) Colloq. C3* **44**, 1–1761 (1983).
7. K. Bechgaard, C. S. Jacobsen, K. Mortensen, H. J. Pedersen, N. Thorup, *Solid State Commun.* **33**, 1119 (1980).
8. T. H. Maugh II, *Science* **222**, 606 (1983); S. S. P. Parkin *et al.*, *Phys. Rev. Lett.* **50**, 270 (1983); E. B. Yagubskii *et al.*, *JETP Lett.* **30**, 12 (1984).
9. P. M. Chaikin, M. Y. Choi, R. L. Greene, *J. Phys. (Paris) Colloq. C3* **44**, 783 (1983).
10. J. F. Kwak, J. E. Schirber, E. M. Engler, R. L. Greene, *Phys. Rev. Lett.* **46**, 1296 (1981).
11. L. P. Gorkov and A. G. Lebed, *J. Phys. (Paris) Lett.* **45**, L433 (1984); P. M. Chaikin, in preparation.
12. P. M. Chaikin *et al.*, *Phys. Rev. Lett.* **51**, 2333 (1983); M. Ribault *et al.*, *J. Phys. (Paris) Lett.* **44**, L953 (1983).
13. C. B. Duke and H. W. Gibson, "Conductive polymers," in *Kirk-Othmer Encyclopedia of Chemical Technology* (Wiley, New York, ed. 3, 1982), vol. 18, p. 755.
14. G. Wegner, *Angew. Chem. Int. Ed. Engl.* **20**, 361 (1982).
15. W. D. Gill, T. C. Clarke, G. B. Street, *Appl. Phys. Commun.* **2**, 211 (1983).
16. M. M. Labes, P. Love, L. F. Nichols, *Chem. Rev.* **79**, 1 (1979).
17. C. K. Chiang, C. R. Fincher, Y. W. Park, A. J. Heeger, H. Shirakawa, E. J. Louis, S. C. Gau, A. G. MacDiarmid, *Phys. Rev. Lett.* **39**, 1098 (1977).
18. D. Baeriswyl, G. Harbeke, H. Keiss, W. Meyer, in *Electronic Properties of Polymers*, J. Mort and C. P. Fisher, Eds. (Wiley, New York, 1982), pp. 267–326.
19. A. G. MacDiarmid and A. J. Heeger, *Synth. Met.* **1**, 101 (1979).
20. F. Garnier and G. Tourillon, *J. Electroanal. Chem.* **148**, 299 (1983).

21. M. Gazard, J. C. Dubois, M. Champagne, F. Garnier, G. Tourillon, *J. Phys. (Paris) Colloq. C3* **44**, 537 (1983).
22. A. G. MacDiarmid, R. B. Kaner, R. J. Mammone, A. J. Heeger, *ibid.*, p. 543.
23. A. F. Diaz and B. Hall, *IBM J. Res. Dev.* **27**, 477 (1983).
24. R. A. Bull, F. R. Fan, A. J. Bard, *J. Electrochem. Soc.* **130**, 1636 (1983).
25. T. Inabe, J. F. Lomax, J. W. Lyding, C. R. Kannewurf, T. J. Marks, *Synth. Met.* **9**, 303 (1984).
26. M. Hanack, *Mol. Cryst. Liq. Cryst.* **105**, 133 (1984).
27. G. B. Street, S. E. Lindsey, A. I. Nazzal, K. J. Wynne, *ibid.*, in press.
28. J. E. Frommer, R. L. Elsenbaumer, R. R. Chance, in *Polymers in Electronics*, T. Davidson, Ed. (ACS Symposium Series No. 242, American Chemical Society, Washington, D.C., 1984), p. 447.
29. S. T. Wellinghoff, T. Kedrowski, S. Lenekhe, T. Tshida, *J. Phys. (Paris) Colloq. C3* **44**, 677 (1983).
30. J. H. Edwards, W. J. Feast, D. C. Bott, *Polymer* **25**, 395 (1984).
31. G. Leising, *Polym. Bull.* **11**, 401 (1984).
32. F. E. Karasz, J. Capistran, D. Gagnon, R. W. Lenz, *Mol. Cryst. Liq. Cryst.*, in press; I. Murase, T. Ohnishi, T. Noguchi, M. Hirooka, S. Murakami, *ibid.*, in press.
33. S. Ikehata, J. Kaufer, T. Woerner, A. Pron, M. A. Druy, A. Sivak, A. J. Heeger, A. G. MacDiarmid, *Phys. Rev. Lett.* **45**, 1123 (1980).
34. W. J. Moore, *Seven Solid States* (Benjamin, New York, 1967).
35. S. A. Brazovskii and N. Kirova, *JETP Lett.* **33**, 4 (1981).
36. A. R. Bishop, D. K. Campbell, K. Fesser, *Mol. Cryst. Liq. Cryst.* **77**, 253 (1981).
37. J. L. Brédas, R. R. Chance, R. Silbey, *ibid.*, p. 319; *Phys. Rev. B* **26**, 5843 (1982).
38. J. L. Brédas, J. C. Scott, K. Yakushi, G. B. Street, *Phys. Rev. B* **30**, 1023 (1984).
39. W. A. Little, *Sci. Am.* **214**, 21 (February 1965).
40. J. C. Scott, J. L. Brédas, K. Yakushi, P. Pfluger, G. B. Street, *Synth. Met.* **9**, 165 (1984).
41. J. H. Kaufman, N. Colaneri, J. C. Scott, G. B. Street, *Phys. Rev. Lett.*, in press.
42. A. J. Heeger, G. Blanchet, T. C. Chung, C. R. Fincher, *Synth. Met.* **9**, 173 (1984).
43. M. S. Wrighton, H. S. White, G. P. Kittlesen, *J. Am. Chem. Soc.* **106**, 5375 (1984).
44. W. H. Meyer, H. Keiss, B. Binggeli, E. Meier, G. Harbeke, in preparation.
45. We thank J. L. Brédas for many helpful comments.

Until only a few years ago, the idea that a traditional organic or metal-organic substance could exhibit the electrical, optical, and magnetic properties of a metal seemed a complete contradiction in terms. Among other features, such substances lack the partially filled, spatially delocalized electronic energy levels (bands) which are an essential characteristic of a metal (*1*). This picture has, however, changed dramatically in the past several years, and the art of chemical synthesis has given rise to whole new classes of molecular (*2–5*) and polymeric (*2–6*) materials with properties analogous to those of metals having restricted dimensionality. The culmination is a new condensed matter field of study at the interface of chemistry, physics, and materials science that is stimulating breakthroughs in synthetic chemical strategy and methodology, in spectroscopic, structural, and transport analysis, and in the fundamental theoretical descriptions of how electrical charge is transported in the solid state of matter. Terms such as "molecular metal," "synthetic metal," "organic superconductor," "soliton conductor," and "bipolaron conductor" were unheard of only a few years ago. The application of this new knowledge to sensors, rectifiers, batteries, switching devices, photoresists, solar energy devices, electrophotographic devices, static charge–dissipating materials, electro-magnetic shielding materials, chemoselective electrodes, and video disk coatings has also received a great deal of discussion (*2–6*).

Despite the impressive advances that have been achieved in the fields of molecular and macromolecular electrically conductive materials, it is fair to say that our current level of physical understanding of and chemical control over such

35

Electrically Conductive Metallomacrocyclic Assemblies

Tobin J. Marks

Science 227, 881–889 (22 February 1985)

systems is at a rather rudimentary level. From a synthetic chemical standpoint, the ability to tailor charge-transporting microstructures at the atomic level represents an exciting challenge and a key both to manipulating macroscopic properties and to testing theoretical predictions. A brief overview of these activities is given in this article.

To review in detail the recent chemical progress made by the host of excellent research groups working on conductive molecular and polymeric substances is not possible in the space available here. Thus, we have chosen to be selective and to illustrate current strategies, approaches, and problems by focusing on work performed in this laboratory on a new family of materials which is at the crossroads between molecular and polymeric conductors and which embodies characteristics of each. Our long-range goal has been to develop rational, flexible syntheses of low-dimensional, metal-like molecular assemblies and, through correlated physical and theoretical investigations, to understand the properties of these assemblies as a function of architecture and electronic structure. This approach leads quite logically from rather simple systems composed of aggregated single molecules to more elaborate, structure-enforced polymeric arrays of covalently linked molecular subunits.

A Molecular Approach and
Model Systems

Two general features are emerging as prerequisites for converting an unorganized collection of molecules into an electrically conductive array (7, 8). First, the molecules must be arranged in close spatial proximity and in similar crystallographic and electronic environments so that an energetically flat (that is, with a minimum of "hills and valleys"), extended pathway exists for electronic charge movement. Such a situation is frequently realized when planar, conjugated molecules crystallize in a "stack" (for example, 1 or 2). The progression by

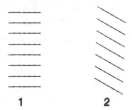

1   2

which a band structure is built up from the highest occupied molecular orbitals (HOMO's) of the molecular components of such a stack is illustrated in Fig. 1A. In a simple, Hückel-like ("tight-binding") description of the electronic structure (1), the transfer integral t (analogous to the Hückel resonance integral β) is a measure of how strongly the component molecular orbitals interact; in an extended solid, the magnitude of this interaction is expressed as the bandwidth, $4t$. An important goal in this area is to measure t accurately and, ultimately, to "tune" it.

The second requisite for the "molecular metallic" state is that the arrayed molecules must be in formal fractional oxidation states ["mixed valence," "partial oxidation (or reduction)," "incomplete charge transfer"]. That is, the molecules must formally have fractionally occupied valence shells. Part of this requirement is analogous to the classical dictate that the highest occupied band of a metal be only partially filled (1). A valence bond depiction is presented in Fig. 1B. However, as was first noted for certain metal oxides (9), situations may

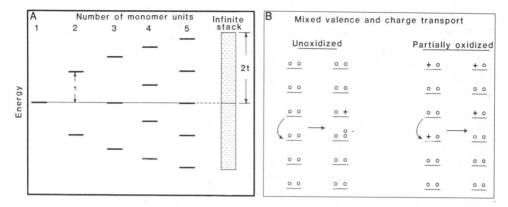

Fig. 1. (A) Schematic illustration of the energetics of arraying increasing numbers of molecular subunit HOMO's, resulting finally in band formation. The parameter t is the tight-binding transfer integral, analogous to the Hückel β integral. (B) Schematic depiction of how partial oxidation enhances charge mobility in a simple molecular stack.

arise in which the coulombic repulsion (denoted by the parameter U) of having two charge carriers on a single site (for example, as expressed in Eq. 1)

$$\bullet \quad \bullet \quad \longrightarrow \quad \underline{} \quad \bullet\bullet \qquad (1)$$

may approach or exceed the bandwidth. In the extreme case ($U \gg 4t$), a material with a formally half-filled band could actually be an insulator (10). For molecular systems, the degree to which molecular, crystal, and electronic structure influence the coulombic interactions is by no means clear, and learning to manipulate U remains an important goal.

In addition to the parameters mentioned above, the conductivity properties of molecular materials are highly sensitive to interactions between the electronic systems and various lattice vibrations (phonons). In the extreme case, a lattice distortion (a Peierls transition) accompanied by an opening of a gap in the conduction band can induce a metal-to-insulator transition at low temperatures (1). Such behavior is analogous to a Jahn-Teller instability and is predicted theoretically for an ideally uni-

dimensional metal. However, metal-to-insulator transitions are not observed in a number of cases, and the most plausible explanation (presently qualitative) appears to be that ideal unidimensionality is not fulfilled. Other electron-phonon effects are believed to be important in the nonclassical temperature dependence of the conductivity of many molecular metals ($\sigma \sim T^{-2}$) (11) and doubtless play a role in the properties of superconducting organic systems (2, 12). Electron-electron effects can also be important (2–5).

Synthetically, mixed valence is usually achieved by chemical or electrochemical oxidation (for example, Eq. 2) or

$$ (2) $$

reduction of the components of the delocalized stack. The success of such a

519

tactic is crucially dependent on a complex, largely uncontrollable variety of Madelung, exchange, polarization, bandwidth, van der Waals, core repulsion, crystallization kinetic, and ionization potential–electron affinity factors, which determine the lattice architecture (segregated arrays of donors and acceptors are essential) and the degree of incomplete charge transfer. Of course, such variables are not readily manipulated by conventional synthetic methodology, and serendipity has been a major ingredient in the synthesis of most molecular metals. Also, obstacles such as the growth of crystals having suitable dimensions, crystallinity, and purity for meaningful measurements are by no means trivial.

Among the donor molecules that have received the greatest attention, the tetrachalcogenafulvalenes (3) in their myriad modifications, have given rise to some of the most spectacular results. Thus, TTF (3; R = H, E = S), in combination with the electron acceptor TCNQ [4; X = H, Y = C(CN)$_2$], gave rise to the first true molecular metal, TTF-TCNQ (13). More recently, salts of TMTSF (3; R = CH$_3$, E = Se) were the first organic supercon-

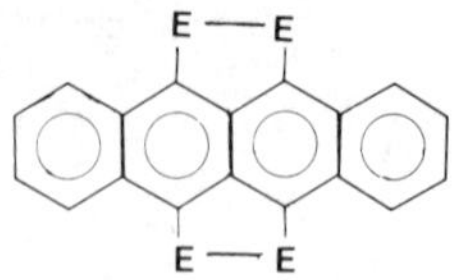

E = S, Se, Te

5

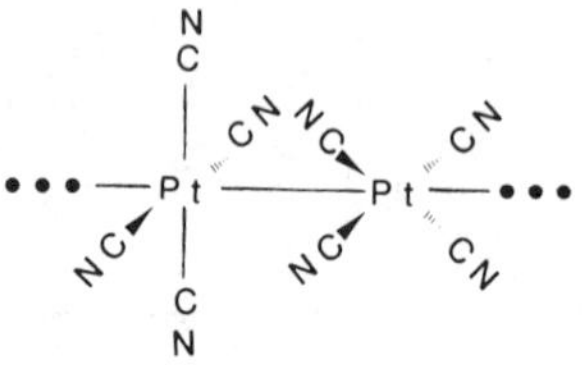

6

Initial efforts in this laboratory (7) focused on synthesizing and understanding new types of stacked molecular conductors composed of glyoxime (7, 8) (7, 16), phthalocyanine [9; M(Pc)] (7, 17), dibenzotetraazaannulene (10) (7, 18), and hemiporphyrazine (11) (19) metallo-

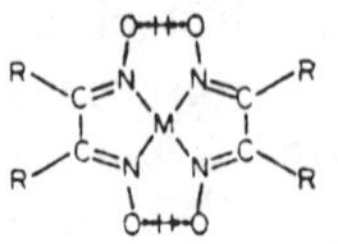

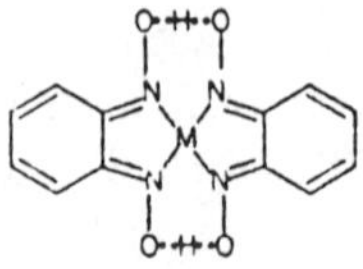

7

8

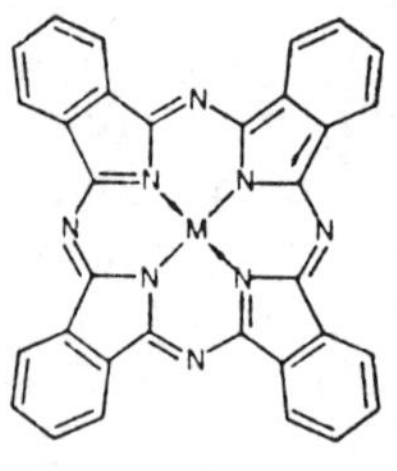

9

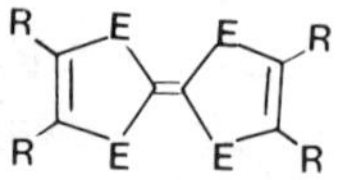

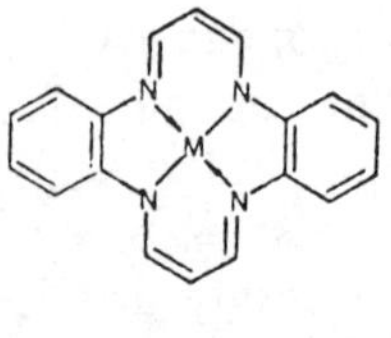

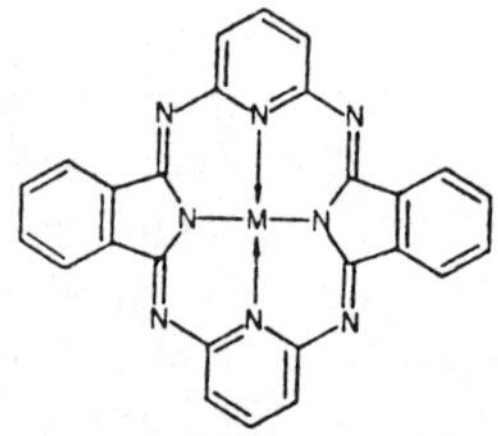

R = organic group
E = S, Se, Te

3

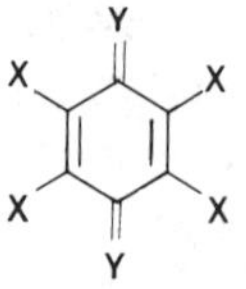

X = H, halogen
Y = O, C(CN)$_2$

4

ductors (12). Other highly conductive systems are based on the tetrachalcogenatetracene (5) (14) and tetracyanoplatinate [6; for example, K$_2$Pt(CN)$_4$Br$_{0.3}$] families. The latter material has a band structure composed primarily of partially filled platinum d_{z^2} orbitals (1, 15).

10

11

macrocycles. As early as 1975, it was shown using resonance Raman and ^{129}I Mössbauer spectroscopy (to identify the form of iodine present in the crystal) that a metallomacrocyclic donor could be co-crystallized with I_2 (Eq. 2) to yield a mixed-valence, low-dimensional conductor (20). The details and import of this molecular cocrystallization–Raman analysis approach were further elaborated in 1976 (21) and 1978 (22). Extension to Br_2 as an acceptor was subsequently described (23), and this general approach has been skillfully applied to numerous other metallomacrocyclic systems (24). In optimum cases, the metallomacrocycle-halogen cocrystallization approach yields segregated arrays of partially oxidized donor ions and polyhalide counterions. The form of the halogen, hence the degree of incomplete charge transfer, can be readily discerned with resonance Raman and iodine Mössbauer spectroscopy (even in polycrystalline or severely disordered samples) (7, 25).

A particularly impressive example of the molecular cocrystallization approach is based on the robust, technologically important phthalocyanine (9) macrocycle (26). The material $H_2(Pc)I$ (9; M = two hydrogen atoms) illustrates conclusively that a central metal ion is not required for high electrical conductivity and is an ideal model material for more elaborate systems (see below) (27). The diffraction-derived crystal structure of this material [tetragonal, space group $P4/mcc$, with crystallographic axes $a = 13.979(6)$ and $c = 6.502(3)$ Å at 25°C; $R_F = 0.041$] consists of stacks of $H_2(Pc)^{+0.33}$ cations arrayed at 3.251(2) Å separations and parallel chains of disordered off-axis I_3^- counterions (Fig. 2A). The staggering angle between adjacent Pc rings is 40.0°. The electrical conductivity ($\sigma_{||}$) of a typical (0.65 by 0.07 by

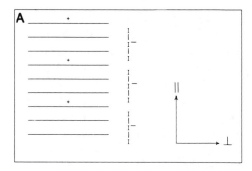

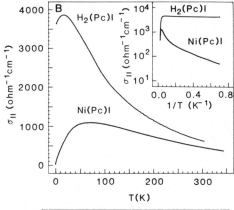

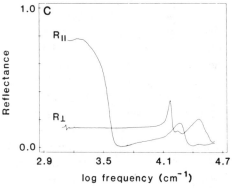

Fig. 2. (A) Schematic illustration of the $H_2(Pc)I$ crystal structure, viewed transverse to the stacking direction (the crystallographic c axis). The symbols || and ⊥ denote directions parallel and perpendicular to the stacking direction, respectively. (B) Variable-temperature electrical conductivity data for typical $Ni(Pc)I$ and $H_2(Pc)I$ single crystals in the macrocycle stacking direction (||). The inset emphasizes the behavior at very low temperatures. (C) Reflectance spectrum of an $H_2(Pc)I$ single crystal; $R_{||}$ and $R_\perp$ denote reflectance parallel and perpendicular to the macrocycle stacking direction, respectively.

0.07 mm!) $H_2(Pc)I$ crystal in the macrocycle stacking direction (the crystallographic c axis) is compared to that of the somewhat less conductive nickel analog in Fig. 2B (28). At room temperature, $\sigma_{||}$ for $H_2(Pc)I$ is approximately 700 ohm^{-1} cm^{-1}, which, adjusted for the high cross-sectional area of an $H_2(Pc)$ stack (that is, the carrier mean free path), is among the highest reported for an organic conductor. Furthermore, the temperature dependence of $\sigma_{||}$ is "metal-like" ($d\sigma/dT < 0$) until very low temperatures and exceeds 3500 ohm^{-1} cm^{-1} even at 1.5 K. The first measurements of the electrical anisotropy of a metallomacrocyclic conductor underscore the pronounced unidimensionality: $\sigma_{||}/\sigma_{\perp} \gtrsim 500$ for a typical $H_2(Pc)I$ crystal.

Optical reflectivity measurements are another powerful method for studying the electronic structure of molecular metals (29). New instrumental techniques have recently yielded the first quantitative, complete far-infrared to ultraviolet polarized reflectance data for a single crystal of a metallomacrocyclic conductor (Fig. 2C) (27). The high anisotropy of this material is evident in the differences in reflectivity parallel to ($R_{||}$) and perpendicular to ($R_{\perp}$) the macrocycle stacking direction. The perpendicularly polarized features at 14,800 and 29,000 cm^{-1} and the parallel-polarized feature at 18,200 cm^{-1} are straightforwardly assigned, on the basis of our earlier work, to Pc-centered ($\pi \rightarrow \pi^*$) and I_3^--centered electronic transitions, respectively (7). Most significant, however, is the parallel-polarized edgelike feature which begins at ~4500 cm^{-1} and extends into the far-infrared. This feature is reminiscent of the classical plasma edge in the reflectance of a metal (1)—clearly the $H_2(Pc)I$ band structure is

highly anisotropic. Detailed numerical analysis of the plasma edge line shape using an "electron gas" (Drude) model yields the plasma frequency (ω_p), the electron relaxation time, the effective mass of the carriers, and the tight-binding bandwidth, $4t = 1.3(1)$ eV (27).

Although less well understood, the magnetic properties of molecular metals are also informative (30). The susceptibility of $H_2(Pc)I$ is only weakly paramagnetic ($\chi_s = 2.21(5) \times 10^{-4}$ emu mol^{-1}) and is relatively independent of temperature. This behavior is reminiscent of the Pauli susceptibility of metals (1), and a simple tight-binding analysis yields a bandwidth (unrealistically small) of 0.38(7) eV. The questionable reliability of susceptibility-derived bandwidths is probably due to large on-site Coulomb repulsions (10, 31). Electron paramagnetic resonance (EPR), thermoelectric power, and solid-state nuclear magnetic resonance (NMR) measurements on H_2-(Pc)I all indicate that the conduction pathway is predominantly if not exclusively through the macrocyclic radical cation stacks and not through the polyiodide chains (27, 32).

Although the molecular cocrystallization method was a useful first-generation approach, attempts to significantly modify the metallomacrocycle molecules (for example, alkylation) or the acceptor moieties (nohalogen oxidizing agents such as TCNQ) soon revealed the limitations of such a simplistic synthetic approach. Without rigorous structural enforcement of the metallomacrocyclic stacking, it is impossible to effect controllable and informative variations in stacking architecture or in donor-acceptor spatial-electronic relationships. It has been evident for some time that the packing forces operative in most molecu-

A

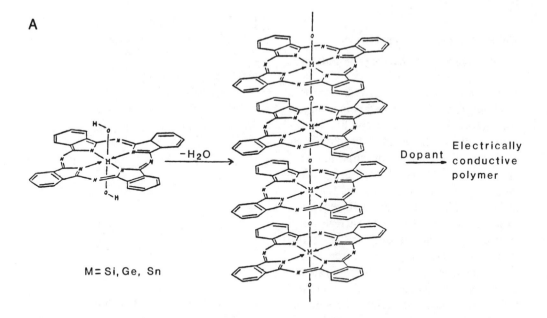

M = Si, Ge, Sn   −H₂O →   Dopant →   Electrically
                                                    conductive
                                                    polymer

B

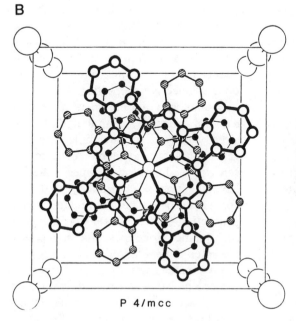

P 4/mcc

Fig. 3. (A) Strategy for the assembly of
structure-enforced, electrically conductive arrays of group IVA metallophthalocyanines. (B) Schematic depiction of the
similar crystal structures of halogen- and
nitrosyl-doped $[Si(Pc^{\rho+})O]_n$ salts.

523

lar conductors are both delicate and ca-
pricious.

Cofacial Assembly Properties of
Conductive Metallomacrocyclic Polymers

One attractive approach to enforce
stacking architecture in metallomacrocylic
conductors is by covalently linking already proven, charge-carrying molecular
subunits in a cofacial orientation. This
approach is illustrated for group IVA
phthalocyanines, $[M(Pc)O]_n$, in Fig. 3A,
and the result is a conceptually generaliz-
able class of extremely robust (the polymerization is carried out at 400°C; the
M = Si polymer can be recovered essentially unchanged after dissolution in strong
acids) macromolecules (33, 34). Characterization studies by a variety of chemical,
radiochemical, spectroscopic (infrared,
Raman, optical, solid-state NMR), diffrac-
tometric, and electron microscopic techniques indicate highly regular, crystalline
structures consistent with the cofacial architecture depicted in Fig. 3A and with
$n = 50$ to 200. Importantly, the Pc-Pc
interplanar separation increases with increasing ionic radius of M: 3.33(2) Å,
M = Si; 3.53(2) Å, M = Ge; 3.82(2) Å,
M = Sn (33). Furthermore, recent scanning and transmission electron microscopic studies have revealed the microscopic
details of how $M(Pc)(OH)_2$ monomer crystals are topotactically transformed upon
heating into $[M(Pc)O]_n$ polymer crystals
(35) and have provided direct images,
within the crystal lattice, of the polymer
chains and the halogen doping process for
M = Ge (36).

The $[M(Pc)O]_n$ polymers can be converted into covalently linked, mixed-valence chain compounds by a variety of
chemical and electrochemical methods

(for example, Eqs. 3 to 5 for M = Si and
Ge). This process has been traditionally

$$[M(Pc)O]_n + 0.55nX_2 \rightarrow$$
$$\{[M(Pc)O]X_{1.1}\}_n \qquad (3)$$
$$X = Br, I$$

$$[M(Pc)O]_n + 0.35nNO^+Y^- \rightarrow$$
$$\{[M(Pc)O]Y_{0.35}\}_n + 0.35nNO(g) \qquad (4)$$
$$Y = BF_4, PF_6, SbF_6$$

$$[M(Pc)O]_n + nyQ \rightarrow \{[M(Pc)O]Q_y\}_n \qquad (5)$$

TCNQ   DDQ

termed "doping" for conductive polymers. In contrast to the simple molecular
materials, M(Pc) stacking is rigorously
enforced, and for the first time it is
possible to meaningfully probe donor-
acceptor relationships for a wide range
of structurally and electronically dissimi-
lar dopants as well as to examine the
effects of differing interplanar spacings
on the collective properties.

The halogenated polymers were studied first since a substantial data base
already existed for the molecular con-
geners and was available for comparisons (33). The halogen-doped materials
have the important property that they
appear to be stable in air for years.
Diffraction studies reveal that the halogen doping process is inhomogeneous;
incremental oxidation yields progressively greater amounts of a new, mixed-
valence phase of discrete stoichiometry
and structure, and the dopant is not
distributed evenly throughout the lattice
as doping progresses. The crystal struc-

tures of the halogenated polymers are illustrated in Fig. 3B. Within experimental error, the structures of Ni(Pc)I, $\{[Si(Pc)O]I_{1.1}\}_n$, and $\{[Ge(Pc)O]I_{1.1}\}_n$ differ only in the ring-ring interplanar spacing, namely 3.244(2), 3.30(2), and 3.48(2) Å, respectively. The form of the halogen in all of these materials is X_3^-, so the formal degree of incomplete charge transfer, $\rho = 0.33$ to 0.37, is rather insensitive to the nature of the halogen and the ring-ring interplanar separation.

The response of the $\{[M(Pc)O]X_{1.1}\}_n$ collective properties to an increase in the key structural parameter of interplanar spacing is beautifully illustrated by a fall in the electrical conductivity ($\sigma_{Ni} > \sigma_{Si} > \sigma_{Ge}$), an increase in the Pauli-like magnetic susceptibility ($\chi_{Ni} < \chi_{Si} < \chi_{Ge}$), and a decrease in the optical reflectivity plasma frequency ($\omega_{Ni} > \omega_{Si} > \omega_{Ge}$), hence the tight-binding bandwidth. The latter two observables clearly indicate a structurally correlated diminution in the bandwidth with increased spacing, while the former shows the close connection to macroscopic charge transport. Relevant data are set out in Table 1. As is usually the case for conductive polymers, it has not yet proved possible to obtain sufficiently large crystals of the $\{[M(Pc)O]X_{1.1}\}_n$ materials for single-crystal conductivity (or optical) studies. Hence, measurements have been on isotropic, compacted, polycrystalline samples, recognizing that for the conductivity of such materials, $\sigma_{\parallel}$ (single crystal) $\sim 100\ \sigma$ (polycrystalline). Thus, in the $[M(Pc^{\rho+})O]_n$ stacking direction, $\sigma_{\parallel}$ (Si) ~ 100 ohm^{-1} cm^{-1} and $\sigma_{\parallel}$ (Ge) ~ 10 ohm^{-1} cm^{-1} (33). Two other observations are of interest with regard to $\{[M(Pc)O]X_{1.1}\}_n$ charge transport. The increase in the electrical conductivity as a function of doping level (a steep rise

followed by a leveling off; Fig. 4A) is explicable in terms of a simple percolation model, that is, a statistical picture of how many randomly dispersed conductive particles must be introduced into a host of nonconductive particles to produce a contiguous conducting pathway (at the "percolation threshold") (33, 37). Second, the temperature dependence of the conductivity can be described by a model (fluctuation-induced tunneling) in which conduction is modulated by small, nonconducting junctions separating the larger, high-conductivity regions (38). This is indicated by a linear relation between $-1/\ln(\sigma/\sigma_0)$ and temperature (Fig. 4B), where σ_0 is a constant characteristic of the particular material. Both of these properties are in accord with the aforementioned doping-structural model for the $\{[M(Pc)O]X_{1.1}\}_n$ materials.

We can now ask a fascinating question. If it were possible to hold the architecture of the donor array constant, how would the halogen-related phenomenology vary as a function of dopant? The rigorously enforced stacking architecture of the $[M(Pc)O]_n$ polymers affords a unique opportunity to address this question. For example, the nitrosyl doping route of Eq. 3 involves oxidants which are different from halogens, and the resulting BF_4^-, PF_6^-, SbF_6^- counterions differ from Br_3^- and I_3^- in spatial demands, charge distributions, and polarizability (39). Nevertheless, the degree of incomplete charge transfer achieved ($\rho \sim 0.35$), the inhomogeneity of the oxidation process, and the geometries (interplanar spacings, ring-ring staggering angles) of the partially oxidized $[Si(Pc^{\rho+})O]_n$ chains (Fig. 3B and Table 1) are almost independent of dopant. As assessed by the magnetic susceptibility and optical reflectivity (Fig. 4C), the

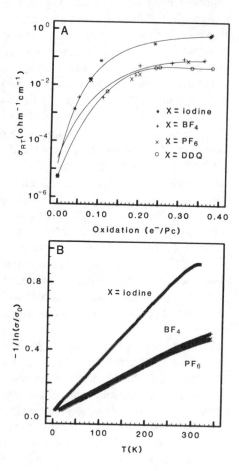

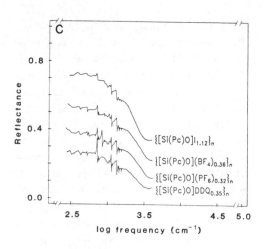

Fig. 4. (A) Electrical conductivity data (polycrystalline samples, 300 K) as a function of doping level for $[Si(Pc^{\rho+})O]_n$ salts prepared with a range of doping agents. (B) Variable temperature electrical conductivity data (polycrystalline samples) for several $[Si(Pc^{\rho+})O]_n$ salts prepared with different doping agents. The specific functional dependence of the conductivity on temperature can be associated with a fluctuation-induced carrier tunneling model for charge transport (see text). (C) Optical reflectance spectra of polycrystalline samples of $[Si(Pc^{\rho+})O]_n$ salts prepared with different doping agents. The plasma-like edges can be discerned in all cases beginning at ~ 3000 cm^{-1}. Successive plots are displaced vertically by $+0.10$ reflectance unit for ease of viewing and, as shown, are uncorrected for the anisotropic nature of the samples.

band structure does not perceptibly "feel" the counterions. It might have been expected that localization of the counterion negative charges (for instance, $I_3^- \rightarrow BF_4^-$) would have caused localization of the $[M(Pc^{\rho+})O]_n$ wave functions and band narrowing. As for the halogenated polymers, the dependence of electrical conductivity on doping level suggests percolation behavior (Fig. 4A) and the temperature dependence of the conductivity agrees with the fluctuation-

induced carrier tunneling mechanism (Fig. 4B). In addition, like the halogenated polymers, the nitrosyl-doped polymers appear to be indefinitely stable in air.

Although the crystallography is considerably more complicated and experiments are still in progress, results with DDQ (Eq. 4) as a $[Si(Pc)O]_n$ dopant are remarkably similar to those for the halogen- and nitrosyl-oxidized polymers (*39, 40*). Evidence for a limiting degree of

Table 1. Structural, transport, magnetic, and bandwidth parameters for electrically conductive phthalocyanine materials.

Compound	Tetragonal lattice (Å)	Charge transfer (ρ)	d-c electrical conductivity at 300 K (ohm^{-1} cm^{-1})	Pauli magnetic susceptibility (emu mol^{-1} $\times 10^4$)	Tight-binding bandwidth $4t^*$ (eV)
$H_2(Pc)I$	$a = 13.979(6)$ $c = 6.502(3)$	0.33	700† 8.0‡	2.21(5)	1.3(1)† 1.1(2)‡
$Ni(Pc)I$	$a = 13.936(6)$ $c = 6.488(3)$	0.33	500† 7.5‡	1.90(10)	0.99(9)‡
$\{[Si(Pc)O]I_{1.12}\}_n$	$a = 13.98(5)$ $c = 6.60(4)$	0.37	0.58‡	2.35(10)	0.60(6)‡
$\{[Si(Pc)O](BF_4)_{0.36}\}_n$	$a = 13.82(7)$ $c = 6.58(4)$	0.36	0.10‡	2.22(6)	0.64(6)‡
$\{[Si(Pc)O](PF_6)_{0.36}\}_n$	$a = 13.99(6)$ $c = 6.58(4)$	0.36	0.10‡	2.49(7)	0.63(6)‡
$\{[Si(Pc)O](SbF_6)_{0.36}\}_n$	$a = 14.31(4)$ $c = 6.58(4)$	0.36	0.15‡	2.22(3)	0.64(6)‡
$\{[Ge(Pc)O]I_{1.12}\}_n$	$a = 13.96(5)$ $c = 6.96(4)$	0.37	0.11‡	2.70(10)	0.48(5)‡

*From analysis of reflectivity data.   †Single-crystal sample.   ‡Polycrystalline sample.

incomplete charge transfer near $\rho \approx 0.40$ under normal doping conditions, percolation behavior (Fig. 4A), and a nearly invariant plasma frequency (Fig. 4C) are again observed. It is conceivable that more drastic oxidizing conditions may effect further depletion of the $[M(Pc)O]_n$ conduction band. Of particular interest would be the properties of the integral oxidation state ($\rho = +1$) material. Would it be a "Mott-Hubbard insulator" (9, 10)? Initial indications obtained with DDQ under more vigorous conditions or electrochemically doping at very high potentials suggest the exciting possibility that more highly oxidized cofacially linked arrays are indeed accessible (41).

The doped $[M(Pc)O]_n$ materials, of course, represent only one component of a rapidly growing field. The past several years have witnessed a veritable explo-sion in the number of other classes of electrically conductive polymers that have been prepared and characterized. Some of the more extensively studied examples are cis- and trans-polyacetylene (12) (6, 42), polypyrrole (13) (6, 43), poly-p-phenylene (14) (6, 44), poly-p-phenylenesulfide (15) (6, 45), and polythiophene (16) (6, 46). In common with the $[M(Pc)O]_n$ polymers, all of these materials have highly conjugated, extended structures that must be oxidized or reduced to create mobile charge carriers. However, it appears likely that completely new kinds of spinless charge carriers are involved in charge transport by oxidatively doped trans-polyacetylene (solitons) (42, 47) and by oxidatively doped polypyrrole (bipolarons) (48). To date, polyacetylene and polypyrrole are probably the best characterized of these

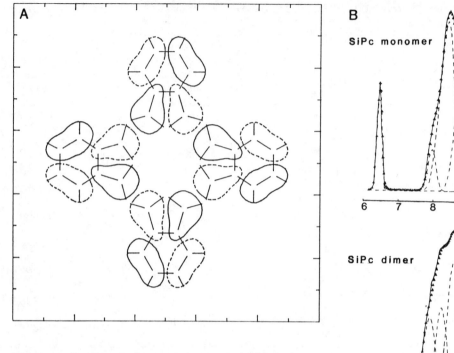

Fig. 5. (A) Graphic representation of the Si(Pc)(OH)$_2$ highest occupied molecular orbital calculated by the first-principles DV-Xα formalism. (B) Gas phase He-I photoelectron spectra of the monomeric [Si(Pc)-(OR)$_2$] and dimeric [ROSi(Pc)OSi(Pc)OR] silicon phthalocyanine molecules; R = Si[C(CH$_3$)$_3$](CH$_3$)$_2$.

12a

12b

13

14

15

16

conductive polymers. Nevertheless, for most of the known materials there are great uncertainties about the structures of the undoped and doped phases, polymer molecular weights, doping homogeneity, and the relative roles of intrachain versus interchain charge transport. As useful polymeric materials, which can be readily processed into films, foils, fibers, and other structures (*49*), many of these substances suffer in the doped state from being intractable, infusible, poorly flexible, and unstable in air. Efforts to improve the characteristics of the known conductive polymers include examples of ingenious alloying (*50*), doping (*51*), solvation (*52*), and functionalization (*53*) approaches, as well as the synthesis of completely new types of conducting polymers (*54*).

Cofacial Assembly: Electronic Structure of Metallomacrocyclic Polymers

Traditionally, quantum chemical approaches to understanding the electronic structures of molecular conductors have included either rigorous calculations for the individual molecular subunits together with imaginary cofacial dimers (**17**; see Fig. 1A) (*55*), or necessarily less rigorous but computationally more efficient (for instance, "crystal orbital") calculations for entire stacks (*56*). One very attractive feature of the $[M(Pc)O]_n$ architecture is that dimers which are actual fragments of the cofacially connected metallomacrocyclic chain (**18**) are no longer imaginary, but can actually be synthesized, isolated, structurally characterized, and compared to the corresponding monomers (**19**) by a variety of physical methods (*57*). The ultimate goal is to understand in depth the relationship between the collective properties of a multimolecular array and the nature of the local π-π and other electronic interactions as a function of various ground-

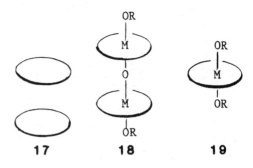

17   **18**   **19**

R = capping group

state geometries and phonon-like excursions therefrom.

As a first step (*57, 58*) in probing $[M(Pc)O]_n$ electronic structure, calculations have been carried out for model phthalocyanines [for example, $Si(Pc)$-$(OH)_2$], using the first-principles discrete variational local exchange (DV-Xα) technique (*59*). In the case of simple arenes and cyclophanes, we have shown this method to reproduce molecular orbital energy orderings determined by photoelectron spectroscopy (which measures the energies of electrons ejected from the various molecular orbitals by an intense monochromatic light source) and details of optical spectra (*60*). The calculated amplitudes for the HOMO of $Si(Pc)$-$(OH)_2$ are shown in Fig. 5A. From this result and the composition of the other molecular orbitals it can be seen that the conduction band of a partially oxidized $[Si(Pc)O]_n$ conductor will be composed principally of carbon $p\pi$ orbitals, with a great deal of the amplitude located near the macrocycle core. These results have several interesting ramifications. First, when it is considered that each $Si(Pc)$ unit is ringed by a phalanx of hydrogen atoms, it can be seen that in the oxidized state, counterions are held at a considerable distance from the atoms and orbitals composing the conduction band structure (see Fig. 3B). This situation is considerably different from that in TMTSF (*12, 61*) and TTF (*7, 8*) salts, where chalcogen-counterion distances can be quite short and are thought to have significant effects on collective properties. The marked insensitivity of the $[Si(Pc^{\rho+})O]_n$ collective properties to the nature of the counterions must reflect this metrical and electronic structural situation. The DV-Xα calculations also indicate little mixing of carbon π orbitals with either silicon or oxygen orbitals in the HOMO. Hence, it is unlikely that the $(-Si-O-)_n$ chains play any direct role in the conduction process.

The cofacial dimer ROSi(Pc)OSi(Pc)OR $\{R = Si[C(CH_3)_3](CH_3)_2\}$, which has OSi(Pc)OSi(Pc)O structural parameters essentially identical to those of the corresponding $[Si(Pc^{\rho+})O]_n$ polymers (*57*), provides important new information on π-π electronic interactions. As can be seen in Fig. 1A, the HOMO-HOMO interaction energy in a dimer ($2t$) should be directly related to the tight-binding bandwidth in the corresponding polymer ($4t$). Such a dimer "through-space" MO-MO splitting should be directly measurable by gas phase photoelectron spectroscopy. This would represent a unique, ex-situ method for assessing the bandwidth divorced from solid-state effects. Computer-analyzed, high-resolution gas phase He-I (21.2 eV) photoelectron spectra of the Si(Pc)(OR)$_2$ monomer and the corresponding dimer are shown in Fig. 5B. The lowest energy features correspond to electron ejection from the respective HOMO's, and it is clear that the dimer band is split. The splitting ($2t$) is 0.29(3) eV, which, as already explained, translates into a bandwidth of 0.58(6) eV. This result is in excellent agreement with a value

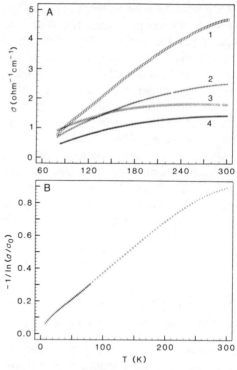

Fig. 6. (A) Variable-temperature four-probe electrical conductivity data for phthalocyanine-Kevlar hybrid fibers of composition: 1, $[Ni(Pc)(K)_{0.43}I_{1.56}]_n$; 2, $[Ni(Pc)(K)_{0.67}I_{1.07}]_n$; 3, $[Ni(Pc)(K)_{1.58}I_{1.27}]_n$; 4, $[Ni(Pc)(K)_{4.36}I_{1.66}]_n$; K = Kevlar monomer unit **20**. (B) Variable-temperature four-probe electrical conductivity data for a $[Ni(Pc)(K)_{0.86}I_{1.71}]_n$ fiber plotted according to the fluctuation-induced tunneling model of charge transport.

(Table 1) of 0.60(6) eV obtained from a Drude analysis of the $\{[Si(Pc)O]I_{1.1}\}_n$ reflectance spectrum (*33*). Moreover, a DV-Xα calculation for the model dimer HOSi(Pc)OSi(Pc)OH yields a bandwidth of 0.76 eV, which is in very good agreement with the above two experimental values (*57*). These conclusions, and recent successes (*62*) in using ab initio results to parameterize the electronic structure calculations for computational efficiency, indicate that truly meaningful connections between quantum chemical theory, chemical synthesis, and physical measurements will be possible in the future.

20

Processable, Conductive Polymeric Materials from Metallomacrocycles

The results discussed above provide a foundation from which to address a major problem in the area of electrically conductive polymeric materials. As already noted, many members of the current generation of substances exhibit severe limitations with regard to chemical-structural control at the molecular level, mechanical stability, air, water, and thermal stability, solubility, and processability. Most of the molecular and macromolecular phthalocyanine compounds discussed in this article have high environmental and thermal stability. Moreover, many are soluble without decomposition in strong acids; indeed, precipitation from acid solution is a common purification technique for phthalocyanines (26). These characteristics suggested a means of producing compositionally diverse, highly oriented conductive materials by a well-known processing technique: wet fiber spinning.

The tendency of rigid-rod polymers to spontaneously form oriented, liquid crystalline solids (63) suggested that $[Si(Pc)O]_n$ might be processed by extruding acid solutions into a precipitation medium. Fibers of the high-performance aramid polymer Kevlar (**20**) are produced by just such a processing technique, and weight for weight have six times the modulus of steel (64). Indeed, it was found that CF_3SO_3H solutions of $[Si(Pc)O]_n$ alone or in combination with Kevlar could be extruded to form relatively strong, oriented $[(--Si-O--)_n$ and

Kevlar chains are aligned along the extrusion direction] fibers. Doping can be carried out prior to or after fiber formation to yield air-stable, electrically conductive fibers (65). Even more interesting is the observation that environmentally stable, electrically conductive, molecular-macromolecular hybrid materials can be produced in an analogous fashion from molecular phthalocyanines and Kevlar (66). For $Ni(Pc)$-Kevlar-I_2, the fibers consist of very small $Ni(Pc)I$ crystalline regions dispersed in the Kevlar host. Both the $Ni(Pc)I$ c axis and the Kevlar chain are preferentially oriented in the extrusion direction. Electrical conductivities are high (Fig. 6A) and variable-temperature data suggest that fluctuation-induced carrier tunneling is again operative (Fig. 6B). Studies of conductivity versus percentage $Ni(Pc)I$ implicate percolation in the transport process (67), and measurements of mechanical properties evidence little degradation of the Kevlar tensile modulus at moderate $Ni(Pc)I$ loadings (68). Further studies of these and related hybrid systems with other polymer hosts are in progress (67, 69). However, it is already evident that the strategy of "marrying" the properties of well-characterized molecular metals to those of a robust, processable, orientable polymer host with complementary solubility characteristics offers an intriguing direction for the design of new electrically conductive, processable polymeric materials.

References and Notes

1. For excellent introductions to condensed-matter science, see J. S. Miller and A. J. Epstein, *Prog. Inorg. Chem.* **20**, 1 (1976); H. M. Rosenberg, *The Solid State, Second Edition* (Clarendon, Oxford, 1983); C. Kittel, *Introduction to Solid State Physics* (Wiley, New York, ed. 5, 1976).
2. C. Pecile, G. Zerbi, R. Bozio, and A. Girlando (Eds.), *Proceedings of the International Conference on the Physics and Chemistry of Low-Dimensional Synthetic Metals* (ICSM 84), Abano Terme, Italy, June 17–22, 1984, *Mol. Cryst. Liq. Cryst.* 117–121 (1985).
3. J. S. Miller, Ed., *Extended Linear Chain Compounds* (Plenum, New York, 1982), vols. 1–3.
4. A. J. Epstein and E. M. Conwell, Eds., "Proceedings of the International Conference on Low-Dimensional Conductors," Boulder, Colo., 9–14 August 1981, *Mol. Cryst. Liq. Cryst.*, parts A–F (1981–1982).
5. L. Alcácer, Ed., *The Physics and Chemistry of Low-Dimensional Solids* (Reidel, Dordrecht, 1980).
6. K. J. Wynne and G. B. Street, *Ind. Eng. Chem. Prod. Res. Dev.* **21**, 23 (1982); R. H. Baughman, J. L. Bredas, R. R. Chance, R. L. Elsenbaumer, L. W. Shacklette, *Chem. Rev.* **82**, 209 (1982); G. Wegner, *Angew. Chem. Int. Ed. Engl.* **20**, 361 (1981); C. G. Duke and H. W. Gibson, in *Kirk-Othmer Encyclopedia of Chemical Technology* (Wiley, New York, ed. 3, 1982), vol. 18, pp. 755–793; R. B. Seymour, Ed., *Conductive Polymers* (Polymer Science and Technology Series, Plenum, New York, 1981), vol. 16.
7. T. J. Marks and D. W. Kalina, in (3), vol. 1, pp. 197–331; T. J. Marks, *Science* **227**, 881 (1985).
8. J. B. Torrance, *Acc. Chem. Res.* **12**, 79 (1979); Z. G. Soos, *Ann. Rev. Phys. Chem.* **25**, 121 (1974).
9. N. F. Mott, *Metal-Insulator Transitions* (Taylor & Francis, London, 1974), chapters 1–5.
10. J. B. Torrance, Y. Tomkiewicz, R. Bozio, O. Pecile, C. R. Wolfe, K. Bechgaard, *Phys. Rev. B* **26**, 2267 (1982); J. B. Torrance, J. J. Mayerle, K. Bechgaard, B. D. Silverman, Y. Tomkiewicz, *ibid.* **22**, 4960 (1980); S. Mazumdar and A. N. Bloch, *Phys. Rev. Lett.* **50**, 207 (1983); S. Mazumdar, S. N. Dixit, A. N. Bloch, *Phys. Rev. B* **30**, 4842 (1984).
11. M. Weger, M. Kaveh, H. Gutfreund, *Solid State Commun.* **37**, 421 (1981); E. M. Conwell, *Phys. Rev. B* **22**, 1761 (1980).
12. K. Bechgaard and D. Jérome, *Sci. Am.* **247**, 52 (July 1982); D. Jérome, *Adv. Phys.* **31**, 299 (1982); F. Wudl, *Acc. Chem. Res.* **17**, 227 (1984).
13. A. J. Heeger, in *Highly Conducting One-Dimensional Solids*, J. T. Devrees, R. P. Evrard, V. E. van Doren, Eds. (Plenum, New York, 1979), chapter 3; A. J. Berlinsky, *Contemp. Phys.* **17**, 331 (1976).
14. I. F. Schegolev and E. B. Yagubskii, in (3), vol. 2, pp. 385–434.
15. J. M. Williams, A. J. Schultz, A. E. Underhill, K. Carneiro, in (3), vol. 1, pp. 73–118; T. W. Thomas and A. E. Underhill, *Chem. Soc. Rev.* **1**, 99 (1982); K. Krogmann, *Angew. Chem. Int. Ed. Engl.* **8**, 35 (1969).

16. M. A. Cowie *et al.*, *J. Am. Chem. Soc.* **101**, 2921 (1979); L. D. Brown *et al.*, *ibid.*, p. 2937; A. S. Foust and R. Soderberg, *ibid.* **89**, 5507 (1967); H. J. Keller and K. Seibold, *ibid.* **93**, 1309 (1971).
17. T. J. Marks, *J. Coatings Technol.* **48**, 53 (1976); J. L. Petersen, C. S. Schramm, D. R. Stojakovic, B. M. Hoffman, T. J. Marks, *J. Am. Chem. Soc.* **99**, 286 (1977); C. S. Schramm *et al.*, *ibid.* **102**, 6702 (1980); J. Martinsen, R. L. Greene, S. M. Palmer, B. M. Hoffman, *ibid.* **105**, 677 (1983); D. R. Stojakovic, thesis, Northwestern University (1978).
18. L.-S. Lin *et al.*, *J. Chem. Soc. Chem. Commun.* (1980), p. 954; L.-S. Lin *et al.*, in preparation.
19. C. W. Dirk and T. J. Marks, *Inorg. Chem.* **23**, 4325 (1984).
20. A. Gleizes, T. J. Marks, J. A. Ibers, *J. Am. Chem. Soc.* **97**, 3545 (1975).
21. T. J. Marks, D. F. Webster, S. L. Ruby, S. Schultz, *J. Chem. Soc. Chem. Commun.* (1976), p. 444.
22. T. J. Marks, *Ann. N.Y. Acad. Sci.* **313**, 594 (1978).
23. D. W. Kalina, J. W. Lyding, M. T. Ratajack, C. R. Kannewurf, T. J. Marks, *J. Am. Chem. Soc.* **102**, 7854 (1980).
24. H. Endres *et al.*, *Ann. N.Y. Acad. Sci.* **313**, 633 (1978); J. S. Miller and C. H. Griffiths, *J. Am. Chem. Soc.* **99**, 749 (1977); W. E. Hatfield, in *Conductive Polymers*, R. B. Seymour, Ed. (Polymer Science and Technology Series, Plenum, New York, 1981), vol. 16, p. 57; A. J. Jircitano, M. C. Colton, K. B. Mertes, *Inorg. Chem.* **20**, 890 (1981); B. M. Hoffman and J. A. Ibers, *Acc. Chem. Res.* **16**, 15 (1983); L. W. ter Haar, W. E. Hatfield, M. Tsutsui, *Mol. Cryst. Liq. Cryst.* **120**, 433 (1985); M. Mossoyan-Deneux, D. Benlian, M. Pierrot, A. Fournel, J. B. Sorbier, *ibid.* **120**, 437 (1985).
25. R. C. Teitelbaum, S. L. Ruby, T. J. Marks, *J. Am. Chem. Soc.* **102**, 3322 (1980); *ibid.* **101**, 7568 (1979).
26. F. H. Moser and A. L. Thomas, *The Phthalocyanines* (CRC Press, Boca Raton, Fla., 1983), vols. I and II.
27. J. W. Lyding, T. Inabe, R. L. Burton, C. R. Kannewurf, T. J. Marks, *Bull. Am. Phys. Soc.* **29**, KH3 (1984); T. Inabe, J. W. Lyding, C. R. Kannewurf, T. J. Marks, *Mol. Cryst. Liq. Cryst.* **118**, 353 (1985); T. Inabe *et al.*, *Solid State Commun.* **54**, 501 (1985). Cell parameters at $-100°C$: $a = 13.928(5)$ and $c = 6.410(4)$ Å.
28. Crystals are necessarily grown with scrupulously purified reagents in a thermostatted, vibration-free enclosure. Abrupt discontinuities in σ versus T data usually indicate poor integrity of electrical contact during measurements. Thermoelectric power measurements confirm the G versus T behavior.
29. D. B. Tanner, in (3), vol. 2, pp. 205–258; M. R. Madison, L. B. Coleman, R. B. Somoano, *Solid State Commun.* **40**, 979 (1981); B. A. Weinstein, M. L. Slade, A. J. Epstein, J. S. Miller, *ibid.* **37**, 643 (1981); J. B. Torrance, B. A. Scott, B. Welber, F. B. Kaufman, P. E. Seiden, *Phys. Rev. B* **19**, 730 (1979).
30. H. Shiba, *Phys. Rev. B* **6**, 930 (1972); J. B. Torrance, Y. Tomkiewicz, B. D. Silverman, *ibid.* **15**, 4738 (1977).

31. K. Mortensen, E. M. Conwell, J. M. Fabre, *ibid.* **28**, 5856 (1983); P. Delhaes, *Mol. Cryst. Liq. Cryst.* **96**, 229 (1983); C. Coulon *et al., J. Phys.* **43**, 1059 (1982).

32. P. J. Toscano and T. J. Marks, *Mol. Cryst. Liq. Cryst.* **118**, 337 (1985); *J. Am. Chem. Soc.*, in press.

33. B. N. Diel *et al., J. Am. Chem. Soc.* **105**, 1551 (1983); C. W. Dirk, K. F. Schoch, Jr., T. J. Marks, *ibid.*, p. 1539; C. W. Dirk, K. F. Schoch, Jr., T. J. Marks, in *Conductive Polymers*, R. B. Seymour, Ed. (Polymer Science and Technology Series, Plenum, New York, 1981), vol. 15, pp. 209–226; C. W. Dirk, E. A. Mintz, K. F. Schoch, Jr., T. J. Marks, *J. Macromol. Sci. Chem.* **16**, 275 (1981); T. J. Marks, K. F. Schoch, Jr., B. R. Kundalkar, *Synth. Met.* **1**, 337 (1980); K. F. Schoch, Jr., B. R. Kundalkar, T. J. Marks, *J. Am. Chem. Soc.* **101**, 7071 (1979); R. D. Joyner and M. E. Kenney, *ibid.* **82**, 5790 (1960).

34. For studies of isoelectric group IIIA [M(Pc)F]$_n$ materials, see R. S. Nohr, P. M. Kuznesof, K. J. Wynne, M. E. Kenney, P. G. Siebenman, *J. Am. Chem. Soc.* **103**, 4371 (1981); for studies of [M(Pc)L]$_n$ materials where L is a bridging organic ligand and M is a transition metal, see O. Schneider and M. Hanack, *Mol. Cryst. Liq. Cryst.* **81**, 273 (1982); B. N. Diel *et al., J. Am. Chem. Soc.* **106**, 3207 (1984); M. Hanack, *Chimia* **37**, 238 (1983).

35. S. H. Carr, X. Zhou, T. Inabe, T. J. Marks, *Polym. Mater. Sci. Eng.* **49**, 94 (1983); X. Zhou, T. J. Marks, S. H. Carr, *ibid.* **51**, 651 (1984); *J. Polym. Sci. Polym. Phys. Ed.*, in press.

36. X. Zhou, T. J. Marks, S. H. Carr, *Mol. Cryst. Liq. Cryst.* **118**, 357 (1985); *J. Polym. Sci., Polym. Phys. Ed.* **23**, 305 (1985).

37. S. Kirkpatrick, *Rev. Mod. Phys.* **45**, 574 (1973); C. H. Seager and G. E. Pike, *Phys. Rev. B* **10**, 1435 (1974).

38. P. Sheng, *Phys. Rev. B* **21**, 2180 (1980); Y.-W. Park, A. J. Heeger, M. A. Druy, A. G. MacDiarmid, *J. Chem. Phys.* **73**, 946 (1980), and references therein.

39. T. Inabe, J. W. Lyding, M. K. Moguel, C. R. Kannewurf, T. J. Marks, *Mol. Cryst. Liq. Cryst.* **93**, 355 (1983); T. Inabe, J. W. Lyding, M. K. Moguel, T. J. Marks, *J. Phys. (Paris)* **C3**, 625 (1983); T. Inabe, M. K. Moguel, T. J. Marks, R. L. Burton, J. W. Lyding, C. R. Kannewurf, *Mol. Cryst. Liq. Cryst.* **118**, 349 (1985); T. Inabe, J. W. Lyding, M. K. Moguel, R. L. Burton, C. R. Kannewurf, T. J. Marks, in preparation.

40. T. J. Marks, C. W. Dirk, K. F. Schoch, Jr., J. W. Lyding, in *Molecular Electronic Devices*, F. L. Carter, Ed. (Dekker, New York, 1982), pp. 195–210; M. K. Moguel and T. J. Marks, *Synth. Met.*, in press.

41. M. K. Moguel, J. G. Gaudiello, W. J. Pietro, T. J. Marks, *Synth. Met.*, in press.

42. J. C. W. Chien, *Polyacetylene* (Academic Press, New York, 1984); S. Etemad, A. J. Heeger, A. G. MacDiarmid, *Annu. Rev. Phys. Chem.* **33**, 443 (1982).

43. A. F. Diaz and K. K. Kanezawa, in (*3*), vol. 3, pp. 417–442; G. Tourillon and F. Garnier, *J. Electroanal. Chem.* **135**, 173 (1982).

44. L. W. Schacklette *et al., J. Chem. Phys.* **73**, 4098 (1980).

45. J. F. Rabolt, T. C. Clark, K. K. Kanazawa, J. R. Reynolds, G. B. Street, *J. Chem. Soc. Chem. Commun.* (1980), p. 347; R. R. Chance *et al., ibid.* (1980), p. 348.

46. K. Kaneto, K. Yoshino, Y. Inuishi, *Solid State Commun.* **46**, 389 (1983).

47. W. P. Su, J. R. Schrieffer, A. J. Heeger, *Phys. Rev. B* **22**, 2099 (1980).

48. J. C. Scott, P. Pfluger, M. T. Krounbi, G. B. Street, *ibid.* **28**, 2140 (1983).

49. H. S. Kaufman and J. J. Falcetta, Eds., *Introduction to Polymer Science and Technology* (Wiley, New York, 1977).

50. K. I. Lee and H. Jopson, *Makromol. Chem. Rapid Commun.* **4**, 375 (1983); M. E. Galvin and G. E. Wnek, *J. Polym. Sci. Polym. Chem. Ed.* **21**, 2727 (1983); O. Niwa and T. Tamamura, *J. Chem. Soc. Chem. Commun.* (1984), p. 817; M.-A. DePaoli, R. J. Waltman, A. F. Diaz, J. Bargon, *ibid.*, p. 1015.

51. W. Wernet, M. Monkenbusch, J. Wegner, *Makromol. Chem. Rapid Commun.* **5**, 157 (1984).

52. M. Aldissi and R. Liepins, *J. Chem. Soc. Chem. Commun.* (1984), p. 255.

53. E. T. Kang, P. Ehrlich, A. P. Bhatt, W. A. Anderson, *Appl. Phys. Lett.* **41**, 1136 (1982).

54. See, for example, S. F. Wellinghoff, T. Kedrowski, S. Jeneke, H. Ishida, *J. Phys. (Paris)* **C3**, 677 (1983); S. E. Tunney, J. Suenaga, J. K. Stille, *Macromolecules* **16**, 1398 (1983).

55. P. M. Grant, *Phys. Rev. B* **27**, 3934 (1983); *ibid.* **26**, 6888 (1982); F. Herman, *Phys. Scr.* **16**, 303 (1977); F. Herman, D. R. Salahub, R. P. Messmer, *Phys. Rev. B* **16**, 2453 (1977).

56. M.-Y. Whangbo, in (*3*), vol. 2, pp. 127–158; C. L. Honeybourne, *J. Chem. Soc. Chem. Commun.* (1982), p. 744; M. C. Böhm, *Solid State Commun.* **45**, 117 (1983); M.-H. Whangbo and K. R. Stewart, *Isr. J. Chem.* **23**, 133 (1983); S. Alvarez and E. Canadell, *Solid State Commun.* **50**, 141 (1984); E. Canadell and S. Alvarez, *Inorg. Chem.* **23**, 573 (1984).

57. K. A. Doris, M. A. Ratner, D. E. Ellis, B. Delley, T. J. Marks, *Bull. Am. Phys. Soc.* **28**, 245 (1983); E. Ciliberto *et al., J. Am. Chem. Soc.* **106**, 7748 (1984); K. A. Doris *et al.*, submitted for publication; N. S. Hush and A. S. Cheung, *Chem. Phys. Lett.* **47**, (1977).

58. W. J. Pietro, D. E. Ellis, T. J. Marks, M. A. Ratner, *Mol. Cryst. Liq. Cryst.* **105**, 273 (1984).

59. E. J. Baerends, D. E. Ellis, P. Ros, *Chem. Phys.* **2**, 41 (1973); Z. Berkovitch-Yellin, D. E. Ellis, M. A. Ratner, *ibid.* **62**, 21 (1981); B. Delley and D. E. Ellis, *J. Chem. Phys.* **76**, 1949 (1982); A. Rosen *et al., ibid.* **65**, 3269 (1975).

60. K. A. Doris, M. A. Ratner, D. E. Ellis, T. J. Marks, *J. Phys. Chem.* **88**, 3157 (1984); K. A. Doris, D. E. Ellis, M. A. Ratner, T. J. Marks, *J. Am. Chem. Soc.* **106**, 2491 (1984).

61. J. M. Williams *et al., J. Am. Chem. Soc.* **105**, 643 (1983).

62. W. J. Pietro, M. A. Ratner, T. J. Marks, *J. Am. Chem. Soc.* **107**, 5387 (1985).

63. P. J. Flory, *Proc. R. Soc. London* **234**, 73 (1956); *J. Polym. Sci.* **49**, 105 (1961).

64. J. R. Schaefgen *et al.*, in *Ultra-High Modulus Polymers*, A. Cifferi and I. M. Ward, Eds. (Applied Science, London, 1979), pp. 173–201; E. E. Magat and R. E. Morrison, *ChemTech* **6**, 702 (1976); G. B. Carter and V. T. J. Schenk, in *Structure and Properties of Oriented Polymers*,

533

I. M. Ward, Ed. (Wiley, New York, 1976), chapter 13; M. G. Dobb and J. D. McIntyre, *Adv. Polym. Sci.* **60–61**, 61 (1984).

65. T. Inabe, J. W. Lyding, M. K. Moguel, T. J. Marks, *J. Phys. (Paris)* **C3**, 625 (1983); T. Inabe, J. W. Lyding, T. J. Marks, *J. Chem. Soc. Chem. Commun.* (1983), p. 1084.

66. T. Inabe, J. F. Lomax, J. W. Lyding, C. R. Kannewurf, T. J. Marks, *Synth. Met.* **9**, 303 (1984); T. Inabe et al., *Macromolecules* **17**, 262 (1984).

67. T. Inabe et al., *Synth. Met.* **13**, 219 (1986).

68. K. J. Wynne, A. E. Zachariades, T. Inabe, T. J. Marks, *Polym. Commun.* **26**, 162 (1985).

69. W.-B. Liang and T. J. Marks, unpublished results.

70. This research on conductive polymers was supported by the National Science Foundation through the Northwestern Materials Research Center (grant DMR82-16972) and by the Office of Naval Research. It is a sincere pleasure to acknowledge the invaluable contributions of R. L. Burton, S. H. Carr, C. W. Dirk, K. A. Doris, D. E. Ellis, I. L. Fragalà, F. H. Herbstein, T. Inabe, C. R. Kannewurf, W.-B. Liang, J. W. Lyding, W. J. McCarthy, M. K. Moguel, W. J. Pietro, M. A. Ratner, G. M. Reisner, K. F. Schoch, Jr., K. J. Wynne, and X. Zhou to this endeavor.

Because of their widespread use as electrical insulators, synthetic polymers have generally come to be regarded as electrically passive. However, in the past two decades, polymeric materials have been developed that can be used as active elements of an electrical circuit; most of these are piezoelectric, although, very recently, polymers that can be made electrically conductive have also appeared. Piezoelectric polymers have elicited much interest because of their many realized and potential applications as electromechanical transducers, for which they are particularly well suited as a result of their easy processability into thin, light, tough, and flexible films.

By the simplest and broadest definition, a piezoelectric material is one that undergoes a change in electrical polarization in response to mechanical stress (or vice versa). From a solid-state viewpoint, piezoelectricity is defined in a much more rigorous and restricted sense as an intrinsic property of only certain crystal classes. The criteria can be made successively more specific to distinguish two important related propertie',—pyroelectricity and ferroelectricity. Of the 32 crystal classes into which all crystalline materials may be categorized, some have symmetry elements that render them nonpolar; other, noncentrosymmetric, crystal classes can exhibit polarization, and it is these that are rigorously defined as piezoelectric. Some such piezoelectric crystals develop polarization only when stressed, while others are permanently polar; the latter crystals will respond not only to stress but also to changes in temperature, and are therefore termed pyroelectric. A final subdivision yields the ferroelectric crystals: these also have a unique polar axis,

36

Ferroelectric Polymers

Andrew J. Lovinger

Science 220, 1115–1121 (10 June 1983)

whose direction can be reoriented by application of an electric field. A ferroelectric crystal is thus a pyroelectric one that exhibits reversible polarization.

In view of the complex molecular, crystalline, and morphological structure of polymers, it may appear surprising that any exist which comply with the very restrictive requirements of ferroelectricity. Nevertheless, in the past few years at least one polymer, poly(vinylidene fluoride), and some of its copolymers have been shown to be ferroelectric, while other piezoelectric polymers are also likely candidates. Poly(vinylidene fluoride) (abbreviated PVF_2), whose molecular repeat formula is $(CH_2-CF_2)_n$, exhibits by far the strongest piezoelectric and pyroelectric activity of all known polymers. These two properties were first reported in 1969 and 1971 (1), respectively, and many actual and potential applications of PVF_2 have been described since then.

The primary goal of this article is to explain the extraordinary characteristics of structure, at all its levels, that must be combined to render a polymer ferroelectric. These are illustrated by investigating in detail the monomeric, macromolecular, crystalline, and morphological structure of PVF_2 and distinguishing it at all these levels from that of other polymers; copolymers of PVF_2 and other piezoelectric (and potentially ferroelectric) polymers are covered more briefly. Macroscopic polarization of samples and their consequent electrical behavior are also discussed; the article then concludes with a brief description of applications of PVF_2. The treatment is necessarily condensed, but extensive reviews are available (2–6). No attempt is made to discuss related materials, such as those in the important categories of (i)

piezoelectric and pyroelectric biopolymers, and (ii) polymeric transducers ("electrets") whose piezoelectric behavior is not of inherent dipolar origin but results from injection of charges; comprehensive reviews exist for both categories [(7) and (8), respectively].

Structure

Monomeric. Macromolecular chains consist of many elementary repeating units, called monomers, that have been chemically linked during polymerization. Many monomers contain polar chemical groups. To yield useful piezoelectric polymers, their constituents should not be so bulky as to prevent crystallization of the macromolecules or to force them into shapes (such as helical) that cause extensive internal compensation of polarization; the resulting macromolecules should also be chemically stable and not cross-linked into infusible and insoluble solids. In light of these considerations, fluorocarbons are highly appropriate monomers to yield piezoelectric polymer crystals. The fluorine atom is very small, its van der Waals radius (1.35 angstroms) being only slightly larger than that of hydrogen (1.2 Å), and it forms highly polar bonds with carbon, having a dipole moment $\mu = 6.4 \times 10^{-30}$ coulomb-meter [the traditional unit of μ—the debye—is expressed in the metric system as 3.34×10^{-30} C-m]. Common resulting polyfluorocarbons are PVF_2, plus poly(vinyl fluoride) [PVF, molecular formula $(CH_2-CHF)_n$] and polytrifluoroethylene [PF_3E, $(CF_2-CHF)_n$]. Other polar groups that can be expected to lead to useful piezoelectric polymers include C–Cl ($\mu = 7.0 \times 10^{-30}$ C-m) and C–CN ($\mu = 12.9 \times 10^{-30}$ C-m); typical poly-

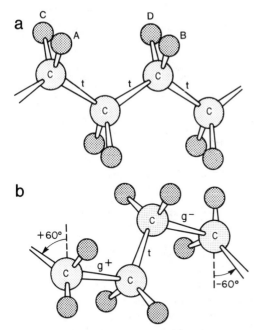

a

C   D
A   B
C   C
t   t   t
C   C

b

+60°
C
g⁻
C
g⁺   t
C   C
C
|−60°

Fig. 1. Schematic representation of polymer chain segments in different conformations: (a) *ttt* and (b) *g⁺tg⁻* .

mers incorporating these groups are poly(vinyl chloride) [PVC, $(CH_2–CHCl)_n$], poly(vinylidene chloride) [PVC₂, $(CH_2–CCl_2)_n$], and polyacrylonitrile [PAN, $(CH_2–CHCN)_n$]. A further possibility is associated with the highly polar hydrogen bond, C=O---H–N ($\mu = 12.0 \times 10^{-30}$ C-m) found in polyamides. We will see how successively higher levels of structure affect all these possibilities and how PVF₂ has emerged as the strongest piezoelectric and pyroelectric polymer with clearly demonstrated ferroelectric behavior.

Chain configuration. This term pertains to the stereochemical manner in which monomers are linked together; it thus describes characteristics of a macromolecule that can be altered only by cleavage of chemical bonds and that are, in effect, immutable after polymeriza-

tion. The monomers listed above have a directionality (9); if we use PVF₂ as an example and denote $-CH_2$ as the "head" of the monomeric unit and $-CF_2$ as its "tail," we usually have regular head-to-tail sequences but can also have reversed monomeric addition leading to head-to-head and tail-to-tail defects. Evidence from nuclear magnetic resonance spectroscopy shows that most monomers are added "isoregically" (that is, head to tail in the same direction) during polymerization of PVF₂ (95 to 97 percent), but less regularly so in the case of some other polymers—for example, 88 to 90 percent in PVF and 87 to 89 percent in PF₃E (10). Moreover, head-to-head defects are generally followed by tail-to-tail addition (10), which causes the average dipole moment of the chain per monomeric unit to be reduced by only 6 to 10 percent in PVF₂ but 20 to 24 percent in PVF and 22 to 26 percent in PF₃E.

A second type of configurational defect commonly seen in vinyl polymers and trifluoroethylene results from their stereochemically asymmetric carbon atoms. Using PVC as an example, we can see with the aid of Fig. 1a that the chlorine atoms could be either always on the same side of an extended carbon chain (positions A and B or C and D), or regularly alternating (A and D or B and C), or randomly located. Although special techniques and catalysts exist that promote stereoregular addition, vinyl polymers are most commonly available only in "atactic," or random, configurations that render them (with few exceptions, such as PVF) incapable of entering into crystallographic lattices and thus excluded from possible ferroelectric behavior. On the other hand, vinylidene polymers, not having asymmetric carbon atoms, are immune from such considerations of tacticity.

Chain conformation. In the melt or in solution, polymer chains have randomly coiled shapes (conformations), which configurationally disordered macromolecules are, to a great extent, forced to retain in their solid state as well; however, chains free from configurational defects can crystallize into regular conformations when cooled from the melt. This is accomplished by rotation about single bonds in a manner that minimizes the potential energy of the chains arising from internal steric and electrostatic interactions. The most favorable torsional bond arrangements have substituents at 180° to each other (called *trans* or *t*) or at ± 60° (*gauche*$^\pm$ or *g*$^\pm$); actual torsional angles commonly deviate somewhat from these values. A sequence of three *t* bonds is illustrated in Fig. 1a and a *g*$^+$*tg*$^-$ in Fig. 1b.

The molecular chains of most polymers are restricted by steric and electrostatic intramolecular interactions into one regular conformation of lowest potential energy. In this regard PVF$_2$ stands out from its counterparts, in that it can adopt at least three regular conformations (of similar potential energies) for reasons associated with the van der Waals radii of its constituents. PVF$_2$ is in the middle of a family of polyfluoroethylenes, whose general repeat formula is (CH$_a$F$_{2-a}$–CH$_b$F$_{2-b}$)$_n$ with *a* and *b* = 0, 1, or 2. Macromolecules of that class that are rich in hydrogen (the smallest possible chain substituent)—that is, polyethylene, (CH$_2$–CH$_2$)$_n$, and PVF, (CH$_2$–CHF)$_n$—will encounter only low rotational barriers separating their possible conformations; as a result, they will easily reach the lowest-energy conformation (the all-*trans*) and adopt only that. At the

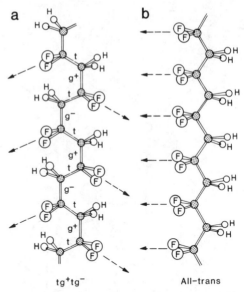

tg$^+$tg$^-$ All-trans

Fig. 2. Schematic depiction of the two most common crystalline chain conformations in PVF$_2$: (a) *tg*$^+$*tg*$^-$ and (b) all-*trans*. The arrows indicate projections of the −CF$_2$ dipole directions on planes defined by the carbon backbone. The *tg*$^+$*tg*$^-$ conformation has components of the dipole moment both parallel and perpendicular to the chain axis, while the all-*trans* conformation has all dipoles essentially normal to the molecular axis.

other end of the spectrum are members rich in the somewhat bulkier fluorine—that is, PF$_3$E, (CF$_2$–CHF)$_n$, and polytetrafluoroethylene, (CF$_2$–CF$_2$); in these, rotation is now sterically quite hindered, so that they are forced to adopt the one conformation of least discomfort (such a conformation is invariably helical or quasi-helical). PVF$_2$ molecules, containing two hydrogen and two fluorine atoms per repeat, are intermediate between these two extremes: they have a choice of multiple conformations, as do their hydrogen-rich counterparts, yet because rotational barriers are now high, the chains can also be stabilized into favorable conformations other than that of lowest energy.

The three known conformations of PVF_2 (*11–13*) are all-*trans*, tg^+tg^-, and $tttg^+tttg^-$ (in all cases there are slight deviations from the 180° and ±60° torsional angles). The first two conformations are by far the most common and important ones and are depicted schematically in Fig. 2. Because of the alignment of all its dipoles in the same direction normal to the chain axis, the all-*trans* is the most highly polar conformation in PVF_2 (7.0×10^{-30} C-m per repeat). The tg^+tg^- conformation is also polar, but because of the inclination of dipoles to the molecular axis (Fig. 2), it has components of the net moment both perpendicular (4.0×10^{-30} C-m) and parallel (3.4×10^{-30} C-m) to the chain; approximately the same values also characterize the $tttg^+tttg^-$ conformation.

It is interesting to note that, for the reason discussed above, PVF differs from other stereoirregular macromolecules in that its atacticity does not prevent it from crystallizing with a regular conformation (all-*trans*); while its dipole moment per repeat is about 50 percent that of PVF_2, regicity defects reduce its overall value for the chain still further.

Chain packing. The ability of molecules of PVF_2 and of other polymers to adopt polar conformations is not sufficient to ensure polarity of their resulting crystals, for these molecules may be crystallographically packed in a lattice so as to cancel each other's moment. For instance, the most common polymorph of PVF_2, the α-phase, which may be obtained by melt-solidification at all temperatures, suffers from exactly such cancellation. As may be seen in Fig. 3a, the unit cell of the lattice of α-PVF_2 consists of two chains in the tg^+tg^- conformation, whose dipole components normal

to the chain axes are antiparallel, thus neutralizing each other (*11*). The disposition of axial components of the dipole moment has been controversial, with suggestions of a regular antiparallel packing (*11*) (that is, one chain as in Fig. 2a, adjacent to one whose C–F dipoles point up) or of a statistical arrangement (*12*); however, recent evidence (*14*) lends support to both by showing that α-chains are generally under statistical packing, which becomes regularly antiparallel upon heat treatment at high temperatures.

Although the dipole moments of tg^+tg^- chains in the α-unit cell are internally compensated, a polar analog (called the δ-phase) can be obtained by application of a high electric field to films of this polymorph (*15*); in effect, this involves rotation of every second chain by 180° about its axis, so that molecules are now packed with the transverse components of their dipole moments pointing in the same direction (Fig. 3b). It is unlikely that chains rotate as physically rigid units; dipole reversal may rather involve propagation of a twist wave along the chains (*16*) or small intramolecular rearrangements (*17*).

The most highly polar phase of PVF_2 is the β-phase, whose unit cell consists of two all-*trans* chains packed with their dipoles pointing in the same direction (*11*) (Fig. 3c); an antipolar analog of this phase is not known. Packing of chains in β-PVF_2 is such that fluorine and hydrogen atoms of neighboring chains are approximately at the same level parallel to the *a*-axis of the unit cell (Fig. 3c); this very favorable intermolecular contribution to the potential energy plays a major role in stabilizing the crystalline structure of β-PVF_2. Chains of the $tttg^+tttg^-$ conformation are packed in a polar fashion to yield the γ-phase of PVF$_2$ (*13*). Its unit cell has the same base dimensions as that of its α-counterpart (Fig. 3a), a fact that enables a solid-state transformation from α-PVF$_2$ to the thermodynamically more stable γ-phase to occur at high temperatures (~ 160°C) solely through limited intramolecular motions (*18*). A nonpolar analog of γ-PVF$_2$ has also been reported but may exist only within a mixture of phases obtained at very high temperatures (*14*). This wealth of crystalline phases for PVF$_2$ is very unusual among polymers (where one or two polymorphs are common) and is another reflection of its unique molecular structure. PVF, on the other hand, crystallizes with only one unit cell that is essentially identical to that of β-PVF$_2$.

Morphology. We have seen so far that polarization of individual chemical groups can survive in a noncompensatory manner through successively higher levels of structure to yield individually polar crystals. However, polymer crystals are extremely small and, when grown from the melt, are arranged into essentially spherically symmetric polycrystalline aggregates that have no net polarization. These aggregates are called spherulites and result from nucleation of primary crystals within the melt, followed by radial growth outward from these nuclei in spherical envelopes (*19*). PVF$_2$ melts at ~ 170° to 200°C (depending on polymorphic form and crystallization temperature) and when cooled from the melt crystallizes in the form of spherulites of the nonpolar α-phase; at high temperatures of crystallization, γ-spherulites are also obtained (*18*). The β-phase is not usually produced from the melt since that requires high pressures (*20*) or epitaxial techniques (*6*), but is obtained by mechanical deformation or electrical

poling as described in the next section. Typical spherulites of α- and γ-PVF$_2$ are seen in Fig. 4a, and a schematic representation of their microstructure is depicted in Fig. 4b. This microstructure is such that what appear in the optical microscope as radial fibers are, in fact, stacks of very thin, platelet-like crystals called lamellae (about 10 nanometers thick and several micrometers in lateral dimensions). These lamellae consist of macromolecular segments that are packed crystallographically, while the intervening amorphous regions contain chain segments in disordered conformations; this two-phase structure of the solid state is typical of crystallizable polymers. In PVF$_2$, crystalline lamellae represent about 50 percent of the total mass, the other half being amorphous. In summary, our review of structure has shown that, while some polymers can form polar crystals, their three-dimensional arrangement in macroscopic specimens results in internal electrical compensation. To overcome this, such specimens must then be externally polarized.

Polarization

The most common technique for obtaining macroscopically polar films in PVF$_2$ involves first mechanical extension and then electrical poling (see Fig. 5). Mechanical drawing causes a breakdown of the original spherulitic structure into an array of crystallites whose molecules are oriented in the direction of the force. When such deformation takes place at high temperatures (for instance, 140° to 150°C), the original tg^+tg^- chains are free to slide past each other without altering their conformation, so that the resulting structure is still of the nonpolar

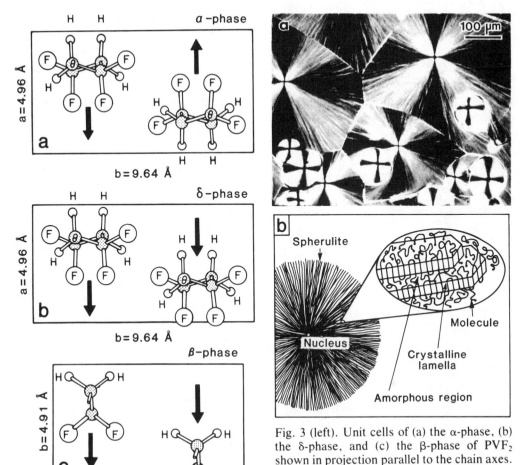

Fig. 3 (left). Unit cells of (a) the α-phase, (b) the δ-phase, and (c) the β-phase of PVF₂ shown in projection parallel to the chain axes. Arrows indicate dipole directions normal to the molecular axes.   Fig. 4 (right). (a) Typical appearance in the polarizing microscope of spherulites of PVF₂ crystallized from the melt at 160°C. The large spherulites are of the antipolar α-phase; the small ones belong to the polar γ-phase. The dark crosses correspond to the optical polarization directions of the microscope. (b) Schematic representation of the structure of polymer spherulites.

α-phase. However, when the deformation is conducted below ~ 90°C, where the polymer is much stiffer, the molecular chains are also forced into their most extended possible conformation, which is the all-*trans*: its monomeric repeat corresponds to 2.56 Å compared to 2.31 Å for the tg^+tg^- (see also Fig. 2). Therefore, mechanical extension at low temperatures has the advantages of producing not only a molecularly oriented morphology, but also one belonging to the polar β-phase. However, as may be seen with the aid of Fig. 5, the dipole vectors are still not uniquely oriented but lie randomly in planes normal to the molecular chains. The final step that is therefore required to produce a macroscopi-

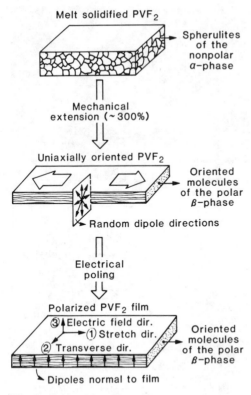

Melt solidified PVF$_2$

Spherulites of the nonpolar α-phase

Mechanical extension (~300%)

Uniaxially oriented PVF$_2$

Oriented molecules of the polar β-phase

Random dipole directions

Electrical poling

Polarized PVF$_2$ film

③ Electric field dir.
① Stretch dir.
② Transverse dir.

Oriented molecules of the polar β-phase

Dipoles normal to film

Fig. 5. Schematic representation of the processes commonly employed to obtain piezoelectrically and pyroelectrically active films of PVF$_2$.

cally polar specimen is to align these dipoles in the direction of an externally applied field normal to the film. This is usually accomplished by evaporating electrodes on the sample and connecting them to a high-voltage source supplying a field of at least 0.5 megavolt per centimeter (*21*), or by using plasma or corona poling (*22*). The facility of field-induced dipole reorientation in β-PVF$_2$ is related to the pseudohexagonal packing of its chains (*23*) (a check of the unit cell base dimensions in Fig. 3c will confirm that they differ only by about 1 percent from hexagonal); this is obviously a factor of major importance in facilitating ferro-

electric switching in β-PVF$_2$, and suggests that this phenomenon takes place in 60° increments (*23*). Films of the α-phase can also be poled: at intermediate fields (1 to 3 MV/cm), the tg^+tg^- chains preserve their conformation, but are packed in a polar unit cell ($\alpha \rightarrow \delta$ transformation), whereas at higher fields (~ 5 MV/cm) they are further transformed both intra- and intermolecularly to the β-phase.

Ferroelectricity in PVF$_2$

Whether PVF$_2$ is a true ferroelectric rather than a trapped-charge electret was a controversial issue for about a decade after the discovery of its strong piezoelectricity; despite its clearly dipolar crystalline structure, there was evidence that polarization is inhomogeneous across the thickness of PVF$_2$ films (being much higher at the side facing the positive electrode), so that piezoelectricity might simply have been a result of trapped charges injected by the electrodes (*24*). It has now been shown that this anisotropy disappears at high poling fields, that dipoles are, in fact, reoriented during application of an electric field, and that other typical phenomena accompanying ferroelectricity—hysteresis loops and Curie transitions—are also seen in PVF$_2$.

Dipolar reorientation has been proved by x-ray (*4*) and infrared (*25*) techniques; results from the latter are depicted in Fig. 6a. Here, the ratio of infrared intensity from a β-PVF$_2$ sample parallel to the molecular chains ($I_{\parallel}$) and perpendicular to them ($I_{\perp}$), the latter parallel to the $-CF_2$ dipole moment, is seen to vary with electric field and to be reversible in a hysteresis-like fashion typical of a ferroelectric. Simultaneously obtained curves of dielectric constant, ϵ, versus poling field show similar behavior (Fig. 6b). What has broadly been considered the most characteristic experimental manifestation of ferroelectricity is the hysteresis behavior of polarization as a function of electric field. Such a curve is also observed for β-PVF$_2$ in Fig. 6c, which shows that polarization rises with applied field, that a "remanent polariza-tion" (P_r in Fig. 6c) persists after the field has been returned to zero, that a negative "coercive field" (E_c in Fig. 6c) is required to depolarize the sample, and that polarization can be repeatedly re-versed by cycling the field between ± 2 MV/cm.

At this stage we may inquire as to what renders PVF$_2$ a ferroelectric in-stead of simply a pyroelectric material whose dipole directions are stable rather

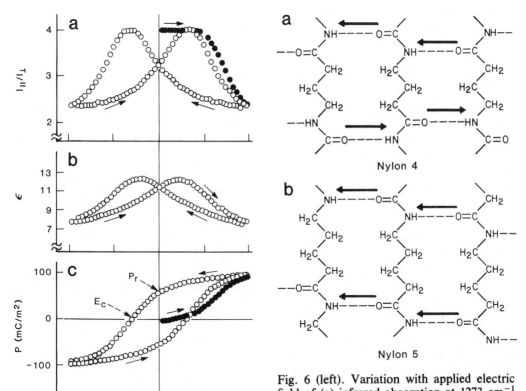

Fig. 6 (left). Variation with applied electric field of (a) infrared absorption at 1273 cm^{-1} corresponding to the $-$CF$_2$ dipole moment of β-PVF$_2$, (b) dielectric constant, and (c) polar-ization of a film of β-PVF$_2$. All curves show the hysteresis behavior typical of ferroelectric materials. (●) Initial poling results, (○) results from subsequent poling cycles. [Reprinted with permission of T. Takahashi, M. Date, E. Fukada, *Applied Physics Letters* **33**, 132 (1978). © 1978 American Institute of Physics.]   Fig. 7 (right). Schematic depiction of hydrogen-bonded sheets in the crystal lattices of (a) even and (b) odd polyamides, showing the dipole directions associated with the hydrogen bonds in each case.

than reversible. For the β-phase, the pseudohexagonal character of its lattice clearly plays a major role (23). However, the exceptional molecular structure of PVF_2 detailed above is also a critical factor, for chains of highly anisotropic cross section, with bulky protrusions or pendant groups, would most likely be locked in place and not allow facile inter- or intramolecular dipole reorientation. The importance of chain structure in this regard is seen in the ferroelectric switching of the δ-phase, whose unit cell is far from hexagonal (Fig. 3, a and b), and where dipole reversal occurs almost certainly through an intramolecular mechanism (16, 17).

Ferroelectric behavior similar to that of β-PVF_2 has also been demonstrated in copolymers of vinylidene fluoride with trifluoroethylene or tetrafluoroethylene; these are randomly added copolymers containing commonly 60 to 80 mole percent VF_2. Since they contain a greater proportion of the comparatively bulky fluorine atoms than PVF_2, their molecular chains cannot accommodate the tg^+tg^- conformation and are therefore forced to crystallize directly with the more extended all-*trans* conformation (26). Copolymers of vinylidene fluoride and trifluoroethylene also exhibit another important aspect of ferroelectricity that so far has not been convincingly demonstrated in PVF_2. This is the Curie temperature at which a ferroelectric crystal undergoes reversibly a solid-state transformation to a nonpolar (paraelectric) state. In these copolymers, the Curie transition was found (27) to involve primarily intramolecular changes of dipole directions through introduction of $g^{\pm}$ bonds that alter the polar all-*trans* conformation to a somewhat disordered arrangement of $tg^{\pm}$ and tt sequences.

Moreover, by extrapolating the observed Curie temperatures as functions of VF_2 content to 100 percent VF_2, the expected ferroelectric-to-paraelectric transition in β-PVF_2 would be in the vicinity of 205°C (27)—about 20°C above its melting point—a fact that explains its experimental elusiveness; however, as the conformational changes characterizing these transitions in VF_2 copolymers were found (27) to span a wide temperature range (30°C or more), the earliest manifestations of such a transition in β-PVF_2 should occur in the vicinity of its own melting point.

Other Piezoelectric Polymers

Our discussion of structure has shown that most potentially ferroelectric polymers have been eliminated primarily for reasons associated with their chain conformation and packing. Poly(vinylidene chloride), while potentially similar in electrical activity to PVF_2, suffers from the presence of bulky chlorine atoms (van der Waals radii, 1.80 Å) which prevent its chains from adopting the highly polar all-*trans* conformation, forcing them instead into one approximating tg^+tg^- or 2_1-helical. Another polyfluorocarbon, PVF, although atactic, adopts a structure similar to that of β-PVF_2 and is expected to be ferroelectric (although data are lacking). Other polar atactic polymers (such as PVC) are amorphous, yet some piezoelectric response is obtained if an electric field is applied above their glass-transition temperatures (at which molecular motions and rotation of dipoles become possible) and held while the sample is cooled to room temperature to freeze the now oriented dipoles in place. Among the cyanide-containing polymers, polyacrylonitrile suffers from intra- and intermolecular association at its $-C{\equiv}N$ groups and poly(vinylidene cyanide) is chemically unstable, although a copolymer of the latter with vinyl acetate has been found to exhibit considerable piezoelectric activity (28).

However, there exists another class of polymers from the above, whose recently discovered piezoelectric activity is second only to that of PVF_2 and which may, in fact, be ferroelectric in nature. This is the class of polyamides (or nylons) of molecular repeat $-HN(CH_2)_{2n}CO-$. Polyamides crystallize in all-*trans* conformations and are packed so as to maximize the number of hydrogen bonds possible between adjacent amine and carbonyl groups, as seen schematically in Fig. 7 for two polyamides containing an even and an odd number of carbon atoms per repeat. The dipoles are clearly associated with the resulting hydrogen bonds, but, as seen in Fig. 7, cancel their net moments in the case of even nylons; it is therefore only the odd members—of which nylon 11, $[HN(CH_2)_{10}CO]_n$, is the most common—that possess a net polarization in their hydrogen-bonded sheets. These sheets are arranged in a polar manner with triclinic symmetry in the common α-phase of such polyamides (29); however, at high temperatures ($> 90°C$) or in rapidly quenched samples the symmetry becomes pseudohexagonal through what is believed to be a randomization of hydrogen bonding, yielding the γ-phase. It has been found (30) that films of γ-nylon 11 display high piezo- and pyroelectric activity when poled, while the response of the α-phase is significantly lower. These phenomena are interpreted by breaking of the existing hydrogen bonds and their re-formation in (at least partial) alignment with the field, a process which is indeed more favorable in the pseudohexagonal γ-phase and may explain its higher electrical response (30). It should be noted that there exists another category of polyamides having molecular repeats of the type $-HN(CH_2)_xNHOC-(CH_2)_yCO-$ which also crystallize in much the same manner; of these, only the "odd-odd" members (those with x and y odd) have their dipoles arranged in a polar unit cell and should thus be expected to behave similarly to nylon 11.

Piezoelectric and Pyroelectric Properties

Although the main purpose of this article is to explain the structural characteristics needed to render a polymer ferroelectric, the piezoelectric and pyroelectric properties of synthetic crystalline polymers are also briefly presented and compared with those of the common ceramics and single crystals.

The piezoelectric and pyroelectric effects are commonly described in terms of compliances between electrical polarization (P) and stress (σ) or temperature (T), respectively, yielding the piezoelectric constant, d, and the pyroelectric constant, p

$$d = \left(\frac{\partial P}{\partial \sigma}\right)_{E,T} \quad \text{and} \quad p = \left(\frac{\partial P}{\partial T}\right)_{E,\sigma}$$

where E is the electric field. The d constant is a tensor whose major components for polymers such as PVF_2 are d_{31}, d_{32}, and d_{33}, the first subscript defining the electric field direction and the second that of mechanical stress; as seen in Fig. 5, in an oriented sample direction 1 is the stretch direction, 2 the transverse, and 3 the film normal.

Realizable values of d and p in PVF_2 depend on a number of parameters, most important of which is the poling field strength; poling temperature plays a small role, while the effects of poling time are significant only up to fields of about 2 MV/cm (21). Typical values of the piezoelectric and pyroelectric coefficients, as well as of other physical properties, of materials that have been used as transducers are given in Table 1. The three-dimensional anisotropy of the piezoelectric effect is seen clearly for oriented PVF_2. The d constant associated with the molecular direction (d_{31}) is an order of magnitude greater than that transverse to the polymer chains (d_{32}); both are positive because a stress in the film plane reduces the specimen thickness, thus increasing the surface charge, whereas d_{33} is negative because a stress normal to the film increases its thickness.

In comparing the piezoelectric and pyroelectric strengths of the various materials in Table 1, we see that although PVF_2 and its copolymers stand out among other polymeric materials, their activity is an order of magnitude lower than that of the traditional piezoelectric ceramics. However, if we calculate the electromechanical coupling coefficient k_{31} (a quantity reflecting the transducing capability of a material) from $k_{31} = d_{31}\sqrt{c/\epsilon}$ (where c is the stiffness and ϵ the permittivity), we see in Table 1 that, as a result of their very low dielectric constant, PVF_2 and its copolymers appear comparably efficient to ceramics. Moreover, PVF_2 films can sustain about 100 times higher fields than ceramics, so that their power output per unit volume and their maximum elongation per unit field strength (quantities that are proportional to d^2cE^2) are about four to five

times greater than in ceramics. Nevertheless, PVF_2 and other polymers are inferior to ceramics in regard to highest temperature of use: piezoelectric constants in PVF_2 reach a maximum at $\sim 80°C$, where activity begins to fall as a result of increased molecular motions and consequent depolarization; this can be retarded to $\sim 110°C$ by cross-linking the molecules with about 40 megarads of γ-radiation (31). However, at even higher temperatures, chemical degradation sets in with loss of HF that increases from α- to γ- to β-PVF_2 as a result of greater steric and electrostatic intramolecular repulsions within trans-segments (6, 32).

Theoretical models have been developed to describe the macroscopic polarization in films of PVF_2 and to predict values of d and p. The two most applicable models treat PVF_2 as a two-phase system (based on its crystallinity of approximately 50 percent) containing either spherical, charged particles (33) or platelet-like polar crystals (34) dispersed within an amorphous, nonpolar matrix. Macroscopic piezoelectricity was found to be due not only to the contributions of polar crystallites, but also to a major contribution from the mechanical and electrical heterogeneity between crystalline and amorphous phases (33).

Applications

Piezoelectric and pyroelectric applications of polymers are for the moment confined to PVF_2 (6) and take advantage of its extraordinary combination of electrical, mechanical, and acoustical properties. Primary among them are its thinness, light weight, flexibility, toughness, and ability to be formed into intricate shapes. In addition to these and to its

Table 1. Typical physical, piezoelectric, and pyroelectric properties of various materials.

Material	Piezoelectric coefficient, d (pC/N)	Pyroelectric coefficient, p (μC/K·m^2)	Density, ρ (g/cm^3)	Elastic modulus, c (GN/m^2)	Dielectric constant, ϵ/ϵ_0	Electromechanical coupling coefficient, k (%)	Acoustic impedance (Gg/m^2-s)
PVF$_2$ (β-phase)	d_{31} = 20–30 d_{32} = 2–3 d_{33} = –30	30–40	1.8	1–3	10–15	11	2–3
PVF$_2$ (δ-phase)	d_{31} = 10–17 d_{32} = 2–3 d_{33} = 10–15	10–15					
Other polymers							
VF$_2$-trifluoroethylene copolymers	d_{31} = 15–30	30–40	~ 1.9		15–20	~ 20	
Poly(vinyl fluoride)	d_{31} = 1	10	1.4	~ 1			
Poly(vinyl chloride)	d_{31} = 1	1–3	1.5	~ 4	3		
Nylon 11 (γ-phase)	d_{31} = 3	3	1.1	1.5	4		
Ceramics and single crystals							
Lead zirconate titanate	d_{31} = 100–300	50–300	7.5	80	1200	30	25
Barium titanate	d_{31} = 80	200	5.7	110	1700	21	25
Quartz	d_{11} = 2		2.7	80	5	10	14

good electromechanical coupling discussed above, Table 1 shows that PVF_2 has a much lower acoustical impedance than ceramics (this quantity is proportional to the product of density and stiffness) and is therefore very suitable for acoustical applications in such media as air or water. Such applications have been reviewed (5, 6) and include microphones, loudspeakers, headphones, and omnidirectional tweeters, all having essentially flat responses over a broad range of frequencies (35). Useful similar applications have also been described in ultrasonics (36), particularly as hydrophones (that is, underwater transducers) for which PVF_2 is most appropriate because it consists of light atoms and has therefore a tenfold better acoustical coupling with water than do ceramics. Elec-

tromechanical applications include pressure switches, detectors in printing equipment, coin sensors, variable-focus mirrors, impact detectors, medical probes, as well as push buttons and keyboards. The structure of one such keyboard (37) is seen in Fig. 8: it consists of a metallized polar PVF_2 film in a holder; films as thin as 16 μm were able to withstand 15 million pushes without failure or deterioration (37). Finally, the pyroelectric properties of PVF_2 may be utilized in such devices as infrared detectors, vidicon cameras, security and alarm systems, and electrostatic copiers (38).

All of the above demonstrate the usefulness and versatility of PVF_2 and show how the study of structure in polymers can aid in understanding and exploiting the properties of this new class of ferroelectric materials.

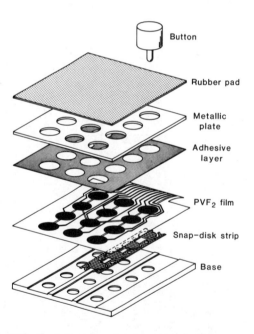

Button
Rubber pad
Metallic plate
Adhesive layer
PVF_2 film
Snap-disk strip
Base

Fig. 8. Schematic drawing of an electrical keyboard with a metallized and polarized PVF_2 film. [Reprinted with permission of G. T. Pearman, J. L. Hokanson, T. R. Meeker, *Ferroelectrics* **28**, 311 (1980).]

References and Notes

1. H. Kawai, *Jpn. J. Appl. Phys.* **8**, 975 (1969); J. G. Bergman, Jr., J. H. McFee, G. R. Crane, *Appl. Phys. Lett.* **18**, 203 (1971).
2. Y. Wada and R. Hayakawa, *Jpn. J. Appl. Phys.* **15**, 2041 (1976).
3. M. G. Broadhurst and G. T. Davis, in *Topics in Modern Physics—Electrets*, G. M. Sessler, Ed. (Springer-Verlag, Berlin, 1980), chapter 5.
4. R. G. Kepler and R. A. Anderson, *CRC Crit. Rev. Solid State Mater. Sci.* **9**, 399 (1980).
5. G. M. Sessler, *J. Acoust. Soc. Am.* **70**, 1596 (1981); _____ and J. E. West, in *Topics in Modern Physics–Electrets*, G. M. Sessler, Ed. (Springer-Verlag, Berlin, 1980), chapter 7.
6. A. J. Lovinger, in *Developments in Crystalline Polymers*, D. C. Bassett, Ed. (Applied Science, London, 1982), vol. 1, chapter 5.
7. E. Fukada, *Adv. Biophys.* **6**, 121 (1974); H. Athenstead, *Ann. N.Y. Acad. Sci.* **238**, 68 (1974); S. Mascarenhas, in *Topics in Modern Physics—Electrets*, G. M. Sessler, Ed. (Springer-Verlag, Berlin, 1980), chapter 6.
8. G. M. Sessler and J. E. West, *J. Acoust. Soc. Am.* **53**, 1589 (1973); in *Topics in Modern Physics—Electrets*, G. M. Sessler, Ed. (Springer-Verlag, Berlin, 1980), chapters 1, 2, and 7.
9. R. E. Cais and N. J. A. Sloane, *Polymer*, in press.
10. C. W. Wilson, III, and E. R. Santee, Jr., *J. Polym. Sci. Part C* **8**, 97 (1965); A. E. Tonelli, F. C. Schilling, R. E. Cais, *Macromolecules* **15**, 849 (1982).

11. R. Hasegawa, Y. Takahashi, Y. Chatani, H. Tadokoro, *Polym. J.* **3**, 600 (1972).
12. M. A. Bachmann and J. B. Lando, *Macromolecules* **14**, 40 (1981).
13. S. Weinhold, M. H. Litt, J. B. Lando, *ibid.* **13**, 1178 (1980); Y. Takahashi and H. Tadokoro, *ibid.*, p. 1317; A. J. Lovinger, *ibid.* **14**, 322 (1981).
14. A. J. Lovinger, *ibid.* **15**, 40 (1982).
15. G. T. Davis, J. E. McKinney, M. G. Broadhurst, S. C. Roth, *J. Appl. Phys.* **49**, 4998 (1978).
16. H. Dvey-Aharon, P. L. Taylor, A. J. Hopfinger, *ibid.* **51**, 5184 (1980).
17. A. J. Lovinger, *Macromolecules* **14**, 225 (1981).
18. _____, *Polymer* **21**, 1317 (1980); *J. Appl. Phys.* **52**, 5934 (1981); *J. Polym. Sci. Polym. Phys. Ed.* **18**, 793 (1980).
19. H. D. Keith and F. J. Padden, Jr., *J. Appl. Phys.* **35**, 1270, 1286 (1964).
20. J. Scheinbeim, C. Nakafuku, B. A. Newman, K. D. Pae, *ibid.* **50**, 4399 (1979).
21. J. M. Kenney and S. C. Roth, *J. Res. Natl. Bur. Stand.* **84**, 447 (1979).
22. J. E. McKinney, G. T. Davis, M. G. Broadhurst, *J. Appl. Phys.* **51**, 1676 (1980); P. D. Southgate, *Appl. Phys. Lett.* **28**, 250 (1976).
23. R. G. Kepler and R. A. Anderson, *J. Appl. Phys.* **49**, 1232 (1978); H. Dvey-Aharon, T. J. Sluckin, P. L. Taylor, A. J. Hopfinger, *Phys. Rev. B* **21**, 3700 (1980).
24. H. Sussner and K. Dransfeld, *J. Polym. Sci. Polym. Phys. Ed.* **16**, 529 (1978).
25. D. Naegele and D. Y. Yoon, *Appl. Phys. Lett.* **33**, 132 (1978); T. Takahashi, M. Date, E. Fukada, *ibid.* **37**, 791 (1980).
26. J. B. Lando and W. W. Doll, *J. Macromol. Sci. Phys.* **B2**, 205 (1968).
27. A. J. Lovinger, G. T. Davis, T. Furukawa, M. G. Broadhurst, *Macromolecules* **15**, 323 and 329 (1982); *Polymer*, in press.
28. S. Miyata, M. Yoshikawa, S. Tasaka, M. Ko, *Polym. J.* **12**, 857 (1980).
29. W. P. Slichter, *J. Polym. Sci.* **36**, 259 (1959).
30. B. A. Newman, P. Chen, K. D. Pae, J. I. Scheinbeim, *J. Appl. Phys.* **51**, 5161 (1980); J. I. Scheinbeim, *ibid.* **52**, 5939 (1981); V. Gelfandbein and D. Katz, *Ferroelectrics* **33**, 111 (1981).
31. T. T. Wang, *J. Polym. Sci. Polym. Lett. Ed.* **19**, 289 (1981); *Ferroelectrics* **41**, 213 (1982).
32. A. J. Lovinger and D. J. Freed, *Macromolecules* **13**, 989 (1980).
33. K. Tashiro, M. Kobayashi, H. Tadokoro, E. Fukada, *ibid.*, p. 691.
34. M. G. Broadhurst, G. T. Davis, J. E. McKinney, R. G. Collins, *J. Appl. Phys.* **49**, 4992 (1978).
35. M. Tamura, T. Yamaguchi, T. Oyaba, T. Yoshimi, *J. Audio Eng. Soc.* **23**, 21 (1975).
36. H. Sussner and K. Dransfeld, *Colloid Polym. Sci.* **251**, 591 (1979).
37. G. T. Pearman, J. L. Hokanson, T. R. Meeker, *Ferroelectrics* **28**, 311 (1980).
38. A. M. Glass, J. H. McFee, J. G. Bergman, *J. Appl. Phys.* **42**, 5219 (1971); J. H. McFee, J. G. Bergman, G. R. Crane, *Ferroelectrics* **3**, 305 (1972); J. G. Bergman, G. R. Crane, A. A. Ballman, H. M. O'Bryan, Jr., *Appl. Phys. Lett.* **21**, 497 (1972).
39. I am very grateful to H. D. Keith for many helpful discussions and for a critical review of the manuscript, and to E. Fukada and T. R. Meeker for permission to use their published figures.

In the early stages of the development of the integrated circuit, solid-state physics and materials science were viewed as the fundamental basis of microelectronics. Chemistry, in particular heterogeneous chemistry, which involves two different material phases such as a gas and a solid, now plays a major role in the fabrication of microelectronic circuits. Recently, laser-initiated, heterogeneous chemical reactions have been used in a variety of novel processing operations for the fabrication of microelectronics components. The study of these reactions has, in turn, led to new insights into the nature of light-driven chemical reactions at the surfaces of solids.

The development of laser-induced chemical processing for microelectronics drew on a considerable body of research in laser-induced chemistry. In general, the goal of early research was to use laser light to initiate specific chemical reactions in order to fabricate specific chemicals or to separate isotopes. In microelectronics fabrication, the specificity of laser-induced chemistry is important; however, the ability of laser light to confine reactions to submicrometer-scale regions is an equally important feature.

The ability to localize chemical reactions on solid surfaces has led to a number of techniques for semiconductor processing (1, 2). In addition, chemical reactions confined to such small regions of a solid-gas interface are sufficiently different from previously studied reactions that a new form of chemistry, laser-induced microchemistry, is evolving (3).

Historical Background

The fabrication of an integrated circuit, containing many transistors on a

37

Laser-Induced Chemistry for Microelectronics

R.M. Osgood
and T.F. Deutsch

Science 227, 709–714 (15 February 1985)

single silicon chip, can be viewed as the assembly of a patterned, multilayer, thin-film structure (4). In most semiconductor processing, photolithography is used extensively to pattern these layers of dielectrics or conductors. A photographic emulsion, or photoresist, is spun onto a wafer, exposed to light through a mask (akin to a photographic negative), and developed to reveal a patterned structure whose features may be as small as one micrometer. This photoresist layer serves to confine subsequent thin-film processing steps to the uncovered regions on the wafer, illustrated in Fig. 1. Insulating or conducting films can be grown by a variety of techniques, including condensation of vaporized metals, deposition using gas-phase chemical reactions (chemical vapor deposition), and oxidation of the silicon substrate. The films can also be etched, either by immersion in appropriate acids (wet chemistry) or by gases excited with an electrical discharge (plasma etching), transferring patterns in the photoresist to the underlying films. Doping involves the controlled introduction of electrically active impurities to portions of the wafer in order to change its electrical conductivity, again using an overlying film to determine the regions the dopant atoms can enter.

Since the current trend in microelectronics is to increase the number of components on a chip, it is desirable to make each component as small as possible. Whereas the features on commercial integrated circuits typically are 3 to 5 μm long, the features on chips that incorporate very large scale integration (VLSI), loosely defined as 65,000 to 2,000,000 components per chip, will be about 1 μm long. Ultradense circuits have put new demands on electronic materials and

processing technology. For example, the unique electrical requirements of small devices have required the development of new insulating and conducting materials. The high component densities also put a premium on yield per component or transistor and require, furthermore, that strategies be developed to overcome the effect of defects in individual components.

Semiconductor processing has increasingly moved to dry processing, in part to realize the high resolution needed for VLSI circuits. Dry processing includes such gas-phase operations as plasma etching and chemical vapor deposition. While most of these processes are well developed for silicon technology, they have features that are sometimes undesirable. For example, the high temperatures involved in some chemical vapor depositions, 900° to 1200°C, limit the kind and number of thin films that can be used, since those temperatures can cause unwanted reactions between layers or material degradation. Some compound semiconductors, such as gallium arsenide (GaAs) which is being used increasingly in fast circuits, decompose at these temperatures and require low-temperature processing.

With these considerations in mind, researchers have turned to lasers for the development of new methods of chemical processing for microelectronics. In comparison with other light sources, lasers offer several special capabilities. One of the better known ones is that a laser beam can be focused to dimensions comparable to the wavelength of the laser light. Thus, even a low-power laser can produce highly intense spots of light with micrometer dimensions. The localized heating generated by focused pulsed-laser beams has previously been used to produce a number of purely thermal effects useful in some areas of semiconductor processing, such as the trimming of thin-film resistors to achieve specified resistance values (5). Such a focused beam can also be used to initiate a chemical reaction that is confined to a region with dimensions comparable to or smaller than the laser spot. By focusing the laser beam onto the substrate and moving the substrate in a plane perpendicular to the beam axis, patterns can be written on a wafer using the laser-initiated doping, deposition, or etching processes described in more detail below. This form of microchemical processing has been termed "direct writing" (6) since it allows one to pattern a wafer directly in one or two steps, without need for the masking steps so important in standard photolithography. Because the patterning is direct, this technique is particularly useful for making discretionary changes in microelectronic components.

Lasers also have a high degree of temporal coherence; this makes it possible to use lasers to perform interferometric imaging. For example, by dividing the output of a laser into two beams and then recombining them at a substrate surface, an interference pattern with extremely fine features can be made. As in the case of direct writing, interferometric processing does not require prior photolithographic masking.

The monochromatic or near-monochromatic output of lasers offers another advantage. In many photochemical reactions, a specific excitation wavelength leads to a specific set of molecular fragments and, ultimately, reaction products. For photochemical reactions, there is a threshold photon energy, generally corresponding to a wavelength in the ultraviolet region, below which no reaction occurs. As a result, the comparatively recent development of high-power ultraviolet lasers has greatly enhanced the ability to perform laser-controlled chemistry. One class of ultraviolet lasers, pulsed excimer lasers, is capable of producing high average power (>40 W) and high peak power (>10 MW) from 150 to 350 nm (7). These lasers are able to dissociate a variety of molecular gases and produce large amounts of specific free radicals, atoms, or molecules that

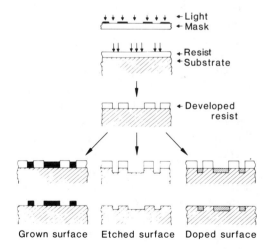

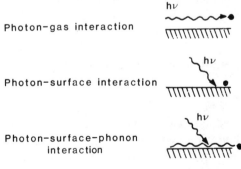

Fig. 1 (left). The main steps in conventional photolithographic processing of semiconductors.   Fig. 2 (right). Possible interactions of a laser beam with a gas-solid interface.

can then deposit on or etch a semiconductor wafer. In this case, laser-induced chemistry performs some of the same processing operations as conventional fabrication methods. However, the monochromatic light produced by the laser allows a specific or, at least, a limited number of reaction channels to be addressed, while conventional techniques, such as plasma etching, generally initiate complex chemical reactions. Hence, laser-induced chemistry has the potential of being both simpler and more controllable than the conventional techniques.

A final property of laser radiation is that it can often be generated in the form of very short pulses, $<10^{-8}$ second long. Laser heating using these short pulses allows an irradiated surface layer to melt and then resolidify before there is substantial conduction of heat into the bulk of the material. At ultraviolet wavelengths, many materials absorb so strongly that light penetrates only a very thin layer at the solid surface. The rapid heating and cooling of a surface region by these lasers has been used in the annealing of semiconductors (8) and in the formation of new kinds of metastable compounds (9), as well as in the doping of semiconductors, discussed below.

Laser-Initiated Surface Chemistry

When a laser beam initiates a chemical reaction near a solid surface, the reaction can occur on or above the surface and may be due to either thermal processes, photochemical processes, or a combination of both (Fig. 2).

Reactions in a gaseous medium above a solid surface are typically based on primary or secondary photochemical phenomena such as:

$$AB \xrightarrow{h\nu} A + B\downarrow$$

or

$$AB \xrightarrow{h\nu} A + B$$
$$B + CD \longrightarrow BC\downarrow + D$$

where $\downarrow$ indicates a species that interacts with the surface, such as an etchant or an atom that deposits on the surface, and $h\nu$ indicates irradiation by photons of energy $h\nu$. Gas-phase transport influences the resolution of the process, since atoms produced far away from the surface spread laterally in the gas as they diffuse toward the surface. Surface-controlled nucleation processes, as well as gas-phasing scavenging processes, minimize this problem, confining the deposit to the region illuminated by the laser beam (10).

Reactions on the surface can be due to either direct or photochemical dissociation of adsorbed molecules or to surface excitation followed by chemical reactions. Even in the case of direct photochemical phenomena, the surface may play a major role in the dissociation chemistry. For example, the yields from photochemical reactions on metal surfaces are expected to be low because of the transfer of molecular excitation to the surface. However, suitably altering the morphology of the metal surface in order to increase the surface excitation can result in a dramatic enhancement in the rate of a photoreaction on or near a surface (11).

Recent experiments using crossed molecular and laser beams have clarified some of the dynamics of energy transfer to the surface in nonreactive gas-surface collisions (12), but studies of the dynamics of surface reactions, with or without light, have not yet been made.

Laser excitation of the surface can also initiate chemical reactions indirectly, through surface heating or through the creation of electron-hole pairs. In the case of surface heating, the reaction mechanism is similar to that of conventional chemical vapor deposition (13). However, in laser-initiated chemical vapor deposition, the heating of the substrate can be localized to a thin surface layer or to a small spot by using a pulsed or a focused laser beam, respectively. Since heating occurs because of absorption of the laser beam by the solid substrate, the reaction rate is sensitive to the optical as well as the thermal properties of the surface. Initiation of reactions via the production of electron-hole pairs requires that the reaction involve an ionic species or a charge complex; thus, a liquid-solid interface is the ideal environment in which to observe these effects (14).

Laser-Assisted Dry Processing

The high average power produced by excimer lasers has led to numerous demonstrations of photon-assisted dry processing wafer-size areas. For example, one application of considerable importance, which can only be touched in here, is the use of excimer lasers as ultraviolet light sources for photolithography. Ultraviolet light can be focused to submicrometer dimensions before the limits imposed by diffraction are reached, so photolithography with ultraviolet rather than visible light offers a decrease in the size of features that can be produced by the lithographic process. The output of ultraviolet light from con-

ventional lamp sources is so weak that exposure times on the order of minutes are required to expose photoresists that are relatively insensitive to ultraviolet light. However, the ultraviolet emission from excimer lasers is 100 to 1000 times higher than from conventional sources, which reduces exposure times to seconds. In fact, the output from excimer lasers can be so intense that new forms of lithography have been demonstrated. Several groups have used excimer lasers to produce a dry photoresist technology based on directly ablating away the resist photochemically. This eliminates the usual wet-chemical development needed in photolithography (15).

Lasers have also been used to initiate specific chemical reactions over large areas at low substrate temperatures with purely gas-phase reagents. Excimer-laser chemistry has been used to dope, deposit, and etch electronic materials. The high average power available from some excimer lasers can allow the reaction zone to be large enough to process an entire wafer with a 2- to 3-inch diameter without scanning it. Several research groups have reported etching silicon and GaAs by dissociating halogen-bearing gases, such as CH_3Br, with excimer-laser radiation (16); and at least one industrial group is actually using laser etching in the fabrication of 64-Kbit memory chips. This approach to etching uses the projection of a patterned laser beam to etch the silicon surface directly, eliminating the need for masking the wafer.

Metals, insulators, and semiconductors have been deposited using an apparatus such as that shown in Fig. 3. The laser beam passes several millimeters above a substrate while decomposing one or more photosensitive gases. The

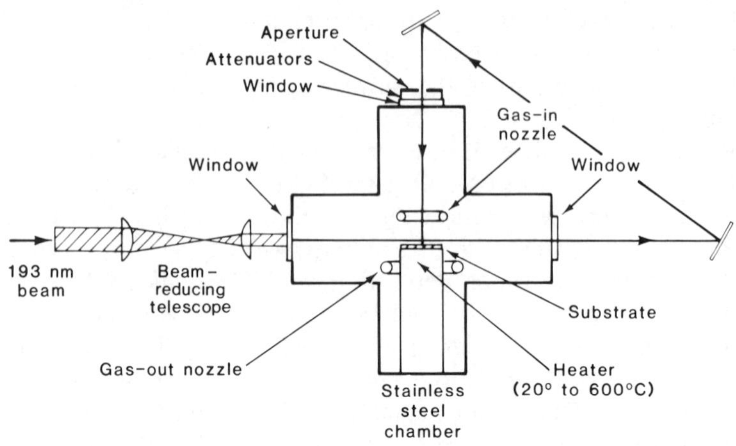

Fig. 3. Schematic diagram of a system for depositing materials using a pulsed excimer laser.

resultant photodecomposition products then either react with each other or move as atoms to the substrate and form a film. The substrate may be heated to about 300°C to increase the density of the deposited film by thermally enhanced adatom mobility or by desorption of impurity gases. Several groups have reported that irradiating the surface with a low dosage of ultraviolet light enhances the film quality, but the mechanism involved is not yet understood.

Metal films have been deposited by dissociating metal carbonyls. For example, when $Mo(CO)_6$ is irradiated with 193-nm radiation, CO ligands are released from the molybdenum atoms, and molybdenum films form (17). Tungsten films have been deposited on silicon and SiO_2 using ArF laser radiation to initiate the reaction of WF_6 with H_2 (18). Figure 4 shows micrographs of a laser-deposited tungsten film over SiO_2 steps on a silicon substrate. The film has excellent conformal coverage, a property needed in depositing metal conductors over sharp edges on a patterned silicon chip. Deposition of compound films, such as SiO_2 or Si_3N_4, requires the initiation of a gasphase chemical reaction (19). For example, SiO_2 films have been deposited by irradiating a mixture of SiH_4 and N_2O with excimer-laser radiation having a wavelength of 193 or 248 nm. The ultraviolet photons dissociate the N_2O, releasing oxygen atoms that then react with the SiH_4 to form SiO_2. The deposited SiO_2 films cover the sharp vertical walls found on some microstructures without breaks, a feature not always achieved with conventional techniques such as plasma deposition. Finally, laser-enhanced growth of single-crystal films has been demonstrated for deposition of germanium from GeH_4 (20).

Lasers have been used to dope semiconductors with impurities that control the electrical properties of the doped region. Figure 5 illustrates the approach used in pulsed-laser doping. The laser serves both to release dopant atoms from an appropriate molecule and to heat the surface to the melting point, allowing the atoms to diffuse into the substrate. The substrate is translated perpendicular to the beam axis in Fig. 5, and the doping is accomplished in a single step. By contrast, the technology often used for doping semiconductors, implantation of high-energy ions, requires a subsequent

556

annealing step to remove damage to the crystal lattice created during the implantation.

Chemical doping with lasers has been performed on both elemental and compound semiconductors, and it has been used to make low-resistance electrical contacts and to form *p-n* junctions (*21*). Solar cells have been made from the *p-n* junctions in silicon that result from doping it with boron. In this process, the boron is produced by irradiating BCl_3 with ArF laser radiation. The solar cells have shown efficiencies of 10.6 percent without any antireflection coating to increase the efficiency of their light collecting, and higher efficiencies should be possible with process optimization. The electrical properties of laser-doped silicon are comparable to those of thermally diffused silicon, and several groups are attempting to apply the technique to the fabrication of transistors.

Direct Writing with Lasers

Direct writing is the production of patterned features on a substrate using localized, laser-initiated chemical reactions. This contrasts with the conventional techniques for producing patterned features described earlier, in which a thin patterned film serves to confine deposition or etching to specific regions. The design and fabrication of the masks used to pattern the films is an expensive and time-consuming process, which is justified for mass production of integrated-circuit chips. While conven-

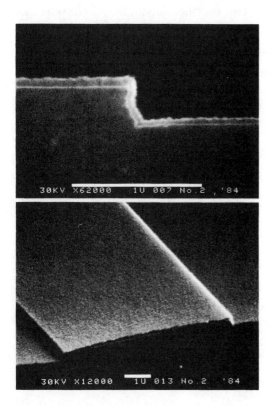

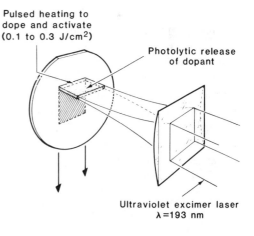

Pulsed heating to
dope and activate
(0.1 to 0.3 J/cm^2)

Photolytic release
of dopant

Ultraviolet excimer laser
$\lambda = 193$ nm

Fig. 4 (left). Micrographs from a scanning electron microscope showing a side and top view of a tungsten film over SiO_2 steps on a silicon substrate. An ArF laser was used to initiate the reaction of WF_6 with H_2. Illustration from (*18*). Fig. 5 (right). Schematic diagram of the doping of a semiconductor with a pulsed excimer laser.

tional techniques are satisfactory for most production operations, there are cases where it would be convenient to produce the patterned layer directly. These cases include the repair, design, and modification of circuits. Direct writing may even make it possible to monitor a device's performance during fabrication.

In the United States, direct writing is being investigated by groups at the MIT Lincoln Laboratory, the Lawrence Livermore National Laboratory, and Columbia University. Although direct writing is simple in principle, its implementation requires high-quality diffraction-limited optics, translators capable of moving a sample smoothly and positioning it with submicrometer accuracy, and both the hardware and software to allow a computer to control the sample position.

In direct writing, an ultraviolet or visible laser beam is focused onto the surface of a semiconductor wafer that is in a cell containing the appropriate reagent gases. Typically, argon-ion laser radiation at 514 nm is used directly; alternatively, the beam is first passed through a nonlinear crystal to obtain 257-nm radiation. If the laser wavelength is in the ultraviolet region, the laser light causes photochemical reactions such as:

$$CF_3Br \xrightarrow{h\nu} CF_3 + Br$$

or

$$Al(CH_3)_3 \xrightarrow{h\nu} Al + 3CH_3$$

to occur. Atoms that may either etch, deposit on, or dope a solid surface are produced. If the laser wavelength is in the visible region and the substrate absorbs light at the laser wavelength, surface heating occurs and the reactions are initiated by thermal chemistry. Figure 6 shows a silicon line drawn over an SiO_2 step on a silicon wafer. In this case a focused argon-ion laser beam was used to initiate the pyrolysis of SiH_4 by heating the silicon substrate (22).

Several unexpected phenomena occur when tightly focused laser beams are used to initiate chemical reactions. First, the rates of deposition for direct writing can be much greater than for conventional chemical vapor deposition. In the conventional process, the deposition rate is generally limited by the need to transport fresh, gaseous reagents to the surface. In focused-laser chemistry, the distance over which gas transport must occur is only about one beam diameter, so the reaction rates are consequently very fast (22, 23).

Under appropriate conditions in direct-writing processes, the region written on the substrate is considerably smaller than the laser spot. This occurs because the overall rate for the process can have a nonlinear dependence on either laser power or energy. Such nonlinearities allow laser writing to be used to produce the submicrometer features that are important in many microelectronic structures.

One example of such a nonlinearity occurs in the photodeposition of zinc or cadmium from the respective alkyl derivatives (10). The metal atoms produced by gas-phase photolysis only stick to or condense on surface areas that already have metal nuclei in place. The nucleated surface is formed by irradiating adsorbed molecules with an ultraviolet beam at a power intensity above a characteristic threshold value. By choosing the beam's power so that only a small portion of the beam has an intensity above the threshold, a deposit can be produced that has dimensions smaller

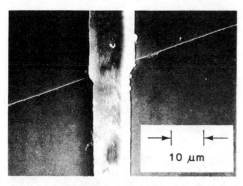

Fig. 6. Micrograph from scanning electron microscope showing a silicon line written by thermochemical dissociation over a SiO_2 step on a silicon substrate. Illustration from (22).

than the Gaussian width of the laser beam.

Another example of a process nonlinearity, which results in improved resolution, occurs in the direct writing of a doping pattern with continuous-wave lasers. As in doping with a pulsed laser, the direct writing of a doping pattern can be based on one of two different methods of releasing the doping atoms, thermally cracking molecules at the surface or dissociating gas-phase molecules with a separate ultraviolet laser beam. With either approach, a focused visible laser beam heats the surface, driving the dopant into a local region of the substrate. Using the thermal-cracking technique, D. J. Ehrlich and J. Y. Tsao at the MIT Lincoln Laboratory have written doped lines 0.2 µm wide in a silicon substrate (24). In this case, the relevant process nonlinearity is the exponential dependence of the diffusion constant of dopant atoms in silicon on the temperature of the substrate.

Direct writing with lasers is just beginning to be applied to the fabrication of complete devices and the modification of actual circuits. Laser writing has been used to fabricate a complete metal-oxide-semiconductor transistor by using laser heating to initiate localized doping, etching, and deposition (25). While the dimensions of that device do not reflect the limits imposed by the laser resolution, the experiment indicated that a working device can be fabricated almost entirely with laser techniques.

At present, the processing rates attainable with lasers are probably too low to produce custom circuits entirely by laser writing. A more attainable goal is to use laser writing to connect standard circuit elements produced by conventional techniques in customized configurations. In an example of this approach, computer-controlled etching and deposition with laser-writing techniques was used to reconfigure a circuit, called a ring oscillator, in a manner that allowed the high-frequency-transmission characteristics of the individual transistors and the metal interconnects between them to be determined (26). Such on-line custom operations, combined with simultaneous testing of the circuit elements, may be useful in shortening the design and test cycle in the development of integrated circuits. Another application amenable to laser writing techniques is the repair or reconfiguration of an integrated circuit. In this case, the defective portion would be disconnected using laser etching and then reconnected by laser writing a conducting link of metal or semiconductor material.

Laser-Assisted Wet Chemistry

Although semiconductor device fabrication relies increasingly on dry processing, liquid-phase chemistry has many powerful features, such as well-established etch rates that are dependent on crystal orientation, low processing temperatures, and, in many cases, the availability of efficient ionic chemical reactions. For example, aqueous solutions can be used to etch crystalline silicon anisotropically and to dissolve relatively inert elements, such as gold. Consequently, wet chemistry is still used in many phases of electronics production, including electroplating, etching of compound semiconductors, and substrate preparation.

Lasers have been used to modify many of these wet chemical techniques. When a solid is placed in an ionic solution, the requirement that the electrochemical potential be the same on both sides of the interface results in charge segregation in a thin layer on both sides of the interface. In those cases where ionic reactions with the surface dominate the chemistry between the solid and the solution, the charged layers near the interface can control the reaction rate. Light, in particular laser light, can alter the dynamics of such interface reactions by several mechanisms (14). The mechanisms include altering the chemical constituents in the solution by photochemical reactions, and, in the case of semiconductor substrates, creating charge carriers by optical absorption.

Thermal alteration of the electrochemical potential is the basis of laser-enhanced electroplating, one of the earliest examples of laser-induced chemical processing. The technique was developed by R. J. Von Gutfeld and co-workers at IBM (27). They used a tightly focused beam from a visible argon-ion laser to illuminate a small region on the electrode of an electrochemical cell. The illuminated region, which could be as small as a few micrometers, then had an

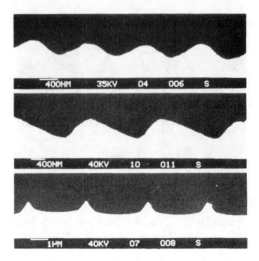

Fig. 7. Profiles of GaAs gratings fabricated using laser-initiated aqueous etching.

electrochemical potential different from that of the rest of the electrode, and the electrode voltage could be adjusted so that plating occurred only in the illuminated region.

This technique has been used for selectively gold plating a variety of conducting substrates. An important application is the localized plating of gold on metal, leaf-spring contacts. Because of the high cost of gold, which is necessary for good electrical contact, it is desirable to apply gold plating only at the point of contact. The complex, three-dimensional shapes of the springs make conventional photomasking techniques difficult and time-consuming.

Laser-enhanced electrochemical processing can also be applied to the plating and etching of semiconductors. In this case, the laser light itself generates the current that drives the electrochemistry, eliminating the need for an externally applied voltage.

Recently a group at Columbia University has used laser sources to produce high-resolution, optical-diffraction gratings on both conducting and semi-insulating GaAs (28). The high electrical resistance of semi-insulating GaAs, important in many of its applications, makes it difficult to etch using standard electrochemical techniques. The laser-based technique uses the interference of two beams from a visible laser to produce a series of alternating light and dark stripes on the semiconductor surface. In the illuminated region, positive charges, created by optical absorption, move to the surface and allow the surface to be oxidized. The oxide then dissolves in the solution, resulting in etching of the illuminated region.

The high-resolution diffraction gratings made this way have groove profiles that are determined by the diffusion of the photogenerated carriers out of the illuminated regions. When ultraviolet light, which is strongly absorbed in GaAs, is used, the charge carriers are primarily created at the surface and their lateral diffusion in the solid is minimized. Gratings with groove spacings as small as 100 nm have been made using ultraviolet laser radiation. Gratings with sinusoidal and cusped profiles have also been made. Figure 7 shows some of the structures.

Conclusion

Research in laser-assisted chemical processing has led to the development of new fabrication techniques for microelectronics. Some of the techniques, such as laser-assisted etching and excimer-laser photolithography, are immediately applicable to manufacturing. This research has also generated an increasing interest in the basic physics and chemistry of light-assisted interface reactions. Among the topics that have been investigated are the role of charged carriers in desorbing molecules from semiconductor surfaces (29), the influence of surface microstructure on photochemical reactions (11), the role of collective surface electromagnetic waves on the deposition process (30), and the effect of surfaces on ultraviolet molecular spectra (31), to name only a few. These studies have not only uncovered unexpected physical and chemical phenomena, but some of them have, in turn, led to novel techniques for microelectronics production. Applications to other disciplines, such as catalysis, can be expected in the future.

References and Notes

1. R. M. Osgood, Jr., *et al.*, Eds., *Laser Diagnostics and Photochemical Processing for Semiconductor Devices* (Elsevier, New York, 1983).
2. A. W. Johnson, D. J. Ehrlich, H. R. Schlossberg, Eds., *Laser-Controlled Chemical Processing of Surfaces* (Elsevier, New York, 1984).
3. D. J. Ehrlich, R. M. Osgood, T. F. Deutsch, *IEEE J. Quantum. Electron.* **QE-16**, 1233 (1980); D. J. Ehrlich and J. Y. Tsao, *J. Vac. Sci. Technol.* **B1**, 969 (1983); J. T. Yardley, in *Laser Handbook*, M. Stich and M. Bass, Eds. (Elsevier, New York, in press), vol. 4.
4. S. Sze, Ed., *VLSI Fabrication* (McGraw-Hill, New York, 1983).
5. S. S. Charschan, Ed., *Lasers in Industry* (Van Nostrand, New York, 1972).
6. R. M. Osgood, D. J. Ehrlich, T. F. Deutsch, D. J. Silversmith, A. Sanchez, in *Laser Processing of Semiconductor Devices*, C. C. Tang, Ed. (Society of Photo-Optical Instrumentation Engineer, Bellingham, Wash., 1983), p. 112; D. J. Ehrlich and J. Y. Tsao, in *VLSI Electronics: Microstructure Science*, N. Einspeuch, Ed. (Academic Press, New York, 1983), vol. 7, p. 129.
7. C. K. Rhodes, H. Egger, H. Pummer, Eds., *Excimer Lasers—1983* (American Institute of Physics, New York, 1983).
8. J. C. C. Fan and N. M. Johnson, Eds., *Energy Beam–Solid Interactions and Transient Thermal Processing* (Elsevier, New York, 1984).
9. S. Oguz, W. Paul, T. F. Deutsch, B. Y. Tsaur, D. V. Murphy, *Appl. Phys. Lett.* **43**, 848 (1983).
10. D. J. Ehrlich, R. M. Osgood, T. F. Deutsch, *ibid.* **38**, 946 (1981).
11. G. M. Goncher and C. B. Harris, *J. Chem. Phys.* **77**, 3767 (1982); C. J. Chen and R. M. Osgood, *Phys. Rev. Lett.* **50**, 1705 (1983).
12. J. Hager and H. Walther, in (2), p. 243.

13. S. D. Allen and M. Bass, *J. Vac. Sci. Tech.* **16**, 431 (1979).
14. S. R. Morrison, *The Chemical Physics of Surfaces* (Plenum, New York, 1972).
15. R. Srinavasan and V. Mayne-Banton, *Appl. Phys. Lett.* **41**, 576 (1982); T. F. Deutsch and M. W. Geis, *J. Appl. Phys.* **54**, 1201 (1983).
16. T. Arikado, M. Sekine, H. Okano, Y. Horiike, in (2), p. 167; P. Brewer, S. Halle, R. M. Osgood, *ibid.*, p. 179.
17. R. Solanki, P. K. Boyer, J. Mahan, G. J. Collins, *Appl. Phys. Lett.* **38**, 572 (1981); D. J. Ehrlich, R. M. Osgood, Jr., T. F. Deutsch, *J. Electrochem. Soc.* **128**, 2041 (1981).
18. T. F. Deutsch and D. D. Rathman, *Appl. Phys. Lett.* **45**, 623 (1984).
19. P. K. Boyer, G. A. Roche, W. H. Ritchie, G. Collins, *Appl. Phys. Lett.* **40**, 716 (1982); T. F. Deutsch, D. J. Silversmith, R. W. Mountain, in (1), p. 129, and (2), p. 67.
20. J. G. Eden, J. E. Greene, J. F. Osmundsen, D. Lubben, C. C. Abele, S. Gorbatkin, H. D. Desai, in (1), p. 185.
21. T. F. Deutsch, in (1), p. 225; M. L. Lloyd and K. G. Ibbs, in (2), p. 35.
22. D. J. Ehrlich, R. M. Osgood, Jr., T. F. Deutsch, *Appl. Phys. Lett.* **39**, 957 (1981).
23. I. P. Herman, R. A. Hyde, B. McWilliams, A. Weisberg, L. Wood, in (1), p. 9; D. J. Ehrlich and J. Y. Tsao, *ibid.*, p. 3.
24. D. J. Ehrlich and J. Y. Tsao, *Appl. Phys. Lett.* **44**, 267 (1984).
25. B. M. McWilliams, I. P. Herman, F. Mitlitsky, R. A. Hyde, L. L. Wood, *ibid.* **43**, 946 (1983).
26. D. J. Ehrlich, J. Y. Tsao, D. J. Silversmith, J. H. C. Sedlacek, W. S. Graber, R. Mountain, *IEEE Electron Device Lett.* **EDL-5**, 32 (1984).
27. R. J. Von Gutfeld, in *Laser Applications*, J. Ready, Ed. (Academic Press, New York, 1984), p. 1.
28. D. Podlesnik, H. H. Gilgen, R. M. Osgood, A. Sanchez, *Appl. Phys. Lett.* **43**, 1083 (1983).
29. F. A. Houle, *J. Chem. Phys.* **79**, 4237 (1983); G. P. Davis, C. A. Moore, R. A. Gottscho, *J. Appl. Phys.*, in press.
30. S. R. J. Brueck and D. J. Ehrlich, *Phys. Rev. Lett.* **48**, 1678 (1982); R. M. Osgood and D. J. Ehrlich, *Opt. Lett.* **7**, 385 (1982).
31. D. J. Ehrlich and R. M. Osgood, Jr., *Chem. Phys. Lett.* **79**, 381 (1981); C. J. Chen and R. M. Osgood, *Chem. Phys. Lett.* **50**, 1705 (1983).
32. Preparation of this manuscript was supported in part by contracts from the National Science Foundation (R.O.), the Department of the Air Force, in part under a specific program sponsored by the Air Force Office of Scientific Research (R.O. and T.D.), and by the Army Research Office (R.O. and T.D.).

The monopoly that the semiconductor silicon has enjoyed for the past 30 years in the area of large-scale electronic device applications will soon be broken by the advent of several compound semiconductor materials composed of crystalline solid solutions of elements from columns IIIa and Va of the periodic table. These III-V compound semiconductor materials have many unique electronic and optical properties that distinguish them from the column IVa semiconductor silicon. For example, many of these alloy semiconductors have a direct band structure; that is, the minimum of the energy for electrons in the "conduction band" and the minimum of the energy for holes in the "valence band" are at the center of the Brillouin zone, where the crystal momentum k is zero. In the case of silicon, the minimum energy in the condition band and the minimum energy in the valence band occur at different points in the Brillouin zone; thus silicon has an indirect band structure.

As a result of the details of their band structure, the alloy semiconductors that have a direct band gap, such as GaAs, are very efficient emitters of photons, while semiconductors having indirect band structures, such as Si, are inefficient sources of light. The use of such direct-band-gap semiconductors for visible light-emitting diode applications is familiar to everyone. An increasingly important application for such light-emitting materials is in the construction of semiconductor lasers that emit coherent light in the infrared portion of the spectrum. Such lasers are in great demand for use in optical data storage and playback applications (for instance, videodisc systems) and for high-speed optical communications systems that use

38

Metalorganic Chemical Vapor Deposition of III-V Semiconductors

R.D. Dupuis

Science 226, 623–629 (9 November 1984)

563

low-optical-loss glass fibers to transmit information coded in a pulsed light beam (for instance, in an undersea communications link across the Atlantic Ocean).

Several materials technologies have been developed for growth of the thin-film epitaxial structures that are required for the realization of such III-V compound semiconductor optoelectronic devices. In general, these growth techniques can be classified in three categories: (i) liquid-phase epitaxy (LPE), (ii) molecular-beam epitaxy (MBE), and (iii) vapor-phase epitaxy (VPE). The LPE process uses heated liquid solutions of the IIIa and Va elements to produce a epitaxial thin crystalline film of the corresponding alloy semiconductor on an appropriate single-crystal substrate or host crystal. This is accomplished by cooling the liquid solution below the saturation temperature while it is in contact with the substrate. The MBE technology employs ultrahigh-vacuum systems in which elemental atomic or molecular beams are produced for each elemental species of the desired III-V compound semiconductor thin film. These molecular beams are formed by heating the corresponding elemental source to a high temperature in vacuum, producing a molecular beam, which then impinges on a heated substrate. Under the proper conditions, an epitaxial layer of the corresponding III-V alloy semiconductor is grown on the substrate (1). The VPE growth technique utilizes chemical reactions that occur between the vapors of certain chemical compounds of the IIIa and Va elements when they are heated together. These reactions produce chemically active species that interact in the vapor phase or at the surface of the substrate to produce the corresponding III-V semiconductor thin film. The particular VPE process that is the subject of this article employs metal-organic compounds as sources of the IIIa elements; the sources of the Va elements are either Va hydrides or mixtures of the Va hydrides and Va metalorganics. This metalorganic chemical vapor deposition (MOCVD) process is becoming widely used in the growth of many important III-V compound semiconductors.

Development of the MOCVD Process

In this article, the term metalorganic denotes the broad class of compounds that contain metal-carbon bonds (known as organometallic compounds), as well as those with metal-oxygen-carbon bonds (the alkoxides) and the coordination compounds of metals and organic molecules. Metalorganic compounds were first identified by Robert Bunsen in 1839 (2). Since that time, there have been extensive studies of their chemical and physical properties as well as their reactions with other compounds, giving rise to a vast area of chemical research, development, and production. In the past 30 years many of these compounds have become of great practical interest. The first American Chemical Society symposium devoted entirely to metalorganics was held in 1957 (3), and numerous international conferences on the chemistry of these compounds have been held since.

Although many of the chemical and physical properties of a large number of metalorganics were established in the 1950's, it was not until 1968 that any work was reported on the growth of epitaxial semiconductor thin films with metalorganic sources (4). In this work, actually begun in 1967, Manasevit (4) showed that vapor-phase mixtures of the metalorganic compound trimethylgallium (CH$_3$)$_3$Ga, and the hydride arsine, AsH$_3$, when pyrolized at 600° to 700°C in an H$_2$ atmosphere, could be used to grow thin single-crystal epitaxial films of GaAs in an open-tube reactor. Manasevit used this MOCVD process to grow heteroepitaxial (5) single-crystal films of GaAs on single-crystal insulating substrates such as sapphire (Al$_2$O$_3$), spinel (MgAl$_2$O$_4$), and beryllium oxide (BeO) (4). His interest in the heteroepitaxial growth of III-V compound semiconductor thin films on insulating substrates such as sapphire (6). This silicon-on-sapphire (SOS) materials technology is applied in certain important specialized device areas.

Other laboratories were also interested in semiconductor-on-insulator epitaxial growth, and much of the early literature on MOCVD research in the growth of compound semiconductors was devoted to the study of heteroepitaxial films on insulating oxide substrates. This work has been reviewed (7–9) and will not be discussed here.

At this same time, research was being done on the application of MOCVD to the growth of homoepitaxial semiconductor thin films and on the growth of heteroepitaxial semiconductor thin films on other semiconductor substrates. However, because of the availability of other, more advanced technologies for the growth of III-V compound semiconductor thin films, such as the LPE process and the halide-transport VPE processes, the MOCVD technology was not extensively used for this purpose.

In the past 5 years, there has been increasing interest throughout the world in MOCVD growth of semiconductors. This is the result of interest in the potential large-scale use of certain compound semiconductor devices based on submicrometer-thick layers [for instance, field-effect transistors (FET's)] and high-quality heterojunctions between two different semiconductor materials (heterojunction solar cells, and heterojunction lasers and light-emitting diodes). The MOCVD materials technology is well-suited to the fulfillment of these requirements and has some advantages over alternative materials technologies. As a result of this interest in device applications, much of the MOCVD materials research is device-oriented, and virtually every major corporation with an interest in compound semiconductor devices has a research program in the MOCVD growth of these materials.

The MOCVD Process

The metalorganic chemical vapor deposition process involves the pyrolysis (10) of vapor-phase mixtures of two or more metalorganic sources, one or more metalorganic compounds, and one or more hydride compounds. Many other names have been used for this process, including "organometallic CVD" (OMCVD), "metal alkyl vapor-phase epitaxy" (MAVPE), "metalorganic VPE" (MOVPE), and "organometallic VPE" (OMVPE). The term "metalorganic chemical vapor deposition" is used here since it is a more general description of the process, encompassing the use of any of the metalorganic compounds (not just the organometallics) and the growth of nonepitaxial (polycrystalline or amorphous) films.

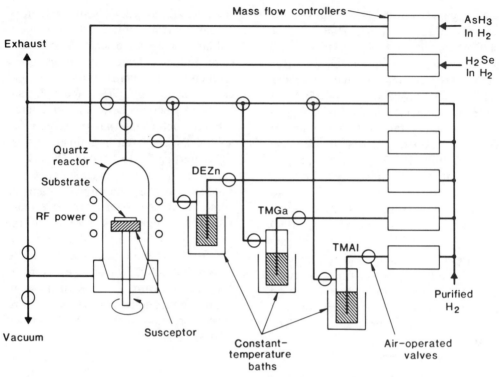

Fig. 1. Schematic diagram of a vertical atmospheric-pressure MOCVD reactor.

The metalorganics of interest for the growth of semiconductor films are typically liquids at room temperature, although some of the useful metalorganic compounds are solids at room temperature and above. These metalorganics generally have relatively high vapor pressures and can be readily transported into the reaction zone by passing a carrier gas, such as H$_2$, through the liquid sources or over the solid sources. The hydrides of interest for semiconductor thin-film growth are gases at room temperature and are generally used as dilute mixtures in an H$_2$ matrix and are contained in high-pressure cylinders. These metalorganic and hydride compounds are mixed in the vapor phase and are usually pyrolyzed in a flowing H$_2$ atmo-

sphere in a open-tube reactor operating at atmospheric pressure or at a reduced pressure of about 70 torr (0.1 atm). Pyrolysis temperatures typically in the range 600° to 800°C are employed. Energy for heating the gaseous source mixtures is usually provided by a high-power radio-frequency (RF) generator operating at about 450 kHz. This RF generator heats a graphite "susceptor" on which the single crystal substrate is placed. The gas mixture at or near the substrate surface is thus heated to high temperatures while the walls of the reactor chamber remain relatively cool, leading to deposition of the semiconductor thin film on the substrate crystal and not much loss of reactants to the surfaces of the reactor chamber. A typical atmospheric-pressure MOCVD reactor is shown schematically in Fig. 1.

The net chemical reaction for the specific case of the MOCVD growth of the semiconductor GaAs from the metalorganic trimethylgallium $(CH_3)_3Ga$ (TMGa), and the hydride arsine, AsH_3, is

$$(CH_3)_3Ga + AsH_3 \xrightarrow[700°C]{H_2} \rightarrow$$

$$GaAs + 3CH_4 \qquad (1)$$

Similar reactions are employed for the growth of other semiconductor binary, ternary, and quaternary compound semiconductor thin films. For example, for the technologically important ternary alloy semiconductor $Al_xGa_{1-x}As$, the following process is typically employed:

$$(1-x)[(CH_3)_3Ga] + x[(CH_3)_3Al] +$$

$$AsH_3 \xrightarrow[\sim 700°C]{H_2} \rightarrow Ga_{1-x}Al_xAs + 3CH_4$$

$$(2)$$

In this case, the alloy composition of the epitaxial thin film is directly related to the relative initial partial pressures of the trimethylgallium and trimethylaluminum (TMAl) in the vapor phase. Unfortunately, this simple relation between the composition of the sources in the vapor phase and the composition of the resulting thin solid film is not always true for the growth of compound semiconductors by MOCVD.

Most of the research on MOCVD growth of compound semiconductors has employed the methyl and ethyl metalorganics. These compounds are relatively simple to prepare and can be readily pyrolyzed in a hydrogen atmosphere to yield the corresponding metal atoms, with methane or ethane produced as a by-product. Table 1 lists the princi-

Table 1. Metalorganics for semiconductor thin film deposition.

Group of metal in periodic table	Compound	Symbol
IIa	Biscyclopentadienyl-magnesium	Cp$_2$Mg
IIb	Dimethylzinc	DMZn
	Diethylzinc	DEZn
	Dimethylcadmium	DMCd
	Dimethylmercury	DMHg
	Diethylmercury	DEHg
IIIa	Trimethylaluminum	TMAl
	Trimethylgallium	TMGa
	Triethylgallium	TEGa
	Diethylgallium chloride	DEGaCl
	Trimethylindium	TMIn
	Triethylindium	TEIn
IVa	Tetramethyltin	TMSn
	Tetraethyltin	TESn
	Tetramethyllead	TMPb
	Tetraethyllead	TEPb
Va	Triethylphosphine	TEP
	Trimethylantimony	TMSb
	Trimethylarsine	TMAs
VIa	Dimethyltelluride	DMTe
	Diethyltelluride	DETe

pal metalorganics that have been utilized in the growth of compound semiconductor thin films. All of these compounds are now commercially available in "electronic grade" purity, typically more than 99.9995 percent pure. However, most applications of high-purity semiconductor materials require epitaxial layers with total impurity levels of less than 0.1 part per million. Thus, reliable methods for producing metalorganics with improved purity are essential in the future large-scale application of MOCVD for the growth of high-purity semiconductors. However, many semiconductor device structures that are being studied today employ "doped" layers, that is, layers in which impurities have been intentionally added to levels of about 100 ppm. Thus, the currently available metalorganic sources are quite useful in the realization of many practical semiconductor devices.

The large variety of metalorganic sources that are listed in Table 1 implies that a wide variety of compound semiconductor materials have been grown by MOCVD. In fact, binary compound semiconductors have been grown in the III-V, II-VI, and IV-VI semiconductor materials families. Also, many of the important ternary and quaternary III-V compound semiconductors have been grown by MOCVD. Table 2 shows the binary and ternary III-V semiconductors that have been produced by MOCVD and the reactants employed. The greatest amount of attention has been devoted to the technologically important alloy $Al_xGa_{1-x}As$, which is of interest for many practical semiconductor devices including solar cells and semiconductor lasers. In addition, a few of the quaternary alloy semiconductors have been grown by MOCVD. The quaternary alloy $In_xGa_{1-x}As_{1-y}P_y$ has been grown with great success by MOCVD and will be discussed below.

Thin films of many of the II-VI and IV-VI compound semiconductors have also been deposited by MOCVD, as indicated in Table 3. Although research in this area has not been as extensive as that on the III-V semiconductors, recent results (*11, 12*) have shown that high-quality films of ZnSe and $ZnSe_{1-x}S_x$ can be produced by MOCVD, indicating that it should be possible to produce similarly high-quality films of other II-VI materials. In addition, the important II-VI ternary alloy

Table 2. Binary and ternary III-V compound semiconductors formed by MOCVD.

Compound	Reactants
Binary	
AlAs	$TMAl$-AsH_3
AlN	$TMAl$-NH_3
GaAs	$TMGa$-AsH_3
	$TEGa$-AsH_3
	$DEGaCl$-AsH_3
GaN	$TMGa$-NH_3
GaP	$TMGa$-PH_3
	TMP-PH_3
	$TEGa$-PH_3
GaSb	$TEGa$-$TMSb$
InAs	$TEIn$-AsH_3
InP	$TEIn$-PH_3
	$TMIn$-PH_3
	$TMIn$-TMP
InSb	$TMIn$-$TMSb$
Ternary	
$GaAs_{1-x}P_x$	$TMGa$-AsH_3-PH_3
($x = 0.0$–1.0)	
$GaAs_{1-x}Sb_x$	$TMGa$-AsH_3-SbH_3
($x = 0.0$–3, 0.6–1.0)	$TMGa$-$TMSb$-AsH_3
	$TMGa$-$TMSb$-$TMAs$
$Ga_{1-x}Al_xAs$	$TMGa$-$TMAl$-AsH_3
($x = 0.0$–1.0)	$TEGa$-$TMAl$-AsH_3
$Ga_{1-x}In_xAs$	$TMGa$-$TEIn$-AsH_3
($x = 0.0$–1.0)	$TMGa$-$TMIn$-AsH_3
	$TMGa$-$TMIn$-$TMAs$
$InAs_{1-x}P_x$	$TEIn$-AsH_3-PH_3
($x = 0.0$–0.6)	
$InAs_{1-x}Sb_x$	$TEIn$-$TESb$-AsH_3
($x = 0.0$–1.0)	

$Hg_{1-x}Cd_xTe$ has been produced by MOCVD (*13*), as have thin films of some of the important IV-VI compound semiconductors, including $Pb_{1-x}Sn_xTe$.

It is evident that the MOCVD materials technology can be used to produce thin films of many compound semiconductors that are of interest today for the fabrication of technologically important semiconductor devices. There are, however, additional features of the MOCVD process that make it attractive for the growth of such semiconductor thin-film structures. These features are described below.

Process control is inherent. The deposition process occurs by passing a homogeneous gas-phase mixture of reactants and carrier gases over a heated substrate. Only the temperature of the substrate need be carefully controlled. Most of the properties of films grown by MOCVD are only weakly dependent on temperature, so that slight variations in substrate temperature are inconsequential. The partial pressure of the various gaseous constituents can be controlled by electronic mass flow control of the source flow rates. These features allow all of the critical variables in the growth process to be controlled with extremely high accuracy, ensuring reproducibility.

Multilayer, multicomponent epitaxial structures can be deposited in a single growth sequence. The reactors typically used for MOCVD growth have facilities for several metalorganic and hydride sources. In addition, process involves no etching species, so that abrupt interfaces between various materials can be formed. The absence of an etching species also helps to achieve the growth of films with uniform thickness and composition, since the growth process is not the result of competing deposition and etching reactions, as is the case for some other vapor-phase epitaxy processes. Such reactions could lead to the introduction into the gas phase of uncontrollable amounts of III-V compound substrate components, which would subsequently upset the desired film stoichiometry and impurity level.

The process is scalable to high volume. Large-area uniform growth of epitaxial structures can be achieved by MOCVD with equipment similar to that currently used in large-scale commercial epitaxial growth of semiconductor silicon. MOCVD systems are available with the capacity for growing simultaneously on 20 substrates that are 2 inches in diameter. While this is still far from the capacity of large silicon epitaxy reactors, it is much larger than the capacity of growth systems based on other III-V materials technologies such as LPE and MBE. In addition, when desirable, high growth rates compatible with production applications are achievable.

Materials Properties of MOCVD Semiconductors

The materials properties of III-V compound semiconductors are of great concern for all device applications. Many applications for high-speed electronic information systems require the growth of epitaxial layers of binary and ternary compound semiconductor thin films with extremely low concentrations of unwanted chemical impurities. The goal is to achieve the highest possible speed for electrons in these devices, and additional impurities can slow the electrons through scattering interactions, thus reducing the speed of response of the device to the electrical signal applied to it.

The purity of a semiconductor material can be determined by measuring the mobility of the electrons and holes through the crystal lattice of the epitaxial layer. High mobilities indicate that the number of impurities in the semiconductor crystal is low, since the electrons are not slowed down by scattering from impurity atoms. Because most useful devices are made from n-type material in which the number of electrons n is greater than the number of holes p, the important measurement is the mobility of the electrons, which is dependent on the total number of impurity atoms in the epitaxial layer. Theoretically, a high-quality GaAs thin film will have an electron mobility of about 125,000 cm^2/V-sec at 77 K when the total impurity concentration is 5×10^{14} cm^{-3}. For lower concentrations of impurities the mobility is higher, and an epitaxial layer having a mobility of 200,000 cm^2/V-sec at 77 K is considered excellent since the impurity concentration should be below 1×10^{14} cm^{-3}.

The effects of metalorganic and hydride source purity, growth temperature, and reactor pressure on residual impurity incorporation have been shown to be important in determining the purity of epitaxial GaAs layers grown by MOCVD (14, 15). In this work, far-infrared photoconductivity measurements and low-temperature photoluminescence measurements showed that C, Si, and Zn are

Table 3. II-VI and IV-VI semiconductors formed by MOCVD.

Compound	Reactants
II-VI	
ZnS	DEZn-H$_2$S
ZnSe	DEZn-H$_2$Se
ZnTe	DEZn-DMTe
CdS	DMCd-H$_2$S
CdSe	DMCd-H$_2$Se
CdTe	DMCd-DMTe
Hg$_{1-x}$Cd$_x$Te	Hg-DMCd-DETe
ZnS$_x$Se$_{1-x}$	DEZn-H$_2$S-H$_2$Se
IV-VI	
SnTe	TESn-DMTe
PbTe	TMPb-DMTe
	TEPb-DMTe
Pb$_{1-x}$Sn$_x$Te	TMPb-TESn-DMTe
PbS	TMPb-H$_2$S
PbSe	TMPb-H$_2$Se

the dominant residual impurities in unintentionally doped MOCVD GaAs films. The highest mobilities were obtained for films grown in the temperature range 600° to 625°C. Mobilities in these undoped GaAs films varied greatly with the trimethylgallium and arsine sources used, as routinely observed by other workers. Purification of the trimethylgallium by redistillation resulted in GaAs films with improved mobilities and fewer unwanted impurities. Also, higher mobilities could be obtained with a given set of sources if the reactor pressure during growth was lowered from about 760 torr to about 70 torr (14, 16). Total impurity concentrations as low as 5×10^{14} cm^{-3} were measured for films with mobilities as high as 125,000 cm^2/V-sec at 77 K. As noted above, this is about the theoretical maximum for GaAs films with this impurity concentration. Other workers (17) have also reported growth of high-purity GaAs films by MOCVD. The films were grown at atmospheric pressure by using specially refined trimethylgallium with mobilities as high as 139,000 cm^2/V-sec. The total impurity concentration was about 4×10^{14} cm^{-3}. More recently, mobilities as high as 150,000 cm^2/V-sec have been reported for MOCVD GaAs epitaxial layers (18).

For comparison, other materials technologies such as VPE and MBE have produced GaAs films with electron mobilities greater than 200,000 cm^2/V-sec and total impurity levels less than 7×10^{13} cm^{-3}. These results indicate that the metalorganic and hdyride sources still need some improvement before MOCVD reaches the level of purity achieved by some of the other III-V materials technologies. However, most devices of interest do not require GaAs of this purity and consequently MOCVD can meet the needs of most device applications.

Another III-V semiconductor that is of great interest is the ternary alloy $Al_xGa_{1-x}As$, which is a solid solution of the two binary compounds GaAs and AlAs. Application of MOCVD for the growth of epitaxial films of this ternary alloy is currently of great interest because of the increasing commercial importance of a wide variety of devices that employ thin layers of this material. One of the unique features of the MOCVD process is that it is the only vapor-phase epitaxy process capable of easily handling the growth of compounds containing aluminum.

Because of the extreme reactivity of aluminum, care must be taken to ensure that the reactor system is leak-tight. It is also critically important that the hydride gas sources be free of even small concentrations (less than 1 ppm) of water vapor. Because of problems with arsine sources and also because reactor systems were not constructed as carefully as they are today, early reports of the properties of MOCVD $Al_xGa_{1-x}As$ indicated excessive incorporation of oxygen. However, results published in 1977 and 1978 (*19, 20*) showed that high-quality $Al_xGa_{1-x}As$ films and $Al_xGa_{1-x}As$-GaAs heterojunctions could be grown by this process. The optical properties of this material are of interest since $Al_xGa_{1-x}As$ thin films are widely used in semiconductor laser structures. These devices, which are described below in more detail, are becoming extremely important in optical communications and information transmission systems.

At present, GaAs and $Al_xGa_{1-x}As$ are the principal III-V compounds grown by MOCVD. However, there is great interest in the application of this materials technology to the growth of other important compound semiconductors, for instance, InP and $In_{1-x}Ga_xAs_yP_{1-y}$ epitaxial thin films. A number of different MOCVD processes are being studied for this purpose. Trimethylindium and triethylindium are primarily used for the indium sources. The phosphorus sources are phosphine, triethylphosphine, and trimethylphosphine. One approach that is being extensively studied is the use of adduct compounds that are formed by reacting a metalorganic containing indium with another metalorganic containing phosphorus, for instance, the trimethyl-indium-trimethylphosphine adduct (*21*). This compound is stable near room temperature and can be handled with somewhat greater safety than the simple metalorganics since the adduct is much less reactive and is not pyrophoric. One of the attractive features of this approach is that it should be possible to grow the quaternary solid $In_{1-x}Ga_xAs_yP_{1-y}$ with a given composition from a single metalorganic adduct source containing the right amounts of In, Ga, As, and P. These III-V semicondcutor materials containing indium are of increasing importance in a variety of device application areas. One specific application to an injection laser device structure is discussed further below.

Heterojunction Structures and Devices

In many of the important semiconductor device structures being developed today, it is critically important that highly perfect interfaces be made between two different single-crystal semiconductor materials. Such an interface between two materials having different chemical compositions is called a heterojunction or heterostructure. If the two different

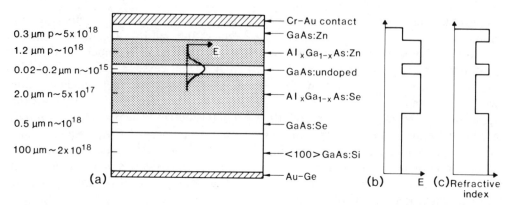

0.3 μm p~5×10^18
1.2 μm p~10^18
0.02–0.2 μm n~10^15
2.0 μm n~5×10^17
0.5 μm n~10^18
100 μm ~2×10^18

Cr–Au contact
GaAs:Zn
Al$_x$Ga$_{1-x}$As:Zn
GaAs:undoped
Al$_x$Ga$_{1-x}$As:Se
GaAs:Se
<100>GaAs:Si
Au–Ge

(a)   (b) E_g (c)Refractive index

Fig. 2. (a) Schematic cross section of a conventional Al$_x$Ga$_{1-x}$As-GaAs double-heterojunction laser. (b) Variation of the energy gap, E_g, in the structure. (c) Variation of the index of refraction.

crystalline materials have the same spacing between corresponding planes of atoms, the heterojunction is called a lattice-matched heterojunction.

One of the most important device structures of this class is the semiconductor double-heterojunction (DH) injection laser. These devices were first grown by LPE (22, 23) and are the light sources used in most high-speed optical communications systems today (24). Typically, these devices have five separate layers of different materials, with the layers being as thin as 0.1 μm or less. The chemical composition of these layers must be carefully controlled in order to achieve the desired device perform-

ance. A schematic diagram of a standard DH laser made in the GaAs-Al$_x$Ga$_{1-x}$As materials system is shown in Fig. 2a. The epitaxial structure consists of five layers, which are, in order of growth, (i) an n-type GaAs layer doped with Se having a carrier concentration $n = 1 \times 10^{18}$ cm^{-3} and a thickness of 0.5 μm; (ii) an n-type Al$_x$Ga$_{1-x}$As layer doped with Se having $n = 5 \times 10^{17}$ cm^{-3} and 2.0 μm thick; (iii) an undoped GaAs layer (the active region) 0.02 to 0.2 μm thick having a background carrier concentration $n = 1 \times 10^{15}$ cm^{-3}; (iv) a p-type Al$_x$Ga$_{1-x}$As layer doped with Zn to a level of $p = 1 \times 10^{18}$ cm^{-3} and 1.2 μm thick; and (v) a p-type GaAs layer doped with

Fig. 3. Schematic cross section of a typical conventional In$_{1-x}$Ga$_x$As$_y$P$_{1-y}$ - InP double - heterostructure laser. The "active layer" is made of the quaternary In$_{1-x}$Ga$_x$As$_y$P$_{1-y}$.

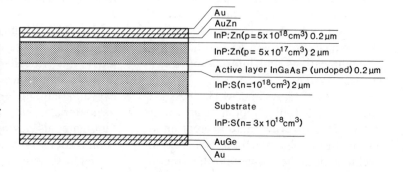

Au
AuZn
InP:Zn(p=5×10^18 cm^3) 0.2 μm
InP:Zn(p= 5×10^17 cm^3) 2 μm
Active layer InGaAsP (undoped) 0.2 μm
InP:S(n=10^18 cm^3) 2 μm
Substrate
InP:S(n= 3×10^18 cm^3)
AuGe
Au

Zn having $p = 5 \times 10^{18}$ cm^{-3}. The epitaxial layers are grown on an n-type GaAs substrate that is doped with silicon. This single-crystal substrate is sawed from a large ingot and is polished and etched before epitaxial crystal growth.

As discussed above, the MOCVD process has many inherent features that permit precise control of the growth of laser structures such as that shown in Fig. 2a. The first room-temperature operation of a semiconductor laser grown by MOCVD was reproted in 1977 (19), and since that time the application of MOCVD to the growth of such device structures has become a subject of intense study, in particular for lasers made of materials in the Al$_x$Ga$_{1-x}$As-GaAs system (25, 26).

Another III-V compound semiconductor system that is of great importance for the fabrication of semiconductor lasers is In$_{1-x}$Ga$_x$As$_y$P$_{1-y}$-InP (27, 28). A typical laser structure composed of materials in this system is shown in Fig. 3. Note that in this case all but one of the layers is composed of InP. The active region is made of In$_{1-x}$Ga$_x$As$_y$P$_{1-y}$. The alloy composition of this layer determines the energy of the photons that the laser will emit. This quaternary system is important because the energies of the photons that are emitted cover the range where the optical losses in silica-based optical fibers reach a minimum (0.8 to 0.9 eV, photon wavelengths of 1.33 to 1.55 μm). Lasers in this III-V system were first reported in 1976 (27) and were grown by LPE. The first In$_{1-x}$Ga$_x$As$_y$P$_{1-y}$-InP lasers grown by MOCVD were reported in 1980 (29). Since then, great progress has been made in the MOCVD growth of these materials (30, 31) and the performance characteristics of these lasers are

equal to those of the best comparable devices grown by LPE and of similar lasers grown recently by MBE (32).

Compared to the more conventional LPE process for growth of these device structures, MOCVD offers a higher degree of uniformity of devices fabricated from a wafer (33, 34) and also a much greater production capacity because of the much larger substrates that can be utilized in an MOCVD reactor. In addition, MOCVD has some advantages over the MBE materials technology in scale-up to large-area device production and in total system cost.

Quantum Well Structures

The operating characteristics of most conventional double-heterostructure devices are determined by the bulk properties of the materials that form their structure. However, if the active region of the device has a thickness less than about 500 Å, the device characteristics are influenced by quantum mechanical effects in addition to the properties of the materials themselves. These "quantum size effects" (35) result from the confinement of electrons and holes to the extremely thin active region of such a "quantum well" double-heterostructure device. The energy levels of these electrons and holes are quantized, in much the same way as the classical "particle in a box." As a result of this quantization of the electron and hole energy levels, the electronic properties of the active region of the device are modified. The effect of this quantization on the density of states for electrons in the conduction band and holes in the valence band is shown schematically in Fig. 4. The dashed curves show the density of states, $g(E)$, as a function of energy for

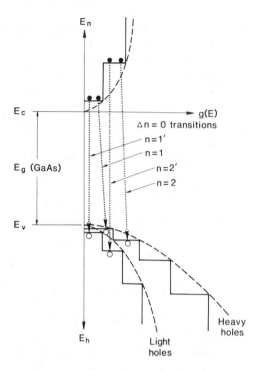

Fig. 4. Density of states $g(E)$ as a function of energy for electrons, E_n, and holes, E_h, in an $Al_xGa_{1-x}As$-GaAs quantum well double heterostructure. The stair-step solid lines show the density of states for the quantum well. The dashed curves show the corresponding density of states for thick layers of GaAs (thickness greater than 500 Å).

electrons, E_n, and holes, E_h, in bulk GaAs. The solid "stair-step" curves are the new densities of states for a quantum well made of GaAs and bounded on both sides by layers of $Al_xGa_{1-x}As$. Because of quantum mechanical effects, radiative recombination of electrons and holes can occur only between electron and hole states having the same quantum number (that is, $\Delta n = 0$); thus the luminescence spectra of such a structure will consist of emission peaks characteristic of the quantum well energy levels of the electrons and holes. The electronic conduction characteristics of thin quantum well

layers are also very different from those of "bulk" layers. Quantum effects have been used to produce extremely high mobility thin GaAs layers in $Al_xGa_{1-x}As$-GaAs double heterostructures (36, 37).

The first room-temperature operation of a quantum well semiconductor laser was reported in 1976 for epitaxial $Al_xGa_{1-x}As$-GaAs quantum well structures grown by MBE (38), but these lasers were optically pumped and had relatively high thresholds for laser operation. In 1977 quantum well lasers were grown by LPE in the $In_{1-x}Ga_xAs_yP_{1-y}$-InP system (39). These lasers were also optically pumped and exhibited a relatively high threshold for laser operation. The first quantum well lasers grown by MOCVD were reported in 1978 (40). They were made in the $Al_xGa_{1-x}As$-GaAs materials system and were the first injection-pumped quantum well lasers; they were also the first that were capable of operation at low thresholds at room temperature. These lasers show pronounced quantum effects in their spectral output and other operating characteristics. Extensive studies of optically pumped and injection quantum well MOCVD lasers have shown that high-performance (that is, low-threshold, high-quantum efficiency, and long life) quantum well lasers can be grown by MOCVD (33, 34, 41).

Quantum well lasers are unique in many respects and typically have low thresholds, high-energy spectral emission (greater than the energy gap of the material), and high external differential quantum efficiencies. These features have been utilized to achieve levels of performance that far exceed those of semiconductor lasers with a more conventional double-heterostructure geometry. In 1983, MOCVD quantum well lasers were reported in the $Al_xGa_{1-x}As$ system that operate in the visible portion of the spectrum at about 7200 Å (42). These were the first low-threshold semiconductor injection lasers that emitted visible light. In addition, such MOCVD quantum well structures have been used to fabricate visible semiconductor lasers emitting 100 mW of optical power at about 7300 Å (43) at a drive current of only 180 mA. They are very much more efficient than the He-Ne gas laser and will probably replace it in many applications, since they require much less space and less power to operate.

Another recent development in the area of MOCVD quantum well lasers is the use of monolithic arrays of such lasers to make a structure capable of emitting greater optical powers than previously reported for any semiconductor laser structure. Optical powers greater than 2.5 W have been reported for continuously operating arrays of 40 individual lasers (44). These high optical powers are emitted from a laser device that is only 400 μm wide; 250 μm long, and 100 μm thick, and they are achievable only because of the high efficiency of the quantum well laser structure and the quality of MOCVD $Al_xGa_{1-x}As$. It is possible that such arrays could replace much larger krypton and argon gas lasers (which require water cooling) in some high-power laser applications requiring high efficiency and small weight, as in certain space-based laser communication systems.

Other Device Structures and
Recent Developments

Several other III-V compound semiconductor device structures have been grown by MOCVD. Many of them are heterojunction devices consisting of $Al_xGa_{1-x}As$ and GaAs. Among these are heterojunction solar cells (45), heterojunction phototransistors (46), heterojunction field-effect transistors (47), and heterojunction photocathodes (48).

As the requirements for heterojunction devices increase MOCVD will become more widely used for a variety of compound semiconductor materials and devices. For example, the MOCVD process has been used to make device structures in the $In_{1-x}Ga_xAs_yP_{1-y}$-InP system. Most of this work has been in the growth of laser structures. However, selectively doped heterojunction structures in the $In_{1-x}Ga_xAs_yP_{1-y}$-InP system have been reported recently (49).

MOCVD has also been used in novel ways to produce thin films of several metals and some of the III-V semiconductors by using laser excitation to either thermally or photolytically decompose the metalorganics. Thin films of such metals as aluminum and cadmium have been produced by laser enhancement of the chemical decomposition reactions, using metalorganic sources (50, 51). Thin epitaxial films of GaAs have been produced by laser-induced pyrolysis of mixtures of $(CH_3)_3Ga$ and AsH_3 (52). More recently, InP epitaxial films have been deposited by photolytic dissociation of In- and P-containing metalorganic compounds (53). The use of laser-induced chemistry in the MOCVD growth of semiconductor films and for the deposition of metals opens vast new areas of research, since this process permits selective growth of thin films, potentially with very high resolution. It may be possible to use the laser light to "write" epitaxial structures on substrates and thus to build complex electronic and optical circuits with this technology. Such integrated circuits are the next step in the application of III-V compound semiconductors to high-speed information processing.

Conclusions

The metalorganic chemical vapor deposition process has been used to grow a wide variety of III-V, II-IV, and IV-VI compound semiconductors. Much research on the application of the MOCVD materials technology has concentrated on the epitaxial growth of III-V compound semiconductor heterojunction structures in the $Al_xGa_{1-x}As$-GaAs and the $In_{1-x}Ga_xAs_yP_{1-y}$-InP systems. The rather brief description has presented some of the results on the application of the MOCVD process to the growth of injection lasers in these two materials systems.

In the future, MOCVD will be used to grow films of other compound semiconductors, and novel approaches to the growth of complex three-dimensional integrated circuits will be developed. MOCVD promises to be an extremely important research tool as well as an important commercial process for the production of compound semiconductor materials and devices.

References and Notes

1. Molecular beam epitaxy was reviewed in M. B. Panish, *Science* **208**, 916 (1980).
2. R. Bunsen, *Ann. Pharm. (Justus Leibigs Ann. Chem.)* **31**, 175 (1839). See also E. G. Rochow, D. T. Hurd, R. N. Lewis, *The Chemistry of Organometallic Compounds* (Wiley, New York, 1957), pp. 1 and 207.
3. *Metal-Organic Compounds* (American Chemical Society, Washington, D.C., 1959).
4. H. M. Manasevit, *Appl. Phys. Lett.* **12**, 156 (1968).
5. _____ and W. I. Simpson, *J. Electrochem. Soc.* **116**, 1725 (1969).
6. _____, *J. Appl. Phys.* **35**, 1349 (1964).
7. H. M. Manasevit, *J. Cryst. Growth* **13/14**, 306 (1972).
8. _____, ibid. **22**, 125 (1974).
9. C. C. Wang and S. H. McFarlane III, *Thin Solid Films* **31**, 3 (1976).
10. Pyrolysis is the transformation of a compound into another substance or substances by heat alone. Although in common usage this term is used to describe a process that results in the decomposition of a chemical species, many pyrolysis reactions lead to the union of chemical species. Pyrolysis is thus a more general term than thermal decomposition.
11. W. Stutius, *J. Cryst. Growth* **59**, 1 (1982).
12. P. J. Dean, A. D. Pitt, P. J. Wright, M. L. Young, B. Cockayne, *Physica Sects. B and C* **116**, 508 (1983).
13. W. E. Hoke and R. Traczewski, *J. Appl. Phys.* **54**, 5087 (1983).
14. P. D. Dapkus, H. M. Manasevit, K. L. Hess, T. S. Low, G. E. Stillman, *J. Cryst. Growth* **55**, 10 (1981).
15. K. L. Hess, P. D. Dapkus, H. M. Manasevit, T. S. Low, B. J. Skromme, G. E. Stillman, *J. Electron. Mater.* **11**, 1115 (1982).
16. For a review of low-pressure MOCVD, see S. D. Hersee and J. P. Duchemin, *Annu. Rev. Mater. Sci. 1982* **12**, 65 (1982).
17. T. Nakanisi, T. Udagawa, A. Tanaka, K. Kamei, *J. Cryst. Growth* **55**, 255 (1981).
18. For a review of MOCVD growth of high-purity GaAs, see T. Nakanisi, ibid., in press.
19. R. D. Dupuis and P. D. Dapkus, *Appl. Phys. Lett.* **31**, 466 (1977).
20. _____, ibid., p. 839.
21. For a review of this work, see R. H. Moss, *J. Cryst. Growth*, in press.
22. Zh. I. Alferov, V. M. Andreev, V. I. Korol'kov, E. L. Portnoi, D. N. Tret'yakov, *Fiz. Tekh. Poluprovodn* **2**, 1545 (1968); *Sov. Phys. Semicond.* **2**, 1289 (1969).
23. I. Hayashi, M. B. Panish, P. W. Foy, *IEEE J. Quantum Electron.* **QE-5**, 211 (1969).
24. See M. B. Panish and I. Hayashi [*Sci. Am.* **225**, 32 (July 1971)] for a discussion of the basic features of semiconductor injection lasers. A more advanced treatment is given by H. C. Casey, Jr., and M. B. Panish, *Heterostructure Lasers*, part A, *Fundamental Principles* (Academic Press, New York, 1978).
25. For a review of early results on MOCVD lasers, see R. D. Dupuis, *Jpn. J. Appl. Phys.* **B19** (Suppl. 19-1), 415 (1980).
26. For a review of recent results on MOCVD quantum well lasers, see R. D. Burnham, T. L. Paoli, W. Streifer, N. Holonyak, Jr., *J. Cryst. Growth*, in press.
27. J. J. Hsieh, J. A. Rossi, J. P. Donnelly, *Appl. Phys. Lett.* **28**, 709 (1976).
28. For a review, see Y. Suematsu, K. Iga, K. Kishino, in *GaInAsP Alloy Semiconductors*, T. P. Pearsall, Ed. (Wiley, New York, 1982), pp. 341–411.
29. J. P. Hirtz, J. P. Duchemin, P. Hirtz, B. de Cremoux, T. P. Pearsall, M. Bonnet, *Electron. Lett.* **16**, 275 (1980).
30. For a review of MOCVD growth of $In_{1-x}Ga_xAs_yP_{1-y}$, see J. P. Hirtz, M. Razeghi, M. Bonnet, J. P. Duchemin, in *GaInAsP Alloy*

Semiconductors, T. P. Pearsall, Ed. (Wiley, New York, 1982), pp. 61–86.

31. M. Razeghi, J. P. Duchemin, B. de Cremoux, *J. Cryst. Growth*, in press.

32. M. B. Panish and H. Temkin, *Appl. Phys. Lett.* **44**, 785 (1984).

33. R. D. Burnham, D. R. Scifres, W. Streifer, *ibid.* **40**, 118 (1982).

34. R. D. Dupuis, R. L. Hartman, F. R. Nash, *IEEE Electron Device Lett.* **EDL-4**, 286 (1983).

35. For a discussion of quantum size effects in III-V semiconductor heterostructures see R. Dingle, in *Festkorperprobleme XV, Advances in Solid State Physics*, H. J. Queisser, Ed. (Pergamon, Vieweg Braunschweig, Germany, 1975), pp. 21–48.

36. R. Dingle, H. L. Stormer, A. C. Gossard, W. Wiegmann, *Appl. Phys. Lett.* **33**, 665 (1978).

37. For a review see R. Dingle, M. D. Feuer, C. W. Tu, in *VLSI Electronics Microstructure Science*, vol. 8, *Gallium Arsenide*, N. G Einspruch, Ed. (Academic Press, New York, in press).

38. R. C. Miller, R. Dingle, A. C. Gossard, R. A. Logan, W. A. Nordland, Jr., W. Wiegmann, *J. Appl. Phys.* **47**, 4509 (1976).

39. E. A. Rezek, N. Holonyak, Jr., B. A. Vojak, G. E. Stillman, J. A. Rossi, D. L. Keune, J. D. Fairing, *Appl. Phys. Lett.* **31**, 288 (1977).

40. R. D. Dupuis, P. D. Dapkus, N. Holonyak, Jr., E. A. Rezek, R. Chin, *ibid.* **32**, 295 (1978).

41. N. Holonyak, Jr., R. M. Kolbas, R. D. Dupuis, P. D. Dapkus, *IEEE J. Quantum Electron.* **QE-16**, 170 (1980).

42. C. Lindstrom, R. D. Burnham, D. R. Scifres, *Appl. Phys. Lett.* **42**, 134 (1983).

43. R. D. Burnham, C. Lindstrom, T. L. Paoli, D. R. Scifres, W. Streifer, N. Holonyak, Jr., *ibid.*, p. 937.

44. D. R. Scifres, C. Lindstrom, R. D. Burnham, W. Streifer, T. L. Paoli, *Electron. Lett.* **19**, 169 (1983).

45. R. D. Dupuis, P. D. Dapkus, R. D. Yingling, L. A. Moudy, *Appl. Phys. Lett.* **31**, 201 (1977).

46. R. A. Milano, T. H. Windhorn, E. R. Anderson, G. E. Stillman, R. D. Dupuis, P. D. Dapkus, *ibid.* **34**, 562 (1979).

47. J. Hallais, J. P. Andre, P. Baudet, D. Boccon-Gibod, in *Gallium Arsenide and Related Compounds, 1978*, C. M. Wolfe, Ed. (Institute of Physics, Bristol, England, 1979), pp. 361–370.

48. J. P. Andre, M. Boulou, P. Guittard, E. Roaux, in *Gallium Arsenide and Related Compounds, 1980*, H. W. Thim, Ed. (Institute of Physics, Bristol, England, 1981), pp. 413–422.

49. R. J. Nicolas, M. A. Brummeil, J. C. Portal, M. Razeghi, M. A. Poisson, *Solid State Commun.* **43**, 825 (1982).

50. T. F. Duetsch, D. J. Erlich, R. M. Osgood, Jr., *Appl. Phys. Lett.* **35**, 175 (1979).

51. D. J. Erlich, T. F. Duetsch, R. M. Osgood, Jr., in *Laser and Electron Beam Processing of Materials*, C. W. White and P. S. Peercy, Eds. (Academic Press, New York, 1980), pp. 671–677.

52. A. Krings and H. Beneking, paper I6.2, presented at the Materials Research Society Symposium, Boston, November 1982.

53. V. M. Donnelly, M. Geva, J. Long, R. F. Karlicek, *Appl. Phys. Lett.* **44**, 951 (1984).

Hans C. Andersen is a professor of chemistry at Stanford University, Stanford, California 94305.

Roger C. Baetzold is a member of the Chemistry Division of the Research Laboratories at Eastman Kodak Company, Rochester, New York 14650.

Allen J. Bard is a professor in the Department of Chemistry at the University of Texas, Austin, Texas 78712.

Robert G. Bergman is a professor in the Department of Chemistry at the University of California, Berkeley, California 94720.

John I. Brauman is the J.G. Jackson-C.J. Wood Professor of Chemistry at Stanford University, Stanford, California 94305.

Ronald Breslow is the S.L. Mitchill Professor of Chemistry at Columbia University, New York, New York 10027.

K.L. Busch, formerly an assistant research scientist in the Department of Chemistry at Purdue University, is now an assistant professor of chemistry at Indiana University, Bloomington, Indiana 47405.

David Chandler, formerly a professor of chemistry at the University of Illinois in Urbana, is now a professor of chemistry at the University of California, Berkeley, California 94720.

Paul B. Comita is a postdoctoral research associate in the chemistry department at Stanford University, Stanford, California 94305.

R.G. Cooks is a professor of chemistry at Purdue University, West Lafayette, Indiana 47907.

Donald J. Cram is a professor of chemistry at the University of California, Los Angeles, California 90024.

T.F. Deutsch, formerly a member of the technical staff of Lincoln Laboratory at the Massachusetts Institute of Technology, is now at Wellman Laboratory at Massachusetts General Hospital, Boston, Massachusetts 02114.

R.D. Dupuis is a member of the technical staff at AT&T Bell Laboratories, Murray Hill, New Jersey 07974.

About the Authors

Kenneth B. Eisenthal is a professor of chemistry at Columbia University, New York, New York 10027.

Kenichi Fukui is professor in the Department of Hydrocarbon Chemistry at Kyoto University, Sakyo-ku, Kyoto 606, Japan.

G.L. Glish is on the research staff of the Analytical Chemistry Division at Oak Ridge National Laboratory, Oak Ridge, Tennessee 37830.

William A. Goddard III is the Charles and Mary Ferkel Professor of Chemistry and Applied Physics at the California Institute of Technology, Pasadena, California 91125.

R.L. Greene is manager of the Condensed Matter Physics Group at the IBM Research Laboratory, San Jose, California 95193.

Jack Halpern is the Louis Block Distinguished Service Professor of Chemistry at the University of Chicago, Chicago, Illinois 60637, and an External Scientific Member of the Max-Planck-Institut für Kohlenforschung, Mülheim-Ruhr, West Germany.

Clayton H. Heathcock is a professor of chemistry at the University of California, Berkeley, California 94720.

Larry L. Hench is a professor of materials science in the College of Engineering and director of the Bioglass Research Center at the University of Florida, Gainesville, Florida 32611.

Roald Hoffmann is John A. Newman Professor of Physical Science at Cornell University, Ithaca, New York 14853.

E.T. Kaiser is professor of chemistry and head of the Laboratory of Bioorganic Chemistry and Biochemistry at The Rockefeller University, New York, New York 10021-6399.

F.J. Kézdy is a professor in the Department of Biochemistry at the University of Chicago, Chicago, Illinois 60637.

Warren D. Lawrance is an Adolf C. and Mary Sprague Miller Institute postdoctoral fellow in the Chemistry Department at the University of California, Berkeley, and a guest scientist at the Materials and Molecular Research Division of the Lawrence Berkeley Laboratory, Berkeley, California, 94720.

D.S. Lawrence is a postdoctoral associate in the chemistry and biochemistry department at The Rockefeller University, New York, New York 10021-6399.

Jean-Marie Lehn directs research groups at the Institut Le Bel, Université Louis Pasteur, 67000 Strasbourg, France, as well as at the Collège de France, 75005 Paris, where he is a professor of chemistry.

Stephen R. Leone, a staff member of the Quantum Physics Division of the National Bureau of Standards and an adjoint professor of chemistry at the University of Colorado, is with the Joint Institute for Laboratory Astrophysics of the National Bureau of Standards and the University of Colorado, Boulder, Colorado 80309.

Andrew J. Lovinger is a member of the technical staff in the Plastics Research and Development Department at Bell Laboratories, Murray Hill, New Jersey 07974.

Neil E. Mackenzie is an associate professor of veterinary microbiology and parasitology in the Center for Biological NMR of the Department of Chemistry at Texas A&M University, College Station, Texas 77843.

J. Paul G. Malthouse is an associate research scientist at the Center for Biological NMR of the Department of Chemistry at Texas A&M University, College Station, Texas 77843.

Tobin J. Marks is a professor of chemistry and a member of the Materials Research Center at Northwestern University, Evanston, Illinois 60201.

Fred W. McLafferty is a professor in the chemistry department at Cornell University, Ithaca, New York 14853.

Terry A. Miller is at Bell Laboratories, Murray Hill, New Jersey 07974.

C. Bradley Moore is faculty senior scientist with the Materials and Molecular Research Division of the Lawrence Berkeley Laboratory, and professor and chairman of the Chemistry Department at the University of California, Berkeley, California 94720.

James D. Morrison is a professor in the Department of Chemistry, University of New Hampshire, Durham, New Hampshire 03824.

Harry S. Mosher is professor emeritus in the Department of Chemistry at Stanford University, Stanford, California 94305.

Robert A. Moss is a professor of chemistry at Rutgers, State University of New Jersey, New Brunswick, New Jersey 08903.

R.M. Osgood is a professor of electrical engineering and applied physics at Columbia University, New York, New York 10027.

Leo A. Paquette is Charles H. Kimberly Professor of Chemistry at Ohio State University, Columbus, Ohio 43210.

Hrvoje Petek is a National Science Foundation predoctoral fellow in the Chemistry Department at the University of California, Berkeley, California 94720.

Stuart L. Schreiber is an associate professor of chemistry at Yale University, New Haven, Connecticut 06511.

Richard R. Schrock is a professor in the Department of Chemistry at the Massachusetts Institute of Technology, Cambridge, Massachusetts 02139.

A.I. Scott is director and Davidson Professor of Science in the Center for Biological NMR of the Department of Chemistry at Texas A&M University, College Station, Texas 77843.

C.V. Shank is head of the Quantum Physics and Electronics Department at Bell Laboratories, Holmdel, New Jersey 07733.

Evgeny Shustorovich is a member of the Chemistry Division of the Research Labora-

tories at Eastman Kodak Company, Rochester, New York 14650.

Robert M. Simon is senior staff officer on the Board on Chemical Sciences and Technology at the National Research Council, Washington, D.C. 20418.

Gabor A. Somorjai is a faculty senior scientist and director of the Catalysis Program of the Center for Advanced Materials at the Lawrence Berkeley Laboratory, and a professor in the Department of Chemistry at the University of California, Berkeley, California 94720.

William Spindel is staff director of the Board on Chemical Sciences and Technology at the National Research Council, Washington, D.C. 20418.

G.B. Street is manager of the Polymer Surfaces and Interfaces Group at the IBM Research Laboratory, San Jose, California 95193.

Henry Taube is Marguerite Blake Wilbur Professor of Chemistry at Stanford University, Stanford, California 94305.

Barry M. Trost is Vilas and Helfaer Professor of Chemistry in the McElvain Laboratories of Organic Chemistry of the Department of Chemistry at the University of Wisconsin, Madison, Wisconsin 53706.

Nicholas J. Turro is a professor of chemistry at Columbia University, New York, New York 10027.

John D. Weeks is a member of the technical staff at Bell Laboratories, Murray Hill, New Jersey 07974.

June Wilson is a research associate in the Department of Surgery of the College of Medicine and is program manager of the Bioglass Research Center at the University of Florida, Gainesville, Florida 32611.

Richard N. Zare is Shell Distinguished Professor of Chemistry in the Department of Chemistry at Stanford University, Stanford, California 94305.

INDEX

Anions
coordination chemistry of, 397–399
dicarboxylate, coreceptors for, 403–404
nucleophilicities of, 70
Anisole, 412–413
Anomeric effect, 314
Anthracene, orbital interpretation of, 295–296
Antibiotics, 354
Antifungals, 354
Antitumor agents, 114
Apolipoproteins, 444–446
Arachidonic acid cascade, 306–307
Argon, liquid, 147, 153
Aromaticity, 294–297, 314
Arylhalocarbenes, 42–44
L-Aspartic acid, production of, 336
Asteltoxin, 357–359
Asymmetric synthesis, 320, 323–336
See also Selectivity; Stereocontrol
Atactic configuration, 537–539
Avenociolide, synthesis of, 354–357
Avidin, 438
Azaaromatics, 104
Azulene, 80

Bardeen-Cooper-Schrieffer (BCS) theory, 299
Barker-Henderson (BH) theory, 152
Beckmann rearrangement, 103
Benzene, 23–24, 153, 219, 296
Benzene chromiumtricarbonyl, 207
2-Benzylsuccinate, 462
Beta decay, double, 120–121
Bimetallic clusters, 215, 226
Bioglass, 490–494, 496–500
Bioinertness, 489
Biomaterials, surface-active, 489–500
Biomolecules, ionization of, 105
Bipolarons, 512–515, 517, 527
Bis(ethylenedithiolo)tetrathiafulvalene (BEDT-TTF), 505–506, 508
Blastomycinone, synthesis of, 344–345
Boltzmann factor, 150–151
Bond-order conservation, 169–171
Botrydiplodin, 365
Branching, photochemical, 68
Breakdown curve, 103–104
t-Butoxide anion, 66–67

Calcitonin, model of, 447–449
Calixarenes, 388

Calutrons, 107
Carbanion chemistry, 316
Carbenes, 33–44, 245–246, 310, 370
Carbinolamine, 462
Carbinols, chiral secondary, synthesis of, 334–335
Carbon-13 nuclear magnetic resonance, 455–468
Carbon-14, sensitivity of analysis for, 94, 102
Carbon dating, 94, 102
Carbon, divalent, 33, 36
Carbonium ion rearrangements, 479–480
Carbon monoxide, 165, 216, 218–219, 222–223
Carbon tetrachloride, 156
Carboxypeptidase, 456, 462
L-Carnitine, synthesis of, 321
Caryophyllene, 349
Catalysis
heterogeneous, 138–140, 163, 224, 245
homogeneous, 245
stereoselective, 229–230
supramolecular, 405–407
surface science and, 213–226
Ziegler-Natta, 205, 245, 255–256
Catalysts
chiral, 229
design of, 225–226
heterogeneous, 138
high-surface-area, 224–225
in olefin metathesis, 135–138
photoelectrosynthesis of, 281
semiconductor powder, 274, 280
Sharpless allylic epoxidation, 329
transition metal, 320
Catalytic reaction, 214, 220–224, 271
Cavitands, 308–309, 379–391
Cedrol, synthesis of, 339–340
Cembranolides, synthesis of, 360
Ceramics, piezoelectric, 546–547
Ceravital, 493–495, 498
$C_6F_6^+$, emission spectrum of, 53–55
CH_2. *See* Methylene
Charge transfer, 10–12, 518
Charge transfer complexes, 503–508
Cheerios, 153
Chemical profiling, 97
Chemical timing, 25
Chemical vapor deposition, 555, 558, 563–576
Chemiluminescence, 4, 10

Chemisorption, 138–139, 163–173, 298
Chemoselectivity, 316–317
Chiral molecules, 323–325, 336 (n.1), 346
   (n.1), 349–361
Chloromethylketones, 458–461
Chorands, 380–381
α-Chymotrypsin, 426
CLT inhibitor, 142
Cluster ions, 72, 106
   See also Ionic clusters
CN free radical, 59
Coadsorption, 219–220
Coal liquids, 104, 107, 110
Cobaltammine self-exchange reaction, 179–180,
   183
Cobalt-carbon bond energy, 472–477, 480, 483
Co(III)/Co(II) couple, 191
Coenzyme B_{12}, 471–484
Coenzymes, 414–416, 426–427, 464–467
Cofactors, ^{13}C-enriched, 464–467
Coherent anti-Stokes Raman spectroscopy
   (CARS), 6
Collisional activation, 66
Collisional quenching, 20
Collisional relaxation, 5
Collision-induced dissociation, 71
Complexes
   activated, 182–184
   catalytic, 229–231
   dihydridometal, 261–263
   d^3 triple-bonded, 208
   hydridoalkyl, 261–268
   ion clustering in, 105–106
   organometallic, 197–210
   transition metal, 166, 197–200, 260–272,
      316–317, 320, 396
Composites, bioactive, 490, 494–495
Computer graphics, 315–316
Conductors
   molecular, 518–524
   organic, 503–515, 517
Conformational analysis, 314–315
Conformers, 314
Conjugation, 295–297
Coordination number, 205–206
Coreceptors, 400–404
Coriolis coupling, 22
Corticosteroids, synthesis of, 311–312
Corticotropin releasing factor, 449–450
Coulomb interactions, 293

Cram's rules, 330
Cr(III)/Cr(II) couple, 189–190
Creutz-Taube ion, 192
$Cr(H_2O)_6^{3+}$-Cr^{2+} (aq) self-exchange, 183
Crown compounds, 308–309, 314, 380
Cryoenzymology, 456, 462–464, 468
Cryptands, 314, 380–381, 394–397
Cryptates, 394–403
Crystal field theory, 205
Crystallography, of surfaces, 215–216
Cycloaddition
   of butadiene and ethylene, 290–291
   to five-membered rings, 310–311
   of olefins, 312
   [1+2], 36
   [2+2], 349–361
Cyclodextrins, 379–380, 412–413, 415–423
Cyclometallation, 261
Cyclopentadiene, 208

Daunomycin, 310, 312
Delocalization interactions, 293
Dewar-Chatt-Duncanson model of olefin bond-
   ing, 205
Diarylphosphines, 239-240
Diastereomeric adducts, 239–240
Diastereoselectivity, 319–320
Dicarboxylates, coreceptors for, 403–404
Dieckmann ester condensation, 102–103
Diels-Alder reactions, 289–291, 294, 309–310
Diene-dienophile reactions, 289–291
Diffusion, facilitated, 407
Dihydridometal complexes, 261–263
Dihydronicotinamides, 428–437
Diol dehydrase rearrangement, 472, 480–481
Dioxin, 94, 108, 110
Dioxygen, 483–484
Diphenylcarbene, 36–42
Diphenyldiazomethane, 38
Diradicals, 36–37
Dispersion interactions, 146
Diterpenes, 361
Dithiols, 435–437
DNA, methylated, 114
Dodecahedrane, 308
L-Dopa, synthesis of, 229, 325
Doping
   of polymers, 509, 511–514, 524–527, 531
   of semiconductors, 552, 554, 556–557,
      559

Glass ceramics, bioactive, 490, 493–496
Glasses, bioactive, 490–493, 495–500
Glassy materials, 159–160
Glucagon, models for, 452–453
Group theory, 203–204
Growth hormone releasing factor, 451

Halomethanes, 23–24
Hammett rule, 294
Hard sphere model, 145, 148–151, 154–157
Harmonic approximation, 21
Hemoglobin, human, 107–108
Hexatic phase, 161 (n.67)
L-Hexoses, synthesis of, 330
Highest occupied molecular orbital (HOMO), 289, 518–519, 528–530
High-performance liquid chromatography (HPLC), 123–125
HNO free radical, 61–62
HOMO-LUMO interaction, 291
"Hopping" mechanism, 187–188
Host-guest chemistry, 308–309, 380, 391
Hydration number, 182
α-Hydride elimination, 249–250, 255
β-Hydride elimination, 249–251, 255
Hydridoalkyl complexes, 261–268
Hydrocarbon conversion reactions, 222–223
Hydrogenation, asymmetric, 229–242, 325–327
Hydrogen atoms
    abstraction of, 246–250, 252, 264–265, 289
    fast, reactions of, 10, 14
Hydrogen bonds, 153
Hydroxylapatite, 490–491, 495, 499
Hyperconjugation, 297

Implants, 489–500
Infrared multiple-photon activation, 66–68
Inhibitors, enzyme, 458–459
Inner-sphere mechanism, 183–187, 194 (n.9), 483
Insertion reactions, 7–9
Interaction diagram, 203–204
Intermediates, chemical, 47–48, 63 (n.2)
    in free jet expansions, 47, 51–62
    long-lived, 7
    organocobalt, 480–482
    photochemistry of, 40–41
    reactive, 34, 43–44, 311
Ion beams, 74, 113

Ion cyclotron resonance, 66–67, 71, 74, 118
Ionic clusters, 55–57, 62, 65, 71–73
    See also Cluster ions
Ionization
    desorption, 99–100, 104–107
    electrohydrodynamic, 104
    electron impact, 53–54
    gas-phase, 104–107
    by lasers, 53–55
    multiphoton, 54, 94, 117–122
    single-photon, 117–118
Ion-molecule complexes, 66–71
Ion-molecule reactions, 67–70, 105, 112
Ions
    energetics of, 72–73
    gas-phase chemistry of, 65–76, 111–112
    laser production and detection of, 53–55
Ion trap, 111
Iridium anomaly, 102
Iron tetracarbonyl ethylene, 200–204
Isotope ratios, 102, 119, 121–122
Isotopes, stable, abundance determinations, 101–102
Isotopic tracer methods, 178

Jahn-Teller effect, 55

Ketene (CH_2CO), 27–28, 57, 68
Ketones, asymmetric reductions of, 334–335
Kevlar, 531
Kinetics, chemical, 3
Krypton, isotope selection in, 121–123

Lasalocid, synthesis of, 343
Laser photolysis, 6
Lasers
    chemical reaction probing by, 3–15, 17–18
    colliding pulse mode-locked, 80–81
    continuous-wave, 559
    double-heterojunction, 570–573
    excimer, 553, 555–557
    passively mode-locked, 80–81
    picosecond, 96
    pulsed tunable dye, 119
    in semiconductor processing, 551–561, 575
Lennard-Jones potential, 150
    See also Liquids, Lennard-Jones; Solids, Lennard-Jones
Leukotrienes, 114, 306–307

587

Ligands, 199–202, 229
Light, conversion to energy, 274–275
Liquid chromatography-mass spectrometry (LC-MS), 96, 100, 107, 110–111
Liquid crystals, 157–158
Liquid junction cell, 278
Liquid metals, 154, 156–157, 159
Liquid-phase epitaxy, 564–565, 569
Liquids
 Lennard-Jones, 147–149, 152, 158
 mixtures of, 154
 molecular, 154–156
 nonassociated, 153–154, 156
 van der Waals theory of, 145–160
Lithium aluminum hydride, 334–335
Lock-and-key concept, 234, 242, 393
Low-energy electron diffraction (LEED), 214–215
Lowest unoccupied molecular orbital (LUMO), 289
Lutetium/ytterbium mixtures, 121
Lysocellin, 315

Macromycin, 101
Macropolycyclic structures, 394–397
Magnetic resonance, laser, 18–20, 73
 See also Nuclear magnetic resonance
Mass spectrometers, 94, 102, 111–113
Mass spectrometry
 analytical capabilities of, 99–103
 angle-resolved, 104
 desorption ionization, 105
 energy resolved, 104
 field desorption, 107
 Fourier transform, 111–112
 high-pressure, 66, 71
 high-resolution, 95
 ion-beam, 72
 laser desorption, 124
 medical diagnostic use of, 101
 MS-MS, 96–97, 100–101, 104–113
 MS-MS-MS, 112
 resonance ionization, 110–111
 time of flight, 118, 120
Matrix effects, 120, 127
Matrix isolation technique, 106, 311
Mechanisms, reaction, 29, 43, 70–72, 102–103
Melittin, model of, 446–447
Melting, 158–159

Metabolic profiling, 101, 114
Metabolites, drug, 101
Metal-carbon bond, 135–136, 245–247, 256, 564
Metal carbonyls, 556
Metal-hydrogen halide reactions, 6–7
Metal-ligand bond, 72
 See also Complexes
Metallocenes, 205
Metalloenzymes, analogs of, 405
Metallophthalocyanines, 521, 523–531
Metalloreceptors, 405
Metalorganic chemical vapor deposition, 564–571, 573–576
Metalorganics, 564, 566–568, 570, 575
Metal oxo bonds, 136–137
Metals
 molecular, 518, 522
 quasi-one-dimensional, 504–506
Metathesis reactions, 135–141, 246, 250–254
Methane, 259–260, 269–270
Methanol, 138–140
Methylene (CH_2)
 bent versus linear, 131–132
 chemistry of, 34–39
 flash kinetic spectra of, 18–20
 in free jets, 57
 free radical, 57
 photoelectron spectrum of, 134
 singlet-triplet gap in, 132–134
Methylenecyclopentanes, 310
α-Methyleneglutarate mutase reaction, 479, 483
Methyl iodide, 29–30
Methylmalonyl-coenzyme A mutase rearrangement, 471, 475, 478–479, 483
Methyl transfer reactions, 70–71
Micellar structures, 444
Microelectronics, 551–561
Microprobe techniques, 96
Mixed valence compounds, 191–193
Mixtures, analysis of, 100, 107–111, 123–124
Molecular beam epitaxy, 564, 569–570
Molecular energy transfer, 5
Molecular orbital theory, 200–205
 See also Frontier orbitals; Orbitals
Molecular weight determination, 106
Molybdenum catalysts, 223, 225
Molybdenum-oxygen bonds, 136–139
Mössbauer spectra, iron-57, 95

in chemisorption theory, 164–166, 172
cryptand complexation of, 396
electronic fluctuations in, 221
as liquid metals, 157
organometallic complexes of, 197–200, 207, 260–272, 316–317, 320
Transition states, 4, 12, 14, 63 (n.2), 341–342
Transport, carrier-mediated, 407–408
Trichlorophenol, 110
Trimethylenemethane iron tricarbonyl, 207
Trimethylsilyl reagents, 371–373
Triple bonds, metathesis-like reactions of, 253–254
Trypsin, 456–461, 464–465
TTF-TCNQ, 504–506, 508, 520
Tungsten alkylidyne complexes, 252–253

Uncertainty principle, 88
Unimolecular reactions, 7, 17, 21, 66–68, 114, 210
Urinary acids, 101

Valence, mixed, 518–519, 521, 524
Van der Waals molecules, 51, 57
Van der Waals theory, 145–160
Vapor-phase epitaxy, 564–565, 570
Vibrational excitation, 7–11, 28–29
Vibrational relaxation, 21–26, 157
Vibration-rotation interactions, 22, 28
Vinoxy free radical, 60–61
Vinylsilanes, 366–370
Vitamin B_{12}, 110
Vitamin D metabolites, 313
Vomitoxin, 110

Wagner-Meerwein rearrangement, 103
WCA (Weeks-Chandler-Anderson) theory, 149–152
"Wet" chemistry, 93–94, 559–561
Woodward-Hoffmann rules, 290
Work function, 163, 167–169

Zeolites, 224–225
Ziegler-Natta catalysts, 205, 245, 255–256